July 29–August 1, 2013
Bangalore, India

I0060982

**Association for
Computing Machinery**

Advancing Computing as a Science & Profession

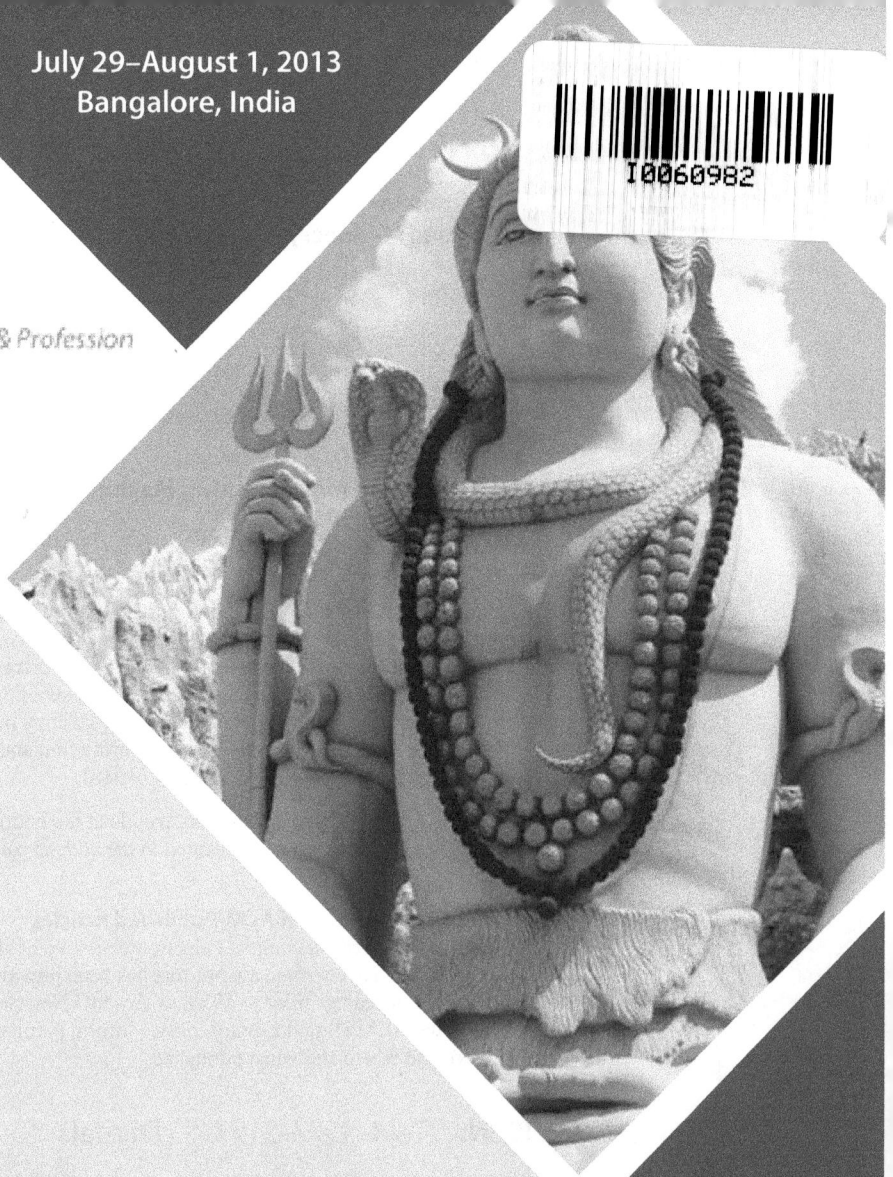

MobiHoc'13

Proceedings of the Fourteenth ACM International Symposium on
Mobile Ad Hoc Networking and Computing

Sponsored by:
SIGMobile

**Association for
Computing Machinery**

Advancing Computing as a Science & Profession

The Association for Computing Machinery
2 Penn Plaza, Suite 701
New York, New York 10121-0701

ISBN: 978-1-4503-2193-8 (Digital)

ISBN: 978-1-4503-2520-2 (Print)

Additional copies may be ordered prepaid from:

ACM Order Department
PO Box 30777
New York, NY 10087-0777, USA

Phone: 1-800-342-6626 (USA and Canada)
+1-212-626-0500 (Global)
Fax: +1-212-944-1318
E-mail: acmhelp@acm.org
Hours of Operation: 8:30 am – 4:30 pm ET

Printed in the USA

Table of Contents

Session 1: Applications

Session 2: Sensor Networks

Session 3: Localization in Sensor Networks

Session 4: Scheduling

Session 5: Routing

MobiHoc 2013 Organization

General Chairs: A. Chockalingam *(IISc Bangalore)*
D. Manjunath *(IIT Bombay)*

Program Chairs: Massimo Franceschetti *(University of California, San Diego)*
Leandros Tassiulas *(University of Thessaly)*

Local Arrangements Chair: T V Prabhakar *(IISc Bangalore)*

Publicity Chairs: Kameswari Chebrolu *(IIT Bombay)*
RangaRao Venkatesha Prasad *(Delft University of Technology)*

Steering Committee Chair: P. R. Kumar *(Texas A&M University)*

Steering Committee: Mario Gerla *(University of California Los Angeles)*
Jean-Pierre Hubaux *(EPFL)*
Joseph Macker *(Naval Research Lab)*
Sergio Palazzo *(University of Catania)*
Charles Perkins *(WiChorus)*
Martha Steenstrup *(Clemson University and Stow Research L.L.C)*

Technical Program Committee: Animashree Anandkumar *(University of California Irvine)*
Stefano Avallone *(University of Naples)*
Randall Berry *(Northwestern University)*
Srikrishna Bhashyam *(IIT Madras)*
Sem Borst *(Technishche Universiteit Eindhoven)*
Ioannis Broustis *(AT&T Labs Research)*
Prasanna Chaporkar *(IIT Bombay)*
Zainul Charbiwala *(IBM India Research Lab)*
Carla-Fabiana Chiasserini *(Politecnico di Torino)*
Song Chong *(KAIST)*
Onkar Dabeer *(TIFR Mumbai)*
Alex Dimakis *(University of Southern California)*
Vijay Erramilli *(Telefonika)*
Atilla Eryilmaz *(The Ohio State University)*
Radha Krishna Ganti *(IIT Madras)*
Jason J. Haas *(Sandia National Labs)*
Jianwei Huang *(The Chinese University of Hong Kong)*
I-Hong Hou *(Texas A&M University)*
Tara Javidi *(UCSD)*

Technical Program Committee (continued):

Ravi Kokku *(IBM Research)*
Koushik Kar *(Rensselaer Polytechnic Institute)*
Hemant Kowshik *(IBM India Research Lab)*
Nikhil Karamchandani *(USC/UCSD)*
Can Emre Koksal *(The Ohio State University)*
Iordanis Koutsopoulos *(CERTH)*
Douglas Leith *(Hamilton)*
Jorg Liebeherr *(University of Toronto)*
Jun Luo *(Nanyang Technological University)*
Peter Marbach *(University of Toronto)*
Athina Markopoulou *(University of California, Irvine)*
Cecilia Mascolo *(University of Cambridge)*
Eytan Modiano *(MIT)*
Anand Muralidhar *(Bell Labs, Alcatel-Lucent)*
Jayakrishnan Unnikrishnan Nair *(California Institute of Technology)*
Samuel C. Nelson *(BBN)*
Max Ott *(NICTA)*
George Paschos *(MIT)*
Konstantinos Pelechrinis *(University of Pittsburgh)*
Chiara Petrioli *(University of Rome "La Sapienza")*
Vinod Prabhakaran *(TIFT Mumbai)*
Alexandre Proutiere *(KTH Royal Institute of Technology)*
Gaurav Raina *(IIT Madras)*
Venkatesh Ramaiyan *(IIT Madras)*
Bhaskaran Raman *(IIT Bombay)*
Ramachandran Ramjee *(Microsoft Research)*
Vinay Ribeiro *(IIT Delhi)*
Theodoros Salonidis *(Technicolor Paris Research Center)*
Saswati Sarkar *(University of Pennsylvania)*
Ivan Seskar *(Rutgers University)*
Srinivas Shakkottai *(Texas A&M University)*
Vinod Sharma *(IISc Bangalore)*
Shreyas Sundaram *(University of Waterloo)*
Vassilis Tsaoussidis *(Democritus University of Thrace, Greece)*
Rahul Urgaonkar *(BBN)*
Rahul Vaze *(TIFR Mumbai)*
Xinbing Wang *(Shanghai Jiao Tong University)*
Xinzhou Wu *(Qualcomm)*
Guoliang Xing *(Michigan State University)*
Yung Yi *(KAIST)*
Lei Ying *(Arizona State University)*
Zheng Zeng *(Apple)*
Yanchao Zhang *(Arizona State University)*

Workshop Chairs:	Aditya Gopalan (*Technion*) Krishna Jagannathan (*IIT Madras*)
Poster Chair:	Sujay Sanghavi (*UT Austin*)
Student Travel Grants Chair:	Vinay Joseph Ribeiro (*IIT Delhi*)
Visa Chair:	Chandrika Sridhar (*IISC Bangalore*)
Publications Chairs:	Zainul M Charbiwala (*IBM Research*) Nishanth Sastry (*King's College London*)
Web Submissions Chair:	Vasilis Sourlas (*CERTH*)
Web and Registration Chair:	Rakesh Chavan (*IIT Bombay*)

MobiHoc 2013 Sponsor & Supporters

Sponsor:

Supporters:

TATA
CONSULTANCY
SERVICES

Google

Microsoft Research

QUALCOMM®

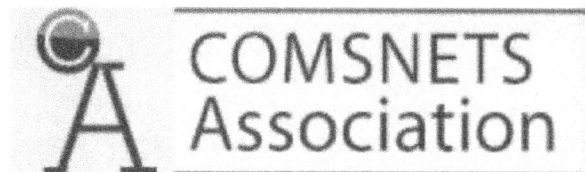

LED-to-LED Visible Light Communication Networks

Stefan Schmid
Disney Research & ETH Zurich
Stampfenbachstrasse 48
8006 Zurich, Switzerland
schmist@disneyresearch.com

Giorgio Corbellini
Disney Research Zurich
Stampfenbachstrasse 48
8006 Zurich, Switzerland
giorgio@disneyresearch.com

Stefan Mangold
Disney Research Zurich
Stampfenbachstrasse 48
8006 Zurich, Switzerland
stefan@disneyresearch.com

Thomas R. Gross
ETH Zurich
Clausiusstrasse 59
8092 Zurich, Switzerland
thomas.gross@inf.ethz.ch

ABSTRACT

Visible Light Communication (VLC) with Light Emitting Diodes (LEDs) as transmitters and receivers enables low bitrate wireless adhoc networking. LED-to-LED VLC adhoc networks with VLC devices communicating with each other over free-space optical links typically achieve a throughput of less than a megabit per second at distances of no more than a few meters. LED-to-LED VLC adhoc networks are useful for combining a smart illumination with low-cost networking. We present and evaluate a software-based VLC physical layer and a VLC medium access control layer that retain the simplicity of the LED-to-LED approach. The design satisfies the requirement that LEDs should always be perceived as on with constant brightness. In each VLC device, in addition to an LED, only a low-cost microcontroller is required for handling the software-based communication protocol. The results of our performance measurements confirm recent claims about the potential of LED-to-LED VLC adhoc networks as a useful technology for sensor networks, smart and connected consumer devices, and the Internet-of-Things.

Categories and Subject Descriptors

C.2 [**Computer-Communication Networks**]: Network Architecture and Design—*Wireless Communications*

Keywords

Visible Light Communication, MAC Protocol, Free-Space Optics

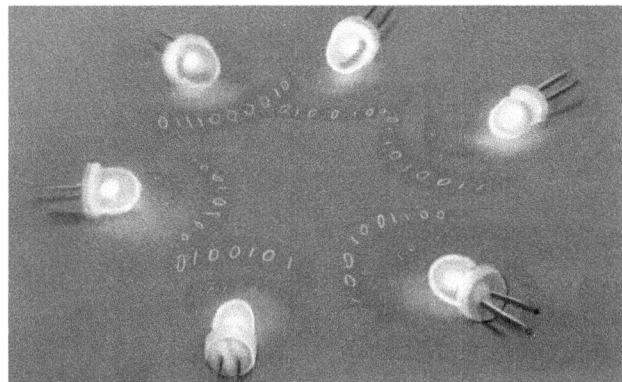

Figure 1: Concept art (©Disney). **LED-to-LED VLC adhoc network with PHY and MAC layers that use LEDs as transmitter and receiver.**

1. INTRODUCTION

Visible Light Communication (VLC) with Light Emitting Diodes (LEDs) as transmitters and receivers provide a novel approach to enable low bitrate wireless adhoc networking for short distances [17]. LED-to-LED VLC networks are formed by VLC devices that communicate with each other over free-space optical line-of-sight channels and typically achieve an overall network throughput of less than a megabit per second at distances of no more than a few meters. Low-complexity LED-to-LED VLC adhoc networks are useful when deploying sensor networks, home networks, smart illumination, or when connecting consumer devices like smart toys. These scenarios are characterized by frequent additions and removals of endpoints (the neighbor's kid may bring over a smart toy) and cost sensitivity. Such networks can exploit the ubiquitous presence of LEDs and leverage their ability to act as cost-effective transceivers (transmitters and receivers) – while allowing the LEDs to continue to operate as a lighting device.

In such low-complexity VLC networks, devices are not configured to operate with photodetectors to receive data; instead the LEDs can be used for data reception [6, 7, 18, 5]. The VLC devices use off-the-shelf 8-bit microcontrollers, powerful enough to operate the required adhoc communication protocols, to coordinate the network and medium ac-

Table 1: Terms and abbreviations.

Term	Description
ACK	Acknowledgement PDU
ADC	Analog-to-Digital Converter
CA	Collision Avoidance
CCDF	Complementary Cumulative Distrib. Fct.
CRC	Cyclic Redundancy Check
CSMA	Carrier Sense Multiple Access
CW	Contention Window with 16 CW slots
CW Slot	Slot used for CA (duration: 16 T)
FCS	Frame Check Sequence used by CRC
IDLE	Idle mode (no TX, no RX)
MAC	Medium Access Control
MPDU	MAC PDU
OFF	Symbol OFF: slot energy < THRS
ON	Symbol ON: slot energy > THRS
PDU	Protocol Data Unit
PHY	Physical
PPM	Pulse Position Modulation
QoS	Quality-of-Service
RX	Reception mode (no IDLE, no RX)
SAP	Service Access Point
SFD	Start Frame Delimiter
T	Target slot duration, 500 μs
THRS	Detection threshold, 10bit ADC reference
TX	Transmission mode (no IDLE, no RX)

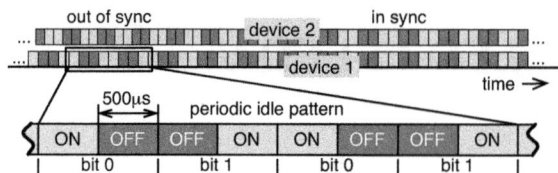

(a) Periodic idle pattern: ON-OFF-OFF-ON (bit 0 1).

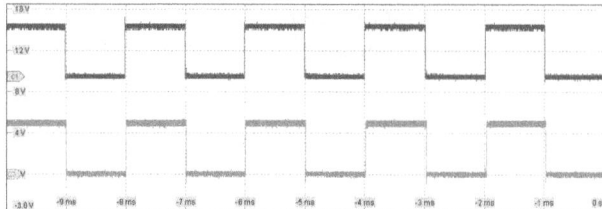

(b) Measurement of two synchronized idle patterns.

Figure 2: Idle pattern transmitted by VLC devices during IDLE mode: LEDs are periodically switched on and off and emit light repeatedly. VLC devices in range of each other synchronize automatically.

cess. The LED-to-LED network obscures the exchange of messages in the existing illumination. The exchange of visible light messages has no effect on the level of brightness (so that an LED appears to be switched on all the time).

The paper describes a novel design of a complete and low-complexity VLC Physical (PHY) layer and a contention-based VLC Medium Access Control (MAC) protocol layer and evaluates the effectiveness of these layers. The novel nature of this kind of adhoc network requires that we present some aspects of the PHY layer. We introduce a time synchronization to improve link reliability. The efficiency of the synchronization protocol is critical for the link throughput especially when the distance between the different devices increases. We also define a method to adapt the sensitivity of the VLC receiver: A threshold adaptation enables VLC devices to operate reliably even when the level of ambient light changes. The adaptation uses the preamble that also serves as start indicator at the beginning of each frame. For the MAC layer, a *Carrier Sense Multiple Access with Collision Avoidance (CSMA/CA)* protocol is defined and its performance evaluated. The MAC protocol relies on a distributed CSMA approach without priority support, i.e., no guarantee of Quality-of-Service (QoS). There is no centralized, contention-free period available for any controlled medium access.

Table 1 summarizes the abbreviations used in this paper. The PHY layer, with its novel synchronization method and adaptive frame reception, and the MAC layer, with its distributed CSMA medium access protocol, are described in Sections 2 and 3, respectively. The system's performance is evaluated in Section 4. Section 5 provides a brief summary of related work and is followed by the conclusions in Section 6.

2. SOFTWARE-DEFINED PHY-LAYER

The PHY layer is responsible for the frame delivery over the free-space optical medium, for synchronization, and for the medium idle/busy sensing. The PHY layer operates with PHY-Protocol Data Units (PHY-PDUs), which are referred to as frames when they are transmitted over the medium.

The VLC system consists of small consumer devices (here referred to as VLC devices) that are equipped with one or more LEDs for light effects and communication. A microcontroller is used to process the transmitted and received signals (PHY layer) and to run the medium access control protocols (MAC layer). In the setup described, VLC devices use the same LEDs for transmission and reception. In case of multiple LEDs embedded in the same VLC device, they can be used in a time multiplexed way [17].

The communication is organized using time slots of target duration T=500 μs. With the given system design, the VLC system could reach a higher data throughput with shorter slot durations. However, we selected this value to ensure that a common low-cost 8-bit microcontroller has sufficient time for processing.

Slots in which at least one VLC device emits light are interpreted to carry the symbol "ON". Slots during which no VLC device emits any light are interpreted as carrying the symbol "OFF". Depending on device locations, ambient light, and network conditions, different VLC devices might interpret the same slot differently at the same time, i.e., as ON or OFF.

2.1 IDLE, TX, and RX Modes

Each VLC device operates in one of three communication modes and changes its mode when required by the communication protocol. A VLC device remains in *IDLE mode* if it does not have any frame to transmit and if it is not receiving any frame. While in IDLE mode, a VLC device periodically transmits the so-called idle pattern to let the LED appear to be switched on at constant brightness, as shown in Figure 2. The device transmits ON and OFF symbols following a predefined pattern with a given alternation frequency ($1/(2 * T)$) where T is the duration of each slot (T=500 μs), so that the light is perceived with constant intensity. The periodic ON-OFF idle pattern during IDLE

Table 2: Bit encoding with 2-PPM.

bit	first symbol	second symbol
0	ON	OFF
1	OFF	ON

remark: ON = light is emitted; OFF = no light

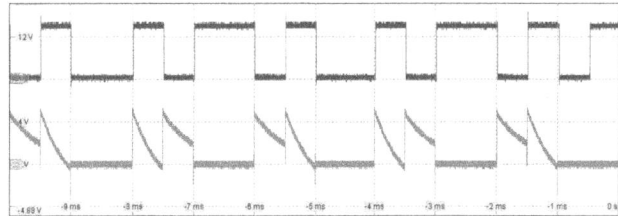

Figure 3: Measurement of incoming light during slots at which the device itself does not emit light. Top: transmitted pattern. Bottom: discharging voltage at the receiving LED.

mode consists of the periodic repetition of four consecutive time slots {ON-OFF-OFF-ON} (see Figure 2(a)), which can be interpreted as a periodic repetition of bit 0 and bit 1.

A VLC device operates in *TX mode* when it transmits a frame and in *RX mode* during the reception of a frame. In TX mode, the device modulates the slots with the 16-slot preamble and the subsequent bitstream, and as a result the ON and OFF symbol pattern changes depending on the transmitted frames (the data bitstream). The bit encoding is a 2-Pulse Position Modulation (2-PPM) with 50% duty cycling and is indicated in Table 2.

Because 2-PPM coding is applied, on average, the same number of ON symbols and OFF symbols are emitted. As a consequence, there is no visual flickering even when bit streams would be correlated with, for example, long streams of only bit 0. Also during the IDLE mode, the LEDs continue to appear switched on without flickering. On the receiver side, in RX mode, a VLC device still transmits ON and OFF symbols as in IDLE mode (a continuation of the IDLE pattern), but it interprets the amount of incoming light during each OFF symbol as incoming symbols, because it intends to receive a frame.

2.2 Receiving with an LED

Often, VLC systems use LEDs for transmission and photodetectors for reception. A photodetector efficiently converts light photons into electrical current. Already with one LED and one photodetector, it is possible to build a one-way VLC system in which frames are transmitted from the LED to the photodetector, but this kind of VLC system does not provide a feedback channel back to the LED. To build a two-way VLC system that allows feedback, two components per device would be required: an LED to transmit, and a photodetector to receive. It is possible to use the LED as photodetector to receive optical messages using the same LED that is used for transmission [6], a set up that reduces the complexity per device. This approach is taken here.

Figure 3 illustrates how LEDs that are charged in reverse bias can be used to receive incoming light. As can be seen in the figure, depending on the intensity of the incoming light, the LED capacitance discharges at different speed. The stronger the incoming light, the faster the discharge. With an adaptive threshold parameter (THRS), the two different symbols ON and OFF can be determined and differentiated by the receiving VLC device at the end of each slot used for measurements.

A drawback of using the LED as a receiver is the fact that an LED is less sensitive than a photodetector; this property negatively affects the achievable communication range. A second drawback is the resulting limited throughput due to the required multiplexing: Whereas with two components per VLC device, transmission and reception can occur at the same time in parallel, with a single LED per VLC device, this LED can either only transmit or only receive but

cannot do both at the same time. To use single LEDs as transceivers, the VLC device needs to alternate transmission and reception periods. And half of the slots are not used for reception, because they are used to periodically emit light, to let the LED appear to be switched on.

2.3 Synchronization

Devices that use a slotted VLC scheme to communicate need to operate in a time-synchronized way: VLC devices in range of each other are synchronized if the beginning of each slot occurs at the same time for all devices. Synchronization errors such as clock shift and jitter are undesirable and reduce the system performance [2]. Using an LED in reverse bias acting as a receiver leads to low sensitivity to optical energy (in comparison with photodetectors). This setup makes obtaining an accurate synchronization more difficult. A possible solution is to leave devices un-synchronized and transmit a dedicated synchronization preamble before each frame [17, 9]. If a portion of time is dedicated to the transmission of the preamble for synchronisation, the overhead increases. In our solution, we achieve synchronization of VLC devices that are in range with each other differently, without the need of transmitting a dedicated synchronization preamble: The periodically repeated idle pattern is continuously used to synchronize the VLC devices. This limits the protocol overhead as no dedicated preamble is required.

To obtain synchronization, all devices follow the same idle pattern ({ON-OFF-OFF-ON}, as indicated in Figure 2) so that all LEDs are perceived as switched on. VLC devices remain in IDLE mode as long as they do not have data to transmit or receive. When powered on, a VLC device performs an initial measurement that typically lasts for a few milliseconds: The device measures the current (ambient) light intensity over several measurement slots. After the initial measurement, every VLC device continuously measures the amount of light detected in measurement slots and compares the values of two consecutive measurements. If the amount of light measured during two consecutive slots is close to each other and if it is also similar to the ambient light measured earlier, the device can conclude that either there is no other device in its vicinity, or that it is synchronized to the other devices near by (within communication range). However, if the device detects that the two consecutive measurement values deviate (with some hysteresis), this might mean that there is another light-emitting device in the vicinity. In this case, for example, if a first VLC device detects that the incoming light in slot n is larger than slot n+1, this means that it is sampling too early and its

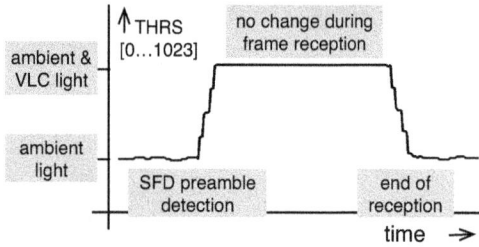

Figure 4: THRS adaption before and after frame reception, logged at microcontroller.

local clock must be shifted forward, that is, it must sample later. On the other side, at the second VLC device, this means that the incoming light in slot n is smaller than slot n+1. Therefore the second VLC device will need to shift its clock backward, i.e., it needs to advance the sampling. By applying a small shift (repeatedly if needed), VLC devices synchronize their time slots during IDLE mode, without the explicit use of a synchronization preamble. This method is not limited to the synchronization of only a pair of devices, it also works with a larger number of devices that are in range of each other.

Speed and accuracy of the synchronization can be increased by separating the measurement for synchronization from the two measurement slots that are used for symbol reception (two consecutive symbols form one bit). Such a separation would introduce two extra measurement slots of short duration: One before and one after the two original measurement slots. This setup allows stable synchronization even in TX and RX mode, because dedicated time is allocated to synchronization during all modes of operation.

2.4 Adaptive Bit Detection Threshold (THRS)

Light intensity modulation is used to differentiate and detect symbols ON and OFF. At the receiver side, the ability of detecting and differentiating symbols ON or OFF is affected by the attenuation of the transmitted light from the source VLC device and the intensity of the ambient light. Assuming that transmitters and receivers are synchronized, an ideal detection threshold is helpful to reliably decode the received bits by distinguishing the ON and OFF symbols. Such a threshold is difficult to determine because the level of ambient light changes over time. An optimal detection threshold is the average of the amount of light corresponding to the reception of a symbol ON and the amount of light corresponding to reception of a symbol OFF (see Figure 4).

Consider two devices that are out of communication range of each other. At the end of the first period of the synchronization, the adaptive bit detection thresholds THRS at both devices are set to the average level of the ambient light, independently. Then, each device continuously updates its THRS value averaging the amount of light detected during slots that were used for measuring incoming light (using a sliding window). Since they are out of range, the two devices remain un-synchronized and stay in IDLE mode as long as there is no data to transmit. When the two devices are placed close to each other with the LEDs pointing towards each other's field of view, they try to synchronize. After the VLC devices achieve synchronization, the devices emit and receive light at the same time: They follow the

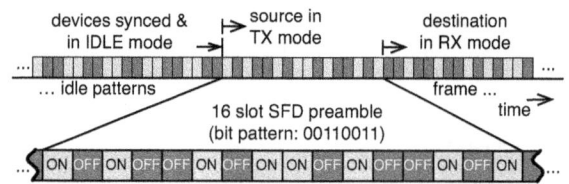

(a) Start Frame Delimiter (SFD)

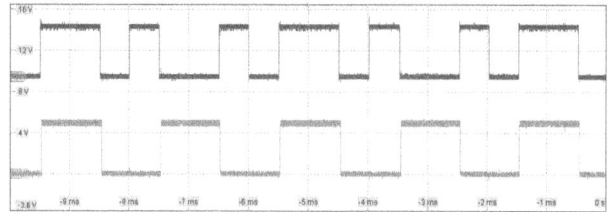

(b) SFD measurement

Figure 5: The SFD preamble: DATA and ACK frames start with the SFD, to achieve reliable frame detection. In addition, the SFD serves as preamble so that each receiver updates its detection threshold (THRS) if needed.

same idle pattern. With ambient light, THRS remains at its minimum level. This changes as soon as the Start Frame Delimiter (SFD) is detected, as described in Section 2.5.

2.5 Start Frame Delimiter (SFD) Preamble

To set THRS at each VLC device to the optimal values, receivers should receive patterns consisting of an even number of symbols ON and OFF. Therefore, each time a VLC device needs to transmit a frame, it precedes the transmission by a so-called SFD preamble (sixteen slots), which is used to alert the receiver that a frame will follow. The SFD preamble is chosen so that the receiver can increase the detection threshold up to the correct value. At the end of the SFD, at any receiving VLC device, the value of THRS is adapted to the current ambient light situation and closer to the optimal (but unknown) value for THRS so that the receiver is able to reliably receive the subsequent frame. As soon as the SFD is decoded, the receiver stops updating THRS and keeps it constant until the end of the current frame reception. After the end of the reception, THRS adapts again and converges to the value related to the ambient light. In case the level of ambient light did not change during the frame exchange, this is a value similar to THRS before the frame exchange. Figure 4 illustrates this adaptive process.

2.6 Carrier Sensing

A VLC device interprets the medium as busy or idle depending on ongoing frame transmissions and/or the level of incoming ambient light. If the MAC layer requests a carrier sensing with the PhySense_req() service primitive via the PHY-Service Access Point (PHY-SAP) to the PHY layer, the PHY reports back the information about whether the optical channel is busy or clear. The PHY layer reports that the channel is clear if the current value of THRS is comparable to the instantaneous level of ambient light. In our testbed, THRS is always updated using the measured light of the last 16 slots. Therefore, if during these last slots

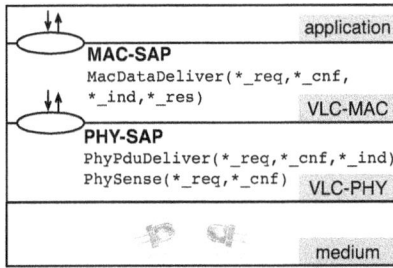

Figure 6: Reference model of the VLC system.

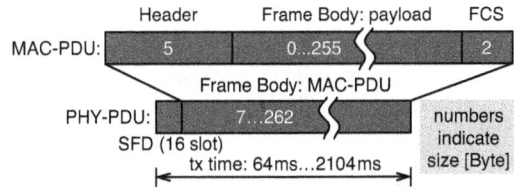

Figure 7: MAC- and PHY-PDU definition.

all VLC devices are in *IDLE mode*, the deviation between the values observed is small; otherwise a large deviation is detected. If the deviation is higher than the difference between THRS and the instantaneous level of ambient light, the PHY layer reports that the channel is busy. This situation is reported with the `PhySense_cnf(ON/OFF)` service primitive at the PHY-SAP. This information is then used by the MAC protocol to coordinate the medium access, as described in detail in Section 3. Figure 6 illustrates how the layer entities communicate via SAPs and service primitives.

3. LOW-COMPLEXITY MAC-LAYER

A reference model for communication networks defines layers, functions, services, SAPs, and service primitives exchanged by the layer entities via these SAPs. A layer entity can be interpreted as a service provider that serves the higher layer entity. The VLC system reference model is shown in Figure 6. This model is an abstraction of the real implementation. The figure illustrates how MAC and PHY interact via the PHY-SAP with so-called service primitives that are abstractions of discrete, instantaneous events (`*_req` for requests, `*_cnf` for confirmations, `*_ind` for indications, and `*_res` for responses). A MAC layer entity serves the application layer by providing the data delivery service, which is requested via the `MacDataDeliver_req(data)` primitive at the MAC-SAP. The MAC layer relies on PHY services at the PHY-SAP, with primitives like, for example, the `PhyPduDeliver_req(MAC-PDU)` primitive.

Two types of frames are used: DATA and Acknowledgement (ACK). The DATA frame body contains the application payload of variable size (0...255 Byte). The ACK frames are control frames used to acknowledge a successful MAC-PDU reception and are transmitted by the destination device back to the source device, right after the DATA frame (after a short transceiver turnaround time of four slots). The structure of the two frames is given by the Protocol Data Unit (PDUs) illustrated in Figure 7 and defined in Table 3. The bit pattern of a frame is determined by the PHY-PDU, which encapsulates a MAC-PDU (MPDU). Each MAC-PDU is composed of various elements that are used to manage communication and error handling. The MAC-PDU of a DATA frame contains the payload of the application. The MAC-PDU of an ACK frame is of a similar structure, but without payload.

A Carrier Sense Multiple Access with Collision Avoidance (CSMA/CA) protocol is used to coordinate how VLC devices access the optical medium. Figure 8(a) shows an example of a DATA/ACK frame exchange with the preceding Contention Window (CW). After the medium became idle,

an initial first wait time of 16 slots is followed by the CW of 16 SFD preamble durations, before the start of a frame exchange. The initial waiting time is needed to ensure that ACK frames are not interfered with. There is a short idle time of four slots between a DATA and ACK frame which must not be misinterpreted as idle channel. Within the CW, there is a maximum of 16 CW slots; each CW slot has the duration of 16 T=16×500 μs (i.e., the SFD preamble duration).

A VLC device that intends to transmit a DATA frame selects one of the sixteen possible CW slots for its transmission: Each CW slot is assigned the same probability of 1/16 to be selected for the transmission start (uniform distribution). This random selection ensures a minimal collision probability. However, before this target CW slot is reached, the VLC device senses the optical medium for each of the preceding CW slots. The device starts transmitting the DATA frame at the target slot only if the medium did not get used by any other VLC device at one of the earlier CW slots. When the medium is detected as busy at such an earlier CW slot (another VLC device has selected an earlier CW slot for its own medium access), the VLC device stops the contention process and switches to RX mode. Then, upon detecting the medium to be idle for the duration of one CW slot (16×500 μs), the device starts again sensing the medium for each following CW slot to eventually start transmitting the DATA frame when the selected target CW slot has been reached.

In unicast, with one unique VLC device as destination, DATA frames that are completely received are acknowledged immediately with an ACK frame from the destination back to the source VLC device.

The ACK is transmitted back to the source device right after the successfully received DATA frame, after a short waiting time of four slots, as indicated in Figure 8(b). The waiting time is needed for the system to remain synchronized. Each message begins and ends during the second ON slot of the idle pattern. When a device receives a message,

Table 3: DATA & ACK Frame Structure

	Element	Byte (Slots)	Time [ms]
1.	SFD	1 (16)	8.0
2.	Header:		
	2.1 frame body size	1 (16)	8.0
	2.2 frame control	1 (16)	8.0
	2.3 destination address	1 (16)	8.0
	2.4 source address	1 (16)	8.0
	2.5 sequence number	1 (16)	8.0
3.	Frame Body:		
	- for DATA	0 - 255 (0 - 4080)	0.0 - 2040.0
	- for ACK	0 (0)	0.0
4.	FCS	2 (32)	16.0

(a) Collision avoidance and DATA / ACK

(b) Timeout and retry after missing ACK

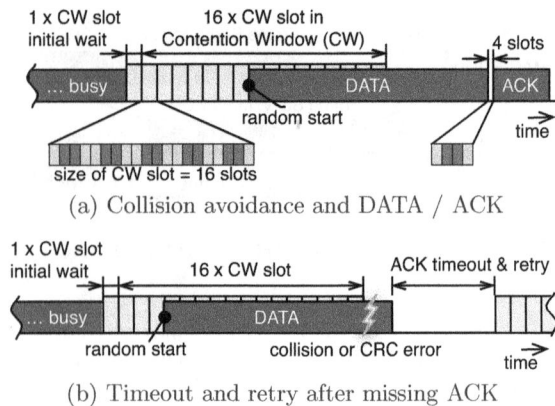

Figure 8: Contention-based medium access.

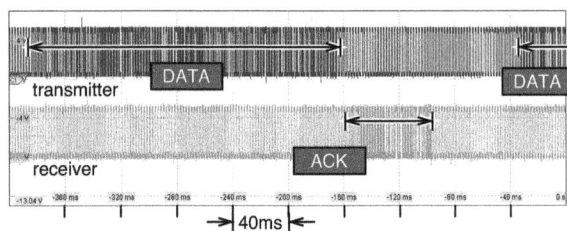

Figure 9: DATA/ACK frame exchange.

it computes the final CRC and transmits the ACK starting at the second following ON slot. With the CSMA/CA approach, multiple VLC devices might select the same CW slot and hence a so-called collision might occur. Collisions are not directly detected by the transmitting devices.

Instead, ACK timeouts indicate a missing ACK frame and lead to retransmission. As indicated in Figure 8(b), in case the ACK is not received within the timeout interval, a retransmission can be initiated. A timeout interval of 134 slots is applied. This interval helps to maintain a synchronized contention avoidance protocol behavior when some devices detect and receive the ACK, but other devices miss the SFD preamble of the ACK frame and therefore time out. When retransmitting a previously failed DATA frame, a new collision avoidance process with the same CW size as before follows. After a number of transmission attempts (four in our testbed, the initial attempt followed by three retransmissions), the DATA frame is discarded and in case there is another MPDU waiting, this next MPDU is processed. In addition to unicast, broadcast transmission is possible. No ACK is transmitted by the receiving VLC devices back to the broadcasting device upon successful reception of the broadcast frame. The address 255 is used to identify a broadcast frame. Figure 9 shows a measured DATA / ACK frame exchange. Devices remain synchronized during the frame exchange. Hence, the receiving device (while in mode RX) can transmit compensating ON symbols periodically to appear on at the same brightness as it was during mode IDLE before. During DATA and during ACK transmission (while in mode TX) the 2-PPM bit encoding (two slots per bit) maintains this constant brightness and mitigates visible flickering at the transmitter.

LED-to-LED VLC networking might be used in consumer electronics of low complexity with limited need for data security (for example, connected toys). Data is not encrypted

at MAC, and anything related to advanced security is handled by higher layers if feasible. VLC devices authenticate against each other and synchronize as soon as they are in communication range of each other. There is no need for a protocol that manages association and authentication, as long as devices can be identified without ambiguity. In our testbed, the address range is relatively small (1 Byte), which allows to realize a small number of unique device addresses.

4. EVALUATION

4.1 Testbed

The testbed consists of Atmel ATmega328P evaluation boards [3] connected to a computer. The computer generates the traffic for all boards and collects the measurement results. Each evaluation board operates one transceiver LED to mimic one VLC device. All VLC devices operate the PHY and MAC software on their microcontrollers without relying on the connection to the computer. Unless stated otherwise, payload is generated so that the VLC devices operate in saturation.

This setup is the interesting point of operation that we look at when evaluating the contention-based access protocol. In saturation, all VLC devices have always at least one packet to transmit. Therefore, the distribution of the inter-arrival time is not relevant. Payload of random content (uncorrelated random bit pattern) with different sizes is generated. The LEDs are located so that they can all communicate with each other: there are no hidden VLC stations, as indicated in Figure 10.

In all evaluation steps, the same type of LED is used: a 5 mm red LED, type Kingbright L-7113SEC-J3, with transparent case. This LED has a peak wavelength of 640 nm, a 20° field of view (radiation angle), and a brightness of 12000 mcd. This type was selected because it is common and widely available. Each board connects an LED to the microcontroller. We are interested in counting successful frame transmissions and measuring the time it takes to deliver them (including waiting time and retransmission time, if any). The tests were performed in an office indoor space with windows but no direct sunlight, during the day, and with normal office artificial lighting.

4.2 Single Link

Figure 11 shows the achievable maximum throughput, i.e., the saturation throughput, in a scenario in which one VLC

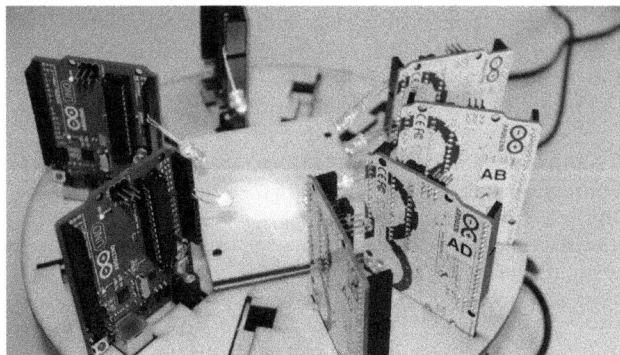

Figure 10: Testbed with boards and LEDs.

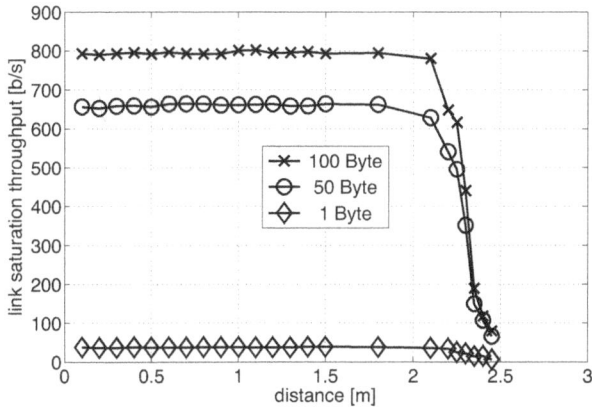

Figure 11: Throughput over distance for a single link. The system operates reliably at distances of up to 2 m.

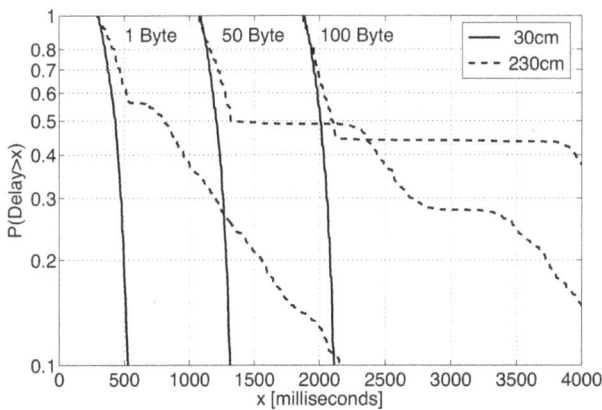

Figure 12: Data transmission delays, including waiting, contention, retransmission times. Single link with different payload sizes.

device continuously attempts to transmit to another VLC device. Except the single transmitter and single receiver, there is no other VLC device in the system. The two LEDs are pointed to each other, and the distance between them is varied between 10 cm and 250 cm. Throughput results for different payload sizes are shown. At distances larger than 2.3 m, the achievable link saturation throughput drops with the increasing distance, down to zero at around 2.5 m. Because of the CSMA/CA protocol applied, the saturation throughput depends considerably on the payload size: With a small payload size of 1 Byte, the saturation throughput is much smaller than with 100 Byte (around 40 b/s instead of 800 b/s). CSMA/CA is not needed when only one transmitter is active. However, the configuration is kept as defined in Section 3, to be able to compare the results to the network scenarios (below and Figure 13).

Figure 12 illustrates the Complementary Cumulative Distribution Function (CCDF) for the delivery delay, for three different payload sizes. For all payload sizes, distributions for two distances (30 cm and 230 cm) are shown. The delivery delay is the time between (1) the packet generation event with the arrival at the transmitting VLC device, and (2) the event when the receiver received the associated data frame successfully. The figure illustrates the time that it takes to

transmit a frame. The minimum is the transmission time over the medium, which can be taken from the curves at probability 1, at the top of the graph. The prolonged delays result from additional waiting and retransmission times. At closer distance (30 cm), retransmissions do not occur and the CCDFs of the three payload sizes show similar behaviors (solid lines). At the larger distance (230 cm), the CCDFs indicate a considerable increase of delivery delay (dashed lines). This behavior is due to the high number of retransmissions that occur because of the many CRCs failing at such a long distance. According to the figure, around 60 % of the frame transmissions are unsuccessful and require at least one retransmission. When retransmissions occur, the resulting delivery delay increases towards delays that might be unacceptable: In particular, very long delivery delays can be seen in the case of the long payload, 100 Byte. Keeping this in mind, since such delays might be undesirable, VLC should be applied when either a smaller number of retransmissions is allowed before discarding a packet, or the payload sizes are small. For example, in a VLC application for geolocation and indoor tracking, broadcasting IPv6 addresses from LED light bulbs requires only a payload size of less than 20 Byte.

4.3 Network

The achievable system saturation throughput in the scenario of networked devices is indicated in Figure 13. Multiple VLC devices that are all in communication range of each other are active (up to six transmitters). Figure 10 shows the testbed setup used for this measurement. Because of the CSMA/CA protocol applied here, the throughput decreases with increasing number of contending VLC devices. This is expected, because colliding frames can occur in contention-based protocols. Collisions consume time, and this waste of time is undesirable. In addition, collided frames lead to retransmissions, which consume additional resources. Because of the large expected increase of the delay due to retransmissions, higher collision probabilities should be avoided. However, because of the limited field of view of the used LEDs, we can assume that only a small number of VLC devices, say, two to three VLC devices, will be in communication range of each other and will compete for medium access. This is the reason why we focus on setups with a moderate number of networked VLC devices. In the figure, the case of one transmitter is comparable to the single link scenario that was discussed earlier.

4.4 Power Consumption

Table 4 shows the average power consumption for different device modes. The first row indicates the power consumption of the evaluation board and the microcontroller running an empty program without attached LEDs. During each slot of duration T, each VLC device processes the signal that has been received in the preceding slot. The processing cost changes from slot to slot and depends on the current mode (idle, receiving, transmitting) of the device. Slot duration is long enough to guarantee that when a device begins a new slot it is not processing any signal. After the end of the processing, the microcontroller is put to sleep mode to save energy until the start of the new slot. The power consumption for the three device modes are only slightly different. The difference of around 1 mW is due to increased processing cost when receiving and transmitting a packet.

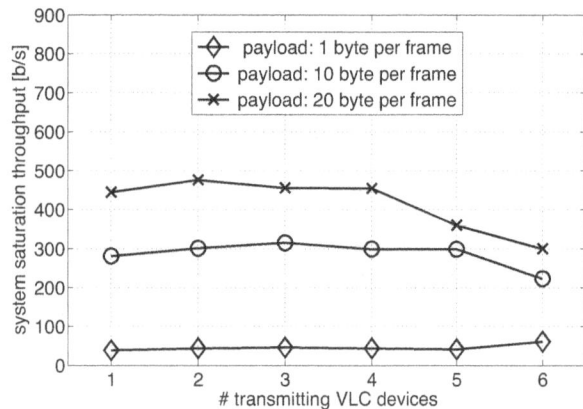

Figure 13: System saturation throughput in the network scenario. CSMA/CA causes throughput to decrease with increasing number of contending VLC devices.

Table 4: Average power consumption.

device mode	power consumption [mW]
board (no LED)	221.134
IDLE mode	288.136
RX mode	288.601
TX mode	289.020

(a) IDLE mode

(b) RX mode

(c) TX mode

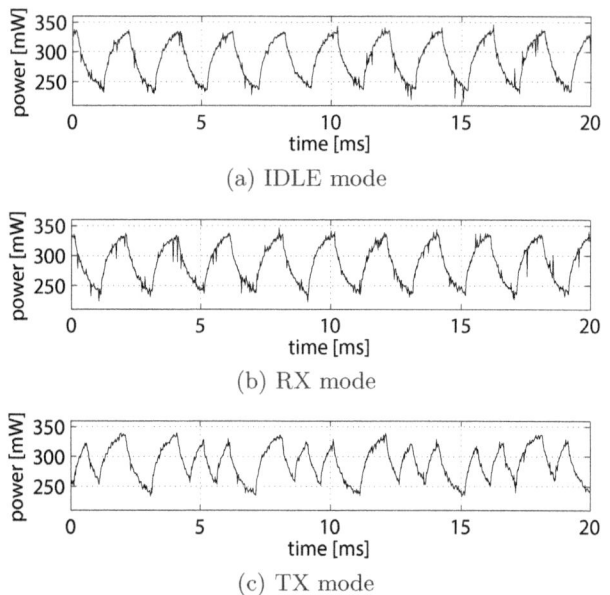

Figure 14: Power consumption over time for the three different device modes.

Table 4 indicates increasing power consumption for receiving and transmitting which correlates with the processing costs of the implementation.

The 67 mW difference between an empty program and idle mode is due to operating the LED. In the measurements, the LED's forward current is not limited to achieve enough brightness to cover large transmitting distances. The brightness can be adapted in software with duty-cycling or in hardware with resistors to fit a specific application and its power consumption constraints.

Figure 14 shows the power consumption over time for the three modes. Figure 14(a) indicates the consumption in IDLE mode. The behavior is regular and the periodicity correlates with the physical layer. The consumption increases while switching on the LED for two slots (2 T=1 ms) and decreases again for the following two slots where the LED is off and the incoming light is measured. Figure 14(b) shows a similar behavior for RX mode. During reception, the LED follows the same pattern, which can be seen in the energy consumption. The additional processing cost is so small that it is not visible in this graph. Figure 14(c) refers to the power consumption in TX mode. Transmitting requires the LED to follow the 2-PPM encoded message which can lead to more mode changes of the LED. These changes and the additional processing cost lead to an increase of the overall power consumption.

5. RELATED WORK

IEEE 802.15.7 standard [1, 16], which was published in 2011, has influenced various aspects of recent VLC research. The document specifies a PHY and MAC protocol stack with different optional transmission schemes and protocol configurations for supporting a variety of use cases. Dimming of light is realized by modifying PPM pulse durations. The standard defines user data communication as well as management functions such as authentication, encryption, and the distribution of data across the 802 architecture. The approach described here follows a different direction; our approach aims towards highest simplicity and the re-use of existing hardware components embedded in the VLC devices. The simplicity of this design is reflected by the fact that we rely on LEDs and microcontrollers instead of dedicated communication systems. Security other than simplest device identity handling is not addressed. This approach can be referred to as steganographic networking [11]: The LED-to-LED VLC network obscures the exchange of messages in the visible light. The objective is to hide the communication into already existing illumination.

The idea to use LEDs as receivers is described by Dietz et al [6, 7]. Applications scenarios have been demonstrated at conferences [18, 5]. This paper goes further and introduces and evaluates MAC protocol, synchronization, and system aspects such as communication range, performance, and device power consumption.

Research challenges in the domain of VLC are summarized in [14, 8, 12]. One challenge is to increase the communication throughput. Increasing the throughput by what is referred to as Multiple-Input-Multiple-Output (MIMO) transceivers is discussed by Azhar et al. and O'Brien et al. [4, 15]. The approach taken in [10] to increase throughput is to operate with multiple colors, such as red, green, and blue for RGB-LEDs. Further, [13] explores the use of an array of diodes to transmit data in parallel, again with the objective to achieve increased throughput. Most research focuses on isolated optical links and ignores the need for multiple access or networking. Also, much of the related work is based on deploying photodiodes instead of LEDs for detecting incoming light. The field of VLC is highly dynamic and there are efforts to investigate link layers for VLC without applying LEDs or photodetectors as receivers: some VLC systems address the communication with a display and to a camera with multiple light emitters [19]. The design presented here is simpler, built with low-cost ubiquitous devices and provides a base for VLC adhoc networking.

6. CONCLUSIONS

LED-to-LED VLC networks use LEDs as transmitters and receivers. We evaluated a novel low-complexity software-based PHY and MAC layer that are designed under the premise that the LEDs should always be perceived (by human observers) as switched on at constant brightness. The VLC protocols discussed here are designed so that only a microcontroller is required for handling the software-defined PHY and the MAC protocol. The evaluation results of our performance measurements confirm recent claims about the potential of LED-to-LED VLC networking. The achievable throughput is limited by the 8-bit microcontroller and in the order of 800 b/s at a remarkable distance of more than 2 m. Collisions have a negative impact as retransmissions consume undesirable additional time. The VLC adhoc network is designed for smaller packets such as broadcasting or forwarding of IPv6 addresses. LEDs consume only relatively small energy, correlating with their brightness, which can be adapted to the needs of the application. The VLC devices evaluated here consume almost equal amount of power in idle mode, and during receiving or transmitting.

VLC is a novel approach that enables low bitrate wireless adhoc networking based on consumer electronic equipment that is often already present in many environments. The simplicity of this approach lies in the re-use of existing components, and the steganographic approach in which data security is deliberately compromised. The LED-based VLC adhoc network obscures the communication within the visible light and hides the communication in the illumination. Such an LED-based VLC adhoc network, in which VLC devices communicate with each other via free-space optics, might in the future achieve a performance so that this approach will be useful for combining smart illumination with low-cost networking, to eventually become a candidate technology for the Internet-of-Things.

7. REFERENCES

[1] 802.15.7. IEEE Standard for Local and Metropolitan Area Networks. Part 15.7: Short-Range Wireless Optical Communication Using Visible Light, Sept. 2011.

[2] S. Arnon. The effect of clock jitter in visible light communication applications. *Lightwave Technology, Journal of*, 30(21):3434–3439, 2012.

[3] Atmel. 8-bit Microcontroller with 4/8/16/32KBytes In-System Programmable Flash. www.atmel.com, 2012.

[4] A. Azhar, T.-A. Tran, and D. O'Brien. Demonstration of high-speed data transmission using MIMO-OFDM visible light communications. In *Globecom Workshops (GC Wkshps), 2010 IEEE*, pages 1052–1056, Dec. 2010.

[5] G. Corbellini, S. Schmid, S. Mangold, T. R. Gross, and A. Mkrtchyan. LED-to-LED Visible Light Communication for Mobile Applications. In *Demo at ACM SIGGRAPH Mobile 2012*, Aug. 2012.

[6] P. Dietz, W. Yerazunis, and D. Leigh. Very low-cost sensing and communication using bidirectional leds. In *UbiComp 2003: Ubiquitous Computing*, pages 175–191. Springer, 2003.

[7] P. Dietz, W. Yerazunis, and D. Leigh. Very Low-Cost Sensing and Communication Using Bidirectional LEDs. In *TR2003-35*, 2003.

[8] H. Elgala, R. Mesleh, and H. Haas. Indoor Optical Wireless Communication: Potential and State-of-the-Art. *IEEE Commun. Mag.*, 49(9):56–62, 2011.

[9] D. Giustiniano, N. Tippenhauer, and S. Mangold. Low-Complexity Visible Light Networking with LED-to-LED Communication. *IFIP Wireless Days 2012*, Nov. 2012.

[10] T. Komiyama, K. Kobayashi, K. Watanabe, T. Ohkubo, and Y. Kurihara. Study of Visible Light Communication System using RGB LED Lights. In *SICE Annual Conference (SICE), 2011 Proceedings of*, pages 1926–1928, Sept. 2011.

[11] J. Lubacz, W. Mazurczyk, and K. Szczypiorski. Voice over IP. *Spectrum, IEEE*, 47(2):42–47, Feb. 2010.

[12] S. Mangold. Visible Light Communications for Entertainment Networking. In *Photonics Society Summer Topical Meeting, 2012 IEEE*, pages 100–101, July 2012.

[13] J. McKendry, R. Green, A. Kelly, Z. Gong, B. Guilhabert, D. Massoubre, E. Gu, and M. Dawson. High-Speed Visible Light Communications Using Individual Pixels in a Micro Light-Emitting Diode Array. *Photonics Technology Letters, IEEE*, 22(18):1346–1348, Sept.15, 2010.

[14] D. O'Brien. Visible Light Communications: Challenges and Potential. In *Photonics Conference (PHO), 2011 IEEE*, pages 365–366, Oct. 2011.

[15] D. O'Brien, S. Quasem, S. Zikic, , and G. E. Faulkne. Multiple Input Multiple Output Systems for Optical Wireless: Challenges and Possibilities. In *Proceedings of SPIE*, volume 6304, 2006.

[16] S. Rajagopal, R. Roberts, and S.-K. Lim. IEEE 802.15.7 Visible Light Communication: Modulation Schemes and Dimming Support. *Communications Magazine, IEEE*, 50(3):72–82, March 2012.

[17] S. Schmid, G. Corbellini, S. Mangold, and T. Gross. An LED-to-LED Visible Light Communication system with software-based synchronization. In *Optical Wireless Communication. Globecom Workshops (GC Wkshps), 2012 IEEE*, pages 1264–1268, Dec. 2012.

[18] N. Tippenhauer, D. Giustiniano, and S. Mangold. Toys communicating with leds: Enabling toy cars interaction. In *Consumer Communications and Networking Conference (CCNC), 2012 IEEE*, pages 48–49, 2012.

[19] H. Ukida, M. Miwa, Y. Tanimoto, T. Sano, and H. Yamamoto. Visual Communication using LED Panel and Video Camera for Mobile Object. In *Imaging Systems and Techniques (IST), 2012*, pages 321–326, July 2012.

Continuous Scanning with Mobile Reader in RFID Systems: an Experimental Study

Lei Xie[†], Qun Li[‡], Xi Chen[†], Sanglu Lu[†], and Daoxu Chen[†]
[†]State Key Laboratory of Novel Software Technology, Nanjing University, China
[‡]College of William and Mary, Williamsburg, VA, USA
lxie@nju.edu.cn, liqun@cs.wm.edu, hawkxc@163.com, {sanglu,cdx}@nju.edu.cn

ABSTRACT

In this paper, we show the first comprehensive experimental study on mobile RFID reading performance based on a relatively large number of tags. By making a number of observations regarding the tag reading performance, we build a model to depict how various parameters affect the reading performance. Through our model, we have designed very efficient algorithms to maximize the time-efficiency and energy-efficiency by adjusting the reader's power and moving speed. Our experiments show that our algorithms can reduce the total scanning time by 50% and the total energy consumption by 83% compared to the prior solutions.

Categories and Subject Descriptors

C.2.1 [**Network Architecture and Design**]: Wireless Communication

Keywords

RFID; Realistic Settings; Algorithm Design; Experimental Study; Model

1. INTRODUCTION

Mobile RFID reading performance is critical to a number of applications that rely on mobile readers. Scanning books in a library or a bookstore, tracking merchandises in a store, all require a mobile reader to be used for continuous scanning over the tags attached to the physical goods and assets. The mobile reader moves continuously to scan a large number of tags effectively compensating for its limited reading range. In those types of mobile reader systems, two performance metrics are highly pertinent: time efficiency to reduce the total scanning time, and energy efficiency to reduce the total power consumption. Unfortunately, there is no realistic model to characterize the performance for mobile RFID reading for a large scale setting. The factors that affect the mobile reading performance are very complicated. For example, the actual scanning time for a number

of tags in a realistic scenario is much longer than the time computed for free space, as shown in our experiments. In addition, RFID readers have a wide range of power selections, e.g., the Alien-9900 reader has a maximum power 30.7dBm, which is 30 times larger than the minimum power 15.7dBm. There is no guideline, however, in selecting a suitable power. Therefore, we aim to design an efficient solution to continuous scanning problem for a mobile RFID reader based on experimental study.

Although there have been some experimental studies on reading performance in a stationary RFID system [4, 1, 7], the previous studies have the following limitations. First, previous experiments were usually conducted in a small scale (fewer than 20 tags), which does not capture the complication for a large number of tags. Second, previous work has been focused on reading performance in a close to free space scenario. In reality, path loss, multi-path effect and mutual interference are common and have a big impact to RFID reading process. Third, previous work mainly examined how factors such as distance, coding scheme and frequency, affect reading performance. Very important factors, i.e., the reader's power and tag density, were neglected. Therefore, the previous work does not give a model for RFID reading process in a realistic and large scale setting; in particular, it does not include the power and tag density. Indeed, before we started our work, there was no realistic model which can guide us in designing an efficient tag identification solution in our setting.

We have, thus, conducted comprehensive measurements over a large number of tags in realistic settings by varying various parameters. Surprisingly, we have a few important new findings from the experiments. For example, we have found that the probabilistic backscattering is a ubiquitous phenomenon of the RFID system in realistic settings, i.e., during every query cycle each tag randomly responds with a certain probability, which has an important effect on the reading performance. This observation is contrary to the previous belief that tags respond to a reader with either probability 1 or 0. We have also found it is not wise to blindly increase the reader's power for tag identification, which can degrade the overall performance including the effective throughput and energy consumption. These findings are essential to improving reading performance for a mobile RFID system. Most importantly, we can (1) model the patterns of reading a large number of tags by giving a probabilistic model to capture the major and minor detection region, and (2) model how the reading power and tag density affect the reading performance by proving an empirical mapping.

Based on the effective models, we can then design efficient algorithms, which can dramatically improve the performance, as shown in our real experiments.

We make the following contributions in this paper. (1) We are the first to conduct an extensive experimental study over a relatively large number of tags (up to 240 tags) and a rather high tag density (up to 90 tags per square meter) in realistic settings. To the best of our knowledge, this is the first work to propose a model for investigating how the important parameters including reader's power, moving speed and tag density jointly affect the reading performance. (2) This is also the first work to give a framework of optimizing reading performance based on experimental study. We apply our model to solve the problem of continuous scanning with mobile reader. By carefully adjusting the power and moving speed, we design efficient algorithms to optimize time-efficiency and energy-efficiency. (3) We have a number of novel techniques in making our algorithms practical. For example, our tag density estimation method is extremely simple and accurate. Our algorithm extension to nonuniform tag density is also effective. (4) Being compatible with RFID standard (with no changes to the C1G2 protocols or low-level parameters for commercial RFID readers), our solutions can deliver significant performance gain. Experiment results indicate that, while achieving the same coverage ratio, our practical solutions respectively reduce scanning time by 50% and energy consumption by 83% compared to the prior solutions.

2. RELATED WORK

In RFID systems, a reader needs to receive data from multiple tags. These tags are unable to self-regulate their radio transmissions to avoid collisions. In light of this, a series of slotted ALOHA-based anti-collision protocols [19, 14, 9, 22], as well as tree-based anti-collision protocols [12, 13, 2], are designed to resolve collisions in RFID systems. In order to deal with the collision problems in multi-reader R-FID systems, scheduling protocols for reader activation are explored in [18], [21]. Recently, a number of polling-based protocols [10, 5, 23, 3] are proposed, aiming to collect information from RFID tags in a time/energy efficient approach. In order to estimate the number of tags without collecting tag IDs, a number of protocols are proposed [8, 11, 15, 6, 17] to leverage the information gathered in slotted ALOHA protocol for fast estimation of the number of tags.

In order to verify the impact of the physical layer's unreliability, Buettner et al. [4] examine the performance of the C1G2 RFID system in a realistic setting. Aroor et al. [1] identify the state of the technical capability of passive UHF RFID systems using a simple, empirical, experimental approach. Jeffery et al. [7] conduct experiments in realistic settings and find that within each reader's detection range, a large difference exists in reading performance. In order to efficiently identify RFID tags in mobile settings, Xie et al. propose a probabilistic model to set optimized parameters for mobile tag identification [20]. Sheng et al. develop efficient schemes for continuous scanning operations [16], aiming to utilize the information gathered in the previous scanning operations to reduce the scanning time of the succeeding ones. Being different from the previous work, this paper conducts an extensive experimental study over a large scale RFID deployment, and proposes an effective model to depict the regularities of reading performance in realistic settings.

3. PROBLEM FORMULATION

We consider a typical scenario of continuous scanning in realistic settings, i.e., using a mobile reader to identify a large volume of tags deployed over a wide area. We respectively consider a situation where the tags are continuously placed with a uniform/nonuniform density, we seek to execute continuous scanning over the tags along a certain direction. The performance metrics in our consideration are as follows:

- *Time-efficiency*: considering it is time-consuming to identify a large volume of tags in realistic settings, the overall scanning time should be as small as possible.

- *Energy-efficiency*: considering the mobile reader is conventionally battery powered, e.g., a typical battery for the mobile reader has a capacity of 3200mAh with output voltage 3.7v, if we scan the tags with a maximum radiation power 36 dBm, the mobile reader can execute continuous scanning for only 3 hours, therefore, the overall energy used should be as small as possible.

- *Coverage ratio*: due to various issues like path loss in realistic settings, it is difficult to identify all tags with a high probability for one single scanning cycle, therefore, the coverage ratio, i.e., the ratio of the number of identified tags to the total number of tags, should be guaranteed, while each tag should have a uniform probability to be identified.

In regard to the continuous scanning, we define the scanning time as T, the overall energy used as E, and the coverage ratio as C. Assuming the tag density is ρ and the length of the scanning area is l, then the total number of tags is $n = l \cdot \rho$, we denote the overall tag set as S. We assume that each tag $t_j \in S$ is successfully identified with probability of p_j after the continuous scanning. The reader's antenna is deployed towards the tags with a distance of d. We can adjust the parameters including the reader's power p_w and the moving speed v to improve the reading performance. Therefore, during the continuous scanning, the problem is how to efficiently set the parameters p_w and v such that the following objectives can be achieved:

Time-efficiency:

$$\text{minimize } T \qquad (1)$$

$$\text{subject to}$$

$$E \leq \alpha \qquad \text{energy constraint} \qquad (2)$$

$$Pr[C \geq \theta] \geq \beta \qquad \text{coverage constraint} \qquad (3)$$

$$\forall t_j \in S \quad p_j = p \qquad \text{coverage constraint} \qquad (4)$$

Energy-efficiency:

$$\text{minimize } E \qquad (5)$$

$$\text{subject to}$$

$$T \leq \gamma \qquad \text{time constraint} \qquad (6)$$

$$Pr[C \geq \theta] \geq \beta \qquad \text{coverage constraint} \qquad (7)$$

$$\forall t_j \in S \quad p_j = p \qquad \text{coverage constraint} \qquad (8)$$

According to the above formulation, in regard to the time-efficiency, the objective is to minimize the overall scanning time T while the energy constraint and the coverage constraint should be satisfied. The energy constraint requires the energy used should be no greater than a certain threshold α. In regard to the coverage constraint, due to the random factors in the anti-collision scheme and the communication environment, the coverage ratio C cannot guarantee to be deterministically equal or greater than a threshold θ, hence we use the probabilistic approach to denote the requirement. The probability for the coverage ratio C to be equal or greater than θ should be no less than β. Moreover, there could exist multiple feasible solutions to guarantee the coverage constraint, in some of the solutions the tags are detected with nonuniform probabilities. In fairness, we require that each tag t_j in the set S should be detected with a uniform probability p, i.e., the detection probability p_j should be equal to p. Similarly, in regard to the energy-efficiency, the objective is to minimize the overall energy E, while the time constraint and the coverage constraint should be satisfied. The time constraint requires that the scanning time should be no greater than a certain threshold, γ.

4. DERIVING A MODEL FROM REALISTIC EXPERIMENTS

In order to understand how the reader's power and tag density affect the reading performance, while dealing with issues like the path loss, energy absorption, and mutual interference, we illustrate several original findings from our realistic experiments. In our experiments, we use the Alien-9900 reader and Alien-9611 linear antenna with a directional gain of 6dB. The 3dB beamwidth is 40 degrees. The RFID tags used are Alien 9640 general-purpose tags which support the EPC C1G2 standards. We attach the RFID tags onto the books which are placed in a large bookshelf. Each tag is attached onto a distinct book with a unique ID. The bookshelf is composed of 12 grids with 4 columns and 3 rows, the height and width of each grid are respectively 60cm and 75cm. The RFID reader is statically deployed by facing its antenna towards the book shelf. Note that in order to set an appropriate value for the distance between the reader and the bookshelf, it is difficult to directly derive the optimal distance from geometry according to the beamwidth, due to issues like the multipath effect. Therefore, we vary the distance from 0.5m to 3m and measure the number of effectively identified tags while scanning 160 tags uniformly distributed on the shelf. We find that the reader achieves the maximum coverage when the distance is 1.5m. Thus, we set the distance to 1.5m to guarantee the reading performance. This setting is close to a typical noisy condition, which is distinct from the free space condition, since the issues in the realistic applications like the path loss, multi-path effect and energy absorption all exist. Considering that we deploy a relatively large number of tags (up to 240 tags) and a rather high tag density (up to 90 tags per square meter) in realistic settings, the experimental findings from the high tag density deployment can be highly scalable and generalized to rather large scale settings.

On the whole, it took us over 300 hours to conduct an extensive experimental study of up to 240 tags in realistic settings. In order to sufficiently understand how the parameters separately/jointly affect the actual reading performance, we conduct up to 100 various experiments, carrying out lots of experimental comparisons and analysis on the ob-

tained results. In order to keep the statistical characteristics, all results are the averaged results of 500 independent trials. We finally summarize these original findings in 12 figures. In the following experiments, we vary the tag density, ρ, from 10 to 40 tags/grid, while adjusting the reader's power from 20.7dBm to 30.7dBm for performance evaluation. Unless otherwise specified, by default we fix the reader towards the center of the bookshelf, set the reader's power to 30.7 dBm, and repetitively scan the tags for 50 query cycles.

4.1 Experimental findings

4.1.1 Probabilistic backscattering

During the query cycles, each tag responds to the reader with a certain probability between 0 and 1. We uniformly deploy 96 tags in the bookshelf with 8 tags in each grid. The grids on the left/middle/right side are respectively numbered $(1,2,3)/(4,5,6,7,8,9)/(10,11,12)$. In Fig.1(a), we respectively compute the read ratios of each tag in the 12 grids, i.e., the ratio of successful number of responses to the expected number of responses for each tag, and illustrate them in histogram grouped by grid ID. We note that the tags respond to the reader with various probabilities between 0 and 1, although basically no parameters are changed during the repetitive scanning. This observation is contrary to the popular idea that each tag either responds thoroughly or does not respond at all. We think this is probably due to the randomness in the backscattering factors, like the power scattering, multi-path propagation. Furthermore, we vary the reader's power, p_w, from 22.7 dBm to 30.7dBm and obtain the probability density functions for the read ratio. According to Fig.1(b), we note that as the reader's power varies, the distribution of the read ratio also varies. The above observation further implies that, due to the probabilistic backscattering, multiple query cycles are essential to successfully identify a typical tag in the tag set, which may cause massive duplicated readings over other tags in the scanning area.

4.1.2 Major vs minor detection region

Within each reader's detection range, there are two distinct regions: the major detection region where the tags can be identified with high probability, and the minor detection region where the tags can be identified with low probability. We uniformly deploy the tags in a row with 4 grids in the bookshelf, where the tag IDs are sequentially numbered from left to right. The reader's power is set to 30.7dBm. Fig.1(c)-Fig.1(f) show the histogram of each tag's read ratio in the order of tag ID, while varying the tag density, i.e., the number of tags per grid. In order to see the two distinct regions, we use red window to depict the boundary of the major detection region. We observe that within each reader's detection range, the major detection region is the area directly in front of the reader, giving high detection probability, and the minor detection region extends from the end of the major detection region to the edge of the detection range, where the read ratio drops off to zero at the end of the detection range. As the tag density increases, the major detection region gradually shrinks. Note that due to issues like the multipath effect and energy absorption in realistic settings, it is difficult to directly derive the detail parameters of the major/minor detection region from geometrical

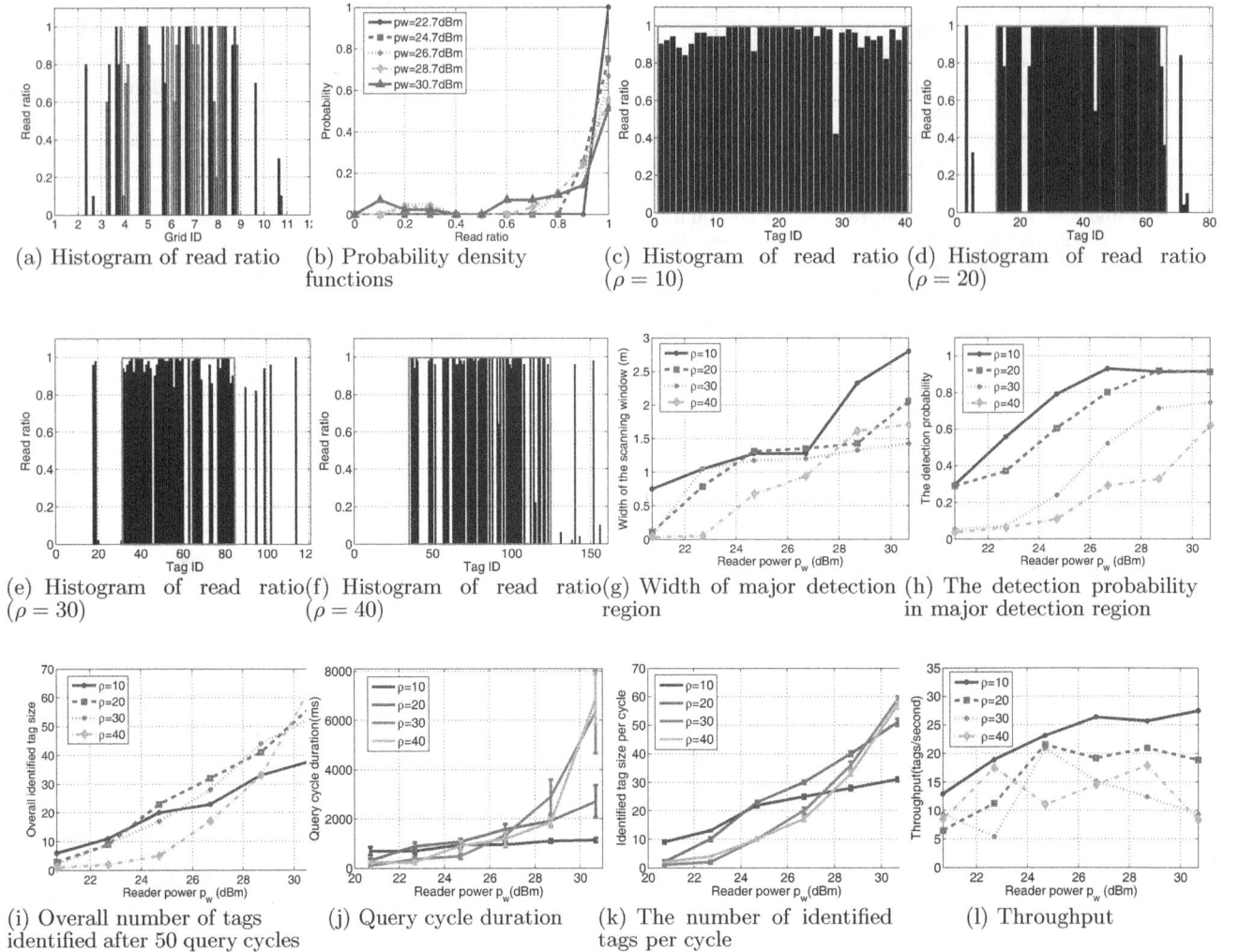

(a) Histogram of read ratio (b) Probability density functions (c) Histogram of read ratio ($\rho = 10$) (d) Histogram of read ratio ($\rho = 20$)

(e) Histogram of read ratio ($\rho = 30$) (f) Histogram of read ratio ($\rho = 40$) (g) Width of major detection region (h) The detection probability in major detection region

(i) Overall number of tags identified after 50 query cycles (j) Query cycle duration (k) The number of identified tags per cycle (l) Throughput

Figure 1: Observations from the realistic experiments

principles, since they are also related to other parameters like the tag density.

4.1.3 Marginal decreasing effect

As the reader's power is increasing, the exact read efficiency including the scanning range, the detection probability, as well as the number of identified tags, is not increasing equally with the power. In Fig.1(g)-1(i), we respectively measure the width of the major detection region, the average detection probability (i.e., read ratio) in the major detection region, as well as the overall number of identified tags, while varying the reader's power from 20.7dBm to 30.7dBm. All of the above three variables are increasing while the reader's power increases. However, as the power is increased by 2dBm (i.e, 1.58 times in watt), they mainly increase with a much smaller speed on average. This observation implies that the read efficiency cannot be sufficiently enhanced by purely increasing the reader's power.

4.1.4 Query cycle duration vs the number of identified tags per cycle

As the reader's power increases, the query cycle duration does not increase linearly with the number of identified tags

per cycle, causing the variation of the throughput. According to the theoretical analysis in the ideal situation, if the frame size is optimally selected, the expected number of slots as well as the query cycle duration should be linearly increasing with the number of identified tags per cycle. However, in realistic settings that doesn't follow at all. Fig.1(j) and Fig.1(k) respectively show the value of cycle duration τ_c and the number of identified tags per cycle n_c while varying the reader's power. Note that the standard deviation of τ_c is much larger than n_c, which is mainly due to the randomness in the anti-collision scheme. As the reader's power increases, the values of τ_c and n_c are both increasing, however, at different rates. Therefore, the ratio of n_c to τ_c, i.e., the throughput, is also varying.

Fig.1(l) shows the throughput variation with 4 different tag densities. We find that in all cases the throughput achieves the peak value when the reader's power is set to an appropriate value between the minimum and maximum power. The reason is as follows: When the reader's power is set to a small value, the number of activated tags is small, then due to the fairly large inter-query cycle overhead, the throughput is fairly small. As the reader's power increases, more tags are involved in the query cycle, the inter-query cycle overhead is sufficiently amortized, thus the throughput is gradually increased. When the reader's power increases to a fairly large value, the number of collisions in the query cycle is greatly increased, resulting in a large value for the cycle duration, thus the throughput is further decreased. This observation implies that it is neither time-efficient nor energy-efficient to blindly increase the reader's power, an optimal value for the reader's power should be determined.

The above experiment results and observations are obtained from the static situation where the reader is statically deployed. In the mobile situation where the reader is continuously moving, since the moving speed cannot be too large due to the large number of tags to be identified, all the above properties should be preserved. In order to verify this statement, we conduct experiments in mobile situations while varying the moving speed from 0.3m/s to 3m/s. We find that all the obtained results, including the width of major detection region, the detection probability, and the query cycle duration are very close to the static situation. Besides, these experiment results are currently obtained from the settings constructed by the Alien reader and antennas, since the Alien reader and antennas are designed and manufactured according to industrial standard, these results can be applicable to other kinds of commercial readers conforming to the standards. Therefore, it is feasible to apply these parameters to the continuous scanning algorithm.

4.2 Model

Based on the above findings, it is essential to build a model to effectively depict the regularities in reading performance. We first propose a model for probabilistic backscattering, and then a model of the effective scanning window to evaluate the reading performance over multiple tags.

4.2.1 Probabilistic backscattering

Suppose an arbitrary tag is separated from the reader at a distance of d. In order for the tag to successfully backscatter the ID message, the reader needs to send a continuous wave to activate the tag. As the tag has a sensitivity threshold t, which is the minimum power required to activate the tag, the incident power to the tag's antenna should be larger than t. It is known that the power budget of conventional RFID systems is *forward-link limited*, which implies that well-designed passive RFID systems are always limited by the tag's sensitivity. Therefore, as long as the reader's power p_w is large enough to activate the tag, the reader is able to resolve the backscattered signal from the tag. We have conducted experiments to evaluate the threshold t. We find that the value of t basically remains unchanged among a certain type of tags.

In the reader's read zone, i.e, the region in which the incident power exceeds the threshold t, it is found that the range is longest along the center and falls off towards the edges. In regard to a plane at a fixed distance from the reader, the incident power varies from the center towards the edges. Besides, the values of incident power has variances since the continuous wave issued from the reader has fluctuations in terms of power. Therefore, assume the reader's power is p_w, in regard to a two dimensional plane at a distance of d from the reader, we respectively use $f_{p_w,d}(x,y)$ and $g_{p_w,d}(x,y)$ to denote the average value and the variance of the incident power in the coordinate (x,y). In the settings where the tags are deployed in a row, we respectively simplify them to

$f_{p_w,d}(x)$ and $g_{p_w,d}(x)$. Fig.2 shows a schematic diagram of the average value and variance of the incident power $p'_w(x)$ in the one-dimensional space. Note that conventionally the incident power achieves the maximum value in the center of the read zone, and gradually decreases towards the edges. Meanwhile, in regard to multiple tags deployed in the plane, the incident power is also affected by tag density and multipath effect.

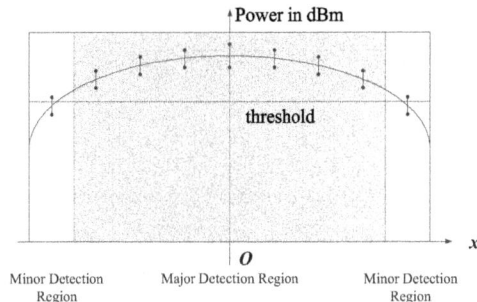

Figure 2: The average value and variance of the incident power in the one-dimensional space: a schematic diagram

In regard to an arbitrary tag in the row, the tag can be successfully identified if and only if the incident power is above the tag's sensitivity threshold t. Due to the fluctuation of the incident power, the tag is successfully identified with some probability, i.e., $Pr[p'_w(x) \geq t]$. In regard to a position x in the effective scanning region, note that once the average value is relatively larger than the threshold t, as the variance is usually relatively small, the detection probability $Pr[p'_w(x) \geq t]$ will be close to 1; similarly, once the average value is relatively smaller than the threshold t, the detection probability $Pr[p'_w(x) \geq t]$ will be close to 0. This property divides the scanning region into two distinct regions, i.e., the major detection region and the minor detection region.

4.2.2 Effective scanning window over multiple tags

As we have observed, the reader's effective scanning region can be divided into a major detection region as well as a minor detection region. In the major detection region, most tags can be detected with a probability close to 100%. As the tag density increases, the diffused power cannot guarantee to activate all tags in the major detection region, each tag has a probability to be detected in a random approach. Therefore, we can use the average detection probability to depict the reading performance in this region. The minor detection region is extending from the end of the major detection region to the edge of the effective range, with the detection probability quickly drops off to 0. Based on the above analysis, we use a trapezoidal curve to denote the expected detection probability of tags in the scanning region, as illustrated in Fig.3. In fact, due to the narrow width of the minor detection region, the average probability for a tag to be detected in this region can be negligible. Therefore, in consideration of the actual reading performance, we only need to focus on the major detection region. In the rest of this paper, we use the term *effective scanning window* to denote the major detection region. We use w and p' to denote the width and the average detection probability of the effective scanning window, respectively.

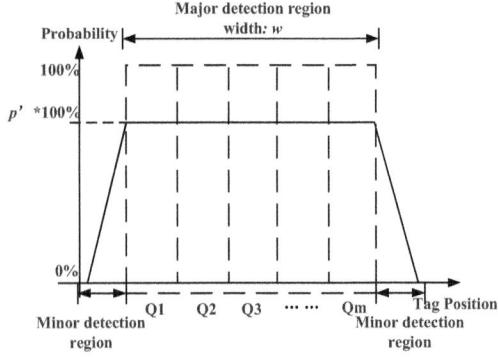

Figure 3: The model of effective scanning window

During continuous scanning, the effective scanning window is continuously moving forward with the mobile reader. Note that there exist overlapping areas between the contiguous scanning windows. During the continuous scanning, each tag gradually enters a minor detection region, then an effective scanning window, finally exits from a minor detection region. While within the effective scanning window, each tag has a probability to be detected for each query cycle. Therefore, in order to guarantee the coverage constraint, multiple query cycles should be issued over each tag while it is within the effective scanning window. Assume that the tags are uniformly deployed along the scanning area, then the number of tags within the effective scanning window is always constant. This infers that the number of tags involved in a query cycle mostly remains unchanged. If the mobile reader is set to a constant power and a constant moving speed, then, after multiple query cycles, each tag has a uniform probability to be detected. This conforms to the requirement in the coverage constraint.

Suppose an arbitrary tag is expected to be queried for m cycles while it is within the effective scanning window, we denote the detection probability in the m query cycles as $p_i (i = 1...m)$. Then, the probability for an arbitrary tag to be identified at least once is as follows:

$$p = 1 - \prod_{i=1}^{m} (1 - p_i) \tag{9}$$

As the reader's power is set to a constant value, due to the uniform tag density, the probability $p_i (i = 1...m)$ in the m query cycles should be uniform. If we use p' to denote the uniform detection probability, then Eq.(9) is further simplified as follows:

$$p = 1 - (1 - p')^m \tag{10}$$

In particular, the value of m is equal to $\frac{\tau_w}{\tau_c}$, here τ_w is the duration in the effective scanning window, and τ_c is the average duration of a query cycle. Moreover, τ_w is equal to the ratio of the window width w to the moving speed v, i.e., $\frac{w}{v}$, hence $m = \frac{w}{v \cdot \tau_c}$. Therefore, in order to increase the detection probability p for an arbitrary tag, it is essential to (1) increase the number of query cycles m as much as possible; (2) increase the detection probability p' as much as possible.

In Fig.1(g), Fig.1(h) and Fig.1(j), we illustrate the value of w, p' and τ_c with various power, p_w. In regard to a fixed tag density, we note that as the value of p_w increases, the value

of p', w and τ_c are all monotonically increasing. Moreover, since the value of w increases much more slowly than τ_c, the value of $m = \frac{w}{v \cdot \tau_c}$ is monotonically decreasing with the value of p_w. Therefore, we reach the following conclusion: as the moving speed v decreases, the value of m is monotonically increasing, while the value of p' remains unchanged. As the reader's power p_w increases, the value of p' is monotonically increasing, while the value of m is monotonically decreasing. Thus the value of p_w should be appropriately selected to optimize the performance.

In regard to the coverage constraints in Eq.(3) and Eq.(4), we use the parameter p to denote the probability that a tag is successfully identified after the continuous scanning. Then, according to the binomial distribution, after the overall scanning procedure, the probability for the reader to identify at least $\theta \cdot 100\%$ percent of the overall tags (i.e., $Pr[C \geq \theta]$), is computed as follows:

$$Pr[C \geq \theta] = \sum_{i=\lceil \theta \cdot n \rceil}^{n} C_n^i \cdot p^i \cdot (1-p)^{n-i} \tag{11}$$

Then it is essential to compute the solution of p to guarantee $Pr[C \geq \theta] \geq \beta$. As $\sum_{i=\lceil \theta \cdot n \rceil}^{n} C_n^i \cdot p^i \cdot (1-p)^{n-i} = \beta$ is an equation of higher degree, it is rather difficult to directly solve the variable p from the above equation.

In fact, the constraint $Pr[C \geq \theta] \geq \beta$ is equivalent to $Pr[C \leq \theta] \leq 1 - \beta$. In particular, according to *Hoeffding's inequality*, for $\theta \leq p$, it yields the bound

$$Pr[C \leq \theta] \leq \exp(-2 \frac{(n \cdot p - n \cdot \theta)^2}{n}).$$

In order for $Pr[C \leq \theta] \leq 1 - \beta$, it is essential to guarantee

$$\exp(-2 \frac{(n \cdot p - n \cdot \theta)^2}{n}) \leq 1 - \beta.$$

The solution of p can be directly solved from the above inequality, that is $p \geq \theta^*$, here $\theta^* = \theta + \sqrt{\frac{\ln(1-\beta)}{-2n}}$. This shows that, as long as the detection probability p is no less than θ^* for any tag, the coverage constraint is guaranteed.

5. CONTINUOUS SCANNING WITH MOBILE READER

5.1 Baseline solution

For both the uniform and nonuniform tag distribution, in order to effectively identify all the tags with the mobile reader, conventionally the reader's power is set to maximum and the moving speed is set to a constant value which is small enough. This baseline solution is very straightforward, which however, is neither time-efficient nor energy-efficient since excessive power is used up and the moving speed is slowed down. Besides, a number of tags are interrogated multiple times during continuous scanning, which is unnecessary as each tag only needs to be identified once.

5.2 Solution for uniform tag density

5.2.1 Solution

Without loss of generality, we first propose an optimized solution for the situation with uniform tag density. Considering the objective as well as the energy/time constraint, we need to figure out the optimized value of p_w and v such that the objective is achieved while the coverage constraints are satisfied.

In regard to the coverage constraint, since we need to guarantee $p = 1 - (1 - p')^m \geq \theta^*$, i.e., $1 - (1 - p')^{\frac{w}{v \cdot \tau_c}} \geq \theta^*$, it is equivalent to ensure $v \leq \frac{1}{|\ln(1 - \theta^*)|} \cdot \frac{w \cdot |\ln(1 - p')|}{\tau_c}$. As the value of w, p' and τ_c all depends on the value of p_w, let $w(p_w)$, $p'(p_w)$ and $\tau_c(p_w)$ respectively denote the mapping function from p_w to w, p' and τ_c, then

$$v^* = \frac{1}{|\ln(1 - \theta^*)|} \cdot \frac{w(p_w) \cdot |\ln(1 - p'(p_w))|}{\tau_c(p_w)}, \quad (12)$$

then, v^* is the maximum allowable moving speed to satisfy the coverage constraint.

Since the length of the scanning area is l, the overall scanning time $T = \frac{l}{v}$, and the overall used energy $E = T \cdot p_w = \frac{p_w \cdot l}{v}$. Therefore, considering the time-efficiency, in order to minimize T, it is equivalent to maximize v. Then, according to Eq.(12), it is essential to maximize $\frac{w \cdot |\ln(1 - p')|}{\tau_c}$. It is known that as the value of p_w increases, the value of $w \cdot |\ln(1 - p')|$ and τ_c are both monotonically increasing, thus an optimized value of p_w should be selected to minimize T. Considering the energy constraint $E \leq \alpha$, the optimal value p_w^* can be computed according to the following formulation:

$$\text{maximize } y_T = \frac{|\ln(1 - p'(p_w))| \cdot w(p_w)}{\tau_c(p_w)} \quad (13)$$

subject to

$$\frac{|\ln(1 - p'(p_w))| \cdot w(p_w)}{p_w \cdot \tau_c(p_w)} \geq \frac{l \cdot |\ln(1 - \theta^*)|}{\alpha} \quad (14)$$

Considering the energy-efficiency, in order to minimize E, it is equivalent to minimize $\frac{p_w}{v}$, then according to Eq.(12), it is essential to maximize $\frac{|\ln(1-p')| \cdot w}{p_w \cdot \tau_c}$. Therefore, considering the time constraint $T \leq \gamma$, the optimal value p_w^* can be computed according to the following formulation:

$$\text{maximize } y_E = \frac{|\ln(1 - p'(p_w))| \cdot w(p_w)}{p_w \cdot \tau_c(p_w)} \quad (15)$$

subject to

$$\frac{|\ln(1 - p'(p_w))| \cdot w(p_w)}{\tau_c(p_w)} \geq \frac{l \cdot |\ln(1 - \theta^*)|}{\gamma} \quad (16)$$

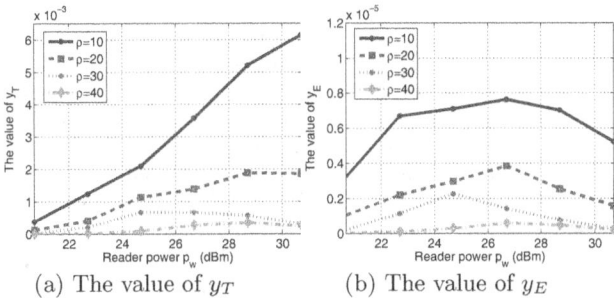

(a) The value of y_T (b) The value of y_E

Figure 4: Compute the value of y_T and y_E with various values of p_w

In regard to a certain tag density ρ, by enumerating the candidate values of the power p_w, we can compute the value of y_T and y_E. Fig.4(a) and Fig.4(b) respectively illustrate the value of y_T and y_E while varying the reader's power p_w. We note that there exist a maximum value of y_T and y_E for each tag density. In regard to a specified tag density ρ, while satisfying the time/energy constraint, we can use the power p_w^* for the maximum value of y_T or y_E as the optimal parameter and compute the corresponding moving speed v^* according to Eq.(12). In this way, the optimal solution (p_w^*, v^*) for time/energy efficiency can be generated. Therefore, in regard to various tag densities ρ, we can collect the performance parameters like w, p' and τ_c in advance, pre-compute the optimal pairs of (p_w^*, v^*), and store them in a table. When dealing with an arbitrary tag density, we can directly use the optimal pair of (p_w^*, v^*) to achieve the time/energy efficiency.

5.2.2 Estimate the tag density

According to the measured data in realistic settings, it is known that the tag density ρ has an important effect on the performance metrics. In situations where the tag density cannot be pre-fetched or the tag density varies along the forwarding direction, it is essential to accurately estimate the current tag density, such that the optimized parameters (p_w^*, v^*) can be effectively computed. Due to the probabilistic backscattering property, it is difficult to directly estimate the tag density according to the observed number of empty/singleton/collision slots [8, 11]. Furthermore, current commercial RFID readers do not expose these low-level data to upper-layer applications. Therefore, it is essential to estimate the tag density in a more practical way.

According to Fig.1(k), we note that if the reader's power p_w is set to a certain value, the number of identified tags per cycle n_c is varying as the tag density ρ varies, with a very small standard deviation. Table 1 shows further details for the average values of n_c. These are obtained through 50 repetitive experiments with various values of ρ and p_w. Due to the small variance of n_c, there is a very stable pattern between n_c and ρ that varies with p_w. Therefore, given a reference tag density ρ_i, we can depict the values of n_c with various powers as a vector $V_i = \{n_{i,1}, n_{i,2}, ..., n_{i,s}\}$, here s is the number of power levels. Then, in regard to an unknown tag density ρ, assume the corresponding vector is $V = \{n_1, n_2, ..., n_s\}$, we can estimate the value of ρ by comparing V with the vectors of reference tag densities. Therefore, we propose an algorithm to estimate the tag density, by leveraging the k-nearest neighbor method, as shown in Algorithm 1.

$p_w =$	20.7	22.7	24.7	26.7	28.7	30.7
$\rho = 10$	9	13	22	25	28	31
$\rho = 20$	2	10	23	30	40	51
$\rho = 30$	1	2	10	20	36	59
$\rho = 40$	2	4	10	17	33	57

Table 1: The number of identified tags per cycle

In Algorithm 1, the similarity $sim(V, V_i)$ is actually calculated by using the cosine value of the angle between the two vectors, hence the value of similarity is between 0 and 1. We use the k-nearest neighbor method to estimate the tag density based on k-nearest reference tag densities. The estimated tag density ρ is computed using an inverse distance weighted average with the k-nearest multivariate neighbors, here the distance is defined as $1 - sim(V, V_i)$. Since the value of n_c has a rather small variance, the accuracy of the estimated tag density can be guaranteed if the number of samplings m is fairly large. In the algorithm, the mobile

Algorithm 1 Tag density estimation algorithm
1: INPUT: $V_i = \{n_{i,1}, n_{i,2}, ..., n_{i,s}\}$: the vectors for reference tag densities $\rho_i (i = 1...h)$.
2: PROCEDURE
3: Set the reader's power to various levels $p_{w,1}, ..., p_{w,s}$. In regard to each power level $p_{w,j} (j \in [1, s])$, issue m query cycles to get the average value of the number of identified tags per cycle as n_j. Assemble them as a vector $V = \{n_1, n_2, ..., n_s\}$.
4: **for** $i \in [1, h]$ **do**
5:     Compute the similarity between V and V_i as follows:

$$sim(V, V_i) = \frac{V \cdot V_i}{|V| \cdot |V_i|} = \frac{\sum_{j=1}^{s} n_{i,j} \cdot n_j}{\sqrt{\sum_{j=1}^{s} n_j^2} \cdot \sqrt{\sum_{j=1}^{s} n_{i,j}^2}}.$$

6: **end for**
7: Sort the value of $sim(V, V_i)$ in decreasing order. Find the first k items of ρ_i according to $sim(V, V_i)$, say $\rho_1', ..., \rho_k'$.
8: Compute $\rho = \sum_{i=1}^{k} \rho_i' \cdot w_i$, here

$$w_i = \frac{1/(1 - sim(V, V_i) + \epsilon)}{\sum_{i=1}^{k} 1/(1 - sim(V, V_i) + \epsilon)} (\epsilon > 0).$$

9: OUTPUT: The estimated tag density ρ.

reader is required to obtain the value of n_c in multiple power levels, which increases overhead in both time and power usage. In regard to the uniform tag density, since the tag density is only necessary to estimate once, this overhead can be effectively amortized by the following multiple query cycles. In regard to the nonuniform tag density, the tag density needs to be continuously estimated. The algorithms for fast tag size estimation [8, 11, 15] can be used to reduce the overhead. In regard to the selection of k, k should be set to neither a too small value nor a too large value for accurate estimation, the optimal value depends on the exact deployment, conventionally k should be set to 2 or 3 for performance consideration.

This algorithm is very practical and fully compatible with the EPC C1G2 standard, and does not require to obtain any low-level parameters for commercial RFID readers.

5.3 Extensions for nonuniform tag density

In the above solution, we assume that the tag density is always uniform. In some applications, the tags are not uniformly deployed. While the mobile reader is continuously scanning the tags, the tag density may always change along the forward direction. In this situation, the constant moving speed and power for the mobile reader is no longer suitable to improve performance. Note that in conventional situations, the tag density changes slowly along the forward direction. Therefore, in regard to each query cycle, we can assume the tag density within the effective scanning window is close to uniform, since the cycle duration is usually small. Therefore, we can reduce the situation with nonuniform tag density into multiple snapshots with fairly uniform tag density. In each query cycle, the mobile reader can be reset with the optimal values of p_w and v according to the nearby tag density. In this way, the reading performance can be effectively improved by dynamically adjusting the reader's power p_w and the moving speed v.

6. PERFORMANCE EVALUATION

We evaluate the performance in realistic settings. The experiment settings are the same as the realistic settings in Section IV, except that in this experiment we use the Alien-9900 reader as the mobile reader to move forward for continuous scanning.

6.1 Evaluate the performance in unform tag density

In order to evaluate the performance in unform tag density, we deploy the tags in a row with 4 grids in the shelf, while varying the tag densities from 10 tags/grid to 40 tags/grid, the length of the scanning area is 3m.

6.1.1 The accuracy of tag density estimation

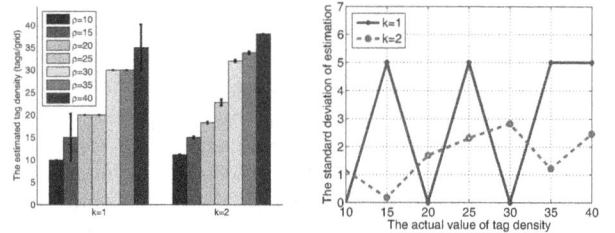

(a) The average value and (b) The estimation error of standard deviation for esti- tag density estimation mated tag density

Figure 5: Evaluate the accuracy of tag density estimation

In order to perform tag density estimation, we utilize Table 1 as the reference set. Then, while varying the tag density from 10 to 40 tags/grid in increments of 5 tags/grid, we estimate the tag densities based on the number of identified tags per cycle, i.e., n_c. For each power level, we collect 10 samples of n_c for estimation. As the k-nearest neighbor method is used in the estimation, we respectively set k to 1 and 2 for the estimation. Fig.5(a) illustrates the estimated values as well as the standard deviations for various tag densities. We find that both the 1-nearest neighbor method (1NN) and the 2-nearest neighbor method (2NN) achieve fairly good performance in terms of estimation accuracy. As 1NN can only select one tag density with the nearest property, the estimate accuracy declines when the exact tag density is between two reference tag densities in Table 1. In comparison, 2NN achieves a much higher estimation accuracy since it can effectively select two closest reference tag densities to estimate the tag densities. Fig.5(b) further compares the estimation error of the two methods. We use the standard deviation as the metric for the estimation error. We note that the standard deviation for 1NN is fluctuating between 0 and 5 tags/grid while the standard deviation for 2NN is relatively stable below 3 tags/grid. This infers that in conventional cases 2NN achieves a much better performance than 1NN in terms of estimation accuracy.

6.1.2 The coverage ratio

According to the analysis in Section IV, in regard to the coverage constraint $Pr[C \geq \theta] \geq \beta$, we need to guarantee $p \geq \theta^*$, here $\theta^* = \theta + \sqrt{\frac{\ln(1-\beta)}{-2n}}$. Without loss of generality, we set θ^* to 90% in our experiments. This means, on average 90% of the tags should be identified after continuous scanning. According to Fig.4, the optimal values of the

(a) The coverage of various powers (time-efficiency) (b) The coverage of various moving speed (time-efficiency) (c) The coverage of various powers (energy-efficiency) (d) Coverage of various moving speed (energy-efficiency)

(e) The energy consumption with various power and moving speed (f) Coverage ratio for time-efficient, energy-efficient and baseline solution (g) Scanning time for time-efficient, energy-efficient and baseline solution (h) Energy consumption for time-efficient, energy-efficient and baseline

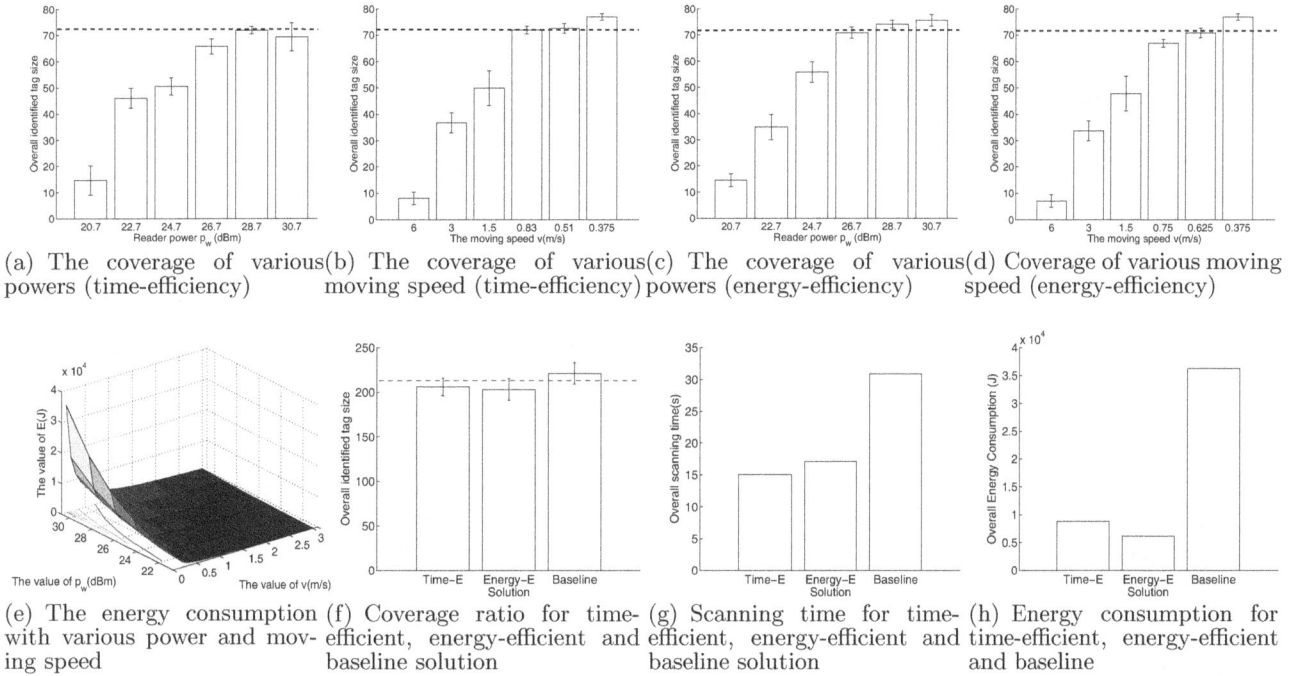

Figure 6: Experiment results in realistic settings

power p_w^* and moving speed v^* for the mobile reader are computed in Table 2. The corresponding scanning time and energy are also illustrated.

TE	$\rho = 10$	$\rho = 20$	$\rho = 30$	$\rho = 40$
p_w^*	30.7dBm	28.7dBm	26.7dBm	28.7dBm
v^*	2.65m/s	0.83m/s	0.29m/s	0.15m/s
T^*	1.13s	3.6s	10.3s	20s
E	1326.62J	2667.6J	4820.4J	14820J
EE	$\rho = 10$	$\rho = 20$	$\rho = 30$	$\rho = 40$
p_w^*	26.7dBm	26.7dBm	24.7dBm	26.7dBm
v^*	1.57m/s	0.625m/s	0.29m/s	0.12m/s
T	1.91s	4.8s	10.3s	25s
E^*	893.88J	2243.75J	3038.5J	11700J

Table 2: Optimal parameters for time-efficiency (TE) and energy-efficiency (EE)

We vary the tag densities ρ from 10 tags/grid to 40 tags/grid and verify the coverage ratio with various configurations for the parameters p_w and v. Due to the limitation of space, we only illustrate the results for $\rho = 20$ tags/grid, the other results are very similar to them. We run each experiment 20 times to obtain the average value and the standard deviation of the number of identified tags. In regard to the time-efficiency, Fig.6(a) shows the coverage of various power levels, while fixing the moving speed to the optimal value $v^* = 0.83$ m/s, the dashed line denotes the threshold for 90% coverage. We note that as the power increases, the number of identified tags gradually increases to the threshold and then further decreases. The threshold is only achieved while the power is set to the optimal value 28.7dBm. Fig.6(b) shows the coverage of various moving speed, while fixing the power to optimal value $p_w^* = 28.7$ dBm. We note that as the moving speed decreases, the number of identified tags

gradually increases to cross the threshold for 90% coverage, the coverage is right at the threshold when the speed is set to the optimal value 0.83 m/s. The above results show that, while guaranteeing the coverage ratio, the optimal settings (p_w^*, v^*) can achieve much better time-efficiency than other settings.

Similarly, in regard to the energy-efficiency, Fig.6(c) and Fig.6(d) respectively show 1) the coverage of various power levels while fixing the moving speed to the optimal value $v^* = 0.625$ m/s; 2)the coverage of various moving speed, while fixing the power to optimal value $p_w^* = 26.7$ dBm. It is found that, among various (p_w, v) parameter pairs, the optimal solution (p_w^*, v^*) achieves the coverage right at the required threshold. These results infer that, while guaranteeing the coverage ratio, the optimal settings (p_w^*, v^*) can achieve much better energy-efficiency than other settings.

6.1.3 The time/energy-efficiency

In regard to the time-efficiency, since the length of the scanning area is 3m, the scanning time $T = \frac{3m}{v}$. Therefore, while guaranteeing the coverage ratio, a high speed v is preferred. Note that as the value of v increases from 0 to 1m/s, the scanning time rapidly decreases from $+\infty$ to 3s; as the value of v further increases, the scanning time decreases rather slowly from 3s to 0, while the coverage ratio can be decreased rapidly. Considering the marginal decreasing effect, the moving speed should be appropriately selected for cost-effective consideration. In regard to the energy efficiency, it is known that the overall energy consumption E is proportional to the power p_w and inverse to the moving speed v. Fig.6(e) illustrates the value of E while varying the value of p_w and v. Moreover, in order to guarantee the coverage ratio, the values of p_w and v are mutually restricted. Recall that Fig.4(a) actually illustrates the maximum allowable moving speed $v = \frac{yT}{|\ln(1-\theta^*)|}$ for reader's power p_w

19

with various tag densities. Integrating with both figures, we can effectively derive the minimum energy to satisfy the coverage constraint.

6.2 Evaluate the performance in non-unform tag density

In order to evaluate the performance in non-uniform tag density, we nonuniformly deploy 240 tags in a row with 12 grids, while varying the tag density from $\rho = 10$ to $\rho = 30$ tags/grid, the length of the scanning area is 9m. We set θ^* to 90% in our experiments, which infers that, on average 90% of the tags should be identified after the continuous scanning. We compare our time-efficient solution (*Time-E*) and energy-efficient solution (*Energy-E*) with the baseline solution (*Baseline*). In regard to the baseline solution, we set the reader's power to its maximum value, i.e., 30.7dBm, which is also the standard configuration for conventional commodity readers. The moving speed is set according to the optimal value $v^* = 0.29$m/s when $\rho = 30$ to tackle the worst case. In Fig.6(f), we evaluate the coverage ratio for the three solutions. Both *Time-E* and *Energy-E* achieve the coverage which is very close to the 90% coverage ratio. *Baseline*'s coverage is slightly more than this threshold, since *Baseline* uses the maximum power and lowest speed. In Fig.6(g) and Fig.6(h), we respectively evaluate the overall scanning time and energy consumption for continuous scanning. Both *Time-E* and *Energy-E* achieves much better performance than *Baseline* in regard to the two metrics. Here, *Time-E* saves more than 50% of the scanning time compared with *Baseline*, and *Energy-E* saves more than 83% of the energy consumption compared with *Baseline*.

7. CONCLUSION

This paper considers how to efficiently identify RFID tags with a mobile reader, from the experimental point of view. We conduct measurements over a large volume of tags in realistic settings, and propose efficient algorithms for continuous scanning. Our experiments show that our algorithms can reduce the total scanning time by 50% and the total energy consumption by 83% compared to the prior solutions. We believe this work gives much insight and inspiration for devising optimized algorithms for reading a large number of tags in realistic settings.

Acknowledgments

This work is supported in part by National Basic Research Program of China (973) under Grant No. 2009CB320705; National Natural Science Foundation of China under Grant No. 61100196, 61073028, 61021062; JiangSu Natural Science Foundation under Grant No. BK2011559. Qun Li is supported in part by US National Science Foundation grants CNS-1117412 and CAREER Award CNS-0747108.

8. REFERENCES

[1] S. R. Aroor and D. D. Deavours. Evaluation of the state of passive uhf rfid: An experimental approach. *IEEE Systems Journal*, 1(2):168–176, 2007.

[2] D. Benedetti, G. Maselli, and C. Petrioli. Fast identification of mobile rfid tags. In *Proc. of IEEE MASS*, 2012.

[3] K. Bu, B. Xiao, Q. Xiao, and S. Chen. Efficient misplaced-tag pinpointing in large rfid systems. *IEEE Transactions on Parallel and Distributed Systems*, 23(11):2094–2106, 2012.

[4] M. Buettner and D. Wetherall. An empirical study of uhf rfid performance. In *Proc. of MobiCom*, 2008.

[5] S. Chen, M. Zhang, and B. Xiao. Efficient information collection protocols for sensor-augmented rfid networks. In *Proc. of INFOCOM*, 2011.

[6] H. Han, B. Sheng, C. C. Tan, Q. Li, W. Mao, and S. Lu. Counting rfid tags efficiently and anonymously. In *Proc. of INFOCOM*, 2010.

[7] S. R. Jeffery, M. Garofalakis, and M. J. Franklin. Adaptive cleaning for rfid data streams. In *Proc. of VLDB*, 2006.

[8] M. Kodialam and T. Nandagopal. Fast and reliable estimation schemes in rfid systems. In *Proc. of ACM MobiCom*, 2006.

[9] S. Lee, S. Joo, and C. Lee. An enhanced dynamic framed slotted aloha algorithm for rfid tag identification. In *Proc. of MobiQuitous*, 2005.

[10] T. Li, S. Chen, and Y. Ling. Identifying the missing tags in a large rfid system. In *Proc. of ACM Mobihoc*, 2010.

[11] T. Li, S. Wu, S. Chen, and M. Yang. Energy efficient algorithms for the rfid estimation problem. In *Proc. of INFOCOM*, 2010.

[12] J. Myung and W. Lee. Adaptive splitting protocols for rfid tag collision arbitration. In *Proc. of ACM MobiHoc*, 2006.

[13] T. L. Porta, G. Maselli, and C. Petrioli. Anti-collision protocols for single-reader rfid systems: temporal analysis and optimization. *IEEE Transactions on Mobile Computing*, 10(2):267–279, 2011.

[14] C. Qian, Y. Liu, H.-L. Ngan, and L. M. Ni. Asap: Scalable identification and counting for contactless rfid systems. In *Proc. of ICDCS*, 2010.

[15] C. Qian, H.-L. Ngan, and Y. Liu. Cardinality estimation for large-scale rfid systems. In *Proc. of PerCom*, 2008.

[16] B. Sheng, Q. Li, and W. Mao. Efficient continuous scanning in rfid systems. In *Proc. of INFOCOM*, 2010.

[17] B. Sheng, C. C. Tan, Q. Li, and W. Mao. Finding popular categoried for RFID tags. In *Proc. of ACM Mobihoc*, 2008.

[18] S. Tang, J. Yuan, X. Y. Li, G. Chen, Y. Liu, and J. Zhao. Raspberry: A stable reader activation scheduling protocol in multi-reader rfid systems. In *Proc. of ICNP*, 2009.

[19] H. Vogt. Efficient object identification with passive rfid tags. In *Proc. of Pervasive*, 2002.

[20] L. Xie, B. Sheng, C. C. Tan, H. Han, Q. Li, and D. Chen. Efficient tag identification in mobile rfid systems. In *Proc. of INFOCOM*, 2010.

[21] L. Yang, J. Han, Y. Qi, C. Wang, T. Gu, and Y. Liu. Season: Shelving interference and joint identification in large-scale rfid systems. In *Proc. of INFOCOM*, 2011.

[22] Y. Yin, L. Xie, S. Lu, and D. Chen. Efficient protocols for rule checking in rfid systems. In *Proc. of ICCCN*, 2013.

[23] Y. Zheng and M. Li. Fast tag searching protocol for large-scale rfid systems. In *Proc. of ICNP*, 2011.

Informative Counting: Fine-grained Batch Authentication for Large-Scale RFID Systems

Wei Gong, Kebin Liu, Xin Miao, Qiang Ma, Zheng Yang, Yunhao Liu
School of Software, TNLIST, Tsinghua University
{gongwei, kebin, miao, maq, yang, yunhao}@greenorbs.com

ABSTRACT

Many algorithms have been introduced to deterministically authenticate Radio Frequency Identification (RFID) tags, while little work has been done to address the scalability issue in batch authentications. Deterministic approaches verify tags one by one, and the communication overhead and time cost grow linearly with increasing size of tags. We design a fine-grained batch authentication scheme, INformative Counting (INC), which achieves sublinear authentication time and communication cost in batch verifications. INC also provides authentication results with accurate estimates of the number of counterfeiting tags and genuine tags, while previous batch authentication methods merely provide 0/1 results indicating the existence of counterfeits. We conduct detailed theoretical analysis and extensive experiments to examine this design and the results show that INC significantly outperforms previous work in terms of effectiveness and efficiency.

Categories and Subject Descriptors

H.4 [**Information Systems Applications**]: Miscellaneous; C.2.1 [**Network Architecture and Design**]: Wireless Communication

Keywords

RFID tags, batch authentication, informative counting

1. INTRODUCTION

Counterfeiting products are growing dramatically in terms of quantity, sophistication, category of goods, and countries affected in the late years [1]. The Counterfeiting Intelligence Bureau of the International Chamber of Commerce estimates that the share of counterfeiting commodities in international trade is about 5% to 7% [2]. The overall economic loss around the world amounts to more than $600 billion and is growing steadily over years [3]. Radio Frequency Identification (RFID) is one of the most promising technologies to help distinguish the genuines from the counterfeits. The RFID technology has several advantages over traditional methods , e.g. bar codes. First, the tag can be read inside containers and covers. Second, the identification information on the tag is unique for each affixed object. Of greater importance, hundreds of tags can be read at a time while bar codes can only be read one by one.

The problem of authenticating tags in large-scale RFID systems can be easily reduced to verifying each tag one by one. A number of different authentication protocols have been proposed to address this issue. Let N be the number of tagsIDs stored in the authentication server. Weis et al. [4] introduce a hash-based approach named Hash Lock. The search complexity of this method is $\mathcal{O}(N)$. In order to optimize searching efficiency, several tree-based approaches are proposed [5][6]. While tree-based data structures effectively reduce search complexity to $\mathcal{O}(\log N)$, they also additionally associate each tag with $\mathcal{O}(\log N)$ items in trees. By reviewing previous methods, we discover three major challenging issues affecting the effectiveness and efficiency of batch authentication in RFID system. First, the scanning time is not scalable. As all tags share communication channel, the reader has to introduce proper anti-collision scheme to receive authentication information from different tags. As a result, the scanning time is $\mathcal{O}(n)$, where n is the number of tags to be verified. Second, the communication overhead is not scalable. In those tree-based approaches, $\mathcal{O}(logN)$ hash values are expected to be exchanged between reader and server for each tag. Thus, the overall communication cost is $\mathcal{O}(n \log N)$. Third, the result of previous batch authentication is coarse-grain. The most recent work SEBA [7] can only provide 0/1 authentication result with probabilistic guarantees for a batch of tags. In particular, it only tells whether there exist counterfeits in verified tags, no further information.

In practical RFID systems, for those very expensive brands and valuable objects such as diamonds and art works, high cost of the deterministic authentication is worthwhile. However, it may not suit well for large amounts of fast-moving consumer goods such as fashion clothes, accessories and wines, which are the leading industries significantly affected by counterfeits. As a matter of fact, authenticating each tag of large quantities of fast-moving consumer goods is not necessary. Instead, knowing the approximate count of counterfeits and genuines with accuracy and error probability guarantees is desired in many large-scale RFID system applications.

In this paper, we propose a fine-grained batch authentication scheme, INformative Counting (INC), which provides a scalable and reliable batch authentication solution for large-scale RFID system. In particular, we geometrically divide the tag set and construct its Authentication Synopsis (AS). With the help of AS, we develop authentication algorithms to approximate both the number of counterfeits and genuines for any given (ε, δ) requirement. Based on authentication algorithms, fine-grained authentication protocols are provided to support fast batch authentication in practical large-scale RFID systems. To the best of our knowledge, we are the first to propose fine-grained batch authentication scheme for large-scale RFID systems. The major contributions of this work are as follows.

1. We propose a fine-grained batch authentication scheme INC. Compared with previous methods that produce only 0/1 authentication results, the result of INC additionally consists of accurate approximations of both the number of counterfeits and genuines in tested tags.

2. Using AS data structure, INC achieves sublinear authentication time and sublinear communication overhead in batch operation while existing work are linear with the number of tags.

3. We validate the proposed algorithms through theoretical analysis and conduct extensive simulations to verify the effectiveness and performance of our scheme.

The rest of this paper is organized as follows. Section 2 discusses related work. We introduce system model in Section 3. The INC design and theoretical analysis are presented in Section 4. We have more detailed discussions on this design in Section 5, and then show simulations results in Section 6. The conclusion is in Section 7.

2. RELATED WORK

Identifying counterfeiting goods serves the main purpose of the RFID tag usage. In early work, much of interests is centered on how to identify a single tag in deterministic and secure ways. A number of hardware-based approaches and security protocols are designed for system anonymity and anti-cloning [8][9][10]. As wireless medium is shared among all tags, the problem of resolving collisions arises when operating a batch of tags. Most of anti-collision schemes can be classified into two categories: ALOHA-based [11][12][13][14][15] and tree-based [16][17]. ALOHA-based protocols have been implemented in EPCGlobal Generation-2 RFID standard [18] and many commercial RFID system. Sheng et al. [12] introduce an efficient continuous scanning scheme which effectively uses the information collected in the previous scanning. While another popular standard, ISO 18000-6 [19], adopts tree-based schemes. The collision resolving processes are dramatically boosted, since all tags are itemized into query tree according to *tagID*s. Recently many methods aiming at providing private authentication are proposed. Weis et al. [4] propose Hash Lock scheme to protect tags from being attacked. But its searching complexity is linear with the size of keys in database. In order to speed up searching process, tree-based data structures are introduced to achieve logarithmic scale in [5][6][20]. Lu et al. even achieve $\mathcal{O}(1)$ authentication efficiency for a single tag, based

on their own weak privacy model [21]. However, even combing state-of-the-art anti-collision methods and tree-based authentication schemes, the maximum identification throughput is still linear with the number of tags. Therefore they do not scale well when the size of tags quickly increases.

Besides above deterministic approaches, several probabilistic schemes are proposed to efficiently estimate the cardinality of tags. Kodialam et al. [22] propose Unified Simple Estimator (USE) and Unified Probabilistic Estimator (UPE) using linear counting technique. Qian et al. [23] introduce geometric distribution hashes to quickly estimate the cardinality of tags and the proposed LOF algorithm achieves $\mathcal{O}(\log n)$ time complexity. Zheng et al [24] propose Probabilistic Estimation Tree (PET), which advances estimation efficiency to $\mathcal{O}(\log \log n)$. A new scheme, Average Run based Tag estimation (ART), is 7x faster than UPE in the most recent work [25]. Li et al. [26] first propose energy-efficient RFID estimation algorithms for active tags. Zheng et al propose Zero-One Estimator (ZOE) protocol which rapidly converges to optimal parameter settings and achieves high estimation efficiency [27]. Although those schemes can estimate the number of distinct tags in RFID systems, they do not discriminate genuine ones from counterfeit ones.

In [7], a batch authentication scheme SEBA is proposed. It is able to detect counterfeits with probabilistic guarantee if the percentage of counterfeit tags is above predefined threshold. Nevertheless, SEBA has several drawbacks. First, its result is binary, merely indicating the existence of counterfeits in batch tags. Second, the frame size is $\mathcal{O}(N)$ and so it is not scalable in large-scale RFID systems. In contrast, INC can authenticate a batch of tags with accurate estimates of the number of counterfeits and genuines. And the authentication time and communication overhead of INC are sublinear with the cardinality of tags.

3. SYSTEM MODEL

In our system model, an RFID system consists of three main parts: one or more servers, several readers and hundreds of tags. Each tag is associated with a unique key, or called *tagID*. And it is attached to the object as an exclusive identity. Through wireless access medium, the reader can interrogate and receive responses from tags. The server usually plays a role of managing all keys of tags, including creation, authentication, and revocation of keys. If the key of a tag is actually stored in the server, we call this tag is *genuine*, otherwise *counterfeit*. Similar to most prior tag authentication schemes, *we do not discuss the issue that the genuine tag is attached to counterfeiting goods, vice versa.* The reader connects to the server through high speed wire or wireless networks. Let N be the number of keys maintained in the server, and n be the cardinality of batch tags to be verfied.

We adopt Listen-before-Talk [28] as the communication model between tag and reader in which the tag listens to the reader's interrogation and then replies. We also assume framed slotted ALOHA model as in [11][12][29][30]. In ALOHA model, the reader first broadcasts frame size f to all tags. Then each tag generates hash value $h(tagID)$ as its slot number. The reader then initializes time slot by sending "slot start" command. If the tag's slot number equals zero, it replies a bit-string to reader, otherwise it decreases slot number by one. This process repeats until f time slots are finished. In addition, we assume the bit-string contains some

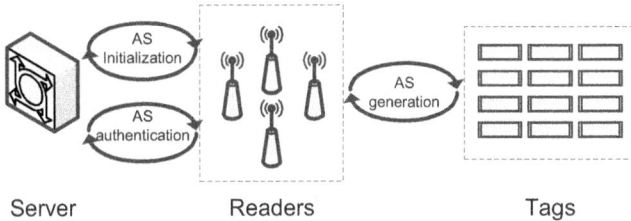

Figure 1: System architecture.

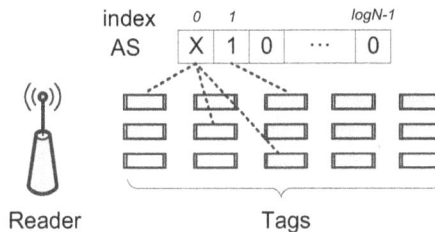

Figure 2: AS construction.

error-detecting codes like CRC. Note that this bit-string is not the identification information for each tag and is just used to detect whether there exist collisions in the time slot. Therefore, 10 bits should be fairly enough [22]. We classify time slots into three categories: *zero slot* means that there is no transmission in this slot while *singleton slot* denotes that only one tag transmits in this slot, and *collision slot* indicates that collisions happen in this slot.

There are some essential requirements of a good batch authentication scheme for large-scale RFID systems. First, the authentication scheme should be efficient, i.e., the authentication time and communication overhead should sub-linearly or near-constantly increase with the cardinality of tags. Second, the authentication result should be informative to support various application demands. Knowing that whether there exist counterfeits in a batch of tags or not is far from adequate, since the administrator of RFID systems may still resort to per-tag authentication to count how many counterfeits in a number of tags. The per-tag verification is rather time-consuming for batch processing [7]. Third, if the authentication scheme can output estimates of the number of counterfeits and genuines, these estimates should be arbitrarily accurate. The last but not the least, the frame size in batch authentication cannot be infinitely large. As stated in [18], the frame size should be set no more than 512 for practical reasons.

We use two parameters as accuracy requirements of the estimation result: relative error ε between 0 and 1, and error probability δ between 0 and 1. Let n_c to be the actual number of counterfeits in a batch of tags, the output result $\hat{n}_c$ of an estimator with required (ε, δ) should satisfy $Pr[|n_c - \hat{n}_c| \leq \varepsilon n_c] \geq 1 - \delta$. For example, if the exact number of counterfeits in batch tags is 1000, and $\varepsilon = 0.05$, $\delta = 0.05$, then the output estimate is between 950 and 1050 with probability at least 0.95. In our design, these two parameters can be arbitrarily small.

4. THE DESIGN OF INFORMATIVE COUNTING

In this section, we first outline the framework of our authentication system in Section 4.1. Then we present how to generate authentication synopsis between reader and tag in Section 4.2. Authentication algorithms of estimating the number of counterfeits and genuines are detailed in Section 4.3.

4.1 System Architecture

The INC authentication scheme consists of three steps: AS initialization, AS generation and AS authentication. As shown in Figure 1, during AS initialization phase, after ac-

quiring the accuracy requirement (ε, δ) from the user, the reader sends authentication request to the server, e.g., the number of hash functions. The server replies authentication parameters to the reader. Then in the AS generation phase, the reader interrogates batch tags and waits for responses. Through several rounds, collected ASes are transmitted to the server for the next AS authentication step. The server runs informative counting algorithms to give authentication results with accurate estimates of the number of counterfeits and genuines.

4.2 Authentication Synopsis

Our AS data structure is an extension of FM-Sketch [31]. In particular, as shown in Figure 2, the first position marked as "X" that denotes collisions happened in this time slot. The second position marked as "1" indicates there is only one tag transmitting in this slot. And the third position marked as "0" denotes that there is no response in this slot. The length of AS is equal to the frame size f. Similar to FM-Sketch, we do assume an ideal uniformly random hash function h_u generating hash values between 0 and $2^f - 1$. Then the second hash H, which counts the number of leading zeros (leftmost) of former hash value, is used to acquire geometric distribution as in [23]. It is expected that there are $\frac{n}{2^t}$ tags responses in the t-th position. For example, if the frame size is 8, one tag with $H(h_u(tagID_1)) = H(7) = H(00000111)_2 = 5$ should reply at slot 5. According to the geometric distribution characteristic, the frame size is $\log N$. So the frame of size 32 can support RFID system of which the cardinality is less than $4,294,967,295$. Thus, in this paper, we set the frame size at 32 which is fairly large for practical use.

The pseudocode of AS generation algorithms for tag and reader are given in Algorithm 1 and Algorithm 2 respectively. In Algorithm 1, the tag will generate the reply slot number k according to the two hashes H and h after receiving the probing message which might contain frame size and random seed. When each time slot starts, the tag responds instantly to the reader if its k is equal to 0. Otherwise it decreases k by 1 and keep silent. In Algorithm 2, the reader first broadcasts a request to all tags. And then the reader listens the status of tag responses in each time slot and sets the flag of slot according to different type of tag responses (0,1,X).

4.3 Algorithms

As in Figure 3, we assume that S is the set of tags on the server, and T is the set of tags to be tested. Accordingly, M is the union of S and T, i.e., $S \cup T$. C is the set of counterfeits which are in T but not in S, i.e., $T - S$. Therefore, we can easily define the number of counterfeits

Algorithm 1 AS generation algorithm for the tag

1: Receive a probing message from reader, compute slot number $k = H(h_u(tagID))$.
2: **while** TRUE **do**
3:    wait-for-slot-start().
4:    **if** $k == 0$ **then**
5:       respond instantly.
6:    **else**
7:       $k \leftarrow k - 1$, keep silent.
8:    **end if**
9: **end while**

Algorithm 2 AS generation algorithm for the reader

1: Initialize $AS[i] \leftarrow 0 (0 \leq i \leq \log N - 1)$;
2: Broadcast a request to tags.
3: **for** $i = 0$ to $\log N - 1$ **do**
4:    wait-for-tags-response().
5:    **if** there is no response in this slot **then**
6:       $AS[i] \leftarrow 0$.
7:    **else**
8:       **if** there is one tag response in this slot **then**
9:          $AS[i] \leftarrow 1$.
10:      **else**
11:         $AS[i] \leftarrow X$.
12:      **end if**
13:    **end if**
14: **end for**

as $|C| = |T - S|$. Similarly, we use G to denote the set of genuines which are both in S and T, i.e., $T \cap S$. And the number of genuines is denoted as $|G| = |T \cap S|$. If there is no counterfeits in the tested tags, then $M = \emptyset$ and $|C| = 0$. The basic idea of our approach is to get an (ε, δ) approximation to $|M|$, and then estimate the PG which is the $|G|$ to $|M|$ ratio and PC which is that $|C|$ to $|M|$. Finally, combining the accurate approximation of $|M|$ and above two ratios, the (ε, δ) approximations for both $|G|$ and $|C|$ are deduced. Note that as these two ratios are constants during the authentication process, custom probabilistic algorithms can be introduced to provide arbitrarily accurate estimates. To simplify the exposition, we treat δ as $\Theta(\delta)$ in following algorithm descriptions.

4.3.1 Estimating $|M| = |S \cup T|$

Estimating $|M|$ mainly consists of four steps. First, the server needs to collect enough number of independent AS_i^T from the reader and then generates its own $AS_i^S (1 \leq i \leq d$ using the same hash function respectively. Second, we construct virtual ASes slot by slot based on the criteria that only if both the corresponding slots of AS_i^T and AS_i^S are empty slots, the slot of virtual ASes would be empty. Otherwise, it is non-empty slot. Third, we find the smallest index level r which satisfying the appropriate threshold and obtain the non-empty probability of M of this level, p_r. Finally, as the expectation of this probability is the function of $|M|$, inverting of this function provides a good estimate, $\hat{M}$.

As shown in Figure 4(a), T_i is an AS from the reader and S_i is the corresponding AS generated by the server. 0 denotes empty slot and 1 denotes non-empty slot. The virtual AS construction is to combine non-empty slots. The slot of

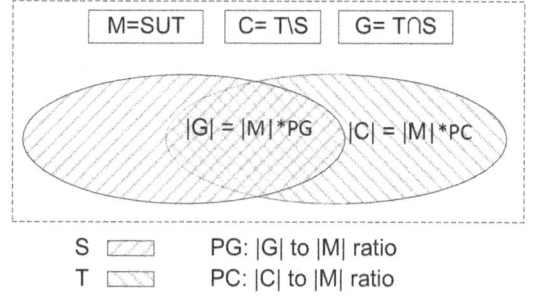

Figure 3: Basic idea of estimating the numbers of counterfeits and genuines.

virtual AS V_i would be non-empty if any one corresponding slot of S_i or T_i is non-empty. The index level search process is in Figure 4(b), NCount is the count of non-empty slots in specific index level and assumed threshold is 3. Therefore, starting from lowest index level (leftmost), the first index level of which the NCount is below the threshold would be the appropriate estimating level. And here we find the qualified level is the third index.

Definition 1. For a specific index r $(0 \leq r \leq 31)$. We define

$$x(r) = \begin{cases} 1 & \text{the index } r \text{ of } AS^{S \cup T} \text{ is non-empty} \\ 0 & \text{otherwise} \end{cases}$$

Therefore the $x(r)$ takes 1 with probability $p_r = 1 - (1 - \frac{1}{U_r})^{|M|}$, where $U_r = 2^{r+1}$.

From above definition, we can invert the probability formula of p_r to deduce the estimate $\hat{M}$

$$\hat{M} = \frac{\ln(1 - \hat{p_r})}{\ln(1 - \frac{1}{U_r})}, \tag{1}$$

where $\hat{p_r}$ is the observed non-empty probability of index level r.

Now, we have the $|M|$ estimating function but there are still two major problems. 1. d_m. How many pairs of independent ASes from reader and server are enough to provide an accurate estimate $\hat{p_r}$? 2. λ. How accurate of $\hat{p_r}$ is adequate to produce the user-specified (ε, δ) approximation of $|M|$? The following lemmas answer those questions.

LEMMA 1. *If* $d_m \geq \frac{96}{7} \lambda^{-2} \ln \frac{2}{\delta}$, r *is the smallest index satisfying* $NCount_r \leq \frac{(1+\lambda)d_m}{4}$, *and* $\hat{p_r} = \frac{NCount_r}{d_m}$, *then* $Pr[|\hat{p_r} - p_r| \leq \lambda p_r] \geq 1 - \delta$.

PROOF. Note that the probability that the index r is non-empty in $AS^{S \cup T}$ is the same as the probability that the index r is non-empty in either AS^S or AS^T. So we fix r to a positive value such that $\frac{1}{8} \leq \frac{M}{U_r} \leq \frac{1}{4}$. By binomial expansion, we know that $p_r = 1 - (1 - \frac{1}{U_r})^M = \sum_{i=1}^{M} (-1)^{i+1} \binom{M}{i} U_r^{-i}$, then $(\frac{M}{U_r} - \frac{1}{2}(\frac{M}{U_r})^2) \leq p_r \leq \frac{M}{U_r}$. Therefore, we obtain that $\frac{7}{32} \leq p_r \leq \frac{1}{4}$. Also remember that $x(r)$ is a binomial random variable, hence we can apply Chernoff bound [32]. Using slightly worse case bound expressions, we know that as long as $d_m p_r \geq \frac{3}{\lambda^2} \ln \frac{2}{\delta}$, i.e,

$$d_m \geq \frac{96}{7\lambda^2} \ln \frac{2}{\delta}, \tag{2}$$

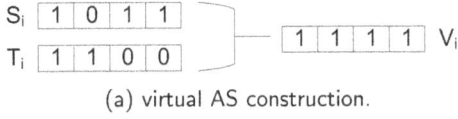

(a) virtual AS construction.

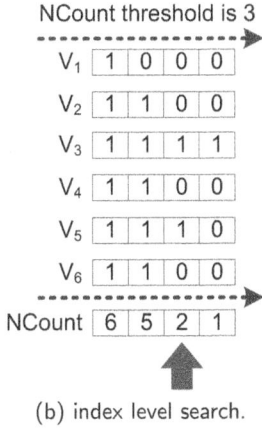

(b) index level search.

Figure 4: An illustration of estimating $|M| = |S \cup T|$.

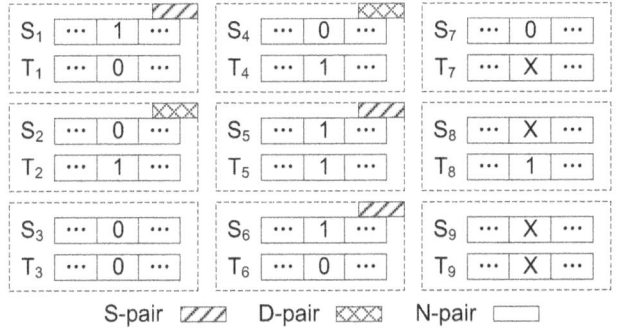

Figure 5: Distinguish different types of pairs for estimating the number of counterfeits.

as $p_r \geq \frac{7}{32}$, then the estimate $\hat{p}_r = \frac{NCount_r}{d_m}$ can satisfy $Pr[|\hat{p}_r - p_r| \leq \lambda p_r] \geq 1 - \delta$. Therefore at index level r, we can find $|\hat{p}_r - p_r| \leq \lambda p_r$ which ensures that $count_r \leq \frac{(1+\lambda)d_m}{4}$ with probability at least $1 - \delta$. □

LEMMA 2. *If $p_r < \frac{1}{4}$, $\lambda = \frac{\varepsilon}{2}$ and $|\hat{p}_r - p_r| \leq \lambda p_r$, then a new estimate $\hat{M}$, defined as $\hat{M} = \frac{\ln(1-\hat{p}_r)}{\ln(1-\frac{1}{U_r})}$, satisfies $|\hat{M} - M| \leq \varepsilon M$.*

PROOF. For a continuous function $f(x) = \ln(1-x)$, we know that $\ln(1-x) < -x$ when $0 < x < 1$. And also there is $|f(x) - f(\bar{x})| \leq \varepsilon |sup_{y \in (x,\bar{x})} f'(y)|$ if $\bar{x}$ is close to x, hence we can get $|\ln(1-x) - \ln(1-\bar{x})| \leq \frac{|x-\bar{x}|}{1-max\{x,\bar{x}\}}$. Therefore,

$$
\begin{aligned}
|\hat{M} - M| &= \frac{|\ln(1-p_r) - \ln(1-\hat{p}_r)|}{-\ln(1-\frac{1}{U_r})} \\
&\leq -\ln(1-\frac{1}{U_r}) \cdot \frac{|p_r - \hat{p}_r|}{1 - max\{p_r, \hat{p}_r\}} \\
&\leq -\ln(1-\frac{1}{U_r}) \cdot \frac{\frac{\varepsilon}{2}p_r}{1 - (1+\frac{\varepsilon}{2})p_r} \\
&\leq -\ln(1-\frac{1}{U_r}) \cdot \varepsilon p_r \\
&\leq -\ln(1-\frac{1}{U_r}) \cdot \varepsilon \cdot (-\ln(1-p_r)) \\
&= \varepsilon M.
\end{aligned}
$$

□

Combining lemma 1 and lemma 2, we can establish following theorem.

THEOREM 1 ($|M|$ ESTIMATE). *If $d_m \geq \frac{384}{7\varepsilon^2} \ln \frac{2}{\delta}$, equation 1 outputs an (ε, δ) estimate $\hat{M}$ for $|M|$.*

4.3.2 Estimating the number of counterfeits

Given the set S and T, the cardinality of the set-difference $|C| = |T - S|$ is the number of distinct tags that are in T but not in S, i.e., the counterfeit ones. Here, we introduce an (ε, δ) approximation scheme for estimating $|C|$ based on the former accurate estimate $\hat{M}$. It is worth noting that a "naive" method may be proposed to use $|T - S| = |T \cup S| - |S|$ formula and corresponding (ε, δ) estimates of both $|T \cup S|$ and $|S|$ to estimate the cardinality of this set-difference. However, this "naive" approach does not give any accuracy guarantees of the approximation for $|T - S|$, i.e., it is not an (ε, δ) estimation scheme as needed.

Estimating $|C|$ procedure is composed of three essential parts. First, assume that we get an estimate $\hat{M}$ with $(\frac{\varepsilon}{3}, \delta)$. And then we choose the level $r_c = \lceil \log(\frac{\alpha \hat{M}}{1-\varepsilon}) \rceil$, such that the number of elements that can be hashed into this level, 2^{r_c+1}, is slightly greater than $\hat{M}$, where α is a constant parameter greater than 1 (line 3). Second, in order to obtain difference ratio (p_d), we seek to distinguish different types of pairs and record their counts. Third, we combine the observed difference ratio and the estimate $\hat{M}$ to compute the approximation of set difference $\hat{C}$.

The core of estimating $|C|$ is the second part. As shown in Figure 5, there are 9 AS pairs. And if one slot is "1" and the corresponding slot is "0" or "1", the pair is called singleton pair, S-pair for short. Further if the slot of S_i is '0' and slot T_i is '1', the pair is denoted as difference pair, D-pair for short. The other pairs are called N-pair. Obviously, all D-pairs are S-pairs. And the count of S-pair is 5 (singletonCount) and its of D-pair is 3 (diffCount). Therefore, we can roughly estimate that the number of counterfeiting ones is about $\frac{3}{5}|M|$.

Definition 2. For a specific index r_c ($0 \leq r_c \leq 31$), let $U_c = 2^{r_c+1}$, we define

$$
x_M(r_c) = \begin{cases} 1 & \text{the index } r_c \text{ of } AS^{S \cup T} \text{ is singleton} \\ 0 & \text{otherwise.} \end{cases}
$$

We can have that the $x_M(r_c)$ takes 1 with probability $p_M = |S \cup T| \cdot \frac{1}{U_c}(1-\frac{1}{U_c})^{|S \cup T|-1}$, as the probability of a given element mapped to level r_c is $\frac{1}{U_c}$, and the probability of a given element being singleton is $\frac{1}{U_c}(1-\frac{1}{U_c})^{|S \cup T|-1}$. And we also define

$$
x_C(r_c) = \begin{cases} 1 & \text{the index } r_c \text{ of } AS^T \text{ is "1" and } AS^S \text{ is "0"} \\ 0 & \text{otherwise.} \end{cases}
$$

25

Likewise, $x_C(r_c)$ takes 1 with probability $p_C = |T - S| \cdot \frac{1}{U_c}(1 - \frac{1}{U_c})^{|S \cup T|-1}$. Therefore, we can define the conditional probability p_d as

$$p_d = \frac{p_C}{p_M} = \frac{|T - S| \cdot \frac{1}{U_c}(1 - \frac{1}{U_c})^{|S \cup T|-1}}{|S \cup T| \cdot \frac{1}{U_c}(1 - \frac{1}{U_c})^{|S \cup T|-1}} = \frac{|T - S|}{|S \cup T|}. \quad (3)$$

We also define $\acute{d}_c = \sum_{i=0}^{d_c-1} x_M(r)_i$ over d_c AS pairs, the number of S-pairs.

From above definitions, we can deduce the formula for estimating $|C|$ as

$$\hat{C} = \hat{p}_d \cdot \hat{M}, \quad (4)$$

where $\hat{p}_d = diffCount/singletonCount$.

In order to produce an (ε, δ) estimate $\hat{C}$, the following lemmas deal with three questions. 1. d_c. How many pairs of independent ASes are enough to produce specified number of S-pairs? 2. $\acute{d}_c$. How many S-pairs are needed to provide accurate estimate of p_d? 3. How accurate of $\hat{M}$ and $\hat{p}_d$ are adequate to give an (ε, δ) estimate $\hat{C}$?

LEMMA 3. Let $\alpha > 1$ and $r_c = \lceil \log(\frac{\alpha|S \cup T|}{1-\varepsilon}) \rceil$. For any constant γ between 0 and 1, if $d_c \geq \frac{3\alpha^2}{\gamma^2(\alpha-1)} \ln \frac{2}{\delta}$, then $\acute{d}_c > \frac{(1-\gamma)(\alpha-1)}{\alpha^2} d_c$ with probability $1 - \delta$.

PROOF. As $r_c = \lceil \log(\frac{\alpha|S \cup T|}{1-\varepsilon}) \rceil$, we have that $\frac{|S \cup T|}{U_c} < \frac{1}{\alpha}$. By Bernoulli inequality [33], we can obtain the lower-bound of p_M

$$p_M > \frac{|S \cup T|}{U_c}(1 - \frac{|S \cup T|}{U_c}) > \frac{\alpha - 1}{\alpha^2}.$$

Again by Chernoff bound [32], we can obtain an estimate $\hat{p_M} = \frac{\acute{d}_c}{d_c}$ such that $Pr[|\hat{p_M} - p_M| \leq \gamma p_M] \geq 1 - \delta$ as long as $d_c \geq \frac{3\alpha^2}{\gamma^2(\alpha-1)} \ln \frac{2}{\delta} \geq \frac{1}{p_M} \frac{3}{\gamma^2} \ln \frac{2}{\delta}$. Thus

$$\hat{p_M} \geq (1-\gamma)p_M \Leftrightarrow \acute{d}_c \geq (1-\gamma)p_M \cdot d_c > \frac{(1-\gamma)(\alpha-1)}{\alpha^2} d_c.$$

□

LEMMA 4. If $\acute{d}_c \geq \frac{|S \cup T|}{|T - S|} \frac{3}{\eta^2} \ln \frac{2}{\delta}$, $\hat{p}_d$ is an (η, δ) estimate.

PROOF. By Chernoff bound [32], we know that as long as $\acute{d}_c p_d \geq \frac{3}{\eta^2} \ln \frac{2}{\delta}$, $\hat{p}_d$ is within a relative error η with probability $1 - \delta$. And combing equation (3), it gives that $\acute{d}_c \geq \frac{|S \cup T|}{|T - S|} \frac{3}{\eta^2} \ln \frac{2}{\delta}$. □

LEMMA 5. If we have an $(\frac{\varepsilon}{3}, \delta)$ estimate $\hat{M}$ for $|M|$, and an $(\frac{\varepsilon}{3}, \delta)$ estimate $\hat{p}_d$ for p_d, then $\hat{p}_d\hat{M}$ is an (ε, δ) estimate of p_dM, i.e., C.

PROOF. $|\hat{p}_d\hat{M} - p_dM| = |p_d(1 \pm \frac{\varepsilon}{3})M(1 \pm \frac{\varepsilon}{3}) - p_dM| \leq p_dM(\pm \frac{2\varepsilon}{3} + \frac{\varepsilon^2}{9}) \leq \varepsilon p_dM$. □

By lemma 3 and 4, we know that in order to get an (η, δ) estimate $\hat{p}_d$, d_c should be at least $\frac{\alpha^2}{(1-\gamma)(\alpha-1)} \frac{|S \cup T|}{|T - S|} \frac{3}{\eta^2} \ln \frac{2}{\delta}$. And also by lemma 3, d_c should be at least $\frac{3\alpha^2}{\gamma^2(\alpha-1)} \ln \frac{2}{\delta}$. Combing those two conditions, hence we get

$$d_c \geq \frac{\alpha^2}{\min\{1-\gamma, \gamma^2\}(\alpha-1)} \frac{|S \cup T|}{|T - S|} \frac{3}{\eta^2} \ln \frac{2}{\delta} = d_c^1. \quad (5)$$

Algorithm 3 Estimating the number of counterfeits

Input: relative error ε, error probability δ.
Output: estimate $\hat{C}$.
1: compute d_c by theorem 2.
2: generate d_c independent AS pairs of S and T as in Algorithm 1 and 2.
3: $\hat{M}$=**GetEstimateM**$(AS_{dc}, \frac{\varepsilon}{3}, \delta)$.
4: $singletonCount \leftarrow 0$, $diffCount \leftarrow 0$.
5: $\alpha = 2$, $r_c \leftarrow \lceil \log(\frac{\alpha\hat{M}}{1-\varepsilon}) \rceil$.
6: **for** $i = 0$ to $d_c - 1$ **do**
7:     **if** $(AS_i^S[r_c] == 0$ and $AS_i^T[r_c] == 1$) or $(AS_i^S[r_c] == 1$ and $AS_i^T[r_c] == 0$) or $(AS_i^S[r_c] == 1$ and $AS_i^T[r_c] == 1$) **then**
8:         $singletonCount \leftarrow singletonCount + 1$.
9:         **if** $AS_i^S[r_c] == 0$ and $AS_i^T[r_c] == 1$ **then**
10:             $diffCount \leftarrow diffCount + 1$.
11:         **end if**
12:     **end if**
13: **end for**
14: **return** $\hat{C} = \frac{diffCount}{singletonCount} \cdot \hat{M}$.
15: **GetEstimateM**$(AS_{dc}, \varepsilon, \delta)$
16: $index \leftarrow 0$, $found \leftarrow false$, $\lambda \leftarrow \frac{\varepsilon}{2}$, $minC \leftarrow \frac{(1+\lambda)d_m}{4}$.
17: $d_m \leftarrow \frac{384}{7\varepsilon^2} \ln \frac{2}{\delta}$, randomly select d_m pairs AS_{dm} from AS_{dc}.
18: **while** $index \leq 31$ and $!found$ **do**
19:     $NCount \leftarrow 0$.
20:     **for** $i = 0$ to $d_m - 1$ **do**
21:         **if** $AS_i^S[index] \neq 0$ or $AS_i^T[index] \neq 0$ **then**
22:             $NCount \leftarrow NCount + 1$.
23:         **end if**
24:     **end for**
25:     **if** $NCount > minC$ **then**
26:         $index \leftarrow index + 1$.
27:     **else**
28:         $found \leftarrow true$.
29:     **end if**
30: **end while**
31: $\hat{p} \leftarrow \frac{NCount}{d_m}$, $\hat{M} \leftarrow \frac{\ln(1-\hat{p})}{\ln(1 - \frac{1}{2^{index+1}})}$.
32: **return** $\hat{M}$.

In order to get minimum d_c, we can compute the optimal values for γ and α. The results are $\gamma = \frac{\sqrt{5}-1}{2}$, $\alpha = 2$. The first lower-bound of d_c is denoted as d_c^1. And by lemma 5, let $\eta = \frac{\varepsilon}{3}$, we can get an $(\frac{\varepsilon}{3}, \delta)$ estimate of p_d. By theorem 1, in order to get an $(\frac{\varepsilon}{3}, \delta)$ estimate for $|M|$, we get

$$d_c \geq \frac{3456}{7\varepsilon^2} \ln \frac{2}{\delta} = d_c^2. \quad (6)$$

The second lower-bound of d_c is denoted as d_c^2. Based on the above analysis, we can state the following theorem.

THEOREM 2 (ESTIMATING THE NUMBER OF COUNTERFEITS). If $d_c = \max\{d_c^1, d_c^2\}$, equation 4 outputs an (ε, δ) estimate $\hat{C}$ for $|C|$.

Algorithm 3 shows the pseudocode of estimating the number of counterfeits.

4.3.3 Estimating the number of genuines

According to the concept of genuineness, the number of genuines in T can be defined as $|G| = |S \cap T|$, i.e., the number of distinct keys that are in both T and S. Fortunately, the (ε, δ) estimation algorithm of G is very similar to Algorithm 3. The only difference is that the condition of line 9 should be changed into "$AS_i^S[r_c] == 1$ and $AS_i^T[r_c] == 1$". Similarly, we can obtain the formula for estimating $|G|$ as

$$\hat{G} = \hat{p_s} \cdot \hat{M}, \qquad (7)$$

where $\hat{p_s}$ is the identical ratio. Likewise, we derive the lower-bound of d_g as follows

$$d_g \geq \frac{\alpha^2}{\min\{1-\gamma, \gamma^2\}(\alpha-1)} \frac{|S \cup T|}{|S \cap T|} \frac{3}{\eta^2} \ln \frac{2}{\delta} = d_g^1 \qquad (8)$$

$$d_g \geq \frac{3456}{7\varepsilon^2} \ln \frac{2}{\delta} = d_g^2 \qquad (9)$$

Thus, we can establish the following theorem.

THEOREM 3 (ESTIMATING THE NUMBER OF GENUINES). If $d_g = \max\{d_g^1, d_g^2\}$, equation 7 outputs an (ε, δ) estimate $\hat{G}$ for $|G|$.

5. DISCUSSION

In this section, we discuss several important issues of our proposed algorithms.

5.1 t-wise Independent Hash Functions

So far, the design and analysis of our schemes assume that there exist ideal uniformly random hash functions. In fact this assumption is unrealistic for practical use. Therefore, we can employ t-wise independent hash functions to ensure our former analysis still hold. Actually, if $t = \Theta(\log(\varepsilon^{-1}))$, then using h_i from t-wise independent hash family $\mathcal{H}_t$ as alternatives of h_u in AS constructions, we can still provide (ε, δ) estimates for the number of counterfeits and genuines. Due to limited space, we omit the details here.

5.2 Singleton Slot Observation

In the line 5 of Algorithm 4, "$AS_i^S == 1$ and $AS_i^T == 1$" is one of conditions that singleton is found in $(S \cup T)$. This condition may fail if two elements are "luckily" enough to be mapped into the same slot. Fortunately, we can prove that the possibility of this "luck" is rather small and negligible in practical use. This possibility, denoted as p_l, is that two different elements are hashed into $index$ level and also are singletons in their respective AS. Therefore, we have $p_l = \frac{1}{U_c}(1 - \frac{1}{U_c})^{|S \cup T|-2} \approx \frac{1}{M}(1 - \frac{1}{M})^{M-2}$ where $U_c = \frac{\alpha \hat{M}}{1-\varepsilon}$. Therefore, we know that even if the server only has $M = 10,000$ keys, then p_l is about $0.000,037$. And with the increasing number of keys on the server, p_l would be even smaller. As a matter of fact, we can also use an extra round to determine whether the undecided slot is the singleton slot in M using rolling schemes similar in [30].

5.3 The Size of Counterfeits/genuines

Another practical problem is that the size of counterfeits/genuines may be too small. For example, if $\frac{|S \cup T|}{|T-S|}$ is quite large, estimation problem would become difficult as in equation 5. To address this issue, we know that the share of counterfeiting commodities in real life is about 5% to 7% [2]. Therefore, we can provide the "sanity" lower bound $B = f(M, d_c, \varepsilon, \delta)$ determined by our theorem. Then our earlier statements about (ε, δ) estimate can be formed like this: "...outputs (ε, δ) estimate of $|C|$ as long as $|C| \geq B$".

Table 1: Scheme comparison

Scheme	Scanning Cost	Communication Cost	Authentication Cost
Hash Lock	$\mathcal{O}(n)$	$\mathcal{O}(n)$	$\mathcal{O}(nN)$
ACTION	$\mathcal{O}(n \log N)$	$\mathcal{O}(n \log N)$	$\mathcal{O}(n\log N)$
SEBA [1]	$\mathcal{O}(N)$	$\mathcal{O}(N)$	$\mathcal{O}(N)$
INC	$\mathcal{O}(\frac{N}{n} \log N)$	$\mathcal{O}(\frac{N}{n} \log N)$	$\mathcal{O}(\frac{N}{n} \log N)$

However, as shown later in evaluation part we know that this sanity bound is rather pessimistic, it may due to that fact that our analysis are based on the worst-case analysis. Furthermore, we may divide keys on the server into tree or other organized data structures in which the initial matching space $\mathcal{O}(N)$ might be effectively reduced. For example, keys of tagged items are group by different categories such as wine and clothes. And so when authenticating wines, S should be S_{wine}, not $S_{wine} \cup S_{clothes}$.

5.4 Time-efficient Optimization

Although the communication cost and AS construction time are sublinear with the cardinality of RFID system, there are possible solutions to further boost authentication efficiency. Remember in Algorithm 3 that the estimation are performed on some specific level, e.g., $index$, because the elements of ASes below this level are almost 'X' or '1'. Therefore, if we can obtain the scale of $|S \cup T|$ as a prior knowledge, AS can be compressed from $\log N$ slots into $(\log N - index)$ slots or $\log \log N$, where $index$ is the estimation start level. Thus, compressed AS may significantly speed up the authentication process.

5.5 Energy-efficient Optimization

The energy cost of the tag is one important issue we should carefully cope with. For example, in a large warehouse equipped with RFID system, active tags are usually used to label commodities [26]. Since active tags are battery-powered, recharging batteries for thousands of tags is really a heavy work, and even in some cases the tags are not easily reachable. As our early analysis shown, although this number of ASes is sublinear with the cardinality of RFID system, it still imposes heavy burdens for resource constraint tags since all tags are required to respond to each interrogation from reader during authentication. To further reduce energy consumption of tags, mechanisms that may shift the energy consumption from tag side to reader side are necessary. One possible solution is to build an energy-efficient AS data structure of which the organization is based on a hash function (e.g., h_g) and each tag contributes to all ASes sequentially. The design this hash function is to divide tags into several groups and minimize the number of data transmissions of tags.

5.6 Comparison

Table 1 compares INC with three state-of-the-art authentication schemes: Hash lock [4], ACTION [6] and SEBA

---

[1] As optimal frame length functions are complex and implicit, therefore we use simple function as $\delta = 0.99$. See details in [7].

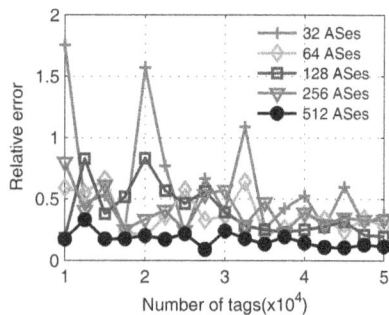

Figure 6: Relative error of estimate $\hat{C}$. Figure 7: Standard deviation of estimate $\hat{C}$. Figure 8: Normalized std deviation of estimate $\hat{C}$.

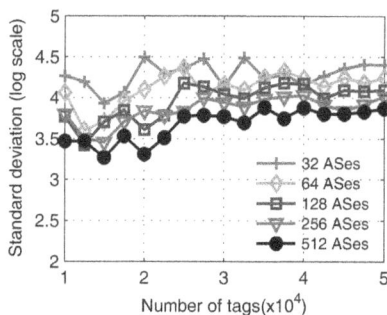

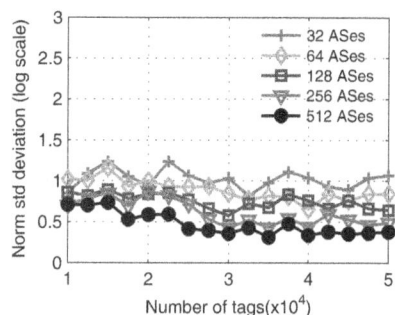

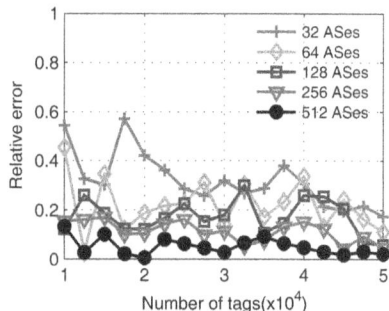

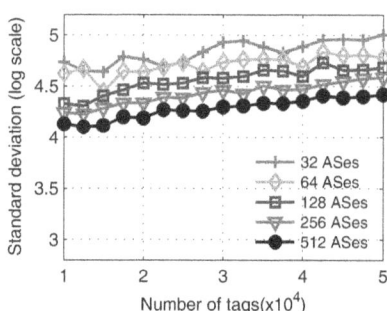

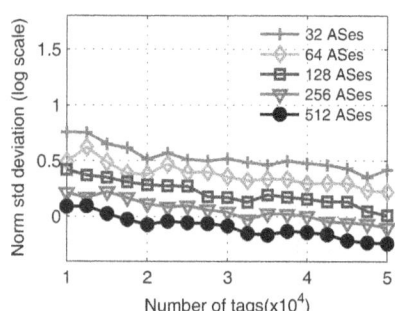

Figure 9: Relative error of estimate $\hat{G}$. Figure 10: Standard deviation of estimate $\hat{G}$. Figure 11: Normalized std deviation of estimate $\hat{G}$.

[7]. Refresh [21], which achieves better authentication efficiency over ACTION, is not included in this comparison in terms of fairness. Since it is based on the weak privacy model in which the improvement of the authentication efficiency is at the cost of the privacy degradation. The comparison mainly consists of three metrics: (i) scanning cost that is the complexity of acquiring the authentication information; (ii) communication cost which is the amount of data transmissions between reader and server; and (iii) authentication cost which is the time of verifying n tags in N server-stored keys. As Hash lock and ACTION are per-tag based deterministic authentication schemes, we assume an omniscient anti-collision solution is with them in which n tags can transmit identification information in a frame of size n perfectly. As characteristics of deterministic approaches, scanning cost and communication cost of Hash lock are $\mathcal{O}(n)$. With the help of tree-based data structures, ACTION achieves $\mathcal{O}(n\log N)$ efficiency in terms of scanning cost, communication cost and authentication cost, while authentication cost of Hash lock is $\mathcal{O}(nN)$ with linear searching. For SEBA [7], as the optimal frame length is $7(N + n\varepsilon)$, the scanning cost, communication cost and authentication cost are $\mathcal{O}(N)$. From theorem 2 and theorem 3, we know that for a given (ε, δ), the space complexity of ASes is $\mathcal{O}(\frac{\log(\delta^{-1})N}{\varepsilon^2 n}\log N)$. If we treat (ε, δ) as constants, therefore we can easily deduce the corresponding scanning cost, communication cost and authentication cost. We can see that if the size of tested set n is relatively small, ACTION maybe the best solution. But when n is growing larger and

larger, INC wins the contest as its complexity is asymptotically towards $\mathcal{O}(\log N)$.

6. EVALUATION

We evaluate the performance of INC under extensive simulations. First, we study the estimation accuracy with tunable size of ASes under various settings. Then we compare INC with two most recent methods ACTION [6] and SEBA [7] in terms of scanning cost and communication overhead.

6.1 Setup and Metrics

The simulations are implemented on a laptop with Intel i7 CPU at 2.8GHz and 4GB RAM, using C# as programming language. In order to support ACTION of depth 21 with $N = 2^{20} = 1,408,576$ server tags, we have $2^{21} = 2,097,152$ keys stored in SQLExpress 9.00 database. For ACTION, we use MD5 as the uniform hash function. Therefore each tag is associated with 21 key values. According to reports in [2], we set the ratio of counterfeiting tags at 7%. We take 300 runs and report the average.

The estimation accuracy of the number of counterfeits or genuines is the most important metric for our authentication scheme. We use three standard parameters to measure the accuracy of INC. The first parameter is relative error: $RelError = |\frac{\hat{\theta} - \theta}{\theta}|$, where θ is the actual number and $\hat{\theta}$ is the estimate. The second parameter is standard deviation: $\sigma = \sqrt{E[(\hat{\theta} - \theta)^2]}$. The third parameter is normalized standard deviation: $\sigma_n = \frac{\sigma}{E[\theta]}$.

28

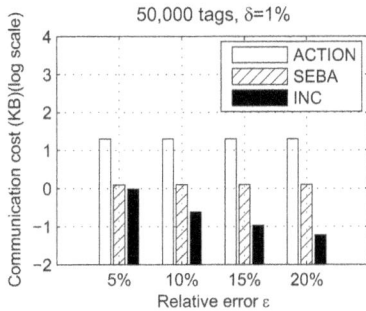

(a) with different relative error ε, 50,000 tags and $\delta = 1\%$.

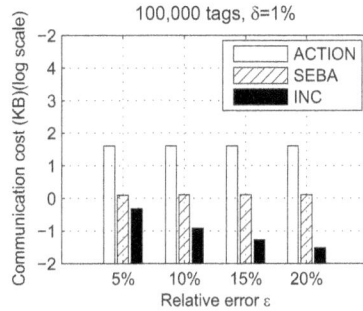

(b) with different relative error ε, 100,000 tags and $\delta = 1\%$.

Figure 12: Communication cost between reader and server (in *log scale*).

Table 2: Total slots to meet $\delta = 1\%$ with different ε

ε	$n = 50,000$		
	ACTION	SEBA	INC
5%	1,050,000	1,051,076	808,992
10%	1,050,000	1,053,576	202,240
15%	1,050,000	1,056,076	89,888
20%	1,050,000	1,058,576	50,560
	$n = 100,000$		
5%	2,100,000	1,053,576	404,480
10%	2,100,000	1,058,576	101,120
15%	2,100,000	1,063,576	44,928
20%	2,100,000	1,068,576	25,280

6.2 INC Investigation

The results in Figures 6, 7, and 8, show that the accuracy of estimating the number of counterfeits in tested tags. From those figures, we make following observations. First, we can improve estimation accuracy of INC by increasing the number of ASes. As illustrated in Figure 6, increasing the number of ASs lead to significant drops of relative error. In particular, when the number of tags is 50,000, the relative error is 0.37 with 32 ASes and drops close to 0.1 with 512 ASes. This nice linearly addible property of INC provides us flexibility of making tradeoffs between estimation accuracy and efficiency according to different application demands. Second, the standard deviation can be greatly reduced by larger size of ASes, which is indicating in Figure 7. Third, normalized standard deviations are getting steadily smaller with increasing the cardinality of tags. As shown in Figure 8, the normalized standard deviation is 5.1 with 10,000 tags and diminishes to 2.4 with 50,000 tags using 512 ASes.

Similar trends can also be observed in Figures 9, 10 and 11. We can see that the estimate $\hat{G}$ of approximating the number of genuines is advantageous over $\hat{C}$ in all three aspects. As we stated in Section 5.3, the main reason is that the number of genuine tags is larger than its of counterfeits, since we set the share of counterfeiting tags at 7%. Relative errors with different size of tags from 10,000 to 50,000 with 512 ASes are below 0.2 and most of them are about 0.01.

From normalized standard deviation perspective, $\hat{G}$ is also an excellent estimator. As shown in Figure 11, for 50,000 tags, the normalized standard deviation is 0.57 with only 512 ASes.

6.3 Performance Comparison

We compare the performance of INC with the two state-of-the-art approaches ACTION and SEBA. We mainly compare the scanning cost and communication cost under different accuracy settings. Here we measure scanning time in terms of the total time slots to acquire authentication information and communication cost in terms of data size of transmissions between reader and server.

We compare three methods given $\delta = 1\%$ and ε changing from 5% to 20% and the size of tags at 50,000 and 100,000. As shown in Table 2, INC significantly outperforms both ACTION and SEBA. For instance, total time slots of INC is 8.5% of SEBA when $\varepsilon = 15\%$, $n = 50,000$ and is 1.2% of ACTION when $\varepsilon = 20\%$, $n = 100,000$. The communication cost results of three schemes are depicted in Figure 12. We assume that the length of identification ID is 96 bits [18] in ACTION and the length of bit-string for each time slot is 10 bits [22] in SEBA and INC. Again, we can see that INC achieves much lower communication cost than both ACTION and SEBA, e.g., the data size of transmissions of INC is merely 0.3% of ACTION and 4.7% of SEBA when $\varepsilon = 20\%$, $n = 50,000$. From a different perspective, above comparison results indicate that given a certain amount of transmission data or scanning time requirement, the authentication accuracy of INC will be much better than ACTION and SEBA. As a matter of fact, since INC achieve $\mathcal{O}(\frac{N}{n} \log N)$ efficiency, when the size of tags quickly scales, the performance gain of INC over ACTION and SEBA is growing larger.

7. CONCLUSION

This paper proposes a probabilistic batch authentication schemes, INC, for large-scale RFID systems. Compared with previous methods. INC not only achieves sublinear authentication efficiency, but also provides accurate estimate of the number of counterfeits and genuines. Both theoretical analysis and extensive simulations are presented to show advantages of INC over prior work. In future work, we plan to examine whether our estimation bounds are tight. And also we intend to extend our framework to multiple readers scenarios, in which each reader has its own operation range and tag set.

8. ACKNOWLEDGMENTS

We would like to thank the anonymous reviewers for valuable and insightful comments. This work is supported in part by the NSFC Young Scholar 61103187, National Basic Research Program of China (973) Grant No. 2011CB302705, NSFC Major Program 61190110 and the NSFC under Grant 61171067. We also acknowledge the support from the codes of USRP2reader from the Open RFID Lab (ORL) project [34].

9. REFERENCES

[1] Thorsten Staake, Frĭedĭęric Thiesse, and Elgar Fleisch. Business strategies in the counterfeit market. *Journal of Business Research*, 65(5):658 – 665, 2012.

[2] ICC Counterfeiting Intelligence Bureau. Countering counterfeiting: A guide to protecting and enforcing intellectual property rights. 1997.

[3] The spread of counterfeiting: Knock-offs catch on. *The Economist*, 2010.

[4] S. Weis, S. Sarma, R. Rivest, and D. Engels. Security and privacy aspects of low-cost radio frequency identification systems. *Security in pervasive computing*, pages 50–59, 2004.

[5] L. Lu, J. Han, L. Hu, Y. Liu, and L.M. Ni. Dynamic key-updating: Privacy-preserving authentication for rfid systems. In *Proc. of IEEE PERCOM*, 2007.

[6] L. Lu, J. Han, R. Xiao, and Y. Liu. Action: breaking the privacy barrier for rfid systems. In *Proc. of IEEE INFOCOM*, 2009.

[7] L. Yang, J. Han, Y. Qi, and Y. Liu. Identification-free batch authentication for rfid tags. In *Proc. of IEEE ICNP*, 2010.

[8] L. Bolotnyy and G. Robins. Physically unclonable function-based security and privacy in rfid systems. In *Proc. of IEEE PERCOM*, 2007.

[9] Y.K. Lee, L. Batina, and I. Verbauwhede. Ec-rac (ecdlp based randomized access control): Provably secure rfid authentication protocol. In *Proc. of IEEE RFID*, 2008.

[10] Y.K. Lee, L. Batina, and I. Verbauwhede. Untraceable rfid authentication protocols: Revision of ec-rac. In *Proc. of IEEE RFID*, 2009.

[11] L.G. Roberts. Aloha packet system with and without slots and capture. *ACM SIGCOMM Computer Communication Review*, 5(2):28–42, 1975.

[12] B. Sheng, Q. Li, and W. Mao. Efficient continuous scanning in rfid systems. In *Proc. of IEEE INFOCOM*, 2010.

[13] D. Benedetti, G. Maselli, and C. Petrioli. Fast identification of mobile rfid tags. In *Proc. of IEEE MASS*, 2012.

[14] T.F. La Porta, G. Maselli, and C. Petrioli. Anticollision protocols for single-reader rfid systems: Temporal analysis and optimization. *IEEE Transactions on Mobile Computing*, 10(2):267 –279, 2011.

[15] R. Kumar, T.F. La Porta, G. Maselli, and C. Petrioli. Interference cancellation-based rfid tags identification. In *Proc. of ACM MSWiM*, 2011.

[16] J. Capetanakis. Tree algorithms for packet broadcast channels. *IEEE Transactions on Information Theory*, 25(5):505–515, 1979.

[17] J. Myung and W. Lee. Adaptive splitting protocols for rfid tag collision arbitration. In *Proc. of ACM MobiHoc*, 2006.

[18] Epcglobal radio-frequency identity protocols class-1 generation-2 uhf rfid protocol for communications at 860 mhz-960mhz, 2008.

[19] Information technology radio frequency identification for item management part 6: Parameters for air interface communications at 860 mhz to 960 mhz, 2010.

[20] T. Dimitriou. A secure and efficient rfid protocol that could make big brother (partially) obsolete. In *Proc. of IEEE PERCOM*, 2006.

[21] Li Lu, Yunhao Liu, and Xiang-Yang Li. Refresh: Weak privacy model for rfid systems. In *Proc. of IEEE INFOCOM*, 2010.

[22] M. Kodialam and T. Nandagopal. Fast and reliable estimation schemes in rfid systems. In *Proc. of ACM MobiCom*, 2006.

[23] C. Qian, H. Ngan, Y. Liu, and L.M. Ni. Cardinality estimation for large-scale rfid systems. *IEEE Transactions on Parallel and Distributed Systems*, 22(9):1441–1454, 2011.

[24] Y. Zheng and M Li. Pet: Probabilistic estimating tree for large-scale rfid estimation. *IEEE Transactions on Mobile Computing*, 11(11):1763–1774, 2012.

[25] M. Shahzad and A. Liu. Every bit counts - fast and scalable rfid estimation. In *Proc. of ACM MobiCom*, 2012.

[26] T. Li, S. S. Wu, S. Chen, and M. C. K. Yang. Generalized energy-efficient algorithms for the rfid estimation problem. *IEEE/ACM Transactions on Networking*, 20(6):1978 –1990, 2012.

[27] Y. Zheng and M. Li. Zoe: Fast cardinality estimation for large-scale rfid systems. In *Proc. of IEEE INFOCOM*, 2013.

[28] P.H. Cole and D.C. Ranasinghe. Networked rfid systems. *Networked RFID Systems and Lightweight Cryptography*, pages 45–58, 2008.

[29] S. Devadas, E. Suh, S. Paral, R. Sowell, T. Ziola, and V. Khandelwal. Design and implementation of puf-based unclonable rfid ics for anti-counterfeiting and security applications. In *Proc. of IEEE RFID*, 2008.

[30] R. Zhang, Y. Liu, Y. Zhang, and J. Sun. Fast identification of the missing tags in a large rfid system. In *Proc. of IEEE SECON*, 2011.

[31] P. Flajolet and G. Nigel Martin. Probabilistic counting algorithms for data base applications. *Journal of computer and system sciences*, 31(2):182–209, 1985.

[32] R. Motwani and P. Raghavan. *Randomized algorithms*. Chapman & Hall/CRC, 2010.

[33] P.S. Bullen. *Handbook of Means and their Inequalities*. Springer, 2003.

[34] Open rfid lab, http://pdcc.ntu.edu.sg/wands/orl.

Balancing Lifetime and Classification Accuracy of Wireless Sensor Networks

Kush R. Varshney and Peter M. van de Ven
Business Analytics and Mathematical Sciences Department
IBM Thomas J. Watson Research Center
1101 Kitchawan Road, Route 134
Yorktown Heights, New York
{krvarshn,pmvandev}@us.ibm.com

ABSTRACT

Wireless sensor networks are composed of distributed sensors that can be used for signal detection or classification. The likelihood functions of the hypotheses are often not known in advance, and decision rules have to be learned via supervised learning. A specific learning algorithm is Fisher discriminant analysis (FDA), the classification accuracy of which has been previously studied in the context of wireless sensor networks. Previous work, however, does not take into account the communication protocol or battery lifetime; in this paper we extend existing studies by proposing a model that captures the relationship between battery lifetime and classification accuracy. To do so, we combine the FDA with a model that captures the dynamics of the carrier-sense multiple-access (CSMA) algorithm, the random-access algorithm used to regulate communications in sensor networks. This allows us to study the interaction between the classification accuracy, battery lifetime and effort put towards learning, as well as the impact of the back-off rates of CSMA on the accuracy. We characterize the tradeoff between the length of the training stage and accuracy, and show that accuracy is non-monotone in the back-off rate due to changes in the training sample size and overfitting.

Categories and Subject Descriptors

C.2 [**Computer Systems Organization**]: Computer-Communication Networks

General Terms

Performance

Keywords

carrier-sense multiple-access, linear discriminant analysis, sensor networks

1. INTRODUCTION

Wireless sensor networks are used for detection or classification, whether for surveillance, environmental monitoring, or any of the myriad other application domains that are emerging in the age of big data. In many such applications, the likelihood functions of the hypotheses, e.g., the presence or absence of a particular physical phenomenon, are not known before the sensor network is deployed; in these applications, the sensor network requires training prior to operation via supervised learning [13, 18]. The resulting classification accuracy improves with the number of measurements taken during training [17], but increasing the length of the training stage further reduces the limited battery capacity for the operational stage. Therefore, the amount of resources expended during training mediates operational lifetime and accuracy of the sensor network.

The energy consumption of sensor nodes, and thus the lifetime of the network, is dominated by energy expended on communication. Node transmissions in wireless sensor networks are commonly regulated by the carrier-sense multiple-access (CSMA) algorithm [1, 16]. This algorithm is implemented in TinyOS, a popular open source operating system for wireless sensor networks, and is part of the IEEE 802.15.4 standard for wireless sensor network communication. Nodes using CSMA access the medium in a distributed manner, and wait some random back-off time between successive transmissions.

In this paper we consider a scenario where a set of measurements and classification is required every time unit. Only nodes that are active at that time perform a measurement and transmit the result, so the number of measurements collected varies over time. We develop and analyze a model of sensor networks that perform supervised classification *in situ*, using the Fisher discriminant analysis (FDA) learning algorithm, with a training stage and an operational stage enabled by CSMA. The specific analysis of focus is the relationship between operational accuracy and lifetime, which we show to be of a fundamentally different character than for the case of detection with known likelihood functions, due to overfitting. In characterizing operational classification accuracy (in contrast to classification accuracy on training samples), we make use of generalization approximations for FDA developed by Raudys et al. [15].

Battery capacity is characterized by the number of transmissions that can be performed, whether they be during training or operation. As every measurement corresponds to one transmission, the expected network lifetime is inversely proportional to the node throughput in our model. The performance measures of interest are the classification accuracy and operational lifetime, which is the lifetime spent in the operational stage, not in the training stage. The two main parameters available for configuring the sensor network are the CSMA back-off rates (the reciprocal of the mean back-off time), and the fraction of the lifetime spent in training.

As the back-off rates of the nodes increase, states with many actively transmitting nodes are more likely. This requires more energy consumption, and also affects classification accuracy. Classification accuracy is not monotonically increasing in the number of active nodes due to the phenomenon of overfitting [17]. We also show that operational accuracy as a function of back-off rate exhibits the hallmarks of overfitting in one regime, but in another regime, has a behavior quite different than any behavior usually encountered in statistical learning.

The analysis of supervised classification for sensor networks in the literature is limited [13]: investigations have been predominantly concerned with the detection case where the likelihood functions are known. Moreover, sensor network research tends to separate learning issues from the communication aspect. There are several works that model CSMA communication in sensor networks generally, e.g. [8] and references therein, but not with the supervised classification application as part of the formulation. Cross-layer work that does consider the networking issues together with a detection or estimation application, e.g., the correlation-based collaborative MAC protocol [20], is again focused on the case with known likelihoods. So although FDA and the performance of CSMA-like algorithms have been widely studied, we are the first to jointly consider classification accuracy and communication aspects of wireless sensor networks.

We consider both the case of statistically independent and identically distributed (i.i.d.) measurements from different nodes, and the case of measurements exhibiting correlation that depends on the spatial distance between the nodes. Having i.i.d. measurements is a common simplifying assumption in wireless sensor network detection [6]. A model with spatially-correlated measurements is much closer to reality in most applications [10]. The spatial correlation is encoded via a Gauss–Markov random field (GMRF) model.

The CSMA model under consideration was first introduced in the 1980s in the context of packet radio networks [5] and was later applied to networks based on the IEEE 802.11 standard [7, 19]. More recently, it has been used to study so-called adaptive CSMA algorithms, where the back-off rate of the nodes changes with their congestion level [14, 4, 9]. Although the representation of binary exponential back-off mechanism in the above-mentioned models is far less detailed than in the landmark work of Bianchi [3] and similar results focusing on sensor networks, e.g., [11], the general interference graph offers greater versatility and covers a broad range of topologies.

The remainder of the paper is organized as follows. In Section 2, we describe the setup of the sensor network system from the FDA supervised classification perspective and in Section 3, we describe the setup of the sensor network system from the CSMA communication perspective. In Section 4 we derive the relationship between operational lifetime and accuracy, and Section 5 presents numerical results of lifetime and accuracy for two special cases, illustrating

the complicated balancing act that is involved. Section 6 provides a discussion and several ideas for future research.

2. FISHER DISCRIMINANT ANALYSIS

Consider a sensor network consisting of n sensor nodes each taking a scalar measurement combined into a joint measurement vector $\mathbf{x}_j \in \mathbb{R}^n$. In the general supervised classification problem, we are given m sample pairs $\{(\mathbf{x}_1, y_1), \ldots, (\mathbf{x}_m, y_m)\}$ known as the training set, with measurement $\mathbf{x}_j$ and class label or hypothesis $y_j \in \{0, 1\}$. The training samples are acquired by the network after deployment and before the operational stage. Once the training set is acquired, the samples are used to learn a classification function or decision rule $\hat{y}(\cdot)$ that will accurately classify new unseen and unlabeled samples $\mathbf{x}$ from the same distribution from which the training set was drawn.

In this paper, we focus on a simple, classical decision rule $\hat{y}$, the Fisher discriminant analysis classifier:

$$\hat{y}(\mathbf{x}) = \text{step}(\mathbf{w}^T \mathbf{x} + \theta), \quad (1)$$

where $\mathbf{w} = \left(\hat{\mathbf{\Sigma}}_0 + \hat{\mathbf{\Sigma}}_1\right)^{-1} (\hat{\boldsymbol{\mu}}_1 - \hat{\boldsymbol{\mu}}_0)$, $\theta = -\frac{1}{2} \mathbf{w}^T (\hat{\boldsymbol{\mu}}_0 + \hat{\boldsymbol{\mu}}_1)$, and $\hat{\boldsymbol{\mu}}_0$, $\hat{\boldsymbol{\mu}}_1$, $\hat{\mathbf{\Sigma}}_0$, and $\hat{\mathbf{\Sigma}}_1$ are the conditional sample means and covariances of the m training samples. The FDA rule is a plug-in classifier that follows from the likelihood ratio test for optimal signal detection between Gaussian signals with the same covariance and different means.

Given the FDA decision rule (1), we would like to characterize its performance, specifically its classification accuracy as it generalizes to new unseen samples in the operational stage. Generalization accuracy, however, is a functional of the underlying data distribution $f_{\mathbf{x},y}(\mathbf{x}, y)$, and we must first specify a probability distribution of the sensor measurements. We employ the same GMRF statistical model for sensor measurements as [17, 2]. That is, the n sensor nodes are deployed on the plane with spatial locations $\mathbf{v}_i \in \mathbb{R}^2$, $i = 1, \ldots, n$. The likelihoods of the two hypotheses are Gaussian: $f_{\mathbf{x}|y}(\mathbf{x}|y = 0) \sim \mathcal{N}(\boldsymbol{\mu}_0, \mathbf{\Sigma})$ and $f_{\mathbf{x}|y}(\mathbf{x}|y = 1) \sim \mathcal{N}(\boldsymbol{\mu}_1, \mathbf{\Sigma})$. The prior probabilities of the hypotheses are equal: $\Pr[y = 0] = \Pr[y = 1] = 1/2$. For simplicity of exposition $\boldsymbol{\mu}_0 = \mathbf{0}$ (the vector of all zeroes) and $\boldsymbol{\mu}_1 = \mathbf{1}$ (the vector of all ones).

The covariance structure is based on the Euclidean nearest neighbor graph of the sensors: the (undirected) nearest neighbor graph contains an edge between sensor i and sensor i' if sensor i is the nearest neighbor of sensor i' or if sensor i' is the nearest neighbor of sensor i. The set of edges in the nearest neighbor graph is denoted $\mathcal{E}$. The diagonal elements of $\mathbf{\Sigma}$ are all equal to σ^2. The elements of $\mathbf{\Sigma}$ corresponding to edges in the nearest neighbor graph are:

$$\{\mathbf{\Sigma}\}_{ii'} = \sigma^2 g(d(\mathbf{v}_i, \mathbf{v}_{i'})), \quad (i, i') \in \mathcal{E}, \quad (2)$$

where $g(\cdot) : \mathbb{R}^+ \to (0, 1)$ is a decreasing function that encodes correlation decay with distance. The inverse covariance matrix $\mathbf{J} = \mathbf{\Sigma}^{-1}$ is used to specify the remaining elements. The off-diagonal elements of $\mathbf{J}$ corresponding to sensor pairs (i, i') that do not have an edge in the nearest neighbor graph are zero, i.e.

$$\{\mathbf{J}\}_{ii'} = 0, \quad i \neq i', (i, i') \notin \mathcal{E}. \quad (3)$$

We also consider the case of i.i.d. observations in the paper, in which case $g(d) = 1$ for $d = 0$ and $g(d) = 0$ otherwise.

A highly accurate approximation of generalization accuracy $A = \Pr[\hat{y}(\mathbf{x}) = y]$ for FDA with GMRF is found in [17]. Based on [15], this approximation is

$$A \approx \Phi\left(\frac{\delta}{2}\left[\left(1 + \frac{4n}{m\delta^2}\right)\frac{m}{m-n}\right]^{-\frac{1}{2}}\right), \quad m > n, \quad (4)$$

where $\Phi(\cdot)$ is the Gaussian cumulative distribution function and δ is known as the Mahalanobis distance

$$\delta^2 = \frac{n}{\sigma^2} - \frac{2}{\sigma^2}\sum_{(i,i')\in\mathcal{E}}\frac{g(d(\mathbf{v}_i,\mathbf{v}_{i'}))}{1 + g(d(\mathbf{v}_i,\mathbf{v}_{i'}))}. \quad (5)$$

In case $m \leq n$ there are insufficient training samples for accurate classification and we have $A = 0.5$. In the i.i.d. case, δ^2 simplifies to $\frac{n}{\sigma^2}$.

3. CARRIER-SENSE MULTIPLE-ACCESS

The CSMA algorithm is a random-access algorithm where nodes decide for themselves when to transmit based only on local information. We assume that the n nodes share the wireless medium according to a CSMA-type protocol.

The network is described by an undirected conflict graph $(\mathcal{V}, \mathcal{E})$, where the set of vertices $\mathcal{V} = \{1, \ldots, n\}$ represents the nodes of the network and the set of edges $\mathcal{E} \subseteq \mathcal{V} \times \mathcal{V}$ indicates which pairs of nodes cannot activate simultaneously. For ease of presentation we assume that the conflict graph is the same as the nearest neighbor graph introduced in Section 2. Nodes that are neighbors in the conflict graph are prevented from simultaneous activity by the carrier-sensing mechanism. An inactive node is said to be blocked whenever any of its neighbors is active, and unblocked otherwise.

The nodes continually monitor the phenomenon of interest, i.e. they are not event-triggered. The transmission times of node i are independent and exponentially distributed with unit mean. When node i is blocked, it remains silent until all its neighbors are inactive, when it tries to activate after an exponentially distributed back-off time with mean $1/\nu_i$.

The set Ω of all feasible joint activity states of the network in this case corresponds to the incidence vectors of all independent sets of the conflict graph. Let the network state at time t be denoted by $\mathbf{Y}(t) = (Y_1(t), Y_2(t), \ldots, Y_n(t)) \in \Omega$, with $Y_i(t)$ indicating whether node i is active at time t ($Y_i(t) = 1$) or not ($Y_i(t) = 0$). Then $\{\mathbf{Y}(t)\}_{t\geq 0}$ is a Markov process which is fully specified by the state space Ω and the transition rates

$$r(\omega, \omega') = \begin{cases} \nu_i, & \text{if } \omega' = \omega + e_i \in \Omega, \\ 1, & \text{if } \omega' = \omega - e_i \in \Omega, \\ 0, & \text{otherwise.} \end{cases} \quad (6)$$

Since $\mathbf{Y}(t)$ is reversible (see [5]), the following product-form stationary distribution π exists:

$$\pi(\boldsymbol{\omega}) = \begin{cases} Z^{-1}\prod_{i=1}^{n}\nu_i^{\omega_i}, & \text{if } \boldsymbol{\omega} \in \Omega, \\ 0, & \text{otherwise,} \end{cases} \quad (7)$$

where Z is the normalization constant that makes π a probability measure.

We assume that nodes attempt to send out data constantly for streaming classification, so the output of a node is limited only by its back-off process. The rate θ_i at which sensor node i makes and transmits an observation is referred

to as the throughput of this node, and may be written as

$$\theta_i = \sum_{\boldsymbol{\omega}\in\Omega}\pi(\boldsymbol{\omega})\mathbb{I}_{\{\omega_i=1\}}. \quad (8)$$

Sensor nodes rely on batteries for energy, and we assume that all nodes have a battery that allows them to make l transmissions each before their battery is drained. Consequently, the expected lifetime of a node can be written as

$$T_i = \frac{l}{\theta_i}. \quad (9)$$

The activity process in the training stage is the same as in the operational stage. We denote by $0 \leq \alpha \leq 1$ the fraction of the battery capacity that is dedicated to training the sensor network. So the testing lifetime of node i is αT_i, and the operational lifetime is:

$$U_i = (1-\alpha)T_i = (1-\alpha)\frac{l}{\theta_i}. \quad (10)$$

The model we have specified is fully general for any n-node conflict graph. We work with this general model throughout the remainder of the paper, but also focus on two illustrative special cases. The two special cases of the CSMA network we consider are an n-node network where all networks are disjoint and a three-node linear network.

3.1 Independent Nodes

First, consider an n-node network where all nodes can be active simultaneously. This corresponds to an interference graph with an empty edge set $\mathrm{E} = \phi$ and $\Omega = \{0,1\}^n$. Note that although in this case the lack of interference eliminates the need for a MAC protocol like CSMA, it is nevertheless interesting because it allows us to derive closed-form results. Moreover, deployment of sensor networks is typically random, and such disjoint network may arise in practice. We set $\nu_i \equiv \nu$ so the stationary distribution (7) simplifies to

$$\pi(\boldsymbol{\omega}) = \frac{1}{(\nu+1)^n}\nu^{\|\boldsymbol{\omega}\|_1}. \quad (11)$$

The stationary probability of any particular state only depends on the number of active nodes in that state and on the back-off rate ν. Thus, for notational convenience, we introduce $\pi(k)$ as the stationary probability of being in any state with k active nodes, and we write

$$\pi(k) = \frac{1}{(\nu+1)^n}\binom{n}{k}\nu^k, \quad (12)$$

which follows since there are $\binom{n}{k}$ different activity states with k nodes transmitting.

With equal back-off rates and disjoint nodes, the stationary throughput (8) is the same for all nodes

$$\theta_i \equiv \theta = \frac{\nu}{\nu+1}. \quad (13)$$

Moreover, all nodes have the same lifetime, and the operational lifetime of the network may be written as

$$U_i \equiv U = (1-\alpha)l\frac{\nu+1}{\nu}. \quad (14)$$

3.2 A Three-Node Linear Network

Consider the three-node network where the nodes are positioned such that the carrier-sensing mechanism prevents node 2 from activating while either node 1 or node 3 is active. Nodes 1 and 3 can be active simultaneously, but their observations are correlated. The network can take five possible states

$$\Omega = \{\mathbf{0}, \mathbf{e}_1, \mathbf{e}_2, \mathbf{e}_3, \mathbf{e}_1 + \mathbf{e}_3\}. \qquad (15)$$

Using (7) we compute the following stationary probabilities:

$$\pi(\mathbf{0}) = Z^{-1},$$
$$\pi(\mathbf{e}_i) = Z^{-1}\nu_i, \quad i = 1, 2, 3,$$
$$\pi(\mathbf{e}_1 + \mathbf{e}_3) = Z^{-1}\nu_1\nu_3. \qquad (16)$$

In order to make sure that all nodes have the same throughput and lifetime, we fix some parameter $\eta > 0$ and choose $\nu_1 = \nu_3 = \eta$ and $\nu_2 = \eta(\eta + 1)$. So node 2 has a shorter mean back-off time in order to compensate for its disadvantageous position in the network, and all nodes have throughput (see [19])

$$\theta_i \equiv \theta = \frac{\eta}{2\eta + 1} \qquad (17)$$

and operational lifetime

$$U_i \equiv U = (1 - \alpha)l\frac{2\eta + 1}{\eta}. \qquad (18)$$

The normalization constant with these back-off rates is

$$Z = 2\eta^2 + 3\eta + 1. \qquad (19)$$

4. RELATIONSHIP BETWEEN LIFETIME AND ACCURACY

We are now in position to combine the FDA model from Section 2 and the CSMA model presented in Section 3 to derive the relationship between generalization accuracy and operational lifetime. This is mediated by two parameters: the back-off rate ν or η and the fraction of the lifetime spent in the training stage α.

Due to the interference constraints and the intermittent nature of CSMA communications, not all nodes produce and validly communicate measurements at all times. So the training samples are acquired under different activity states $\boldsymbol{\omega} \in \Omega$. Thus studying the relationship between accuracy and lifetime is not simply a matter of joining the corresponding expressions (4) and (10).

This issue of incomplete data due to the activity process can be addressed in several ways, including data imputation. Although various elaborate schemes are available, they come at the cost of additional computation, communication, and coordination that are at a premium in the sensor network setting. Instead, we choose to model the classification by having separately learned classifiers for different activity states. In the operational stage the appropriate classifier is used for prediction based on the activity state of the measurements. In this setup, we associate with each state $\boldsymbol{\omega}$ a number of training samples

$$m_{\boldsymbol{\omega}} = \alpha T \pi(\boldsymbol{\omega}). \qquad (20)$$

Then we compute the overall generalization accuracy as the weighted sum of the individual generalization accuracies for each pattern according to their stationary probabilities:

$$A \approx \sum_{\boldsymbol{\omega} \in \Omega} \pi(\boldsymbol{\omega}) \Phi \left(\frac{\delta}{2} \left[\left(1 + \frac{4\|\boldsymbol{\omega}\|_1}{m_{\boldsymbol{\omega}}\delta^2} \right) \frac{m_{\boldsymbol{\omega}}}{m_{\boldsymbol{\omega}} - \|\boldsymbol{\omega}\|_1} \right]^{-\frac{1}{2}} \right), \qquad (21)$$

with π the stationary distribution (7) and $m_{\boldsymbol{\omega}}$ as in (20).

We now compute the generalization accuracies for the two special cases introduced in Section 3 with the GMRF of the measurements having the same graph structure as the CSMA network.

4.1 Independent Nodes

As discussed in Section 3 for a set of n disjoint nodes, all patterns with k active nodes have the same stationary probability $\pi(k)$ given in (12), and all nodes have equal throughput (13) and lifetime (14). We denote by m_k the number of training samples for patterns with k active nodes, and by summing (20) over all states with k active nodes, we write

$$m_k = \alpha l \binom{n}{k} \frac{\nu^{k-1}}{(\nu + 1)^{n-1}}. \qquad (22)$$

As discussed in Section 2, with i.i.d. measurements from n sensors, the squared Mahalanobis distance is $\frac{n}{\sigma^2}$. Thus, with k active sensors, the squared Mahalanobis distance is $\frac{k}{\sigma^2}$. Substituting the expression for the stationary distribution (12) and the number of training samples (22) into the expression for the generalization accuracy (21) we obtain

$$A \approx \frac{1}{(\nu + 1)^n} \sum_{k=0}^{n} \binom{n}{k} \nu^k \Phi \left(\frac{\sqrt{k}}{2\sigma} \left[\left(1 + \frac{4\sigma^2}{m_k} \right) \frac{m_k}{m_k - k} \right]^{-\frac{1}{2}} \right). \qquad (23)$$

4.2 A Three-Node Linear Network

Recall from Section 3.2 that the three-node network has 5 feasible states. The four non-empty states have squared Mahalanobis distance

$$\delta_{\mathbf{e}_i}^2 = \frac{1}{\sigma^2}, \quad i = 1, 2, 3,$$
$$\delta_{\mathbf{e}_1 + \mathbf{e}_3}^2 = \frac{2}{\sigma^2} \cdot \frac{1}{1 + g(d(\mathbf{v}_1, \mathbf{v}_2))g(d(\mathbf{v}_2, \mathbf{v}_3))}. \qquad (24)$$

Note that since $g < 1$, the Mahalanobis distance of the larger state is larger than that of the states with only one node active, and is more valuable.

Evaluating (20) we obtain an expression for the number of training samples for each state:

$$m_{\mathbf{0}} = \alpha l \frac{1}{\eta^2 + \eta}$$
$$m_{\mathbf{e}_i} = \alpha l \frac{1}{\eta + 1}, \quad i = 1, 3,$$
$$m_{\mathbf{e}_2} = \alpha l,$$
$$m_{\mathbf{e}_1 + \mathbf{e}_3} = \alpha l \frac{\eta}{\eta + 1}. \qquad (25)$$

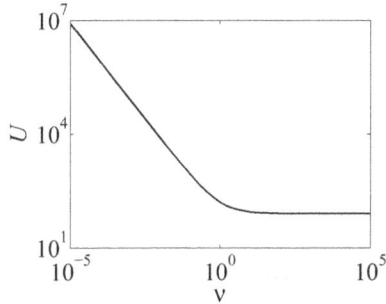

Figure 1: Operational lifetime as function of back-off rate.

By weighting the individual generalization accuracies (21),

$$A \approx$$

$$\frac{1}{2\eta^2 + 3\eta + 1} \left[\frac{1}{2} + \eta\Phi\left(\frac{1}{2\sigma} \left[\left(1 + \frac{4\sigma^2}{m_{e_1}}\right) \frac{m_{e_1}}{m_{e_1} - 1} \right]^{-\frac{1}{2}} \right) + \right.$$

$$(\eta^2 + \eta)\Phi\left(\frac{1}{2\sigma} \left[\left(1 + \frac{4\sigma^2}{m_{e_2}}\right) \frac{m_{e_2}}{m_{e_2} - 1} \right]^{-\frac{1}{2}} \right) +$$

$$\eta\Phi\left(\frac{1}{2\sigma} \left[\left(1 + \frac{4\sigma^2}{m_{e_3}}\right) \frac{m_{e_3}}{m_{e_3} - 1} \right]^{-\frac{1}{2}} \right) +$$

$$\left. \eta^2\Phi\left(\frac{\delta_{e_1+e_3}}{2} \left[\left(1 + \frac{8}{m_{e_1+e_3}\delta^2_{e_1+e_3}}\right) \frac{m_{e_1+e_3}}{m_{e_1+e_3} - 2} \right]^{-\frac{1}{2}} \right) \right]. \tag{26}$$

5. EXAMPLES

In Section 4 we derived the operational lifetime U, the number of training samples m_ω and the operational classification accuracy A for a wireless sensor network with random-access communication as a function of the back-off rate ν and the fraction of the lifetime spent in training α. Here we numerically evaluate these quantities for the special cases of independent nodes and the three-node linear network. We include a comparison to the Bayes optimal detector with known likelihood functions and see that the accuracy behavior is markedly different. Additionally, we see that there are two different regimes in the accuracy behavior as a function of the back-off rate, the second regime different than that usually seen in statistical learning. The overall behavior is unique due to the combination of CSMA and FDA.

5.1 Independent Nodes

We consider a network of $n = 8$ independent nodes with $l = 100$ transmissions allowed by the battery per node. The sensor measurement noise variance is set to $\sigma^2 = 1$. Other parameter settings produce qualitatively similar results. First, in Fig. 1, we plot the operational lifetime U as a function of the back-off rate ν for a fixed lifetime fraction devoted to training: $\alpha = 0.2$. The operational lifetime is very high with low back-off rate because the system is mostly in the state with $k = 0$ active sensors, which does not drain sensor batteries at all. Once states with more active sensors become more probable with increasing back-off rate, the lifetime drops rapidly to $U = (1 - \alpha)l$, the lifetime

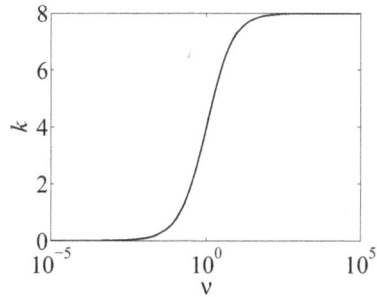

Figure 2: Expected number of active sensors as function of back-off rate.

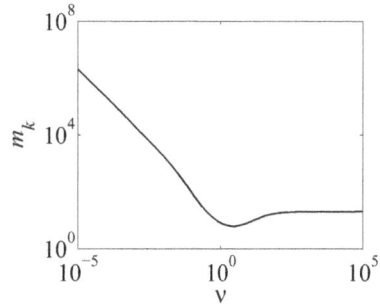

Figure 3: Expected number of training samples per state classifier as function of back-off rate.

in the case where all nodes are always active. Fig. 2 shows the expected number of active sensors $\bar{k}$ as a function of ν.

One of the components of the generalization accuracy expression (23) is the number of active sensors k; the other is m_k, the number of training samples. In Fig. 3 we set $\alpha = 0.2$, and plot $\bar{m}_k$, the weighted average over k of m_k:

$$\bar{m}_k = \sum_{k=0}^{n} \pi(k)m_k. \tag{27}$$

Interestingly, this number is not monotonically decreasing as a function of ν like we see with the operational lifetime. This is because when several different states all have non-negligible probability, the acquired training samples get divided to all of the different states. Initially the number of training samples is very high because almost all of the training samples are for the state with no active sensors. For large ν the number of training samples approaches αl.

Now that we have looked at $\bar{k}$ and $\bar{m}_k$, we now examine the accuracy A as a function of ν, plotted in Fig. 4 for $\alpha = 0.2$. The figure also shows the detection accuracy of the Bayes optimal decision rule with known likelihood functions. The Bayes optimal accuracy is monotonically increasing in ν, following the expected value of k. On the other hand, the FDA classification accuracy first increases in ν, starts decreasing with local bumps, and then increases. The local bumps arise from the generalization accuracy behavior for different states, which are shown in Fig. 5. Specifically, the figure shows the different $\Phi(\cdot)$ components of A given in (23); these are functions of ν because the m_k are.

The phenomenon of overfitting, demonstrated in [17], is that for a fixed number of training samples and an increasing number of sensors, the generalization accuracy first increases and then decreases. Conversely, for a fixed number of sensors and an increasing number of training samples, the

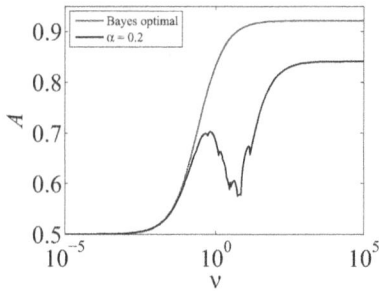

Figure 4: Operational classification accuracy as function of back-off rate.

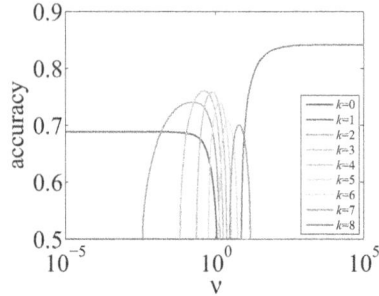

Figure 5: Operational classification accuracy of different state classifiers as function of back-off rate.

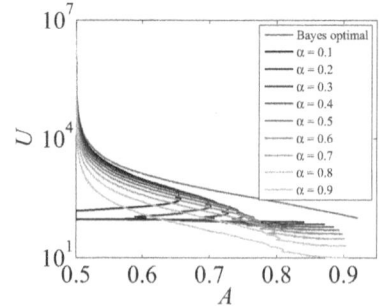

Figure 6: Relationship between operational lifetime and operational classification accuracy.

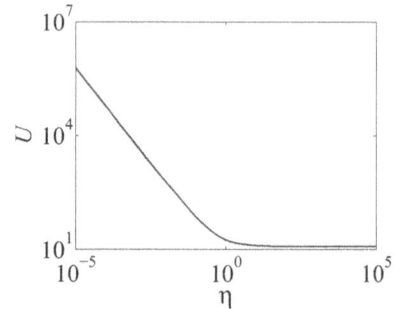

Figure 7: Operational lifetime as function of back-off rate.

generalization accuracy monotonically increases. With the wireless sensor network with CSMA communication, both of these effects intermingle as a function of ν because both the number of active sensors and the number of training samples changes. The initial increase and decrease in A is the manifestation of overfitting, where the generalization accuracy is best around $k = 3$ and $k = 4$. For large ν, the number of active sensors is essentially fixed at $k = n$ (seen in Fig. 2) and the number of training samples increases (seen in Fig. 3), resulting in improving classification accuracy.

Finally, we examine the relationship between lifetime and accuracy in Fig. 6. For comparison, the figure shows the relationship for the Bayes optimal decision rule, in which there is no lifetime devoted to training, only to operation. All curves represent parametric functions of ν, and correspond to different values of α ranging from 0.1 to 0.9. Different values of α contribute to the frontier of the relationship, the parts of the curve closest to the Bayes decision rule and closest to the top right corner of the plot. At the extreme of random guessing, i.e. $A = 0.5$, lifetime is maximized by not doing any training, i.e. $\alpha = 0$. Small but increasing values of α then contribute to the frontier until a point when smaller values of α abruptly again become part of the frontier. Very large values of α contribute to the frontier only when the very best accuracies possible are desired.

5.2 A Three-Node Linear Network

Having seen quite interesting behaviors for independent nodes, we now turn to a three-node linear network with correlated measurements and conflict graph preventing sensors 1 and 2, and sensors 2 and 3 from transmitting simultaneously. We present similar plots as in Section 5.1, with $l = 10$ and $\sigma^2 = 1$. We present results for $g(d(\mathbf{v}_1, \mathbf{v}_2)) = g(d(\mathbf{v}_2, \mathbf{v}_3)) = \frac{1}{4}$ as the distance-based correlations. For the

dependent linear network case, we see more or less the same behavior as for the independent nodes in Fig. 7–Fig. 12. Fig. 7, Fig. 9, and Fig. 11 are given for $\alpha = 0.4$. One difference from the independent nodes case is that for the e_2 state, m_ω is constant and not a function of η.

6. CONCLUSION AND OUTLOOK

In this paper we proposed a model to investigate the interaction between generalization accuracy and operational lifetime in wireless sensor networks. We demonstrate that this relationship is highly nontrivial, due to the joint effects of overfitting, the number of training samples and the changing weights of the various states. The two special cases for which we provide result plots are qualitatively similar, and changing the conflict graph and spatial correlation does not affect the general behavior.

For small increasing back-off rates, the accuracy improves until peaking. At intermediate back-off rates, the accuracy gets worse due to overfitting, and then improves again for large back-off rates due to increasing training samples per state. Due to these different regimes of increasing and decreasing accuracy in the back-off rate, along with different values of the fraction of lifetime to spend in training affecting both the operational lifetime and the accuracy, setting the parameters ν and α to achieve certain target performance is not straightforward. The parameterized curves in Fig. 6 and Fig. 12 give (not necessarily intuitive) recommendations for balancing lifetime and accuracy as a function of the back-off rate and training fraction parameters.

The classification and communication models we have used, i.e. FDA with GMRF-dependent sensor measurements and binary exponential back-off mechanism, are certainly simplified, but are general and amenable to analysis. The guidelines and behaviors we see will transfer over in a broad sense

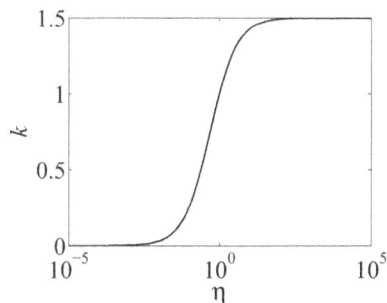

Figure 8: Expected number of active sensors as function of back-off rate.

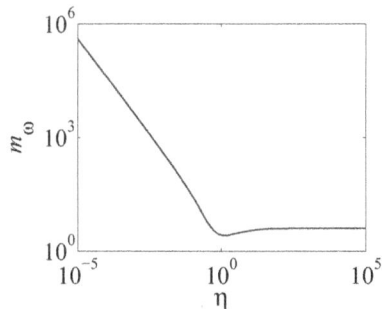

Figure 9: Expected number of training samples per state classifier as function of back-off rate.

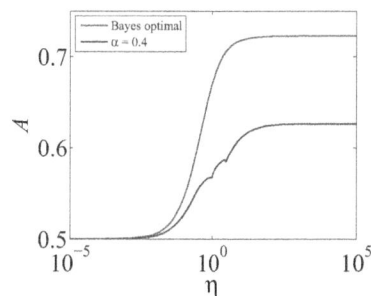

Figure 10: Operational classification accuracy as function of back-off rate.

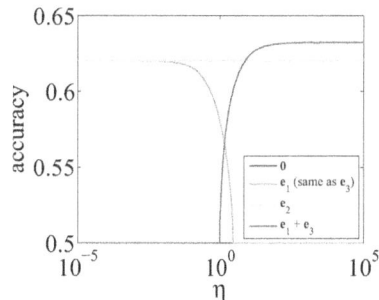

Figure 11: Operational classification accuracy of different state classifiers as function of back-off rate.

to other classifiers and other similar random-access communication protocols.

6.1 Outlook

Having made the connection between lifetime and accuracy in Fig. 6 and Fig. 12, the next step is to find the values of α and ν_i that achieve a certain target performance. For example, we may want to maximize the lifetime of the network, subject to certain accuracy constraints $\beta \in (0, 1)$:

$$(\alpha^*, \nu^*) = \operatorname{argmax} U(\alpha, \nu_1, \ldots, \nu_n)$$
$$\text{s.t. } A(\alpha, \nu_1, \ldots, \nu_n) \geq \beta. \qquad (28)$$

The optimization problem (28) is non-convex (as illustrated in Fig. 6), and we may approximate its solution using numerical methods. Some preliminary results are shown in Figs. 13 and 14, where we plot the solution to (28) for increasing β, in the model with n independent nodes with the parameters as in Section 5. Fig. 13 shows that the ν increases almost monotonically, and jumps to infinity around $\beta = 0.772$. In practice we see that the back-off rate is constrained by physical limitations and by the communication protocol, so ν is bounded from above. Fig. 14 shows a more irregular behavior for α, with a sharp drop when ν jumps to infinity.

The non-monotonicity of the classification accuracy in the back-off rates makes an analytic approach to optimization difficult, and an alternative solution would be to approximate the expression for the detection accuracy (21) with some convex function. This would reduce the complexity of numerical optimization, and may even allow for analytical results.

The effect of overfitting for medium back-off rates can be mitigated by choosing different back-off rates of the training stage and operational stage. For example, choosing larger

back-off rates during training should increase the number of samples for states with many active nodes, thus reducing the risk of overfitting. Although this would simultaneously reduce the number of samples for smaller states, the risk of overfitting is not as high there due to the smaller number of active nodes.

Another direction for future research is to model temporal correlation in the sensor measurements in addition to spatial correlation. In the present work, successive measurements in time are assumed independent, but including temporal correlation is more realistic. If temporal correlation is part of the sensing and classification model, its interaction with the temporal back-off mechanism may produce quite interesting phenomena. Moreover, many other properties and parameters of the network such as link quality, physical properties measured in joules, watts and volts, network diameter, and multihop routing can be analyzed.

Finally, we also mention that asymptotic analysis is of interest in the future study of this cross-layer supervised learning and random-access communication setup. Developing expressions for the three-node dependent network, e.g., (25) and (26), requires us to keep track of many details; larger networks will require us to keep track of many more. By performing an asymptotic analysis of an increasing number of randomly placed sensor nodes with constant density we are able to eliminate many such details in the sensor network generalization error using geometric probability [12]. Having now set forth this extended model with CSMA communication, similar asymptotic analysis using geometric probability is certainly warranted. Such asymptotics can also inform classifier design in that a separate classifier for each specific state not be required.

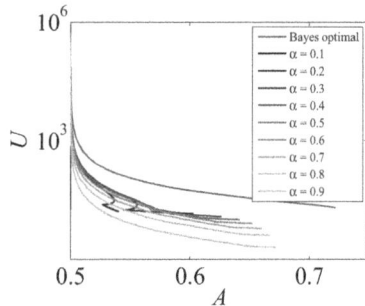

Figure 12: Relationship between operational lifetime and operational classification accuracy.

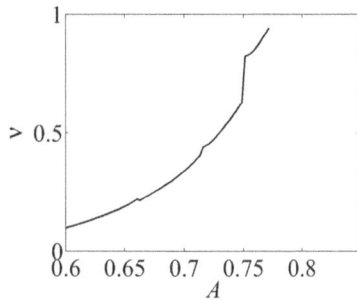

Figure 13: ν^* as function of desired accuracy.

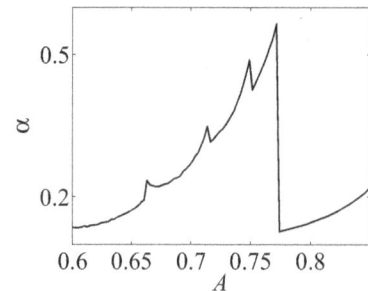

Figure 14: α^* as function of desired accuracy.

7. REFERENCES

[1] I. F. Akyildiz, W. Su, Y. Sankarasubramaniam, and E. Cayirci. A survey on sensor networks. *IEEE Commun. Mag.*, 40(8):102–114, Aug. 2002.

[2] A. Anandkumar, L. Tong, and A. Swami. Detection of Gauss–Markov random fields with nearest-neighbor dependency. *IEEE Trans. Inf. Theory*, 55(2):816–827, Feb. 2009.

[3] G. Bianchi. Performance analysis of the IEEE 802.11 distributed coordination function. *IEEE J. Sel. Areas Commun.*, 18(3):535–547, Mar. 2000.

[4] T. Bonald and M. Feuillet. On the stability of flow-aware CSMA. *Perform. Evaluation*, 67(11):1219–1229, Nov. 2010.

[5] R. Boorstyn, A. Kershenbaum, B. Maglaris, and V. Sahin. Throughput analysis in multihop CSMA packet radio networks. *IEEE Trans. Commun.*, COM-35(3):267–274, Mar. 1987.

[6] J.-F. Chamberland and V. V. Veeravalli. Decentralized detection in sensor networks. *IEEE Trans. Signal Process.*, 51(2):407–416, Feb. 2003.

[7] M. Durvy, O. Dousse, and P. Thiran. Self-organization properties of CSMA/CA systems and their consequences on fairness. *IEEE Trans. Inf. Theory*, 55(3):931–943, Mar. 2009.

[8] E. Feo and G. A. Di Caro. An analytical model for IEEE 802.15.4 non-beacon enabled CSMA/CA in multihop wireless sensor networks. Technical Report 05-11, Istituto Dalle Molle di Studi sull'Intelligenza Artificiale, Lugano, Switzerland, May 2011.

[9] L. Jiang and J. Walrand. A distributed CSMA algorithm for throughput and utility maximization in wireless networks. *IEEE/ACM Trans. Netw.*, 18(3):960–972, June 2010.

[10] A. Jindal and K. Psounis. Modeling spatially correlated data in sensor networks. *ACM Trans. Sensor Netw.*, 2(4):466–499, Nov. 2006.

[11] T. Park, T. Kim, J. Choi, S. Choi, and W. Kwon. Throughput and energy consumption analysis of IEEE 802.15.4 slotted CSMA/CA. *Electron. Lett.*, 41(18):1017–1019, Sept. 2005.

[12] M. D. Penrose and J. E. Yukich. Weak laws of large numbers in geometric probability. *Ann. Appl. Prob.*, 13(1):277–303, Jan. 2003.

[13] J. B. Predd, S. R. Kulkarni, and H. V. Poor. Distributed learning in wireless sensor networks. *IEEE Signal Process. Mag.*, 23(4):56–69, July 2006.

[14] S. Rajagopalan, D. Shah, and J. Shin. Network adiabatic theorem: An efficient randomized protocol for content resolution. In *Proc. ACM SIGMETRICS/Performance*, pages 133–144, Seattle, WA, June 2009.

[15] Š. Raudys and D. M. Young. Results in statistical discriminant analysis: A review of the former Soviet Union literature. *J. Multivariate Anal.*, 89(1):1–35, Apr. 2004.

[16] Y. C. Tay, K. Jamieson, and H. Balakrishnan. Collision-minimizing CSMA and its applications to wireless sensor networks. *IEEE J. Sel. Areas Commun.*, 22(6):1048–1057, Aug. 2004.

[17] K. R. Varshney. Generalization error of linear discriminant analysis in spatially-correlated sensor networks. *IEEE Trans. Signal Process.*, 60(6):3295–3301, June 2012.

[18] K. R. Varshney and A. S. Willsky. Linear dimensionality reduction for margin-based classification: High-dimensional data and sensor networks. *IEEE Trans. Signal Process.*, 59(6):2496–2512, June 2011.

[19] P. M. van de Ven, J. S. H. van Leeuwaarden, D. Denteneer, and A. J. E. M. Janssen. Spatial fairness in linear random-access networks. *Perform. Evaluation*, 69(3–4):121–134, Mar.–Apr. 2012.

[20] M. C. Vuran and I. F. Akyildiz. Spatial correlation-based collaborative medium access control in wireless sensor networks. *IEEE/ACM Trans. Netw.*, 14(2):316–329, Apr. 2006.

FAVOR: Frequency Allocation for Versatile Occupancy of spectRum in Wireless Sensor Networks

Feng Li*   Jun Luo*   Gaotao Shi†   Ying He*

*School of Computer Engineering, Nanyang Technological University, Singapore
†School of Computer Science and Technology, Tianjin University, China
{fli3, junluo, yhe}@ntu.edu.sg, shgt@tju.edu.cn

ABSTRACT

While the increasing scales of the recent WSN deployments keep pushing a higher demand on the network throughput, the 16 orthogonal channels of the ZigBee radios are intensively explored to improve the parallelism of the transmissions. However, the interferences generated by other ISM band wireless devices (e.g., WiFi) have severely limited the usable channels for WSNs. Such a situation raises a need for a spectrum utilizing method more efficient than the conventional multi-channel access. To this end, we propose to shift the paradigm from discrete channel allocation to continuous frequency allocation in this paper. Motivated by our experiments showing the flexible and efficient use of spectrum through continuously tuning channel center frequencies with respect to link distances, we present FAVOR (Frequency Allocation for Versatile Occupancy of spectRum) to allocate proper center frequencies in a continuous spectrum (hence potentially overlapped channels, rather than discrete orthogonal channels) to nodes or links. To find an optimal frequency allocation, FAVOR creatively combines location and frequency into one space and thus transforms the frequency allocation problem into a spatial tessellation problem. This allows FAVOR to innovatively extend a spatial tessellation technique for the purpose of frequency allocation. We implement FAVOR in MicaZ platforms, and our extensive experiments with different network settings strongly demonstrate the superiority of FAVOR over existing approaches.

Categories and Subject Descriptors

C.2.1 [**Computer-Communication Networks**]: Network Architecture and Design—*Network topology*

General Terms

Algorithms, Design, Performance

Keywords

Continuous Frequency Allocation, Wireless Sensor Networks, CC2420 Radio, Spatial Tessellation.

1. INTRODUCTION

Deeply exploited as an emerging technology, *Wireless Sensor Networks* (WSNs) have the potential to be widely deployed to support a variety of applications. In many recent applications, both the scale of WSN deployments and the demand in data rate for individual sensor nodes keep increasing [3,6]. However, such a development is severely hampered by the co-channel interference produced by the ever increasing wireless transmissions and their intensity. Moreover, co-channel interference may come from not only the ZigBee [2] devices involved in a WSN, but also other 2.4GHz ISM band occupants such as WiFi and Bluetooth [11,12].

As ZigBee compatible radios (e.g., CC2420 of MicaZ Motes [1]) may operate on up to 16 channels, common wisdom suggests that one can make use of the multi-channel ability to prevent WSN links from interfering each other. This has led to quite a few research proposals, including prominently multi-channel scheduling for multi-hop transmissions (e.g., [24]) and multi-channel MACs (e.g., [21]). However, the number of channels available to a WSN is much lower than what ZigBee radios can offer, mainly due to the strong interference from other occupants in the 2.4GHz ISM band [11]. In our case (as shown in Figure 1), the experimental field is occupied by many WiFi testbeds, which effectively constrains the "clean" spectrum[1] to 2473–2483MHz where only two ZigBee channels can fit in [2].

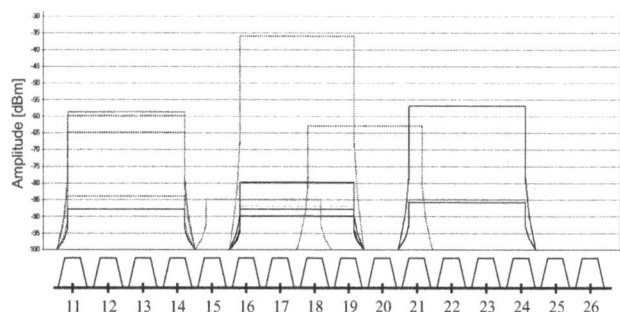

Figure 1: We overlap the WiFi signal strength observed through inSSIDer with the 16 channels of ZigBee. It is easy to spot that, apart from channels 25 and 26, other channels can be heavily interfered by multiple WiFi hotspots.

---

[1]Though avoiding WiFi interference temporally [12] or in frequency [28] is possible, directly incorporating these algorithms may severely complicate the system design, so we leave the ZigBee-Wifi coexistence issue as a future work.

By intensively experimenting (to be reported later) on our MicaZ Motes and their CC2420 radios, we have discovered that it would be more efficient to use the spectrum resource in a **continuous** manner. In other words, we should allocate channels to nodes or links in terms of their center frequencies (rather than a finite set of channel IDs). As it is well known that interference attenuates with distance and the distance between two interfering nodes/links is a **continuous** variable in Euclidean space, it is bounded to be more flexible and efficient (in term of spectrum utilization) to tune the frequencies of the nodes/links in a continuous manner, compared with the conventional graph coloring approach [18] where discrete channels are allocated based on the concept of (discrete) graph distance. Therefore, if one could allocate spectrum resource by jointly taking into account frequency and distance in a continuous manner, there is still a great potential to improve the performance of WSNs. For practical implementation, the ability of CC2420 radio in adjusting its channel center frequency at a granularity of 1MHz [1] does offer a good approximation of frequency fine-tuning.

Based on the above discovery, we first propose to shift the paradigm from *channel allocation* to (center) *frequency allocation* in utilizing frequency spectrum for WSNs. Although continuously allocating channel center frequencies to different links may lead to partially overlapped channels being operating simultaneously, we can combat the resulting interference by spacing them with a proper distance. In order to find the optimal frequency allocation for a set of spatially distributed nodes (or links), we further propose *Frequency Allocation for Versatile Occupancy of spectRum* (FAVOR) as a novel framework for frequency allocation in WSNs. FAVOR consists of two main components: i) a metric that unifies frequency and distance, which allows us to transform the frequency allocation problem into a spatial tessellation problem in a higher dimensional space, and ii) an algorithm that innovates on the *Centroidal Voronoi Tessellation* (CVT) method [8] to search for the local but nearly optimal frequency allocations.

Roughly speaking, FAVOR results in a frequency allocation such that nodes/links that are closer to each other in distance are further away from each other in frequency, while those far from each other in distance are allowed to be close in frequency. While FAVOR can be viewed as a continuous version of the conventional graph coloring approach, it offers a much greater freedom due to the relaxation from discrete sets to continuous spaces. The improved spectrum efficiency is bounded to favor WSN performance in throughput. Moreover, the FAVOR algorithm can be performed in a distributed manner using only local information. In order to verify the efficacy of FAVOR, we perform extensive experiments on a set of arbitrary links, as well as on two data collection trees. Our experiments demonstrate that FAVOR outperforms conventional graph coloring channel allocations and a recent overlapped channel allocation mechanism [29] in terms of throughput, given a stringent spectrum resource. In summary, our main contributions are:

- We propose to replace channel allocation with (center) frequency allocation, representing a paradigm shift in utilizing the scarce frequency spectrum.

- We define an optimization objective unifying frequency and distance; it enables us to formulate a frequency

allocation problem into a spatial tessellation problem in a high dimensional space.

- We propose a new algorithm inspired by CVT to search for the (local) optimal frequency allocations; the algorithm entails an easy distributed implementation with a need for only local information.

- We perform extensive experiments, both to investigate the frequency-distance tradeoff and to verify the efficacy of our proposed FAVOR framework.

In the remaining of our paper, we first report the experiments that lead to our discovery of the distance driving frequency allocation method in Sec. 2. Then we present our FAVOR framework in Sec. 3. Further experiments in demonstrating the superiority of FAVOR are reported in Sec. 4. We also discuss the related work and possible extensions of FAVOR to general wireless networks in Sec. 5, before concluding our paper in Sec. 6.

2. MOTIVATION AND MATHEMATICAL BACKGROUND

In this section, we first motivate the design of FAVOR, by reporting our experiment results that exhibit an almost continuous tradeoff between (center) frequency and (Euclidean) distance for two competing wireless links. We also briefly discuss the mathematical background needed for the development of FAVOR optimal frequency allocation algorithm.

2.1 Frequency–Distance Tradeoff for 2 Links

Our simple experiment setting involves two parallel wireless links, l_1 and l_2, operated by four MicaZ nodes. The configuration is illustrated in Figure 2: we fix the location of transmission link l_1, and change the location of l_2 with the distance to l_1 ranging from 1.2 m to 4.8 m. We make the transmitters of the two links to send packets persistently as fast as possible, and measure the throughput of the two links, i.e., how many packets can be received correctly.

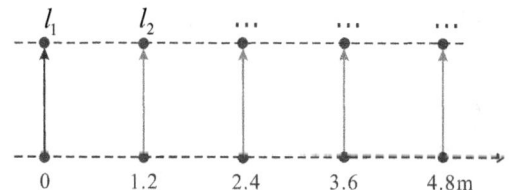

Figure 2: Experiment configuration. Each link has a length (the distance from its sender to its receiver) of 3.6m.

2.1.1 The Anomaly of MicaZ with CSMA

We first report an anomaly of MicaZ motes to better motivate our case.[2] It is well known that, though the *nominal data rate* of CC2420 is 250kbps, the maximum stable data rate one may squeeze from it is much lower (it is only about 50kbps for TelosB [22]). In our experience with MicaZ, we can only push one packet (45 bytes in total including 26 byes

---

[2]This anomaly also exists for other nodes using CC2420 radio, such as TelosB motes. However, we will not report results for other platforms due to the space limit.

pay load and 19 bytes header) through a ZigBee channel every 9ms, which achieves a data rate of 40kpbs (equivalent to a throughput of 0.16 with respect to the nominal data rate). In the following, we say a link achieving *full* data rate (or throughput) if it has a performance comparable to a single link. One would expect that operating two links simultaneously in the same channel would result in much lower (at least halved) data rate for each link. The results in Figure 3, nevertheless, show very different outcome, where we measure the throughput as the ratio between receiving rates at the receivers and the nominal data rate.

An amazing observation is that both links roughly achieve the full throughput of 0.16. The reason accounting for this

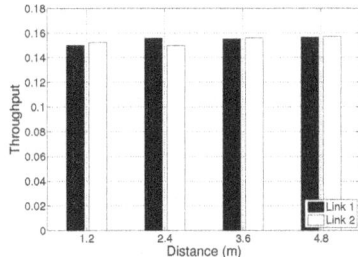

Figure 3: The throughput for two co-channel links with different distances.

anomaly is that the MCU (Atmel AVR Atmega 128L) used by MicaZ spends a certain amount of time moving packets within the protocol stacks and, most prominently, performing CSMA, the radio interface is left in an unsaturated mode. Consequently, CSMA may perfectly coordinate the two links such that they both achieve nearly full throughput. Actually, as we will show later, even adding up to 5 links will not drastically decrease the throughput of individual links. So one important message we have here is the follow:

> *Testing a channel allocation mechanism on CSMA-enabled platforms can lead to wrong conclusion, as even co-channel links may NOT conflict if we observe the outcome only in terms of throughput.*

2.1.2 Frequency vs. Distance without CSMA

In light of the results presented in Sec. 2.1.1, we test the throughput of two links under different (center) frequency and distance with CSMA disabled. While the four distances are shown in Figure 3, we also vary the frequency of l_1 from 2475MHz to 2480MHz but keep the frequency of l_2 at 2480MHz. When CSMA is disabled, a single link can carry a packet as fast as every 2ms, resulting in a data rate of 180kbps and a throughput (against the nominal data rate) of 0.72. Therefore, the full data rate in this case is 4.5 times higher than a CSMA-enabled link. The throughputs for all the frequency–distance combinations are shown in Figure 4. The figure clearly shows a throughput trade-off surface we may achieve by extending from a 1D discrete frequency space (for conventional channel allocations) to a higher dimensional *spatial-frequency space*.

We also plot the two sets of results in 2D figures in Figure 5, with Figure 5(a) fixing the distance at 1.2m and Figure 5(b) fixing the frequency of l_1 at 2479MHz. It is clear from Figure 5(a) that even a frequency interval of only 1MHz is rather usable at the minimum distance: while the higher

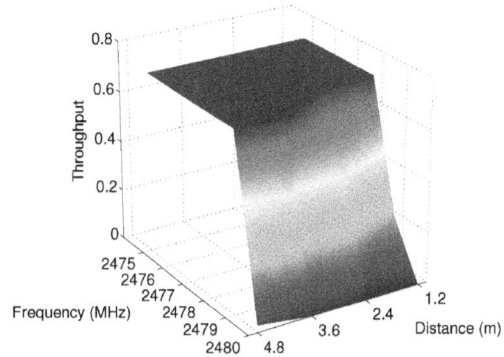

Figure 4: The lower throughput for two links with different frequency–distance combinations. CSMA is disabled for both links.

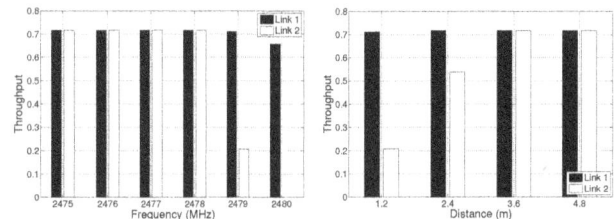

(a) Throughput vs. frequency   (b) Throughput vs. distance

Figure 5: The throughput of both links as functions of frequency interval (a) and distance (b).

throughput is almost full (beyond 0.7), the lower throughput still offers a reasonable data rate of about 50kbps (even higher than the full rate of a single link with CSMA enabled). Figure 5(b) further shows that, by increasing the distance between two links, both links may achieve a nearly full throughput. Remember that, if we follow the IEEE 802.15.4 standard, only two independent channels are available for the spectrum we can use (due to the heavy interference from WiFi hotspots, see Sec. 1), whereas we almost get six usable but overlapped channels by varying center frequencies and distances in a continuous manner.[3] Therefore, another important message we get is this:

> *Continuously tuning frequencies allocated to links with respect to the distances between them allows for a more flexible and efficient use of spectrum.*

Note that a byproduct is that CSMA can be disabled as soon as the center frequencies of all links are slightly misaligned, reducing overhead and hence further improving throughput.

Now the question is, *given a set of wireless nodes or links (whose locations are already fixed), how can we find the optimal (center) frequency allocation for them?* Our FAVOR framework is exactly proposed to address this question. Before diving into the details of our proposal, we need to briefly discuss the mathematical background relevant to our algorithm designed for FAVOR.

---

[3]We are still confined by the current implementation of CC2420 radio, whose center frequency can be tuned only at a granularity of 1MHz. With more flexible radios developed in the future, we will have more channels available.

2.2 Mathematical Background on Spatial Tessellation and CVT

Given a region $\mathcal{A} \subseteq \mathbb{R}^n$, the set $\{A_i\}$ is called a *tessellation* of $\mathcal{A}$ if $A_i \cap A_j = \emptyset$ for $i \neq j$ and $\cup_i A_i = \mathcal{A}$. Let $\|\cdot\|_{\ell^2}$ denote the Euclidean norm on $\mathbb{R}^n$. For a set of points $\{u_i\}$ belonging to $\mathbb{R}^n$, the Voronoi region V_i corresponding to the point u_i is defined by

$$V_i = \left\{ v \in \mathcal{A} \mid \|v - u_i\|_{\ell^2}^2 \leq \|v - u_j\|_{\ell^2}^2, \forall j \neq i \right\}.$$

The set $\{V_i\}$ is termed *Voronoi tessellation* of $\mathcal{A}$, with points $\{u_i\}$ called *generators* and each V_i referred to as the *Voronoi cell* corresponding to u_i.

For a fixed number of generators, varying their locations results in different tessellations of $\mathcal{A}$. One way to identify the "best" tessellation is to define a metric that measures the quality of a tessellation, and a typical metric is the "impact" of the generator u_i to a point v in its cell,[4] represented often by $\|v - u_i\|_{\ell^2}$ for $v \in V_i$. This ends up with an objective that represents the total impact of the generators to their individual cells:

$$\Im(\{A_i\}, \{u_i\}) = \sum_i \int_{A_i} \|v - u_i\|_{\ell^2}^2 \Phi(v) dv, \quad (1)$$

where $\Phi(v)$ indicates the density at location v. This objective is neither convex nor concave, so optimizing this objective may lead to many local minima (or maxima). According to the theory of *Centroidal Voronoi Tessellations* (CVT) [8], we have the following four basic conclusions:

- The Voronoi tessellation is optimal for a fixed set of points $\{u_i\}$.

- A (good) local optimal solution is when every u_i coincides with the gravity center (centroid) of its cell V_i.

- Lloyd's method [14] that moves each u_i to the centroid of its current cell in every iteration terminates at this local optimal solution with a linear convergence rate.

- Lloyd's method outputs $\{u_i\}$ that are uniformly distributed in $\mathcal{A}$ and, if $\mathcal{A} \subseteq \mathbb{R}^2$, the cells that are almost all regular hexagons.

The interesting observation of the outcome of CVT is that $\{u_i\}$ at termination are uniformly distributed in the space, which is intuitively related to our need of "spreading" nodes/links over the available frequency spectrum. In fact, our FAVOR framework is a non-trivial extension of CVT to a space involving both frequency and distance (or location).

3. FAVOR: A LOCATION-AWARE FREQUENCY ALLOCATION SCHEME

In this section, we first discuss our system model, then we present our FAVOR framework in terms of the optimization objective and the algorithm to find a local optimal solution. Finally, we discuss how the algorithm can be implemented in practical scenarios.

---

[4]There are many interpretations to this metric [8], we choose the one that is relevant to our design later.

3.1 System Model

We assume a WSN consisting of a set of sensor nodes $\mathcal{N} = \{n_1, n_2, ..., n_N\}$ with $|\mathcal{N}| = N$, which are deployed on a 2D plane. Although our proposal can be readily extensible to 3D volume deployments in theory, we confine our scenarios to 2D deployments due to the limitations imposed by our experimental conditions. Let $\{u_i\}_{i=1,...,N}$ be the locations of the sensor nodes, where $u_i \in \mathbb{R}^2$. Given a frequency band $\mathcal{B} = [f_{min}, f_{max}]$ and the channel width f_w, we assign each sensor node n_i a channel with center frequency $f_i \in \mathcal{B}'$, where $\mathcal{B}' = [f_{min} + \frac{f_w}{2}, f_{max} - \frac{f_w}{2}]$.[5] Combining the node's location and frequency, a node n_i now has a new "coordinate" $(u_i, f_i) \in \mathbb{R}^2 \times \mathcal{B}'$. We denote by $\mathcal{N}(n_i)$ the one-hop neighbors of node n_i: the nodes with whom n_i can communicate directly given a common channel and a fixed transmit power.

3.2 FAVOR Objective: Balancing Distance and Frequency

According to our observation in Sec. 2.1.2, a good frequency allocation scheme should assign very different frequencies to nodes that are close to each other but arbitrary frequencies to nodes that are far from each other. We illustrate such a possible location-dependent frequency allocation scheme in Figure 6. However, as there are many possi-

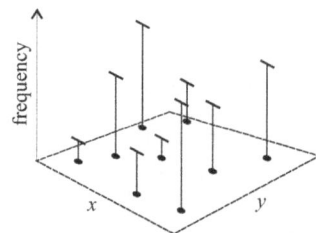

Figure 6: Allocating center frequencies based on the nodes' locations. While the black points indicate the sensor nodes, the "poles" on the points (with their different heights) represent different center frequencies allocated to the nodes.

ble allocations given a certain WSN deployment, we need to find the best possible allocations, for which we need a metric to evaluate different allocations.

To this end, we define the *impact metric* as

$$\|v - u_j\|_{\ell^2} + \beta \|f - f_i\|_{\ell^2},$$

where $v \in \mathbb{R}^2$ and $f \in \mathcal{B}'$, β is a weight to tune the tradeoff between frequency and distance, and we aim at a "tessellation" of the subset $\mathcal{A} = \mathbb{R}^2 \times \mathcal{B}'$ in $\mathbb{R}^3$ such that the following FAVOR objective is minimized:

$$\mathfrak{F}(\{A_i\}, \{f_i\}) = \sum_i \int_{A_i} \left(\|v - u_i\|_{\ell^2}^2 + \beta \|f - f_i\|_{\ell^2}^2 \right) dz, \quad (2)$$

where $z = (u, f) \in \mathcal{A}$. It is obvious that (2) differs from (1) mainly in that part of the coordinates are fixed: we do not get the freedom to move nodes around, only the frequencies allocated to them are variables. Intuitively, $\mathfrak{F}(\{A_i\}, \{f_i\})$ represents the total "impact" from nodes to their cells. Given

---

[5]Channels can also be assigned to links rather than nodes; we will discuss this later in Sec. 3.4.

fixed locations, minimizing the objective has the effect of minimizing interference to some extent: according to what we discussed in Sec. 2.2, an optimal solution tends to spread out the generators (nodes' locations in this case). Since the locations in $\mathbb{R}^2$ are fixed, frequencies for two close-by nodes will be pushed further from each other. Another (relatively minor) difference is that we let $\Phi(v) = 1$ in (2), as both Euclidean and frequency spaces are assumed to be homogeneous for now. We could make use of $\Phi(\cdot)$ to characterize the non-homogeneity in space for our future development.

3.3 FAVOR: A CVT-based Approach

Given the similarity between (1) and (2), it is natural that one would propose to apply CVT to find a local optimal solution. Without loss of generality, we let $\beta = 1$ to simplify our derivation. The problem we face now is twofold: i) as we need to perform CVT at least in a 3D space (it can be 4D if nodes are distributed in a 3D Euclidean space), we need an algorithm more efficient than the Lloyd's method [14]; otherwise nodes with limited computation resource cannot afford it,[6] and ii) part of the coordinates for each $u_i \in \mathcal{A}$ are fixed, whereas CVT requires all the coordinates to be variables. To tackle these issues, we first propose *Approximate CVT* (A-CVT) to transform the problem into a more tractable and implementable form, then we apply gradient projection method to handle the fixed coordinates.

Given a region $\mathcal{A} = \mathbb{R}^2 \times \mathcal{B}' \subseteq \mathbb{R}^3$ and suppose we apply Voronoi tessellations to partition $\mathcal{A}$, we may re-write the objective (2) as the following.

$$\mathfrak{F}(\{A_i\}, \{f_i\}) = \int_A \min_i \left(\|v - u_i\|_{\ell^2}^2 + \|f - f_i\|_{\ell^2}^2 \right) dz \quad (3)$$

The equivalence between (2) and (3) is obvious: for each generator (u_i, f_i), integrating over its own cell V_i implies an integration over all the points in $\mathcal{A}$ that are closer to (u_i, f_i) than to any other generators (by the definition of a Voronoi cell shown in Sec. 2.2). Now we get a global integration over $\mathcal{A}$, eliminating the need for re-computing the Voronoi tessellations in every iteration, but we have to face the non-differentiable function $\min(\cdot)$. To make the problem tractable, we apply an approximation to the $\min(\cdot)$ function to "smooth" it, leading to the following A-CVT objective.

$$\mathfrak{F}(\{A_i\}, \{f_i\}) = \int_A \left[\sum_i \left(\|v - u_i\|_{\ell^2}^2 + \|f - f_i\|_{\ell^2}^2 \right)^\lambda \right]^{\frac{1}{\lambda}} dz \quad (4)$$

Due to page limit, we omit the proof of (4) converging to (3) when $\lambda \to -\infty$. In practice, we take $\lambda \in [-40, -20]$.

As minimizing the A-CVT objective (4) is a typical nonlinear optimization problem, we apply a gradient-descent method with gradient projection to search for a local minimum. The pseudocodes of the algorithm are shown by **Algorithm 1**. Roughly, the algorithm proceeds in rounds and takes the following three steps in each round:

---

[6]Lloyd's method requires to recompute the Voronoi tessellation in every iteration (due to the modified locations of the generators and the need to find the centroids of the cells), entailing high computational cost in $\mathbb{R}^n$ for $n > 2$. In fact, the complicated data structures required to model a $\mathbb{R}^3$ Voronoi tessellation cannot be easily implemented in sensor nodes, and no algorithm for CVT in $\mathbb{R}^n, n > 3$ has been implemented even for common CPUs.

1. Compute gradient for each generator z_i as (line 2)

$$g(z_i) = \int_{\mathcal{A}} 2 \left(\|z - z_i\|_{\ell^2}^2 \right)^{\lambda-1} (z_i - z) \left(\sum_j \|z - z_j\|_{\ell^2}^{2\lambda} \right)^{\frac{1-\lambda}{\lambda}} dz.$$

In order to facilitate localized computation and also to reduce the complexity, the summation in the third term can be applied only for $j : n_j \in \mathcal{N}(n_i)$ and the integration can be done only for z in the neighborhood of z_i. This is possible because the terms introduced by those far-away locations contribute only insignificantly to $g(z_i)$, due to $\lambda \to \infty$. This is also intuitively correct as the change in z_i for CVT is only affected by z_j whose cell shares boundaries with that of z_i. For a WSN, we can use the communication neighborhood to approximate the tessellation neighborhood.

2. As u_i is fixed and we can change only f_i, we take $g_f(z_i)$ as the projection of $g(z_i)$ on the frequency axis (line 3).

3. A tentative update is applied to the frequency by $f_i^+ = f_i - \alpha \cdot g_f(z_i)$ where α is a step size. If both $|g_f(z_i)|$ and $|f_i^+ - f_i|$ become sufficiently small, the algorithm is terminated, returning the optimal frequency allocation (line 8); otherwise, the frequency of each n_i is updated by $f_i \leftarrow f_i^+$, the outcome is exchanged among neighboring nodes (line 6), and further compuation will be conducted during the next round.

Algorithm 1: FAVOR

Input: For each $n_i \in \mathcal{N}$, location u_i, initial frequency $f_i^0 \in \mathcal{B}'$, stopping tolerance ε_1 and ε_2
Output: f_i^* for each n_i
1 For every node $n_i \in \mathcal{N}$ in each round (every τ ms):
2 Compute the gradient $g(z_i)$ of (4)
3 Project $g(z_i)$ on f_i to get $g_f(z_i)$
4 $f_i^+ = f_i - \alpha \cdot g_f(z_i)$ /*α is the step size*/
5 **if** $|g_f(z_i)| > \varepsilon_1 \lor |f_i^+ - f_i| > \varepsilon_2$ **then**
6 | $f_i \leftarrow f_i^+$; BROADCAST(f_i) to nodes in $\mathcal{N}(n_i)$
7 **else**
8 └ $f_i^* \leftarrow f_i^+$

We omit the convergence analysis as it follows directly from the basic theory of gradient-descent methods [4]. The convergence can be even faster if a centralized Quasi-Newton method is used to solve (4). We illustrate the results of our FAVOR algorithm in Figure 7. It is shown that, while the frequency allocation is initially ascending from left to right (with small frequency separations between neighboring nodes), the outcome of FAVOR exhibits much better separation in frequency for nodes close to each other. To facilitate visual illustration, we use only a 1D deployment (nodes on a line) as an example, which leads to easily discernable Voronoi tessellations in $\mathbb{R}^2$. However, our FAVOR algorithm works for any dimension higher than 2, while our experiments in Sec. 4 will be done for 2D deployments.

3.4 FAVOR for Disjoint Link Set

As FAVOR relies on a set of (point) locations $\{u_i\}$ to perform allocation, an obvious difficulty it may face is what if no obvious points exist in a networking scenario. In particular, a network may consist of several point-to-point wireless

43

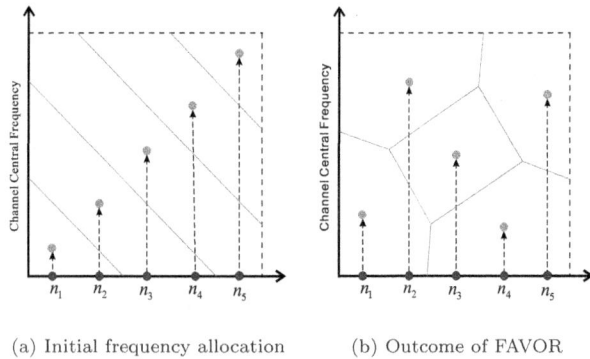

(a) Initial frequency allocation   (b) Outcome of FAVOR

Figure 7: Frequency allocation based on Voronoi tessellations. With a regular (ascending) initial allocation (a), our FAVOR algorithm outputs a much better allocation (b). In both cases, the corresponding Voronoi tessellations are also shown using red line segments as cell boundaries.

links [29]. We propose two possible solutions for centralized and distributed computing separately. If a centralized computing is feasible, we may apply the extended Voronoi diagram where generators are not points but line segments (representing the links) [5]. Whereas this method may result in rather accurate frequency allocation, computing Voronoi tessellation with line segments as generators is quite time consuming. Therefore, we pick one point to represent each link in a distributed computing environment. This point can be either the source or the destination of the link, or it can even be the middle point of the link. If a point is chosen as the source or the destination, the computation is performed by that node. If the middle point is chosen, the computation can be done by either of the two nodes. Our experiments reported in Sec. 4.2 show that replacing line segments by points still leads to very good performance in throughput.

3.5 FAVOR For Tree-based Data Collection

One typical communication pattern in WSNs is the convergecast. More specifically, nodes of a WSN are organized into a data collection tree with a root at sink n_s, and every other node n_i sends data directly or indirectly (through multi-hop routing) to n_s [9]. In this case, frequency allocation itself is not sufficient to tackle the conflicts in media access: it cannot avoid conflicts for either multiple links ending at the same node or one node having both incoming and outgoing links, as such conflicts are the consequence of equipping a node with only one radio. Therefore, we need to perform both the frequency allocation and a TDMA-like time schedule. While a joint frequency allocation and scheduling problem is beyond the scope of our paper, we simply adapt the minimum latency scheduling mechanism [23].

We basically order the nodes into layers according to their hop-distances to the sink on the collection tree T, then use a similar labeling mechanism as proposed in [23]. The idea of this labeling is twofold: i) to guarantee that the number of labels assigned to an outgoing link of node n_i should be 1 plus the number of descendants of n_i in T, and ii) to assign different labels for links sharing the same node. Whereas the proposal in [23] adopts (orthogonal) channel allocation to enable parallel transmissions, our frequency al-

location potentially allows more parallel transmissions. For the convenience of deriving bounds, channels are allocated using a first-fit distance-$(\rho + 1)$ graph coloring in [23], but we allocate different frequencies using FAVOR. As a node n_j needs to tune to the frequency f_i when transmitting to a receiver n_i, leaf nodes (being receivers to no one) do not need their own frequencies. We will show an example of our frequency allocation and scheduling in Sec. 4.3.

A similar frequency allocation and scheduling method also works for other types of communication patterns, such as broadcast (disseminating commands from the sink to the network) and aggregation (each relay node may send out less data then it receives). However, we focus only on data collection in our paper.

3.6 Time Synchronization

Time synchronization severs as a fundamental infrastructure for the TDMA scheduling used by FAVOR. Unfortunately, the existing synchronization protocols (e.g., FTSP [17]) rely on periodically flooding, resulting in a heavy traffic load that may significantly compromise the network throughput. In this paper, we employ a link-based approach that piggybacks control information with data traffic: for each transmission link, the transmitter synchronizes its transmitting schedule with that of the receiver based on the information piggyback with the acknowledgements sent by the receiver. To further suppress the control traffic, the transmitter may require acknowledgement only for the first transmission in each transmission schedule that consists of several transmissions as shown by Figure 11(b). In effect, non-root nodes report data to their respective parents while getting synchronized with the latter. Therefore, running this protocol within a data collection tree simply forces every node to stay synchronized with the root.

4. EVALUATION

We have implemented FAVOR in our MicaZ platforms and performed extensive experiments on them. The exciting experimental results are reported in this section. We first discuss the basic parameter settings for our experiments, then we describe our experiments on two different network scenarios. Finally, we briefly examine the convergence of FAVOR algorithm.

4.1 Experiment Settings

We apply MicaZ Motes and TinyOS 2.1 as the hardware and software platforms. As explained in Sec. 1, the available frequency spectrum we adopt is 2474–2481MHz, to avoid the "contaminated" spectrum by WiFi devices. According to the ZigBee standard [2], each channel has a 2MHz bandwidth and the center frequencies of two neighboring channels have to be separated by 5MHz. Therefore, only two orthogonal channels (with center frequencies 2475 and 2480) can fit into this spectrum (hence we term this scheme **two-channel**). However, we modify the codes for FAVOR to choose six possible center frequencies: 2475, 2476, 2477, 2478, 2479, and 2480MHz.[7] We will also test the proposal made in [29], where a center frequency interval of 3 or 2MHz has been suggested. This means that, given our available spectrum, three

---

[7] The CC2420 radio used by MicaZ Motes can adjust the frequency with only a granularity of 1MHz, which somewhat limits the potential of FAVOR. FAVOR can perform better if frequencies can be tuned continuously.

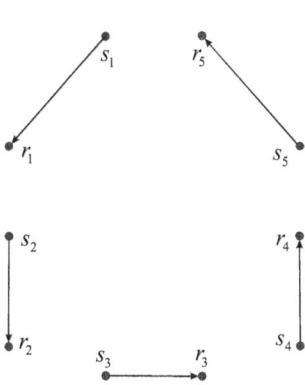

(a) Network deployment   (b) Statistics on TRPs for all links

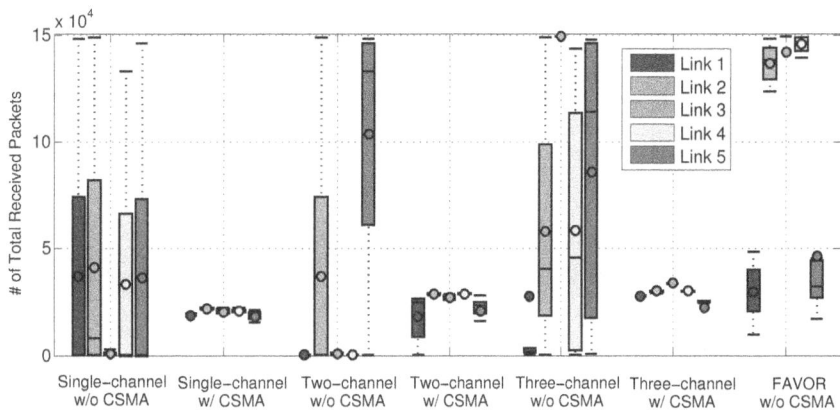

Figure 8: A five-link scenario and the corresponding experimental results.

channels are available: 2475, 2477 (or 2478), and 2480MHz (hence we term this scheme **three-channel**). We apply a greedy graph coloring approach to allocate channels for both two- and three-channel schemes, while FAVOR uses its own algorithm for frequency allocation. In the following, we use only the center frequency to indicate a channel.

If we can avoid assigning the same channel to more than one link, we disable CSMA for all links. This allows packets to be pushed into a channel faster: one packet every 2ms according to Sec. 2.1.2. For cases where the same channel has to be assigned to different links, we test both with and without CSMA: we can only send one packet every 9ms in the former situation (see Sec. 2.1.2 again), whereas collisions may totally ruin some links in the latter situation, as already shown in Figure 5(a). We deploy our WSNs on the ceiling of our laboratory (see Figure 9), in order to emulate an indoor monitoring application scenario. We perform each experiment for 5 minutes, and use the *total received packets* (TRPs) at each destination node as the performance measure; it is actually an indicator of the **throughput**. For each reported data point, we perform 10 experiments, and plot their statistical quantities such as means and/or interquartile ranges.

Figure 9: A MicaZ-based WSN testbed on the ceiling of our research center.

4.2 A Five-Link Scenario

We first test a scenario containing five links $\{l_i = (s_i, r_i), \forall i = 1, ..., 5\}$, as shown in Figure 8(a). We deliberately put the

five links in a relatively small area (about $20m^2$), in order to mimic a small section of a densely deployed WSNs (which we do not have at our disposal). The optimal frequency allocations based on different schemes are shown in Table 1. Obviously, for schemes other than FAVOR, the number of available frequencies (channels) is smaller than the number of links, so some frequency has to be allocated to more than one link; the optimal allocation can simply try to space these co-channel links as far as possible.

Table 1: Frequency allocations for the five-link scenario based on different schemes.

Links	l_1	l_2	l_3	l_4	l_5
Single-channel	2480	2480	2480	2480	2480
Two-channel	2480	2475	2480	2475	2480
Three-channel	2475	2477	2480	2475	2477
FAVOR	2476	2477	2480	2475	2478

We test all the four schemes: single-channel, two-channel, three-channel, and FAVOR; each with CSMA enabled and disabled (except FAVOR, as it does not need CSMA at all). The results in terms of TRPs per link are shown in Figure 8(b), where both interquartile ranges (boxes) and means (circles) are shown. It is obvious that FAVOR operates three out of five links (l_2, l_3, and l_4) much better than other schemes: mean TRPs can be up to three times higher than the one second to it. For another two links, FAVOR also does a relatively good job (better than any CSMA-enabled cases). Consequently, the total throughput of FAVOR is **five** times of the commonly used single channel with CSMA. Although three-channel without CSMA appears to achieve rather high mean TRPs in some links (l_3 and l_5), the huge interquartile ranges indicate a very unstable performance, rendering this scheme useless in practice.

The tradeoff between using CSMA or not (for other schemes) is very evident: CSMA delivers a rather stable performance by sacrificing throughput: the nature of a random access scheme. However, FAVOR achieves both stable and high throughput without the need for CSMA. In fact, FAVOR

45

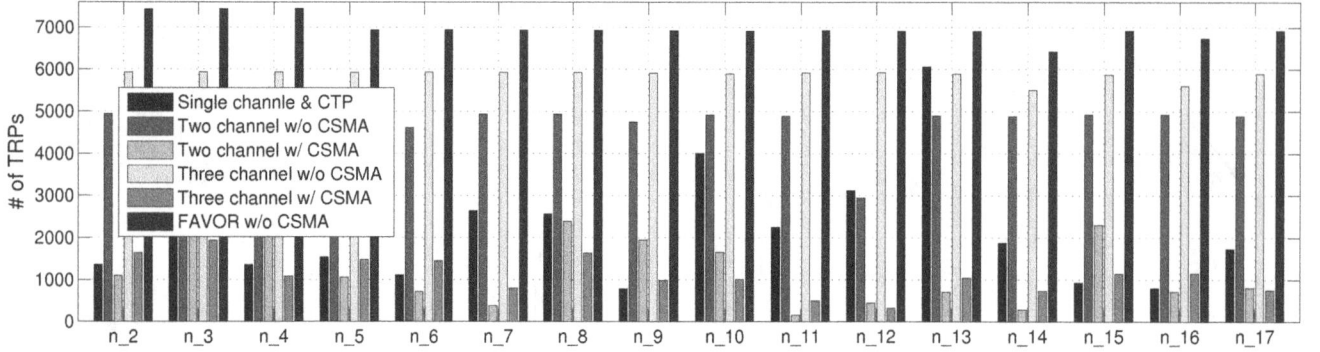

Figure 10: Performance of six different media access schemes in an unbalanced data collection tree.

could perform better if the radio allowed a finer granularity in tuning frequencies: we have to round the frequency allocation done by FAVOR to integer values of MHz, which leads to the relatively bad performance of l_1 and l_2.

4.3 Unbalanced Tree-based Data Collection

In order to demonstrate the benefit of applying FAVOR in a practical situation, we test the performance of different schemes in an arbitrary (unbalanced) data collection tree. The deployment area is roughly 600m^2, and the node locations, as well as the network topology, are shown in Figure 11(a). The frequencies (or channels) are allocated to non-leaf nodes, and we apply the labeling method in [23] for link schedules. Note that the disjoint-tree based scheduling [24] does not apply to this small network.

According to what is shown in Sec. 4.2, the schedule needs also to avoid links with the same frequency transmitting simultaneously. This leads to the obvious consequence that the more frequencies we can allocate, the less time slots we need in one round (during which every node gets a chance to transmit). As a result, two-channel scheme needs at least 29 slots, three-channel scheme needs at least 24 slots, while FAVOR needs only 22 slots. For single channel, we enable CSMA and apply CTP [9] to perform data collection. We also enable CSMA for two- and three-channel schemes, but with a different schedule: it needs only to guarantee a sender and a receiver staying at the same frequency when the link between them is active. Consequently, only 9 time slots are needed. Given limited space, we show only the frequency allocation and schedule for FAVOR in Figure 11.

As expected, the results in Figure 10 show that FAVOR surpasses all other schemes, and it achieves a throughput that is 3.36 times of that achieved by CTP. Apparently, FAVOR beats two- and three-channel schemes due to the less time slots in a round, and it prevails against all CSMA-enabled cases thanks to the elimination of the overhead brought by CSMA. Moreover, FAVOR allows for a more fair sharing of the bandwidth (every node gets roughly the same throughput), which cannot be guaranteed by any CSMA-based schemes.

4.4 Balanced Tree-based Data Collection

One may argue that FAVOR's continuous (hence potentially irregular) frequency allocation cannot offer significant network performance in a regular network topology, e.g., a balanced data collection tree. Therefore, we re-deploy

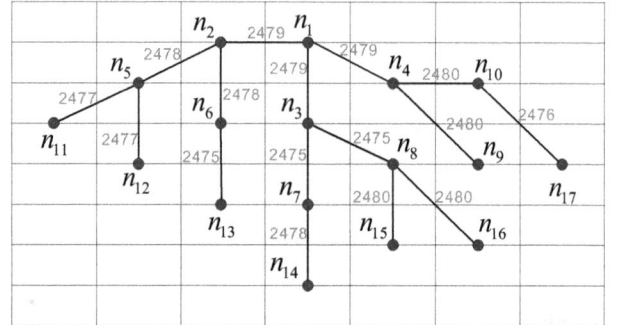

(a) Data collection tree (frequency allocation marked on links).

(b) Minimum delay schedule on the tree.

Figure 11: Unbalanced data collection tree, with FAVOR frequency allocation (a) and min-delay transmission schedule (b). The node locations in (a) are plotted roughly proportional to their actual locations in our ceiling testbed.

our testbed to form a balanced data collection tree with nodes regularly spaced, as shown in Figure 12. All settings are maintained as those in Sec. 4.3, except that different FAVOR frequency allocation and transmission schedule are computed to suit this tree.

The results in term of TRPs are reported in Figure 13. Obviously, the observations made for Figure 10 still hold here, except that the advantage of FAVOR over others has been slightly reduced. This stems from the reduced number of links involved in this tree: as FAVOR surpasses others by offering higher frequency utilization, its advantage becomes more conspicuous if a higher utilization is actually needed.

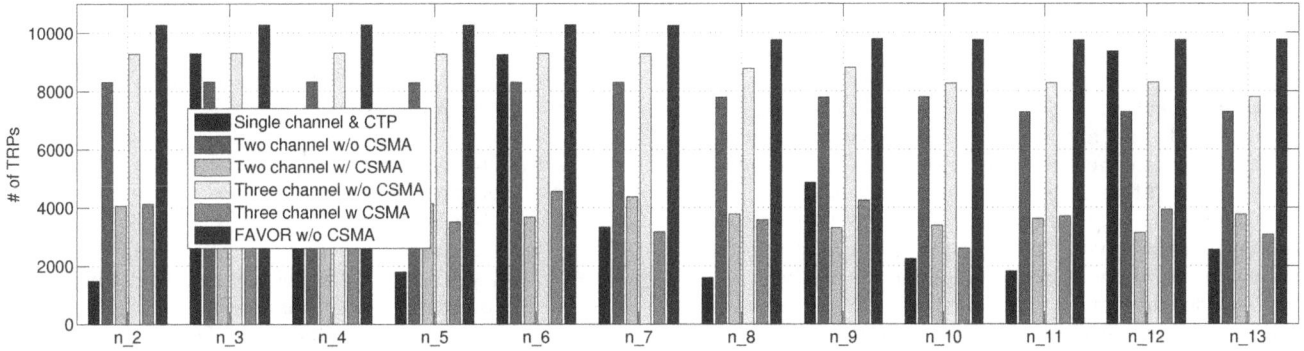

Figure 13: Performance of six different media access schemes in a balanced data collection tree.

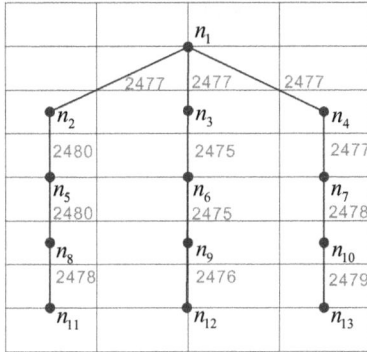

Figure 12: Balanced data collection tree with FA-VOR frequency allocations marked on links. The node locations are plotted roughly proportional to their actual locations in our ceiling testbed.

4.5 Convergence of FAVOR

We briefly verify the complexity (in terms of communication rounds) of FAVOR in a distributed computing scenario: a 30-node WSN. As we unify both frequency and distance into a scale of $[0, 1]$, the A-CVT objective value is always smaller than 1. As shown in Figure 14, the convergence under different step size α's often takes 20–30 rounds. As such message exchanges can piggyback with other transmission activities and frequency (re)allocation does not happen very often, the overhead of FAVOR (in terms of the entailed communication and computation costs) is affordable, given the substantial throughput improvement it can bring.

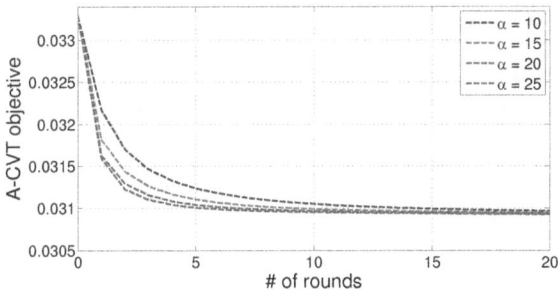

Figure 14: The convergence of FAVOR in 30-node WSN.

5. RELATED WORK AND DISCUSSIONS

While exploiting multi-channel access to improve the performance of general multi-hop wireless network has been intensively studied in the last decade (e.g., [20]), dedicated investigations for WSNs started rather late [13, 24]. Most proposals make use of graph-coloring heuristics to allocate channels in the whole network, but the work of [24] innovated in allocating channels to disjoint trees in WSNs. As the allocated channels are assumed to be orthogonal, the inter-channel interference is often neglected. However, the inter-channel interference does exist and its impact on link capacity and the performance of multi-channel protocols are systematically explored in [24, 27].

Given the limited number of non-overlapping (or orthogonal) channels, partially overlapped channels were later introduced to improve the spectrum utilization in WiFi networks [19]. However, it is only recently that the special inter-channel interference feature of ZigBee radio was identified [29]. In particular, whereas the center frequencies of two partially overlapped channels have to be sufficiently apart from each other for WiFi radios to properly operate [19], the ZigBee radios used by WSNs are far more robust due to their simple design: as we have also shown in Sec. 2.1.2, only a difference of 1MHz is enough for two channels to deliver a reasonable throughput. However, the work in [29] did not discover the advantage of continuous frequency allocation; it focuses only on adjusting CSMA. As demonstrated by our FAVOR, CSMA may not be needed anymore if frequency allocation is applied.

The multi-channel feature of ZigBee radios has also been applied to improve the quality of individual links [7] and to avoid the interference from WiFi devices [28]. However, the scalability of these proposals are still confined by the limited channels. We believe that the flexible spectrum utilization offered by FAVOR can also contribute to tackling these problems, as well as problems under a joint routing and link scheduling framework (e.g., [16]) or for duty-cycled WSNs (e.g., [10]).

Currently, FAVOR cannot be directly applied to WiFi networks due to the significant inter-channel interference between two partially overlapped channels [25]. This may stem from the particular filter design for a WiFi radio receiver, but we may be able to redesign the filter to reduce the inter-channel interference (hence to apply FAVOR for achieving a higher spectrum utilization), at a cost of slightly reduced data rate.

6. CONCLUSION

In this paper, we present FAVOR, a novel framework for efficient spectrum utilization in WSNs. We exploit a continuous frequency allocation to replace the conventional discrete multi-channel allocation. We then combine frequency and location into one space and thus transform the optimal frequency allocation problem into a spatial tessellation problem. Our FAVOR algorithm innovates on the Centroidal Voronoi Tessellation method to search for the nearly optimal frequency allocations. Finally, we perform extensive experiments to demonstrate the feasibility and superiority of FAVOR. For future work, we plan to apply FAVOR to broader scenarios including data aggregation WSNs (e.g., [26]) and WSNs with mobile elements (e.g., [15]).

Acknowledgments

We would like to thank the anonymous reviewers for their insightful comments and constructive suggestions. This work was supported in part by AcRF Tier 2 Grant ARC15/11 and Tianjin Research Program of Application Foundation and Advanced Technology (China) No. 12JCQNJC00200.

7. REFERENCES

[1] Chipcon's CC2420 2.4G IEEE 802.15.4/ZigBee-ready RF Transceiver.

[2] IEEE Standard 802.15.4, 2011.

[3] TOES: Terrestrial Ecology Observing Systems.

[4] D. Bertsekas. *Nonlinear Programming*. Athena Scientific, Belmont, MA, 2 edition, 1999.

[5] C. Burnikel, K. Mehlhorn, and S. Schirra. How to Compute the Voronoi Diagram of Line Segments: Theoretical and Experimental Results. In *Proc. of the 2nd ESA (LNCS 855)*, 1994.

[6] S. Dawson-Haggerty, S. Lanzisera, J. Taneja, R. Brown, and D. Culler. @scale: Insights from a Large, Long-Lived Appliance Energy WSN. In *Proc. of the 11th ACM/IEEE IPSN*, 2012.

[7] M. Doddavenkatappa, M. Chan, and B. Leong. Improving Link Quality by Exploiting Channel Diversity in Wireless Sensor Networks. In *Proc. of the 33rd IEEE RTSS*, 2011.

[8] Q. Du, V. Faber, and M. Gunzburger. Centroidal Voronoi Tessellations: Applications and Algorithm. *SIAM Review*, 41(4):637–676, 1999.

[9] O. Gnawali, R. Fonseca, K. Jamieson, D. Moss, and P Levis. Collection Tree Protocol. In *Proc. of the 7th ACM SenSys*, 2009.

[10] K. Han, Y. Liu, and J. Luo. Duty-Cycle-Aware Minimum-Energy Multicasting in Wireless Sensor Networks. *IEEE/ACM Trans. on Netw.*, 21(3):910–923, 2013.

[11] J.-H. Hauer, V. Handziski, and A. Wolisz. Experimental Study of the Impact of WLAN Interference on IEEE 802.15.4 Body Area Networks. In *Proc. of the 6th EWSN*, 2009.

[12] J. Huang, G. Xing, G. Zhou, and R. Zhou. Beyond Co-Existence: Exploiting WiFi White Space for ZigBee Performance Assurance. In *Proc. of the 18th IEEE ICNP*, 2010.

[13] H.K. Le, D. Henriksson, and T. Abdelzaher. A Control Theory Approach to Throughput Optimization in Multi-Channel Collection Sensor Networks. In *Proc. of ACM/IEEE IPSN*, 2007.

[14] S. Lloyd. Least Square Quantization in PCM. *IEEE Trans. on Info. Theory*, 28:129–137, 1982.

[15] J. Luo and J.-P. Hubaux. Joint Sink Mobility and Routing to Increase the Lifetime of Wireless Sensor Networks: The Case of Constrained Mobility. *IEEE/ACM Trans. on Netw.*, 18(3):871–884, 2010.

[16] J. Luo, C. Rosenberg, and A. Girard. Engineering Wireless Mesh Networks: Joint Scheduling, Routing, Power Control and Rate Adaptation. *IEEE/ACM Trans. on Netw.*, 18(5):1387–1400, 2010.

[17] M. Maróti, B. Kusy, G. Simon, and Á. Lédeczi. The flooding time synchronization protocol. In *Proceedings of the 2nd ACM SenSys*, pages 39–49, 2004.

[18] A. Mishra, S. Banerjee, and W. Arbaugh. Weighted Coloring Based Channel Assignment for WLANs. *ACM Mobile Computing and Communication Review*, 9(3):19–31, 2005.

[19] A. Mishra, V. Shrivastava, S. Banerjee, and W. Arbaugh. Partially Overlapped Channels Not Considered Harmful. In *Proc. of the ACM/IFIP SIGMetrics/Performance*, 2006.

[20] J. So and N. Vaidya. Multi-Channel MAC for Ad Hoc Networks: Handling Multi-Channel Hidden Terminals Using a Single Transceiver. In *Proc. of the 5th ACM MobiHoc*, 2004.

[21] L. Tang, Y. Sun, O. Gurewitz, and D.B. Johnson. EM-MAC: A Dynamic Multichannel Energy-Efficient MAC Protocol for Wireless Sensor Networks. In *Proc. of the 12th ACM MobiHoc*, 2011.

[22] P. Trenkamp, M. Becker, and C. Goerg. Wireless Sensor Network Platforms – Datasheets versus Measurements. In *Proc. of IEEE SenseApp*, 2011.

[23] P.-J. Wan, Z. Wang, Z. Wan S.C. Huang, and H. Liu. Minimum-Latency Scheduling for Group Communication in Multi-channel Multihop Wireless Networks. In *Proc. of WASA*, 2009.

[24] Y. Wu, J. Stankovic, T. He, and S. Lin. Realistic and Efficient Multi-Channel Communications in Wireless Sensor Networks. In *Proc. of the 27th IEEE INFOCOM*, 2008.

[25] L. Xiang and J. Luo. Joint Channel Assignment and Link Scheduling for Wireless Mesh Networks: Revisiting the Partially Overlapped Channels. In *Proc. of the 21st IEEE PIMRC*, pages 2063–2068, 2010.

[26] L. Xiang, J. Luo, and C. Roserberg. Compressed Data Aggregation: Energy Efficient and High Fidelity Data Collection. *IEEE/ACM Trans. on Netw.*, 2013 (accepted to appear).

[27] G. Xing, M. Sha, J. Huang, G. Zhou, X. Wang, and S. Liu. Multi-Channel Interference Measurement and Modeling in Low-Power Wireless Networks. In *Proc. of IEEE RTSS*, 2009.

[28] R. Xu, G. Shi, J. Luo, Z. Zhao, and Y. Shu. MuZi: Multi-channel ZigBee Networks for Avoiding WiFi Interference. In *Proc. of the 4th IEEE CPSCom*, pages 323 – 329, 2011.

[29] X. Xu, J. Luo, and Q. Zhang. Design of Non-Orthogonal Multi-Channel Sensor Networks. In *Proc. of the 30th IEEE ICDCS*, 2010.

Barrier Coverage in Bistatic Radar Sensor Networks: Cassini Oval Sensing and Optimal Placement

Xiaowen Gong
Arizona State University
Tempe, AZ 85287, USA
xgong9@asu.edu

Junshan Zhang
Arizona State University
Tempe, AZ 85287, USA
junshan.zhang@asu.edu

Douglas Cochran
Arizona State University
Tempe, AZ 85287, USA
cochran@asu.edu

Kai Xing
University of Science and Technology of China
Hefei, Anhui 230027, China
kxing@ustc.edu.cn

ABSTRACT

By taking advantage of active sensing using radio waves, *radar* sensors can offer several advantages over passive sensors. Although much recent attention has been given to multistatic and MIMO radar concepts, little has been paid to understanding the performance of radar networks (i.e., multiple individual radars working in concert). In this context, we study the optimal placement of a *bistatic radar* (BR) sensor network for barrier coverage. The coverage problem in a bistatic radar network (BRN) is challenging because: 1) in contrast to the disk sensing model of a traditional passive sensor, the sensing region of a BR depends on the locations of both the BR transmitter and receiver, and is characterized by a *Cassini oval*; 2) since a BR transmitter (or receiver) can potentially form multiple BRs with different BR transmitters (or receivers, respectively), the sensing regions of different BRs are coupled, making the coverage of a BRN highly non-trivial.

This paper considers the problem of deploying a network of BRs in a region for maximizing the *worst-case intrusion detectability*, which amounts to minimizing the *vulnerability* of a *barrier*. We show that the *shortest barrier*-based placement is optimal if the shortest barrier is also the shortest line segment connecting the region's two boundaries. Based on this observation, we study the optimal placement of the BRs on a line segment for minimizing its vulnerability, which is a non-convex optimization problem. By exploiting some specific structural properties pertaining to the problem (particularly an important structure of detectability), we find the optimal *placement order* and the optimal *placement spacing* of the BR nodes, both of which exhibit elegant *balanced* structures. Our findings give valuable insight for the placement of BRs for barrier coverage. To our best knowledge, this is the first work to explore the coverage of a network of BRs.

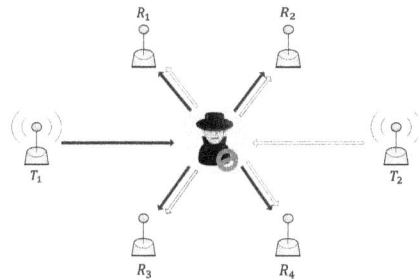

Figure 1: A bistatic radar network consisting of 2 radar transmitters T_1, T_2 and 4 radar receivers R_1, R_2, R_3, R_4. Each transmitter-receiver pair operates as a bistatic radar.

Categories and Subject Descriptors

C.2.1 [**Computer-Communication Networks**]: Network Architecture and Design—*Network Topology*

Keywords

Barrier Coverage, Bistatic Radar Sensor Network, Optimal Placement, Worst-case Intrusion

1. INTRODUCTION

Wireless sensor networks have received tremendous attention over the past decade. Typically, it is assumed that a sensor network is composed of *passive sensors* (e.g., thermal, acoustics, optic sensors) which detect radiation that is emitted or reflected by an object. In contrast, an active *radar* (RAdio Detection And Ranging) purposefully emits radio waves with the objective of collecting echoes. The ability to design the structure and power of the transmitted radio signal imbues active radars with performance advantages over passive sensors in many application scenarios, though this is typically at the expense of additional system complexity.

Thanks to recent technological advances, radars are becoming less expensive and more compact, making it feasible to deploy a network of radars working in concert. Indeed, the application scale and scope of networked radar sensors[1] are

---

[1]For brevity, we may use "radar" and "radar sensor" interchangeably.

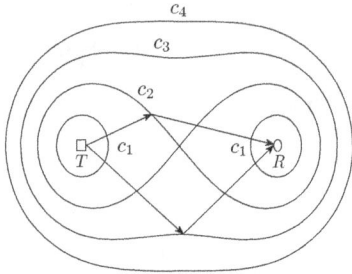

Figure 2: Bistatic radar SNR contours as Cassini ovals with foci at BR transmitter T and receiver R for distance products: $c_1 < c_2 < c_3 < c_4$.

expected to expand significantly. Due to radars' advantages over traditional passive sensors, radar networks have great potential for many applications, such as border security [1] and traffic monitoring [2]. Nevertheless, to fully exploit this potential, radar networks should be judiciously designed.

Coverage, which defines how well the object of interest is monitored, is a critical performance metric for sensor networks. *Barrier coverage* has recently emerged as an efficient coverage strategy for numerous sensor network applications centered around *intruder detection*, such as border monitoring and drug interdiction, and has drawn a surge of research interest [3–6]. Despite tremendous research progress on coverage problems for sensor networks [7], those pertaining to radar sensors remain largely unexplored, and this is the main subject of this study.

In this paper, we consider the problem of deploying a network of bistatic radars (BRs) for intrusion detection. Due to the flexibility to deploy the radar transmitter and receiver separately, a BR is more favorable than a monostatic radar (MR) for coverage. Our goal is to *build a fundamental understanding of a bistatic radar network (BRN) for coverage*. In particular, a central question we ask here is: *Where should the BRs be placed to achieve the optimal coverage quality?*

The coverage problem of a BRN is dramatically different and more challenging than that of a network of traditional passive sensors, because 1) departing from the disk sensing model of a passive sensor, the sensing region of a BR depends on the locations of both transmitter and receiver, and is characterized by a Cassini oval. Formally, a Cassini oval is a locus of points for which the distances to two fixed points (foci) have a constant product (as illustrated in Figure 2); 2) the sensing regions of different BRs are coupled with each other, since each BR transmitter[2] (or receiver) can potentially pair with different BR receivers (or transmitters, respectively) to form multiple BRs such that its location would impact multiple BRs.

We summary the main contributions of this paper as follows.

- We consider the problem of deploying a network of BRs in a region for maximizing the *worst-case intrusion detectability*, which is equivalent to minimizing the *vulnerability* of a *barrier*. We show that the *shortest barrier*-based placement is not optimal in general, and it is optimal if the shortest barrier is also the shortest line segment connecting the region's two boundaries.

- We focus on characterizing the optimal placement of the BRs on a line barrier for minimizing its vulnerability, *which is a highly non-trivial optimization problem due to its non-convexity*. To tackle the challenges herein, we reformulate the problem as finding the optimal *placement order* with the optimal *placement spacing* of the BR nodes. Based on an important structure of detectability, we characterize *balanced* placement spacing and show that it is optimal. Based on the optimal placement spacing, we then characterize the optimal placement orders, which also present balanced structures. These findings provide valuable insight for the placement of BRs for barrier coverage.

Although it is somewhat idealized, the Cassini oval sensing model (see SNR equation (1)) used in this paper can capture the essential feature of a BR, compared to a passive sensor or MR. Furthermore, the coverage problem of a BRN corresponding to the Cassini oval sensing model gives rise to significant technical difficulties (as will be seen later). Needless to say, future work is needed to generalize this study to more complex and realistic situations. In short, we believe that this study will open a new door to explore radar sensor networks.

The rest of this paper is organized as follows. Section 2 introduces the model of our work and the problem definition. In Section 3, we investigate our problem based on the barrier coverage strategy. We study the optimal placement of BRs on a line barrier in Section 4. Section 5 provides numerical results and Section 6 reviews the related works. Section 7 concludes this paper and discusses future works.

2. MODEL AND PROBLEM DEFINITION

In this section, we first describe the model of our work, including bistatic radar network and network coverage, and then introduce the problem definition.

2.1 Bistatic Radar Network

The radar transmitter and receiver are placed at different locations for a BR, whereas they are co-located for a MR. Intuitively, a BR can achieve enhanced coverage by appropriate placement of the transmitter and receiver, such that an object is more likely to be physically closer to either a transmitter or receiver, and thus attains a high signal-to-noise ratio (SNR). This advantage of BR will be illustrated by a concrete example in Section 3.

One important metric of a BR's capability for target detection is its received SNR: the strength of the received radar signal indicates whether the target is present. Let $\|AB\|$ and $\overline{AB}$ denote the (Euclidean) distance and the line segment between two points A and B, respectively. For convenience, we also use T_i or R_j to denote the location (point) of a BR transmitter node T_i or receiver node R_j, respectively. For a BR $T_i - R_j$, the received SNR from the target at a point X is given by [8]:

$$\text{SNR} = \frac{K}{\|T_i X\|^2 \|R_j X\|^2} \qquad (1)$$

where K denotes a *bistatic radar constant* that reflects physical characteristics of the BR, such as transmit power, *radar*

---

[2]For brevity, we may say "transmitter" and "receiver" instead of "BR transmitter" and "BR receiver", respectively.

cross section[3], transmitter and receiver antenna power gains. The SNR contours of a BR are characterized by the Cassini ovals with foci at the transmitter and receiver.

In a network of BRs, we assume that all transmitters operate on orthogonal radio resources (e.g., different frequencies or time slots, orthogonal waveforms) to avoid interferences at receivers. While multiple receivers can pair with the same transmitter to form multiple BRs, one receiver can also pair with multiple transmitters (e.g., by tuning to different transmitters in different time slots periodically). Typically, a BRN has more receivers than transmitters, mainly because that a transmitter incurs higher cost than a receiver (e.g., since signal transmission consumes much more energy than other sensor activities including signal reception and processing). In addition, the number of transmitters can also be limited by the available radio resources (e.g., the number of different frequencies).

We consider the deployment of a BRN consisting of M transmitters $T_i \in \mathcal{T}$, $i \in \mathcal{M} \triangleq \{1, \cdots, M\}$ and N receivers $R_j \in \mathcal{R}$, $j \in \mathcal{N} \triangleq \{1, \cdots, N\}$ where $M \leq N$. For ease of exposition, we assume that transmitters and receivers, respectively, have homogeneous physical characteristics such that all BRs have the same bistatic radar constant. We also assume that a receiver can potentially pair with all transmitters to form multiple BRs. However, our results in Section 4 will show that it suffices for a receiver to pair with *at most two* transmitters. We further assume that transmitted and reflected radar signals are *omni-directional*[4].

2.2 Network Coverage

The BRN is deployed in a 2D geographical *region of interest* F for detecting an intruder traversing through the region. The region F is defined by an entrance side, a destination side, a left boundary F_l, and a right boundary F_r (as illustrated in Figure 3). The intruder can choose any *intrusion path* P in region F connecting the entrance to the destination.

The existing studies on sensor network coverage [9–11] have used the distance from a point to its *closest* sensor to measure the coverage of that point (also known as the *closest sensor observability*). In the same spirit, we measure the coverage of a point by the *highest* SNR received by all BRs when the target is present at that point. Considering (1), we have the following definition.

DEFINITION 1 (**Detectability**). *The detectability[5] $I(X)$ of a point X is the minimum distance product from X to all BRs:*

$$I(X) \triangleq \min_{T_i \in \mathcal{T}, R_j \in \mathcal{R}} \|T_i X\| \|R_j X\|. \tag{2}$$

In other words, the detectability of a point is determined by the point's *closest BR* consisting of its closest transmitter and closest receiver. As in [9–11], we use the *worst-case intrusion* to quantify the coverage of the intruder.

---

[3]Radar cross section measures the amount of radar signal energy reflected by an object depending on its physical characteristics (e.g., shape, material).

[4]The propagation direction of the reflected radar signal depends on the object's physical characteristics (e.g., shape, material), and is not omni-directional in general.

[5]With a little abuse of notation, we use $I(X)$ to denote the detectability of X, but X's detectability changes inversely with the value of $I(X)$.

DEFINITION 2 (**Worst-case Intrusion [9]**). *The worst-case intrusion path P^* is the intrusion path with the minimum detectability:*

$$P^* \triangleq \arg\max_{P \in \mathcal{P}} D(P) \tag{3}$$

where $\mathcal{P}$ is the set of all possible intrusion paths, and the detectability $D(P)$ of an intrusion path P is the maximum detectability of all points in P:

$$D(P) \triangleq \min_{X \in P} I(X). \tag{4}$$

2.3 Problem Definition

We are interested in finding the optimal placement of the BRN (i.e., optimal locations of M transmitters and N receivers) in region F for maximizing the worst-case intrusion detectability:

$$\underset{T_i \in F, R_j \in F}{\text{minimize}} \quad D(P^*). \tag{5}$$

Based on the notion of worst-case intrusion, problem (5) is of great interest for the intruder detection problem. In particular, solving problem (5) allows us to answer an important question: How many transmitters and receivers are needed and where do we place them for providing the required coverage such that *at least one* BR will receive an SNR above some predefined threshold, regardless of the intruder's path?

It is worth noting that problem (5) is difficult to solve in general (even for a sensor network under the disk sensing model). This is because 1) the shape of region F can be arbitrary and 2) the feasible solution space, which includes any placement in region F, is large. In the next section, we will investigate problem (5) based on the strategy of barrier coverage.

3. PLACEMENT FOR BARRIER COVERAGE

In this section, we consider problem (5) by placing the BRs for barrier coverage. As in [3–6], a *barrier* is defined as a curve lying in region F such that *any* intrusion path intersects with the curve. We next define the coverage metric of a barrier.

DEFINITION 3 (**Vulnerability**). *The vulnerability $V(U)$ of a barrier U is the minimum detectability of all points in U:*

$$V(U) \triangleq \max_{X \in U} I(X). \tag{6}$$

The rationale of using such a coverage metric is that, since a barrier intersects with any possible intrusion path, the vulnerability of a barrier serves as an *upper bound* on the worst-case intrusion detectability. This bound becomes *tight* when all barriers are considered:

$$D(P^*) = \min_{U \in \mathcal{U}} V(U) \tag{7}$$

where $\mathcal{U}$ is the set of all barriers.

Using (7), problem (5) boils down to finding the optimal barrier that has the minimum *achievable vulnerability* $V^*(U)$:

$$\underset{U \in \mathcal{U}}{\text{minimize}} \quad V^*(U), \tag{8}$$

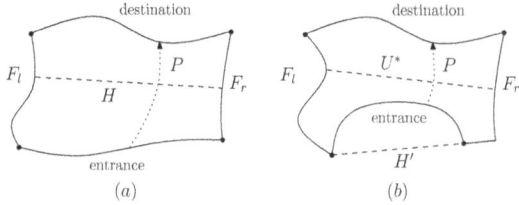

Figure 3: (a) H is the shortcut barrier; (b) H' is the shortest line segment connecting F_l and F_r but not a barrier, so the shortcut barrier does not exist; the shortest barrier U^* (also a line barrier) is not the shortcut barrier.

where $V^*(U)$ is the vulnerability of U under the optimal placement of the BRs in region F for minimizing U's vulnerability, i.e., the optimal value of the following problem:

$$\underset{T_i \in F, R_j \in F}{\text{minimize}} \quad V(U). \tag{9}$$

For problem (8), it seems plausible to select the *shortest barrier* U^*, which is the barrier (possibly multiple) with the minimum length. However, this strategy is not optimal in general because $V^*(U^*)$ is not necessarily lower than $V^*(U)$ for a barrier U with a *larger* length than U^*. We give a simple counterexample as illustrated in Figure 4. For a line barrier[6] $\overline{AB}$ in Figure 4(a), one can easily figure out that the optimal placement of a BR $T-R$ is to set $\|AT\| = \|RB\| = \sqrt{2} - 1$ such that $V^*(\overline{AB}) = \|AT\|\|AR\| = 1$. For a barrier $\widetilde{CF}$[7] consisting of $\overline{CD}$, $\overline{DE}$, $\overline{EF}$ in Figure 4(b), which has a larger length than $\overline{AB}$, we have $V^*(\widetilde{CF}) \leq V(\widetilde{CF}) = \|TE\|\|ER\| = \sqrt{5}/4 < 1$ where T and R are placed at the midpoints of $\overline{CD}$ and $\overline{EF}$, respectively.

Before proceeding further, we use a simple example to illustrate the advantage of a BR over a MR for barrier coverage. If we place a MR (co-located transmitter and receiver) on $\overline{AB}$ in Figure 4(a) to minimize $V(\overline{AB})$, the optimal placement is clearly at the midpoint G of $\overline{AB}$ such that $V(\overline{AB}) = \|AG\|^2 = 2$, which is greater than $V^*(\overline{AB}) = 1$.

Although the shortest barrier-based placement is not optimal in general, it is optimal if the shortest barrier is also the *shortcut barrier*.

DEFINITION 4 (**Shortcut Barrier**). *The shortcut barrier exists and is the shortest barrier if and only if the shortest barrier is the shortest line segment connecting F_l and F_r (i.e., its length is the minimum distance between a point in F_l and a point in F_r).*

Although there must exist the shortest line segment connecting F_l and F_r, it does not necessarily lie in region F as a barrier (as illustrated in Figure 3(b)). The shortcut barrier exists for a large class of shapes of region F. For example, any convex region falls in this class. Worth noting is that if the shortest barrier is not the shortcut barrier, it can still be a line barrier (as illustrated in Figure 3(b)).

THEOREM 1. *If the shortcut barrier H exists, then it is the optimal barrier for problem (8), indicating that it suffices to solve problem (9) for H.*

---

[6]We refer a barrier as a line barrier if it is a line segment.
[7]We use $\widetilde{PQ}$ and $\|\widetilde{PQ}\|$ to denote a curve with end points P and Q, and its length, respectively.

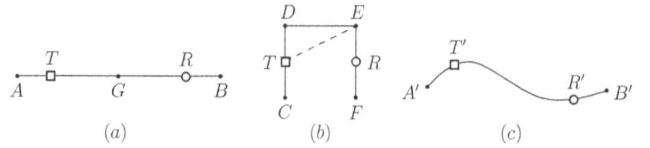

Figure 4: (a) $\|AB\| = 2\sqrt{2}$; (b) $\|CD\| = \|DE\| = \|EF\| = 1$, $\overline{CD} \perp \overline{DE}$, $\overline{EF} \perp \overline{DE}$; (c) $\|\widetilde{A'B'}\| = 2\sqrt{2}$.

We can show that the optimal placement for problem (9) for a line barrier is *on* that barrier. In view of the optimality of the shortcut barrier-based placement, in Section 4, we will focus on finding the optimal placement of the BRs on a line barrier for minimizing its vulnerability.

If the shortest barrier U^* is not the shortcut barrier, it can be an arbitrary curve and it is not necessarily the optimal barrier for problem (8). Furthermore, it is in general difficult to find the optimal placement for problem (9) for an arbitrary barrier (even for a sensor network under the disk sensing model). In this situation, we can take an *approximate* approach for U^* by applying the optimal placement for problem (9) for a line barrier (will be obtained in Section 4). Specifically, we can treat U^* as a line barrier $\overline{U^*}$ (if U^* is not) with the same length, and place the BR nodes on U^* according to the optimal placement for $\overline{U^*}$. An appealing property of this approach is that $V(U^*)$ under such placement must be *no greater than* $V^*(\overline{U^*})$, and hence $V^*(\overline{U^*})$ serves as an upper bound for the worst-case intrusion detectability $D(P^*)$. For example, as $\overline{AB}$ in Figure 4(a) has the same length with $\widetilde{A'B'}$ in Figure 4(c), we can apply the optimal placement of $\{T, R\}$ for minimizing $V(\overline{AB})$ to $\widetilde{A'B'}$ such that $\|\widetilde{A'T'}\| = \|AT\|$ and $\|\widetilde{T'B'}\| = \|TB\|$. Then we have that $V(\widetilde{A'B'})$ under $\{T', R'\}$ is no greater than $V^*(\overline{AB})$ under $\{T, R\}$. Since we can show that $V^*(U)$ for a line barrier U is increasing with U's length and U^* has the minimum length, applying the approximate approach to U^* can achieve the *minimum* upper bound $V^*(\overline{U^*})$ for $D(P^*)$ compared to applying it to other barriers.

4. OPTIMAL PLACEMENT ON A LINE BARRIER

In this section, we study the optimal placement of the BRs on a line barrier (line segment) H for minimizing its vulnerability $V(H)$.

4.1 Problem Reformulation

Let H_l and H_r be the end points of H and h be its length. Also let $t_i \triangleq \|H_l T_i\|$ and $r_j \triangleq \|H_l R_j\|$. Mathematically, the problem can be written as

$$\underset{t_i, r_j}{\text{minimize}} \quad \max_{0 \leq x \leq h} \min_{i \in \mathcal{M}, j \in \mathcal{N}} |x - t_i||x - r_j| \tag{10}$$

$$\text{subject to} \quad 0 \leq t_i \leq h, \forall i \in \mathcal{M}$$
$$0 \leq r_j \leq h, \forall j \in \mathcal{N}$$

where $\min_{i \in \mathcal{M}, j \in \mathcal{N}} |x - t_i||x - r_j|$ represents the detectability of a point $X \in H$ with $\|H_l X\| = x$. Since we can show that problem (10) is *non-convex* in general, standard optimization methods would not work well here.

To gain more insight of problem (10), we reformulate it as follows. We first treat H_l and H_r as two (virtual) nodes and ignore the constraint $\|H_l H_r\| = h$. Then we place H_l, H_r, and all the BR nodes on a horizontal line such that H_l and H_r are the *leftmost* and *rightmost* nodes, respectively.

DEFINITION 5 (**Placement Order and Spacing**). *A placement order ("order" for short) $\mathbf{S}$ is an order of all the nodes on the line from left to right:*

$$\mathbf{S} \triangleq (H_l, S_1, \cdots, S_J, H_r)$$

where $J \triangleq M + N$ and $(S_1, \cdots, S_J)$ is a permutation of the BR nodes such that $\|H_l H_l\| \leq \|H_l S_1\| \leq \cdots \leq \|H_l S_J\| \leq \|H_l H_r\|$. The placement spacing ("spacing" for short) $\mathbf{D_S}$ of a placement order $\mathbf{S}$ consists of the distances between neighbor nodes in $\mathbf{S}$:

$$\mathbf{D_S} \triangleq (\|H_l S_1\|, \cdots, \|S_J H_r\|).$$

A local placement order ("local order" for short) $(S_{i+1}, \cdots, S_{i+j})$ is an order of some neighbor nodes in $\mathbf{S}$, and its placement spacing is

$$\mathbf{D}_{(S_{i+1}\cdots,S_{i+j})} \triangleq (\|S_{i+1}S_{i+2}\|, \cdots, \|S_{i+j-1}S_{i+j}\|).$$

Any order $\mathbf{S}$, together with any spacing $\mathbf{D_S}$, correspond to some placement of the BRs on a line segment with length $\|H_l H_r\|$; conversely, any placement of the BRs on the line segment H has a corresponding order $\mathbf{S}$ with corresponding spacing $\mathbf{D_S}$ subject to $\|H_l H_r\| = h$. Therefore, problem (10) can be recast as

$$\underset{\mathbf{S,D_S}}{\text{minimize}} \quad V(\overline{H_l H_r}) \tag{11}$$

$$\text{subject to} \quad \|H_l H_r\| = h.$$

For any order $\mathbf{S}$, we note that the vulnerability $V(\overline{H_l H_r})$ is *non-decreasing* as any distance in $\mathbf{D_S}$ increases. In light of this, we can formulate a related problem of problem (11) as

$$\underset{\mathbf{S,D_S}}{\text{maximize}} \quad \|H_l H_r\| \tag{12}$$

$$\text{subject to} \quad V(\overline{H_l H_r}) \leq c.$$

Let l_c denote the optimal value of problem (12) under the constraint $V(\overline{H_l H_r}) \leq c$. We can verify that l_c is strictly increasing in c and, in particular, $l_c \to 0$ when $c \to 0$ and $l_c \to \infty$ when $c \to \infty$. Therefore, if we can solve problem (12) for any $c > 0$, we can also solve problem (11) by a *bisection search* described in Algorithm 1.

We make two general observations to be used later. First, since all BRs are homogeneous, swapping the locations of any pair of transmitters (or receivers, respectively) results in an equivalent placement. Second, transmitters and receivers are *reciprocal* in the sense that replacing all transmitters by receivers while replacing all receivers by transmitters results in an equivalent placement.

4.2 Optimal Placement Order and Spacing

In this subsection, we focus on characterizing the optimal order and the optimal spacing for problem (12). We outline the major steps as follows.

1) We identify an important structure of detectability on $\overline{H_l H_r}$ (Lemma 1), based on which we define *balanced spacing*.

Algorithm 1: Compute the optimal solution to problem (11)

input : the line barrier length h, tolerance ϵ
output: the optimal order $\mathbf{S}^*$, the optimal spacing $\mathbf{D}_{\mathbf{S}^*}^*$, the optimal value c^*

1 $c_1 \leftarrow 0$, $c_2 \leftarrow h^2$, $c \leftarrow \frac{c_1+c_2}{2}$;
2 **repeat**
3 $\quad$ compute the optimal order $\mathbf{S}^*$ with the optimal spacing $\mathbf{D}_{\mathbf{S}^*}$ for problem (12) subject to $V(\overline{H_l H_r}) \leq c$;
4 $\quad$ **if** $l_c > h + \epsilon$ **then**
5 $\quad\quad$ $c_2 \leftarrow c$; $c \leftarrow \frac{c_1+c_2}{2}$;
6 $\quad$ **end**
7 $\quad$ **if** $l_c < h - \epsilon$ **then**
8 $\quad\quad$ $c_1 \leftarrow c$; $c \leftarrow \frac{c_1+c_2}{2}$;
9 $\quad$ **end**
10 **until** $|l_c - h| \leq \epsilon$;
11 **return** $\mathbf{S}^*, \mathbf{D}_{\mathbf{S}^*}^*, c^* \leftarrow c$;

---

2) We define *independent local orders*. Then we characterize the balanced spacing for an independent local order (Lemma 2) and show that it is optimal (Lemma 3).

3) We show that the optimal order consists of independent local orders with *disjoint* sets of spacing distances (Lemma 4), based on which we show that the balanced spacing is optimal for the optimal order (Theorem 2).

4) Based on the optimal spacing, we characterize the optimal orders (Theorem 3).

Let Y_{AB} denote the midpoint between two points A and B. We first show an important structure of detectability on $\overline{H_l H_r}$.

LEMMA 1. *For any order $\mathbf{S}$ with any spacing $\mathbf{D_S}$, the detectability on $\overline{H_l H_r}$ attains local maximums at the end nodes and at the midpoints of all pairs of neighbor nodes in $(S_1, \cdots, S_J)$ (as illustrated in Figure 5):*

$$\arg \max_{X \in \overline{H_l S_1}} I(X) = H_l, \quad \arg \max_{X \in \overline{S_J H_r}} I(X) = H_r,$$

$$\arg \max_{X \in \overline{S_i S_{i+1}}} I(X) = Y_{S_i S_{i+1}}, \quad \forall i \in \{1, \cdots, J-1\}.$$

DEFINITION 6 (**Local Vulnerable Point**). *A local vulnerable point is a local maximum point of detectability on $\overline{H_l H_r}$ (described in Lemma 1), and a local vulnerable value is its detectability.*

By Lemma 1, it suffices to focus on the local vulnerable values for determining the vulnerability of $\overline{H_l H_r}$.

DEFINITION 7 (**Independent Local Order**). *A local order $\mathbf{S}_i$ is an independent local order if it exhibits one of the following orders of node types (H_l, H_r, transmitter type T, or receiver type R):*

$$(T, R^k, H_r), (R, T^k, H_r), (H_l, R^k, T), (H_l, T^k, R), k \geq 1;$$

$$(T, R^k, T), (R, T^k, R), k \geq 1; \quad (T, R), (R, T);$$

where we use T^k or R^k to denote k consecutive T or R, respectively. The independent local zone $Z_{\mathbf{S}_i}$ of an independent local order $\mathbf{S}_i$ is the line segment between the two end nodes in $\mathbf{S}_i$, with its length denoted by $L_{\mathbf{S}_i}$.

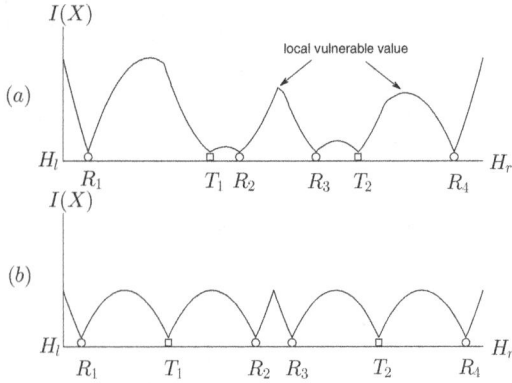

Figure 5: The local vulnerable values are (a) unequal with some placement spacing and (b) equal with balanced placement spacing.

An independent local order $\mathbf{S}_i$ has an *independent* property as described below. Note that the *closest* transmitter and *closest* receiver for any local vulnerable point on $Z_{\mathbf{S}_i}$ are nodes in $\mathbf{S}_i$. For example, for $\mathbf{S}_i = (T_1, R_1)$, the closest transmitter and closest receiver for $Y_{T_1 R_1}$ are T_1 and R_1, respectively; for $\mathbf{S}_i = (T_1, R_1, \cdots, R_k, H_r)$, the closest transmitter for any of $Y_{T_1 R_1}$, $\cdots$, $Y_{R_{k-1} R_k}$, and H_r, is T_1, while the closest receiver for any of $Y_{T_1 R_1}$, $\cdots$, $Y_{R_{k-1} R_k}$, and H_r, is some node in $\{R_1, \cdots, R_k\}$. Therefore, *all the local vulnerable values on the independent local zone $Z_{\mathbf{S}_i}$, and hence the vulnerability $V(Z_{\mathbf{S}_i})$, are determined by the spacing $\mathbf{D}_{\mathbf{S}_i}$ (i.e., independent of any distance not in $\mathbf{D}_{\mathbf{S}_i}$).*

DEFINITION 8 (**Balanced Spacing**). *The spacing $\mathbf{D}_\mathbf{S}$ (or $\mathbf{D}_{\mathbf{S}_i}$) of an order $\mathbf{S}$ (or an independent local order $\mathbf{S}_i$, respectively) is balanced if all the local vulnerable values on $\overline{H_l H_r}$ (or the independent local zone $L_{\mathbf{S}_i}$, respectively) are equal (as illustrated in Figure 5).*

Using the independent property and Lemma 1, we can characterize the balanced spacing of an independent local order as follows.

LEMMA 2. *For any $c > 0$, let $e_c^0 \triangleq 2\sqrt{c}$ and e_c^j be the unique positive value of x such that $(\sum_{i=0}^{j-1} e_c^i + x/2)(x/2) = c$ for any $j \in \mathbb{N}^+$. For any $c > 0$ and any independent local order $\mathbf{S}_i$, there exists a unique balanced spacing $\mathbf{D}_{\mathbf{S}_i}$ such that $V(Z_{\mathbf{S}_i}) = c$. Furthermore, it is given by, e.g., for $\mathbf{S}_i = (T_1, R_1)$, $\mathbf{D}_{\mathbf{S}_i} = (e_c^0)$; for $\mathbf{S}_i = (T_1, R_1, \cdots, R_k, H_r)$,*

$$\mathbf{D}_{\mathbf{S}_i} = (e_c^0, e_c^1, \cdots, e_c^{k-1}, \frac{e_c^k}{2});$$

for $\mathbf{S}_i = (T_1, R_1, \cdots, R_k, T_2)$ and even k,

$$\mathbf{D}_{\mathbf{S}_i} = (e_c^0, e_c^1, \cdots, e_c^{\frac{k}{2}-1}, e_c^{\frac{k}{2}}, e_c^{\frac{k}{2}-1}, \cdots, e_c^1, e_c^0)$$

or odd k,

$$\mathbf{D}_{\mathbf{S}_i} = (e_c^0, e_c^1, \cdots, e_c^{\frac{k-1}{2}}, e_c^{\frac{k-1}{2}}, \cdots, e_c^1, e_c^0).$$

Similar results can be obtained for any other independent local order.

By definition, given c, the value of e_c^i, $i \in \mathbb{N}^+$ can be found iteratively and it decreases as i becomes larger (as shown in Table 1).

Table 1: Values of balanced spacing

c	e_c^0	e_c^1	e_c^2	e_c^3	e_c^4
1	2.0000	0.8284	0.6357	0.5359	0.4721
5	4.4721	1.8524	1.4214	1.1983	1.0557
10	6.3246	2.6197	2.0102	1.6947	1.4930
20	8.9443	3.7048	2.8428	2.3966	2.1115

Based on the independent property, we can cast a problem similar to problem (12) but for a *given* independent local order $\mathbf{S}_i$ as

$$\underset{\mathbf{D}_{\mathbf{S}_i}}{\text{maximize}} \quad L_{\mathbf{S}_i} \tag{13}$$

$$\text{subject to} \quad V(Z_{\mathbf{S}_i}) \leq c.$$

The balanced spacing proves to be the optimal spacing for problem (13).

LEMMA 3. *For any $c > 0$ and any independent local order $\mathbf{S}_i$, the balanced spacing $\mathbf{D}_{\mathbf{S}_i}$ such that $V(Z_{\mathbf{S}_i}) = c$ is the optimal spacing for problem (13).*

Next we turn to show that the optimal order has a *dividable* structure.

DEFINITION 9 (**Dividable Order**). *An order $\mathbf{S}$ is dividable if it consists of independent local orders $\mathbf{S}_1, \cdots, \mathbf{S}_m$ such that 1) each node in $\mathbf{S}$ is included in some $\mathbf{S}_i$; 2) the last node of $\mathbf{S}_i$ is the first node of $\mathbf{S}_{i+1}$ for all $i = 1, \cdots, m-1$. Therefore, $\overline{H_l H_r}$ under $\mathbf{S}$ consists of independent local zones $Z_{\mathbf{S}_1}, \cdots, Z_{\mathbf{S}_m}$, and $\mathbf{D}_\mathbf{S}$ consists of disjoint sets of spacing distances $\mathbf{D}_{\mathbf{S}_1}, \cdots, \mathbf{D}_{\mathbf{S}_m}$.*

For example,

$$\mathbf{S} = (\overbrace{H_l, R_1, R_2, \underbrace{T_1}_{}, R_3, R_4, R_5, \overbrace{T_2}, R_6, T_3, T_4, H_r}^{\mathbf{S}_1}) \tag{14}$$

is a dividable order.

LEMMA 4. *The optimal order $\mathbf{S}^*$ is dividable.*

It is worth mentioning that Lemma 4 provides a *necessary* condition for the optimal order. In other words, a non-optimal order (such as (14)) can also have this structure.

The structure of a dividable order allows us to break down the problem. Consider the following problem that is related to problem (12) but for a *given* order $\mathbf{S}$:

$$\underset{\mathbf{D}_\mathbf{S}}{\text{maximize}} \quad \|H_l H_r\| \tag{15}$$

$$\text{subject to} \quad V(\overline{H_l H_r}) \leq c.$$

For a dividable order $\mathbf{S}$, the set of local vulnerable points on $\overline{H_l H_r}$ is the union of *disjoint* sets of local vulnerable points on the independent local zones $Z_{\mathbf{S}_1}, \cdots, Z_{\mathbf{S}_m}$ of independent local orders $\mathbf{S}_1, \cdots, \mathbf{S}_m$. Therefore, *problem (15) for a dividable order $\mathbf{S}$ can be broken down into independent subproblems, each of which is an instance of problem (13)* for one of $\mathbf{S}_1, \cdots, \mathbf{S}_m$, such that each can be solved using Lemma 3. Note that if the optimal order $\mathbf{S}^*$ is known, then problem (12) boils down to problem (15) for $\mathbf{S} = \mathbf{S}^*$. Since the optimal order $\mathbf{S}^*$ is dividable by Lemma 4, the next result directly follows.

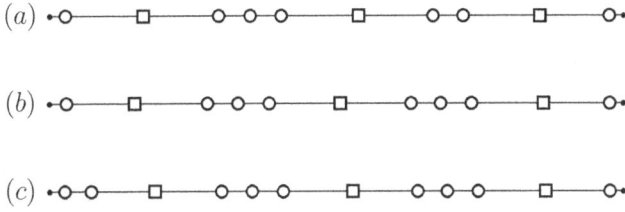

(a)
(b)
(c)

Figure 6: Optimal placement of transmitters (square) and receivers (circle) for (a) $M = 3$, $N = 7$; (b) $M = 3$, $N = 8$; (c) $M = 3$, $N = 9$.

THEOREM 2. *For the optimal order $\mathbf{S}^*$ and any $c > 0$, the balanced spacing $\mathbf{D}_{\mathbf{S}^*}$ such that $V(\overline{H_l H_r}) = c$ exists and is the optimal spacing for problem (15) for $\mathbf{S} = \mathbf{S}^*$.*

Next we characterize the optimal order $\mathbf{S}^*$. Let $f_c^{\mathbf{S}}$ denote the optimal value of problem (15) for an order $\mathbf{S}$ under the constraint $V(\overline{H_l H_r}) \leq c$. Then $f_c^{\mathbf{S}}$ achieves the maximum when $\mathbf{S} = \mathbf{S}^*$. Since all transmitters are homogeneous, we index the transmitters from left to right such that $0 \leq \|H_l T_1\| \leq \cdots \leq \|H_l T_M\| \leq \|H_l H_r\|$. Define

$$\mathbf{N}_{\mathbf{S}} \triangleq (n_1, n_2, \cdots, n_M, n_{M+1})$$

where n_i, n_1, n_{M+1} denote the number of receivers in $\mathbf{S}$ between T_{i-1} and T_i for $i \in \{2, \cdots, M\}$, between H_l and T_1, between T_M and H_r, respectively. Since all receivers are homogeneous, it suffices to determine $\mathbf{N}_{\mathbf{S}^*}$ for characterizing the optimal order $\mathbf{S}^*$.

THEOREM 3. *An order $\mathbf{S}$ is optimal if and only if*

$$|n_i - n_j| \leq 1, \forall i, j \in \{2, \cdots, M\}$$
$$|n_i - 2n_1| \leq 1, |n_i - 2n_{M+1}| \leq 1, \forall i \in \{2, \cdots, M\}.$$

Using Theorem 3, we can describe the optimal order $\mathbf{S}^*$ as follows. Let two integers q and r be the quotient and remainder of N/M, respectively. If q is even, we have

$$\mathbf{N}_{\mathbf{S}^*} = (\frac{q}{2}, \overbrace{q+1, \cdots, q+1}^{r}, \overbrace{q, \cdots, q}^{M-1-r}, \frac{q}{2});$$

if q is odd and $r = 0$, we have

$$\mathbf{N}_{\mathbf{S}^*} = (\frac{q+1}{2}, \overbrace{q, \cdots, q}^{M-1}, \frac{q-1}{2});$$

if q is odd and $r \geq 1$, we have

$$\mathbf{N}_{\mathbf{S}^*} = (\frac{q+1}{2}, \overbrace{q+1, \cdots, q+1}^{r-1}, \overbrace{q, \cdots, q}^{M-r}, \frac{q+1}{2}).$$

In addition, any order obtained from the above optimal order by swapping the values of n_1 and n_{M+1}, or the values of n_i and n_j for $i, j \in \{2, \cdots, M\}$, also satisfies the conditions in Theorem 3, and hence is optimal.

Given the optimal order $\mathbf{S}^*$ for problem (12), we can find the optimal spacing $\mathbf{D}_{\mathbf{S}^*}$ for problem (12) by using Theorem 2 to solve problem (15) for $\mathbf{S}^*$.

4.3 Discussions

The optimal order and the optimal spacing both present elegant *balanced* structures (as illustrated in Figure 6). As noted earlier, the optimal spacing is balanced in the sense that all the local vulnerable values on $\overline{H_l H_r}$ are *equal* under

the optimal spacing. The optimal orders are balanced in a more subtle sense: for the optimal order $\mathbf{S}^*$, n_i is *as equal as possible* across all $i = 2, \cdots, M$, and it is *as equal as possible to twice* the value of n_1, n_{M+1}, respectively. The balanced structures are partly due to that all the BRs are homogeneous.

The optimal placement results imply that one assumption made in Section 2 can be greatly relaxed without losing the optimality. Since the detectability of a point is determined by the closest transmitter and closest receiver, it suffices for a receiver to pair with a transmitter only if they are the closest receiver and closest transmitter, respectively, for some point on the line barrier. Thus, under the optimal placement of a BRN consisting of more receivers than transmitters (such as in Figure 6), one can see that a receiver only needs to pair with its closest transmitter(s), the number of which is *one or two*.

5. NUMERICAL RESULTS

In this section, we present numerical results to illustrate the effectiveness of the optimal placement of BRs on a line barrier H.

As none of the existing works has studied the placement of BRs for barrier coverage, we compare the optimal placement strategy (OPT) with two heuristic strategies. The heuristics are motivated by the rationale of the optimal placement strategy for a network of homogeneous sensors under the disk sensing model, which is to minimize the maximum distance from a point on H to its closest sensor.

One heuristic (HEU-1) is to place transmitters (or receivers, respectively) with uniform spacing such that the maximum distance from a point on H to its closest transmitter (or receiver, respectively) is minimized (as illustrated in Figure 7(a)):

$$2\|H_l T_1\| = \|T_1 T_2\| = \cdots = \|T_{M-1} T_M\| = 2\|T_M H_r\|$$

$$2\|H_l R_1\| = \|R_1 R_2\| = \cdots = \|R_{N-1} R_N\| = 2\|R_N H_r\|.$$

We can see from Figure 6 that neither the transmitters nor the receivers in OPT follow the placement of HEU-1. Compared to OPT, the drawback of HEU-1 is mainly due to that it places transmitters and receivers *independently*.

Another heuristic (HEU-2) is to place transmitters and receivers according to the optimal order $\mathbf{S}^*$, but with uniform spacing such that the maximum distance from a point on H to its closest BR node (transmitter or receiver) is minimized (as illustrated in Figure 7(b)):

$$2\|H_l S_1\| = \|S_1 S_2\| = \cdots = \|S_{M-1} S_M\| = 2\|S_M H_r\|.$$

Although HEU-2 follows the optimal order, its main drawback lies in that it treats transmitters and receivers *equivalently*.

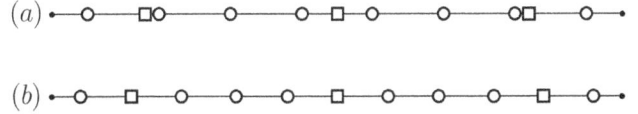

(a)
(b)

Figure 7: Heuristic placement of transmitters (square) and receivers (circle) for $M = 3$, $N = 8$: (a) HEU-1; (b) HEU-2.

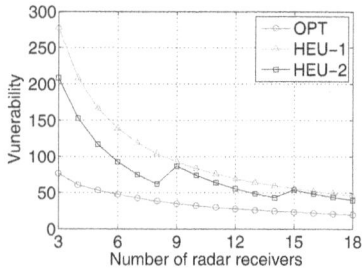

Figure 8: Vulnerability for 3 transmitters.

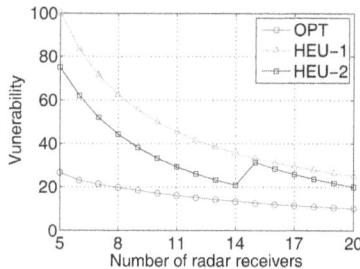

Figure 9: Vulnerability for 5 transmitters.

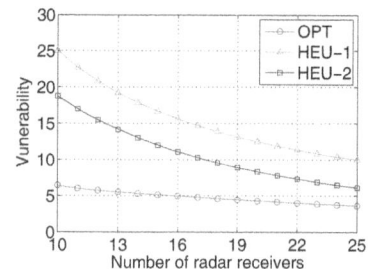

Figure 10: Vulnerability for 10 transmitters.

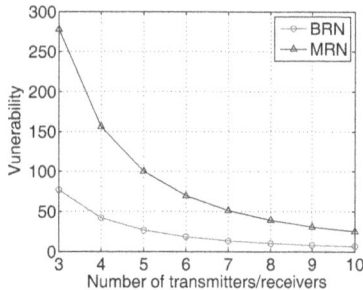

Figure 11: Minimum vulnerability: BRN vs MRN.

Figure 8, Figure 9 and Figure 10 depict the vulnerability of H under OPT, HEU-1, and HEU-2 for a varying number of receivers and 3, 5, 10 transmitters, respectively. We set the length of H to 100m. We observe that HEU-2 results in considerably lower vulnerability than HEU-1, and OPT further outperforms HEU-2 significantly. This shows that OPT is highly advantageous for improving the barrier coverage, which is essentially due to that the design rationale for a BRN under the Cassini oval sensing model is quite different and more complicated than that for a network of passive sensors (or MRs) under the disk sensing model. Therefore, the optimal placement of a BRN requires judicious design of transmitters and receivers as we do in this paper.

In Figure 11, we compare the vulnerability of H under the optimal placement of a BRN to that of a monostatic radar network (MRN) for a varying number of transmitters and receivers. The optimal placement strategy for a MRN is to minimize the maximum distance from a point on H to its closest MR. For fair comparison, we set the number of transmitters equal to that of receivers in the BRN, and also equal to the number of MRs in the MRN. We observe that the advantage of a BRN is significant, which demonstrates that the flexibility to place transmitters and receivers separately is highly beneficial for barrier coverage.

6. RELATED WORK

Radar has been extensively studied for decades [12]. However, radar sensor networks have garnered attention only in the past few years, largely driven by the emergence of cheaper and more compact radar sensors replacing conventionally expensive and bulky radar systems. For example, in [13], a platform has been successfully designed and built to integrate ultrawideband radars with mote-class sensor de-

vices. The existing literature has studied different problems for radar sensor networks, including waveform design and diversity [14], radar scheduling [15], data management [16], for a variety of objectives, such as target detection [17] and localization [18]. In particular, BRs have also been considered in [18]. However, coverage problems of a radar sensor network have received very little attention. Recently, a novel Doppler coverage model has been introduced in [19] for a radar sensor network that exploits the Doppler effect. To our best knowledge, our work is the first to explore the coverage of a network of BRs.

Numerous studies on sensor network coverage can be found in the literature [7]. Worst-case intrusion was first introduced in [9]. [9,10,20] have studied how to find the worst-case intrusion path for arbitrary deployed sensors. [21,22] have considered adding sensors to improve the coverage of the worst-case intrusion path. Along another avenue, barrier coverage was first introduced in [3] and has attracted much research interests recently. [3,5] have studied the critical sensor density for barrier coverage under random deployment. The coverage of a barrier has been investigated using a quantitative metric in [4]. Barrier coverage of sensors with mobility have been considered in [6,23]. Barrier coverage for camera sensor networks has also been studied recently based on a novel full-view coverage model [24,25].

While most aforementioned studies are concerned with how to find the worst-case intrusion path or a barrier covered by sensors (if such a barrier exists) under an *existing* deployment of sensors, this work focuses on *where* we should deploy sensors to cover a barrier such that the worst-case intrusion detectability is maximized. More importantly, the existing sensing models (particularly the widely used disk sensing model) are quite different from the Cassini oval sensing model of a BR, and the latter is further complicated by the coupling of sensing regions across multiple BRs.

7. CONCLUSION AND FUTURE WORK

Radar sensor networks have great potential in many applications, such as border surveillance and traffic monitoring. In this paper, we studied the problem of deploying a BRN for barrier coverage. The optimal placement of BRs is highly non-trivial, since 1) the coverage region of a BR is characterized by a Cassini oval that presents complex geometry; 2) the coverage regions of different BRs are coupled and the network coverage is intimately related to the locations of all BR nodes. We show that it is in general not optimal to place the BRs on the shortest barrier, but it is optimal if the shortest barrier is also the shortcut barrier. Further, we characterized the optimal placement order and the optimal placement spacing of the BRs on a line barrier, both of which present elegant balanced structures.

Although the models are built upon some idealized assumptions, we believe that this work provides some initial steps for understanding the coverage of networked BRs. There are still many questions remaining open. For example, while the Cassini oval sensing model used in this work is based on SNR, it would be interesting to also take into account the *Doppler effect*. As the Doppler effect is intimately related to the motion of objects, it would give rise to a number of challenging problems in the context of networked radars. Furthermore, as this work assumes that all BRs are homogeneous, an interesting future direction is to consider a network of heterogeneous BRs.

ACKNOWLEDGEMENT

This research was supported in part by the NSF Grant CNS-0901451, U.S. AFOSR project FA9550-10-1-0464, DoD MURI project No. FA9550-09-1-0643, and NSFC Grant 61170267.

8. REFERENCES

[1] C. J. Baker and H. D. Griffiths, "Bistatic and multistatic radar sensors for homeland security," *Advances in Sensing with Security Applications*, vol. 2, pp. 1–22, Feb. 2006.

[2] B. Donovan, D. J. McLaughlin, and J. Kurose, "Principles and design considerations for short-range energy balanced radar networks," in *IGARSS 2005*.

[3] S. Kumar, T.-H. Lai, and A. Arora, "Barrier coverage with wireless sensors," in *ACM MOBICOM 2005*.

[4] A. Chen, T.-H. Lai, and D. Xuan, "Measuring and guaranteeing quality of barrier-coverage in wireless sensor networks," in *ACM MOBIHOC 2008*.

[5] B. Liu, O. Dousse, J. Wang, and A. Saipulla, "Strong barrier coverage of wireless sensor networks," in *ACM MOBIHOC 2008*.

[6] A. Saipulla, B. Liu, G. Xing, X. Fu, and J. Wang, "Barrier coverage with sensors of limited mobility," in *ACM MOBIHOC 2010*.

[7] B. Wang, "Coverage problems in sensor networks: A survey," *ACM Computing Surveys*, vol. 43, Oct. 2011.

[8] N. Willis, *Bistatic Radar*. SciTech Publishing, 2005.

[9] S. Meguerdichian, F. Koushanfar, M. Potkonjak, and M. Srivastava, "Coverage problems in wireless ad-hoc sensor network," in *IEEE INFOCOM 2001*.

[10] S. Meguerdichian, S. Slijepcevic, V. Karayan, and M. Potkonjak, "Localized algorithms in wireless ad-hoc networks: Location discovery and sensor exposure," in *ACM MOBIHOC 2001*.

[11] X.-Y. Li, P.-J. Wan, and O. Frieder, "Coverage in wireless ad hoc sensor networks," *IEEE Transactions on Computers*, vol. 52, pp. 753–763, Jun. 2003.

[12] M. Skolnik, *Introduction to radar systems*. McGraw-Hill, 2002.

[13] A. A. Dutta, P. and S. Bibyk, "Towards radar-enabled sensor networks," in *ACM/IEEE IPSN 2006*.

[14] Q. Liang, "Waveform design and diversity in radar sensor networks: Theoretical analysis and application to automatic target recognition," in *IEEE SECON 2006*.

[15] T. Hanselmann, M. Morelande, B. Moran, and P. Sarunic, "Constrained multi-object Markov decision scheduling with application to radar resource management," in *Infomation Fusion 2010*.

[16] M. Li, T. Yan, D. Ganesan, E. Lyons, P. Shenoy, A. Venkataramani, and M. Zink, "Multi-user data sharing in radar sensor networks," in *ACM SenSys 2007*.

[17] S. Bartoletti, S. Conti, and A. Giorgetti, "Analysis of UWB radar sensor networks," in *IEEE ICC 2010*.

[18] E. Paolini, A. Giorgetti, M. Chiani, R. Minutolo, and M. Montanari, "Localization capability of cooperative anti-intruder radar systems," *EURASIP Journal on Advances in Signal Processing*, 2008.

[19] X. Gong, J. Zhang, and D. Cochran, "When target motion matters: Doppler coverage in radar sensor networks," in *IEEE INFOCOM 2013*.

[20] S. Meguerdichian, F. Koushanfar, G. Qu, and M. Potkonjak, "Exposure in wireless ad-hoc sensor network," in *ACM MOBICOM 2001*.

[21] R.-H. Gau and Y.-Y. Peng, "A dual approach for the worst-case-coverage deployment problem in ad-hoc wireless sensor networks," in *IEEE MASS 2006*.

[22] C. Lee, D. Shin, S.-W. Bae, and S. Choi, "Best and worst-case coverage problems for arbitrary paths in wireless sensor networks," in *IEEE MASS 2010*.

[23] G.-Y. Keung, B. Li, and Q. Zhang, "The intrusion detection in mobile sensor network," in *ACM MOBIHOC 2010*.

[24] Y. Wang and G. Cao, "Barrier coverage in camera sensor networks," in *ACM MOBIHOC 2011*.

[25] H. Ma, Y. Meng, D. Li, Y. Hong, and W. Chen, "Minimum camera barrier coverage in wireless camera sensor networks," in *IEEE INFOCOM 2012*.

[26] X. Gong, J. Zhang, and D. Cochran, "Optimal placement for barrier coverage in bistatic radar sensor networks," Tech. Rep., http://informationnet.asu.edu/bistatic-radar.pdf.

APPENDIX

In this section, due to space limitation, we only provide the main ideas of the proofs of the results presented in this paper. The detailed proofs can be found in our online technical report [26].

PROOF SKETCH OF THEOREM 1

For any barrier $\widetilde{AB}$ and the optimal placement $\{T_i, R_j\}$ for minimizing $V(\widetilde{AB})$, we can construct a placement $\{T_i', R_j'\}$ for $\overline{AB}$ by moving each T_i (or R_j) to its projection T_i' (or R_j', respectively) on the line passing through A and B (not necessarily falls on $\overline{AB}$), as illustrated in Figure 12(a). Then for any point $X' \in \overline{AB}$, there exists a point $X \in \widetilde{AB}$ whose projection on $\overline{AB}$ is X', and we can show that $I'(X')^8 \leq I(X)$. This implies that $V(\overline{AB})$ under $\{T_i', R_j'\}$ is no greater than $V^*(\widetilde{AB})$, and hence $V^*(\overline{AB}) \leq V^*(\widetilde{AB})$. Since $V^*(U)$ for a

---

[8] We will use $I'(X)$ and $I^-(X)$ to denote the detectability of a point X under a placement of nodes with superscripts $'$ and $^-$, respectively.

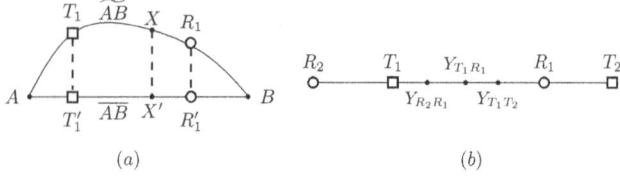

Figure 12: Examples of proofs of (a) Theorem 1 and (b) Lemma 1.

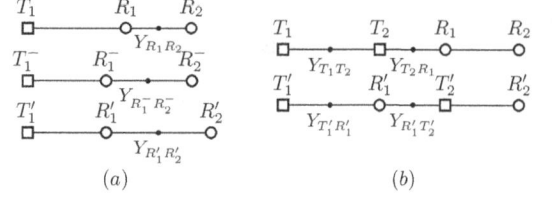

Figure 13: Examples of proofs of (a) Lemma 3 and (b) Lemma 4.

line barrier U is increasing with U's length and the short-cut barrier H is no longer than $\overline{AB}$, we have $V^*(H) \leq V^*(\overline{AB}) \leq V^*(\widetilde{AB})$.

PROOF SKETCH OF LEMMA 1

The main idea of this proof is to divide the line segment between each pair of neighbor nodes into intervals such that all points on an interval have the *same* closest BR, and then we examine the detectability structure on each interval. For example, suppose T_1 and R_1 are neighbor nodes as illustrated in Figure 12(b). Then suppose there exist $R_2 \in \overline{H_l T_1}$ and $T_2 \in \overline{R_1 H_r}$ such that the closest BRs for a point on $\overline{T_1 Y_{R_2 R_1}}$, $\overline{Y_{R_2 R_1} Y_{T_1 T_2}}$, $\overline{Y_{T_1 T_2} R_1}$ are $T_1 - R_2$, $T_1 - R_1$, $T_2 - R_1$, respectively. We can show that $I(X)$ increases as $X \in \overline{T_1 Y_{R_2 R_1}}$ is closer to $Y_{R_2 R_1}$; $I(X)$ increases as $X \in \overline{Y_{R_2 R_1} Y_{T_1 T_2}}$ is closer to $Y_{T_1 R_1}$; $I(X)$ decreases as $X \in \overline{Y_{T_1 T_2} R_1}$ is closer to R_1. Therefore, $I(X)$ attains maximum on $\overline{T_1 R_1}$ when $X = Y_{T_1 R_1}$.

PROOF SKETCH OF LEMMA 2

The main idea of this proof is to successively determine the distances between neighbor nodes. For example, suppose $\mathbf{S}_i = (T_1, R_1, \cdots, R_k, H_r)$ and $\mathbf{D}_{\mathbf{S}_i}$ is balanced such that $V(\overline{T_1 H_r}) = c$. Since $I(Y_{T_1 R_1}) = (\|T_1 R_1\|/2)^2 = c$, we obtain $\|T_1 R_1\| = 2\sqrt{c} = e_c^0$. Then since $I(Y_{R_1 R_2}) = \|T_1 Y_{R_1 R_2}\| \|Y_{R_1 R_2} R_2\| = (\|T_1 R_1\| + \|R_1 R_2\|/2)(\|R_1 R_2\|/2) = c$, using $\|T_1 R_1\| = 2\sqrt{c}$, we obtain a unique value e_c^1 for $\|R_1 R_2\|$. Following the same argument recursively, using the values of $\|T_1 R_1\|$, $\cdots$, $\|R_{i-1} R_i\|$, we obtain a unique value e_c^i for $\|R_i R_{i+1}\|$ such that $I(Y_{R_i R_{i+1}}) = c$ ... until we obtain a unique value $e_c^k/2$ for $\|R_k H_r\|$ such that $I(H_r) = c$.

PROOF SKETCH OF LEMMA 3

The proof is based on contradiction. For example, suppose $\mathbf{S}_i = (T_1, R_1, \cdots, R_k, H_r)$ with balanced spacing $\mathbf{D}_{\mathbf{S}_i}$ such that $V(Z_{\mathbf{S}_i}) = c$. Suppose there exists another placement $\mathbf{S}_i' = (T_1', R_1', \cdots, R_k', H_r')$ with spacing $\mathbf{D}_{\mathbf{S}_i'}$ such that $V(Z_{\mathbf{S}_i'}) \leq c$ and $\|T_1' H_r'\| > \|T_1 H_r\|$. Since $I(Y_{T_1 R_1}) = (\|T_1 R_1\|/2)^2 \geq I'(Y_{T_1' R_1'}) = (\|T_1' R_1'\|/2)^2$, we have $\|T_1' R_1'\| \leq \|T_1 R_1\|$. Then, using this and $I(Y_{R_1 R_2}) \geq I'(Y_{R_1' R_2'})$, we can show that $\|T_1' R_2'\| \leq \|T_1 R_2\|$: As illustrated in Figure 13(a), if $\|T_1' R_2'\| > \|T_1 R_2\|$, we can find $\{T_1^-, R_1^-, R_2^-\}$ with $\|T_1^- R_1^-\| = \|T_1' R_1'\|$ and $\|T_1^- R_2^-\| = \|T_1 R_2\|$ such that $I'(Y_{R_1' R_2'}) > I^-(Y_{R_1^- R_2^-}) > I(Y_{R_1 R_2})$, which is a contradiction. Following the same argument recursively, using $\|T_1' R_{i-1}'\| \leq \|T_1 R_{i-1}\|$ and $I(Y_{R_{i-1} R_i}) \geq I'(Y_{R_{i-1}' R_i'})$, we can show that $\|T_1' R_i'\| \leq \|T_1 R_i\|$ ... until we can show that $\|T_1' H_r'\| \leq \|T_1 H_r\|$, which is a contradiction.

PROOF SKETCH OF LEMMA 4

The proof is based on the following result: The optimal order does not have a local order with one of these node types: (T, T, R, R), (R, R, T, T), (H_l, R, T, T), (R, R, T, H_r), (T, T, R, H_r). We can use a *swapping* argument to show this result. For example, suppose $\mathbf{S} = (\cdots, T_1, T_2, R_1, R_2, \cdots)$ with any spacing $\mathbf{D}_{\mathbf{S}}$. We can construct a new order $\mathbf{S}' = (\cdots, T_1', R_1', T_2', R_2', \cdots)$ from $\mathbf{S}$ with the same *values* of spacing $\mathbf{D}_{\mathbf{S}'}$ by swapping the *locations* of nodes T_2 and R_1. Then we can show that any local vulnerable value (and hence the vulnerability) under $\mathbf{S}'$ and $\mathbf{D}_{\mathbf{S}'}$ must be *no greater than* that under $\mathbf{S}$ and $\mathbf{D}_{\mathbf{S}}$. For example, as illustrated in Figure 13(b), since the closest BRs to $Y_{T_2 R_1}$, $Y_{R_1' T_2'}$ are $T_2 - R_1$, $T_2' - R_1'$, respectively, we have $I(Y_{T_2 R_1}) = I'(Y_{R_1' T_2'})$; since the closest transmitters to $Y_{T_1 T_2}$, $Y_{T_1' R_1'}$ are T_1, T_1', respectively, and the distance from $Y_{T_1 T_2}$ to $Y_{T_1 T_2}$'s closest receiver is no less than that from $Y_{T_1' R_1'}$ to $Y_{T_1' R_1'}$'s closest receiver, we have $I(Y_{T_1 T_2}) \geq I'(Y_{T_1' R_1'})$. This implies that $\mathbf{S}$ is not the optimal order.

Based on the above result, we next show that the optimal order $\mathbf{S}^*$ must be dividable. We construct a *super order* $\mathbf{S}^+$ from an order $\mathbf{S}$ by combining neighbor nodes of the same type in $\mathbf{S}$ into a *super node*. For the example in (14), the super order is given by $\mathbf{S}^+ = (H_l, R_{1,2}^+, T_1, R_{3,4,5}^+, T_2, R_6, T_{3,4}^+, H_r)$. By construction, two neighbor nodes in $\mathbf{S}^+$ (excluding H_l and H_r) are of different types (transmitter or receiver type). Using the previous result, for $\mathbf{S}^+$ constructed from the optimal order $\mathbf{S}^*$, two neighbor nodes in $\mathbf{S}^+$ cannot be both super nodes, and a super node cannot be the third or third-to-last node in $\mathbf{S}^+$. Then we can see that $\mathbf{S}^*$ must have the dividable structure.

PROOF SKETCH OF THEOREM 3

We first can show that the optimal order $\mathbf{S}^*$ must have $n_i \geq 1$ for all $i \in \{1, \cdots, M+1\}$. Based on this, we can see that $\mathbf{S}^*$ consists of independent local orders $(H_l, \cdots, T_1)$, $\cdots$, $(T_{i-1}, \cdots, T_i)$, $\cdots$, $(T_M, \cdots, H_r)$. Let $g_c(n)$ denote the optimal value of problem (13) for such an independent local order $(T_{i-1}, \cdots, T_i)$ containing n receivers under the constraint $V(Z_{\mathbf{S}_i}) \leq c$. Then we have $f_c^{\mathbf{S}^*} = \sum_{i=2}^{M} g_c(n_i) + g_c(2n_1)/2 + g_c(2n_{M+1})/2$. By Lemma 2 and Lemma 3, we have $g_c(n) = 2\sum_{i=0}^{\frac{n}{2}-1} e_c^i + e_c^{\frac{n}{2}}$ if n is even or $2\sum_{i=0}^{\frac{n+1}{2}-1} e_c^i$ if n is odd. Then it follows that $g_c(n+1) - g_c(n) = e_c^{\frac{n}{2}}$ if n is even or $e_c^{\frac{n+1}{2}}$ if n is odd, and $g_c(2n+2) - g_c(2n) = e_c^{n+1} + e_c^n$. Using these, we can show that if $\mathbf{S}^*$ does not satisfy the desired conditions, there exists some $\mathbf{S}'$ constructed from $\mathbf{S}^*$ by adding n_i by 1 while subtracting n_j by 1 for some $i \neq j$ such that $f_c^{\mathbf{S}^*} < f_c^{\mathbf{S}'}$, which contradicts that $\mathbf{S}^*$ is optimal.

A Distributed Delaunay Triangulation Algorithm Based on Centroidal Voronoi Tessellation for Wireless Sensor Networks

Hongyu Zhou
zhou.hongyu@me.com

Miao Jin
mjin@cacs.louisiana.edu

Hongyi Wu
wu@cacs.louisiana.edu

The Center for Advanced Computer Studies (CACS)
University of Louisiana at Lafayette
Lafayette, LA 70503

ABSTRACT

A wireless sensor network can be represented by a graph. While the network graph is extremely useful, it often exhibits undesired irregularity. Therefore, special treatment of the graph is required by a variety of network algorithms and protocols. In particular, many geometry-oriented algorithms depend on a type of subgraph called *Delaunay triangulation*. However, when location information is unavailable, it is nontrivial to achieve Delaunay triangulation by using connectivity information only. The only connectivity-based algorithm available for Delaunay triangulation is built upon the property that the dual graph for a Voronoi diagram is a Delaunay triangulation. This approach, however, often fails in practical wireless sensor networks because the boundaries of Voronoi cells can be arbitrarily short in discrete sensor network settings. In a sensor network with connectivity information only, it is fundamentally unattainable to correctly judge neighboring cells when a Voronoi cell boundary is less than one hop. Consequently, the Voronoi diagram-based Delaunay triangulation fails. The proposed algorithm employs a distributed approach to perform centroidal Voronoi tessellation, and constructs its dual graph to yield Delaunay triangulation. It exhibits several distinctive properties. First, it eliminates the problem due to short cell boundaries and thus effectively avoids crossing edges. Second, the proposed algorithm is proven to converge and succeed in constructing a Delaunay triangulation, if the CVT cell size is greater than a constant threshold. Third, the established Delaunay triangulation consists of close-to-equilateral triangles, benefiting a range of applications such as geometric routing, localization, coverage, segmentation, and data storage and processing. Extensive simulations are carried out under various 2D network models to evaluate the effectiveness and efficiency of the proposed CVT-based triangulation algorithm.

Categories and Subject Descriptors

C.2.1 [**Computer Systems Organization**]: Computer-Communication Networks—*Network Architecture and Design*

Keywords

Delaunay, Triangulation, Centroidal Voronoi Tessellation, Wireless Sensor Networks

1. INTRODUCTION

This work aims to develop a distributed and efficient algorithm to construct Delaunay triangulation for wireless sensor networks. This section first introduces the motivations, then discusses the state-of-the-art Delaunay triangulation algorithms, followed by a summary of the main contributions of this paper.

1.1 Motivation

A wireless sensor network can be represented by a graph, where a node corresponds to a sensor and an edge indicates the communication link between two adjacent sensors with radio transmission range. While the network graph itself is extremely useful, it often exhibits undesired randomness and irregularity. Therefore, special treatment of the graph is required by a variety of network algorithms and protocols. In particular, many geometry-oriented algorithms, such as geometric routing [1–4], autonomous localization [5,6], sensor coverage [7], network segmentation [8], and distributed data storage and processing [9], all depend on a type of subgraph called *triangulation*.

In general, triangulation is a subdivision of a geometric object into simplices. The triangulation of a discrete set of points is a subdivision of the convex hull of the points into simplices such that any two simplices intersect in no more than one common face and the vertices of the subdividing simplices coincide with the points [10]. Several algorithms have been proposed for triangulation in wireless sensor networks [1,8,11]. Obviously, while a network graph is usually nonplanar (see the crossing edges in Fig. 1(a)), a triangulation is a planar graph, where no edges cross each other (as illustrated in Fig. 1(b)).

Delaunay triangulation is a special triangulation defined as follows.

Definition 1. *A Delaunay triangulation for a set of points on a plane is a triangulation such that no point is inside the circumcircle of any triangle [12].*

An example of Delaunay triangulation is shown in Fig. 1(c), where none of the circumcircles of the triangles contain a node. On the other hand, the triangulation in Fig. 1(b) does not meet the above definition. For instance, the circumcircle of $\triangle ABD$ contains Node C. Thus it is not a Delaunay triangulation. The Delaunay triangulation is preferred in many network algorithms, because it

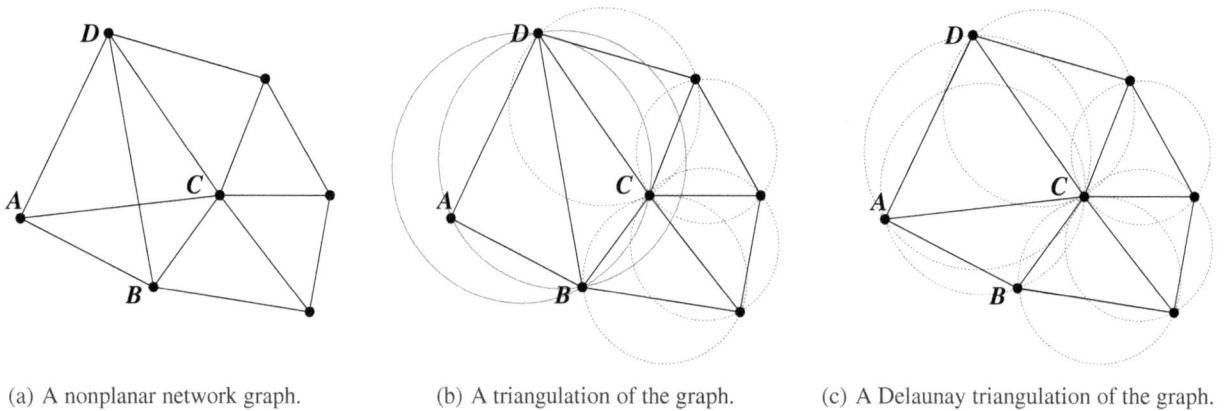

| (a) A nonplanar network graph. | (b) A triangulation of the graph. | (c) A Delaunay triangulation of the graph. |

Figure 1: Triangulation and Delaunay triangulation of a nonplanar network graph.

tends to avoid skinny triangles by maximizing the minimum angle in the triangulation. For instance, greedy forwarding is guaranteed to succeed in a Delaunay triangulation except at network boundaries [1–4]. And better localization can be achieved based on a Delaunay triangulation than that based on non-Delaunay triangulation [5,6]. The Delaunay triangulation with all edges equal is call an equilateral Delaunay triangulation.

1.2 State-of-the-Art for Connectivity-Based Delaunay Triangulation

If the location information or distance measurement is available, a Delaunay triangulation can be straightforwardly constructed [4]. However, it is nontrivial to achieve Delaunay triangulation by using connectivity information only. Note that the triangulation method proposed in [1,8] is not Delaunay-guaranteed.

The only connectivity-based algorithm for Delaunay triangulation in practical wireless sensor networks is built upon the property that the dual graph for a Voronoi diagram[1] is a Delaunay triangulation. To this end the planarization algorithm proposed in [13] is employed to establish an approximate Voronoi diagram. More specifically, a node is randomly chosen as a generating point, which claims its K-hop neighbors to form a cell. Then a node is randomly chosen among the rest nodes as the next generating point. The process repeats until every node in the network is either selected as a generating point or associated with a generating point. If a node is claimed by multiple generating points, it chooses the closest one (in term of hop count) and joins its cell. The distance between any two adjacent generating points is greater than K hops but no more than $2K+1$ hops.

Then, the dual graph of the approximate Voronoi diagram, called combinatorial Delaunay graph (CDG), is constructed as follows. If two cells are adjacent to each other, i.e., have at least one pair of neighboring nodes (one in each cell), a virtual edge is established to connect the corresponding generating points. Note that, while the precisely computed dual graph of a Voronoi diagram is a Delaunay triangulation as discussed earlier (e.g., as shown in Fig. 2(a)), the same result no longer holds for CDG. As a matter of fact, CDG is not even necessarily planar, as illustrated in Fig. 2(b) that shows crossing edges. This is because the boundary of two adjacent ap-

proximate Voronoi cells can be shorter than one hop, and thus two non-neighboring cells may be mistakenly considered as neighbors. For example, as illustrated in Fig. 2(b), the shared boundary of Cells A and B is virtually zero (with only one pair of nodes connected). At the same time, a node in Cell C is directly connected to a node in Cell D. Hence Cells C and D are deemed adjacent, resulting in crossing edges. Practically, there are less crossing edges under a larger K. However no matter how large the K (i.e., the cell) is, the approximate Voronoi diagram cannot guarantee every cell boundary to be greater than one hop. Therefore this scheme does not ensure the success of Delaunay triangulation. Moreover, even a Delaunay triangulation is successfully constructed, the formed triangles are often nonuniform in terms of edge length (see Fig. 2(a)).

The crossing edges in CDG can be removed, yielding a Combinatorial Delaunay Map (CDM), which is planar but has polygon holes [13]. Therefore it is not a triangulation. The algorithms proposed in [1, 11] try to fill the holes to form triangles. However, the resulting triangulation is not guaranteed to hold the Delaunay property.

1.3 Contribution of This Work

The problem of the Voronoi diagram-based (or VD-based) algorithm stems from the fact that the boundaries of Voronoi cells are nonuniform. Particularly, some cell boundaries can be arbitrarily short. In a sensor network with connectivity information only, it is fundamentally unattainable to correctly judge neighboring cells when a Voronoi cell boundary is less than one hop. Consequently, some generating points are mistakenly connected, forming crossing edges. As a result, the VD-based Delaunay triangulation fails. It is worth clarifying that Delaunay triangulation can aways be achieved if accurate location information is available. With the location information, a Voronoi diagram can be precisely computed, and accordingly its dual graph can be readily obtained by connecting the generating points whose cells share a boundary (no matter how short the boundary is). Such a dual graph is proven to be a Delaunay triangulation.

It is obviously desired to build a Voronoi diagram with uniform cell boundaries, so as to maximize the minimum boundary length. This observation motivates a new triangulation algorithm based on centroidal Voronoi tessellation (CVT).

Definition 2. *A centroidal Voronoi tessellation (CVT) is a special Voronoi diagram, where the generating point of each Voronoi cell is also its mean (i.e., center of mass) [14].*

---

[1]Given a finite set of generating points on a plane, the Voronoi diagram is a partitioning of the plane with points into convex polygons (or cells) such that each polygon contains exactly one generating point and every point in a given polygon is closer to its generating point than to any other [12].

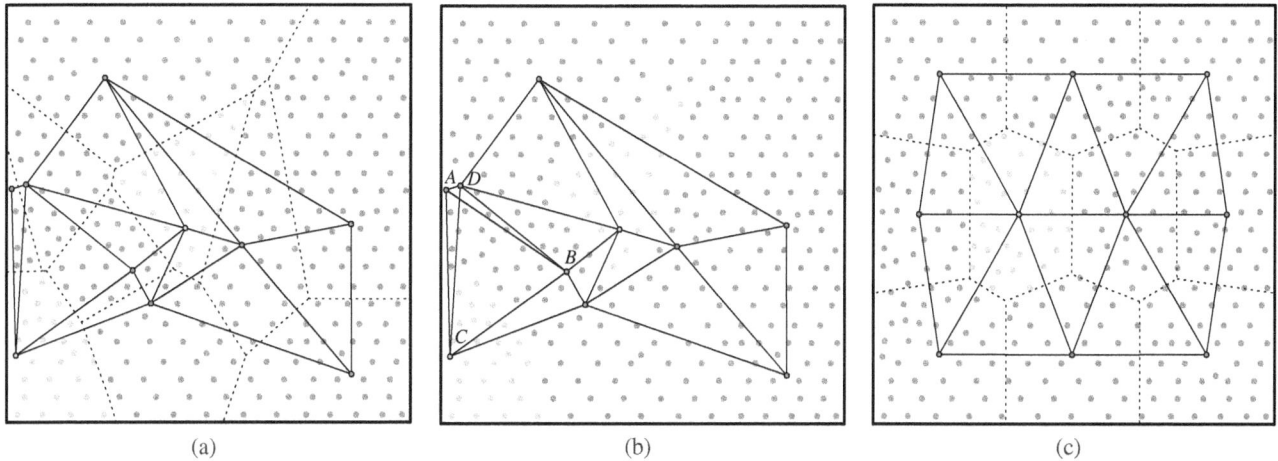

Figure 2: Comparison between the dual graphs of Voronoi diagram and CVT. (a) The dual graph of a precisely computed Voronoi diagram. (b) The dual graph with crossing edges of an approximate Voronoi diagram under discrete sensor network settings. (c) CVT has uniform cell boundaries and thus its dual graph forms a Delaunay triangulation with close-to-equilateral triangles.

The proposed algorithm employs a distributed approach to perform centroidal Voronoi tessellation, and constructs its dual graph to yield Delaunay triangulation. It exhibits several distinctive properties. First, it eliminates the problem due to short cell boundaries and thus effectively avoids crossing edges, significantly improving the probability to successfully establish a Delaunay triangulation in comparison with its VD-based counterpart. Second, as shown by Theorems 1 and 2, the proposed algorithm can always converge and succeed in constructing a Delaunay triangulation, if the CVT cell size is greater than a constant threshold. Third, the established Delaunay triangulation consists of close-to-equilateral triangles, benefiting a range of applications such as geometric routing [1–4], autonomous localization [5,6], sensor coverage [7], network segmentation [8], and distributed data storage and processing [9].

The rest of this paper is organized as follows: Sec. 2 introduces the proposed CVT-based Delaunay triangulation algorithm and proves its convergence and correctness when the cell size is greater than a constant threshold. Sec. 3 presents simulation results. Finally, Sec. 4 concludes the paper.

2. PROPOSED CVT-BASED TRIANGULATION ALGORITHM

This section presents the proposed CVT-based triangulation algorithm and proves its convergence and correctness when the cell size is greater than a constant threshold.

The proposed algorithm consists of three phases. First, it samples the generally distributed network to yield a uniform nodal density. Second, it iteratively builds an approximate CVT. Finally it obtains the dual graph of the CVT to construct the Delaunay triangulation.

2.1 Single-Hop Voronoi Diagram Sampling

As shown in Definition 2, the generating point of a CVT cell is the centroid of the cell. In order to build uniform cells, which induce a Delaunay triangulation with equilateral triangles, the network must have a uniform density function [14]. However, a general sensor network can be non-uniformly distributed, due to the lack of precise nodal deployment and the nondeterministic sensor failures and link dynamics. To this end, a sampling process is employed. It essentially builds a Voronoi diagram with small, constant

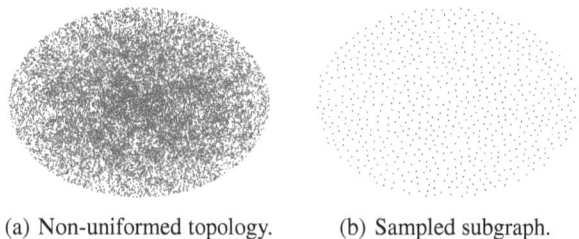

(a) Non-uniformed topology.   (b) Sampled subgraph.

Figure 3: Single-hop Voronoi diagram sampling that yields uniform nodal density.

cell size (e.g., one hop). The generating points of the Voronoi diagram have a uniform density, and thus serving as the ideal input for CVT construction.

More specifically, let an undirected graph $G = \{V, E\}$ represent a wireless sensor network, where V is the set of sensor nodes and E is the set of communication links. A node can be in one of the three possible states, i.e., a cell generating point, a cell member, or undetermined. Each node is initialized as the undetermined state. It starts a random back-off timer. If it is in the undetermined state when the timer expires, it changes its state to a generating point, and informs its undetermined one-hop neighbors to change their states to cell member. The process terminates when every node in the network becomes either a generating point or a cell member. Two generating points are adjacent if a pair of their cell members (one in each cell) are one-hop neighbors. The generating points are largely uniformly distributed, where any two adjacent generating points are at least one hop and at most three hops away from each other. Let $G' = \{V', E'\}$ be the sampled subgraph, where V' consists of the generating points and E' are the virtual links that connect adjacent generating points.

An example of sampling is shown in Fig. 3(b), where the nodes are uniform, in comparison with the nonuniform original graph Fig. 3(a).

2.2 CVT Construction

Based on the sampled subgraph G', the CVT is constructed via a distributed, iterative process. It starts with an arbitrary Voronoi diagram with a cell size of K-hops, which can be established according to the method introduced in Sec. 1.2. Let $T = \{T_1, T_2, ..., T_m\}$ denote the set of tessellations (or cells) and $L = \{L_1, L_2, ..., L_m\}$ the corresponding generating points of the initial Voronoi diagram. They will be updated by repeating the following two steps.

(1) Cells construction: The nodes in G' are associated with their closest generating points to form tessellations. Given Node n_i, its closest generating point is

$$L(n_i) = \arg \min_{L_j \in L} D(n_i, L_j), \quad (1)$$

where $D(n_i, L_j)$ denotes the hop distance between n_i to L_j. The nodes associated with the same generating point form a tessellation $T_j = \{n_i \in G' | L(n_i) = L_j\}$. Let $|T_j|$ denote the number of nodes in T_j. The colored cells in Fig. 4(a) show the initial tessellation.

(2) Centroid Calculation: The generating point of each cell is updated to the current centroid of the cell. More specifically, every node learns its hop distance to every other node in the same cell via localized flooding (within its cell). Then each node calculates the standard deviation of such distances. For example, given Node n_i in Cell T_j, its standard deviation σ_i is determined by the following equations:

$$\sigma_i = \sqrt{\frac{1}{|T_j| - 1} \sum_{n_k \in T_j} (D(n_i, n_k) - \overline{D_j})^2}, \quad (2)$$

where $\overline{D_j}$ is the mean of distances between any two nodes in Cell T_j, i.e.,

$$\overline{D_j} = \frac{2}{|T_j|(|T_j| - 1)} \sum_{n_p, n_q \in T_j} D(n_p, n_q). \quad (3)$$

The node with the minimal standard deviation is the approximate centroid of the tessellation and selected as the new generating point, i.e,

$$L_j = \arg \min_{n_k \in T_j} \sigma_k. \quad (4)$$

The above calculation involves all nodes in the cell (i.e., T_j). In fact, given the uniformly distributed nodes (after sampling), the centroid can be determined according to the cell boundary that depicts the shape of the cell. The boundary nodes of the cell can be easily identified. If a node is on the network boundary or has a neighbor node that belongs to a different cell, it is a cell boundary node. The centroid calculation can be carried out with reduced computing time, communication overhead and energy consumption by simply replacing T_j in Eqs. (2)-(4) with the set of cell boundary nodes. The boundary-based approximation yields similar CVT construction as to be demonstrated in Sec. 3.

Once the new generating points are determined, the cells are updated accordingly as discussed in the previous step.

Fig. 4 illustrates the evolution of the cells while the above steps repeat. As can be seen, the cells become more uniform after more iterations. The process terminates when the generating points remain unchanged. Fig. 4(d) shows the final CVT.

The iterative process always converges as shown by the following theorem.

Theorem 1. *The proposed CVT construction algorithm converges.*

PROOF. For a given cell j, both $|T_j|$ and $\overline{D_j}$ are constant. Thus, CVT construction intrinsically aims to minimize the following optimal function $J(T, L)$:

$$J(T, L) = \sum_{i \in G'} (D(n_i, L(n_i)))^2. \quad (5)$$

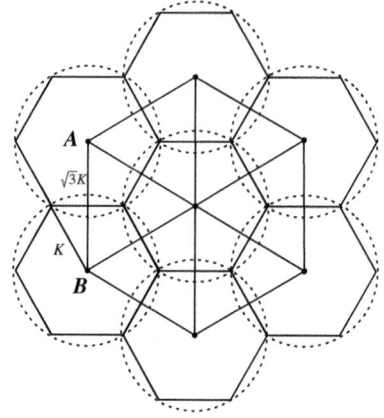

Figure 5: An ideal CVT consists of regular hexagons.

Assume J is not minimal, one can either fix generating points L to adjust the nodes' association to reduce $J(T, L)$ just like cells construction, or fix cell nodes C to select new generating points as the centroids. Both processes are monotonically decreasing $J(T, L)$. When $J(T, L)$ reaches the minimal value, the cells T and generating points L converge at the same time. As a matter of fact, the CVT construction algorithm is a coordinate descent algorithm with guaranteed convergence. □

While it is difficult to derive a theoretic bound for the number iterations in order to reach the convergence, the algorithm converges very fast in practice. For example, merely four to five iterations are needed in most cases according to the simulation results to be presented in Sec. 3.

2.3 Delaunay Triangulation

After CVT is constructed, its dual graph is established by connecting every two adjacent generating points with a virtual edge. Two generating points are adjacent if at least one pair of their cell members (one in each cell) are neighbors in G'. An example of the dual graph is illustrated Fig. 4(d).

The following theorem formally shows that the proposed algorithm can always construct a Delaunay triangle mesh, if the cell size is greater than a constant threshold. Without loss of generality, the maximum radio transmission range is normalized to one in the following discussions.

Theorem 2. *Given an asymptotically deployed wireless sensor network, the centroidal Voronoi tessellation (CVT) with a cell size greater than 15 always yields a Delaunay triangulation.*

PROOF. As introduced in Sec. 2.1, a sampling process (that results in a constant density function) is employed before the centroidal Voronoi tessellation. Accordingly, every cell of CVT is a regular hexagon (see Fig. 5) if the nodal density approaches infinity (i.e., the continuous case) and the network is asymptotically deployed with no boundaries [15]. The distance between any two neighboring centroids is $\sqrt{3}K$, where K is the hexagon edge length or the cell radius. The dual of the CVT is a Delaunay triangle mesh that consists of equilateral triangles [14]. For convenience, such a CVT is referred as the *ideal CVT* and the corresponding Delaunay triangle mesh is referred as the *ideal triangulation* in the following discussions.

Under a discrete sensor network setting, the proposed algorithm yields an approximate CVT, where the tessellation is performed according to hop counts instead of real distance. Thus, the actual cell

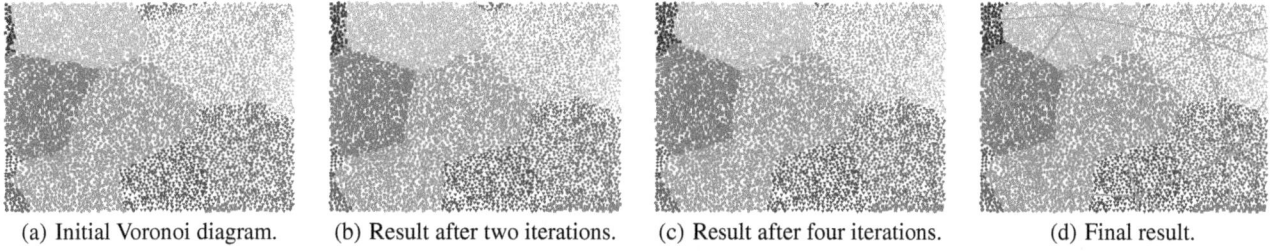

| (a) Initial Voronoi diagram. | (b) Result after two iterations. | (c) Result after four iterations. | (d) Final result. |

Figure 4: CVT-Based Delaunay Triangulation Algorithm. It starts with an arbitrary Voronoi diagram (depicted in (a)), and refines the cells by a small number of iteration (see (b) and (c)) to yield the final CVT cells (shown in (d)). The dual graph of the CVT cells is a close-to-equilateral Delaunay triangulation.

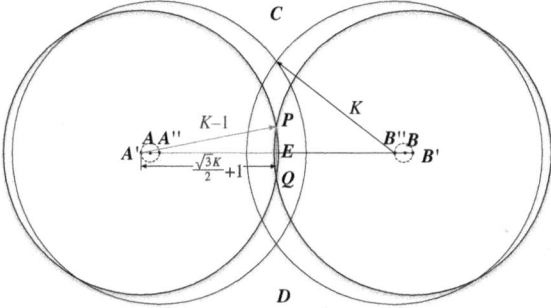

Figure 6: Illustration for the proof of Theorem 2. The length of PQ reaches minimum when the two generating points are farthest away from each other (i.e., at A' and B') and the cell size is the smallest (i.e., $K-1$).

size is in the range of $(K-1, K]$. Given such an approximate CVT cell, a node is not always available at its real centroid. For example, Fig. 6 illustrates two adjacent cells, with their real centroids at A and B, respectively. The actual generating point of a cell could be anywhere within 1-hop range of the real centroid, as delineated by the small dashed circle around A or B. Similar to the continuous setting, the proposed algorithm obtains the dual of the approximate CVT, aiming to produce a triangle mesh.

The rest of the proof shows that, when K is greater than a constant, the approximate CVT induces the same triangle mesh as the ideal triangulation, with neither extra edges nor missing edges.

First, if an extra edge is added into the ideal triangulation, it must result in a crossing edge [12]. To ensure free of crossing edges in the dual of the approximate CVT, the boundary between the two cells must be greater than the radio transmission range (i.e., 1). Otherwise, the cells on the two sides of AB (i.e., Cells C and D) might be mistakenly considered as neighboring cells since their nodes can be connected, leading to a crossing edge CD (see a similar example in Fig. 2(b)). To this end, the proof intends to examine the worst case scenario that results in the shortest boundary between Cells A and B. The worst case occurs when the two generating points are farthest away from each other (i.e., at A' and B' in Fig. 6) and the cell size reaches minimum (i.e., $K-1$). Let P and Q denote the intersections of the two cells under the worst case, and E denote the intersection of AB and PQ. Obviously, $\|B'E\| = \sqrt{3}K/2 + 1$ and $\|B'P\| = K-1$. Let the boundary greater than one, i.e.,

$$\|PQ\| = 2\sqrt{\|B'P\|^2 - \|B'E\|^2} > 1. \qquad (6)$$

The inequality holds when and only when K is greater than 15.06. Therefore, if $K > 15$, the dual graph of the approximate CVT includes no extra edges compared with the ideal triangulation.

Second, since PQ is greater than one, Cells C and D are disconnected, i.e., there does not exist a path between C and D that involves the nodes in the two cells only. If Cells A and B are not connected either, there must exist a void (i.e., a hole) in the middle of the four cells. The hole forms a network boundary [16], which contradicts to the assumption of asymptotic sensor deployment with no boundaries. Hence, Cells A and B must be neighbors, inducing Edge AB in the dual of the approximate CVT.

Therefore, if $K > 15$, the dual graph of the approximate CVT is the same as the ideal triangulation, i.e., a Delaunay triangle mesh. $\square$

Note that Theorem 2 only shows a provable sufficient condition for Delaunay triangulation. It is not always necessary to have $K > 15$ to establish a Delaunay triangle mesh. When $K \leq 15$, although without a proof, it is intuitively obvious that the proposed algorithm yields better triangulation in comparison with its VD-based counterpart. More specifically, it results in close-to-equilateral triangles. The uniform edge length reduces the probability of crossing edges as discussed in Sec. 1. This observation is verified by simulation results to be presented in the next section.

2.4   Time Complexity and Communication Cost

The proposed Delaunay triangulation algorithm has a linear time complexity and communication cost (measured by messages sent) with respect to the size of the network. A brief analysis is summarized below.

First, the single-hop Voronoi diagram sampling has a time complexity of $O(n)$, where n is the total number of nodes in the network. As each node communicates with its neighbors only, the communication cost is also $O(n)$.

In every iteration of CVT construction, each cell has a time complexity and communication cost of $O(m^2)$, where m is the number nodes in a cell. Since all cells are processed simultaneously, the network-wide time complexity remains $O(m^2)$, but the overall communication cost becomes $O(cm^2)$, where c is the number of cells. Obviously, $O(cm) = O(n)$, hence $O(cm^2) = O(nm)$. For a given cell size (i.e., K) and nodal density, m is bounded by a constant. Therefore, $O(nm) = O(n)$. Moreover, the proposed algorithm adopts a constant number of iterations (e.g., four iterations that are enough to yield satisfied results).

Finally, the time complexity and communication cost for dual graph construction are both $O(k)$.

In summary, the overall time complexity and communication cost of the Delaunay triangulation algorithm are dominated by $O(n)$.

63

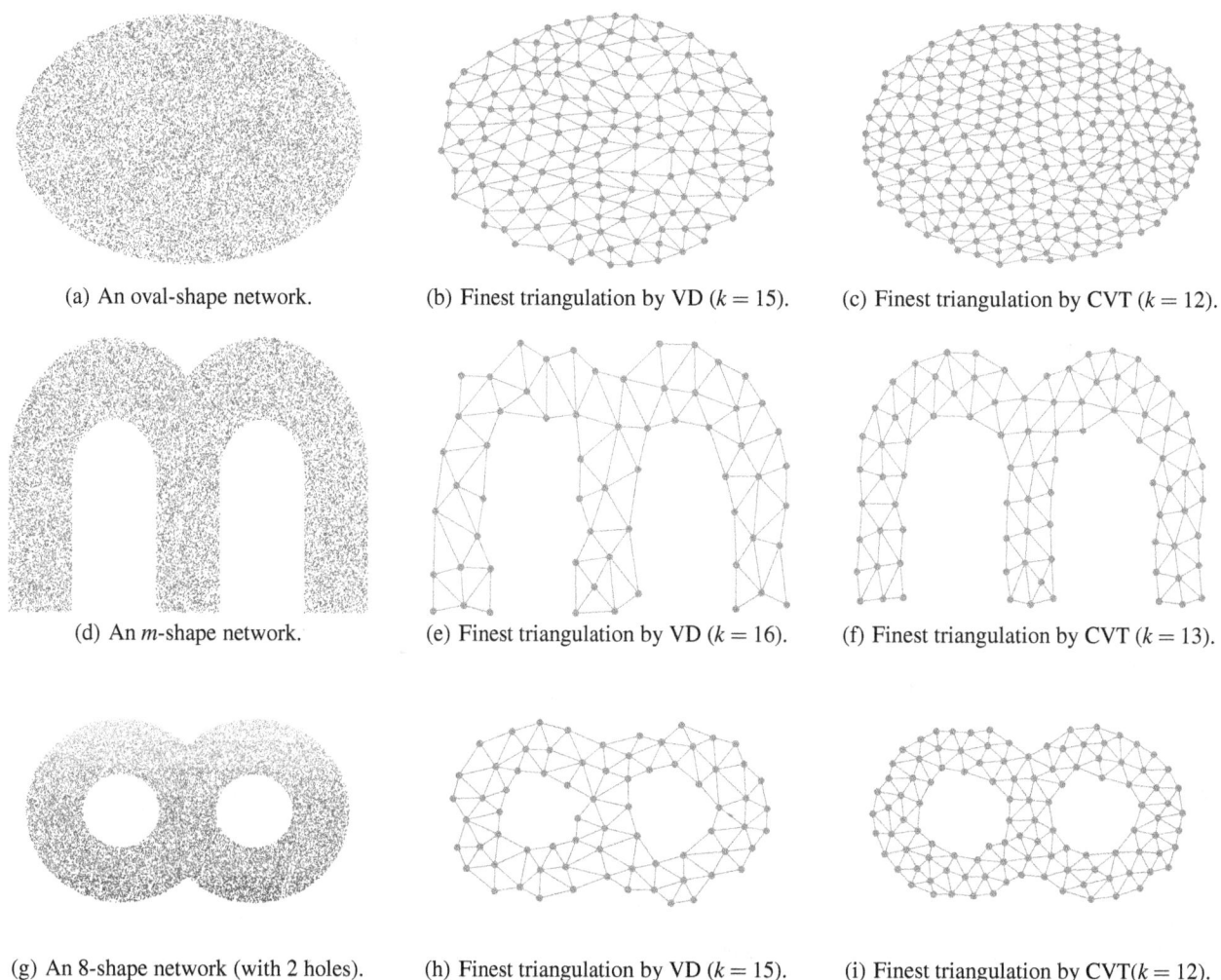

(a) An oval-shape network. (b) Finest triangulation by VD ($k = 15$). (c) Finest triangulation by CVT ($k = 12$).

(d) An *m*-shape network. (e) Finest triangulation by VD ($k = 16$). (f) Finest triangulation by CVT ($k = 13$).

(g) An 8-shape network (with 2 holes). (h) Finest triangulation by VD ($k = 15$). (i) Finest triangulation by CVT($k = 12$).

Figure 7: Example of 2D networks that demonstrate the proposed CVT-based algorithm produces finer triangulations than the VD-based algorithm does.

3. SIMULATIONS RESULTS

In order to evaluate the effectiveness and efficiency of the proposed CVT-based triangulation algorithm, extensive simulations have been carried out under various 2D network models. The sensor nodes are randomly deployed in each network. The proposed algorithm does not depend on any specific communication model. This simulation adopts a general model, with merely a constraint on the maximum radio transmission range, which is normalized to one. It is similar to the most general Quasi-UDG model. Quasi-UDG determines connectivity according to a parameter $\alpha < 1$. Two nodes are disconnected if they are separated by a distance greater than one, or connected if their distance is less than α, or connected with a probability if their distance is between α and one. In this simulation, α is set to 0. The performance of the proposed CVT-based algorithm is compared against its VD-based counterpart, in terms of crossing edge rate, the deviation of triangle edge length, and the success rate of Delaunay triangulation construction.

3.1 Triangulation Granularity

First, simulations are performed to compare the granularity of triangulations produced by the proposed CVT-based algorithm and its VD-based counterpart. Several sample networks are illustrated

in Fig. 7, in oval, *m*-shape, and 8-shape, respectively. As discussed in Sec. 2, the cell size K is crucial for both algorithms, dictating the possibility of having crossing edges. K must be large enough to ensure the free of crossing edge, and thus a valid Delaunay triangulation.

The simulation intends to find the minimum K under both algorithms, which induce a successful triangulation. A small K is highly desired because it represents the original network with finer granularity. The simulation results show that the CVT-based algorithm always yields a finer triangulation (with smaller K) than the VD-based algorithm does (see the second and the third columns of Fig. 7).

As proven in Sec. 2, CVT with a cell size greater than $K = 15$ ensures a Delaunay triangulation. Note that, $K = 15$ is the theoretical bound. A small K is often sufficient in practice. However, the VD-based algorithm does not have such nice property. As a matter of fact, it is interesting to observe that even when the VD-based algorithm successfully produces a Delaunay triangulation under certain K, it may still fail when K further increases. This is evidenced in Fig. 8(a) that shows the crossing edge rate. The crossing edges are the edges that intersect, i.e., are non-planar. The crossing edge rate is defined to be the ratio between the number of crossing edges to the total number of edges in the dual graph of VD or CVT. Obviously, the smaller the crossing edge rate, the better. A successful Delaunay triangulation is constructed if the crossing edge rate is zero. As can be seen, the crossing edge rate reaches and stays to be zero after $K = 12$ under the CVT-based algorithm. The VD-based algorithm, however, offers no guarantee for achieving free of crossing edges. The crossing edge rate exhibits significant fluctuations even when K is large.

It is also observed in simulation that the boundary will have some impact on the CVT construction. The shape of the boundary cells are more likely irregular (non-hexagon) compared with inner cells, because the shape of boundary cells are constrained by the network shape. Crossing edges occur more often around boundary cells.

3.2 Triangulation Regularity

In addition to granularity, it is obvious that the triangulation produced by the CVT-based algorithm is more uniform. This is because the CVT cells are generally more regular and uniform. Fig. 8(b) compares the deviation of cell size under the CVT and VD-based algorithms. As can be seen, the former achieves a consistently lower deviation in cell size than the latter. This is a natural result of CVT cell construction, which intends to improve the regularity of cells over the Voronoi diagram. Consequently, the edges of the dual graph under the former always have a smaller deviation than that of the latter.

The regularity of triangulation benefits a range of applications such as geometric routing, localization, coverage, segmentation, and data storage and processing. For example a connectivity-based localization method is introduced in [6]. It takes triangulations as input and applies a Ricci flow algorithm to computes the optimal flat metric of the triangulation in order to embed the network to plane (and accordingly determining the locations of sensors). It has been shown in [6] that the algorithm achieves the highest localization accuracy compared with other competing methods including multi-dimensional scaling (MDS) [17, 18] and neural network based methods [19, 20] under various representative network shapes.

This simulation intends to demonstrate the benefits of CVT-based triangulation in support of localization. To this end, the same cell size (i.e., K) is employed to build both VD and CVT cells to construct triangulations. Two network examples are depicted in Fig. 9, where nodes are randomly deployed with representative boundary shapes. The triangulations serve as inputs for the same localization algorithm. The localization result are shown in the second row of Fig. 9, where a blue line segment is drawn for each node, starting from its real coordinates marked with blue dot and ending at the computed coordinates. The gray triangulation is based on the computed coordinates. The longer the line segment, the lower accuracy of the localization results. It is obviously noticed that the CVT-based triangulation helps yield more accurate localization results. The quantitative average localization errors are summarized in Table 1. The localization error is computed as the ratio of the average node distance error (based on all edges in the network) and the averaged transmission range. As can be seen, a significant reduction of error (up to 50%) is achieved by using the CVT-based triangulation.

3.3 Comparison Cell Centroid Calculation

As discussed in Sec. 2, the calculation of cell centroid can be performed according to all nodes in the cell or only boundary nodes of the cell. Fig. 10 compares their performance. As can be seen, they achieve similar triangulation results. However, the latter involves only 24% of the nodes in comparison with the former, thus signifi-

Table 1: Localization errors of [6] with CVT-based and VD-based triangulations as inputs.

	C-shape Model (see Fig. 9(a)-9(b))	Rectangle Model (see Fig. 9(c)-9(d))
VD	0.46	0.24
CVT	0.21	0.18

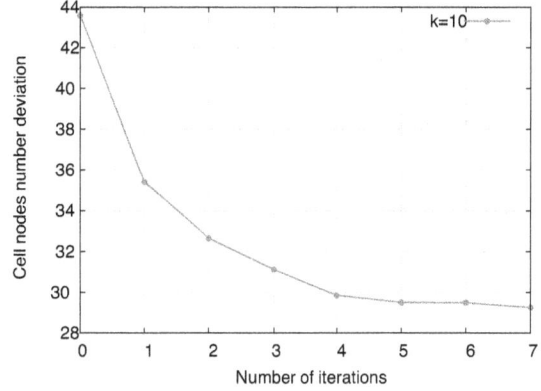

Figure 11: CVT triangulation convergence (K=10).

cantly reducing the communication overhead and energy consumption.

3.4 Convergence

The convergence of the proposed algorithm has been proven in Sec. 2. Fig. 11 shows the convergence speed under practical implementation. The standard deviation of cell size reaches minimum and keeps constant after about five iterations. The fast convergence is highly desired, because it means not only shorter algorithm running delay but also lower communication overhead. In each iteration, every nodes in the network must communicate with other nodes in the same cell to determine the new generating point and construct the CVT cell. As discussed in Sec. 2, the overall communication cost of one iteration is $O(nm)$, where m is the number nodes in a cell (which can be deem as a constant for a given K) and n is the number of nodes in the entire network. Apparently, a smaller number of iterations can significantly reduce the communication overhead and energy consumption in practical sensor network settings. The result also justifies the assumption of constant number of iterations used in complexity analysis in Sec. 2.

4. CONCLUSION

A wireless sensor network can be represented by a graph. While the network graph is extremely useful, it often exhibits undesired randomness and irregularity. Therefore, special treatment of the graph is required by a variety of network algorithms and protocols. In particular, many geometry-oriented algorithms depend on a type of subgraph called *Delaunay triangulation*. However, when location information is unavailable, it is nontrivial to achieve Delaunay triangulation by using connectivity information only. The only connectivity-based algorithm available for Delaunay triangulation is built upon the property that the dual graph for a Voronoi diagram is a Delaunay triangulation. This approach, however, often fails in practical wireless sensor networks because the boundaries of Voronoi cells can be arbitrarily short in discrete sensor network settings. In a sensor network with connectivity information only, it

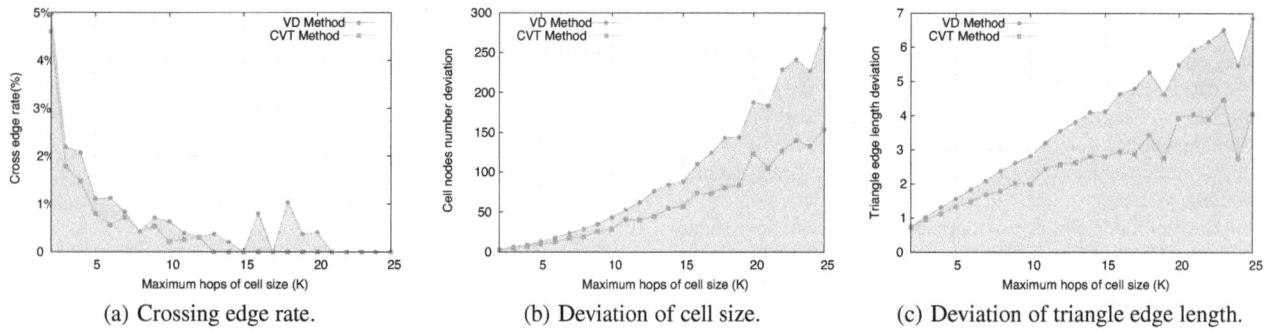

(a) Crossing edge rate.

(b) Deviation of cell size.

(c) Deviation of triangle edge length.

Figure 8: Quantitative comparison of CVT-based and VD-based triangulation algorithms. As depicted in (a), the crossing edge rate reaches and stays to be zero after $K = 12$ under the CVT-based algorithm, but exhibits significant fluctuations under the VD-based algorithm even when K is large. As shown in (b) and (c), the CVT-based triangulation achieves a consistently lower deviation in cell size and edge length than the VD-based counterpart does.

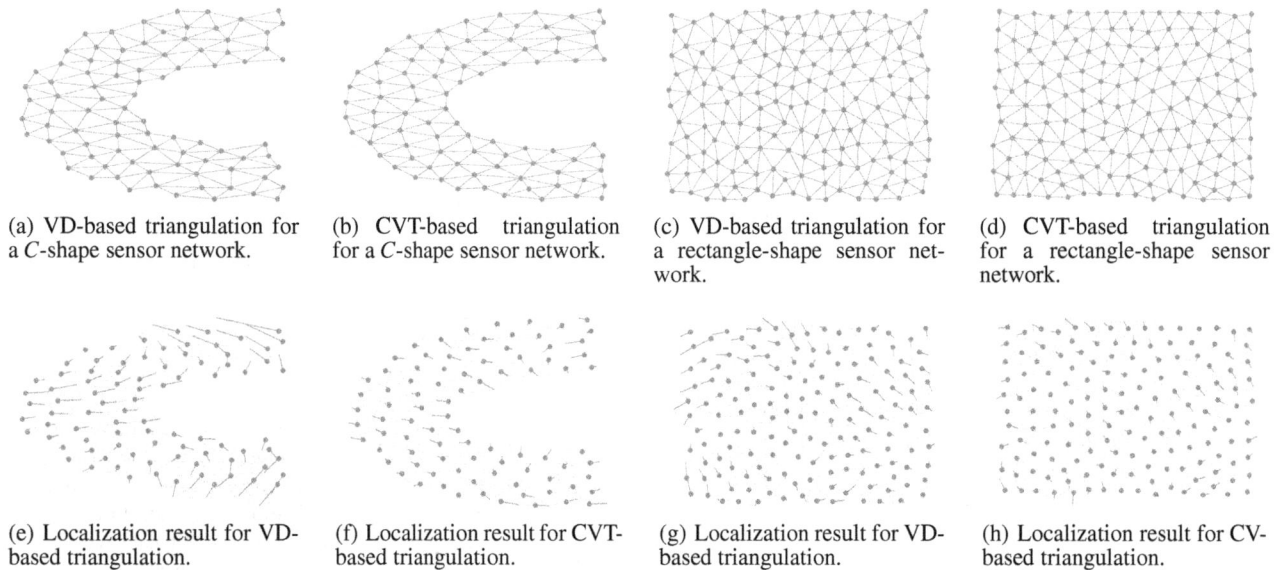

(a) VD-based triangulation for a C-shape sensor network.

(b) CVT-based triangulation for a C-shape sensor network.

(c) VD-based triangulation for a rectangle-shape sensor network.

(d) CVT-based triangulation for a rectangle-shape sensor network.

(e) Localization result for VD-based triangulation.

(f) Localization result for CVT-based triangulation.

(g) Localization result for VD-based triangulation.

(h) Localization result for CV-based triangulation.

Figure 9: Comparison of localization results with CVT-based and VD-based triangulations as inputs. A blue line segment is drawn for each node, starting from its real coordinates marked with blue dot and ending at the computed coordinates. The gray triangulation is based on the computed coordinates. The CVT-based triangulation supports more accurate localization than its VD-based counterpart does.

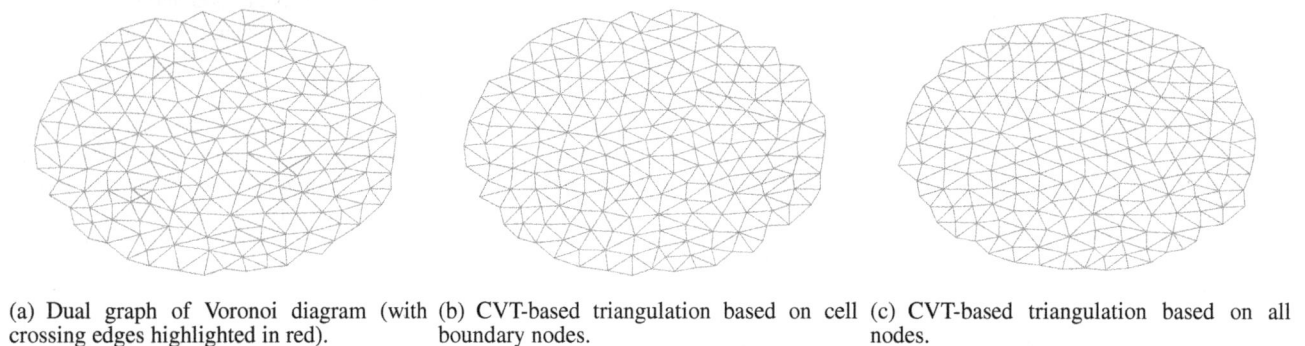

(a) Dual graph of Voronoi diagram (with crossing edges highlighted in red).

(b) CVT-based triangulation based on cell boundary nodes.

(c) CVT-based triangulation based on all nodes.

Figure 10: Similar triangulation result are achieved by the proposed CVT-based algorithm with either cell boundary nodes or all nodes (K=13).

is fundamentally unattainable to correctly judge neighboring cells when a Voronoi cell boundary is less than one hop. Consequently, the Voronoi diagram-based Delaunay triangulation fails.

This paper has proposed a distributed algorithm that performs centroidal Voronoi tessellation and constructs its dual graph to yield Delaunay triangulation. It exhibits several distinctive properties. First, it eliminates the problem due to short cell boundaries and thus effectively avoids crossing edges. Second, the proposed algorithm has been proven to converge and succeed in constructing a Delaunay triangulation, if the CVT cell size is greater than a constant threshold. Third, the established Delaunay triangulation consists of close-to-equilateral triangles, benefiting a range of applications such as geometric routing, localization, coverage, segmentation, and data storage and processing. Extensive simulations have been carried out under various 2D network models to evaluate the effectiveness and efficiency of the proposed CVT-based triangulation algorithm.

ACKNOWLEDGEMENT

M. Jin is partially supported by NSF CCF-1054996 and CNS-1018306. H. Wu is partially supported by NSF CNS-1018306, CNS-0831823, and CNS-0821702.

5. REFERENCES

[1] R. Sarkar, X. Yin, J. Gao, F. Luo, and X. D. Gu, "Greedy routing with guaranteed delivery using ricci flows," in *Proc. of the 8th ACM/IEEE International Conference on Information Processing in Sensor Networks (IPSN)*, pp. 121–132, April 2009.

[2] R. Flury, S. V. Pemmaraju, and R. Wattenhofer, "Greedy Routing with Bounded Stretch," in *Proc. of the 28th IEEE International Conference on Computer Communications (INFOCOM)*, pp. 1737–1745, 2009.

[3] W. Zeng, R. Sarkar, F. Luo, X. D. Gu, and J. Gao, "Resilient Routing for Sensor Networks using Hyperbolic Embedding of Universal Covering Space," in *Proc. of the 29th IEEE International Conference on Computer Communications (INFOCOM)*, pp. 1–9.

[4] P. Bose and P. Morin, "Competitive Online Routing in Geometric Graphs," *Theoretical Computer Science*, vol. 324, no. 2, pp. 273–288, 2004.

[5] S. Lederer, Y. Wang, and J. Gao, "Connectivity-based Localization of Large Scale Sensor Networks with Complex Shape," in *Proc. of the 27th IEEE International Conference on Computer Communications (INFOCOM)*, pp. 789–797, 2008.

[6] M. Jin, G. Rong, H. Wu, L. Shuai, and X. Guo, "Optimal Surface Deployment Problem in Wireless Sensor Networks," in *Proc. of IEEE Conference on Computer Communications (INFOCOM)*, pp. 2345–2353, 2012.

[7] M.-C. Zhao, J. Lei, M.-Y. Wu, Y. Liu, and W. Shu, "Surface Coverage in Wireless Sensor Networks," in *Proc. of the 28th IEEE International Conference on Computer Communications (INFOCOM)*, pp. 109–117, 2009.

[8] H. Zhou, H. Wu, S. Xia, M. Jin, and N. Ding, "A Distributed Triangulation Algorithm for Wireless Sensor Networks on 2D and 3D Surface," in *Proc. of the 30th IEEE International Conference on Computer Communications (INFOCOM)*, pp. 1053–1061, 2011.

[9] R. Sarkar, W. Zeng, J. Gao, and X. Gu, "Covering space for in-network sensor data storage," in *Proc. of the 9th ACM/IEEE International Conference on Information Processing in Sensor Networks (IPSN)*, pp. 232–243, 2010.

[10] E. L. Lloyd, "On Triangulations of a Set of Points in the Plane," in *Proc. of the 18th Annual IEEE Symposium on Foundations of Computer Science*, pp. 228–240, 1977.

[11] H. Zhou, S. Xia, M. Jin, and H. Wu, "Localized algorithm for precise boundary detection in 3d wireless networks," in *Proc. in the 30th IEEE International Conference on Distributed Computing Systems (ICDCS)*, pp. 744–753, 2010.

[12] M. De Berg, O. Cheong, M. Van Kreveld, and M. Overmars, *Computational geometry: algorithms and applications.* Springer, 2008.

[13] S. Funke and N. Milosavljevi, "How Much Geometry Hides in Connectivity? - Part II," in *Proc. of the 8th Annual ACM-SIAM Symposium on Discrete Algorithms (SODA)*, pp. 958–967, 2007.

[14] Q. Du, V. Faber, and M. Gunzburger, "Centroidal voronoi tessellations: Applications and algorithms," *Society for Industrial and Applied Mathematics (SIAM) review*, vol. 41, no. 4, pp. 637–676, 1999.

[15] D. Newman, "The hexagon theorem," *IEEE Transactions on Information Theory*, vol. 28, no. 2, pp. 137–139, 1982.

[16] D. Dong, Y. Liu, and X. Liao, "Fine-grained boundary recognition in wireless ad hoc and sensor networks by topological methods," in *Proc. of the 10th ACM International Symposium on Mobile Ad Hoc Networking and Computing (MOBIHOC)*, pp. 135–144, 2009.

[17] Y. Shang, W. Ruml, Y. Zhang, and M. P. J. Fromherz, "Localization from Mere Connectivity," in *Proc. of ACM Int'l Symposium on Mobile Ad hoc Networking and Computing (MobiHOC)*, pp. 201–212, 2003.

[18] Y. Shang and W. Ruml, "Improved MDS-based Localization," in *Proc. of IEEE International Conference on Computer Communications*, pp. 2640–2651, 2004.

[19] G. Giorgetti, S. Gupta, and G. Manes, "Wireless Localization Using Self-Organizing Maps," in *Proc. of The International Symposium on ACM/IEEE International Conference on Information Processing in Sensor Networks (ACM/IEEE International Conference on Information Processing in Sensor Networks)*, pp. 293 – 302, 2007.

[20] L. Li and T. Kunz, "Localization Applying An Efficient Neural Network Mapping," in *Proc. of The 1st International Conference on Autonomic Computing and Communication Systems*, pp. 1–9, 2007.

Cut-and-Sew: A Distributed Autonomous Localization Algorithm for 3D Surface Wireless Sensor Networks

Yao Zhao
yxz4655@louisiana.edu

Hongyi Wu
wu@cacs.louisiana.edu

Miao Jin
mjin@cacs.louisiana.edu

Yang Yang
yxy6700@louisiana.edu

Hongyu Zhou
zhou.hongyu@me.com

Su Xia
suxia.ull@gmail.com

The Center for Advanced Computer Studies
University of Louisiana at Lafayette
Lafayette, USA

ABSTRACT

Location awareness is imperative for a variety of sensing applications and network operations. Although a diversity of GPS-less and GPS-free solutions have been developed recently for autonomous localization in wireless sensor networks, they primarily target at 2D planar or 3D volumetric settings. There exists unique and fundamental hardness to extend them to 3D surface. The contributions of this work are twofold. First, it proposes a theoretically-proven algorithm for the 3D surface localization problem. Seeing the challenges to localize general 3D surface networks and the solvability of the localization problem on single-value (SV) surface, this work proposes the *cut-and-sew* algorithm that takes a divide-and-conquer approach by partitioning a general 3D surface network into SV patches, which are localized individually and then merged into a unified coordinates system. The algorithm is optimized by discovering the minimum SV partition, an optimal partition that creates a minimum set of SV patches. Second, it develops practically-viable solutions for real-world sensor network settings where the inputs are often noisy. The proposed algorithm is implemented and evaluated via simulations and experiments in an indoor testbed. The results demonstrate that the proposed cut-and-sew algorithm achieves perfect 100% localization rate and the desired robustness against measurement errors.

Categories and Subject Descriptors

C.2.1 [**Network Architecture and Design**]: Wireless communication

Keywords

3D surface, autonomous localization, wireless sensor networks

1. INTRODUCTION

Location awareness is of significant importance to wireless sensor networks. It is imperative for a variety of sensing applications

and network operations, ranging from position-aware sensing to sensor deployment and geometric routing. While location service can be readily provided by the global navigation system (GPS), it is often unaffordable to integrate a GPS receiver in every single sensor for large-scale deployment, due to its high cost and lavish energy consumption. Moreover, part or all of the sensors (e.g., deployed underground or underwater) may be prohibited from receiving line-of-sight satellite signals, rendering it infeasible to solely rely on GPS for sensor localization. To this end, a diversity of GPS-less and GPS-free solutions have been developed recently for *autonomous* localization in wireless sensor networks [1–26].

1.1 Challenges in 3D Surface Localization

A wireless sensor network may be deployed on a 2D plane (e.g., for crop sensing in fields or wildlife tracking on plains), or in a 3D volume (for underwater or space reconnaissance), or on a 3D surface (such as for seismic monitoring on ocean floors or in mountainous regions). Most autonomous localization algorithms are based on 2D sensor networks [1–24]. They take Euclidean distance information as input, and search the solution space to discover optimal sensor coordinates that minimize the average distance error (which is defined as the average difference between the real distance and the distance under the established coordinates system). The real Euclidean distance between two adjacent sensors can be approximately measured by received signal strength (RSS) or time difference of arrival (TDOA) or simply assumed as a constant radio range, while the real distance between two remote nodes is often estimated by their shortest path. A diversity of approaches have been proposed for distance error minimization in order to determine the coordinates of sensors, with different localization accuracy and time complexity [15–21]. Generally, distance information is sufficient to localize sensor nodes on a 2D plane (except for non-rigid shapes [27]). For example, Fig. 1(b) illustrates the localization result of the network shown in Fig. 1(a) by using multi-dimensional scaling (MDS) [15, 16].

It is straightforward to extend the 2D localization algorithms to 3D volume. Introducing the third dimension does not substantially increase the hardness of the problem. For instance, based on the estimated pair-wise Euclidean distances in a 3D volumetric network (shown in Fig. 1(c)), the MDS algorithm can be readily applied to establish the coordinates of the sensors (see Fig. 1(d)).

However, similar extension is not applicable to 3D surface networks as reported in [28]. While a 3D surface appears to be a special case of 3D volume or a generalization of 2D plane, surprising challenges exist in efforts to apply 2D planar or 3D volumetric lo-

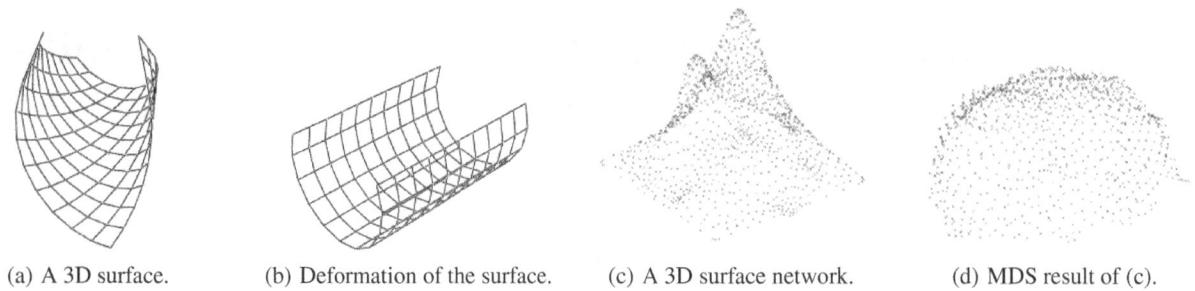

(a) A 3D surface.　　(b) Deformation of the surface.　　(c) A 3D surface network.　　(d) MDS result of (c).

Figure 2: A general 3D surface network is not localizable based on surface distances only, since it can be deformed to another surface without changing the surface distance between any pair of points as shown in (a) and (b). When a distance-based localization algorithm (e.g., MDS) is applied on 3D surface, it simply fails as illustrated in (c) and (d).

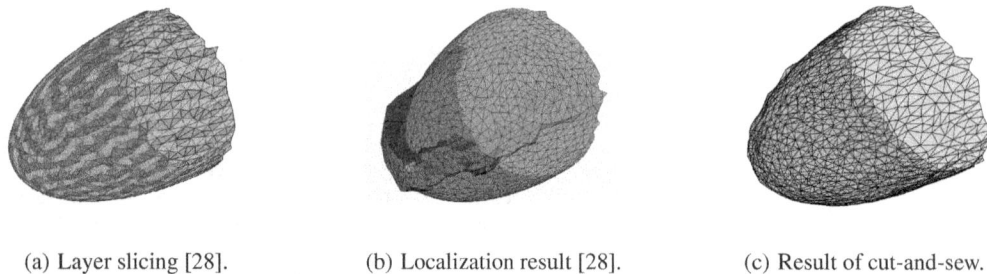

(a) Layer slicing [28].　　(b) Localization result [28].　　(c) Result of cut-and-sew.

Figure 3: The layer slicing approach proposed in [28] divides the surface into layers (see (a)). But it often fails in localizing some layers because they are NSV, thus resulting in large distortions in its localization result illustrated in (b). The proposed cut-and-sew algorithm produces much more accurate localization results as shown in (c).

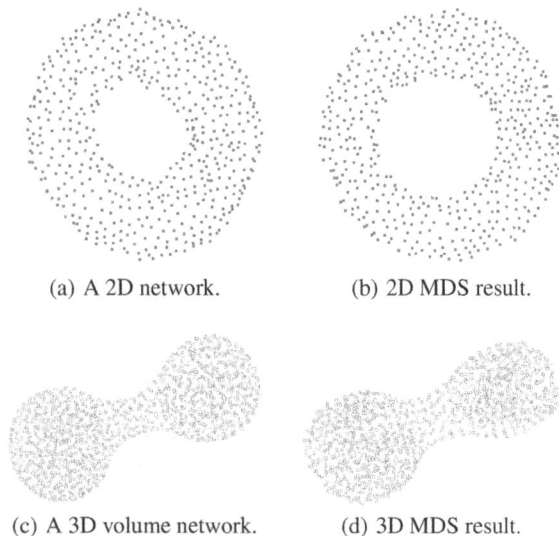

(a) A 2D network.　　(b) 2D MDS result.

(c) A 3D volume network.　　(d) 3D MDS result.

Figure 1: Localization is feasible based on distance information in 2D plane and 3D volume sensor networks.

calization techniques to a 3D surface network. The hardness of the problem stems from the lack of Euclidean distance as input. More specifically, with short radio range, the measurable distance between two remote sensors on a 3D surface is their surface distance, which is essentially the length of geodesic, i.e., the (locally) shortest path between them on the surface. Such surface distance is often dramatically different from the corresponding 3D Euclidean distance. As revealed by Theorem 1 of [28], *a general 3D surface*

network is not localizable, given surface distance constraints only. An intuitive explanation is depicted in Fig. 2, where the surface in Fig. 2(a) can be deformed to another surface (see Fig. 2(b) for example) without changing any surface distance. Thus, it is obviously impossible to determine a unique 3D embedding merely based on surface distances. When a distance-based localization algorithm (e.g., MDS) is applied to a 3D surface network, it simply fails as illustrated in Fig. 2(c)-2(d).

1.2　Contribution of This Work

Seeing the fundamental challenges in 3D surface localization based on surface distances only, a practical setting with augmented input information has been considered in [28], where not only surface distances but also nodal height measurements are assumed to formulate the localization problem. The height (or altitude) of a sensor is measurable via atmospheric pressure. Such measurement is of extremely low cost. As a matter of fact, many sensors have integrated barometer for gauging their altitude. For example, the Crossbow MTS400/MTS420 sensor board is equipped with Intersema MS55ER pressure sensor with an error margin of about 1.5% and thus able to determine its height with high accuracy. Similarly, the underwater height (or depth) may be measured via water pressure. Therefore, the height information can be taken along with surface distances as inputs to formulate the 3D surface localization problem.

At the first glimpse, the problem seems to become trivially easy with the given height (i.e., Z-coordinates) of sensors. For example, a naive approach is to project the sensors to X-Y plane and then apply 2D localization algorithms [15–17, 19–21] to determine their X-Y coordinates. The sensors are thus localized by putting X, Y and Z coordinates together. However this naive approach often fails because the projection of a general 3D surface on the X-Y plane is non-planar. For instance, when the 3D surface network shown in Figs. 3 or 4 is projected to X-Y plane, the upper and lower parts of the surface will overlap, yielding a folded (i.e., non-planar) graph. The 2D localization algorithms either fail or result in extraordinarily large errors when they are applied to a significantly non-planar graph.

DEFINITION 1. *A 3D surface network is* localizable *if a unique embedding can be determined according to surface distance and height information.*

In general, a sensor network deployed on a single-value (SV) 3D surface is localizable. The formal definition of SV is to be given in Sec. 2.1. Briefly, a SV surface is a surface on which any two points have different projections on the X-Y plane. The definition is in reference to X-Y plane since sensors' heights are given as input. A more general definition of SV surface can be made according to any arbitrary plane, but is not considered in this research. Obviously, a network on a SV surface has a planar projection on X-Y plane, converting the problem to a 2D setting. Once the nodes are localized on 2D (i.e., X-Y coordinates are determined), they are mapped back to 3D by adding the height as Z-coordinate.

Seeing the challenges to localize general 3D surface networks and the solvability of the localization problem on SV surface, *this paper proposes a divide-and-conquer approach, dubbed cut-and-sew, by partitioning a general 3D surface network into SV patches, which are localized individually and then merged into a unified coordinates system.* The contributions of this work are twofold, including the design of theoretically-proven algorithm and the development of practically-viable solution.

(1) Theoretically-Proven Algorithm: There are obviously many options to partition a network. As a matter of fact, there is a theoretically infinite solution space to be explored. For example, a simple heuristic has been discussed in [28]. Given Z-coordinates of sensors, it is natural to divide a 3D surface network into short (e.g., one-hop high) horizontal layers, which are more likely to be SV and thus localizable. This approach does not guarantee all layers are SV. The non-single-value (NSV) layers are simply marked non-localizable. The localized layers are then combined together by least square alignment. Its localizable rate and location error highly depend on the percentage of NSV layers. Given a NSV surface, it is generally unavoidable to have NSV layers. Hence the localization result often exhibits significant distortion (as shown in Fig. 3(b)). Moreover, the algorithm may fail completely in a 3D surface network if most of its layers are NSV. This is just an example that an arbitrary partition does not yield the desired localization result.

This research aims to discover the minimum SV partition, an optimal partition that creates a set of patches satisfying two conditions. First, all patches must be SV to ensure their localizability. Second, the number of patches should be minimized to avoid unnecessary partitioning and merging, which are subject to linear transformation errors. This is obviously different from the trivially arbitrary partition adopted in several early works (such as [16,28]). The proposed approach is to identify NSV edges (as to be elaborated in Sec. 2.1) to guide the division of a 3D surface network into SV patches. Once the network is partitioned, the individual patches can be readily projected to 2D plane, where various algorithms are available for localization. Finally, the damped least-square algorithm [29] is employed to combine the localized patches by minimizing average distance error.

(2) Practically-Viable Solutions: Under practical sensor network settings, the inputs are often noisy. For example, both surface distances and sensors heights are subject to measurement errors. Al-

though the basic ideas still apply, the noisy inputs obviously lead to inaccurate localization results or even a total failure of the localization algorithm. Practically-viable solutions must be developed to filter out input noise, aiming to improve the robustness and reliability and minimize localization errors. More specifically, the inaccurate distance and height measurement directly affects the identification of NSV edges, which are often deviated from the ground truth and become isolated. It is apparently impossible to partition the network according to such noisy NSV edges. The proposed idea is to fuse nearby NSV edges to form a band and then cut the network along the medial axis of the band. This approach effectively minimizes the impact of input errors on network partition and localization.

The propose algorithm is implemented and evaluated via simulations and experiments in an indoor testbed. The results demonstrate that the proposed cut-and-sew algorithm achieves nearly perfect 100% localization rate and the desired robustness against measurement errors. The rest of this paper is organized as follows: Sec. 2 introduces the proposed localization algorithm. Secs. 3 and 4 present testbed experiments and simulation results, respectively. Finally, Sec. 5 concludes the paper.

2. THE CUT-AND-SEW ALGORITHM

The inputs of the 3D surface localization problem include the height of the sensor and the connectivity and distance between neighboring nodes. The proposed distributed localization algorithm, named *cut-and-sew*, consists of three components as outlined below. First, it identifies NSV edges. Then, according to the NSV edges, the network is partitioned into a minimum set of SV patches that can be readily localized. Finally, the patches are merged together to produce a unified coordinates system. For a lucid exposition of the proposed scheme, the distance and height are first assumed free of errors. The problem due to measurement inaccuracy will be discussed in Sec. 2.4.

2.1 Identification of NSV Edges

To facilitate network partitioning and localization, a distributed algorithm [30] is employed to establish a triangular mesh structure (or triangulation) based on local connectivity and distance information (see Fig. 4(a) and Fig. 4(b) for the original sensor network graph and the corresponding triangular mesh, respectively). If an edge in the triangular mesh is not on the boundary, it must be shared by two and only two triangles. For example, Fig. 5(a) illustrates two neighboring triangles, $\triangle ABC$ and $\triangle BCD$, which share a common edge, i.e., BC. Without loss of generality, the 3D surface network is oriented such that Edge BC is on the Y-Z plane. Let $\triangle abc$ and $\triangle bcd$ denote the projected triangles. Obviously, the projection of Edge BC, i.e., bc, is on the Y-axis. The length of each edge on the projected plane is determined according to the following equation:

$$L_{ij} = \sqrt{L_{IJ}^2 - (Z_I - Z_J)^2},\qquad(1)$$

where Z_I is the height of Node I, L_{IJ} is the length of Edge IJ, and L_{ij} is the length of Edge IJ's projection.

DEFINITION 2. *The local distance information of a node includes the Euclidean distances of the node to its one-hop and two-hop neighbors.*

The absolutely accurate distance measurement is generally unattainable in practice. However the Euclidean distance between two neighboring nodes can be estimated by RSS or TDOA. It often consumes higher power to measure the distance to a two-hop neighbor. But such measurement is required only once for a static network.

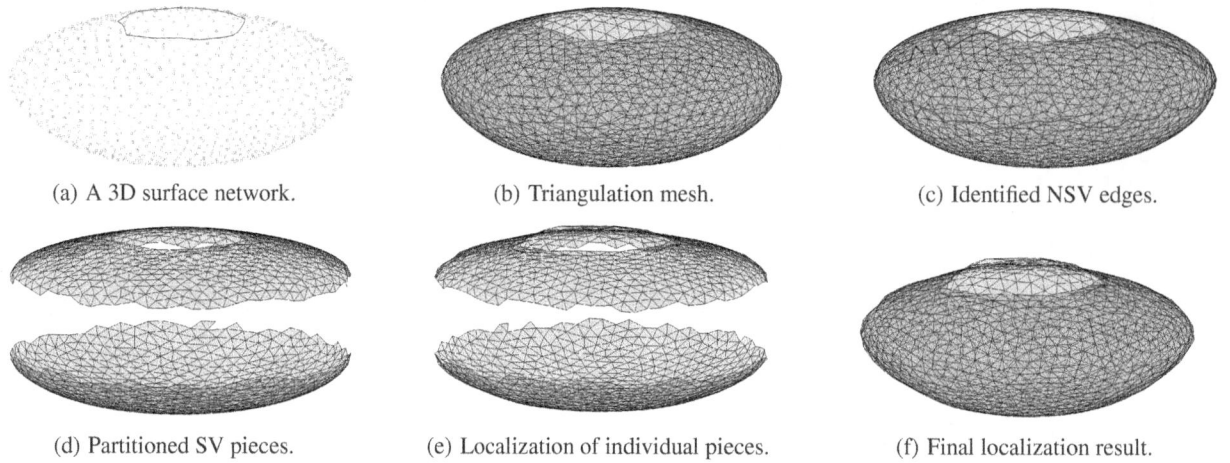

(a) A 3D surface network.   (b) Triangulation mesh.   (c) Identified NSV edges.

(d) Partitioned SV pieces.   (e) Localization of individual pieces.   (f) Final localization result.

Figure 4: An overview of the proposed cut-and-sew algorithm for 3D surface localization.

This is very different from using long links for communication, which results in significant power consumption.

DEFINITION 3. *In the triangulation of a 3D surface network, an edge is locally NSV (or NSV for short) if the projection of its two associated triangles overlap on the X-Y plane.*

It has been assumed in the above definition that neither triangles are vertical (i.e., the projection of a triangle is not colinear). The exception of vertical triangle will be discussed in Sec. 2.3. It can be checked by a simple, local test and in fact leads to a trivial problem that is readily solvable.

DEFINITION 4. *A 3D surface sensor network is called a NSV network if it contains NSV edges.*

DEFINITION 5. *A 3D surface sensor network is called a SV network if it does not contain NSV edges.*

As discussed earlier, the NSV network, particularly the NSV edges, introduce problem in localization. The proposed algorithm, however, exploits them to partition the network into SV patches that can be readily localized. The rest of this subsection shows that NSV edges can be identified by using local information.

LEMMA 1. *Given an edge in the triangular mesh of a 3D surface sensor network, its associated local distance information is sufficient to determine whether it is a NSV edge.*

PROOF. Consider Edge BC shared by $\triangle ABC$ and $\triangle BCD$, which are projected to $\triangle abc$ and $\triangle bcd$ on the 2D plane. Obviously, to check if $\triangle abc$ and $\triangle bcd$ overlap is equivalent to examine whether Nodes a and d are on the same side of Edge bc. If Nodes a and d are on the same side of Edge bc, $\triangle abc$ and $\triangle bcd$ overlap; otherwise, they do not.

Without loss of generality, Node a is assumed on an arbitrary side of Edge bc. Now, the problem is reduced to check if Node d is on the same side. Assume the coordinates of Nodes b and c are (X_b, Y_b) and (X_c, Y_c). Given the edge length information, i.e., L_{bd} and L_{cd}, the basic triangulation can be applied to derive the coordinates of Node d. Clearly, there are two possible solutions symmetric about Y-axis, denoted as d' and d'' with coordinates (X_d, Y_d) and $(-X_d, Y_d)$, respectively, as illustrated in Fig. 5(b).

So far it still unknown on which side Node d is located. However, it can be shown by contradiction that the distances between

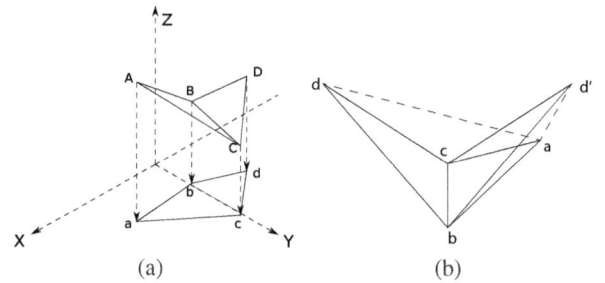

(a)   (b)

Figure 5: (a) Projections of neighboring triangles. (b) Hypothetic nodal positions become deterministic with 2-hop distance.

Node a and the two hypothetic positions of Node d (i.e., d' and d'') are always different. If $L_{ad'} = L_{ad''}$, then

$$(X_a - X_d)^2 + (Y_a - Y_d)^2 = [X_a - (-X_d)]^2 + (Y_a - Y_d)^2. \quad (2)$$

The solution of the equation is $X_a = 0$ or $X_d = 0$, i.e., either Node a or Node d must be collinear with Nodes b and c. This contradicts the fact that Nodes a, b, c and d form two triangles. Therefore, if L_{ad} is known, one can readily determine whether Node a and Node d are on the same side of Edge bc. More specifically, one can first embed $\triangle abc$ on the 2D plane according to the local distances L_{ab}, L_{bc}, and L_{ac}. Then, based L_{bd} and L_{cd}, two hypothetic positions of Node d are computed. Finally, one of them is chosen according to L_{ad}. If Nodes a and d are on the same side, Edge BC is NSV; otherwise, it is not. Thus the lemma is proven. □

The proof of Lemma 1 clearly suggests a simple and localized scheme to examine if an edge is NSV. Every node in the network can perform such local calculation for its associated edges and mark the ones that are NSV. Fig. 4(c) illustrates the identified NSV edges.

2.2 Network Partition

This subsection shows that the minimum SV partition is achieved by dividing the network along NSV edges.

DEFINITION 6. *Given a triangular mesh, a boundary edge is an edge contained in one and only one triangle.*

A boundary edge is obviously not NSV because it is contained in one triangle only.

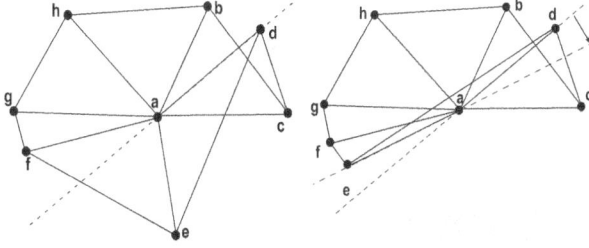

(a) Node *e* on right side of *ad*.   (b) Node *e* on left side of *ad*.

Figure 6: Illustration of Lemma 2, where Edge *ac* is NSV.

LEMMA 2. *A NSV edge must connect to other NSV edges or boundary edges.*

PROOF. Consider a NSV edge, denoted as Edge *AC*. Its projection on the 2D plane is *ac*. First, Edge *AC* must be shared by two triangles according to Definition 3. Therefore it is not a boundary edge. If either Node *A* or Node *C* is on boundary, Edge *AC* must connect to a boundary edge, and thus the lemma is proven.

Now assume Nodes *A* and *C* are non-boundary nodes. A non-boundary node must be surrounded by a set of triangles on the 3D surface. Fig. 6 illustrates the projection of the set of triangles around Node *A* on the 2D plane. Since Edge *AC* is NSV, the projection of its associated triangles must overlap. In other words, Nodes *b* and *d* must be on the same side of *ac* on the projected plane. Thus either *ad* is located within ∠*bac* or *ab* is in ∠*dac*. Since the two cases are symmetric, only the first one is discussed here. Given *ad* is within ∠*bac*, Nodes *b* and *c* are obviously on the two sides of Line *ad* (see the dashed line in Fig. 6). To facilitate the discussion, a directions is defined as follows. Assume one stand at Node *a* and face Node *d*. Node *c* is on the right hand side, thus it is said on the right side of *ad*. Similarly, Node *b* is on the left side of *ad*. Besides left and right, the following discussion also uses the clockwise order around Node *a*.

Next the proof shows by contradiction that besides Edge *AC*, there must exists another NSV edge associated with Node *A*. To construct the contradiction, all edges that connect to Node *A*, except *AC*, are assumed SV. Consequently, the neighboring nodes of Node *A* (excluding *C*) must be in a clockwise order on the projected plane. More specifically, let's begin with Node *d* as shown in Fig. 6(a). Without loss of generality, assume *d* is on the left side of *ae*. Since *ae* is SV, Node *f* must be on its right side. Thus Nodes *d*, *e* and *f* must be in a clockwise order around Node *a*. Similarly, since *af* is SV, Nodes *e*, *f*, and *g* must follow the clockwise order. Thus Nodes *d* to *g* are all ordered. By deduction, all nodes around Node *a*, except *c*, must be clockwise ordered (see Nodes *d*, *e*, ..., *b* in the figure).

Now the proof shows that either Edge *AD* or Edge *AE* must be NSV. First, examine △*dae*, which shares *ad* with △*cad*. As discussed earlier, Node *c* is on the right side of *ad*. If Node *e* is also on the right side of *ad* (as illustrated in Fig. 6(a)), then Edge *AD* must be NSV. If Node *e* is on the left side of *ad* (as illustrated in Fig. 6(b)), let the dashed line *ad* rotate clockwise around Node *a*. Since Nodes *d*, *e*, ..., *b* are clockwise ordered, the dashed line must meet Node *e* first. Obviously, both *d* and *f* are on the same side of *ae*. Accordingly, Edge *AE* is NSV. The above results contradict the earlier assumption that all edges that connect to Node *A*, except *AC*, are SV. Thus besides Edge *AC*, there must exists another NSV edge associated with Node *A*.

The above discussions focuses on Node *A* and its surrounding triangles. Similar results can be obtained by analyzing Node *C* and its neighboring edges. Therefore, the proof concludes that the two ends of an NSV edge must connect to other NSV edges or boundary edges. The lemma is proven.   □

In fact, the above lemma is intuitively understandable, because the NSV edge is where the surface is folded when it is projected to the *X-Y* plane. Given a 3D surface, the folding line must be continuous until it extends to the boundary of the surface or meets other folding lines.

THEOREM 1. *The minimum SV partition is achieved by dividing the network along NSV edges.*

PROOF. Let the partition process start from any node on an arbitrary NSV edge and cut the network along all of its connected NSV edges. Since the NSV edges must connected to each other or to boundary edges as shown by Lemma 2, the cutting process will either form a loop or stop at the boundaries of the network. In either case, the network is partitioned into two or more separated patches. The NSV edges used in such partition become boundary edges of the newly created patches. As a result, they are no longer NSV edges, because a boundary edge is contained in one triangle only. The process repeats until no NSV edges exist in the entire network. It is clear that none of the patches contains an NSV edge. Therefore they are all SV.

On the other hand, it is straightforward to show the partition is minimum, because all NSV edges must be cut open, otherwise a patch that contains NSV edges must be an NSV patch. The theorem is thus proven.   □

Fig. 4(d) illustrates the SV patches by partitioning the 3D surface sensor network along NSV edges.

2.3 Localization and Combination

As discussed above, each edge decides if it is NSV by using local information only. A set of NSV edges form the boundary of an SV patch. At least one randomly elected node in each patch initiates projection and 2D localization by constructing a local flooding packet that contains a patch ID (which, e.g., can be simply its own node ID). The packet is dropped when it reaches the NSV edges. Obviously the local flooding packet is limited within the given patch. If multiple nodes flood at the same time, the one with the highest ID wins. All nodes in the patch thus begin projection and 2D localization.

The projection of a SV patch is planar. It remains a triangulation on the X-Y plane, with only changes in edge length that can be determined by surface edge length and height information (i.e., by using Eq. (1)). Given the height information, Z-coordinates are already known. Thus only X-Y-coordinates are yet to be determined. To this end, several distributed 2D localization algorithms can be applied [15–17,31]. The localization result is then mapped back to 3D by including the known Z-coordinates (see Fig. 4(e)).

There is a rare exceptions of Definition 3 that may occur when a triangle is vertical. A vertical triangle can be checked by a simple, local test since it is colinear on the X-Y plane. If one and only one triangle associated with an edge is vertical, the edge is considered a NSV edge. However, if both triangles are vertical, the edge is not treated as NSV.

Similarly, an entire patch may be vertical. Its projection becomes a line, which is not localizable by 2D algorithms. However, the patch is in fact already planar (without a projection), and thus can be readily localized on the X-Y plane and mapped to 3D as discussed above.

Finally, since neighboring patches share common vertexes and edges, they can be readily "sewed" together by using a distributed least square algorithm [28,32], yielding a unified coordinate system for the entire 3D surface sensor network as illustrated in Fig. 4(f).

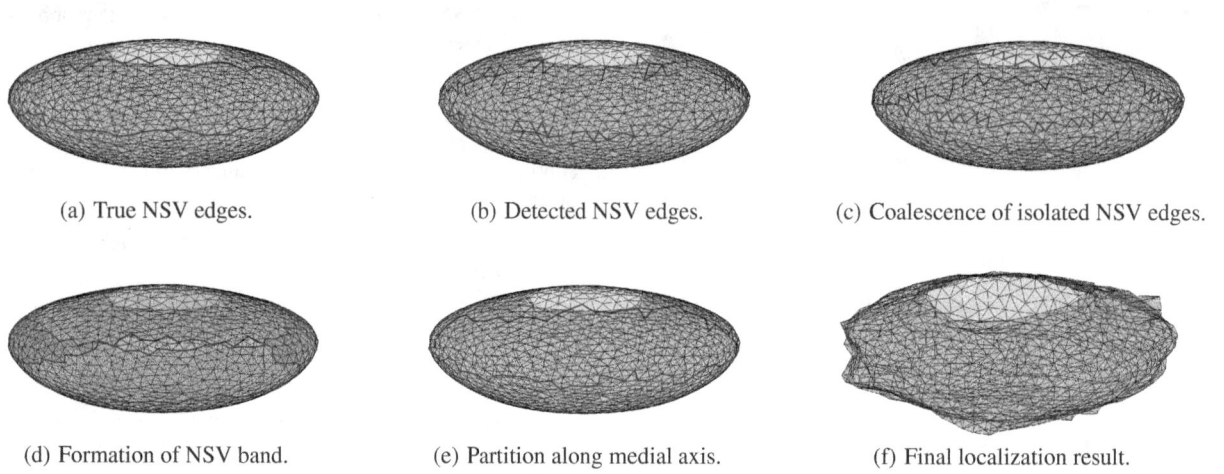

(a) True NSV edges. (b) Detected NSV edges. (c) Coalescence of isolated NSV edges.

(d) Formation of NSV band. (e) Partition along medial axis. (f) Final localization result.

Figure 7: Network partition under noisy measurements.

2.4 Practical Solution with Noisy Inputs

So far accurate distance and height measurements are assumed as inputs of the 3D surface localization problem. Both of them can be noisy under practical sensor network settings. The proposed algorithm still applies but needs further treatment on network partitioning.

The inaccurate inputs directly affect the identification of NSV edges. Fig. 7(b) illustrates an example of the detected NSV edges with 10% measurement errors. As can be seen, they become isolated and many of them are deviated from the true NSV edges shown in Fig. 7(a). It is obviously impossible to partition the network according to such noisy NSV edges directly. The proposed idea is to fuse the nearby NSV edges to form a band and then cut the network along the medial axis of the band. More specifically, the algorithm consists of the following three steps.

(1) Coalescence of Isolated NSV Edges. The algorithm fuses the detected NSV edges by expanding and connecting them. First, if a triangle contains an NSV edge, the entire triangle, i.e., all of its edges, are marked NSV. Second, If two NSV triangles are one-hop away from each other, the edges between them are marked as NSV too. The NSV edges are now better connected, forming a number of clusters. The NSV edges in each cluster are connected, but different clusters are still isolated. The two closest clusters are connected by marking all the edges on their shortest path as NSV edges. They are thus merged into one cluster. The process repeats until all clusters are connected. Fig. 7(c) shows the result after coalescence of isolated NSV edges.

(2) Formation of NSV Band. The above step has created an expanded set of NSV edges. The algorithm further marks the edges within 1-hop of existing NSV edges as NSV, forming a NSV band. To smoothen the band, an edge is marked NSV if it is included in a triangle with two NSV edges. An example of the resulting NSV band is depicted in Fig. 7(d).

(3) Partition along Medial Axises of NSV Band. Finally, the medial axis of the NSV band is identified for partitioning. A NSV band may have two or more boundaries. The basic idea is to shrink the boundaries of the NSV band until they meet, forming the approximate medial axis. This is achieved by a distributed process initiated by boundary edges of the NSV band. Each boundary edge is involved in a triangle inside the band. It is replaced by two other edges of the triangle, which become new boundary edges. Thus the

boundaries grow inward into the band. The process repeats until convergence (which is obviously guaranteed because the band has a finite size), yielding the medial axis.

Once the network is partitioned, the patches are localized and combined as discussed earlier. Note that the partition along medial axis is an approximation. It does not guarantee every patch is SV. Thus, the 2D projection of a patch may have minor overlap between its edges (especially at the boarder of the patch), resulting in localization errors. This is evident from the minor distortion shown in Fig 7(f). Such errors will be quantitatively studied and discussed in Sec. 4.

2.5 Complexity and Overhead

The NSV edge identification, network partitioning and projection are all done locally by the individual nodes, thus resulting in a constant computation complexity and communication overhead. However, to initiate projection, a local flooding is required in all individual patches, which together yield a communication overhead of $O(n)$ in the entire network, where n is the total number of nodes in the network.

When inputs are noisy, the NSV edges are further processed to identify the medial axis of the NSV band. This process consumes a total time of $O(n)$ and the corresponding communication overhead is also $O(n)$.

The computation time and communication overhead for localizing individual patches depend on the 2D localization algorithm employed. For example, MDS results in a linear communication overhead and a computation complexity of $O(m^3)$ where m is the number of nodes in a patch. The computation is carried out in all patches simultaneously.

The combination of two patches can be completed in $O(p)$ time and results in $O(p)$ communication overhead, where p is the set of nodes involved in alignment (usually the common nodes shared by the two patches). To merge all patches, the total computational complexity and communication overhead are both $O(n)$.

In summary, the proposed cut-and-sew algorithm results in an overall communication overhead of $O(n)$ and time complexity of $O(Max\{m^3,n\})$.

3. PROTOTYPING AND EXPERIMENTS

Several indoor testbed models have been built in this research for prototyping 3D surface sensor networks. An example is shown in

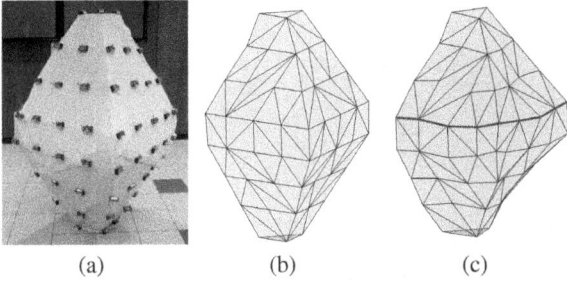

Figure 8: Experimental setup and result. (a) Indoor 3D surface network testbed. (b) Input triangulation. (c) Localization result (with an average location error of 14%).

Table 1: Localizable Rate

	Stadium	Sea Cave	Mine Pit
Slice-Based [28]	98.8%	90.8%	81.3%
Cut-and-Sew	100%	100%	100%

Fig. 8(a), which is $5^1/_4$ feet high, 3 feet long and 3 feet wide. Forty eight Crossbow MICAz motes are attached to its surface. The algorithmic codes are implemented in Tiny OS and run on the Crossbow sensor nodes. The sensors are configured to use close to minimum radio transmission power (Level 2), with a communication range between 25 to 55 cm. The short radio range avoids undesired connections through volume.

Every sensor periodically broadcasts a beacon message that contains its node ID to its neighbors. Based on received beacon messages, a node builds a neighbor list with the RSSI of corresponding links. RSSI is used to estimate the length of links by looking up a RSSI-distance table established by experimental training data. The preliminary test shows that, under low transmission power, such estimation has an error rate about 20%. At the same time, the ground truth of surface distances and sensor coordinates are manually measured. A triangular mesh structure is constructed by using the distributed algorithm [30]. Fig. 8(b) illustrates the triangulation based on ground truth inputs.

The localization result is depicted in Fig. 8(c). As can be seen, the NSV edges are identified correctly (as highlighted in the figure). The network is thus partitioned into two SV patches. Then MDS is applied to localize each of them. The combined patches largely restore the original 3D surface network, with an average location error of about 14%. The largest errors are observed at the middle of the network (around the NSV edges), while the top and bottom are localized more accurately. This is because the triangles right below the NSV edges are almost vertical, resulting in extremely skinny projections on the X-Y plane (where some edges of the projected triangles are extremely short). Note that localization errors are inevitable in MDS, due to inaccurate distance inputs. Given the same amount of errors, they apparently have larger impact on short edges, which lead to significant distortion on 3D surface, thus producing the aforementioned localization errors.

The layer slicing approach proposed in [28] largely fails in the above experimental setting, because it cannot locate about 50% of sensor nodes (around the NSV edges at the middle of the network).

4. SIMULATION RESULTS

Besides the indoor testbed experiments, the proposed cut-and-sew algorithm is implemented and evaluated via simulations under a variety of practical 3D surface sensor networks. For example,

the results under three representative network models (for monitoring stadium, sea cave, and mine pit) are presented in this section. Note that the localization results obviously depend on individual network settings. It would mix up the performance trend or even become misleading if data from different networks are combined and averaged.

The simulation adopts a general communication model with merely a constraint on the maximum radio transmission range, which is normalized to one. Two nodes are connected with a probability if their distance is less than one. Each node is assumed to measure its own height and local distances. A range of measurement errors are considered in simulations. The proposed algorithm has been proven to work perfectly when the measurements are accurate. Therefore, the simulations focus on studying its tolerance against measurement errors and the impact of sensor densities. There is almost no competing scheme to compare with. The only related work is [28]. However, it is difficult to compare their localization errors, because the approach in [28] does no even localize all nodes in the network. Therefore, the comparison is based on localizable rate only.

4.1 NSV Edge Detection Error

The proposed algorithm heavily depends on NSV edges, which guide the partition of the network. Let S denote the set of perfect NSV edges (identified under precise distance and height measurements). While the measurements are often noisy in practice, the proposed algorithm still yields a set of NSV edges, T. Obviously, the accuracy of network partition and final localization depend on how close T is to S. Hence, the NSV edge detection error is defined as $E_{NSV} = \frac{\sum_{t_k \in T} N(t_k, S)}{|T|}$, where $N(t_k, S)$ is the hop distance between Edge t_k and its closest edge in S, and $|T|$ is the total number of identified NSV edges.

Fig. 9 illustrates the impact of measurement errors and nodal densities on E_{NSV}. Under the same nodal density, E_{NSV} increases sharply with larger measurement errors. While the results show undesired vulnerability of NSV edge detection under measurement errors, the simulation data to be discussed next demonstrates that the proposed algorithm effectively minimizes the impact of the NSV edge errors on network partition and final localization. On the other hand, the nodal density (varying from 1k to 4k in each network model) does not noticeably affect NSV edge detection.

4.2 Network Partition Error

The network partition is guided by the NSV edges. Therefore, NSV edge errors obviously have a negative impact on partitioning. However, the proposed algorithm employs several techniques to minify the effect. More specifically, it fuses the nearby NSV edges to form a band and then identifies the medial axis of the band for network partitioning. The medial axis is expected to approximate the set of true NSV edges. Let U denote the set of edges that constitutes the identified medial axis. The network partition error E_{PAR} is defined as the maximum deviation between U and S (i.e., the set of true NSV edges for ideal partitioning), i.e., $E_{PAR} = \frac{\max\{N(u_i, S)|u_i \in U\}}{M}$, where $N(u_i, S)$ is the shortest hop distance between u_i and S, and M is a normalizer defined as the maximum shortest hop distance between any two nodes in the network. Higher E_{PAR} indicates that, after the network is partitioned, more NSV areas still exist in the patches, leading to potentially worse localization result.

As shown in Fig. 10, E_{PAR} is not sensitive to measurement errors. With the increase of measurement inaccuracy, the network partition error largely remains a constant. Nodal density has a diverse effect

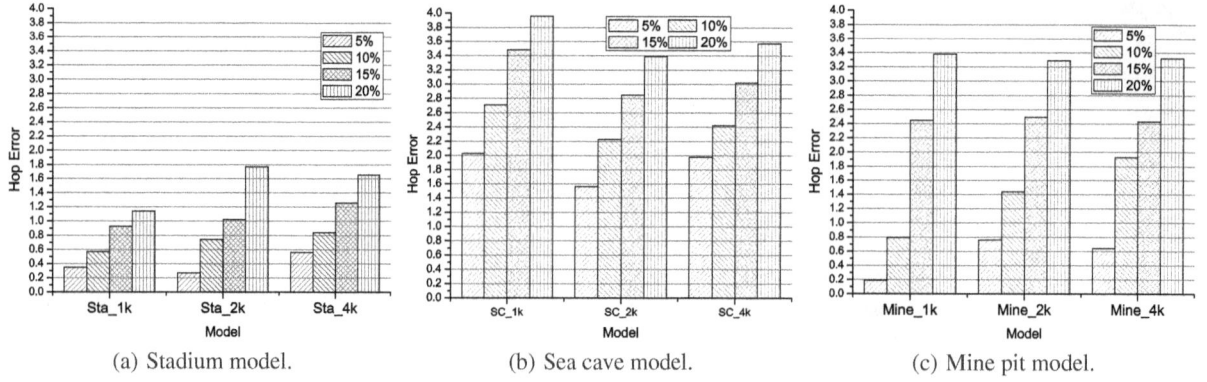

(a) Stadium model. (b) Sea cave model. (c) Mine pit model.

Figure 9: NSV edge detection error increases dramatically with higher measurement errors. Each model is simulated with 1k, 2k and 4k nodes.

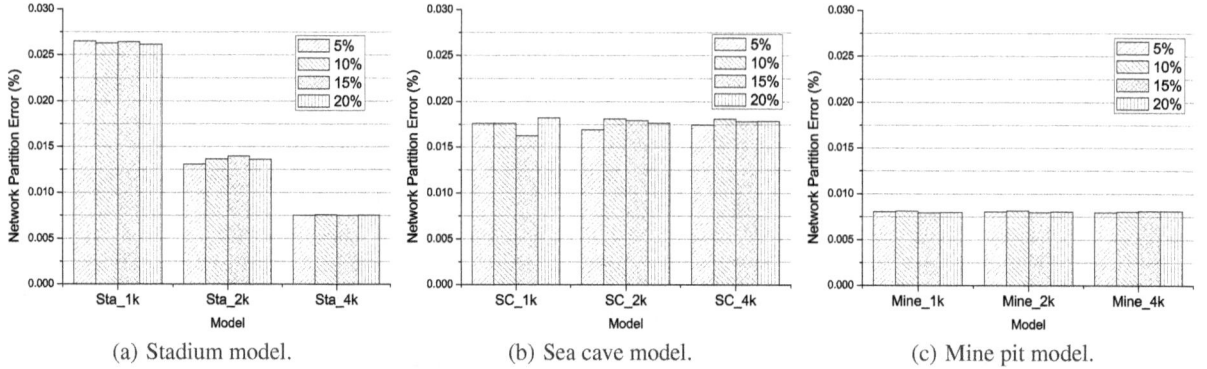

(a) Stadium model. (b) Sea cave model. (c) Mine pit model.

Figure 10: Network partition error is insensitive to measurement errors. Nodal density has a diverse effect on network partition error depending on individual network models.

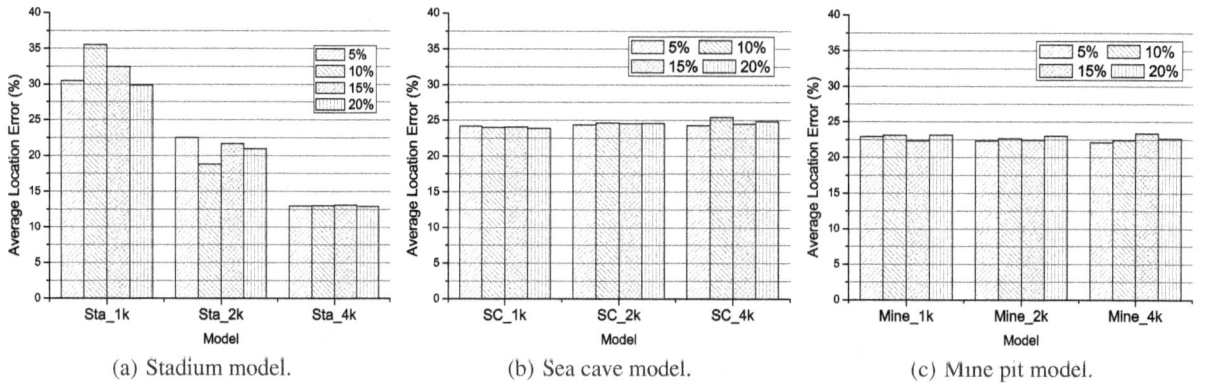

(a) Stadium model. (b) Sea cave model. (c) Mine pit model.

Figure 11: Average localization error is not significantly affected by measurement errors.

on network partition error depending on individual network models. For example, the impact of nodal density is negligible under the sea cave and mine pit models. However, the stadium model noticeably benefits from higher nodal density, owing to the shape of the area surrounding NSV edges.

4.3 Average Location Error

The final output of the algorithm is the coordinates of sensor nodes. Such coordinates can be aligned to any desired coordinates system (e.g., the global positioning system) by using the least square algorithm [32]. However, the alignment to a particular coordinates system is out of the interests of this research. Therefore, the localization error is computed by comparing the edge length under the established coordinates system with the real edge length. More specifically, let $\{l_1, l_2, ..., l_n\}$ denote the edge lengths calculated according to the established coordinates and $\{l'_1, l'_2, ...l'_n\}$ the real edge lengths. Location error E_{LOC} is defined as $E_{LOC} = \frac{\sum_{k=1}^{n} |l_k - l'_k|}{n \times \max\{l'_i | 1 \le i \le n\}}$.

It is not a surprise that the localization error shows similar trend as the network partition error. As a matter of fact, the performance of localization is predominated by the errors in network partitioning. With an average localization error less than 25%, the localization algorithm is not significantly affected by inaccurate distance and height measurements (see Fig. 11). A higher sensor density helps reduce the localization errors in the stadium model. However, the same effect is not observed in the sea cave and mine pit models, in line with the tread in Fig. 10.

Finally, the proposed algorithm achieves 100% localizable rate, which is in a sharp contrast to the earlier heuristic [28]. Table 1 shows a comparison of the localizable rate (averaged over different sensor densities).

5. CONCLUSION

This paper has proposed a divide-and-conquer approach, named cut-and-sew, for autonomous localization of 3D surface wireless sensor networks. Seeing the challenges to localize general 3D surface networks and the solvability of the localization problem on single-value (SV) surface, the proposed cut-and-sew algorithm partitions a general 3D surface network into SV patches, which are localized individually and then merged into a unified coordinates system. The algorithm has been optimized by discovering the minimum SV partition, an optimal partition that creates a minimum set of SV patches. Moreover, the paper has introduced a practically-viable solution for real-world sensor network settings where the inputs are noisy. The proposed algorithm has been implemented and evaluated via simulations and indoor testbed experiments. The results have demonstrated perfect 100% localization rate and the desired robustness against measurement errors.

ACKNOWLEDGEMENT

M. Jin is partially supported by NSF CCF-1054996 and CNS-1018306. H. Wu is partially supported by NSF CNS-1018306, CNS-0831823, and CNS-0821702.

6. REFERENCES

[1] N. Bulusu, J. Heidemann, and D. Estrin, "GPS-less Low Cost Outdoor Localization For Very Small Devices," *IEEE Personal Communications Magazine*, vol. 7, no. 5, pp. 28–34, 2000.

[2] A. Savvides, C. Han, and M. B. Strivastava, "Dynamic Fine-Grained Localization in Ad-Hoc Networks of Sensors," in *Proc. of MobiCom*, pp. 166–179, 2001.

[3] J. Albowicz, A. Chen, and L. Zhang, "Recursive Position Estimation in Sensor Networks," in *Proc. of ICNP*, pp. 35–41, 2001.

[4] L. Doherty, L. Ghaoui, and K. Pister, "Convex Position Estimation in Wireless Sensor Networks," in *Proc. of INFOCOM*, pp. 1655–1663, 2001.

[5] T. He, C. Huang, B. Blum, J. Stankovic, and T. Abdelzaher, "Range-free Localization Schemes in Large Scale Sensor Networks," in *Proc. of MobiCom*, pp. 81–95, 2003.

[6] D. Niculescu and B. Nath, "Ad Hoc Positioning System (APS)," in *Proc. of GLOBECOM*, pp. 2926–2931, 2001.

[7] C. Savarese and J. Rabaey, "Robust Positioning Algorithms for Distributed Ad-Hoc Wireless Sensor Networks," in *Proc. of USENIX Annual Technical Conference*, pp. 317–327, 2002.

[8] A. Nasipuri and K. Li, "A Directionality Based Location Discovery Scheme for Wireless Sensor Networks," in *Proc. of ACM International Workshop on Wireless Sensor Networks and Applications*, pp. 105–111, 2002.

[9] D. Niculescu and B. Nath, "Ad Hoc Positioning System (APS) Using AOA," in *Proc. of INFOCOM*, pp. 1734–1743, 2003.

[10] H. S. AbdelSalam and S. Olariu, "Passive Localization Using Rotating Anchor Pairs in Wireless Sensor Networks," in *Proc. of The 2nd ACM International Workshop on Foundations of Wireless Ad Hoc and Sensor Networking*, pp. 67–76, 2009.

[11] Z. Zhong and T. He, "MSP: Multi-Sequence Positioning of Wireless Sensor Nodes," in *Proc. of SenSys*, pp. 15–28, 2007.

[12] J. Aspnes, T. Eren, D. K. Goldenberg, A. S. Morse, W. Whiteley, Y. R. Yang, B. D. O. Anderson, and P. N. Belhumeur, "A Theory of Network Localization," *IEEE Transactions on Mobile Computing*, vol. 5, no. 12, pp. 1663–1678, 2006.

[13] K. Yedavalli and B. Krishnamachari, "Sequence-Based Localization in Wireless Sensor Networks," *IEEE Transactions on Mobile Computing*, vol. 7, no. 1, pp. 81–94, 2008.

[14] S. Capkun, M. Hamdi, and J. Hubaux, "GPS-Free Positioning in Mobile Ad-Hoc Networks," in *Proc. of The 34th Annual Hawaii International Conference on System Sciences*, pp. 3481–3490, 2001.

[15] Y. Shang, W. Ruml, Y. Zhang, and M. P. J. Fromherz, "Localization from Mere Connectivity," in *Proc. of MobiHoc*, pp. 201–212, 2003.

[16] Y. Shang and W. Ruml, "Improved MDS-based Localization," in *Proc. of INFOCOM*, pp. 2640–2651, 2004.

[17] V. Vivekanandan and V. W. S. Wong, "Ordinal MDS-based Localization for Wireless Sensor Networks," *International Journal of Sensor Networks*, vol. 1, no. 3/4, pp. 169–178, 2006.

[18] H. Wu, C. Wang, and N.-F. Tzeng, "Novel Self-Configurable Positioning Technique for Multi-hop Wireless Networks," *IEEE/ACM Transactions on Networking*, vol. 13, no. 3, pp. 609–621, 2005.

[19] G. Giorgetti, S. Gupta, and G. Manes, "Wireless Localization Using Self-Organizing Maps," in *Proc. of IPSN*, pp. 293 – 302, 2007.

[20] L. Li and T. Kunz, "Localization Applying An Efficient Neural Network Mapping," in *Proc. of The 1st International Conference on Autonomic Computing and Communication Systems*, pp. 1–9, 2007.

[21] M. Jin, S. Xia, H. Wu, and X. Gu, "Scalable and Fully Distributed Localization With Mere Connectivity," in *Proc. of INFOCOM*, pp. 3164–3172, 2011.

[22] J. Wang, M. Tian, T. Zhao, and W. Yan, "A GPS-Free Wireless Mesh Network Localization Approach," in *Proc. of International Conference on Communications and Mobile Computing*, pp. 444–453, 2009.

[23] R. Magnani and K. K. Leung, "Self-Organized, Scalable GPS-Free Localization of Wireless Sensors," in *Proc. of WCNC*, pp. 3798–3803, 2008.

[24] H. Akcan, V. Kriakov, H. Bronnimann, and A. Delis, "GPS-Free Node Localization in Mobile Wireless Sensor Networks," in *Proc. of The 5th ACM International Workshop on Data Engineering for Wireless and Mobile Access*, pp. 35–42, 2006.

[25] C. Wang, H. Wu, and N.-F. Tzeng, "RFID-Based 3-D Positioning Schemes," in *Proc. of INFOCOM*, pp. 1235–1243, 2007.

[26] J. Maneesilp, C. Wang, H. Wu, and N. F. Tzeng, "RFID Support for Accurate 3-Dimensional Localization," *IEEE Transactions on Computers*, vol. 62, no. 7, pp. 1447–1459, 2013.

[27] Z. Yang, Y. Liu, and X.-Y. Li, "Beyond Trilateration: On the Localizability of Wireless Ad-Hoc Networks," in *Proc. of INFOCOM*, pp. 2392–2400, 2009.

[28] Y. Zhao, H. Wu, M. Jin, and S. Xia, "Localization in 3D surface sensor networks: Challenges and solutions," in *Proc. of INFOCOM*, pp. 55–63, 2012.

[29] K. Levenberg, "A Method for the Solution of Certain Non-linear Problems in Least-Squares," *Quarterly of Applied Mathematics*, vol. 2, no. 2, pp. 164–168, 1944.

[30] H. Zhou, H. Wu, S. Xia, M. Jin, and N. Ding, "A Distributed Triangulation Algorithm for Wireless Sensor Networks on 2D and 3D Surface," in *Proc. of INFOCOM*, pp. 1053–1061, 2011.

[31] H. Lim and J. Hou, "Distributed Localization for Anisotropic Sensor Networks," *ACM Transactions on Sensor Networks*, vol. 5, no. 2, pp. 11:1–11:26, 2009.

[32] A. Bjorck, *Numerical Methods for Least Squares Problems*. No. 51, Society for Industrial Mathematics, 1996.

Heavy-Traffic-Optimal Scheduling with Regular Service Guarantees in Wireless Networks

Bin Li
Department of ECE
The Ohio State University
Columbus, OH 43210, USA
lib@ece.osu.edu

Ruogu Li
Department of ECE
The Ohio State University
Columbus, OH 43210, USA
lir@ece.osu.edu

Atilla Eryilmaz
Department of ECE
The Ohio State University
Columbus, OH 43210, USA
eryilmaz@ece.osu.edu

ABSTRACT

We consider the design of throughput-optimal scheduling policies in multi-hop wireless networks that also possess good mean delay performance and provide regular service for all links – critical metrics for real-time applications. To that end, we study a parametric class of maximum-weight type scheduling policies with parameter $\alpha \geq 0$, called Regular Service Guarantee (RSG) Algorithm, where each link weight consists of its own queue-length and a counter that tracks the time since the last service. This policy has been shown to be throughput-optimal and to provide more regular service as the parameter α increases, however at the cost of increasing mean delay.

This motivates us to investigate whether satisfactory service regularity and low mean-delay can be simultaneously achieved by the RSG Algorithm by carefully selecting its parameter α. To that end, we perform a novel Lyapunov-drift based analysis of the steady-state behavior of the stochastic network. Our analysis reveals that the RSG Algorithm can minimize the total mean queue-length to establish mean delay optimality under heavily-loaded conditions as long as α scales no faster than the order of $\frac{1}{\sqrt[5]{\epsilon}}$, where ϵ measures the closeness of the network load to the boundary of the capacity region. To the best of our knowledge, this is the first work that provides regular service to all links while also achieving heavy-traffic optimality in mean queue-lengths.

Categories and Subject Descriptors

C.2 [**Computer-Communication Networks**]: Miscellaneous

General Terms

Performance

Keywords

Wireless scheduling, service regularity, throughput, mean delay, heavy-traffic analysis

MobiHoc'13, July 29–August 1, 2013, Bangalore, India.
Copyright 2013 ACM 978-1-4503-2193-8/13/07 ...$15.00.

1. INTRODUCTION

Real-time applications, such as voice over IP or live multimedia streaming, are becoming increasingly popular as smart phones proliferate in wireless networks. To support real-time applications, network algorithm design should not only efficiently manage the interference among simultaneous transmissions, but also meet the requirements of Quality-of-Service (QoS) including delay, packet delivery ratio, and jitter. Such QoS requirements, in turn, depend on the higher-order statistics of the arrival and service process, which poses significant challenges for effective network algorithm design. For example, the well-known Maximum Weight Scheduling (MWS) Algorithm (e.g., [12], [11]) that prioritizes service of links with the largest backlog levels achieves maximum throughput, but does not provide any guarantees on the regularity of service that most real-time applications demand.

In recent years, there has been an increasing understanding on the algorithm design that target various aspects of QoS, especially packet delivery ratio requirement (e.g., [5], [6], [7]). However, there has been considerably less progress associated with the regularity of service, which is clearly important for real-time applications with stringent jitter requirements. Our work is motivated by the recent advances made in [8] that provides a promising approach for managing this critical QoS metric. In particular, [8] provides a throughput-optimal algorithm that is parametrized with a design variable $\alpha \geq 0$, which improves service regularity as α increases (see Section 3 for more details). Yet, increasing α also has an averse effect on the mean delay performance, which is also vital for most applications.

With this motivation, this paper focuses on the trade-off between the service regularity and the mean delay performance that this class of policies achieves. In particular, we are interested in identifying the range of values for α in which the mean delay performance guarantees can be provided, while the regularity characteristics are preserved. To that end, we build on the recently developed approach of using Lyapunov drifts for the steady-state analysis of queueing networks [2]. *The main result emanating from this analysis is the scaling law of α as the system gets more and more heavily loaded so that the algorithm is mean delay optimal among all feasible scheduling policies, and provides the best service regularity among this class of policies.* Specifically, we show that the heavy-traffic optimality is preserved as long as α scales[1] as $O\left(\frac{1}{\sqrt[5]{\epsilon}}\right)$, where ϵ is the heavy-traffic param-

---

[1] We say $a_n = O(b_n)$ if there exists a $c > 0$ such that $|a_n| \leq c|b_n|$ for two real-valued sequences $\{a_n\}$ and $\{b_n\}$.

eter characterizing the closeness of the arrival rate vector to the boundary of the capacity region.

Our analysis relates to the vast literature on heavy-traffic analysis of queueing networks (for example, [13], [3], [1], [14], [10], [9]), and in particular extends the Lyapunov drift-based approach in [2]. A critical step in most of these results is to establish a *state-space collapse* along a single dimension, and thus relate the multi-dimensional system operation to a *resource-pooled* single dimensional system. Our construction also follows such line of argument in broad strokes. However, the new dynamics of the considered class of algorithms require new Lyapunov functions and techniques in establishing their heavy-traffic optimality.

Note on Notation: We use bold and script font of a variable to denote a vector and a set. Also, let $|\mathcal{A}|$ to denote the cardinality of the set $\mathcal{A}$. We use $\text{Int}(\mathcal{A})$ to denote the set of interior points of the set $\mathcal{A}$. We use $\langle \mathbf{x}, \mathbf{y} \rangle, \mathbf{x} \cdot \mathbf{y}$ to denote the inner product and component-wise product of the vector $\mathbf{x}$ and $\mathbf{y}$, respectively. We use $\mathbf{x}^2$ and $\sqrt{\mathbf{x}}$ to denote the component-wise square and square root of the vector $\mathbf{x}$, respectively. We also use $\preceq, \succeq, \prec, \succ$ to denote component-wise comparison of two vectors, respectively. Let $\|\mathbf{x}\|_1$ and $\|\mathbf{x}\|$ denote the l_1 and l_2 norm of the vector $\mathbf{x}$, respectively.

2. SYSTEM MODEL

We consider a wireless network represented by a graph $\mathcal{G} = (\mathcal{N}, \mathcal{L})$, where $\mathcal{N}$ is the set of nodes and $\mathcal{L}$ is the set of links. A node represents a wireless transmitter or receiver, while a link represents a pair of transmitter and receiver that are within the transmission range of each other. We use $L \triangleq |\mathcal{L}|$ for convenience. We consider the *link-based conflict model*, where links conflicting with each other cannot be active at the same time. We call a set of links that can be active simultaneously as a *feasible schedule* and denote it as $\mathbf{S}[t] = (S_l[t])_{l \in \mathcal{L}}$, where $S_l[t] = 1$ if the link l is scheduled in time slot t and $S_l[t] = 0$, otherwise.

We capture the channel fading over link l in time slot t via a *non-negative-integer-valued* random variable $C_l[t]$, with $C_l[t] \leq C_{\max}, \forall l, t$, for some $C_{\max} < \infty$, which measures the maximum amount of service available in slot t, if scheduled. We use $\mathcal{J}$ to denote the set of global channel states (with finite cardinality). Let $J[t] \in \mathcal{J}$ denote the global state of the channel states of all links in time slot t. We assume that $\{J[t] \in \mathcal{J}\}_{t \geq 0}$ is an independently and identically distributed (i.i.d.) sequence of random variables with $\psi_j \triangleq \Pr\{J[t] = j\}$. Let $\mathcal{S}^j$ denote the set of feasible schedules when the channel is in state $j \in \mathcal{J}$. Then, the *capacity region* is defined as

$$\mathcal{R} \triangleq \sum_{j \in \mathcal{J}} \psi_j \cdot \text{CH}\{\mathcal{S}^j\}, \quad (1)$$

where $\text{CH}\{\mathcal{A}\}$ denotes the convex hull of the set $\mathcal{A}$.

We assume a per-link traffic model[2], where $A_l[t]$ denotes the number of packets arriving at link l in slot t that are independently distributed over links and i.i.d. over time with finite mean λ_l, and $A_l[t] \leq A_{\max}, \forall l, t$, for some $A_{\max} < \infty$. Accordingly, a queue is maintained for each link l with $Q_l[t]$ denoting its queue length at the beginning of time slot t. Let $U_l[t] = \max\{0, C_l[t]S_l[t] - Q_l[t] - A_l[t]\}$ be the unused

---

[2] We note that our algorithm can be extended to serve multi-hop traffic, but the notion of service regularity is clearer in the per-link context.

service for queue l in slot t. Then, the evolution of queue l is described as follows:

$$Q_l[t+1] = Q_l[t] + A_l[t] - C_l[t]S_l[t] + U_l[t], \quad \forall l. \quad (2)$$

We say that the queue l is *strongly stable* if it satisfies

$$\limsup_{T \to \infty} \frac{1}{T} \sum_{t=1}^{T} \mathbb{E}[Q_l[t]] < \infty.$$

We call an algorithm *throughput-optimal* if it makes all queues strongly stable for any arrival rate vector $\boldsymbol{\lambda} = (\lambda_l)_l$ that lies strictly within the capacity region.

Our goal is to design a *throughput-optimal* scheduling algorithm that also possesses the following desirable properties for satisfying the QoS requirements: (i) provides *regular services* in the sense that the *second-moment of the inter-service times* of the links is small; and (ii) achieves *low mean delay* in the sense that the total mean queue-lengths is small, especially in the regime where the system is *heavily-loaded* – when delay effects are most pronounced.

Next, we provide a *regular service scheduler* that possesses throughput-optimality and regular service guarantees, and then investigate its mean-delay performance under the heavy-traffic regime.

3. REGULAR SERVICE SCHEDULER

One of our goals is to provide regular services for each link, which is related to the second moment of the inter-service times. To characterize the inter-service time, we introduce a counter T_l for each link l, namely Time-Since-Last-Service (TSLS), to keep track of the time since link l was last served. In particular, each T_l increases by 1 in each time slot when link l has zero transmission rate, either because it is not scheduled, or because its channel is unavailable, i.e., $C_l[t] = 0$, and drops to 0, otherwise. More precisely, the evolution of T_l is described as follows:

$$T_l[t+1] = \begin{cases} 0 & \text{if } S_l[t]C_l[t] > 0; \\ T_l[t] + 1 & \text{if } S_l[t]C_l[t] = 0. \end{cases} \quad (3)$$

Thus, the TSLS records the link "age" since the last time it received service, and is closely related to the inter-service time. Indeed, the authors in [8] showed that the normalized second moment of the inter-service times of each link is proportional to the mean value of its TSLS for any stabilizing policy. Thus, the TSLS has a direct impact on service regularity: the smaller the mean TSLS value, the more regular the service.

This connection motivates the following maximum-weight type algorithm that uses a combination of queue-lengths and TSLS values as its weights, extending the algorithm in [8] to multi-hop fading networks:

Regular Service Guarantee (RSG) Algorithm:

In each time slot t, select the schedule as

$$\mathbf{S}^*[t] \in \arg\max_{\mathbf{S} \in \mathcal{S}} \sum_{l=1}^{L} (Q_l[t] + \alpha T_l[t])C_l[t]S_l, \quad (4)$$

where $\alpha > 0$ is a design parameter.

Note that the RSG Algorithm coincides with the MWS Algorithm when $\alpha = 0$. Yet, the true significance of the RSG algorithm is observed for large α, since as α increases, the RSG Algorithm prioritizes the schedule with the larger TSLS, hence providing more regular services for each link. We can show that the RSG Algorithm in the multi-hop setup not only achieves throughput optimality but also provides regular service guarantees, which extends the results in [8].

Yet, large values of α may also deteriorate the mean delay performance. We demonstrate this tradeoff in a single-hop non-fading network with 4 links, where the number of packets arriving at each link follows a Bernoulli distribution with the arrival rate of 0.225. Figure 1 shows the mean delay and service regularity performance of the RSG Algorithm with varying α.

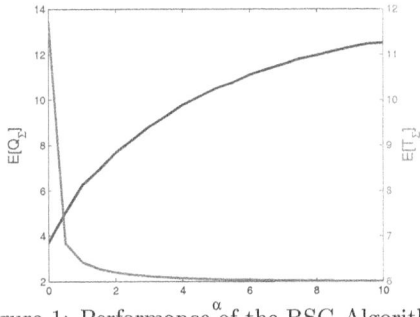

Figure 1: Performance of the RSG Algorithm

Figure 1 reveals that the improved regularity of the RSG Algorithm with increasing α comes at the cost of larger mean delays. We can show that the mean of the total TSLS value is minimized as α goes to ∞ (see [8]). On the other hand, it is known (e.g. [10, 2]) that the mean queue-lengths are minimized under heavily-loaded conditions (cf. Section 4 for more detail) when $\alpha = 0$. In view of the tradeoff observed in the above figure, our objective is to understand whether *both the regularity and the mean-delay optimality characteristics of the RSG Algorithm can be preserved, especially under heavily-loaded conditions, by carefully selecting α.*

In the next section, we answer this question in the affirmative by explicitly characterizing how α should scale with respect to the traffic load in order to achieve the heavy-traffic optimality while also preserving the regularity performance of the RSG Algorithm.

4. HEAVY-TRAFFIC OPTIMALITY RESULT

In this section, we present our main result for the RSG Algorithm in terms of its mean delay optimality under the heavy-traffic limit, where the arrival rate vector approaches the boundary of the capacity region.

We first note that the capacity region $\mathcal{R}$ is a polyhedron due to the discreteness and finiteness of the service rate choices, and thus has a finite number of faces. We consider the exogenous arrival vector process $\{\mathbf{A}^{(\epsilon)}[t]\}_{t\geq 0}$ with mean vector $\boldsymbol{\lambda}^{(\epsilon)} \in \text{Int}(\mathcal{R})$, where ϵ measures the Euclidean distance of $\boldsymbol{\lambda}^{(\epsilon)}$ to the boundary of $\mathcal{R}$ (see Figure 2).

In heavy-traffic analysis, we study the system performance as ϵ decreases to zero, i.e., as the arrival rate vector approaches $\boldsymbol{\lambda}^{(0)}$ belonging to the *relative interior* of a face, referred to as the *dominant hyperplane* $\mathcal{H}^{(\mathbf{c})}$. We denote $\mathcal{H}^{(\mathbf{c})} \triangleq \{\mathbf{r} \in \mathbb{R}^L : \langle \mathbf{r}, \mathbf{c} \rangle = b\}$, where $b \in \mathbb{R}$, and $\mathbf{c} \in \mathbb{R}^L$ is the normal vector of the hyperplane $\mathcal{H}^{(\mathbf{c})}$ satisfying $\|\mathbf{c}\| = 1$ and $\mathbf{c} \succeq \mathbf{0}$.

We are interested in understanding the steady-state queue-length values with vanishing ϵ. To that end, we first provide

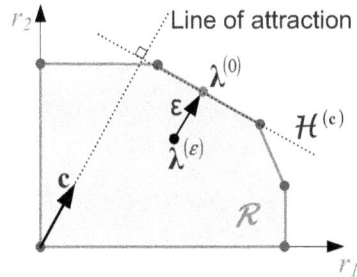

Figure 2: Geometric structure of capacity region

a generic lower bound for all feasible schedulers by constructing a hypothetical single-server queue with the arrival process $\langle \mathbf{c}, \mathbf{A}^{(\epsilon)}[t] \rangle$, and the i.i.d service process $\beta[t]$ with the probability distribution

$$\Pr\{\beta[t] = b_j\} = \psi_j, \qquad \text{for each } j \in \mathcal{J}, \qquad (5)$$

where $b_j \triangleq \max_{\mathbf{s} \in \mathcal{S}^{(j)}} \langle \mathbf{c}, \mathbf{s} \rangle$ is the maximum $\mathbf{c}$-weighted service rate achievable in channel state $j \in \mathcal{J}$. By the construction of capacity region $\mathcal{R}$, we have $\mathbb{E}[\beta[t]] = b$. Also, it is easy to show that the constructed single-server queue-length $\{\Phi[t]\}_{t\geq 0}$ is stochastically smaller than the queue-length process $\{\langle \mathbf{c}, \mathbf{Q}^{(\epsilon)}[t] \rangle\}_{t\geq 0}$ under any feasible scheduling policy. Hence, by using Lemma 4 in [2], we have the following lower bound on the expected limiting queue-length vector under any feasible scheduling policy.

PROPOSITION 1. *Let $\overline{\mathbf{Q}}^{(\epsilon)}$ be a random vector with the same distribution as the steady-state distribution of the queue length processes under any feasible scheduling policy. Consider the heavy-traffic limit $\epsilon \downarrow 0$, suppose that the variance vector $\left(\boldsymbol{\sigma}^{(\epsilon)}\right)^2$ of the arrival process $\{\mathbf{A}^{(\epsilon)}[t]\}_{t\geq 0}$ converges to a constant vector $\boldsymbol{\sigma}^2$. Then,*

$$\lim_{\epsilon \downarrow 0} \epsilon \mathbb{E}\left[\langle \mathbf{c}, \overline{\mathbf{Q}}^{(\epsilon)} \rangle\right] \geq \frac{\zeta}{2}, \qquad (6)$$

where $\zeta \triangleq \langle \mathbf{c}^2, \boldsymbol{\sigma}^2 \rangle + \text{Var}(\beta)$.

This fundamental lower bound of all feasible scheduling policies motivates the following definition of *heavy-traffic optimality* of a scheduler.

DEFINITION 1. *(Heavy-Traffic Optimality) A scheduler is called heavy-traffic optimal, if its limiting queue length vector $\overline{\mathbf{Q}}^{(\epsilon)}$ satisfies*

$$\lim_{\epsilon \downarrow 0} \epsilon \mathbb{E}\left[\langle \mathbf{c}, \overline{\mathbf{Q}}^{(\epsilon)} \rangle\right] \leq \frac{\zeta}{2}, \qquad (7)$$

where ζ is defined in Proposition 1.

It is well-known that the MWS Algorithm, which corresponds to the RSG Algorithm with $\alpha = 0$, is heavy-traffic optimal (e.g., [10, 2]). This is shown by first establishing a *state-space collapse*, i.e., the deviations of queue lengths from the direction $\mathbf{c}$ are bounded, independent of heavy-traffic parameter ϵ. Since the lower bound of mean queue length is of order of $\frac{1}{\epsilon}$, the deviations from the direction $\mathbf{c}$ are negligible compared to the large queue length for a sufficiently small ϵ, and thus the queue lengths concentrate along the normal vector **c**. Because of this, we also call the normal vector **c** the *line of attraction*.

However, as discussed in Section 3, we are interested in large values of α to provide satisfactory service regularity. Yet, it is unknown whether the RSG Algorithm can remain heavy-traffic optimal when α is non-zero, since larger values of α leads to higher mean queue-lengths (cf. Figure 1). Also, the state-space collapse result is not applicable since the deviations from the line of attraction depend on α. This raises the question of how $\alpha(\epsilon)$ should scale with ϵ in order to achieve heavy-traffic optimality while allowing $\alpha(\epsilon)$ to take large values (providing more regular services). We answer this interesting and challenging question by providing the following main result, proved in Section 6.

PROPOSITION 2. *Let $\overline{\mathbf{Q}}^{(\epsilon)}$ be a random vector with the same distribution as the steady-state distribution of the queue length processes under the RSG Algorithm. Consider the heavy-traffic limit $\epsilon \downarrow 0$, suppose that the variance vector $\left(\boldsymbol{\sigma}^{(\epsilon)}\right)^2$ of the arrival process $\{\mathbf{A}^{(\epsilon)}[t]\}_{t \geq 0}$ converges to a constant vector $\boldsymbol{\sigma}^2$. Suppose the channel fading satisfies the mild assumption[3] $\Pr\{C_l[t] = 0\} > 0$, for all $l \in \mathcal{L}$. Then,*

$$\epsilon \mathbb{E}\left[\langle \mathbf{c}, \overline{\mathbf{Q}}^{(\epsilon)} \rangle\right] \leq \frac{\zeta^{(\epsilon)}}{2} + \overline{B}^{(\epsilon)}, \qquad (8)$$

where $\zeta^{(\epsilon)} \triangleq \langle \mathbf{c}^2, \left(\boldsymbol{\sigma}^{(\epsilon)}\right)^2 \rangle + Var(\beta) + \epsilon^2$ and $\overline{B}^{(\epsilon)}$ is defined in (22).

Further, if $\alpha(\epsilon) = O(\frac{1}{\sqrt[5]{\epsilon}})$, then $\lim_{\epsilon \downarrow 0} \overline{B}^{(\epsilon)} = 0$ and thus the RSG Algorithm is heavy-traffic optimal.

This result is interesting in that it provides an explicit scaling regime in which the design parameter $\alpha(\epsilon)$ can be increased to utilize the service regulating nature of the RSG Algorithm without sacrificing the heavy-traffic optimality. Intuitively, if $\alpha(\epsilon)$ scales slowly as ϵ vanishes, each link weight is dominated by its own queue length in the heavy-traffic regime and thus the heavy-traffic optimality may be maintained; otherwise, the heavy-traffic optimality result may not hold, as will be demonstrated in the next section.

5. SIMULATION RESULTS

In this section, we provide simulation results to compare the mean delay and service regularity performance of the RSG Algorithm with the MWS Algorithm. In the simulation, we consider a single-hop non-fading network with 4 links. Its capacity region is $\mathcal{R} = \{\boldsymbol{\lambda} = (\lambda_l)_{l=1}^4 \succeq \mathbf{0} : \sum_{l=1}^4 \lambda_l < 1\}$. We use arrival process where the number of arrivals in each slot follows a Bernoulli distribution. We consider the symmetric case $\boldsymbol{\lambda}^{(\epsilon)} = (1 - \frac{\epsilon}{2}) \times [\frac{1}{4}, \frac{1}{4}, \frac{1}{4}, \frac{1}{4}]$, and the asymmetric case $\boldsymbol{\lambda}^{(\epsilon)} = [\frac{1}{2}, \frac{1}{4}, \frac{1}{8}, \frac{1}{16}] + (1 - \frac{\epsilon}{32}) \times [\frac{1}{64}, \frac{1}{64}, \frac{1}{64}, \frac{1}{64}]$.

From Figure 3a and 4a, we can observe that the RSG Algorithm with both $\alpha = 1$ and $\alpha = \frac{1}{\sqrt[5]{\epsilon}}$, and the MWS Algorithm converge to the theoretical lower bound and thus is heavy-traffic optimal, which confirms our theoretical results. Yet, the RSG Algorithm with $\alpha = \frac{1}{\epsilon}$ has large mean

---

[3] We note that our result holds in single-hop network topologies without this assumption, and its extension to more general settings is part of our future work.

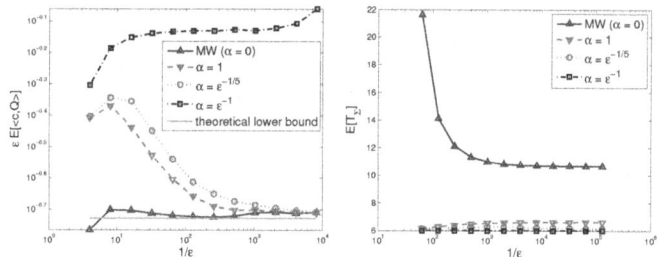

(a) Mean queue length   (b) Service regularity

Figure 3: Symmetric arrivals in a single-hop network

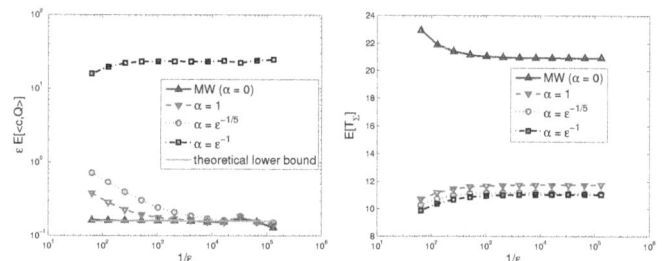

(a) Mean queue length   (b) Service regularity

Figure 4: Asymmetric arrivals in a single-hop network

queue length, which does not match with the theoretical lower bound and thus is not heavy-traffic optimal. Hence, α should scale as slowly as $O(\frac{1}{\sqrt[5]{\epsilon}})$ to preserve heavy-traffic optimality.

From Figure 3b and 4b, we can see that the RSG Algorithm with even $\alpha = 1$ significantly outperforms the MWS Algorithm in terms of service regularity. More remarkably, the RSG Algorithm with $\alpha = \frac{1}{\sqrt[5]{\epsilon}}$ can achieve the lower bound (see [8]) achieved by the round robin policy under symmetric arrivals.

6. DETAILED HEAVY-TRAFFIC ANALYSIS

In this section, we prove Proposition 2 by using the analytical approach in [2], which includes two parts: (i) showing state-space collapse; (ii) using the state-space collapse result to obtain an upper bound on the mean queue lengths. Yet, it is worth noting that the strong coupling between queue length processes and TSLS counters in the RSG Algorithm poses significant challenges in heavy-traffic analysis. In particular, it requires new Lyapunov functions and a novel technique to establish heavy-traffic optimality of the RSG Algorithm.

6.1 State-Space Collapse

We have mentioned in Section 3 that the RSG Algorithm is throughput-optimal, i.e., it stabilizes all queues for any arrival rate vector that are strictly within the capacity region. Let $\{\mathbf{Q}^{(\epsilon)}\}_{t \geq 0}$ and $\{\mathbf{T}^{(\epsilon)}\}_{t \geq 0}$ be queue-length processes and TSLS counters under the RSG Algorithm, respectively. Also, we use $\overline{\mathbf{Q}}^{(\epsilon)}$ and $\overline{\mathbf{T}}^{(\epsilon)}$ to denote their limiting queue-length random vector and limiting TSLS random vector, respectively. Then, by the continuous mapping theorem, we have

$$\mathbf{Q}_\parallel^{(\epsilon)} \Rightarrow \overline{\mathbf{Q}}_\parallel^{(\epsilon)}, \qquad \mathbf{Q}_\perp^{(\epsilon)} \Rightarrow \overline{\mathbf{Q}}_\perp^{(\epsilon)}; \qquad (9)$$

$$\mathbf{T}_\parallel^{(\epsilon)} \Rightarrow \overline{\mathbf{T}}_\parallel^{(\epsilon)}, \qquad \mathbf{T}_\perp^{(\epsilon)} \Rightarrow \overline{\mathbf{T}}_\perp^{(\epsilon)}, \qquad (10)$$

where $\Rightarrow$ denotes convergence in distribution, and we define the projection and the perpendicular vector of any given L-dimensional vector $\mathbf{I}$ with respect to the normal vector $\mathbf{c}$ as:

$$\mathbf{I}_\parallel \triangleq \langle \mathbf{c}, \mathbf{I} \rangle \mathbf{c}, \quad \mathbf{I}_\perp \triangleq \mathbf{I} - \mathbf{I}_\parallel.$$

Next, we will show that under the RSG Algorithm, the second moment of $\|\overline{\mathbf{Q}}_\perp^{(\epsilon)}\|$ is bounded, dependent on $\alpha(\epsilon)$, while the second moment of $\|\overline{\mathbf{T}}^{(\epsilon)}\|$ is bounded by some constant independent of ϵ.

PROPOSITION 3. *If* $\Pr\{C_l[t] = 0\} > 0, \forall l \in \mathcal{L}$, *then, under RSG Algorithm, there exists a constant* $N_{T,2}$, *independent of* ϵ, *such that*

$$\mathbb{E}[\|\overline{\mathbf{Q}}_\perp^{(\epsilon)}\|^2] = O\left((\alpha(\epsilon))^4 (\log \alpha(\epsilon))^2\right), \qquad (11)$$

$$\mathbb{E}[\|\overline{\mathbf{T}}^{(\epsilon)}\|^2] \leq N_{T,2}. \qquad (12)$$

We prove Proposition 3 by first studying the drift of the Lyapunov function

$$V_\perp(\mathbf{Q}^{(\epsilon)}, \mathbf{T}^{(\epsilon)}) \triangleq \|(\mathbf{Q}_\perp^{(\epsilon)}, \sqrt{2\alpha(\epsilon)C_{\max}\mathbf{T}^{(\epsilon)}})\|,$$

and show that when $V_\perp(\mathbf{Q}^{(\epsilon)}, \mathbf{T}^{(\epsilon)})$ is sufficiently large, it has a strictly negative drift independent of ϵ, which is characterized in the following key lemma.

LEMMA 1. *Under the RSG Algorithm, there exist positive constants* d *and* ς, *independent of* ϵ, *such that whenever* $V_\perp(\mathbf{Q}^{(\epsilon)}[t], \mathbf{T}^{(\epsilon)}[t]) > d$, *we have*

$$\mathbb{E}[\Delta V_\perp(\mathbf{Q}^{(\epsilon)}[t], \mathbf{T}^{(\epsilon)}[t]) | \mathbf{Q}^{(\epsilon)}[t], \mathbf{T}^{(\epsilon)}[t]] < -\varsigma, \qquad (13)$$

where $\Delta V_\perp(\mathbf{Q}^{(\epsilon)}[t], \mathbf{T}^{(\epsilon)}[t]) \triangleq V_\perp(\mathbf{Q}^{(\epsilon)}[t+1], \mathbf{T}^{(\epsilon)}[t+1]) - V_\perp(\mathbf{Q}^{(\epsilon)}[t], \mathbf{T}^{(\epsilon)}[t])$.

The proof of Lemma 1 is available in Appendix A.

Note that the TSLS counters have bounded increment but unbounded decrement, since they can at almost increase by 1 and drop to 0 once their corresponding links are scheduled. Due to this characteristic of TLSL, the absolute value of the drift $\Delta V_\perp(\mathbf{Q}^{(\epsilon)}, \mathbf{T}^{(\epsilon)})$ has neither an upper bound nor an exponential tail given the current system state $(\mathbf{Q}^{(\epsilon)}, \mathbf{T}^{(\epsilon)})$. Thus, we cannot directly apply Theorem 2.3 in [4], which requires either boundedness or the exponential tail of the Lyapunov drift to establish the existence of the second moment of the stochastic process. Indeed, for a Markov Chain with a strictly negative drift of Lyapunov function, if its Lyapunov drift has bounded increment but unbounded decrement, its second moment may not exist.

Counterexample: Consider a Markov Chain $\{X[t]\}_{t \geq 0}$ with the following transition probability:

$$P_{j,j+1} = \begin{cases} 1 & \text{if } j = 0; \\ \frac{1}{2} & \text{if } j = 1; \\ \frac{j-1}{j+1} & \text{if } j \geq 2. \end{cases} \quad P_{j,0} = \begin{cases} \frac{1}{2} & \text{if } j = 1; \\ \frac{2}{j+1} & \text{if } j \geq 1. \end{cases}$$

The state transition diagram of Markov Chain $\{X[t]\}_{t \geq 0}$ is shown in Figure 5. Consider a linear Lyapunov function

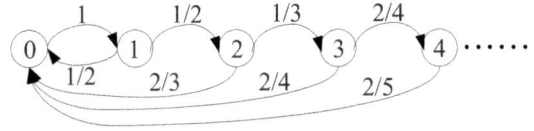

Figure 5: Markov Chain $\{X[t]\}_{t \geq 0}$

X. For any $X \geq 2$, we have

$$\mathbb{E}[X[t+1] - X[t] | X[t] = X] = \frac{X-1}{X+1} - \frac{2X}{X+1} = -1.$$

Thus, the Lyapunov function X has a strictly negative drift when $X \geq 2$ and hence the steady-state distribution of the Markov Chain exists. Recall that its drift increases at almost by 1, but has unbounded decrement, which has similar dynamics with the system under the RSG Algorithm.

Next, we will show that even the first moment of this Markov Chain does not exist, let alone its second moment. Let $\overline{X}$ be the limiting random variable of the Markov Chain and $\pi_j \triangleq \Pr\{\overline{X} = j\}$. According to the global balance equations, we can easily calculate

$$\pi_1 = \pi_0 = \frac{1}{3}, \quad \pi_j = \frac{1}{3j(j-1)}. \qquad (14)$$

Thus, we have

$$\mathbb{E}[\overline{X}] = \sum_{j=1}^\infty j\pi_j = \frac{1}{3} + \sum_{j=2}^\infty \frac{1}{3(j-1)} = \infty.$$

Fortunately, we can establish the boundedness of the second moment of $\|\overline{\mathbf{Q}}_\perp^{(\epsilon)}\|$ under the RSG Algorithm by exploiting its unique dynamics under a mild assumption that $\Pr\{C_l[t] = 0\} > 0, \forall l \in \mathcal{L}$, which leads to the following lemma that all TSLS counters have an exponential tail independent of ϵ.

LEMMA 2. *If* $p_l \triangleq \Pr\{C_l[t] = 0\} > 0, \forall l \in \mathcal{L}$, *then, under the RSG Algorithm, there exists a* $\vartheta \in (0,1)$, *independent of* ϵ, *such that*

$$\Pr\{\overline{T}_l^{(\epsilon)} = m\} \leq 2\vartheta^m, \quad \forall l \in \mathcal{L}. \qquad (15)$$

The proof of Lemma 2 is available in Appendix B.

Remark: We can also show that all TLSL counters still have an exponential tail independent of ϵ in non-fading single-hop network topologies. The extension to the more general setup is left for future search.

Lemma 2 directly implies (12). The rest of proof mainly builds on the analytical technique in [4], while it requires carefully partitioning the space $(\mathbf{Q}_\perp^{(\epsilon)}, \mathbf{T}^{(\epsilon)})$. The detailed outline can be found in Appendix C.

6.2 Proof of Main Result

We first give an upper bound on $\mathbb{E}[\langle \mathbf{c}, \overline{\mathbf{Q}}^{(\epsilon)} \rangle]$ by using the methodology of "setting the drift of a Lyapunov function equal to zero". We will omit the superscript ϵ associated with the queue lengths and TSLS counters for brevity in the rest of proof. To derive an upper bound, we need the following fundamental identity (see Lemma 8 in [2]):

$$\frac{\mathbb{E}\left[\langle \mathbf{c}, \mathbf{U}(\overline{\mathbf{Q}}, \overline{\mathbf{T}}, J)\rangle^2\right]}{2} + \frac{\mathbb{E}\left[\langle \mathbf{c}, \mathbf{A} - \mathbf{S}^*(\overline{\mathbf{Q}}, \overline{\mathbf{T}}, J)\rangle^2\right]}{2}$$

$$+ \mathbb{E}\left[\langle \mathbf{c}, \overline{\mathbf{Q}} + \mathbf{A} - \mathbf{S}^*(\overline{\mathbf{Q}}, \overline{\mathbf{T}}, J)\rangle\langle \mathbf{c}, \mathbf{U}(\overline{\mathbf{Q}}, \overline{\mathbf{T}}, J)\rangle\right]$$

$$= \mathbb{E}\left[\langle \mathbf{c}, \overline{\mathbf{Q}}\rangle\langle \mathbf{c}, \mathbf{S}^*(\overline{\mathbf{Q}}, \overline{\mathbf{T}}, J) - \mathbf{A}\rangle\right], \qquad (16)$$

which is derived through setting $\mathbb{E}[\Delta W_\parallel(\mathbf{Q}, \mathbf{T})] = 0$.

Next, we give upper bounds for each individual term in the left hand side of (16) and a lower bound for the right hand side of (16). Due to the space limitations, we omit the details and directly give results with some simple explanations.

By setting the mean drift of $\langle \mathbf{c}, \mathbf{Q}\rangle$ equal to zero and using the fact that $U_l \le C_{\max}$ for all l, we have

$$\frac{1}{2}\mathbb{E}\left[\langle \mathbf{c}, \mathbf{U}(\overline{\mathbf{Q}}, \overline{\mathbf{T}}, J)\rangle^2\right] \le \frac{\epsilon}{2}\langle \mathbf{c}, C_{\max}\mathbf{1}\rangle. \qquad (17)$$

This means that there is almost no unused services under heavy-traffic conditions.

By observing that the RSG Algorithm selects the schedule $\mathbf{S}$ which maximizes $\langle \mathbf{c}, \mathbf{S}\rangle$ with high probability, we can show

$$\mathbb{E}\left[\langle \mathbf{c}, \mathbf{A} - \mathbf{S}^*(\overline{\mathbf{Q}}, \overline{\mathbf{T}}, J)\rangle^2\right]$$

$$\le \zeta^{(\epsilon)} + 2b\sum_{j=0}^{M}\frac{\epsilon}{\gamma_j}b_j + \sum_{j=0}^{M}\frac{\epsilon}{\gamma_j}\left((b_j)^2 + \langle \mathbf{c}, C_{\max}\mathbf{1}\rangle^2\right), \quad (18)$$

where we recall that $\zeta^{(\epsilon)}$ is defined in Proposition 2, and

$$\pi_j \triangleq \Pr\left\{\langle \mathbf{c}, \mathbf{S}^*(\overline{\mathbf{Q}}, \overline{\mathbf{T}}, J)\rangle = b_j \,\middle|\, J = j\right\},$$

$$\gamma_j \triangleq \min\left\{b_j - \langle \mathbf{c}, \mathbf{r}\rangle : \text{for all } \mathbf{r} \in \mathcal{S}^j \setminus \mathcal{H}^{(\mathbf{c})}\right\}.$$

Inequality (18) indicates that the second moment of $\mathbf{c}$-weighted difference between arrivals and services is dominated by the $\mathbf{c}^2$-weighted variance of the arrival process and the variance of the channel fading process in the heavy-traffic limit.

In addition, by using similar arguments as in the proof for Proposition 4 in [2], we have

$$\mathbb{E}\left[\langle \mathbf{c}, \overline{\mathbf{Q}} + \mathbf{A} - \mathbf{S}^*(\overline{\mathbf{Q}}, \overline{\mathbf{T}}, J)\rangle\langle \mathbf{c}, \mathbf{U}(\overline{\mathbf{Q}}, \overline{\mathbf{T}}, J)\rangle\right]$$

$$\le \sqrt{\epsilon\mathbb{E}[\|\overline{\mathbf{Q}}_\perp\|^2]\frac{C_{\max}}{c_{\min}}}, \qquad (19)$$

where $c_{\min} \triangleq \min_{m \in \{l : c_l > 0\}} c_m$.

Finally, by using the definition of the RSG Algorithm and Proposition 3, we have

$$\mathbb{E}\left[\langle \mathbf{c}, \overline{\mathbf{Q}}\rangle\langle \mathbf{c}, \mathbf{S}^*(\overline{\mathbf{Q}}, \overline{\mathbf{T}}, J) - \mathbf{A}\rangle\right]$$

$$\ge \epsilon\mathbb{E}\left[\|\overline{\mathbf{Q}}_\parallel\|\right] - \cot(\theta)\sqrt{2\left(\mathbb{E}[\|\overline{\mathbf{Q}}_\perp\|^2] + (\alpha(\epsilon))^2 N_{\mathbf{T},2}\right)\epsilon}$$

$$\times \sqrt{\sum_{j=0}^{M}\frac{1}{\gamma_j}\left((b_j)^2 + \langle \mathbf{c}, C_{\max}\mathbf{1}\rangle^2\right)}, \qquad (20)$$

where $\theta \in (0, \frac{\pi}{2}]$ is an angle such that $\langle \mathbf{c}, \mathbf{R}^*(\mathbf{Q}, \mathbf{T})\rangle = b$, for all $\mathbf{Q}$ and $\mathbf{T}$ satisfying $\frac{\|(\mathbf{Q}+\alpha\mathbf{T})_\parallel\|}{\|\mathbf{Q}+\alpha\mathbf{T}\|} \ge \cos(\theta)$, and $\mathbf{R}^*(\mathbf{Q}, \mathbf{T}) \triangleq \mathbb{E}[\mathbf{S}^*(\mathbf{Q}, \mathbf{T}, J) | \mathbf{Q}, \mathbf{T}]$.

By substituting bounds (17), (18), (19) and (20) into identity (16), we have

$$\epsilon\mathbb{E}\left[\|\overline{\mathbf{Q}}_\parallel\|\right] \le \frac{\zeta^{(\epsilon)}}{2} + \overline{B}^{(\epsilon)}, \qquad (21)$$

where

$$\overline{B}^{(\epsilon)} \triangleq \frac{\epsilon}{2}\langle \mathbf{c}, C_{\max}\mathbf{1}\rangle + \sqrt{\epsilon\mathbb{E}[\|\overline{\mathbf{Q}}_\perp\|^2]\frac{C_{\max}}{c_{\min}}}$$

$$+ \frac{1}{2}\sum_{j=0}^{M}\frac{\epsilon}{\gamma_j}\left((b_j)^2 + \langle \mathbf{c}, C_{\max}\mathbf{1}\rangle^2\right)$$

$$+ b\sum_{j=0}^{M}\frac{\epsilon}{\gamma_j}b_j + \cot(\theta)\sqrt{2\left(\mathbb{E}[\|\overline{\mathbf{Q}}_\perp\|^2] + (\alpha(\epsilon))^2 N_{\mathbf{T},2}\right)\epsilon}$$

$$\times \sqrt{\sum_{j=0}^{M}\frac{1}{\gamma_j}\left((b_j)^2 + \langle \mathbf{c}, C_{\max}\mathbf{1}\rangle^2\right)}. \qquad (22)$$

Thus, if $\lim_{\epsilon\downarrow 0}\overline{B}^{(\epsilon)} = 0$, then RSG Algorithm is heavy-traffic optimal. Noting that $N_{\mathbf{T},2}$ is independent of ϵ, to satisfy $\lim_{\epsilon\downarrow 0}\overline{B}^{(\epsilon)} = 0$, it is enough to have

$$\lim_{\epsilon\downarrow 0}\epsilon\mathbb{E}[\|\overline{\mathbf{Q}}_\perp\|^2] = 0 \text{ and } \lim_{\epsilon\downarrow 0}\epsilon\left(\alpha(\epsilon)\right)^2 = 0. \qquad (23)$$

By using Proposition 3, it is easy to see that $\alpha(\epsilon) = O(\frac{1}{\sqrt[5]{\epsilon}})$ meets the above requirements.

7. CONCLUSION

In this paper, we studied the heavy-traffic behavior of the recently proposed maximum-weight type scheduling algorithm, called Regular Service Guarantee (RSG) Algorithm, that not only achieves throughput optimality but also provides regular services through the control parameter $\alpha \ge 0$. We showed that the RSG Algorithm is heavy-traffic optimal as long as $\alpha = O(\frac{1}{\sqrt[5]{\epsilon}})$, where ϵ is the heavy-traffic parameter characterizing the closeness of the arrival rate vector to the boundary of the capacity region. Noting that the service regularity improves with increasing α, our result reveals that the RSG Algorithm with a carefully selected parameter α can achieve the best service regularity performance among the class of the RSG Algorithms without sacrificing the mean delay optimality under heavy-traffic conditions.

8. ACKNOWLEDGMENTS

This work is supported by the NSF grants: CAREER-CNS-0953515 and CCF-0916664. Also, the work of A. Eryilmaz was in part supported by the QNRF grant number NPRP 09-1168-2-455.

9. REFERENCES

[1] M. Bramson. State space collapse with application to heavy traffic limits for multiclass queueing networks. *Queueing Systems*, 30(1):89–140, 1998.

[2] A. Eryilmaz and R. Srikant. Asymptotically tight steady-state queue length bounds implied by drift conditions. *Queueing Systems*, 72:311–359, 2012.

[3] G. Foschini and J. Salz. A basic dynamic routing problem and diffusion. *IEEE Transactions on Communications*, 26(3):320–327, 1978.

[4] B. Hajek. Hitting-time and occupation-time bounds implied by drift analysis with applications. *Advances in Applied Probability*, 14(3):502–525, 1982.

[5] I. Hou, V. Borkar, and P. R. Kumar. A theory of QoS for wireless. In *Proceedings of IEEE INFOCOM*, Rio de Janeiro, Brazil, April 2009.

[6] I. Hou and P. R. Kumar. Scheduling heterogeneous real-time traffic over fading wireless channels. In *Proceedings of IEEE INFOCOM*, San Diego, CA, March 2010.

[7] J. Jaramillo, R. Srikant, and L. Ying. Scheduling for optimal rate allocation in ad hoc networks with heterogeneous delay constraints. *IEEE Journal on Selected Areas in Communications*, 29(5):979–987, 2011.

[8] R. Li, B. Li, and A. Eryilmaz. Throughput-optimal scheduling with regulated inter-service times. In *Proceedings of IEEE INFOCOM*, Turin, Italy, April 2013.

[9] D. Shah and D. Wischik. Switched networks with maximum weight policies: Fluid approximation and multiplicative state space collapse. *Annals of Applied Probability*, 22(1):70–127, 2012.

[10] A. Stolyar. Maxweight scheduling in a generalized switch: State space collapse and workload minimization in heavy traffic. *The Annals of Applied Probability*, 14(1):1–53, 2004.

[11] L. Tassiulas. Scheduling and performance limits of networks with constantly changing topology. *IEEE Transactions on Information Theory*, 43(3):1067–1073, 1997.

[12] L. Tassiulas and A. Ephremides. Stability properties of constrained queueing systems and scheduling policies for maximum throughput in multihop radio networks. *IEEE Transactions on Automatic Control*, 36(12):1936–1948, 1992.

[13] W. Whitt. Weak convergence theorems for priority queues: Preemptive-resume discipline. *Journal of Applied Probability*, 8(1):74–94, 1971.

[14] R. Williams. Diffusion approximations for open multiclass queueing networks: sufficient conditions involving state space collapse. *Queueing Systems*, 30(1):27–88, 1998.

APPENDIX

A. PROOF OF LEMMA 1

We assume $\boldsymbol{\lambda}^{(\epsilon)} \succ \mathbf{0}$. Indeed, if $\lambda_l^{(\epsilon)} = 0$ for some link l, then no arrivals occur in the link l. Thus, we do not need to consider such links. Since normal vector $\mathbf{c} \succeq \mathbf{0}$, we have $\boldsymbol{\lambda}^{(0)} \succ 0$. In addition, since $\boldsymbol{\lambda}^{(0)}$ is a relative interior point of dominant hyperplane $\mathcal{H}^{(\mathbf{c})}$, there exists a small enough $\delta > 0$ such that

$$\mathcal{B}_\delta \triangleq \mathcal{H}^{(\mathbf{c})} \bigcap \left\{ \mathbf{r} \succ \mathbf{0} : \|\mathbf{r} - \boldsymbol{\lambda}^{(0)}\| \leq \delta \right\}, \quad (24)$$

representing the set of vectors on the hyperplane $\mathcal{H}^{(\mathbf{c})}$ that are within δ distance from $\boldsymbol{\lambda}^{(0)}$, lies strictly within the face $\mathcal{F}^{(\mathbf{c})} \triangleq \mathcal{H}^{(\mathbf{c})} \bigcap \mathcal{R}$.

In the rest of proof, we will omit ϵ associated with the queue length processes, the TSLS counters and parameter $\alpha(\epsilon)$ for brevity. Noting the difficulty to directly study the drift of Lyapunov function $V_\perp(\mathbf{Q}, \mathbf{T})$, we relate it with the drift of other proper Lyapunov functions, which is characterized in the following lemma.

LEMMA 3. *Define the following Lyapunov functions:*

$$W(\mathbf{Q}, \mathbf{T}) \triangleq \|(\mathbf{Q}, \sqrt{2\alpha C_{\max} \mathbf{T}})\|^2, \quad (25)$$

$$W_\parallel(\mathbf{Q}, \mathbf{T}) \triangleq \|\mathbf{Q}_\parallel\|^2. \quad (26)$$

Then, given $\mathbf{Q}[t] = \mathbf{Q}$ *and* $\mathbf{T}[t] = \mathbf{T}$*, their one-step drifts denoted by:*

$$\Delta W(\mathbf{Q}, \mathbf{T}) \triangleq [W(\mathbf{Q}[t+1], \mathbf{T}[t+1]) - W(\mathbf{Q}[t], \mathbf{T}[t])],$$

$$\Delta W_\parallel(\mathbf{Q}, \mathbf{T}) \triangleq [W_\parallel(\mathbf{Q}[t+1], \mathbf{T}[t+1]) - W_\parallel(\mathbf{Q}[t], \mathbf{T}[t])],$$

satisfy the following inequality:

$$\Delta V_\perp(\mathbf{Q}, \mathbf{T}) \leq \frac{\Delta W(\mathbf{Q}, \mathbf{T}) - \Delta W_\parallel(\mathbf{Q}, \mathbf{T})}{2\|(\mathbf{Q}_\perp, \sqrt{2\alpha C_{\max} \mathbf{T}})\|}. \quad (27)$$

The proof of Lemma 3 is similar to that in [2] and is omitted here for brevity.

The rest of proof follows from Lemma 3 by studying the conditional expectation of $\Delta W(\mathbf{Q}, \mathbf{T})$ and $\Delta W_\parallel(\mathbf{Q}, \mathbf{T})$. We will omit the time reference $[t]$ without confusion.

We first consider $\mathbb{E}[\Delta W(\mathbf{Q}, \mathbf{T})|\mathbf{Q}[t] = \mathbf{Q}, \mathbf{T}[t] = \mathbf{T}]$. It is not hard to show that

$$\mathbb{E}[\Delta W(\mathbf{Q}, \mathbf{T})|\mathbf{Q}[t] = \mathbf{Q}, \mathbf{T}[t] = \mathbf{T}]$$
$$\leq 2\mathbb{E}[\langle \mathbf{Q}, \mathbf{A} - \mathbf{S}^* \cdot \mathbf{C} \rangle|\mathbf{Q}, \mathbf{T}] + K_1 - 2\alpha \mathbb{E}[\langle \mathbf{T}, C_{\max} \mathbf{S}^* \rangle|\mathbf{Q}, \mathbf{T}], \quad (28)$$

where $K_1 \triangleq L \max\{A_{\max}^2, C_{\max}^2\} + 2\alpha L C_{\max}$, and we use the fact that $\sum_{l=1}^L T_l[t+1] - \sum_{l=1}^L T_l[t] = L - |\mathbf{H}^*| - \sum_{l \in \mathbf{H}^*} T_l[t]$, where $\mathbf{H}^* \triangleq \{l \in \mathcal{L} : S_l^*[t]C_l[t] > 0\}$.

Next, we consider $\mathbb{E}[\langle \mathbf{Q}, \mathbf{A} - \mathbf{S}^* \cdot \mathbf{C} \rangle|\mathbf{Q}, \mathbf{T}]$. By using the definition of projection $\boldsymbol{\lambda}^{(0)}$, we have

$$\mathbb{E}[\langle \mathbf{Q}, \mathbf{A} - \mathbf{S}^* \cdot \mathbf{C} \rangle|\mathbf{Q}, \mathbf{T}]$$
$$= \langle \mathbf{Q}, \boldsymbol{\lambda}^{(0)} - \epsilon \mathbf{c} \rangle - \mathbb{E}[\langle \mathbf{Q}, \mathbf{S}^* \cdot \mathbf{C} \rangle|\mathbf{Q}, \mathbf{T}]$$
$$= -\epsilon \|\mathbf{Q}_\parallel\| + \langle \mathbf{Q}, \boldsymbol{\lambda}^{(0)} \rangle - \mathbb{E}[\langle \mathbf{Q} + \alpha \mathbf{T}, \mathbf{S}^* \cdot \mathbf{C} \rangle|\mathbf{Q}, \mathbf{T}]$$
$$\quad + \alpha \mathbb{E}[\langle \mathbf{T}, \mathbf{S}^* \cdot \mathbf{C} \rangle|\mathbf{Q}, \mathbf{T}]. \quad (29)$$

Given the queue-length vector $\mathbf{Q}[t]$, TSLS vector $\mathbf{T}[t]$ and the global channel state $J[t]$ at the beginning of slot t, according to the definition of the RSG Algorithm, we have

$$\langle \mathbf{Q}[t] + \alpha \mathbf{T}[t], \mathbf{S}^*[t] \cdot \mathbf{C}[t] \rangle = \max_{\mathbf{S} \in \mathcal{S}^{J[t]}} \langle \mathbf{Q}[t] + \alpha \mathbf{T}[t], \mathbf{S} \cdot \mathbf{C}[t] \rangle,$$

which implies

$$\langle \mathbf{Q} + \alpha \mathbf{T}, \mathbb{E}[\mathbf{S}^* \cdot \mathbf{C}|\mathbf{Q}, \mathbf{T}] \rangle = \max_{\mathbf{r} \in \mathcal{R}} \langle \mathbf{Q} + \alpha \mathbf{T}, \mathbf{r} \rangle. \quad (30)$$

Thus, we have

$$\langle \mathbf{Q} + \alpha \mathbf{T}, \mathbb{E}[\mathbf{S}^* \cdot \mathbf{C}|\mathbf{Q}, \mathbf{T}] \rangle = \max_{\mathbf{r} \in \mathcal{R}} \langle \mathbf{Q} + \alpha \mathbf{T}, \mathbf{r} \rangle$$
$$\geq \max_{\mathbf{r} \in \mathcal{B}_\delta} \langle \mathbf{Q} + \alpha \mathbf{T}, \mathbf{r} \rangle \geq \max_{\mathbf{r} \in \mathcal{B}_\delta} \langle \mathbf{Q}, \mathbf{r} \rangle + \langle \alpha \mathbf{T}, \mathbf{r}^* \rangle,$$

where $\mathbf{r}^* \in \arg \max_{\mathbf{r} \in \mathcal{B}_\delta} \langle \mathbf{Q}, \mathbf{r} \rangle$. Since $\boldsymbol{\lambda}^{(0)} \succ \mathbf{0}$, we can find a $\delta > 0$ sufficiently small such that $r_l \geq r_{\min}$ for all $\mathbf{r} = (r_l)_{l \in \mathcal{L}} \in \mathcal{B}_\delta$ and some $r_{\min} > 0$. Hence, we have

$$\langle \mathbf{Q} + \alpha \mathbf{T}, \mathbb{E}[\mathbf{S}^* \cdot \mathbf{C}|\mathbf{Q}, \mathbf{T}] \rangle \geq \max_{\mathbf{r} \in \mathcal{B}_\delta} \langle \mathbf{Q}, \mathbf{r} \rangle + \alpha r_{\min} \|\mathbf{T}\|_1.$$

By substituting above inequality into (29), we have

$$\mathbb{E}[\langle \mathbf{Q}, \mathbf{A} - \mathbf{S}^* \cdot \mathbf{C} \rangle|\mathbf{Q}, \mathbf{T}] \leq -\epsilon \|\mathbf{Q}_\parallel\| + \min_{\mathbf{r} \in \mathcal{B}_\delta} \langle \mathbf{Q}, \boldsymbol{\lambda}^{(0)} - \mathbf{r} \rangle$$
$$- \alpha r_{\min} \|\mathbf{T}\|_1 + \alpha \mathbb{E}[\langle \mathbf{T}, \mathbf{S}^* \cdot \mathbf{C} \rangle|\mathbf{Q}, \mathbf{T}].$$

Since $\boldsymbol{\lambda}^{(0)} - \mathbf{r}$ is perpendicular to the normal vector $\mathbf{c}$ for $\mathbf{r} \in \mathcal{B}_\delta$, we have

$$\min_{\mathbf{r} \in \mathcal{B}_\delta} \langle \mathbf{Q}, \boldsymbol{\lambda}^{(0)} - \mathbf{r} \rangle = \min_{\mathbf{r} \in \mathcal{B}_\delta} \langle \mathbf{Q}_\perp, \boldsymbol{\lambda}^{(0)} - \mathbf{r} \rangle = -\delta \|\mathbf{Q}_\perp\|.$$

Hence, we have

$$\mathbb{E}\left[\langle \mathbf{Q}, \mathbf{A} - \mathbf{S}^* \cdot \mathbf{C} \rangle | \mathbf{Q}, \mathbf{T}\right] \leq -\epsilon \|\mathbf{Q}_\|\| - \delta \|\mathbf{Q}_\perp\|$$
$$- \alpha r_{\min} \|\mathbf{T}\|_1 + \alpha \mathbb{E}[\langle \mathbf{T}, \mathbf{S}^* \cdot \mathbf{C} \rangle | \mathbf{Q}, \mathbf{T}]. \quad (31)$$

Thus, by substituting (31) into (28), we have

$$\mathbb{E}[\Delta W(\mathbf{Q}, \mathbf{T}) | \mathbf{Q}, \mathbf{T}] \leq -2\epsilon \|\mathbf{Q}_\|\| - 2\delta \|\mathbf{Q}_\perp\|$$
$$- 2\alpha r_{\min} \|\mathbf{T}\|_1 + K_1. \quad (32)$$

Using techniques in showing (32) in [2], we have

$$\mathbb{E}[\Delta W_\|(\mathbf{Q}, \mathbf{T}) | \mathbf{Q}, \mathbf{T}] \geq -2\epsilon \|\mathbf{Q}_\|\| - K_2, \quad (33)$$

where $K_2 \triangleq 2LC_{\max}^2$.

By using the bounds (32), (33) and Lemma 3, we have

$$\mathbb{E}[\Delta V_\perp(\mathbf{Q}, \mathbf{T}) | \mathbf{Q}, \mathbf{T}]$$
$$\leq \frac{\mathbb{E}\left[\Delta W(\mathbf{Q}, \mathbf{T}) - \Delta W_\|(\mathbf{Q}, \mathbf{T}) | \mathbf{Q}, \mathbf{T}\right]}{2\|(\mathbf{Q}_\perp, \sqrt{2\alpha C_{\max} \mathbf{T}})\|}$$
$$\leq \frac{-2\delta \|\mathbf{Q}_\perp\| - 2\alpha r_{\min} \sum_{l=1}^{L} T_l + K_1(\alpha) + K_2}{2\|(\mathbf{Q}_\perp, \sqrt{2\alpha C_{\max} \mathbf{T}})\|}.$$

Note that $\alpha T_l \geq \alpha T_l \mathbb{1}_{\{\alpha T_l \geq 1\}} \geq \sqrt{\alpha T_l} \mathbb{1}_{\{\alpha T_l \geq 1\}} = \sqrt{\alpha T_l} - \sqrt{\alpha T_l} \mathbb{1}_{\{\alpha T_l < 1\}} \geq \sqrt{\alpha T_l} - 1$, and $\|\mathbf{Q}_\perp\| \geq \frac{1}{\sqrt{L}} \|\mathbf{Q}_\perp\|_1$, where $\mathbb{1}_{\{.\}}$ is an indicator function. Thus, we have

$$\mathbb{E}[\Delta V_\perp(\mathbf{Q}, \mathbf{T}) | \mathbf{Q}, \mathbf{T}]$$
$$\leq \frac{-\frac{2\delta}{\sqrt{L}} \|\mathbf{Q}_\perp\|_1 - 2r_{\min} \sum_{l=1}^{L} \sqrt{\alpha T_l} + K_1 + K_2 + 2Lr_{\min}}{2\|(\mathbf{Q}_\perp^{(k)}, \sqrt{2\alpha C_{\max} \mathbf{T}})\|}$$
$$\leq -\min\left\{\frac{\delta}{\sqrt{L}}, \frac{r_{\min}}{\sqrt{2C_{\max}}}\right\} + \frac{K_1 + K_2 + 2Lr_{\min}}{2\|(\mathbf{Q}_\perp, \sqrt{2\alpha C_{\max} \mathbf{T}})\|}.$$

Hence, for any $0 < \varsigma < \min\left\{\frac{\delta}{\sqrt{L}}, \frac{r_{\min}}{\sqrt{2C_{\max}}}\right\}$, by taking

$$d \triangleq \frac{K_1 + K_2 + 2Lr_{\min}}{2\left(\min\left\{\frac{\delta}{\sqrt{L}}, \frac{r_{\min}}{\sqrt{2C_{\max}}}\right\} - \varsigma\right)}, \quad (34)$$

we have the desired result.

B. PROOF OF LEMMA 2

If the event $\mathcal{E}_j \triangleq \{C_l[j] > 0, C_i[j] = 0, \forall i \neq l\}$ always happens from $j = t - m + 1$ to t and there is at least one packet arriving at link l in this time duration, then under the RSG Algorithm, link l should be scheduled at least once during the past m slots and thus $T_l^{(\epsilon)}[t] < m$. This implies

$$\Pr\{T_l^{(\epsilon)}[t] = m\}$$
$$\leq \Pr\{\mathcal{E}_j \text{ didn't happen for some } j \in [t - m + 1, t]\}$$
$$+ \Pr\{\text{No packet arrived at link } l \text{ from } t - m + 1 \text{ to } t\}.$$

Hence, we have $\Pr\{T_l^{(\epsilon)}[t] = m\} \leq (1 - (1 - p_l)\Pi_{i \neq l} p_i)^m + q_l^m \leq 2\vartheta_l^m$, where $q_l \triangleq \Pr\{A_l[t] = 0\} < 1, \forall l \in \mathcal{L}$, and $\vartheta_l \triangleq \max\{1 - (1 - p_l)\Pi_{i \neq l} p_i, q_l\} \in (0, 1)$. Hence, we have $\Pr\{\overline{T}_l^{(\epsilon)} = m\} < 2\vartheta_l^m$. Thus, by taking $\vartheta \triangleq \max_{l \in \mathcal{L}} \vartheta_l$, we have the desired result.

C. PROOF OUTLINE OF PROPOSITION 3

In the rest of proof, we will omit ϵ associated with the queue length processes, the TSLS counters and parameter $\alpha(\epsilon)$ for brevity. It is quite challenging to directly give an upper bound on $\mathbb{E}[\|\overline{\mathbf{Q}}_\perp\|^2]$. Instead, we upper-bound the moment generation function of $\|\overline{\mathbf{Q}}_\perp\|$, and use the relationship between the moments of a random variable and its moment generation function to upper-bound $\mathbb{E}[\|\overline{\mathbf{Q}}_\perp\|^2]$ as shown in the following lemma.

LEMMA 4. *For a random variable X with $\mathbb{E}[e^{\eta X}] < \infty$ for some $\eta > 0$, we have*

$$\mathbb{E}[X^n] \leq \frac{1}{\eta^n}\left(\log\left(e^{n-1}\mathbb{E}[e^{\eta X}]\right)\right)^n, \quad (35)$$

for $n = 1, 2, 3 \cdots$.

Please see the Appendix D for the proof of Lemma 4.

Let $\mathbf{Z}[t] \triangleq \left(\mathbf{Q}_\perp[t], \sqrt{2\alpha C_{\max} \mathbf{T}[t]}\right)$. We first give an upper bound on $\mathbb{E}\left[e^{\eta\|\mathbf{Z}[t+1]\|} \big| \mathbf{Q}[t], \mathbf{T}[t]\right]$. To that end, let $l^*[t] \in \arg\max_l T_l[t]$. We partition $(\mathbf{Q}_\perp[t], \mathbf{T}[t])$ into sets $\mathcal{F}_1$, $\mathcal{F}_2$ and $\mathcal{F}_3$, where

$$\mathcal{F}_1 \triangleq \{\|\mathbf{Z}[t]\| \leq d\}; \mathcal{F}_2 \triangleq \left\{\|\mathbf{Z}[t]\| > d, \|\mathbf{Q}_\perp[t]\| > T_{l^*[t]}[t]\right\};$$
$$\mathcal{F}_3 \triangleq \left\{\|\mathbf{Z}[t]\| > d, \|\mathbf{Q}_\perp[t]\| \leq T_{l^*[t]}[t]\right\}.$$

Then, we have

$$\mathbb{E}\left[e^{\eta\|\mathbf{Z}[t+1]\|} \big| \mathbf{Q}[t], \mathbf{T}[t]\right] = \sum_{i=1}^{3} \mathbb{E}\left[e^{\eta\|\mathbf{Z}[t+1]\|}; \mathcal{F}_i \big| \mathbf{Q}[t], \mathbf{T}[t]\right]. \quad (36)$$

Next, we consider each term in (36) individually.

(i) On event $\mathcal{F}_1$, we can show that if $\|\mathbf{Z}[t]\| \leq d$, then

$$\|\mathbf{Z}[t+1]\|^2 \leq d^2 + 2\left(A_{\max} + 2\sqrt{L}C_{\max}\right) d\sqrt{L}$$
$$+ L\left(A_{\max} + 2\sqrt{L}C_{\max}\right)^2 + 2L\alpha C_{\max} \triangleq G_1^2, \quad (37)$$

Hence, we have

$$\mathbb{E}\left[e^{\eta\|\mathbf{Z}[t+1]\|}; \mathcal{F}_1 \big| \mathbf{Q}[t], \mathbf{T}[t]\right] \leq e^{\eta G_1} \quad (38)$$

To consider other two terms in (36), we need the following lemma.

LEMMA 5. *Under the RSG Algorithm, if $\|\mathbf{Z}[t]\| > d$, then*

$$|\|\mathbf{Z}[t+1]\| - \|\mathbf{Z}[t]\||$$
$$\leq 2L \max\{A_{\max}, C_{\max}\} + \frac{2\alpha C_{\max} L}{d}$$
$$+ 2\alpha C_{\max} \frac{\sum_{l \in \mathbf{H}^*} T_l[t]}{\sqrt{\|\mathbf{Q}_\perp[t]\|^2 + 2\alpha C_{\max} \sum_{l=1}^{L} T_l[t]}}, \quad (39)$$

where $\mathbf{H}^* \triangleq \{l : S_l^*[t] C_l[t] > 0\}$.

The proof is omitted due to space limitations.

(ii) On event $\mathcal{F}_2$, we have

$$\frac{\sum_{l \in \mathbf{H}^*} T_l[t]}{\sqrt{\|\mathbf{Q}_\perp[t]\|^2 + 2\alpha C_{\max} \sum_{l=1}^{L} T_l[t]}} \leq \frac{LT_{l^*[t]}[t]}{\|\mathbf{Q}_\perp[t]\|} \leq L.$$

By substituting above inequality into (39), we get

$$|\|\mathbf{Z}[t+1]\| - \|\mathbf{Z}[t]\|| \leq G_2, \quad (40)$$

where $G_2 \triangleq 2L \max\{A_{\max}, C_{\max}\} + \frac{2\alpha C_{\max} L}{d} + 2\alpha C_{\max} L$. Noting that (13) and (40) satisfy conditions of Lemma 2.2 in [4], there exists $\eta_1 > 0$, and $\rho = e^{\eta G_2} - \eta(G_2 + \varsigma) \in (0, 1)$, independent of ϵ, such that

$$\mathbb{E}\left[e^{\eta(\|\mathbf{Z}[t+1]\| - \|\mathbf{Z}[t]\|)}; \mathcal{F}_2 \Big| \mathbf{Q}[t], \mathbf{T}[t]\right] \leq \rho, \forall 0 < \eta < \eta_1.$$

Thus, we have

$$\mathbb{E}\left[e^{\eta\|\mathbf{Z}[t+1]\|}; \mathcal{F}_2 \Big| \mathbf{Q}[t], \mathbf{T}[t]\right] \leq \rho e^{\eta\|\mathbf{Z}[t]\|}. \quad (41)$$

(iii) On event $\mathcal{F}_3$, we have

$$\frac{\sum_{l \in \mathbf{H}^*} T_l[t]}{\sqrt{\|\mathbf{Q}_\perp[t]\|^2 + 2\alpha C_{\max} \sum_{l=1}^{L} T_l[t]}}$$
$$\leq \frac{L T_{l^*[t]}[t]}{\sqrt{2\alpha C_{\max} T_{l^*[t]}[t]}} = \frac{L}{\sqrt{2\alpha C_{\max}}} \sqrt{T_{l^*[t]}[t]}. \quad (42)$$

By substituting (42) into (39), we get

$$\|\mathbf{Z}[t+1]\| - \|\mathbf{Z}[t]\|$$
$$\leq 2L \max\{A_{\max}, C_{\max}\} + \frac{2\alpha C_{\max} L}{d} + L\sqrt{2\alpha C_{\max}} \sqrt{T_{l^*[t]}[t]}.$$

In addition, on event $\mathcal{F}_3$, we have

$$\|\mathbf{Z}[t]\| \leq \sqrt{T_{l^*[t]}^2[t] + 2\alpha C_{\max} L T_{l^*[t]}[t]}. \quad (43)$$

Hence, we have

$$\|\mathbf{Z}[t+1]\| \leq \|\mathbf{Z}[t]\| + \|\mathbf{Z}[t+1]\| - \|\mathbf{Z}[t]\|$$
$$\leq F_1 T_{l^*[t]}[t] + F_2, \quad (44)$$

where $F_1 \triangleq L\sqrt{2\alpha C_{\max}} + \sqrt{1 + 2\alpha C_{\max} L}$ and $F_2 \triangleq \frac{2\alpha C_{\max} L}{d} + 2L\max\{A_{\max}, C_{\max}\}$. Thus, we have

$$\mathbb{E}\left[e^{\eta\|\mathbf{Z}[t+1]\|}; \mathcal{F}_3 \Big| \mathbf{Q}[t], \mathbf{T}[t]\right] \leq e^{\eta F_2} e^{\eta F_1 T_{l^*[t]}[t]}. \quad (45)$$

By substituting (38), (41) and (45) into (36), we have

$$\mathbb{E}\left[e^{\eta\|\mathbf{Z}[t+1]\|} \Big| \mathbf{Q}[t], \mathbf{T}[t]\right] \leq e^{\eta G_1} + \rho e^{\eta\|\mathbf{Z}[t]\|} + e^{\eta F_2} e^{\eta F_1 T_{l^*[t]}[t]}.$$

By taking expectation on both sides, we have

$$\mathbb{E}\left[e^{\eta\|\mathbf{Z}[t+1]\|}\right] \leq e^{\eta G_1} + \rho \mathbb{E}\left[e^{\eta\|\mathbf{Z}[t]\|}\right] + e^{\eta F_2} \mathbb{E}\left[e^{\eta F_1 T_{l^*[t]}[t]}\right]$$
$$\leq e^{\eta G_1} + \rho \mathbb{E}\left[e^{\eta\|\mathbf{Z}[t]\|}\right] + e^{\eta F_2} \sum_{l=1}^{L} \mathbb{E}\left[e^{\eta F_1 T_l[t]}\right]. \quad (46)$$

By Lemma 2, there exist $\eta_2 > 0$ such that $e^{\eta_2 F_1} \vartheta < 1$ and for $0 < \eta < \eta_2$, we have

$$\mathbb{E}\left[e^{\eta F_1 T_l[t]}\right] \leq 2 \sum_{m=0}^{\infty} e^{\eta F_1 m} \vartheta^m = \frac{2}{1 - e^{\eta F_1} \vartheta} \quad (47)$$

By substituting (47) into (46), we have

$$\mathbb{E}\left[e^{\eta\|\mathbf{Z}[t+1]\|}\right] \leq \rho \mathbb{E}\left[e^{\eta\|\mathbf{Z}[t]\|}\right] + G, \quad (48)$$

holding for $0 < \eta < \eta_0 \triangleq \min\{\eta_1, \eta_2\}$, where $G \triangleq e^{\eta G_1} + \frac{2Le^{\eta F_2}}{1 - e^{\eta F_1} \vartheta}$. By using inequality (48) and iterating over t, we have

$$\mathbb{E}\left[e^{\eta\|\mathbf{Z}[t]\|}\right] \leq \rho^t e^{\eta\|\mathbf{Z}[0]\|} + \frac{1 - \rho^t}{1 - \rho} G \leq e^{\eta\|\mathbf{Z}[0]\|} + \frac{G}{1 - \rho},$$

which implies $\mathbb{E}\left[e^{\eta\|\mathbf{Q}_\perp[t]\|}\right] \leq e^{\eta\|\mathbf{Z}[0]\|} + \frac{G}{1 - \rho}$.

Thus, we have

$$\mathbb{E}\left[e^{\eta\|\overline{\mathbf{Q}}_\perp[t]\|}\right] \leq e^{\eta\|\mathbf{Z}[0]\|} + \frac{G}{1 - \rho} \quad (49)$$

where $G \triangleq e^{\eta G_1} + \frac{2Le^{\eta F_2}}{1 - e^{\eta F_1} \vartheta}$, $\rho \triangleq e^{\eta G_2} - \eta(G_2 + \varsigma) \in (0, 1)$, $d = O(\alpha)$, $G_1 = O(\alpha)$, $G_2 = O(\alpha)$, $F_1 = O(\sqrt{\alpha})$ and $F_2 = O(1)$. Note that we need to choose a $\eta > 0$ such that

$$1 - e^{\eta F_1} \vartheta < 1 \quad (50)$$
$$e^{\eta G_2} - \eta(G_2 + \varsigma) < 1. \quad (51)$$

It is not hard to verify that

$$0 < \eta \leq \frac{1}{2} \min\left\{\frac{1}{F_1} \ln \frac{1}{\vartheta}, \frac{1}{G_2} \ln \frac{G_2 + \varsigma}{G_2}\right\} \quad (52)$$

satisfies above requirements. If α is large enough such that $\frac{\varsigma}{G_2} < 1$ and $G_2 \gg F_1$, then we have

$$\frac{1}{G_2} \ln \frac{G_2 + \varsigma}{G_2} \leq \frac{1}{F_1} \ln \frac{1}{\vartheta}. \quad (53)$$

Thus, we can take $\eta^* \triangleq \frac{1}{2G_2} \ln \frac{G_2 + \varsigma}{G_2}$ to meet the above requirements, and hence $\eta^* = O(\frac{1}{\alpha^2})$.

Taking $\eta = \eta^*$ and noting that $\eta^* < \frac{1}{2F_1} \ln \frac{1}{\vartheta}$, we have

$$\frac{G}{1 - \rho} = \frac{e^{\eta^* G_1} + \frac{2Le^{\eta^* F_2}}{1 - \vartheta e^{\eta^* F_1}}}{1 - (e^{\eta^* G_2} - \eta^*(G_2 + \varsigma))}$$
$$\leq \frac{e^{\eta^* G_1} + \frac{2Le^{\eta^* F_2}}{1 - \sqrt{\vartheta}}}{1 - (e^{\eta^* G_2} - \eta^*(G_2 + \varsigma))}$$
$$= \frac{e^{\eta^* G_1} + \frac{2Le^{\eta^* F_2}}{1 - \sqrt{\vartheta}}}{1 - \left(1 + \frac{\varsigma}{G_2}\right)^{\frac{1}{2}} + \frac{1}{2}\left(1 + \frac{\varsigma}{G_2}\right) \ln\left(1 + \frac{\varsigma}{G_2}\right)}$$
$$\overset{(a)}{=} O\left(\frac{1}{1 - \left(1 + \frac{\varsigma}{2G_2}\right) + \frac{1}{2}\left(1 + \frac{\varsigma}{G_2}\right)\frac{\varsigma}{G_2}}\right) = O\left(\alpha^2\right),$$

where the step (a) uses $\eta^* G_1 = O\left(\frac{1}{\alpha}\right)$ and $\eta^* F_2 = O\left(\frac{1}{\alpha^2}\right)$. Thus, we have $\mathbb{E}\left[e^{\eta^*\|\overline{\mathbf{Q}}_\perp\|}\right] = O(\alpha^2)$. By using Lemma 4, we have

$$\mathbb{E}[\|\overline{\mathbf{Q}}_\perp\|^2] \leq \frac{1}{(\eta^*)^2}\left(\log\left(e\mathbb{E}\left[e^{\eta^*\|\overline{\mathbf{Q}}_\perp[t]\|}\right]\right)\right)^2$$
$$= O\left(\alpha^4 (\log \alpha)^2\right).$$

D. PROOF OF LEMMA 4

$$\mathbb{E}[X^n] = \frac{1}{\eta^n} \mathbb{E}\left[\left(\log e^{\eta X}\right)^n\right]$$
$$\overset{(a)}{\leq} \frac{1}{\eta^n} \mathbb{E}\left[\left(\log\left(e^{n-1} e^{\eta X}\right)\right)^n\right]$$
$$\overset{(b)}{\leq} \frac{1}{\eta^n}\left(\log\left(e^{n-1} \mathbb{E}[e^{\eta X}]\right)\right)^n, \quad (54)$$

where the step (a) follows from the fact that $f(y) = (\log y)^n$ is increasing in $y \in [1, \infty)$ for $n = 1, 2, \cdots$; (b) uses the fact that $g(y) = \left(\log\left(e^{n-1}y\right)\right)^n$ is concave in $[1, \infty)$ for $n = 1, 2, \cdots$, and Jensen's Inequality.

Distributed Greedy Approximation to Maximum Weighted Independent Set for Scheduling with Fading Channels

Changhee Joo
ECE, UNIST
UNIST-gil 50
Ulsan, South Korea
cjoo@unist.ac.kr

Xiaojun Lin
ECE, Purdue University
465 Northwestern Ave.
West Lafayette, IN 47907
linx@ecn.purdue.edu

Jiho Ryu
ECE, UNIST
UNIST-gil 50
Ulsan, South Korea
jihoryu@unist.ac.kr

Ness B. Shroff
ECE and CSE, OSU
2015 Neil Ave.
Columbus, OH 43210
shroff@ece.osu.edu

ABSTRACT

Developing scheduling mechanisms that can simultaneously achieve throughput optimality and good delay performance often require solving the Maximum Independent Weighted Set (MWIS) problem. However, under most realistic network settings, the MWIS problem can be shown to be NP-hard. In non-fading environments, low-complexity scheduling algorithms have been provided that converge either to the MWIS solution in time or to a solution that achieves at least a provable fraction of the achievable throughput. However, in more practical systems the channel conditions can vary at faster time-scales than convergence occurs in these lower-complexity algorithms. Hence, these algorithms cannot take advantage of the opportunistic gain, and may no longer guarantee good performance. In this paper, we propose a low-complexity scheduling scheme that performs provably well under fading channels and is amenable to implement in a distributed manner. To the best of our knowledge, this is the first scheduling scheme under fading environments that requires only local information, has a low complexity that grows logarithmically with the network size, and achieves provable performance guarantees (which is arbitrarily close to that of the well-known centralized Greedy Maximal Scheduler). Through simulations we verify that both the throughput and the delay under our proposed distributed scheduling scheme are close to that of the optimal solution to MWIS. Further, we implement a preliminary version of our algorithm in a testbed by modifying the existing IEEE 802.11 DCF. The preliminary experiment results show that our implementation successfully accounts for wireless fading, and attains the opportunistic gains in practice, and hence substantially outperforms IEEE 802.11 DCF.

Categories and Subject Descriptors

C.2.1 [**Network Architecture and Design**]: Wireless communication; G.2.2 [**Graph Theory**]: Network problems

Keywords

Distributed algorithm, wireless scheduling, maximum weighted independent set

1. INTRODUCTION

Scheduling is one of the most fundamental functionalities of wireless networks. It determines which links should transmit at what time and at what data rate. It is well-known that solving the scheduling problem is inherently difficult because the interference relationship is often non-convex and even combinatorial in nature. Further, for large networks it is imperative that the scheduling algorithm is of low complexity and can be implemented in a fully distributed manner. Such requirements make it highly challenging to design easy-to-implement scheduling algorithms.

In the literature, it is well-known that the so-called Max-Weight algorithm is throughput optimal [27]. For graph based interference models, where whether two links interfere or not can be specified by a binary parameter, the MaxWeight algorithm corresponds to the solution to a Maximum Weighted Independent Set (MWIS) problem in the conflict (or interference) graph as follows. In the conflict graph, each link is mapped onto a vertex and two vertices (links) that interfere with each other are connected by an edge. A set of non-connected vertices, which is called an independent set, can transmit data simultaneously. Further, each vertex of the conflict graph is given a weight, which is typically the product of the link rate and its queue length, and which varies across time due to changing queue lengths and time-varying channels. The MaxWeight algorithm then computes an independent set that has the largest total weight (i.e., solution to the MWIS problem). Although the MaxWeight algorithm is throughput optimal, the MWIS problem is NP-Hard in general [13]. Hence, the MaxWeight algorithm incurs high complexity, and further, it is a centralized algorithm that requires global information. Thus, the MaxWeight algorithm is not amenable to practical implementation.

In the literature, there have been many efforts to develop low-complexity and distributed scheduling algorithms with provably good throughput performance [6,8,11,12,14,16,17, 20,21,23]. These algorithms differ in terms of their throughput guarantee, complexity, and delay performance. They can be classified into two categories, depending on whether or not they account for channel fading.

There have been many more scheduling solutions for wireless systems *without fading*. Low-complexity scheduling algorithms have been developed with complexity that grows significantly slower than the network size, and can yet guarantee a non-negligible fraction of the optimal system capacity. As a point of comparison, the *Greedy Maximal Scheduling* (GMS) algorithm (also known as *Longest Queue First* (LQF) algorithm) can provably attain a fraction of the optimal capacity, with complexity that grows linearly with the total number of links L [5]. Other algorithms can reduce the complexity even further. For example, the *Maximal Scheduling* algorithm can attain at least $\frac{1}{\Delta}$ of the optimal capacity, with $O(\log N)$ complexity [29], where Δ denotes the maximum conflict degree (see (2) for the definition) and N denotes the number of nodes. The *Constant-time* scheduling algorithms, instead, can achieve a comparable capacity with $O(1)$ complexity, i.e., the complexity does not grow with the network size [17]. Further, both the *Pick-and-Compare* algorithm [6,21] and *Carrier Sensing Multiple Access (CSMA)* algorithm [11,23] have been shown to achieve the optimal throughput. They incur $O(L)$ and $O(1)$ complexity, respectively. We note that these two algorithms have been observed to lead to poor delay performance [8,20], and hence the utility of the throughput gain may be debatable, especially for delay-sensitive applications. Nonetheless, these results indicate that good throughput performance may be attained for non-fading environments using algorithms with very low complexity.

In practice, however, most wireless systems experience some level of channel fading. When link rates vary across time due to fading, the system throughput can be further improved by scheduling links when their rate are high. This is known as the opportunistic gain [19]. Exploiting opportunistic gain has been extremely popular in cellular systems. For ad hoc wireless networks, the MaxWeight algorithm can exploit this opportunistic gain and in fact achieve the optimal throughput even with fading. However, many of the low-complexity scheduling algorithms described in the previous paragraph cannot exploit the opportunistic gain, and their performance in fading environments will be much worse.

Take *CSMA* and *Pick-and-Compare* algorithms as examples. They reduce complexity by amortizing the computation across many time-slots, and hence need to take many iterations to find a close-to-optimal schedule. In fading environments, the link rates could have changed significantly before these algorithms can find a good schedule. Hence, they will not be able to exploit the opportunistic gain unless the fading is very slow [30]. Similarly, it appears to be difficult for the *Maximal Scheduling* algorithm and the *Constant-time* scheduling algorithm to account for channel fading and still guarantee a provable fraction of the optimal capacity. Recently, there have been a few other low-complexity schemes that are provably efficient with fading channels. However, they are either limited to single-hop networks [16] or their performance guarantees are much lower [12].

An exception is perhaps the GMS scheduling algorithm, which computes an approximation to the MWIS problem by choosing the highest weight vertex first, and can guarantee $\frac{1}{\Delta}$ fraction of the optimal capacity *in both fading and non-fading environment*. Other greedy approximations have also been proposed in [24,28]. However, they require centralized operations and linear complexity $O(L)$. Although distributed greedy approximation algorithms have been developed [4,10], they still incur a worst case time-complexity of $O(L)$. This high complexity has become a major obstacle preventing these algorithms from being used in practical system because the channel conditions can vary at faster time-scales than $O(L)$. Given that fading is a prevalent phenomenon in most modern wireless systems, an interesting open question is how one can develop *distributed* scheduling algorithms with even *lower complexity* and yet *guarantee good performance* .

In this paper, we answer this open question by proposing a novel low-complexity and distributed greedy approximation algorithm, called DistGreedy, for both fading and non-fading environments. In contrast to the known greedy approximations [4,10,24,28], our proposed DistGreedy algorithm incurs a low logarithmic complexity $O(\log L)$ that grows slowly with the network size. Further, it requires only local information (such as queue length and link rates of neighboring links), and can be implemented in a distributed fashion. We analytically show that our low-complexity distributed algorithm produces a schedule that is a $\frac{1}{\Delta}$-approximation to the MWIS problem, and show through simulations that DistGreedy often achieves scheduling performance far better than the provable bounds. Indeed, it empirically achieves throughput and delay performance that is close to that of the MaxWeight scheduler. We also conduct preliminary experiments with implementation in a real testbed. We implement a new MAC protocol that captures the essence of DistGreedy by modifying the IEEE 802.11 DCF. Performance comparison with the IEEE 802.11 DCF under channel fading shows that the DistGreedy algorithm can exploit the opportunistic gains and thus substantially outperform IEEE 802.11 DCF in fading environments.

The rest of the paper is organized as follows. The system model is described in Section 2. The DistGreedy algorithm is proposed and analyzed in Section 3. We numerically evaluate its performance in Section 4, and provide preliminary experiment results based on a testbed implementation in Section 5. Then, we conclude.

2. SYSTEM MODEL

We consider a wireless network with N nodes and L directed links. We assume that time is slotted and that a single frequency channel is shared by all the links. Multiple link transmissions at the same time slot may fail due to wireless interference. We assume that there is no link error, i.e., a link transmission is successful if there is no simultaneous interfering transmission.

The link rate of a successful transmission depends on its channel state. We assume that the channel state is fixed during a time slot, and changes across time slots. We denote the (global) channel state by h. when the channel is in state h, link l can transfer r_l^h unit of data if its transmission is successful. Let $\mathcal{H}$ denote the set of all the channel states.

We assume that the channel state has a finite space with a stationary distribution π^h, with $\sum_{h \in \mathcal{H}} \pi^h = 1$.

In order to account for wireless fading, we employ a channel-dependent interference model as follows. Let $C_{kl}^h \in \{0, 1\}$ denote the interference relationship between link k and link l when the channel state is h. We set $C_{kl}^h = 0$ if link l does not interfere with link k (and therefore they can transmit simultaneously), and $C_{kl}^h = 1$, otherwise. We assume that the interference relationship is symmetric, i.e., $C_{kl}^h = C_{lk}^h$. We note that the dependency on h represents a major departure from existing works for non-fading environments. Specifically, the interference relationship as well as the link rate may change across time in our model. This model is not only a simplified version that captures the fundamental characteristics of the more accurate SINR interference model [9], but also a general model that includes many interference models used in the literature to model FH-CDMA, Bluetooth, and IEEE 802.11 DCF network systems [25, 29], as special cases.

Given a network system, our interference model admits a unique *conflict graph* at each channel state h, which clearly presents the underlying interference constraints. For each link $l \in L$, we draw a vertex in the conflict graph, which is also denoted by the same alphabet l. For every two vertices k, l with $C_{kl}^h = 1$, we connect them with an edge in the conflict graph. Let $G^h = (V, E^h)$ denote the conflict graph with the set V of vertices and the set E^h of edges under channel state h. The conflict graph explicitly shows the interference relationship of any two vertices (i.e., links in the original network). In the sequel, we deal with the conflict graph throughout the paper.

We now formally formulate the Maximum Weighted Independent Set (MWIS) problem. Suppose that the channel state is h at time slot t. We consider the conflict graph G^h constructed from the interference constraints under channel state h. We begin with some definitions. Vertex x is a *neighbor* of vertex v, if they are connected by an edge in the conflict graph. Let $I^h(v)$ denote the set of neighbors of vertex v including v, and let $I^h(A)$ denote the set of neighbors of vertices in A, i.e., $I^h(A) := \cup_{v \in A} I^h(v)$. Let $w_v(t, h)$ denote a weight associated with vertex v. In particular, we define the weight of vertex v as the product of queue length $Q_v(t)$ and transmission rate r_v^h. Let $w^*(t, h)$ denote the largest weight, i.e., $w^*(t, h) := \max_{v \in V} w_v(t, h)$. Further, let $\bar{w}_v(X; t, h)$ denote the largest weight in the neighborhood of vertex v within X, i.e., $\bar{w}_v(X; t, h) := \max_{x \in X \cap I^h(v)} w_x(t, h)$.

We say that a set S of vertices is an *independent set* (or a *feasible schedule*) if no two vertices in the set are neighbors. Further, an independent set is maximal if no extra vertex can be added. Such an independent set is also called a *maximal matching*. Let $\mathbb{S}^h$ denote the collection of all the feasible independent sets that are available in G^h. The MWIS problem can be formulated as finding S^* such that

$$S^* \in \operatorname*{argmax}_{S \in \mathbb{S}^h} \sum_{v \in S} w_v(t, h). \tag{1}$$

It has been known that at each time t, given a channel state h, the solution to the MWIS problem with weight $w_v(t, h) = Q_v(t) \cdot r_v^h$ results in a throughput-optimal MaxWeight scheduling scheme [22]. However, due to the high computational complexity and the requirement of global information, such a MaxWeight algorithm is difficult to implement in practice. On the other hand, it has been shown in [15, 18] that

an imperfect scheduling solution that solves (1) within a factor of γ at every time t achieves at least γ fraction of the optimal throughput. To this end, our goal is to develop practical low-complexity scheduling algorithm that can approximately solve (1) with a provable fraction in a distributed fashion.

Remarks: In the above MaxWeight scheduling scheme, we implicitly assume single-hop traffic, i.e., packets are transmitted over a single link and leave the system immediately after the transmission. For multi-hop traffic, the same MaxWeight algorithm can be used by replacing the queue length with a queue differential. (See [27] for the details.) Similarly, our DistGreedy algorithm described in the next section can be extended to multi-hop scenarios in a straightforward manner.

Finally, we define the vertex degree $\delta(h) := \max_{v \in V} |I^h(v)|$, where $| \cdot |$ denotes the cardinality of the set, and the maximum conflict (or interference) degree $\Delta(h)$ as

$$\Delta(h) := \max_{v \in V, S \in \mathbb{S}} |I^h(v) \cap S|. \tag{2}$$

In the network, the maximum conflict degree represents the maximum number of simultaneous transmissions in the neighborhood of any link, which can be upper bounded by a constant in many practical interference models [7, 9]. Also, we define $\Delta^* = \max_{h \in \mathcal{H}} \Delta(h)$.

3. DISTRIBUTED GREEDY APPROXIMATION

In this section, we describe our distributed approximate solution to (1) and analyze its performance. We emphasize that the algorithm operates in a distributed manner and each vertex (link) requires only local information from its neighbors in the conflict graph. Throughout this section, we consider the conflict graph G^h at time t under channel state h, and omit the subscripts t and h if there is no confusion.

3.1 Algorithm description

We assume that each time slot has two parts: contention and transmission. The contention part has several intervals, and each interval is further divided into mini-slots. We determine a feasible schedule during the contention part, and with the computed schedule, transmits actual data during the transmission part.

At a time slot, let B denote the feasible schedule (independent set of vertices) chosen by our algorithm. We explain our solution, starting with an empty set and add vertices to B by executing an iterative algorithm as shown in Algorithm 1.

At each interval i, some vertices are 'determined' as to whether they belong to set B or not. Specifically, vertices in B_i are 'determined' to be in B at interval i, and vertices in $(I(B_i) \backslash B_i)$ are 'determined' not to be in B at interval i. Let V_i denote the set of vertices that have not been determined yet at the beginning of interval i, i.e.,

$$V_i := V \backslash \left(\cup_{j=1}^{i-1} I(B_j) \right),$$

which can be rewritten in a recursive form as

$$V_i = V_{i-1} \backslash I(B_{i-1}).$$

We say that a vertex in V_i is *eligible* at interval i. Let $A_i \subset V_i$ denote the set of vertices that will be 'determined' during

Algorithm 1 DistGreedy algorithm.

$V_0 \leftarrow V, B_0 \leftarrow \emptyset$
1: **for** $i = 1$ to $\log_\alpha \beta |V|$ **do**
2:    $V_i \leftarrow V_{i-1} \setminus I(B_{i-1})$
3:    $A_i \leftarrow \emptyset$
4:    **for** each $v \in V_i$ **do**
5:       calculate $\bar{w}_v(V_i) := \max_{x \in V_i \cap I(v)} w_x$
6:       **if** $w_v \geq \frac{\bar{w}_v(V_i)}{\alpha}$ **then**
7:          $A_i \leftarrow A_i \cup \{v\}$
8:       **end if**
9:    **end for**
10:   $B_i \leftarrow$ dist_maximal_matching(A_i)
11: **end for**

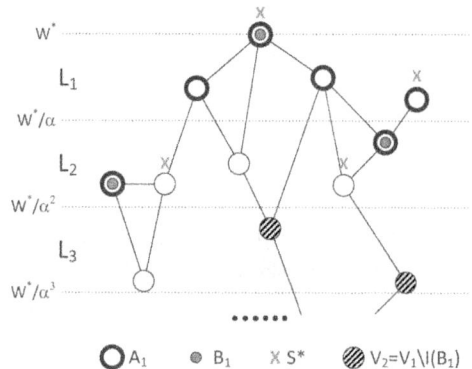

Figure 1: **Conflict graph with vertices and edges in the layered format, where vertices are partitioned into layers according to their weights.**

interval i, from which we will compute B_i. We will soon see how to find A_i and B_i. From the definitions, it is clear that $V_0 = V$ and $B_0 = \emptyset$. Finally, we have a couple of configuration parameters α, β that will be explained later.

The algorithm can be described as follows. We start with the entire set of vertices $V_0 = V$ and an empty set $B_0 = \emptyset$. Suppose that each vertex v knows its neighbors' weights. In wireless networks, this can be obtained by piggybacking/overhearing the information exchange or by explicitly exchanging control messages.

1. At each interval i, the set of eligible vertices V_i is updated by excluding $I(B_{i-1})$ from V_{i-1} (line 2 in Algorithm 1), where B_{i-1} denotes the set of vertices that are chosen during interval $i-1$ and $I(B_{i-1})$ denotes the set of neighbors for B_{i-1} (including B_{i-1} itself). For this purpose, each vertex that belongs to B_{i-1} should notify its neighbors by broadcasting a control message during interval $i - 1$. (See Step 4 below.)

2. Each vertex v in V_i calculates its local maximum weight $\bar{w}_v(V_i) := \max_{x \in V_i \cap I(v)} w_x$ from the weight information of its neighbors (in V_i). Then each vertex v sets itself as one of A_i if $w_v \geq \bar{w}_v(V_i)/\alpha$, where $\alpha > 1$. (Lines $5 - 8$.)

3. On the set A_i, we compute a maximal matching in a distributed manner (line 10), which requires $O(\delta \log^2 |V|)$ complexity [26] or $O(\delta)$ complexity with precomputation [14], where δ is the vertex degree. Let B_i denotes the obtained set.

4. In the process of computing the maximal matching, each neighbor of vertex $v \in B_i$ should be informed that v belongs to B_i. Hence, the vertices in $I(B_i)$ will not participate in the next interval.

5. The above procedure repeats for J times, where $J := \lceil \log_\alpha \beta |V| \rceil$. The set $B(= \cup_i B_i)$ of vertices will be returned as the final result. This is the set of links that will transmit data packets during the time slot.

We have two configuration parameters α and β that will be further discussed in the next section.

Note that our distributed greedy (DistGreedy) algorithm requires $O(\delta)$ complexity at each interval and will run for $O(\log |V|)$ intervals (using the algorithm in [14]). Hence, the worst-case complexity will be $O(\delta \log |V|)$. In some applications, e.g., regular topologies, δ is a fixed constant.

Thus, DistGreedy can be implemented with $O(\log |V|)$ complexity (or with polylogarithmic complexity in random networks[1]), which is much lower than the $O(|V|)$ complexity of the known distributed implementation of GMS [4, 10].

Remarks: Note that the previous greedy approximations to the MWIS problem shown in [24] have a similar iterative procedure as DistGreedy. However, their algorithm works vertex-by-vertex sequentially, which results in linear complexity in the worst case (e.g., consider a ring topology such that, starting from a link, the link weights decrease in a clockwise direction). Further, they have a different rule for selecting a vertex at each interval, e.g., they select vertex v with the largest $\frac{w_v}{|V_i \cap I(v)|}$ or with $w_v \geq \sum_{x \in V_i \cap I(v)} \frac{w_x}{|V_i \cap I(x)|}$. This selection rule is the key to achieve the provable approximation ratio of $\frac{1}{\delta}$. Unlike this previous work [24], DistGreedy reduces the complexity significantly by considering multiple vertices in parallel and do not follow the strict sequential ordering. Further, the procedure stops after a certain number of intervals. At each interval, DistGreedy selects the vertices v with $w_v \geq \max_{x \in V_i \cap I(v)} w_x/\alpha$. The end result is a much better approximation ratio ($\approx \frac{1}{\Delta}$) and a much better complexity ($O(\log |V|)$). However, the parallel processing also makes it more difficult to analyze the performance of DistGreedy. Nonetheless, in the next section, we show that due to the selection rule of DistGreedy, it can achieve the approximation ratio arbitrarily close to $\frac{1}{\Delta}$.

3.2 Performance Analysis

We evaluate the performance of our distributed greedy (DistGreedy) algorithm, and show that it is in fact a $\frac{1}{\Delta}$-approximation to the optimal solution. Motivated by [28], we divide the vertices into layers $L_1, L_2, \ldots$ based on the ratio of their weight to the maximum weight w^*, as [2]

$$L_i = \left\{ v \in V \mid \frac{w^*}{\alpha^i} < w_v \leq \frac{w^*}{\alpha^{i-1}} \right\}. \quad (3)$$

---

[1] In random networks, it is well-known that $\delta \sim O(\log |V|)$ to ensure connectivity [9].

[2] The algorithm in [28] computes a maximal matching for each layer, and thus requires for each node to know which layer it belongs to, or equivalently to know w^*. However, knowing the maximum weight w^* may take $O(|V|)$ time to propagate in the worst case. In contrast, DistGreedy works with local weight information and the layering structure is only for the purpose of analysis.

Fig. 1 illustrates an example conflict graph in the layered format.

We start our analysis with the following lemmas.

Lemma 1. *For* $i \leq \log_\alpha \beta |V|$, *if vertex* $v \in L_i$, *then* $v \in I(\cup_{j=1}^i B_j)$, *and thus*

$$L_i \subset I(\cup_{j=1}^i B_j). \qquad (4)$$

PROOF. If each vertex $v \in L_i$ selects itself for distributed maximal matching no later than the i-th interval, i.e., if $v \in L_i$ implies $v \in \cup_{j=1}^i A_j$, then we can obtain the lemma, since

$$v \in L_i \Rightarrow v \in \cup_{j=1}^i A_j \Rightarrow v \in \cup_{j=1}^i I(B_j) \Rightarrow v \in I(\cup_{j=1}^i B_j), \qquad (5)$$

where the second step comes from the fact $A_j \subset I(B_j)$, since B_j is a maximal matching on A_j.

Now what remains to be shown is that $v \in L_i$ implies $v \in \cup_{j=1}^i A_j$. We show this by induction. It is clear that when $i = 1$, all vertices $v \in L_1$ belong to A_1, because $w_v > \frac{w^*}{\alpha}$. Suppose that the statement is true for all $i \leq c$. Note that all vertices in $\cup_{j=1}^c A_j$ are not eligible at interval $c + 1$ since each vertex in A_j belongs to $I(B_j)$ for $j = 1, 2, \ldots, c$ under our algorithm. Hence, at interval $c + 1$, no vertex in $\cup_{j=1}^c A_j$ is eligible, which immediately implies that no vertex in $\cup_{j=1}^c L_j$ is eligible since $L_i \subset \cup_{j=1}^c A_j$ for all $i \leq c$. Now, if there is a vertex $v \in L_{c+1}$ eligible at interval $c + 1$, i.e., $v \in V_{c+1}$, then vertex v should be included in A_{c+1}, since all vertices e with $w_e > \frac{w^*}{\alpha^c}$ (i.e., $e \in \cup_{j=1}^c L_j$) are not eligible. Hence, the induction hypothesis must hold for $i = c + 1$. This completes the proof. $\square$

Lemma 1 states that under DistGreedy, any vertex in layer L_i will be 'determined' after interval i ends. In the following lemma, we show that each vertex in layer i must have a neighboring vertex that has a similar or higher weight and that is chosen by DistGreedy after interval i ends. Combining these two lemmas, we can show that after interval i, every vertex in L_i or above is a neighbor of a vertex that is already chosen by DistGreedy.

Lemma 2. *For each vertex* $x \in L_i$ *(with* $i \leq \log_\alpha \beta |V|$*), there exists* $v \in \cup_{j=1}^i B_j$ *such that* $x \in I(v)$ *and* $\alpha \cdot w_v \geq w_x$.

PROOF. From Lemma 1, we have $x \in I(\cup_{j=1}^i B_j)$, and thus there exists $v \in \cup_{j=1}^i B_j$ such that $x \in I(v)$. Let $k \leq i$ be the smallest index such that $x \in I(B_k)$. Then, $x \notin I(\cup_{j=1}^{k-1} B_j)$ and there exists $v \in B_k$ with $x \in I(v)$. Since $v \in B_k \subset A_k$, it should satisfy $\bar{w}_v(V_k)/w_v \leq \alpha$ from line 6 of Algorithm 1. Also since $x \in I(v)$ and x is eligible at interval k (because $x \notin I(\cup_{j=1}^{k-1} B_j)$), we have $\bar{w}_v(V_k) \geq w_x$. Hence, we obtain that $\alpha \cdot w_v \geq w_x$. $\square$

Recall that S^* denotes the maximum weighted independent set over V. We define $D_i(v)$ as the set of vertices in $S^* \cap L_i$ that are connected to v by an edge in the conflict graph, i.e.,

$$D_i(v) := \{x \mid x \in S^* \cap L_i, \text{ and } x \in I(v)\}. \qquad (6)$$

Then $|D_i(v)|$ denotes the number of vertices selected by the MWIS solution in layer L_i that conflicts with v. The following lemma shows that the weight sum for S^* within layer L_i can be bounded by the weight sum for the independent set chosen by DistGreedy up to interval i, multiplied by $\alpha \cdot |D_i(v)|$.

Lemma 3. *At each interval* i, *we have*

$$\sum_{v \in \cup_{j=1}^i B_j} \alpha \cdot |D_i(v)| \cdot w_v \geq \sum_{x \in L_i \cap S^*} w_x. \qquad (7)$$

PROOF. From Lemma 2, we have that for each vertex $x \in S^* \cap L_i$, there exists $v \in \cup_{j=1}^i B_j$ such that $x \in I(v)$ and $\alpha \cdot w_v \geq w_x$. However, multiple x may map to the same v. Nonetheless, for each of such v, at most $|D_i(v)|$ vertices in $S^* \cap L_i$ can potentially be neighbor of v in the conflict graph. Therefore, we can obtain the result. $\square$

By summing both sides of (7) for all i, we can bound the maximum weight sum by the weight sum of the vertices chosen by DistGreedy within a constant factor $\alpha\Delta$. (See the proof of Lemma 4 below.) However, if we were to terminate after all vertices are considered, it would have resulted in $O(|V|)$ complexity (e.g., consider a fully connected graph with vertices whose weights are $1, \frac{1}{\alpha+\epsilon}, \frac{1}{(\alpha+\epsilon)^2}, \ldots$). In the next lemma, we show that even if DistGreedy stops after $O(\log |V|)$ intervals, the performance loss would still be negligible.

Lemma 4. *By setting* $\alpha \to 1$ *and* β *sufficiently large, Algorithm 1 is a* $\frac{1}{\Delta(h)}$-*approximation algorithm.*

PROOF. Let $B := \cup_{j=1}^J B_j$, where $J := \lceil \log_\alpha \beta |V| \rceil$. By summing (7) from $i = 1$ to J, we can obtain that

$$\begin{aligned}
\sum_{i=1}^J \sum_{x \in L_i \cap S^*} w_x &\leq \sum_{i=1}^J \sum_{v \in \cup_{j=1}^i B_j} \alpha \cdot |D_i(v)| \cdot w_v \\
&\leq \sum_{v \in B} \sum_{i=1}^J \alpha \cdot |D_i(v)| \cdot w_v \\
&\leq \sum_{v \in B} \alpha \cdot \Delta \cdot w_v.
\end{aligned} \qquad (8)$$

Also, for $i > J$, we can obtain that

$$\sum_{i=J+1}^\infty \sum_{x \in L_i \cap S^*} w_x \leq \sum_{i=J+1}^\infty \sum_{x \in L_i} w_x \leq |V| \cdot \frac{w^*}{\alpha^J} \leq \frac{w^*}{\beta}, \qquad (9)$$

where w^* denotes the largest weight among all the vertices. The last inequality holds since $J = \lceil \log_\alpha \beta |V| \rceil$.

Combining (8) and (9), we can obtain that

$$\alpha\Delta \sum_{v \in B} w_v + \frac{w^*}{\beta} \geq \sum_{x \in S^*} w_x. \qquad (10)$$

Thus our result follows. $\square$

It has been shown in non-fading environments that a scheduling solution that is a γ-approximation to the MWIS problem at each time slot can achieve at least γ fraction of the optimal throughput [15,18]. It is straightforward to extend the result to fading environment: a scheduling solution that is a $\gamma(h)$-approximation to the MWIS problem under channel state h at each time slot can achieve at least $\min_{h \in \mathcal{H}} \gamma(h)$ fraction of the optimal throughput. Combining it with Lemma 4, we can obtain the following Proposition.

Proposition 5. *A scheduling solution that executes Dist-Greedy at each time slot can achieve* $\frac{1}{\Delta^*}$ *fraction of the optimal throughput, where* $\Delta^* = \max_{h \in \mathcal{H}} \Delta(h)$.

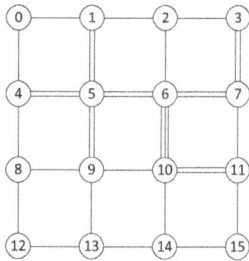

Figure 2: Grid network topology.

Remarks: In developing a distributed low-complexity GMS algorithm, one of the main difficulties lies in the requirement of the strict global ordering of selected vertices. For example, if the conflict graph is a linear graph, where the weights of vertices monotonically decrease from the left to the right, then the selection of the right-most vertex v with the smallest weight can be made only after the selection of its left-side neighbor u, which again can be made only after the selection of the left-side neighbor of vertex u due to the linear topology. This implies that the selection of the right-most vertex v needs to be made after $O(|V|)$ time.

Our result implies that the strict global ordering in the GMS algorithm is not required for high performance. A loose ordering would be sufficient, which can result in significant complexity reduction with negligible performance degradation. We highlight that the state-of-the-art "distributed" $\frac{1}{\Delta(h)}$-approximation algorithm requires $O(|V|)$ complexity [4,10], while our local greedy algorithm significantly lowers the complexity to $O(\delta(h) \log |V|)$.

4. NUMERICAL RESULTS

We evaluate DistGreedy, Greedy, and MaxWeight through simulations, where MaxWeight is the optimal scheduler that solves the MWIS problem at each time slot. We simulate two networks: one with a grid topology and the other with a randomly generated topology.

We first consider a grid topology as shown in Fig. 2. Each link has an average transmission rate of one, two, or three packets per time slot, which are signified in the figures by the number of lines between two nodes, e.g., one line implies one packet per time on average. At each time slot, actual link rate changes and is chosen uniformly at random from the range $[0, 2(\text{Avg. rate})]$. Since DistGreedy approximates the optimal solution to the MWIS problem at each time slot, we focus on the behavior of DistGreedy under static interference models, where the interference relationship does not change across time. In particular, we use one-hop (or primary) interference model, under which two links that share a node cannot transmit at the same time. We impose single-hop traffic of load ρ on every link: at each time slot, each link has a packet arrival with probability ρ. The arrivals are i.i.d. across time slots and links. We set DistGreedy to have $\lceil \log_\alpha \beta |V| \rceil$ intervals at each time slot. We use a link-coloring technique to find a maximal matching, under which $(\delta + 1)$ mini-slots are sufficient [14]. Since $\delta = 6$ in our grid topology, we use 10 mini-slots at each interval. The number of mini-slots are not taken into account in the performance measurements. Each result is an average of 10 simulation runs for 10^6 time slots.

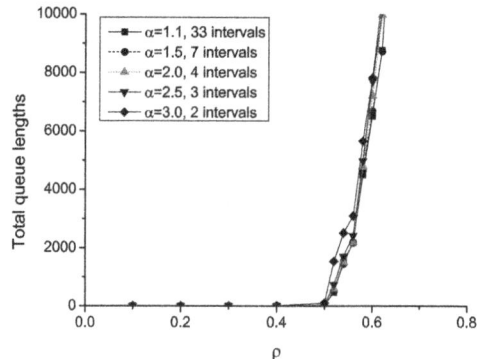

Figure 3: Performance of DistGreedy with different α.

Fig. 3 illustrates the performance of DistGreedy in terms of total queue lengths with different α settings and $\beta = 1$. Sharp increases of queue lengths imply the boundary of the capacity region. Note that a larger α means thicker layers and thus a smaller number of intervals. The results show that the performance is not sensitive to the value of α in a range $[1.1, 3]$ and a small number of intervals (e.g., $\alpha = 2.0$) would be sufficient for high throughput performance.

While running DistGreedy, we also trace the maximum weight sum, the weight sum of the schedule selected by DistGreedy, and the weight sum of the schedule that would be chosen by Greedy, at each time slot. Fig. 4 depicts the ratio of each weight sum (from DistGreedy and Greedy) to the maximum weight sum. It shows that both GMS and DistGreedy typically achieve much higher ratios than the analytical bound, which is $\frac{1}{2}$ in the one-hop interference model and shown in the figure using a dotted line. Also, as α gets closer to 1, DistGreedy algorithm approaches Greedy algorithm because layers become narrower and the number of intervals increases.

In Fig. 5, we compare the performances of MaxWeight, Greedy, DistGreedy (with $\alpha = 2$, $\beta = 1$), and Q-CSMA, where Q-CSMA is a CSMA algorithm known to be throughput-optimal in non-fading environments. For Q-CSMA, we use the version shown in [8], i.e., each link v sets its access probability for the decision vector to $\frac{1}{|I(v)|}$ and sets its weight for link activity to $\log(Q_l(t) \cdot r_l^h(t))$. The results in Fig. 5 illustrate that Q-CSMA, which has the lowest complexity $O(1)$ among the scheduling schemes, has much poor throughput and delay performance than the others. In particular, its delay grows quickly at an offered load much lower than other algorithms, which suggests that it is not throughput optimal in fading environments. In contrast, DistGreedy has similar queue lengths to MaxWeight and Greedy. In other words, it empirically achieves similar throughput and delay performance to the optimal. Further, the simulation results suggest that the actual performance of DistGreedy could significantly outperforms the analytical lower bounds.

Next we consider a network that is randomly generated. We place a total of 32 nodes at random within 1×1 area. We connect two nodes with a link if they are within a distance of 0.25. Each link has a time-varying link rate, which is randomly chosen in the range of $[0, 10]$ packets per slot, and i.i.d. across links and time. We generated single-hop traffic over 24 links, which are chosen at random. Each source

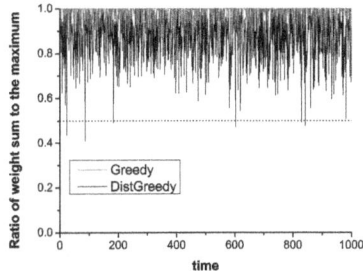

(a) $\alpha = 2.0$

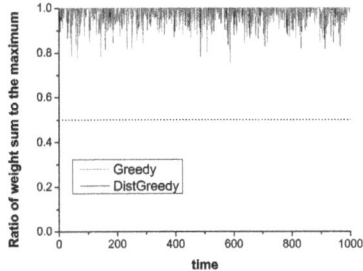

(b) $\alpha = 1.1$

Figure 4: Ratio of the achieved weight sum to the maximum weighted sum.

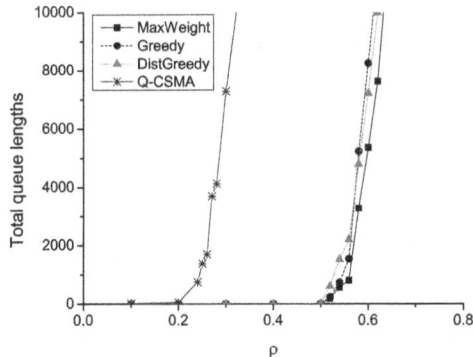

Figure 5: Performance comparison of scheduling schemes.

injects a random number of packets in the range of $[1, 5]$ at each time slot, with probability ρ. The results are similar to the grid network and are omitted due to lack of space. We can observe that the throughput and delay performance of DistGreedy are close to those of MaxWeight and Greedy, even though it has a significantly lower complexity.

5. PRELIMINARY EXPERIMENT RESULTS

In this section, we provide preliminary experimental results and show that the opportunistic gains of wireless fading can be achieved in practice. We have implemented a version of DistGreedy in hardware driver by modifying the medium access control of IEEE 802.11 DCF. Specifically, we use the `ath5k` device driver [2] over Voyage Linux [3] installed into the Alix 2D2 system board [1]. We modified it such that each node maintains information of recent queue lengths and transmission rates (link capacity) of neighboring nodes, and that a header in the data frame includes the queue length information of the transmitter. The transmis-

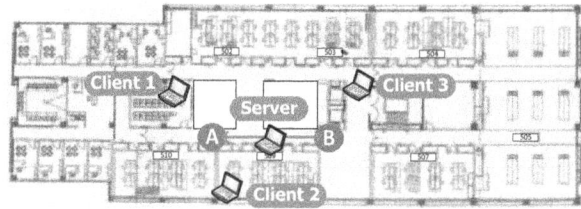

Figure 6: Experiment setup.

sion rate information can also be obtained from the device if the frame is successfully received. Each node who receives or overhears the frame updates the queue and rate information of the transmitter from the header.

Unlike IEEE 802.11 DCF, DistGreedy in Algorithm 1 requires time synchronization between nodes. We avoid the synchronization overheads by implementing an approximation of our algorithm while capturing the essential feature of the algorithm. Further, while the DistGreedy scheme is a link-based algorithm, current implementation of IEEE 802.11 DCF operates on nodes. Due to this difference, we use $w'_n(t)$ and $\bar{w}'_n(t)$ to denote node n's weight and its local maximum weight at time t, respectively. Our DistGreedy implementation works just like IEEE 802.11 DCF, except that when the transmitter n has a frame to send (asynchronously with other nodes) at time t, it calculates the local maximum weight $\bar{w}'_n(t)$ from the information that it has received from its neighbors, and estimates $\theta := \lfloor \frac{\bar{w}'_n(t)}{\alpha \cdot w'_n(t)} \rfloor$. Then, it chooses the contention window size at random within $[0, 7]$ if $\theta = 0$, within $[0, 31]$ if $\theta = 1$, within $[0, 63]$ if $\theta = 2$, and within $[0, 127]$ otherwise[3]. In this way, our implementation effectively has four intervals. Moreover, our implementation does not compute a maximal matching, which reduces the complexity further (compared with Algorithm 1, line 10). We set $\alpha = 10/9$, the maximum buffer size to 100 frames, and the IP packet size to 512 bytes. We remark that finding an optimal number of intervals and the contention window sizes is an interesting open question but beyond the scope of the paper.

Fig. 6 shows our experiment setting at the 5th floor of the ECE department building in UNIST, South Korea. We set three stationary clients and let each client transmit data at 3 Mbps to the single mobile server. All the clients can overhear each other's transmissions. The server moves between two positions A and B as shown in Fig. 6, at every 60 seconds, starting from A. Clearly, Client 1 has a good channel state when the server locates in Position A, and Client 3 performs well when the server locates in Position B. We conduct our experiments for 5 minutes and measure transmission rate, queue lengths, and throughput of each client.

Figs. 7 and 8 illustrate the experiment results. Due to high measurement variations, we show time-averaged values using exponential weighted moving average. Figs. 7(a) and 8(a) show that the variations of the total (average) capacity of three clients is less than the variations of the transmission rate of an individual link, which is significant especially when the server moves. Note that transmission rate of a link has a discrete value of $\{6, 9, 12, 24, 36, 48, 56\}$ Mbps. Instantaneous link rate changes very frequently across time though

---

[3]Non-overlapped windows for each θ seems to be a more intuitive choice. Unfortunately, we are unable to select a window that starts with non-zero value due to configuration restriction in our device firmware.

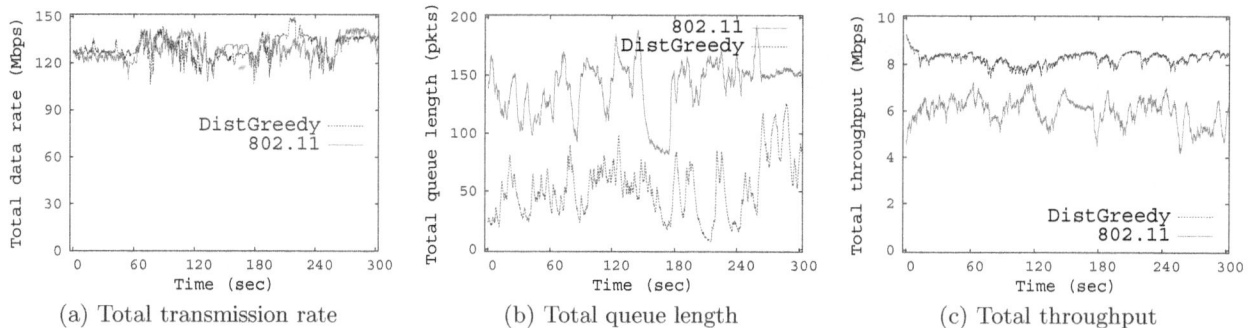

(a) Total transmission rate   (b) Total queue length   (c) Total throughput

Figure 7: Overall performance across all clients. Total transmission rate (a) shows that the total (average) link rate sum is similar. However, total queue lengths (b) and throughputs (c) clearly show that DistGreedy outperforms the IEEE 802.11 DCF.

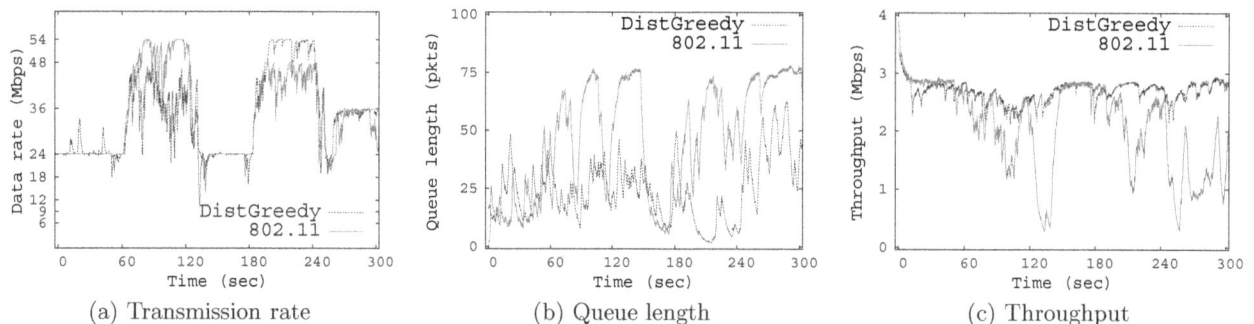

(a) Transmission rate   (b) Queue length   (c) Throughput

Figure 8: Performance of Client 3. Similar results as in Fig. 7 are observed when we focus on the performance of a client.

these changes are not shown in the figures due to averaging. Further the link rate is chosen by the hardware physical layer depending on the channel state at a given time. We have observed that transmission rate changes frequently and often in an unpredictable manner. Since the transmission rate is not under our control, it is difficult to maintain exactly the same channel environment when we compare two different MAC protocols. This is the reason that DistGreedy and IEEE 802.11 DCF have different link rates in Figs. 7(a) and 8(a). Nonetheless, the figures show that overall link rates are similar for both cases. We may force the hardware to use a fixed transmission rate at the transmitter. However, in such a case, we may see frequent transmission failures due to insufficient SINR at the receiver.

Fig. 7 shows that our DistGreedy implementation achieves better network performance in terms of reduced total queue length (Fig. 7(b)) and total throughput performance (Fig. 7(c)) under similar channel states. Indeed, DistGreedy implementation maintains queue lengths low for all the three clients, and IEEE 802.11 DCF results in many drops due to buffer overflows. For example, the queue length of Client 3 frequently increases up to 100 under IEEE 802.11 DCF even through it does not show up in Fig. 8 due to exponential weighted time averaging.

In Figs. 8(a) and 8(b), we can observe that under IEEE 802.11 DCF, high link rate (i.e., in $[60, 120]$ and $[180, 240]$) does not lead to low queue lengths, since IEEE 802.11 DCF does not opportunistically exploit wireless fading. In contrast, it shows that the DistGreedy implementation successfully keeps the queue length low, especially when the link rate is high. This implies that DistGreedy can account

for the channel variations, and thus takes the advantage of the opportunistic gains. Another interesting observation is that in Fig. 8(c), IEEE 802.11 DCF often suffers from poor throughput performance when the server moves, i.e., after 120 and 240 seconds, while DistGreedy implementation maintains high throughput performance during the moves.

6. CONCLUSION

In this paper, we develop a distributed scheduling scheme that is provably efficient under wireless fading. By taking a local greedy approach, we prove that our scheme is a $\frac{1}{\Delta^*}$-approximation to the Maximum Weighted Independent Set problem, where Δ^* is the maximum conflict degree, and has $O(\log |V|)$ complexity (or polylogarithmic complexity), where $|V|$ is the number of links.

We evaluate our scheme through simulations in grid networks and random networks. The results show that our distributed scheduling scheme is insensitive to parameter settings, and achieves throughput and delay performance similar to those of the optimal solution. We also implement our scheme with hardware by modifying the existing IEEE 802.11 DCF. The experimental results show that our modification results in better throughput performance with low queue length by taking into account time-varying link capacities.

7. ACKNOWLEDGMENTS

This work has been partially supported by NSF grants CNS-1012700 and CNS-0643145, and a MURI grant from the Army Research Office W911NF-08-1-0238, and in part by the Basic Science Research Program through the NRF of Korea, funded by the Ministry of Science, ICT, and Future Planning (No. 2011-0008549).

8. REFERENCES

[1] ALIX system boards. http://pcengines.ch/alix.htm.

[2] Atheros Linux Wireless Drivers. http://wireless.kernel.org/en/users/Drivers/ath5k.

[3] Voyage Linux. http://linux.voyage.hk.

[4] S. Basagni. Finding a Maximal Weighted Independent Set in Wireless Networks. *Telecommunication Systems*, 18:155–168, 2001.

[5] B. Birand, M. Chudnovsky, B. Ries, P. Seymour, G. Zussman, and Y. Zwols. Analyzing the Performance of Greedy Maximal Scheduling via Local Pooling and Graph Theory. *IEEE/ACM Trans. Netw.*, 20(1), Feburary 2012.

[6] L. Bui, S. Sanghavi, and R. Srikant. Distributed Link Scheduling with Constant Overhead. *IEEE/ACM Trans. Netw.*, 17(5):1467–1480, October 2009.

[7] P. Chaporkar, K. Kar, X. Luo, and S. Sarkar. Throughput and Fairness Guarantees Through Maximal Scheduling in Wireless Networks. *IEEE Trans. Inf. Theory*, 54(2):572–594, February 2008.

[8] J. Ghaderi and R. Srikant. Effect of Access Probabilities on the Delay Performance of Q-CSMA Algorithms. In *IEEE INFOCOM*, April 2012.

[9] P. Gupta and P. R. Kumar. The Capacity of Wireless Networks. *IEEE Trans. Inf. Theory*, 46(2):388–404, March 2000.

[10] J.-H. Hoepman. Simple Distributed Weighted Matchings. eprint, October 2004.

[11] L. Jiang and J. Walrand. A Distributed CSMA Algorithm for Throughput and Utility Maximization in Wireless Networks. *IEEE/ACM Trans. Netw.*, 18(13):960–972, June 2010.

[12] C. Joo. On Random Access Scheduling for Multimedia Traffic in Multi-hop Wireless Networks, 2012. to appear in IEEE Trans. Mobile Computing.

[13] C. Joo, G. Sharma, N. B. Shroff, and R. R. Mazumdar. On the Complexity of Scheduling in Wireless Networks. *EURASIP Journal of Wireless Communications and Networking*, October 2010.

[14] C. Joo and N. B. Shroff. Local Greedy Approximation for Scheduling in Multi-hop Wireless Networks. *IEEE Trans. Mobile Computing*, 11(3):414–426, March 2012.

[15] E. Leonardi, M. Mellia, F. Neri, and M. A. Marsan. On the Stability of Input-Queued Switches with Speed-Up. *IEEE/ACM Trans. Netw.*, 9(1), Feburary 2001.

[16] B. Li and A. Eryilmaz. A Fast-CSMA Algorithm for Deadline Constraint Scheduling over Wireless Fading Channels. In *Workshop on Research Allocation and Cooperation in Wireless Networks (RAWNET)*, May 2011.

[17] X. Lin and S. Rasool. Constant-Time Distributed Scheduling Policies for Ad Hoc Wireless Networks. *IEEE Trans. Autom. Control*, 54(2):231–242, Feburary 2009.

[18] X. Lin and N. B. Shroff. The Impact of Imperfect Scheduling on Cross-Layer Congestion Control in Wireless Networks. *IEEE/ACM Trans. Netw.*, 14(2):302–315, April 2006.

[19] X. Liu, E. K. P. Chong, and N. B. Shroff. A Framework for Opportunistic Scheduling in Wireless Networks. *Computer Networks*, 41(4):451–474, March 2003.

[20] M. Lotfinezhad and P. Marbach. Delay Performance of CSMA Policies in Multihop Wireless Networks: A New Perspective. In *Information Theory and Application Workshop*, Feburary 2010.

[21] E. Modiano, D. Shah, and G. Zussman. Maximizing Throughput in Wireless Networks via Gossiping. *Sigmetrics Performance Evaluation Review*, 34(1):27–38, 2006.

[22] M. J. Neely, E. Modiano, and C. E. Rohrs. Dynamic Power Allocation and Routing for Time-varying Wireless Networks. *IEEE J. Sel. Areas Commun.*, 23(1), 2005.

[23] J. Ni, B. Tan, and R. Srikant. Q-CSMA: Queue-Length Based CSMA/CA Algorithms for Achieving Maximum Throughput and Low Delay in Wireless Networks. *IEEE/ACM Trans. Netw.*, 20(3), June 2012.

[24] S. Sakai, M. Togasaki, and K. Yamazaki. A Note on Greedy Algorithms for the Maximum Weighted Independent Set Problem. *Discrete Appl. Math.*, 126(2-3):313–322, March 2003.

[25] S. Sarkar and L. Tassiulas. End-to-end Bandwidth Guarantees Through Fair Local Spectrum Share in Wireless Ad-hoc Networks. In *IEEE CDC*, pages 564–569, December 2003.

[26] G. Sharma, C. Joo, N. B. Shroff, and R. R. Mazumdar. Joint Congestion Control and Distributed Scheduling for Throughput Guarantees in Wireless Networks. *ACM Trans. Model. and Comput. Simul.*, 21(1):5:1–5:25, December 2010.

[27] L. Tassiulas and A. Ephremides. Stability Properties of Constrained Queueing Systems and Scheduling Policies for Maximal Throughput in Multihop Radio Networks. *IEEE Trans. Autom. Control*, 37(12):1936–1948, December 1992.

[28] P.-J. Wan, O. Frieder, X. Jia, F. Yao, X. Xu, and S. Tang. Wireless Link Scheduling under Physical Interference Model. In *IEEE INFOCOM*, April 2011.

[29] X. Wu and R. Srikant. Scheduling Efficiency of Distributed Greedy Scheduling Algorithms in Wireless Networks. In *IEEE INFOCOM*, April 2006.

[30] S. Yun, J. Shin, and Y. Yi. Medium Access over Time-varying Channels with Limited Sensing Cost. *CoRR*, abs/1206.5054, 2012.

On the Performance of Largest-Deficit-First for Scheduling Real-Time Traffic in Wireless Networks

Xiaohan Kang*, Weina Wang*, Juan José Jaramillo† and Lei Ying*

*School of Electrical, Computer and Energy
Engineering
Arizona State University
Tempe, AZ 85287, USA
{xkang6, wwang136, lying6}@asu.edu

†Department of Applied Math and Engineering
Universidad EAFIT
Medellín, Colombia
jjaram93@eafit.edu.co

ABSTRACT

This paper considers the problem of scheduling real-time traffic in wireless networks. We consider an ad hoc wireless network with general interference and general one-hop traffic. Each packet is associated with a deadline and will be dropped if it is not transmitted before the deadline expires. The number of packet arrivals in each time slot and the length of a deadline are both stochastic and follow certain distributions. We only allow a fraction of packets to be dropped. At each link, we assume the link keeps track of the difference between the minimum number of packets that need to be delivered and the number of packets that are actually delivered, which we call deficit. The largest-deficit-first (LDF) policy schedules links in descending order according to their deficit values, which is a variation of the largest-queue-first (LQF) policy for non-real-time traffic. We prove that the efficiency ratio of LDF can be lower bounded by a quantity that we call the real-time local-pooling factor (R-LPF). We further prove that given a network with interference degree β, the R-LPF is at least $1/(\beta + 1)$, which in the case of the one-hop interference model translates into an R-LPF of at least $1/3$.

Categories and Subject Descriptors

C.2.1 [**Computer-Communication Networks**]: Network Architecture and Design—*wireless communication*

General Terms

Theory, performance

Keywords

Stability; real-time scheduling; largest-deficit-first; local-pooling factor; fluid limit

1. INTRODUCTION

With the increasing number of real-time applications in wireless networks, scheduling traffic of packets with hard deadlines has become a very important problem. However, the problem is very challenging due to the stochastic nature of the traffic arrivals and deadlines. Hou et al. first proposed a frame-based analytical framework for studying scheduling real-time traffic in wireless networks [6]. In the frame-based framework it is assumed that each frame is a number of consecutive time slots, and all packets arrive at the beginning of a frame and have to be scheduled before the end of the frame. They also characterized the real-time capacity region and developed the optimal scheduling algorithm for collocated networks. Later, the frame-based framework has been generalized to networks with heterogeneous delays, fading, congestion control, etc. [7, 8, 9, 10, 11] In particular, Jaramillo et al. extended the idea to general arrival/deadline patterns within a frame and general network topology, and found the optimal scheduling policy [11], where they assumed that packets can arrive at any time slot during a frame, and the deadline of a packet can be any time after its arrival and before the end of the frame. Their paper assumes that the arrival and deadline information is available at the beginning of the frame, so future knowledge is assumed. Furthermore, the computational complexity of the optimal algorithm is prohibitively high except for some special cases such as collocated networks.

In this paper, we consider the case of general real-time traffic patterns without the assumption of frames and with a general interference model. Under the general settings, the stability region is difficult to characterize, and the optimal policy is unknown. In this paper, we are interested in the performance of a low-complexity greedy policy called the largest-deficit-first (LDF) policy [6], which is the real-time variation of the longest-queue-first (LQF) policy that iteratively selects the link with the largest deficit that does not interfere with those links that are already selected. It has been shown that the largest-deficit-first policy is optimal for scheduling real-time traffic in collocated networks [6, 11] under the frame-based model. The performance of the LDF in general networks has not been studied.

Since LDF can be directly applied to networks with general, non-frame-based real-time traffic, we are interested in characterizing the performance of LDF. Although the capacity region and optimal scheduling algorithm for networks with general real-time traffic remain unknown, we are able to establish the efficiency ratio of LDF by connecting it to the frame-based optimal scheduling algorithm, and obtain a lower bound on the efficiency ratio in terms of a new quantity, called the real-time local-pooling factor (R-LPF). The R-LPF extends the idea of the local-pooling factor for non-real-time traffic [12] and its extension for fading channels [18].

We show using the fluid limit technique [3] that this R-LPF can be successfully used to provide a minimum performance guarantee of LDF under real-time traffic. More interestingly, we are able to connect the R-LPF with the interference degree, and prove that the R-LPF is bounded by $1/(\beta + 1)$, where β is the interference degree [2]. Our contributions are therefore twofold:

1. We formulate the construction of the R-LPF and prove that it is a lower bound of the efficiency ratio of LDF in the presence of general deadline constraints.

2. We show that in a network with interference degree β, the R-LPF is at least $1/(\beta + 1)$. In particular, the R-LPF is at least $1/3$ in a network with one-hop interference model.

We would like to emphasize again that for general (non-frame-based) real-time traffic, to the best of our knowledge, there are no known theoretical results on any scheduling policy, which makes the lower bound obtained in this paper a novel contribution.

2. MODEL

In this paper, we consider a wireless network consisting of K links. The set of links is denoted by $\mathcal{K}$. Assume time is slotted, and at each time slot one packet can be successfully transmitted over a link if no interfering links are transmitting at the same time. We remark that the constant service rate assumption has been widely used in the literature, e.g., [5, 13]. We consider a general interference model. We call a set of links $\mathcal{Z} \subseteq \mathcal{K}$ a maximal schedule if links in $\mathcal{Z}$ can be scheduled at the same time without interfering with each other, but no other link can be further scheduled without interfering with links in $\mathcal{Z}$. We assume that there are R possible maximal schedules and the set of maximal schedules is represented by a maximal schedule matrix M, which is a K-by-R matrix with binary entries such that each column represents a distinct maximal schedule and the set of links that are included in this schedule have value 1 in that column. For example, let M_r be the rth column of matrix M, then the set of links $\{l \in \mathcal{K}: M_{r,l} = 1\}$ is a maximal schedule, where $M_{r,l}$ is the (r, l) entry of the matrix. By abuse of notation we also let $M = \{M_1, M_2, \ldots, M_R\}$. It is easy to see that any subset of a maximal schedule is a feasible schedule (i.e., all links in that set can be scheduled at the same time).

We consider single hop traffic with deadline constraints. Let $a_l(t)$ denote the number of packets that arrive at the beginning of time slot t at link l, where we assume that all packets have the same size and can be transmitted in a single time slot. We assume that $a(\cdot)$ is a stochastic process that is temporally independent and identically distributed (i.i.d.) and independent across links, with probability mass function (p.m.f.) $(f_l(i): i = 0, 1, 2, \ldots)$. We also assume that $f_l(i) = 0$ for $i > N$; i.e., the number of packets arriving on a link at each time slot is at most N. Denote by α_l the rate of arrivals on link l; i.e., $\alpha_l = \mathbb{E}[a_l(t)] = \sum_{i=1}^{N} i f_l(i)$.

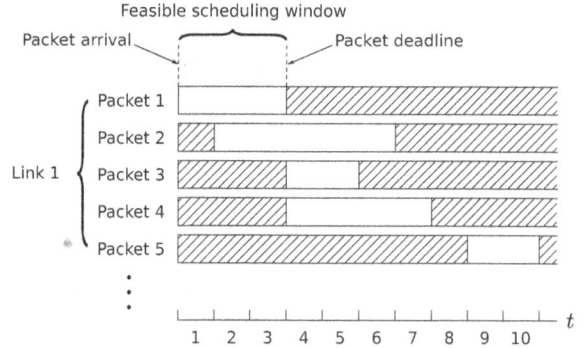

Figure 1: An example of the arrival and maximum delay pattern of packets on a link. For each packet, the beginning of the blank bar is the time slot when that packet arrives, and the end of the blank bar is the deadline associated with that packet. So the feasible scheduling window denoted by the blank bar represents the time slots when the packet is available for transmission, while the shaded part indicates that the packet is not available, either because it has not arrived or because its deadline has passed. Note that here the cumulative numbers of packet arrivals to link 1 are $A_1(\cdot) = (1, 2, 2, 4, 4, 4, 4, 4, 5, 5, \ldots)$ and the maximum delays are $\tau_1(\cdot) = (3, 5, 2, 4, 2, \ldots)$.

Each packet is associated with a maximum delay τ, which is a random variable with integer value between $\tau_{\min}$ and $\tau_{\max}$ and follows a p.m.f. $(\gamma(\tau): \tau_{\min} \leq \tau \leq \tau_{\max})$. Furthermore, let $A_l(t)$ be the cumulative number of packet arrivals to link l up to time slot t for any $l \in \mathcal{K}$ and any nonnegative integer t; i.e., $A_l(t) = \sum_{t'=1}^{t} a_l(t')$. We order the packets arriving on link l according to the arriving time with arbitrary tie-breakings Then we let $b_l(n)$ be the time slot during which the nth packet arrives on link l; i.e., $b_l(n) = \min\{t: A_l(t) \geq n\}$. We also let $e_l(n)$ be the deadline of the nth packet on link l. Note that $e_l(n) = b_l(n) + \tau_l(n)$, where $\tau_l(n)$ is the maximum delay associated with the nth packet on link l. Then the nth arriving packet on link l will be immediately dropped if the deadline is missed. Note that $A(\cdot), \tau(\cdot), b(\cdot)$ and $e(\cdot)$ are all stochastic processes, and $A(\cdot)$ and $\tau(\cdot)$ determine $b(\cdot)$ and $e(\cdot)$. Denote the space of sample paths of the cumulative arrival process $(A_l(\cdot): l \in \mathcal{K})$ and the maximum delay process $(\tau_l(\cdot): l \in \mathcal{K})$ by $\mathcal{A}$. An example of a sample path of the arrival and maximum delay processes on a link during the first 10 time slots is shown in Figure 1.

We assume that each link l is associated with a minimum delivery rate p_l, which is the minimum fraction of packets that should be delivered on link l. The goal of a scheduling policy is to keep the long term delivery rate on link l at least p_l for all $l \in \mathcal{K}$.

Now consider a scheduling policy μ. Denote by $S^\mu(t)$ the cumulative service up to time t, in which $S_l^\mu(t)$ is the service link l received up to time slot t. For any scheduling policy, it is easy to see that the following three conditions hold:

1. (Initialization) $S_l^\mu(0) = 0$ for all $l \in \mathcal{K}$.

2. (Feasibility) The incremental service vector is a feasible schedule; i.e., $0 \preceq S^\mu(t) - S^\mu(t-1) \preceq M_r$ for some $M_r \in M$, for any positive integer t, where $\preceq$ denotes entrywise less than or equal to.

3. (Deadline constraint) All served packets are served before their deadlines. Formally, let $\zeta_l^\mu(n)$ be the time slot in which the nth packet on link l is scheduled by μ if that packet is ever scheduled, and $\zeta_l^\mu(n) = 0$ if that packet is never scheduled by μ. Then the deadline constraint can be stated as follows: For any n and any l with $\zeta_l^\mu(n) > 0$,

$$b_l(n) \leq \zeta_l^\mu(n) \leq b_l(n) + \tau_l(n).$$

In this paper, we will consider a greedy scheduling policy, called Largest-Deficit-First (LDF) [6] based on the following *deficit process* $D^\mu(t)$ (also known as debts or virtual queues)

1. (Initialization) $D_l^\mu(0) = 0$ for all $l \in \mathcal{K}$.

2. (Dynamics) The evolution of the deficit process for link l is given by

$$D_l^\mu(t) = [D_l^\mu(t-1) + (B_l(t) - B_l(t-1)) \\ - (S_l^\mu(t) - S_l^\mu(t-1))]^+,$$

where $(\cdot)^+ = \max\{0, \cdot\}$ and $B_l(t)$ is the cumulative deficit arrival on link l given by

$$B_l(0) = 0$$

and

$$B_l(t) - B_l(t-1) = \sum_{n=A_l(t-1)+1}^{A_l(t)} c_l(n),$$

where by definition $B_l(t) - B_l(t-1) = 0$ if $A_l(t-1) = A_l(t)$, and $c_l(\cdot)$ is an i.i.d. Bernoulli process with mean p_l. Hence $c_l(n)$ determines whether the nth arriving packet on link l is counted as a deficit arrival or not.

Observe from the definition that the deficit process keeps track on the amount of service we owe to a link in order to fulfill the minimum delivery rate. To see that, note that the arrival rate of deficit on link l is $\alpha_l p_l$. The deficit of link l reduces by one when a packet is successfully transmitted over link l before its deadline. So if all deficits are bounded, then the requirements on packet minimum delivery rates are fulfilled.

The LDF scheduling policy is defined as follows. At each time slot, LDF first sorts the links $\mathcal{K}$ according to the current deficits D with arbitrary tie-breaks, and gets the index vector I such that $D_{I_1} \geq D_{I_2} \geq \cdots \geq D_{I_K}$. LDF starts with the *selection* $\mathcal{E} = \{I_1\}$, which only consists of the link with the largest deficit. Then LDF repeatedly considers the link with the next largest deficit I_i for i from 2 to K and adds it into the selection $\mathcal{E}$ if the following two conditions are satisfied:

1. Link I_i does not interfere with any link in $\mathcal{E}$.

2. There is at least one packet available for transmission on link I_i; i.e., $Q_{I_i} > 0$, where Q_l is the number of available packets on link l.

The procedure ends when all links have been considered, and the final selection of links is the desired LDF schedule.

3. PRELIMINARIES

In this section, we introduce basic definitions on stability and efficiency ratio that will be used in the following sections. We first define the stability of the system [17].

DEFINITION 1. *The system is* stable *under a scheduling policy μ if the corresponding deficit process $D^\mu(\cdot)$ satisfies*

$$\limsup_{C\to\infty} \limsup_{t\to\infty} \Pr\left(\sum_{l\in\mathcal{K}} D_l^\mu(t) \geq C\right) = 0.$$

Obviously, the stability of the system depends on the arrival distributions given by $f(\cdot)$, the maximum delay distribution given by $\gamma(\cdot)$, and the required minimum delivery rate $p = (p_l \colon l \in \mathcal{K})$. Without loss of generality, we fix f and γ and consider the stability of the system in terms of the deficit arrival rate $\lambda = (\lambda_l \colon l \in \mathcal{K})$ with $\lambda_l = \alpha_l p_l$. We then have the following definition for characterizing such a relation.

DEFINITION 2. *The deficit arrival rate vector λ is supportable by a scheduling policy if the system is stable under that policy with deficit rate λ_l for each link l.*

DEFINITION 3. *The stability region of a scheduling policy μ is*

$$\Lambda_\mu = \{\lambda \succeq 0 \colon \lambda \text{ is supportable by } \mu\},$$

where $\succeq$ denotes pairwise greater than or equal to.

Let the set of all causal scheduling policies be $\mathcal{M}$, where a causal scheduling policy, also known as an online policy, is one that makes decision on past and statistical information but not future information. We then have the following characterization.

DEFINITION 4. *The stability region of the system is*

$$\Lambda = \bigcup_{\mu\in\mathcal{M}} \Lambda_\mu.$$

That is, the stability region is the set of deficit arrival vectors that can be supported by some causal scheduling policy.

For a given scheduling policy μ, the efficiency ratio of the scheduling policy is defined as follows.

DEFINITION 5. *The efficiency ratio of a scheduling policy μ is*

$$\gamma_\mu^* = \sup\{\gamma \colon \gamma\Lambda \subseteq \Lambda_\mu\}.$$

While refined characterizations of the stability region are possible [14, 15], the efficiency ratio is still a critical metric to evaluate the throughput performance of a scheduling policy.

4. MAIN RESULTS

In this section we present the main results of the LDF policy for scheduling real-time traffic in wireless networks. The first result is Theorem 1, which provides a lower bound on the efficiency ratio of the LDF policy, called the real-time local-pooling factor. The second result is Theorem 2, which states that the efficiency ratio is at least $1/(\beta+1)$ in a network with interference degree β.

We provide a roadmap of the proof of Theorem 1 in Figure 2. The goal of Theorem 1 is to establish the connection between Λ, the stability region of the system, and Λ_{LDF}, the stability region of the LDF policy. However, characterizing Λ turns out to be extremely difficult due to the general arrival and maximum delay distributions. We therefore have to introduce a region called $\Lambda_{\text{NC}}(F)$, which is the stability region by dividing the time into frames with length F and

$$\Lambda_{\mathrm{LDF}} \ - \ - \ - \ - \ - \ \Lambda$$

$$\Lambda_{\mathrm{NC}}(F)$$

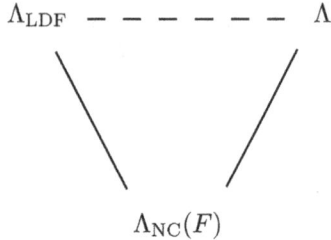

Figure 2: A roadmap of the proof of Theorem 1

assume (i) all information within a frame (arrivals and maximum delays) are known at the beginning of a frame and (ii) at the end of a frame, all packets that have not been transmitted are dropped. The region is denoted by $\Lambda_{\mathrm{NC}}(F)$ since the frame length is F and the system is non-causal because of condition (i).

This novel frame concept was first introduced by Hou et al. [6] for real-time scheduling in wireless networks and provides an analytical framework for understanding real-time communication in wireless networks. The framework has then been extended to general networks and traffic patterns. In particular, the capacity of the non-causal system and heterogeneous deadlines has been characterized by Jaramillo et al. [11]; i.e., $\Lambda_{\mathrm{NC}}(F)$ is known.

We will use $\Lambda_{\mathrm{NC}}(F)$ to bridge Λ and Λ_{LDF}. In the theorem, we will first show that

$$\mathrm{int}(\Lambda) \subseteq \liminf_{F \to \infty} \Lambda_{\mathrm{NC}}(F),$$

where $\mathrm{int}(\Lambda)$ is the interior of the set Λ and $\liminf_{F \to \infty} \Lambda_{\mathrm{NC}}(F)$ is the limit set of $\Lambda_{\mathrm{NC}}(F)$ as F goes to infinity. After that, we will prove that

$$\sigma^* \, \mathrm{int}\left(\Lambda_{\mathrm{NC}}(F)\right) \subseteq \Lambda_{\mathrm{LDF}},$$

where σ^* is the constant, called the real-time local-pooling factor whose definition is presented in Section 4.1. Combining the two results together, we will be able to prove that σ^* is a lower bound on the efficiency ratio. *We remark that the second step is non-trivial since we will compare the time-slot-based, causal LDF (not frame based LDF) with the frame-based, non-causal system.*

As for the proof of Theorem 2, we first show that the real-time local-pooling factor used in Theorem 1 has a lower bound, which is the ratio of the number of links scheduled under the (causal) LDF to the maximum number of links that can be scheduled within a frame. A similar result has been observed by Reddy et al. (Theorem 3 [18]) for characterizing the local-pooling factor for fading channels. We then make use of the fundamental fact that a one-hop neighborhood of a link under the one-hop interference model contains at most two scheduled links in one time slot, and group the scheduled links by LDF and any arbitrary policy in such a way that each link scheduled by LDF corresponds to at most three links by the other policy, with the correspondence covering all the scheduled links by both policies. At that point, the $1/3$ result can be obtained. The proof can be easily generalized to get the $1/(\beta + 1)$ bound where the interference degree is β.

4.1 Real-Time Local-Pooling Factor

We will define a quantity analogous to the local-pooling factor [13] and the fading local-pooling factor [18]. Before we do that, we need the following two definitions.

DEFINITION 6. *A non-causal, frame-based scheduling policy called F-framed for abbreviation is defined as follows: the packet arrivals and deadlines in the kth frame are known at the beginning of the frame and all packets that arrive during the kth frame are dropped at the end of the frame if not transmitted. Formally, for any $l \in \mathcal{K}$ and positive integer n with $\zeta_l^\mu(n) > 0$, there exists a positive integer k such that*

$$kF + 1 \le b_l(n) \le \zeta_l^\mu(n) \le (k+1)F,$$

where $\zeta_l^\mu(n)$ was defined in Section 2 in the deadline constraint condition.

Let the set of all F-framed policies be $\mathcal{M}_{\mathrm{NC}}(F)$. Note that $\mathcal{M}_{\mathrm{NC}}(F)$ is not a subset of $\mathcal{M}$ since policies in $\mathcal{M}_{\mathrm{NC}}(F)$ can be non-causal. The frame concept (alternatively called intervals or periods) has been used in the literature for tractable analytical analysis of delay constrained traffic [7, 8, 9, 10, 11], where packets that arrive in a frame have deadlines in the same frame. In this paper, we adopt this concept to derive the real-time local-pooling factor for the general traffic model.

DEFINITION 7. *The stability region of F-framed policies for a positive integer F is*

$$\Lambda_{\mathrm{NC}}(F) = \bigcup_{\mu \in \mathcal{M}_{\mathrm{NC}}(F)} \Lambda_\mu.$$

We now introduce some notations needed for the main results. Let $\mathcal{J}(F)$ be the set of arrival and maximum delay patterns in a frame of F time slots. We will call an element of $\mathcal{J}(F)$ an F-pattern. An F-pattern is represented by $J = (A, \tau)$ with $A = (A_l(t): l \in \mathcal{K}, 1 \le t \le F)$ and $\tau = (\tau_l(n): l \in \mathcal{K}, 1 \le n \le A_l(F))$, where $A_l(t)$ is the cumulative packet arrival to link l by time slot t in the frame, and $\tau_l(n)$ is the maximum delay associated with the nth packet on link l. Due to the i.i.d. distributions of the arrival and deadline given by f and γ, there is a stationary distribution of the set of F-patterns, denoted by $(\pi(J): J \in \mathcal{J}(F))$.

For a given F-pattern $J = (A, \tau)$, a *schedule* $s = (s_l(n): l \in \mathcal{K}, 1 \le n \le A_l(F))$ specifies the time slot at which each packet is scheduled to be transmitted (if it ever gets scheduled), where $s_l(n)$ is a nonnegative integer that indicates the nth packet on link l is scheduled at time slot $s_l(n)$ if $s_l(n) \in \{1, 2, \ldots, F\}$, and is never scheduled if $s_l(n) = 0$. A schedule s is *feasible* for the F-pattern J if each scheduled packet is scheduled within its feasible scheduling window and no two interfering packets (either two packets on the same link or two packets on two interfering links) are scheduled at the same time slot. We also say that a schedule s is *maximal* for J if no more packets can be further scheduled (i.e., no $s_l(t)$ can be changed from 0 to a positive integer) without breaking feasibility. We denote the maximal feasible schedules for J by $S^*(J)$.

We define the *total service vector of schedule s* to be the column vector $W(s) = (W_i(s): i \in \mathcal{K})$ with $W_i(s) = \sum_{n=1}^{A_i(F)} I(s_i(n) \neq 0)$, where $I(\cdot)$ is the indicator function. Then $W(s)$ is the vector of total number of scheduled packets on each link for the schedule s. Let the *maximal service*

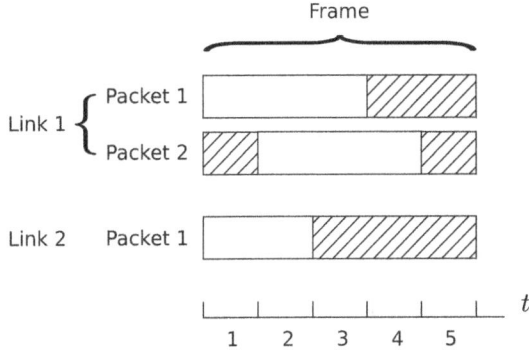

Figure 3: An example of a 5-pattern for two links

matrix for J be

$$M_J = \{W(s) \colon s \in S^*(J)\}.$$

Then the columns of M_J are the total service vectors of the maximal schedules. We note that M_J does not contain all-zero columns if and only if J includes at least one packet arrival on some link, since schedules in $S^*(J)$ are maximal. Similarly, define $M_{J,L}$ to be the maximal service matrix restricted to the set of links L for given pattern J. Then $M_{J,L}$ has no all-zero columns if and only if the pattern J includes at least one packet on some link in L. Also note that $M_{J,L}$ has $|L|$ rows while M_J has $|\mathcal{K}|$ rows.

We use the example in Figure 3 to illustrate the above notations and the concept of the maximal service matrix. As shown in the figure, we consider a frame with size 5 and a 5-pattern J with two packets arriving to link 1 and one packet arriving to link 2, whose arriving times and deadlines are indicated by the blank bars in the figure. The corresponding pattern can be represented by $J = (A, \tau)$, where $A_1(\cdot) = (1, 2, 2, 2, 2)$, $A_2(\cdot) = (1, 1, 1, 1, 1)$, $\tau_1(\cdot) = (3, 3)$, and $\tau_2(\cdot) = (2)$. Assume the two links interfere with each other, so at each time slot only one of them can be scheduled. We can check that there are eight maximal feasible schedules in $S^*(J)$ as follows:

$$s^1 = \begin{pmatrix} 1 & 2 \\ 0 & \end{pmatrix}, s^2 = \begin{pmatrix} 1 & 3 \\ 2 & \end{pmatrix}, s^3 = \begin{pmatrix} 1 & 4 \\ 2 & \end{pmatrix}, s^4 = \begin{pmatrix} 2 & 3 \\ 1 & \end{pmatrix},$$

$$s^5 = \begin{pmatrix} 2 & 4 \\ 1 & \end{pmatrix}, s^6 = \begin{pmatrix} 3 & 2 \\ 1 & \end{pmatrix}, s^7 = \begin{pmatrix} 3 & 4 \\ 1 & \end{pmatrix}, s^8 = \begin{pmatrix} 3 & 4 \\ 2 & \end{pmatrix},$$

where the first row of s^i is the schedule for the two packets on link 1, and the second row is the schedule for the packet on link 2. Then the total service vectors are

$$W(s^1) = \begin{pmatrix} 2 \\ 0 \end{pmatrix} \text{ and } W(s^i) = \begin{pmatrix} 2 \\ 1 \end{pmatrix} \text{ for } 2 \le i \le 8.$$

Hence the maximal service matrix is

$$M_J = \begin{pmatrix} 2 & 2 \\ 0 & 1 \end{pmatrix}.$$

We remark from the above example that unlike in the scenario of non-real-time traffic [5], the total service vector of one maximal schedule could be dominated by that of another in the real-time setting. Thus the maximal service matrix M_J can be huge and hard to compute, especially for large frame size F and complex traffic pattern J.

DEFINITION 8. *The real-time local-pooling factor (R-LPF) for the F-framed scheduling policies for a set of links $L \subseteq \mathcal{K}$ is*

$$\sigma_L^*(F) = \inf\{\sigma \colon \exists \phi_1, \phi_2 \in \Phi_L(F) \text{ such that } \sigma\phi_1 \succeq \phi_2\},$$

where $\Phi_L(F)$ is the stability region restricted to the set of links $L \subseteq \mathcal{K}$ defined by

$$\Phi_L(F) = \{\phi \colon \phi = \sum_{J \in \mathcal{J}(F)} \pi(J)\eta_J, \eta_J \in \mathcal{CH}(M_{J,L})\},$$

and $\mathcal{CH}(M_{J,L})$ defines the convex hull over the columns of the matrix $M_{J,L}$.

DEFINITION 9. *The R-LPF for the F-framed scheduling policies is*

$$\sigma^*(F) = \min_{L \subseteq \mathcal{K}} \sigma_L^*(F).$$

DEFINITION 10. *The R-LPF for the system is*

$$\sigma^* = \liminf_{F \to \infty} \sigma^*(F).$$

We then have the following theorem stating that R-LPF is a lower bound on the efficiency ratio of LDF. The proof is presented in Section 5.1.

THEOREM 1. $\gamma_{\mathrm{LDF}}^* \ge \sigma^*$.

By the definition of R-LPF, we can get the R-LPF by solving the following linear program for each $L \subseteq \mathcal{K}$, as suggested in [5]:

$$\sigma_L^*(F) = \max_{x, (\rho(J)), (\theta(J))} \sum_{J \in \mathcal{J}(F)} \pi(J)\rho(J)$$

$$\text{s.t.} \quad x' M_{J,L} \succeq \rho(J)\mathbf{1}' \quad \forall J \in \mathcal{J}(F)$$

$$x' M_{J,L} \preceq \theta(J)\mathbf{1}' \quad \forall J \in \mathcal{J}(F)$$

$$\sum_{J \in \mathcal{J}(F)} \pi(J)\theta(J) = 1$$

$$x \succeq 0,$$

where x is a column vector of length L, $\rho(J)$ and $\theta(J)$ are scalars for all J, $\mathbf{1}$ is the all-one column vector of length equal to the number of columns of $M_{J,L}$, which we denote by $r_{J,L}$, and x' is the transpose of x. That said, computing the exact R-LPF is usually complex, as it involves roughly

$$\sum_{L \subseteq \mathcal{K}} \left(2 \sum_{J \in \mathcal{J}(F)} r_{J,L} + 1 \right)$$

constraints for each F, which increases exponentially with both the size of the network $\mathcal{K}$ and the frame size F. Thus, we seek lower bounds of the R-LPF in the next subsection.

4.2 A Lower Bound on R-LPF under One-Hop Interference Model

THEOREM 2. *Given a network with interference degree β, which is the maximum number of links in the interference neighborhood of some link that can be scheduled without interference, we have*

$$\sigma^* \ge \frac{1}{\beta + 1}. \tag{1}$$

So under the one-hop interference model, $\sigma^* \ge 1/3$.

Remark. This result for one-hop interference model is related to the well-known result that LQF has efficiency ratio at least 1/2 in packet switches [19, 4] and in wireless networks under the one-hop interference [16, 20]. The result makes use of the fact that a one-hop neighborhood contains at most two scheduled links under such an interference model.

Proof outline. We only give the proof for the one-hop interference model case where $\beta = 2$ since the general case follows from nearly identical arguments. The result is proved by showing that $\sigma_L^*(F) \geq 1/3$ for any L and F. The proof consists of two lemmas. The first lemma gives a lower bound on the R-LPF similar to Theorem 3 in [18]. Basically the lower bound on $\sigma_L^*(F)$ can be found by finding the ratio between the "smallest" maximal schedule and the "largest" maximal schedule, both in terms of the number of links scheduled.

The second lemma uses a *multigraph* representation of the network. A multigraph is a graph that allows multiple edges between two nodes. The idea is that we transform the original graph of the network into a multigraph by replacing a link with multiple packets by multiple links with a single packet on each link. Then we prove that the number of links scheduled under LDF over a frame is at least 1/3 of any scheduling policy over the same frame.

The detailed proofs can be found in Section 5.2.

5. PROOFS

5.1 Proof of Theorem 1

LEMMA 1. *The stability region of the F-framed policies can be characterized by*

$$\overline{\Lambda_{\mathrm{NC}}(F)} = \left\{ \lambda \succeq 0 \colon \lambda F \preceq \sum_{J \in \mathcal{J}(F)} \pi(J)\eta_J, \eta_J \in \mathcal{CH}(M_J) \right\},$$

where $\overline{A}$ denotes the closure of A and $\mathcal{CH}(M_J)$ is the convex hull over the set of columns of M_J.

PROOF OF LEMMA 1. If λ is strictly outside the stability region, it can be proved that the total amount of deficits increase to infinity with probability one using the strictly separating hyperplane theorem [1] and Lyapunov drift arguments. If λ is strictly inside the stability region, then we can find $\eta = (\eta_J \colon J \in \mathcal{J}(F))$ that dominates λ and make the long-term-average of the scheduling process be at least $\eta \succ \lambda$, where $\succ$ denotes strict pairwise greater than. So the system can be stabilized. $\square$

We then have the next lemma.

LEMMA 2. *If $F > \tau_{\max}$, then*

$$\Lambda_{\mathrm{LDF}} \supseteq \sigma^* \cdot \mathrm{int}(\Lambda_{\mathrm{NC}}(F)).$$

PROOF OF LEMMA 2. Let $\lambda' \in \mathrm{int}(\Lambda_{\mathrm{NC}}(F))$ and let $\lambda = \sigma^* \lambda'$. Then by the definition of interior point and the characterization of $\Lambda_{\mathrm{NC}}(F)$ in Lemma 1, there exist $(\xi_J \colon J \in \mathcal{J}(F))$ with $\xi_J \in \mathcal{CH}(M_J)$ for each $J \in \mathcal{J}(F)$ and $\delta > 0$ such that

$$\lambda' + \delta \mathbf{1} \preceq \frac{1}{F} \sum_{J \in \mathcal{J}(F)} \pi(J)\xi_J, \qquad (2)$$

where $\mathbf{1}$ is a vector with all 1's. Let $D(t)$ and $S(t)$ be the cumulative deficit and service processes under LDF (without frame). We sample $D(t)$ and $S(t)$ every F time slots, and let the fluid limits of sampled $D(tF)$ and $S(tF)$ be $\bar{D}(t)$ and $\bar{S}(t)$. The construction of fluid limits follows the standard procedure given in [3]. Let $L_0(t)$ be the set of links with the largest deficit fluids, and let $L(t) \subseteq L_0(t)$ be the set of links in $L_0(t)$ with largest derivatives at time t; i.e.,

$$L_0(t) = \left\{ l \in \mathcal{K} \colon \bar{D}_l(t) = \max_{i \in \mathcal{K}} \bar{D}_i(t) \right\}$$

and

$$L(t) = \left\{ l \in L_0(t) \colon \frac{\mathrm{d}}{\mathrm{d}t} \bar{D}_l(t) = \max_{i \in L_0(t)} \frac{\mathrm{d}}{\mathrm{d}t} \bar{D}_i(t) \right\},$$

where we assume t is a regular point; i.e., the derivatives of the fluid limits exist at t. Then we can construct $\eta_J \in \mathcal{CH}(M_J)$ such that for any $l \in L(t)$, the service fluids satisfy

$$\frac{\mathrm{d}}{\mathrm{d}t} \bar{S}_l(t) \geq \sum_{J \in \mathcal{J}(F)} \pi(J)\eta_{J,l} - \tau_{\max}, \qquad (3)$$

where $\eta_{J,l}$ is the l'th entry of the vector $\eta_J \in \mathcal{CH}(M_J)$ for all $J \in \mathcal{J}(F)$. To understand (3), note that $\bar{S}(t)$ is the fluid limit of $S(t)$ sampled every F time slots, so the derivative of $\bar{S}(t)$ is the average service over F time slots under LDF. Now consider a frame of F time slots with arrival and maximum delay pattern J, and denote by s_J^{LDF} the link schedule under LDF during the F time slots. We next construct another link schedule s_J^F, which is a maximal link schedule under the F-framed policy. The construction is to remove those transmissions in s_J^{LDF}, which serve packets that arrived before the frame started, and then add more transmissions to make it a maximal schedule. So for link l,

$$W(s_J^{\mathrm{LDF}})_l - o_{J,l} + n_{J,l} = W(s_J^F)_l,$$

where $o_{J,l}$ is the number of removed transmissions on link l, $n_{J,l}$ is the number of added transmissions on link l, $W(s_J^F) \in \mathcal{M}_J$, and $(\cdot)_l$ denotes the lth component of the vector. Note that those removed transmissions must occur at the first $\tau_{\max}$ time slots because the maximum delay is $\tau_{\max}$ so none of the packets that arrived before the frame can be transmitted after the first $\tau_{\max}$ time slots. This also implies that the added transmissions must be in the first $\tau_{\max}$ time slots as well. Therefore, $n_{J,l} \leq \tau_{\max}$, and

$$W(s_J^{\mathrm{LDF}})_l \geq W(s_J^F)_l - \tau_{\max}$$

holds for any J.

Now assuming $\bar{D}_l(t) > 0$ for $l \in L(t)$, the derivative of $\bar{D}_l(t)$ is

$$\frac{\mathrm{d}}{\mathrm{d}t}\bar{D}_l(t) = \lambda_l F - \frac{\mathrm{d}}{\mathrm{d}t}\bar{S}_l(t) \tag{4}$$

$$\leq \lambda_l F - \sum_{J \in \mathcal{J}(F)} \pi(J)\eta_{J,l} + \tau_{\max} \tag{5}$$

$$\leq \sigma^* \left(\sum_{J \in \mathcal{J}(F)} \pi(J)\xi_{J,l} - \delta F \right)$$
$$- \sum_{J \in \mathcal{J}(F)} \pi(J)\eta_{J,l} + \tau_{\max} \tag{6}$$

$$= \left[\sigma^* \left(\sum_{J \in \mathcal{J}(F)} \pi(J)\xi_{J,l} \right) \right. \tag{7}$$

$$\left. - \left(\sum_{J \in \mathcal{J}(F)} \pi(J)\eta_{J,l} \right) \right] \tag{8}$$

$$- \sigma^* \delta F + \tau_{\max}, \tag{9}$$

where (5) comes from (3), and (6) holds because $\lambda = \sigma^* \lambda'$ and inequality (2). By the definition of R-LPF and the fact that $L(t)$ has higher scheduling priority over $\mathcal{K}\backslash L(t)$, there exists $i \in L(t)$ such that

$$\sigma^* \left(\sum_{J \in \mathcal{J}(F)} \pi(J)\xi_{J,i} \right) \leq \left(\sum_{J \in \mathcal{J}(F)} \pi(J)\eta_{J,i} \right).$$

Thus by definition of $L(t)$,

$$\frac{\mathrm{d}}{\mathrm{d}t}\bar{D}_l(t) = \frac{\mathrm{d}}{\mathrm{d}t}\bar{D}_i(t) \leq \tau_{\max} - \sigma^* \delta F.$$

We note that for any positive integer k,

$$\Lambda_{\mathrm{NC}}(F) \subseteq \Lambda_{\mathrm{NC}}(kF),$$

since any F-framed policy is a valid but more restrictive kF-framed policy. Then (2) holds with the same δ for any frame size kF. Thus for large enough integer k, the deficit fluid limits associated with the frame size kF satisfy

$$\frac{\mathrm{d}}{\mathrm{d}t}\bar{D}_l(t) \leq \tau_{\max} - \sigma^* \delta kF \leq -\epsilon < 0$$

for some $\epsilon > 0$. Then using the results from [3] we get that the system is positive recurrent under LDF, and hence stable in the sense of Definition 1. $\square$

LEMMA 3. *For any F,*

$$\Lambda_{\mathrm{NC}}(F) \supseteq \mathrm{int}\left(\Lambda \cap \left(\Lambda - \frac{\tau_{\max}}{F}\mathbf{1} \right) \right),$$

where $\Lambda - \frac{\tau_{\max}}{F}\mathbf{1} = \{\lambda - \frac{\tau_{\max}}{F}\mathbf{1} : \lambda \in \Lambda\}$.

PROOF OF LEMMA 3. Note that given the schedules of any causal policy, we can convert them into valid schedules under the F-framed policy by remove those transmissions that serve those packets whose arrival times and transmission times are not in the same frame. For each link, we need to remove at most $\tau_{\max}$ transmissions within a frame of F time slots, which is equivalent to at most $\tau_{\max}/F$ packets per time slot. So the lemma holds. $\square$

We can now proceed to prove Theorem 1.

PROOF OF THEOREM 1. By Lemma 2 and Lemma 3, we have

$$\Lambda_{\mathrm{LDF}} \supseteq \sigma^* \cdot \mathrm{int}\left(\Lambda \cap \left(\Lambda - \frac{\tau_{\max}}{F}\mathbf{1} \right) \right).$$

The theorem holds by letting $F \to \infty$. $\square$

5.2 Proof of Theorem 2

LEMMA 4.

$$\sigma_L^*(F) \geq \frac{\sum_{J \in \mathcal{J}(F)} \pi(J)n(M_{J,L})}{\sum_{J \in \mathcal{J}(F)} \pi(J)N(M_{J,L})},$$

where $n(M) = \min_j \sum_i M_{ij}$ is the minimum number of scheduled links in a maximal schedule of M, and $N(M) = \max_j \sum_i M_{ij}$ is the maximal number of scheduled links in a maximal schedule of M.

The proof is almost the same as the proof of Theorem 3 in [18] and is thus omitted. We bound the ratio between $n(M_{J,L})$ and $N(M_{J,L})$ under the one-hop interference model in the following lemma.

LEMMA 5. *Under the one-hop interference model, for any $J \in \mathcal{J}(F)$ and any $L \subseteq \mathcal{K}$,*

$$\frac{n(M_{J,L})}{N(M_{J,L})} \geq \frac{1}{3}.$$

PROOF OF LEMMA 5. We fix $J \in \mathcal{J}(F)$ and $L \subseteq \mathcal{K}$ and focus on the arrival and maximum delay pattern given by J restricted to the subset of links L. For each link $l \in L$, replace it with n links (each of which has a single packet arrival in the frame) if the total number of packets arriving on l in the frame is $n \geq 2$, leave it along if the total number of packets arriving on l in the frame is 1, and remove it from our consideration if no packet arrives in this frame according to J. We then get a multigraph whose set of links is denoted by $\mathcal{K}'$, where $K' = |\mathcal{K}'|$ equals the total number of packets arriving on L in the original graph according to J, and each link in $\mathcal{K}'$ represents a packet in the original graph with arriving time and deadline given by J. The interference model of $\mathcal{K}'$ inherits from the 1-hop interference model of $\mathcal{K}$; i.e., two links in $\mathcal{K}'$ interference with each other if they share an end node. Let $I(l)$ denote the set of links that interfere with link l in $\mathcal{K}'$, and by convention assume $l \in I(l)$.

A schedule over the multigraph $\mathcal{K}'$ in the frame is represented by a function

$$s \colon \mathcal{K}' \times \{1, 2, \ldots, F\} \to \{0, 1\}$$
$$(i, t) \mapsto s_i(t)$$

with $s_i(t) = 1$ if link $i \in \mathcal{K}'$ is scheduled by s at time slot t, and $s_i(t) = 0$ otherwise. A schedule s is feasible if no two interfering links are scheduled at the same time slot, no link is scheduled before its arriving time or after its deadline, and each link is scheduled at most once during the entire frame. A feasible schedule s is maximal if no more links can be scheduled without breaking the feasibility. We note that a feasible (or maximal, respectively) packet schedule for J over the original set of links $\mathcal{K}$ corresponds to a feasible (or maximal, respectively) schedule over the multigraph $\mathcal{K}'$ given by J. Let $\mathrm{supp}(s)$ be the support of s, i.e., the set of (link, time slot) pairs of scheduled links by s. Let $\|s\| = \sum_i \sum_t s_i(t)$ be the total number of links scheduled by s. We note that $\|s\| = |\mathrm{supp}(s)|$.

Define the one-hop interference neighborhood of the (link, time slot) pair (i,t) to be the interfering links of link i at time slot t, i.e.,

$$I(i,t) = I(i) \times \{t\} \subseteq \mathcal{K}' \times \{1,2,\ldots,F\}.$$

We now consider another maximal schedule u, and the set of (link, time slot) pairs that are in supp(u) but not in the one-hop interference neighborhoods of (link, time slot) pairs in supp(s), i.e., the set

$$P = \text{supp}(u) \Big\backslash \bigcup_{(i,t) \in \text{supp}(s)} I(i,t).$$

We note that for any (link, time slot) pair $(j,t') \in P$, we must have $(j,\tilde{t}) \in \text{supp}(s)$ for some $\tilde{t} \neq t'$; in other words, link j must be scheduled in s at some time slot other than t'. This holds because otherwise, link j can be added to s at time t without interfering any links in s (note that (j,t') is not in any interference neighborhoods of (link, time slot) pairs in supp(s)). We use an example in Figure 4 to illustrate this point.

From the analysis above, we know that

$$|P| \leq |\text{supp}(s)|.$$

Furthermore, schedule u can have at most two active (link, time slot) pairs in $I(i,t)$ for any $(i,t) \in \text{supp}(s)$ because of the one-hop interference [19, 4, 16, 20]. Therefore, we have

$$
\begin{aligned}
|\text{supp}(u)| &\leq |P| + \sum_{(i,t) \in \text{supp}(s)} \Big| \text{supp}(u) \bigcap I(i,t) \Big| \\
&\leq |\text{supp}(s)| + 2|\text{supp}(s)| \\
&= 3|\text{supp}(s)|.
\end{aligned}
$$

Since s and u are any arbitrary maximal schedules, we have

$$\frac{n(M_{J,L})}{N(M_{J,L})} \geq \frac{1}{3}.$$

$\square$

Theorem 2 is an immediate result of Lemmas 4 and 5, given the one-hop interference model. The proof for general interference is almost identical, and therefore omitted.

6. DISCUSSIONS

In this section we illustrate that the bound in Theorem 2 can be tight in the scenario of collocated networks, where at most one link can be scheduled at each time slot. Notice that in collocated networks any two links interfere with each other, so the interference degree of the network is $\beta = 1$. Hence according to Theorem 2, LDF achieves at least half of the stability region. We now show that there exists an adversarial traffic pattern such that LDF cannot achieve a fraction greater than half of the whole stability region.

Consider a collocated network with two links interfering each other. Suppose the traffic pattern is given as in Figure 5. Assume that the deficits for both links are the same at the beginning of time slot 0. Also assume that when there is a packet arriving to each link (time slots 1, 3, 5, 7, ...), the deficits on both links increase by one with probability $1/2+\epsilon$ for some small positive ϵ, and remain unchanged with probability $1/2 - \epsilon$. This results in minimum delivery rates $p_i = 1/2 + \epsilon$ for $i = 1, 2$. We further assume that when the deficits on the two links are equal, the tie-breaking rule of

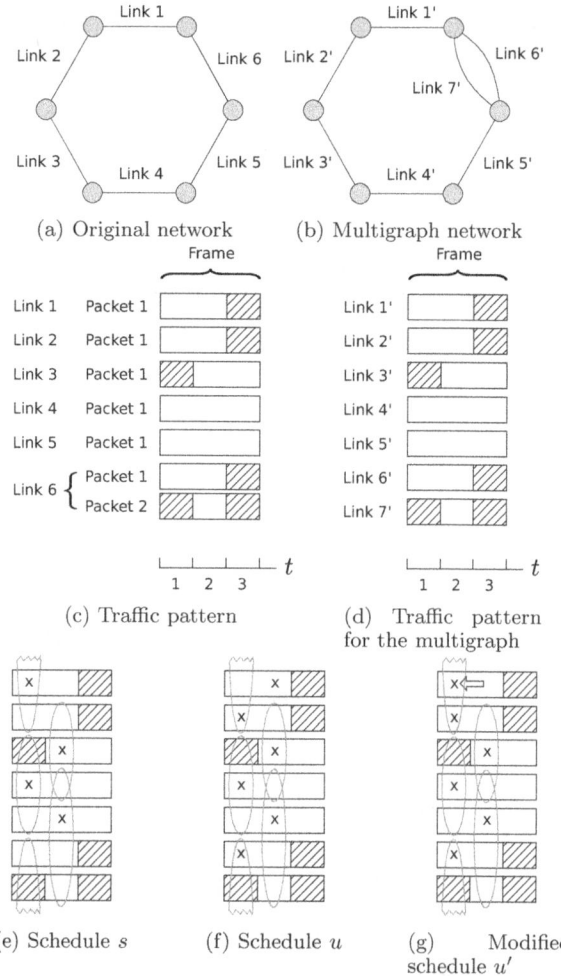

(a) Original network   (b) Multigraph network

(c) Traffic pattern   (d) Traffic pattern for the multigraph

(e) Schedule s (f) Schedule u (g) Modified schedule u'

Figure 4: Consider a six-cycle network with one-hop interference as in (a) and traffic pattern for a frame of 3 time slots as in(c). The network is converted into a multigraph in (b) by dividing the two packets on link 6 into two links with the same end nodes, and the corresponding traffic pattern J is shown in (d). Two maximal schedules, s and u, are given in (e) and (f) with x's denoting the scheduled links, and the one-hop neighborhoods of the scheduled links of s are illustrated by circles. We see that the one-hop neighborhoods of s cover all scheduled links by s, but miss one link scheduled by u. However, as shown in (g), the missed link can be inserted into the one-hop neighborhood in the first time slot so that all scheduled links by u are now covered by the one-hop neighborhoods of s.

LDF gives priority to Link 2. Then one can easily see that LDF schedules Link 2 at time slots 1, 5, 9, ..., and schedules Link 1 at time slots 3, 7, 11, ..., while LDF idles at even time slots. Then the average deficit arrival to each link per time slot is $1/4 + \epsilon/2$, and the average deficit departure from each link per time slot is $1/4$. Hence the deficits are not stable under LDF given this traffic pattern. However, one would notice that the optimal scheduler could schedule Link 1 in time slots $4k$ and $4k + 1$ and schedule Link 2 in time slots $4k + 2$ and $4k + 3$, for all positive integer k. Hence the optimal scheduler can stabilize the system when the mini-

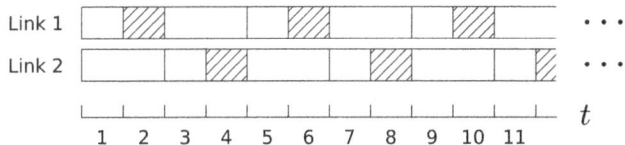

Figure 5: An adversarial traffic pattern for a collocated network with two links. Each blank bar indicates the arriving time and deadline of a real-time packet. The packets arrive on Link 1 at the beginning of time slots 1, 3, 5, 7, ..., and must be scheduled before the end of time slots 1, 4, 5, 8, ... The packets arrive on Link 2 at the beginning of time slots 1, 3, 5, 7, ..., and must be scheduled before the end of time slots 2, 3, 6, 7, ...

mum delivery rates are $p_i = 1$ for $i = 1, 2$. By making ϵ arbitrarily small we can see that the lower bound of $1/2$ on the efficiency ratio of LDF is tight in this two-link collocated network.

7. SIMULATIONS

In this section we use simulations to evaluate the stability performance of LDF. Since to the best of our knowledge, neither the stability region nor an optimal scheduling policy has been obtained in the literature, we do not have a benchmark for the stability performance of LDF. As a result, we compare LDF to two other scheduling policies that do not depend on frames and evaluate the performance using simulations. The first simple scheduling policy we consider is RandMax, which randomly chooses a maximal schedule over the links with packets in each time slot. The other one is MaxWeight, which chooses a maximal schedule with the maximum deficit sum over the links with packets in each time slot.

We first considered a 4-link linear network with one-hop interference. We assumed the packet arrival distribution is binomial with number of trials 2 and success probability 0.5, and the maximum delay distribution is uniform over $\{2, 3, 4\}$. This gives us packet arrival rate $\bar{\alpha} = 1$ and mean maximum delay $\bar{\tau} = 3$. We varied the minimum delivery rate to vary the deficit arrival rate. We compare the average deficit sums of the last 1,000 iterations under the three policies, where each simulation is run for 100,000 iterations. The results are shown in Figure 6. As can be observed from the figure, LDF and MaxWeight have similar stability performance, achieving a maximum deficit arrival rate of roughly 0.5 and significantly outperform the simple RandMax policy, which achieves a maximum deficit arrival rate of roughly 0.33. We further remark that for non-real-time traffic, the maximum deficit arrival rate is 0.5. Thus both LDF and MaxWeight have a near-optimal performance in this case.

We also consider a nine-cycle network with two-hop interference, whose non-real-time local-pooling factor is $2/3$. The arrival and deadline distributions are the same as the previous case, and the number of iterations is 100,000. The results are shown in Figure 7. Note that in this example, RandMax is still the worst of the three, achieving a maximum deficit arrival rate roughly 0.12, while MaxWeight is slightly better than LDF, both of which achieve a maximum deficit arrival rate roughly 0.16. We note that for non-real-time traffic the maximum deficit arrival rate is $1/3$. As we have been trying to convey in this paper, the stability re-

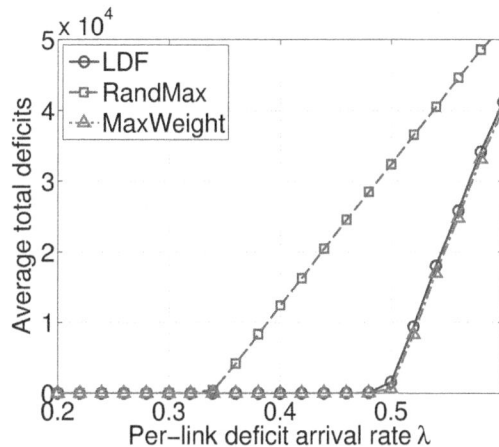

Figure 6: Comparison of the three scheduling policies on a four-linear network with one-hop interference

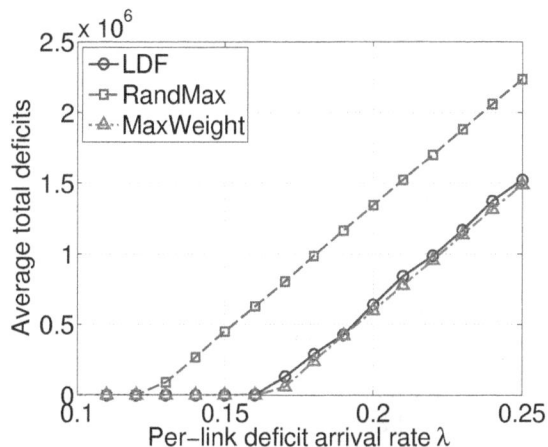

Figure 7: Comparison of the three scheduling policies on a nine-cycle network with two-hop interference

gion for the specific packet arrival and deadline distribution is unknown. We only know that the maximum rate for the real-time traffic is $\bar{\lambda} \leq 1/3$. Note that the nine-cycle has an interference degree of 2, so by Theorem 2, LDF has an efficiency ratio of $1/3$, which agrees with the simulation result since $0.16 > \frac{1}{3} \times \frac{1}{3} \geq \frac{1}{3}\bar{\lambda}$.

Therefore, both simulations imply good throughput performance of LDF and validate our lower bound on the efficiency ratio.

8. CONCLUSIONS

In this paper we considered the problem of scheduling real-time traffic in wireless networks under general stochastic arrivals and deadlines and general interference model. The fraction of delivered packets at a link is required to be no less than a certain threshold. We used deficits to inspect the stability of the system, and studied the stability performance of a scheduling policy that we call the largest-deficit-first (LDF) policy. We proved that the efficiency ratio of LDF can be lower bounded by a quantity that we call the real-time local-pooling factor (R-LPF). Furthermore, we proved that in a network with interference degree β, the R-LPF is at least $1/(\beta+1)$. In particular, for the one-hop interference model, we proved that the R-LPF is at least $1/3$.

9. ACKNOWLEDGMENTS

This work was supported in part by the National Science Foundation (NSF) under Grants CNS-1261429, CNS-1264012, and CNS-1262329, in part by the Defense Threat Reduction Agency (DTRA) under Grant HDTRA1-13-1-0030.

10. REFERENCES

[1] S. Boyd and L. Vandenberghe. *Convex Optimization*. Cambridge Unversity Press, New York, NY, 2004.

[2] P. Chaporkar, K. Kar, and S. Sarkar. Throughput guarantees through maximal scheduling in wireless networks. In *Proc. Annu. Allerton Conf. Communication, Control and Computing*, pages 28–30, Monticello, IL, Sept. 2005.

[3] J. G. Dai. On positive Harris recurrence of multiclass queueing networks: a unified approach via fluid limit models. *Ann. Appl. Probab.*, 5(1):49–77, 1995.

[4] J. G. Dai and B. Prabhakar. The throughput of data switches with and without speedup. In *Proc. IEEE Int. Conf. Computer Communications (INFOCOM)*, volume 2, pages 556–564, Tel Aviv, Israel, Mar. 2000.

[5] A. Dimakis and J. Walrand. Sufficient conditions for stability of longest queue first scheduling: Second-order properties using fluid limits. *Adv. in Appl. Probab.*, 38(2):505–521, June 2006.

[6] I.-H. Hou, V. Borkar, and P. R. Kumar. A theory of QoS for wireless. In *Proc. IEEE Int. Conf. Computer Communications (INFOCOM)*, pages 486–494, Rio de Janeiro, Brazil, Apr. 2009.

[7] I.-H. Hou and P. R. Kumar. Admission control and scheduling for QoS guarantees for variable-bit-rate applications on wireless channels. In *Proc. ACM Int. Symp. Mobile Ad Hoc Networking and Computing (MobiHoc)*, pages 175–184, New Orleans, LA, May 2009.

[8] I.-H. Hou and P. R. Kumar. Scheduling heterogeneous real-time traffic over fading wireless channels. In *Proc. IEEE Int. Conf. Computer Communications (INFOCOM)*, pages 1–9, San Diego, CA, Mar. 2010.

[9] I.-H. Hou and P. R. Kumar. Utility-optimal scheduling in time-varying wireless networks with delay constraints. In *Proc. ACM Int. Symp. Mobile Ad Hoc Networking and Computing (MobiHoc)*, pages 31–40, Chicago, IL, Sept. 2010.

[10] J. J. Jaramillo and R. Srikant. Optimal scheduling for fair resource allocation in ad hoc networks with elastic and inelastic traffic. *IEEE/ACM Trans. Netw.*, 19:1125–1136, Aug. 2011.

[11] J. J. Jaramillo, R. Srikant, and L. Ying. Scheduling for optimal rate allocation in ad hoc networks with heterogeneous delay constraints. *IEEE J. Sel. Areas Commun.*, 29(5):979–987, May 2011.

[12] C. Joo, X. Lin, and N. B. Shroff. Performance limits of greedy maximal matching in multi-hop wireless networks. In *Proc. IEEE Conf. Decision and Control (CDC)*, pages 1128–1133, New Orleans, LA, Dec. 2007.

[13] C. Joo, X. Lin, and N. B. Shroff. Understanding the capacity region of the greedy maximal scheduling algorithm in multihop wireless networks. *IEEE/ACM Trans. Netw.*, 17:1132–1145, Aug. 2009.

[14] B. Li, C. Boyaci, and Y. Xia. A refined performance characterization of longest-queue-first policy in wireless networks. *IEEE/ACM Trans. Netw.*, 19:1382–1395, Oct. 2011.

[15] B. Li, C. Boyaci, and Y. Xia. Performance guarantee under longest-queue-first schedule in wireless networks. *IEEE Trans. Inf. Theory*, 58:5878–5889, Sept. 2012.

[16] X. Lin and N. B. Shroff. The impact of imperfect scheduling on cross-layer rate control in wireless networks. In *Proc. IEEE Int. Conf. Computer Communications (INFOCOM)*, volume 3, pages 1804–1814, Miami, FL, Mar. 2005.

[17] R. M. Loynes. The stability of a queue with non-independent inter-arrival and service times. *Math. Proc. Cambridge Philos. Soc.*, 58(3):497–520, July 1962.

[18] A. A. Reddy, S. Sanghavi, and S. Shakkottai. On the effect of channel fading on greedy scheduling. In *Proc. IEEE Int. Conf. Computer Communications (INFOCOM)*, pages 406–414, Orlando, FL, Mar. 2012.

[19] T. Weller and B. Hajek. Scheduling nonuniform traffic in a packet-switching system with small propagation delay. *IEEE/ACM Trans. Netw.*, 5:813–823, Dec. 1997.

[20] X. Wu and R. Srikant. Regulated maximal matching: A distributed scheduling algorithm for multi-hop wireless networks with node-exclusive spectrum sharing. In *Proc. IEEE Conf. Decision and Control (CDC)*, pages 5342–5347, Seville, Spain, Dec. 2005.

On Traveling Path and Related Problems for a Mobile Station in a Rechargeable Sensor Network

Liguang Xie
Department of Electrical and
Computer Engineering
Virginia Tech
Blacksburg, Virginia, USA
xie@vt.edu

Yi Shi
Intelligent Automation Inc. &
Virginia Tech
Rockville, Maryland, USA
yshi@vt.edu

Y. Thomas Hou[*]
Department of Electrical and
Computer Engineering
Virginia Tech
Blacksburg, Virginia, USA
thou@vt.edu

Wenjing Lou
Department of Computer
Science
Virginia Tech
Falls Church, Virginia, USA
wjlou@vt.edu

Hanif D. Sherali
Department of Industrial and
Systems Engineering
Virginia Tech
Blacksburg, Virginia, USA
hanifs@vt.edu

ABSTRACT

Wireless power transfer is a promising technology to fundamentally address energy problems in a wireless sensor network. To make such a technology work effectively, a vehicle is needed to carry a charger to travel inside the network. On the other hand, it has been well recognized that a mobile base station offers significant advantages over a fixed one. In this paper, we investigate an interesting problem of co-locating the mobile base station on the wireless charging vehicle. We study an optimization problem that jointly optimizes traveling path, stopping points, charging schedule, and flow routing. Our study is carried out in two steps. First, we study an idealized problem that assumes zero traveling time, and develop a provably near-optimal solution to this idealized problem. In the second step, we show how to develop a practical solution with non-zero traveling time and quantify the performance gap between this solution and the unknown optimal solution to the original problem.

Categories and Subject Descriptors

C.2.1 [**Computer-Communication Networks**]: Network Architecture and Design—*Wireless communication*; G.1.6 [**Numerical Analysis**]: Optimization—*Nonlinear programming, Linear programming*

Keywords

Modeling and optimization; nonlinear programming; wireless power transfer; mobile base station; wireless sensor network

---
[*]Please direct all correspondence to Prof. Tom Hou.

1. INTRODUCTION

Recently, wireless power transfer (WPT) has been demonstrated to be a promising technology to address energy problems in a wireless sensor network (WSN) [8, 10]. This new WPT technology was based on the so-called *magnetic resonant coupling* [3, 4], which allows electric energy to be transferred from a source to a number of receivers via a nonradiative magnetic resonant induction. The most attractive features of this WPT technology are high energy transfer efficiency even under omni-direction, not requiring line-of-sight, and being insensitive to the neighboring environment.

In [8, 10], the authors showed how a wireless charging vehicle (WCV) can support WPT by bringing an energy source charger to the proximity of sensor nodes and charging their batteries wirelessly. In those studies, the authors assumed a simple setting where the location of the base station is fixed. On the other hand, it has been well recognized in the sensor network community that a mobile base station (MBS) offers significant advantages over a static one (see, e.g., [1, 5, 7, 13]). Since a base station is the sink node for all data that are collected from the sensor nodes, a mobile base station helps alleviate the traffic relay burden from a fixed set of sensor nodes near the base station to other sensor nodes in the network, thus avoiding energy hot spots and prolonging network lifetime.

Allowing the base station to be mobile adds considerable complexity to the underlying problem. In the most general case, the MBS could travel separately from the WCV, which calls for a separate vehicle to carry the base station. Given that the energy consumption for a vehicle is likely to be the dominant component in the big picture of energy consumption, we do not advocate this approach and defer it to a future study. Instead, in this paper, we consider the case where the MBS is co-located with the WCV. This allows us to explore traveling related questions for only one vehicle while still enjoying the benefits associated with a MBS.

Specifically, we consider the following problem in this paper. Suppose the base station is co-located with the WCV and we call the combines objects simply as WCV when there is no ambiguity. There is a home service station for the WCV (see Fig. 1). The WCV follows a periodic schedule to travel inside the network for charging the sensor nodes. While traveling inside the network, the WCV makes a number of stops and charge sensor nodes near those stops. At any time, data collected from the sensor nodes are relayed

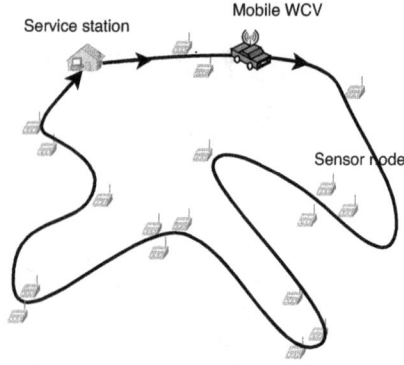

Figure 1: An example sensor network with a mobile WCV.

to the WCV (via multi-hop). By satisfying certain constraints, we hope that none of the sensor nodes in the network will ever run out of energy, i.e., the WSN will remain operational indefinitely.

Apparently, the above problem brings in a number of technical challenges. First of all, the traveling path for the WCV is unknown and needs to be determined. Second, we need to find the optimal stopping points along this path as well as the charging schedule of the WCV (i.e., how long it shall stay at each stopping point). Finally, the data flow routing in the network is dynamic and depends on where the WCV is in the network. Among these challenges, we find that the traveling path problem is most crucial and solutions to the other sub-problems all hinge upon the determination of a traveling path. In this paper, we address these challenges by studying an optimization problem.[1]

The main contributions of this paper are as follows:

- We formulate an optimization problem (TPP) that involves joint optimization of traveling path, stopping points, charging schedule, and data flow routing. This is shown to be a nonlinear program (NLP).

- To tackle TPP, we first consider an idealized problem (OPT-ub) that assumes zero traveling time (i.e., infinite traveling speed) from one point to another. The optimal solution of OPT-ub gives an upper bound to TPP.

- Subsequently, we develop a provably near-optimal solution to OPT-ub for any desired level of accuracy ϵ. Our solution involves several novel techniques, such as discretization of energy reception rate and energy consumption rate, double partitioning of the smallest enclosing disk (SED) into smaller subareas with tight upper energy consumption bounds and lower energy reception bounds, and representation of each subarea by a logical point as its "worst-case" energy reception and energy consumption behavior.

- Based on the near-optimal solution to the idealized problem OPT-ub, we return to the original problem TPP by incorporating non-zero traveling time for the WCV. In particular, we determine the traveling path in TPP by finding the shortest Hamiltonian cycle to connect all the logical points that have

---

[1] A simpler version of bundling WPT and MBS problem was studied in [11], where the traveling path for the WCV was assumed to be given *a priori*. This assumption simplifies the problem considerably and the optimization problem only needs to find solutions to stopping points, charging schedule, and data flow routing. In contrast, this paper considers a much harder problem where the traveling path is unknown.

non-zero stopping time in OPT-ub. Note that this Hamiltonian cycle is fundamentally different from the Hamiltonian cycle that connects all sensor nodes in the network. Based on this traveling path, we can obtain a feasible solution to the original problem TPP. We further quantify the performance gap between this feasible solution and optimal solution to TPP.

The remainder of this paper is organized as follows. In Section 2, we give some essential background and necessary mathematical models for this problem. We also give a formulation for the optimization problem TPP. In Section 3, we study an idealized problem OPT-ub and develop a near-optimal solution. In Section 4, we determine a traveling path based on the solution in Section 3. From this path, we develop a feasible solution to the original problem TPP. We then quantify the performance gap between this solution and the optimal solution. Section 5 presents numerical results. Section 6 concludes this paper.

2. MODELING AND FORMULATION

Suppose that we have a sensor network $\mathcal{N}$ deployed over a two-dimensional area. A WCV is employed to recharge sensor nodes in the network and to collect data from nodes in real time. The WCV follows a periodic schedule: In each cycle, it starts from its home service station, travels inside the network, and returns to the service station. While traveling, the WCV makes a number of stops and charges sensor nodes that are in the vicinity of those stops (see Fig. 1). For the traveling path that we are investigating in this paper, the WCV is allowed to visit *anywhere* over the two-dimensional area, i.e., its traveling path is unconstrained and is part of the optimization problem. At any time, the data generated from the sensor nodes are relayed through multi-hops toward the WCV (in real time).

2.1 Traveling Path and Stopping Schedule

Denote $\mathcal{P}$ as the traveling path and τ as the amount of time for each cycle. Then τ includes three components:

- The total traveling time along path $\mathcal{P}$, $D_{\mathcal{P}}/V$, where $D_{\mathcal{P}}$ is the distance along path $\mathcal{P}$ and V is the traveling speed of the WCV.

- The total sojourn time along path $\mathcal{P}$, which is defined as the sum of all stopping time of the WCV when it travels on $\mathcal{P}$.

- The vacation time for the WCV at its home service station, τ_{vac}, which starts when the WCV returns to its home service station (after traveling path $\mathcal{P}$) and ends when the WCV leaves for the next trip.

Then we have:

$$\tau = \frac{D_{\mathcal{P}}}{V} + \sum_{\substack{p \in \mathcal{P}, \ p \neq p_{\text{vac}}}}^{\omega(p) > 0} \omega(p) + \tau_{\text{vac}}, \qquad (1)$$

where $\omega(p)$ denotes the aggregate amount of time the WCV stays at point $p \in \mathcal{P}$ and p_{vac} denotes the location of the home service station.

2.2 Energy Charging Model

We assume that the WCV can only perform its charging function when it makes a stop along path $\mathcal{P}$ (excluding p_{vac}). Based on the current charging technology [4, 10], the WCV can charge multiple neighboring nodes simultaneously as long as they are within its charging range, although the power transfer rate at a sensor node decreases over the distance.

Denote $U_{iB}(p)$ as the power reception rate at node i when the WCV is located at $p \in \mathcal{P}$. Denote the efficiency of wireless charging by $\mu(D_{iB}(p))$, which is a decreasing function of $D_{iB}(p)$, the distance between node i and the WCV located at p. Then the wireless charging model is as follows [10]:

$$U_{iB}(p) = \begin{cases} \mu(D_{iB}(p)) \cdot U_{\max} & \text{if } D_{iB}(p) \leq D_\delta \\ 0 & \text{if } D_{iB}(p) > D_\delta , \end{cases} \quad (2)$$

where $U_{\max}$ is the maximum output power for a single sensor node and D_δ is the charging range of the WCV, beyond which wireless charging will not occur. In other words, D_δ is defined in a way such that the power reception rate at a sensor node is at least over a threshold value δ.

2.3 Dynamic Data Flow Routing

Recall that the base station is co-located at the WCV and all data generated from the sensor nodes shall be delivered to the base station. To conserve energy, multi-hop data routing is necessary among the sensor nodes in the network. Due to the mobility of the WCV, data flow routing is dynamic, with routing topology changing over time.

Suppose that each sensor node i ($i \in \mathcal{N}$) generates a constant rate R_i. Denote $f_{ij}(p)$ and $f_{iB}(p)$ as flow rates from sensor node i to sensor node j and to the base station when the WCV is at location $p \in \mathcal{P}$, respectively. Then we have the following flow balance at each sensor node i:

$$\sum_{\substack{k \in \mathcal{N} \\ k \neq i}} f_{ki}(p) + R_i = \sum_{\substack{j \in \mathcal{N} \\ j \neq i}} f_{ij}(p) + f_{iB}(p) \quad (i \in \mathcal{N}, p \in \mathcal{P}) \quad (3)$$

The above flow balance equation indicates that we are dealing with real-time flow routing, rather than DTN-like data routing (e.g., data MULEs [6] or message ferry [14]), where data can be delayed and delivered till a later time.

2.4 Sensor Energy Consumption

At a sensor node, we assume data communications (transmission and reception) is the dominant source for energy consumption.[2] Denote C_{ij} as the energy consumption rate for transmitting one unit of data flow from sensor node i to sensor node j. Then C_{ij} (in Joule/bit) can be modeled as [2, 7]:

$$C_{ij} = \beta_1 + \beta_2 D_{ij}^\alpha ,$$

where D_{ij} is the distance between nodes i and j, β_1 and β_2 are constant terms, and α is the path loss index. Given that all sensor nodes are stationary, we have that D_{ij} and C_{ij} are all constants.

Denote $C_{iB}(p)$ as the energy consumption rate for transmitting one unit of data flow from sensor node i to base station B when the WCV is at location $p \in \mathcal{P}$. We have

$$C_{iB}(p) = \beta_1 + \beta_2 \left[\sqrt{(x_p - x_i)^2 + (y_p - y_i)^2} \right]^\alpha , \quad (4)$$

where (x_p, y_p) and (x_i, y_i) are the coordinates of p and node i, respectively. Note that unlike C_{ij}'s, which are all constants, $C_{iB}(p)$ varies with the base station's position p.

[2]Energy consumption for hardware device and information processing can be assumed to be constants and can be easily integrated into total energy consumption without major change of the problem structure.

Then the total energy consumption rate for both transmission and reception at node i when the WCV is at $p \in \mathcal{P}$, denoted as $r_i(p)$, is

$$r_i(p) = \rho \sum_{\substack{k \in \mathcal{N} \\ k \neq i}} f_{ki}(p) + \sum_{\substack{j \in \mathcal{N} \\ j \neq i}} C_{ij} \cdot f_{ij}(p)$$
$$+ C_{iB}(p) \cdot f_{iB}(p) \quad (i \in \mathcal{N}, p \in \mathcal{P}), \quad (5)$$

where ρ is a constant term associated with the rate of energy consumption for receiving one unit of data.

2.5 Energy Cycle at a Sensor Node

We will develop a travel schedule (including charging schedule) for the WCV and data flow routing among the nodes so that no sensor node ever runs out of energy. Such travel schedule follows a periodic cycle, as discussed in Section 2.1, with a cycle time of τ.

Suppose that each sensor node is fully charged initially. Denote $E_{\max}$ as its battery capacity and $E_{\min}$ as the minimum energy threshold for a node to be operational. We offer two energy renewable conditions, and show that once they are met, then the energy level at each sensor node at time t, denoted as $e_i(t)$, never falls below $E_{\min}$.

First, we split energy consumption at node i into two parts:

- Energy consumed whenever the WCV makes any stop (including vacationing at its service station): $r_i(p_{\text{vac}}) \cdot \tau_{\text{vac}} + \sum_{\substack{p \in \mathcal{P}, \ p \neq p_{\text{vac}}}}^{\omega(p)>0} r_i(p) \cdot \omega(p)$,

- Energy consumed when the WCV is moving along $\mathcal{P}$, i.e., $\int_{s \in [0, D_{\mathcal{P}}]}^{\omega(p(s))=0} \frac{1}{V} \cdot r_i(p(s)) \, ds$, where $s \in [0, D_{\mathcal{P}}]$ and the integration is taken over the distance traversed by the WCV along $\mathcal{P}$, and $p(s)$ is the WCV's location corresponding to s.

Following the results in [10], it can be shown that $e_i(t) \geq E_{\min}$ for all $t \geq 0$, $i \in \mathcal{N}$ if the following conditions are satisfied:

$$E_{\max} - \left[r_i(p_{\text{vac}}) \cdot \tau_{\text{vac}} + \sum_{\substack{p \in \mathcal{P}, \ p \neq p_{\text{vac}}}}^{\omega(p)>0, \ D_{iB}(p)>D_\delta} r_i(p) \cdot \omega(p) \right.$$
$$\left. + \int_{s \in [0, D_p]}^{\omega(p(s))=0} \frac{1}{V} \cdot r_i(p(s)) ds \right] \geq E_{\min} , \quad (i \in \mathcal{N}) \quad (6)$$

$$r_i(p_{\text{vac}}) \cdot \tau_{\text{vac}} + \sum_{\substack{p \in \mathcal{P}, \ p \neq p_{\text{vac}}}}^{\omega(p)>0} r_i(p) \cdot \omega(p)$$
$$+ \int_{s \in [0, D_p]}^{\omega(p(s))=0} \frac{1}{V} \cdot r_i(p(s)) ds$$
$$\leq \sum_{p \in \mathcal{P}}^{\omega(p)>0, \ D_{iB}(p) \leq D_\delta} U_{iB}(p) \cdot \omega(p) , \quad (i \in \mathcal{N}) \quad (7)$$

In constraint (6), $\sum_{\substack{p \in \mathcal{P}, \ p \neq p_{\text{vac}}}}^{\omega(p)>0, \ D_{iB}(p)>D_\delta} r_i(p) \cdot \omega(p)$ is the amount of energy consumed at node i when the WCV is making stops near those nodes other than i. Constraint (6) ensures that $e_i(t)$, which starts from $E_{\max}$ at $t = 0$, will not fall below $E_{\min}$ at the end of the first cycle $t = \tau$. In constraint (7), the left hand side is the amount of energy consumed at node i during τ while the right hand side is maximum amount of potential energy received by node i in a cycle. Note that the actual amount of energy received by node i in the first cycle may be less than the right hand side due to battery overflow.[3] Constraint (7) ensures that $e_i(t)$, which starts at full level $E_{\max}$, will be charged back to $E_{\max}$ before the end of the first cycle τ.

2.6 Problem Formulation

Based on the above mathematical models, a number of problems can be formulated and studied. As a case study, we consider an optimization problem involving joint optimization of traveling path, stopping points, charging schedule, and flow routing. For the objective function, we consider minimizing energy consumption of the entire system, which includes power used by the WCV

[3]Once a battery is charged to $E_{\max}$, its energy cannot be further increased.

and the power consumed for wireless power transfer.[4] Since power used by the WCV is the dominant component in the overall energy consumption, our objective function will focus on this component. Specifically, we aim to minimize the fraction of time that the WCV is at work (i.e., away from its service station) in each cycle period, i.e., $\frac{D_{\mathcal{P}}/V + \sum_{p \in \mathcal{P}, \, p \neq p_{\text{vac}}}^{\omega(p)>0} \omega(p)}{\tau}$.[5] Note that by (1), minimizing $\frac{D_{\mathcal{P}}/V + \sum_{p \in \mathcal{P}, \, p \neq p_{\text{vac}}}^{\omega(p)>0} \omega(p)}{\tau}$ is equivalent to maximizing $\frac{\tau_{\text{vac}}}{\tau}$, which is the percentage of time that the WCV is on vacation at its service station. Therefore, we have the following optimization problem.

TPP:
maximize $\frac{\tau_{\text{vac}}}{\tau}$
s.t.   Time constraints: (1);
       Flow routing constraints: (3);
       Energy consumption model: (5);
       Energy renewable constraints: (6), (7).
       $\tau, \tau_{\text{vac}}, \omega(p) \geq 0 \quad (p \in \mathcal{P})$
   $f_{ij}(p), f_{iB}(p), r_i(p) \geq 0 \quad (i, j \in \mathcal{N}, i \neq j, p \in \mathcal{P})$.

In this formulation, V, R_i, ρ, C_{ij}, $E_{\max}$, and $E_{\min}$ are constants, and $U_{iB}(p)$ and $C_{iB}(p)$ can be computed by (2) and (4), respectively. The path $\mathcal{P}$ and $D_{\mathcal{P}}$ are to be determined in problem TPP. The time intervals τ, τ_{vac}, and $\omega(p)$, the flow rates $f_{ij}(p)$ and $f_{iB}(p)$, and the power consumption rate $r_i(p)$ are also optimization variables.[6] Note that problem TPP is a nonlinear program, and is NP-hard in general.

In problem TPP, the WCV can travel anywhere in the two-dimensional plane. It is not hard to see that the WCV's roaming area can be narrowed down to a much smaller area. In particular, it is sufficient for the WCV to roam in the smallest enclosing disk (SED) [9], denoted as $\mathcal{A}$, which covers all the sensor nodes in the network and the home service station. This result is stated in the following lemma.

LEMMA 1. *The optimal traveling path for the WCV must stay inside the SED $\mathcal{A}$.*

A proof sketch: A proof of this lemma can be easily constructed based on contradiction. That is, if there exists an optimal solution that involves the WCV traveling outside of the SED, we can always find a better solution (in terms of objective value) by bringing the WCV inside the SED, which leads to a contradiction. A formal proof is given in [12].

3. A NEAR-OPTIMAL SOLUTION TO AN IDEALIZED PROBLEM

A major difficulty in problem TPP is that the traveling path $\mathcal{P}$ is unknown and is part of the optimization problem. What further complicates this matter is that it takes time for the WCV to travel along the path. In this section, we consider an idealized problem

---

[4]Note that except their initial energy, the energy consumed at all sensor nodes comes from the WCV.

[5]We assume the WCV keeps its engine running as long as it is away from its service station.

[6]Note that variables in this formulation are only dependent on the WCV's location p and are independent of the time when the WCV visits this location. In [7], Shi *et al.* showed that a time-based formulation involving a MBS like ours can be transformed into a location-based formulation. In light of that result, we start directly with a location-based formulation in this paper without going through the details of such transformation, which are similar to those in [7].

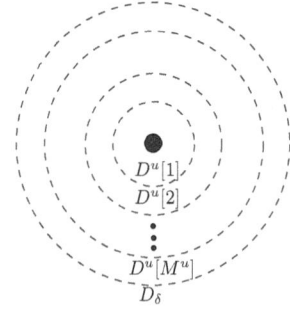

Figure 2: A sequence of circles centered at a node with decreasing energy charging rates.

that ignores the time for the WCV to travel from one point to another along $\mathcal{P}$. We will show that it is possible to develop a provably near-optimal solution to this idealized problem. Based on this result, in Section 4, we address the practical problem which considers non-zero traveling time for the WCV.

3.1 An Idealized Problem with Zero Traveling Time

In the idealized problem, the traveling time of the WCV is assumed to be zero. In this section, we give a formulation for this idealized problem (denoted as OPT-ub) based on our formulation for TPP. Since $V \to \infty$, constraint (1) becomes

$$\tau = \sum_{p \in \mathcal{P}, \, p \neq p_{\text{vac}}}^{\omega(p)>0} \omega(p) + \tau_{\text{vac}} . \qquad (8)$$

Further, since $V \to \infty$ in OPT-ub, energy consumed at a sensor node i when the WCV travels along $\mathcal{P}$ (i.e., $\int_{s \in [0, D_p]}^{\omega(p(s))=0} \frac{1}{V} \cdot r_i(p(s)) \, ds$) degenerates to 0. Hence, (7) can be simplified to

$$r_i(p_{\text{vac}}) \cdot \tau_{\text{vac}} + \sum_{p \in \mathcal{P}, \, p \neq p_{\text{vac}}}^{\omega(p)>0} r_i(p) \cdot \omega(p)$$
$$\leq \sum_{p \in \mathcal{P}}^{\omega(p)>0, \, D_{iB}(p) \leq D_\delta} U_{iB}(p) \cdot \omega(p) \quad (i \in \mathcal{N}) , \qquad (9)$$

and (6) can be simplified to

$$r_i(p_{\text{vac}}) \cdot \tau_{\text{vac}} + \sum_{p \in \mathcal{P}, \, p \neq p_{\text{vac}}}^{\omega(p)>0, \, D_{iB}(p)>D_\delta} r_i(p) \cdot \omega(p)$$
$$\leq E_{\max} - E_{\min} \quad (i \in \mathcal{N}) . \qquad (10)$$

Summarizing all these updates, OPT-ub can be written as follows:

OPT-ub:
maximize $\frac{\tau_{\text{vac}}}{\tau}$
s.t.   Time constraints: (8);
       Flow routing constraints: (3);
       Energy consumption model: (5);
       Energy renewable constraints: (9), (10) .
       $\tau, \tau_{\text{vac}}, \omega(p) \geq 0 \quad (p \in \mathcal{P})$
   $f_{ij}(p), f_{iB}(p), r_i(p) \geq 0 \quad (i, j \in \mathcal{N}, i \neq j, p \in \mathcal{P})$.

Denote ψ_{TPP}^* and $\psi_{\text{OPT-ub}}^*$ as an optimal solution to problem TPP and problem OPT-ub, respectively. Denote η_{TPP}^* as the objective value achieved by ψ_{TPP}^* and $\eta_{\text{OPT-ub}}^*$ as the objective value achieved by $\psi_{\text{OPT-ub}}^*$, respectively. The following lemma shows the relationship between $\eta_{\text{OPT-ub}}^*$ and η_{TPP}^*.

LEMMA 2. *$\eta_{\text{OPT-ub}}^*$ is an upper bound of η_{TPP}^*, i.e., $\eta_{\text{OPT-ub}}^* > \eta_{\text{TPP}}^*$.*

A proof sketch: The proof of Lemma 2 can be constructed as follows. Suppose ψ_{TPP}^* is given. Then we can construct a solution

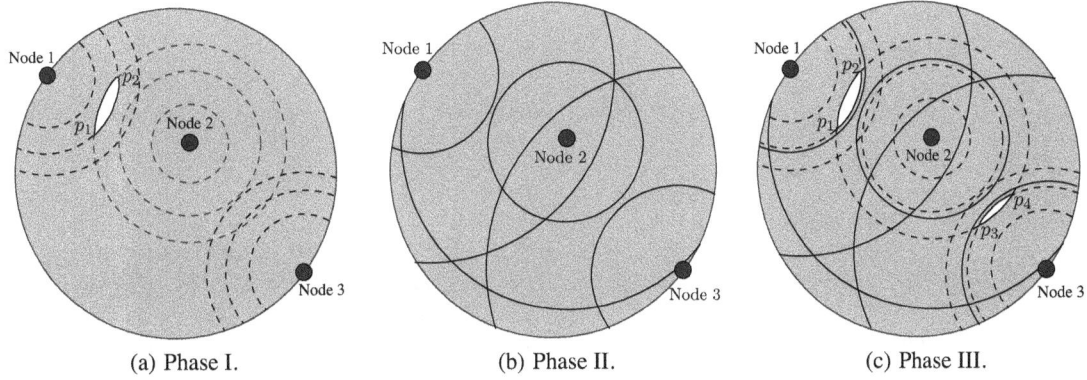

| (a) Phase I. | (b) Phase II. | (c) Phase III. |

Figure 3: An example three-node sensor network illustrating area partition in three phases. The gray region is the SED of the three nodes.

$\psi_{\text{OPT-ub}}$ to problem OPT-ub with strictly greater objective value than η_{TPP}^*. Such construction of $\psi_{\text{OPT-ub}}$ is straightforward as OPT-ub is an ideal (or relaxed) case of TPP assuming zero traveling time. Since the obtained solution $\psi_{\text{OPT-ub}}$ is only feasible to OPT-ub, $\eta_{\text{OPT-ub}}^*$ (for $\psi_{\text{OPT-ub}}^*$) must be greater than η_{TPP}^*. A formal proof is given in [12].

In the rest of this section, we develop a near-optimal solution to OPT-ub. In Section 3.2, we partition the SED $\mathcal{A}$ into a number of smaller subareas. In Section 3.3, we employ the so-called *logical point* as a "worst-case" representation for each subarea in terms of energy reception and energy consumption. The logical point concept that we propose here generalizes the "fictitious cost point" proposed in [7], which only considered energy consumption. In Section 3.4, we propose an approximation algorithm to problem OPT-ub, and prove it near-optimality.

3.2 Area Partitioning

In this section, we partition the SED $\mathcal{A}$ into a number of smaller subareas. The partition is performed in a special way such that some lower and upper bounds can be derived regarding energy charging and consumption at each sensor node.

Phase I: Area Partition based on WPT. First, we discretize energy reception rate. Due to one-to-one mapping between distance and energy reception rate (see (2)), a discretization of energy reception rate also corresponds to a discretization of distance.

We discretize energy reception rates $U[1], U[2], \ldots, U[M^u]$ as follows:

$$U[h] = \begin{cases} U_{\max}(1 - \frac{\epsilon}{W})^h & \text{if } 0 \leq h \leq M^u \\ 0 & \text{otherwise} , \end{cases} \quad (11)$$

where M^u is the largest integer such that $U[M^u] > \delta$, ϵ is the allowed error margin for the near-optimal solution, W is a parameter controlling the granularity of the discretization (i.e., a large W leads to a large M^u).

Corresponding to $U[1], U[2], \ldots, U[M^u]$, we can discretize distance into $D^u[1], D^u[2], \ldots, D^u[M^u]$, and define $D^u[h]$ as follows:

$$D^u[h] = \mu^{-1}\left(\frac{U[h]}{U_{\max}}\right), \quad 1 \leq h \leq M^u ,$$

where $\mu^{-1}(\cdot)$ denotes the inverse function of (2).

To determine M^u, recall that M^u is the largest integer such that $U[M^u] > \delta$. By (11), we have

$$M^u = \left\lfloor \frac{\ln\left(\frac{\delta}{U_{\max}}\right)}{\ln\left(1 - \frac{\epsilon}{W}\right)} \right\rfloor . \quad (12)$$

For each node $i \in \mathcal{N}$, we draw M^u circles centered at node i, with each circle having an increasing radius $D^u[h], 1 \leq h \leq M^u$. Based on (2), there is a circle with a radius of D_δ that cuts off the charging area for node i. That is, when $D_{iB}(p) = D_\delta, U_{iB}(p) = \delta$ (see Fig. 2). Note that this last cut-off circle, along with the M^u-th inner circle (i.e. the second outermost circle) partition $\mathcal{A}$ into three regions:

(i). A disk with a radius of $D^u[M^u]$, where $U[M^u] \leq U_{iB}(p) \leq U_{\max}$.

(ii). A ring bounded by these two circles with radiuses of $D^u[M^u]$ and D_δ, respectively, where $\delta \leq U_{iB}(p) < U[M^u]$.

(iii). The area outside of the cut-off circle, in which $U_{iB}(p) = 0$.

Since we have $(M^u + 1)$ circles for each node i, the intersections of these circles partition disk $\mathcal{A}$ into a number of irregular subareas. As an example, for the 3-node network in Fig. 3(a), suppose that $M^u = 2$. Then we draw 3 circles for each node. Disk $\mathcal{A}$ in Fig. 3(a) is partitioned into 19 irregular subareas. For the subarea of white color with corner points p_1 and p_2, any point p in this subarea satisfies $U[2] \leq U_{1B}(p) \leq U[1]$ with respect to node 1. With respect to nodes 2 and 3, for any point p in this same subarea, we have $\delta \leq U_{2B}(p) \leq U[2]$, and $U_{3B}(p) = 0$.

As shown in the example, the proposed area partition gives tight lower and upper bounds for each subarea. In particular, for a subarea $\mathcal{A}^u$ with $D_{iB}(p) \leq D^u[M^u]$ where $p \in \mathcal{A}^u$, we have

$$U[h_i^u(\mathcal{A}^u)] \leq U_{iB}(p) \leq U[h_i^u(\mathcal{A}^u) - 1] , \quad p \in \mathcal{A}^u , \quad (13)$$

where $h_i^u(\mathcal{A}^u)$ denotes the index of the outer circle (centered at node i), and $h_i^u(\mathcal{A}^u) \leq M^u$. For a given subarea $\mathcal{A}^u$, $h_i^u(\mathcal{A}^u)$ can be determined by (11) and (13). We have

$$h_i^u(\mathcal{A}^u) = \left\lceil \frac{\ln(U_{iB}(p)/U_{\max})}{\ln(1 - \epsilon/W)} \right\rceil , \quad p \in \mathcal{A}^u , \quad (14)$$

Therefore, for any $p \in \mathcal{A}^u$, we have the lower bound for energy reception $U_{iB}(p)$ as follows:

$$U[h_i^u(\mathcal{A}^u)] = \begin{cases} U_{\max}(1 - \frac{\epsilon}{W})^{h_i^u(\mathcal{A}^u)}, & \text{if } D_{iB}(p) \leq D^u[M^u], \\ \delta, & \text{if } D^u[M^u] < D_{iB}(p) \leq D_\delta, \\ 0, & \text{otherwise.} \end{cases}$$
$$(15)$$

where $h_i^u(\mathcal{A}^u)$ is determined by (14).

Phase II: Area Partition based on Energy Consumption. Following a similar token to that in Phase I, we discretize energy consumption rate, which also corresponds to a discretization of distance. Specifically, for each node $i \in \mathcal{N}$, we define a sequence of increasing energy costs $C[1], C[2], \ldots, C[M_i^c]$ as follows:

$$C[h] = \beta_1 \left(1 + \frac{\epsilon}{W}\right)^h , \qquad (16)$$

where M_i^c is the largest number of elements in the sequence of $C[h]$. Corresponding to $C[h]$, $h = 1, 2, \ldots, M_i^c$, we can discretize distance into $D^c[1], D^c[2], \ldots, D^c[M_i^c]$, where

$$D^c[h] = \left(\frac{C[h] - \beta_1}{\beta_2}\right)^{-\alpha} , \quad 0 \le h \le M_i^c + 1 .$$

For each node $i \in \mathcal{N}$, we can draw M_i^c circles centered at node i, with increasing radii $D^c[1], D^c[2], \ldots, D^c[M_i^c]$. To determine M_i^c, denote $O_\mathcal{A}$ and $R_\mathcal{A}$ as the origin and radius of $\mathcal{A}$, respectively. Denote $D_{i,O_\mathcal{A}}$ as the distance from node i to $O_\mathcal{A}$. As $D_{iB}(p) \in [0, D_{i,O_\mathcal{A}} + R_\mathcal{A}]$, by (4), we have $C_{iB} \in [\beta_1, \beta_1 + \beta_2 \cdot (D_{i,O_\mathcal{A}} + R_\mathcal{A})^\alpha]$. Since the WCV can only travel within $\mathcal{A}$, M_i^c is the largest integer such that $C[M_i^c] < \beta_1 + \beta_2 \cdot (D_{i,O_\mathcal{A}} + R_\mathcal{A})^\alpha$. By (16), we have

$$M_i^c = \left\lfloor \frac{\ln\left(1 + \frac{\beta_2}{\beta_1} \cdot (D_{i,O_\mathcal{A}} + R_\mathcal{A})^\alpha\right)}{\ln\left(1 + \frac{\epsilon}{W}\right)} \right\rfloor .$$

Since we have M_i^c circles for $i \in \mathcal{N}$, the intersections of these circles partition $\mathcal{A}$ into a number of subareas (see Fig. 3(b)). For a subarea $\mathcal{A}^c$, we have

$$C[h_i^c(\mathcal{A}^c) - 1] \le C_{iB}(p) \le C[h_i^c(\mathcal{A}^c)] , \quad p \in \mathcal{A}^c , \qquad (17)$$

where $h_i^c(\mathcal{A}^c)$ denotes the index of the outer circle (centered at node i) that contains $\mathcal{A}^c$. Given a subarea $\mathcal{A}^c$, $h_i^c(\mathcal{A}^c)$ can be determined by (16) and (17). Thus, we have

$$h_i^c(\mathcal{A}^c) = \left\lceil \frac{\ln\left(1 + \frac{\beta_2}{\beta_1} \cdot D_{iB}(p)^\alpha\right)}{\ln\left(1 + \frac{\epsilon}{W}\right)} \right\rceil . \qquad (18)$$

Therefore, for any $p \in \mathcal{A}^c$, a tight upper bound of $C_{iB}(p)$ is:

$$C[h_i^c(\mathcal{A}^c)] = \beta_1 \left(1 + \frac{\epsilon}{W}\right)^{h_i^c(\mathcal{A}^c)} , \qquad (19)$$

where $h_i^c(\mathcal{A}^c)$ is determined by (18).

Phase III: Joint Area Partition. By combining the partitions in both Phases I and II, the disk $\mathcal{A}$ is partitioned into smaller subareas $\mathcal{A}_k^{u+c}$, $k = 1, 2, \ldots, K$ (see Fig. 3(c)). For each subarea $\mathcal{A}_k^{u+c}$, both the energy reception and consumption can be tightly bounded.

Now we give an upper bound on the number of subareas K. By (12), we have

$$M^u = \left\lfloor \frac{\ln\left(\frac{\delta}{U_{\max}}\right)}{\ln\left(1 - \frac{\epsilon}{W}\right)} \right\rfloor = O\left(\left\lfloor \frac{1}{\frac{\epsilon}{W}} \right\rfloor\right) = O\left(\frac{W}{\epsilon}\right) ,$$

where the second equality holds since $\ln(\delta/U_{\max})$ is a negative constant and $\ln(1 - \epsilon/W) \approx -\epsilon/W$ for small ϵ/W. Similarly, we have

$$M_i^c = \left\lfloor \frac{\ln\left(1 + \frac{\beta_2}{\beta_1} \cdot (D_{i,O_\mathcal{A}} + R_\mathcal{A})^\alpha\right)}{\ln\left(1 + \frac{\epsilon}{W}\right)} \right\rfloor$$
$$= O\left(\left\lfloor \frac{1}{\frac{\epsilon}{W}} \right\rfloor\right) = O\left(\frac{W}{\epsilon}\right) .$$

For each sensor node, there are $(M^u + 1)$ circles from Phase I and M_i^c circles from Phase II. Putting these circles and one more circle for $\mathcal{A}$ together, the total number of circles is $B = 1 + \sum_{i \in \mathcal{N}} (M^u + M_i^c + 1)$. Given B circles, the maximum number of subareas K is

upper bounded by $K \le B^2 - B + 2$ (which can be easily verified by induction). That is

$$K = O(B^2) = O\left(\left[1 + \sum_{i \in \mathcal{N}} (M^u + M_i^c + 1)\right]^2\right)$$
$$= O\left(\left(\frac{W|\mathcal{N}|}{\epsilon}\right)^2\right) . \qquad (20)$$

3.3 Logical Point Representation

For each subarea $\mathcal{A}_k^{u+c}$, $k = 1, 2, \ldots, K$, we represent it as a *logical point* $\mathcal{L}_k$. Denote $\omega(\mathcal{L}_k)$ as the total stopping time when the WCV is in subarea $\mathcal{A}_k^{u+c}$. Then we have,

$$\omega(\mathcal{L}_k) = \sum_{p \in \mathcal{A}_k^{u+c}}^{\omega(p) > 0} \omega(p) .$$

To characterize a logical point $\mathcal{L}_k$, we use the worst case bounds of energy charging and energy consumption rates within the subarea $\mathcal{A}_k^{u+c}$. Specifically, for a logical point $\mathcal{L}_k$, we use a $|2\mathcal{N}|$-tuple to represent a logical point, where the first $|\mathcal{N}|$ components are for energy charging and the other $|\mathcal{N}|$ components are for energy consumption, i.e., $[U_1(\mathcal{L}_k), U_2(\mathcal{L}_k), \ldots, U_{|\mathcal{N}|}(\mathcal{L}_k), C_{1B}(\mathcal{L}_k), C_{2B}(\mathcal{L}_k), \ldots, C_{|\mathcal{N}|B}(\mathcal{L}_k)]$. In this vector, the first $|\mathcal{N}|$ components are

$$U_i(\mathcal{L}_k) = U[h_i^u(\mathcal{A}_k^{u+c})] , \qquad (21)$$

where $U[h_i^u(\mathcal{A}_k^{u+c})]$ is the lower bound of $U_{iB}(p)$ for any $p \in \mathcal{A}_k^{u+c}$ and is determined by (15), while the next $|\mathcal{N}|$ components are

$$C_{iB}(\mathcal{L}_k) = C[h_i^c(\mathcal{A}_k^{u+c})] , \qquad (22)$$

where $C[h_i^c(\mathcal{A}_k^{u+c})]$ is the upper bound of $C_{iB}(p)$ for any $p \in \mathcal{A}_k^{u+c}$ and is determined by (19).

3.4 A Near-Optimal Solution

Based on these logical points, we can develop a provably near-optimal solution to problem OPT-ub. Recall that $\psi_{\text{OPT-ub}}^*$ is an optimal solution to problem OPT-ub, and $\eta_{\text{OPT-ub}}^*$ is the objective value achieved by $\psi_{\text{OPT-ub}}^*$. Our goal is to find a feasible solution to problem OPT-ub, denoted as $\psi_{\text{OPT-ub}}$, so that $\eta_{\text{OPT-ub}} \ge \eta_{\text{OPT-ub}}^* - \epsilon$.

A Worst-Case Formulation and Its Solution. Note that a logical point is a worst-case representation of the subarea in terms of energy charging and energy consumption. Based on these logical points, we can have a formulation, denoted as OPT-$\mathcal{L}$, that can be used to derive a lower bound to OPT-ub. Problems OPT-$\mathcal{L}$ and OPT-ub are similar to each other except the following differences:

- OPT-$\mathcal{L}$ is based on a finite number of logical points while OPT-ub is based on an infinite number of physical points.

- For $p \ne p_{\text{vac}}$, we have $\omega(\mathcal{L}_k), f_{ij}(\mathcal{L}_k), f_{iB}(\mathcal{L}_k), r_i(\mathcal{L}_k)$ in OPT-$\mathcal{L}$ rather than $\omega(p), f_{ij}(p), f_{iB}(p), r_i(p)$ in OPT-ub.

- We have $U_i(\mathcal{L}_k)$ and $C_{iB}(\mathcal{L}_k)$ in OPT-$\mathcal{L}$ rather than $U_i(p)$ and $C_{iB}(p)$ in OPT-ub.

Through a number of changes of variables, OPT-$\mathcal{L}$ can be reformulated into a linear program (LP), which can be solved in polynomial time (see Appendix A for more details).

Recover a Feasible Solution to OPT-ub. After we obtain an optimal solution to problem OPT-$\mathcal{L}$, denoted as $\psi_{\text{OPT-}\mathcal{L}}^*$, we need to recover a solution to OPT-ub (denoted by $\psi_{\text{OPT-ub}}$). Suppose that $\psi_{\text{OPT-ub}} = (\tau, \tau_{\text{vac}}, \omega(p), f_{ij}(p), f_{iB}(p), r_i(p))$. From $\psi_{\text{OPT-}\mathcal{L}}$, $\psi_{\text{OPT-ub}}$ can be constructed as follows:

- τ and τ_{vac} are the same as their counterparts in $\psi_{\text{OPT-}\mathcal{L}}$.

- For $p = p_{\text{vac}}$, $f_{ij}(p_{\text{vac}}), f_{iB}(p_{\text{vac}}), r_i(p_{\text{vac}})$ are the same as their counterparts in $\psi_{\text{OPT-}\mathcal{L}}$.

- For any $\omega(\mathcal{L}_k) > 0$, choose a point $p_k \in \mathcal{A}_k^{u+c}$ and set $\omega(p_k) = \omega(\mathcal{L}_k)$. Further, set flow routing $f_{ij}(p_k) = f_{ij}(\mathcal{L}_k)$ and $f_{iB}(p_k) = f_{iB}(\mathcal{L}_k)$. Determine $r_i(p_k)$ by (5).

Denote $\eta_{\text{OPT-}\mathcal{L}}$ and $\eta_{\text{OPT-ub}}$ as the objective values achieved by $\psi_{\text{OPT-}\mathcal{L}}$ and $\psi_{\text{OPT-ub}}$, respectively. Since τ and τ_{vac} are unchanged in the foregoing solution construction, we have

$$\eta_{\text{OPT-ub}} = \eta_{\text{OPT-}\mathcal{L}} \ . \tag{23}$$

The following lemma affirms the feasibility of the constructed solution $\psi_{\text{OPT-ub}}$ to problem OPT-ub.

LEMMA 3. *$\psi_{\text{OPT-ub}}$ is a feasible solution to problem OPT-ub.*

A proof of Lemma 3 is given in Appendix B.

Proof of Near-Optimality. The near-optimality of $\psi_{\text{OPT-ub}}$ is stated in the following theorem.

THEOREM 1. *For a given $\epsilon > 0$, $\eta_{\text{OPT-ub}} \geq \eta_{\text{OPT-ub}}^* - \epsilon$.*

A proof sketch: Theorem 1 can be proved based on solution construction. First, we show that given a feasible solution $\hat{\psi}_{\text{OPT-ub}}$ to OPT-ub with an objective value $\hat{\eta}_{\text{OPT-ub}}$, we can construct a solution $\hat{\psi}_{\text{OPT-}\mathcal{L}}$ to OPT-$\mathcal{L}$ with an objective value $\hat{\eta}_{\text{OPT-}\mathcal{L}} \geq \hat{\eta}_{\text{OPT-ub}} - \epsilon$. Second, we consider a special case that the given solution $\hat{\psi}_{\text{OPT-ub}}$ is an optimal solution $\psi_{\text{OPT-ub}}^*$ to OPT-ub with an objective value $\eta_{\text{OPT-ub}}^*$. Based on this construction, we can obtain a solution to OPT-$\mathcal{L}$ with an objective value at least $\eta_{\text{OPT-ub}}^* - \epsilon$. Since this solution is only a feasible solution to OPT-$\mathcal{L}$ while $\psi_{\text{OPT-}\mathcal{L}}$ is an optimal solution, we have $\eta_{\text{OPT-}\mathcal{L}} \geq \eta_{\text{OPT-ub}}^* - \epsilon$. Further, we have $\eta_{\text{OPT-ub}} = \eta_{\text{OPT-}\mathcal{L}} \geq \eta_{\text{OPT-ub}}^* - \epsilon$, where the first equality holds by (23).

4. A PRACTICAL SOLUTION AND PERFORMANCE GAP ANALYSIS

In Section 3, we have developed a near-optimal solution to an idealized problem, in which a WCV's traveling time is assumed to be zero. In this section, we incorporate traveling time and develop a practical solution to our original problem (TPP). We also quantify the performance gap between this solution and optimal (unknown) solution to TPP.

4.1 Fixing a Traveling Path

The near-optimal solution $\psi_{\text{OPT-ub}}$ to the idealized problem OPT-ub in Section 3 offers us several tools in designing a solution to the original problem TPP. First, the objective value $\eta_{\text{OPT-ub}}$ of $\psi_{\text{OPT-ub}}$ is at least $\eta_{\text{OPT-ub}}^* - \epsilon$ while $\eta_{\text{OPT-ub}}^*$ is an upper bound for the unknown objective value η_{TPP}^* of TPP. This can be used as a performance benchmark to measure the quality of the practical solution (which includes traveling time) to the original problem TPP. Second, the logical point concept in the near-optimal solution $\psi_{\text{OPT-ub}}$ offers a hint on where the WCV should make stops and charge the sensor nodes. We will exploit these stops in $\psi_{\text{OPT-ub}}$ to design a solution to problem TPP.

For the stopping points (any physical point within a logical point that has non-zero stopping time) in $\psi_{\text{OPT-ub}}$, we propose to find a

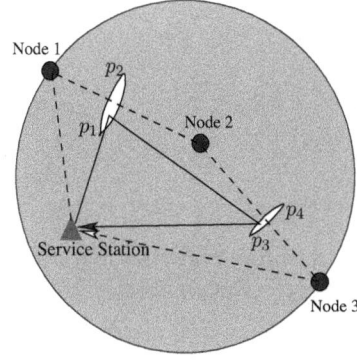

Figure 4: Comparison between a Hamiltonian cycle connecting the logical points and the service station and a Hamiltonian cycle connecting the sensor nodes and the service station.

shortest Hamiltonian cycle to connect them. Note that the service station is also included in this shortest Hamiltonian cycle. Denote this path by $\mathcal{P}_{\text{OPT-lb}}$. The decision of using shortest Hamiltonian cycle is obvious as it corresponds to the least amount of traveling time (not including stopping time).

It is important to realize that the shortest Hamiltonian cycle that we find here is based on the connection of logical points, rather than the actual sensor nodes. A Hamiltonian cycle for the latter would be fundamentally different from the former. As an example, in Fig. 4, the solid triangle is a Hamiltonian cycle connecting two logical points (with corner points (p_1, p_2) and (p_3, p_4)) and the service station while the dotted quadrangle is a Hamiltonian cycle connecting three sensor nodes and the service station.

4.2 Incorporating Traveling Time

Under the selected traveling path $\mathcal{P}_{\text{OPT-lb}}$, denote G as the total number of physical points with positive stopping time. We re-index these points by traveling order under $\mathcal{P}_{\text{OPT-lb}}$, and let $\mathcal{P}_{\text{OPT-lb}} = (p_{\text{vac}}, p_1, \ldots, p_G, p_{\text{vac}})$, with the starting and ending point being the service station (p_{vac}) and the j-th stop along the path being p_j, $1 \leq j \leq G$.

To find flow routing when the WCV travels along $\mathcal{P}_{\text{OPT-lb}}$, we discretize $\mathcal{P}_{\text{OPT-lb}}$ into a sequence of segments based on $\mathcal{P}_{\text{OPT-lb}}$'s intersection with the subareas (see Figs. 3(c) and 4). Based on this path discretization, we can rewrite TPP into a new formulation (denoted as OPT-lb), in which we can obtain flow routing when the WCV is traveling on each segment along path $\mathcal{P}_{\text{OPT-lb}}$.

Note that each segment is contained by one subarea. Among these traversed subareas, the WCV makes stops at some of them while only traversing others without making any stop. Denote $\mathcal{Q}$ as the set of indexes for all the traversed subareas and $\mathcal{Q}_s$ as the set of indexes for those subareas that the WCV makes stops, respectively. For $m \in \mathcal{Q}$, the WCV's traveling time (not including stopping time) in this subarea is $D(\mathcal{L}_m)/V$, where $\mathcal{L}_m$ is the logical point corresponding to this subarea, and $D(\mathcal{L}_m)$ denotes the distance traversed within $\mathcal{L}_m$ along $\mathcal{P}_{\text{OPT-lb}}$.

For $m \in \mathcal{Q}_s$, since the WCV makes a stop only at one point, we have $|\mathcal{Q}_s| = G$. The total time that the WCV spends in $\mathcal{L}_m$ is

$$\frac{D(\mathcal{L}_m)}{V} + \omega(\mathcal{L}_m) \ ,$$

where $\omega(\mathcal{L}_m)$ is the stopping time within $\mathcal{L}_m$. Based on $\mathcal{P}_{\text{OPT-lb}}$ and $\mathcal{L}_m, m \in \mathcal{Q}$, we rewrite problem TPP to OPT-lb as follows:

OPT-lb

maximize $\quad\quad\quad \frac{\tau_{\text{vac}}}{\tau}$

s.t. $\quad \tau = \tau_{\text{vac}} + \sum_{m \in \mathcal{Q}_s} \omega(\mathcal{L}_m) + \sum_{m \in \mathcal{Q}} \frac{D(\mathcal{L}_m)}{V}$ $\quad$ (24)

$\sum_{k \in \mathcal{N}}^{k \neq i} f_{ki}(\mathcal{L}_m) + R_i = \sum_{j \in \mathcal{N}}^{j \neq i} f_{ij}(\mathcal{L}_m) + f_{iB}(\mathcal{L}_m)$

$\quad\quad\quad\quad\quad\quad\quad\quad\quad\quad\quad\quad (i \in \mathcal{N}, m \in \mathcal{Q})$

$r_i(\mathcal{L}_m) = \rho \sum_{k \in \mathcal{N}}^{k \neq i} f_{ki}(\mathcal{L}_m) + \sum_{j \in \mathcal{N}}^{j \neq i} C_{ij} \cdot f_{ij}(\mathcal{L}_m)$

$\quad\quad\quad + C_{iB}(\mathcal{L}_m) \cdot f_{iB}(\mathcal{L}_m) \quad (i \in \mathcal{N}, m \in \mathcal{Q})$

$r_i(p_{\text{vac}}) \cdot \tau_{\text{vac}} + \sum_{m \in \mathcal{Q}_s} r_i(\mathcal{L}_m) \cdot \omega(\mathcal{L}_m)$

$\quad\quad + \sum_{m \in \mathcal{Q}} \frac{D(\mathcal{L}_m)}{V} \cdot r_i(\mathcal{L}_m)$

$\quad\quad \leq \sum_{m \in \mathcal{Q}_s}^{D_{iB}(p) \leq D_\delta, p \in \mathcal{A}_m^{u+c}} U_{iB}(\mathcal{L}_m) \cdot \omega(\mathcal{L}_m) \quad (i \in \mathcal{N})$ $\quad$ (25)

$r_i(p_{\text{vac}}) \cdot \tau_{\text{vac}} + \sum_{m \in \mathcal{Q}_s}^{D_{iB}(p) > D_\delta, p \in \mathcal{A}_m^{u+c}} r_i(\mathcal{L}_m) \cdot \omega(\mathcal{L}_m)$

$\quad\quad + \sum_{m \in \mathcal{Q}} \frac{D(\mathcal{L}_m)}{V} \cdot r_i(\mathcal{L}_m) \leq E_{\max} - E_{\min} \quad (i \in \mathcal{N})$ $\quad$ (26)

$\tau, \tau_{\text{vac}}, \omega(\mathcal{L}_m) \geq 0 \quad (m \in \mathcal{Q}_s)$

$f_{ij}(\mathcal{L}_m), f_{iB}(\mathcal{L}_m), r_i(\mathcal{L}_m) \geq 0 \quad (i, j \in \mathcal{N}, i \neq j, m \in \mathcal{Q})$

$f_{ij}(p_{\text{vac}}), f_{iB}(p_{\text{vac}}), r_i(p_{\text{vac}}) \geq 0 \quad (i, j \in \mathcal{N}, i \neq j)$

In problem OPT-lb, time constraint (24) incorporates traveling time along $\mathcal{P}_{\text{OPT-lb}}$. Also, constraints (25) and (26) incorporate energy consumption when the WCV is traveling.

Through a similar change-of-variable procedure to that in Appendix A for OPT-$\mathcal{L}$, we can reformulate OPT-lb to an LP. By solving this LP, we can obtain a feasible solution to problem TPP. Denote $\psi_{\text{OPT-lb}}$ as this feasible solution and $\eta_{\text{OPT-lb}}$ as the objective value achieved by $\psi_{\text{OPT-lb}}$. The relationship between $\eta_{\text{OPT-lb}}$ and the optimum objective value η_{TPP}^* is given in the following lemma.

LEMMA 4. $\eta_{\text{OPT-lb}}$ is a lower bound of η_{TPP}^*, i.e., $\eta_{\text{OPT-lb}} \leq \eta_{\text{TPP}}^*$.

Clearly, Lemma 4 holds since OPT-lb is based on a specified path $\mathcal{P}_{\text{OPT-lb}}$ and thus $\psi_{\text{OPT-lb}}$ is only a feasible solution to problem TPP.

4.3 Analysis of Performance Gap and Algorithm Complexity

Denote θ as the performance gap between $\eta_{\text{OPT-lb}}$ and the unknown optimal objective η_{TPP}^*. We have the following lemma.

LEMMA 5. $\theta \leq \epsilon + \eta_{\text{OPT-ub}} - \eta_{\text{OPT-lb}}$.

PROOF. By definition, $\theta = \eta_{\text{TPP}}^* - \eta_{\text{OPT-lb}}$, we have

$$\theta \leq \eta_{\text{OPT-ub}}^* - \eta_{\text{OPT-lb}} \leq \epsilon + \eta_{\text{OPT-ub}} - \eta_{\text{OPT-lb}},$$

where the first inequality holds by Lemma 2, and the second inequality holds by Theorem 1. □

In the above solution, solving two LPs (i.e., problems OPT-$\mathcal{L}$ and OPT-lb) has the highest complexity. The problem size of either LP is decided by the maximum number of subareas, which is a polynomial in $|\mathcal{N}|$ (see (20)). Thus, both LPs have polynomial size and the algorithm complexity is polynomial.

5. NUMERICAL RESULTS

Network and Parameter Settings. In the numerical results, the units of distance, time, data rate, and energy are all normalized appropriately. We assume sensor nodes are randomly deployed over a unit square area. The data rate R_i, $i \in \mathcal{N}$, is randomly generated within $[0.1, 1]$. The home service station is assumed to be at $(0.5, 0.5)$, and the WCV travels at a speed $V = 0.1$.

Table 1: Location and data rate R_i for each node in a 50-node network.

Node Index	Location	R_i	Node Index	Location	R_i
1	(0.547,0.644)	0.1	26	(0.833,0.115)	0.2
2	(0.662,0.757)	0.7	27	(0.639,0.658)	0.1
3	(0.037,0.859)	0.4	28	(0.704,0.930)	0.6
4	(0.723,0.741)	1.0	29	(0.977,0.306)	0.8
5	(0.529,0.778)	0.9	30	(0.673,0.386)	0.5
6	(0.316,0.035)	0.4	31	(0.021,0.745)	0.7
7	(0.190,0.842)	0.8	32	(0.924,0.072)	0.6
8	(0.288,0.106)	0.8	33	(0.270,0.829)	0.1
9	(0.040,0.942)	0.2	34	(0.777,0.573)	0.8
10	(0.264,0.648)	0.4	35	(0.097,0.512)	0.9
11	(0.446,0.805)	0.5	36	(0.986,0.290)	0.2
12	(0.890,0.729)	0.5	37	(0.161,0.636)	0.7
13	(0.370,0.350)	0.1	38	(0.355,0.767)	0.9
14	(0.006,0.101)	0.7	39	(0.655,0.574)	0.5
15	(0.393,0.548)	0.1	40	(0.031,0.052)	0.4
16	(0.629,0.623)	0.1	41	(0.350,0.150)	0.3
17	(0.084,0.954)	0.5	42	(0.941,0.724)	0.1
18	(0.756,0.840)	0.2	43	(0.966,0.430)	0.2
19	(0.966,0.376)	0.7	44	(0.107,0.191)	0.3
20	(0.931,0.308)	0.6	45	(0.007,0.337)	0.3
21	(0.944,0.439)	0.1	46	(0.457,0.287)	0.4
22	(0.626,0.323)	0.4	47	(0.753,0.383)	0.1
23	(0.537,0.538)	0.2	48	(0.945,0.909)	0.1
24	(0.118,0.082)	0.3	49	(0.209,0.758)	0.3
25	(0.929,0.541)	0.2	50	(0.221,0.588)	0.8

Table 2: Index of stopping points along the path, location and stopping time at each stopping point for the 50-node network.

Stopping point	Location	τ_k	Stopping point	Location	τ_k
1	(0.575,0.550)	0.3	18	(0.525,0.775)	179.3
2	(0.600,0.575)	77.2	19	(0.375,0.775)	69.8
3	(0.750,0.575)	179.4	20	(0.350,0.775)	117.9
4	(0.675,0.375)	50.9	21	(0.200,0.825)	175.9
5	(0.650,0.350)	66.2	22	(0.075,0.950)	110.0
6	(0.900,0.075)	98.6	23	(0.025,0.775)	166.6
7	(0.925,0.075)	38.8	24	(0.025,0.750)	3.1
8	(0.975,0.300)	12.2	25	(0.200,0.625)	86.5
9	(0.975,0.325)	81.8	26	(0.225,0.625)	43.5
10	(0.975,0.525)	102.9	27	(0.150,0.550)	136.8
11	(0.925,0.525)	42.3	28	(0.100,0.525)	105.1
12	(0.900,0.725)	103.6	29	(0.000,0.325)	64.7
13	(0.950,0.900)	21.8	30	(0.000,0.100)	26.4
14	(0.700,0.925)	124.2	31	(0.050,0.125)	180.6
15	(0.725,0.750)	160.6	32	(0.300,0.100)	173.3
16	(0.700,0.750)	32.4	33	(0.450,0.300)	84.3
17	(0.675,0.750)	18.9	34	(0.400,0.550)	19.8

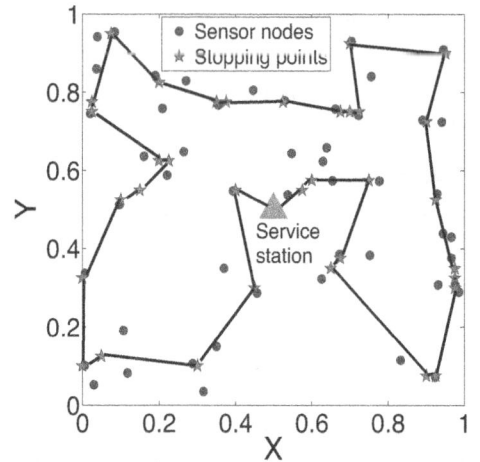

Figure 5: A traveling path for the WCV in the 50-node sensor network.

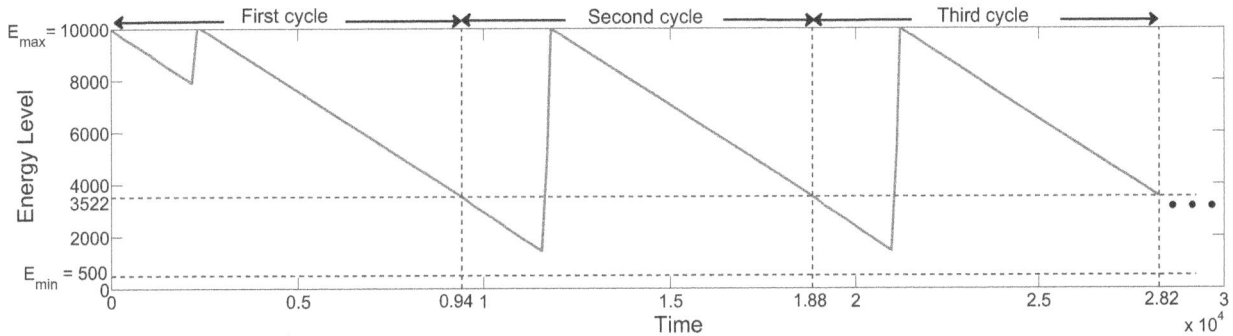

Figure 6: The energy behavior of the 35th sensor node in the 50-node network during the first three cycles.

Suppose that a sensor node uses a rechargeable battery with $E_{\max}$ = 10,000, and $E_{\min}$ = 500. For the charging efficiency function $\mu(D_{iB})$, we assume a decreasing function $\mu(D_{iB}) = -40 \cdot D_{iB}^2 - 4 \cdot D_{iB} + 1.0$. Letting $U_{\max} = 50$ and $\delta = 10$, we have $D_\delta = 0.10$ for a maximum distance of effective charging. The normalized parameters in energy consumption model are $\beta_1 = \beta_2 = \rho = 1$. The path loss index is $\alpha = 4$. We set $W = 3$ and $\epsilon = 0.05$ for the numerical results.

We consider a 50-node network. The normalized location of each node and its data rate are given in Table 1.

Results. Table 2 gives the stopping points (each within a logical point with a non-zero stopping time) along the travel path for the WCV. The traveling path is shown in Fig. 5. For $\mathcal{P}_{\text{OPT-lb}}$, $D_{\mathcal{P}_{\text{OPT-lb}}} = 4.89$ and the traveling time $D_{\mathcal{P}_{\text{OPT-lb}}}/V = 48.9$. Table 2 shows the charging schedule at each stopping point on $\mathcal{P}_{\text{OPT-lb}}$. Following $\mathcal{P}_{\text{OPT-lb}}$ and this charging schedule, our solution ensures that any sensor node never runs out of energy. As an example, Fig. 6 shows the energy behavior of a sensor node (the 35th node) during the first three cycles. During each cycle, this node is charged by the WCV when it makes stops at two stopping points (i.e., the 27th and 28th points). Starting from the second cycle, the node's energy behavior repeats from cycle to cycle.

For the given $\epsilon = 0.05$, we have $\eta_{\text{OPT-ub}} = 68.62\%$. For the obtained practical solution $\psi_{\text{OPT-lb}}$, the cycle time $\tau = 9414$, the vacation time $\tau_{\text{vac}} = 6410$, and the objective value is $\eta_{\text{OPT-lb}} = 68.09\%$. By Lemma 5, the performance gap $\theta \leq \epsilon + \eta_{\text{OPT-ub}} - \eta_{\text{OPT-lb}} = 0.05 + 0.6862 - 0.6809 = 0.0553$, where the given ϵ is the dominant part. This shows that the objective value by the lower bound feasible solution is very close to that by the upper bound solution.

6. CONCLUSIONS

In this paper, we studied the problem of co-locating the MBS on the WCV in a WSN, with a focus on the traveling path problem of the WCV. The goal was to minimize energy consumption of the the entire system while ensuring none of the sensor nodes runs out of energy. We formulated an optimization problem (TPP) that involved joint optimization of traveling path, stopping points, charging schedule, and data flow routing. We first considered an idealized problem (OPT-ub) that assumed zero traveling time. For OPT-ub, we developed a provably near-optimal solution which involves several novel techniques, such as discretization of energy reception rate and energy consumption rate, double partitioning of the SED into smaller subareas with tight energy bounds, and representation of each subarea by a logical point as its "worst-case" energy reception and consumption behavior. Based on the near-optimal solution to the idealized problem OPT-ub, we set the trav-

eling path as the shortest Hamiltonian cycle connecting the logical points and the service station. We then obtained a practical solution (with non-zero traveling time) to the original problem TPP, and quantified the performance gap between this feasible solution and an optimal (unknown) solution to TPP.

7. ACKNOWLEDGMENTS

The authors thank the anonymous reviewers for their constructive comments. This research was supported in part by NSF Grants 0925719 (Hou), 0831865 (Hou), 0969169 (Sherali), 1156311 (Lou), and 1156318 (Lou).

8. REFERENCES

[1] S. Basagni, A. Carosi, C. Petrioli, and C.A. Phillips. Coordinated and controlled mobility of multiple sinks for maximizing the lifetime of wireless sensor networks. *Springer Wirel. Netw.*, 17(3): 759–778, Apr. 2011.

[2] Y.T. Hou, Y. Shi, and H.D. Sherali. Rate allocation and network lifetime problems for wireless sensor networks. *IEEE/ACM Trans. Netw.*, 16(2): 321–334, Apr. 2008.

[3] A. Kurs, A. Karalis, R. Moffatt, J.D. Joannopoulos, P. Fisher, and M. Soljacic. Wireless power transfer via strongly coupled magnetic resonances. *Science*, 317(5834): 83–86, July 2007.

[4] A. Kurs, R. Moffatt, and M. Soljacic. Simultaneous mid-range power transfer to multiple devices. *Appl. Phys. Lett.*, 96(4): article 044102, Jan. 2010.

[5] J. Luo and J.-P. Huabux. Joint mobility and routing for lifetime elongation in wireless sensor networks. In *Proc. IEEE INFOCOM*, pages 1735–1746, March 2005.

[6] R. C. Shah, S. Roy, S. Jain, and W. Brunette. Data MULEs: Modeling a three-tier architecture for sparse sensor networks. In *Proc. IEEE International Workshop on Sensor Network Protocols and Applications (SNPA)*, pages 30–41, May 2003.

[7] Y. Shi, and Y.T. Hou. Theoretical results on base station movement problem for sensor network. In *Proc. IEEE INFOCOM*, pages 376–384, Apr. 2008.

[8] Y. Shi, L. Xie, Y.T. Hou, and H.D. Sherali. On renewable sensor networks with wireless energy transfer. In *Proc. IEEE INFOCOM*, pages 1350–1358, Apr. 2011.

[9] E. Welzl. Smallest enclosing disks. *Lecture Notes in Computer Science (LNCS)*, 555: 359–370, 1991.

[10] L. Xie, Y. Shi, Y.T. Hou, W. Lou, H.D. Sherali, and S.F. Midkiff. On renewable sensor networks with wireless energy transfer: The multi-node case. In *Proc. IEEE SECON*, pages 10–18, June 2012.

[11] L. Xie, Y. Shi, Y.T. Hou, W. Lou, H.D. Sherali, and S.F. Midkiff. Bundling mobile base station and wireless energy transfer: Modeling and optimization. In *Proc. IEEE INFOCOM*, pages 1684–1692, Apr. 2013.

[12] L. Xie, Y. Shi, Y.T. Hou, W. Lou, and H.D. Sherali. On traveling path and related problems for a mobile station in a rechargeable sensor network. Technical Report, Department of Electrical and Computer Engineering, Virginia Tech, Blacksburg, VA, June 2013. Available at http://filebox.vt.edu/users/windgoon/papers/TR13.pdf.

[13] G. Xing, T. Wang, W. Jia, and M. Li. Rendezvous design algorithms for wireless sensor networks with a mobile base station. In *Proc. ACM MobiHoc*, pages 231–240, May 2008.

[14] W. Zhao, M. Ammar, and E. Zegura. A message ferrying approach for data delivery in sparse mobile ad hoc networks. In *Proc. ACM MobiHoc*, pages 187–198, May 2004.

APPENDIX

A. REFORMULATION

We show how to reformulate problem OPT-$\mathcal{L}$ to an LP via a change-of-variable technique. For the fractional objective function $\frac{\tau_{\text{vac}}}{\tau}$, we define $\eta_{\text{vac}} = \frac{\tau_{\text{vac}}}{\tau}$. We also define $\eta(\mathcal{L}_k) = \frac{\omega(\mathcal{L}_k)}{\tau}$ and $q = \frac{1}{\tau}$. For time constraint (8), we divide both sides by τ, and rewrite it as $\eta_{\text{vac}} + \sum_{k=1}^{K} \eta(\mathcal{L}_k) = 1$.

For (3) and (5), we consider logical points $\mathcal{L}_k, k = 1, 2, \cdots, K$, and p_{vac} separately. First, for $\mathcal{L}_k$, we multiple both sides of (3) by $\eta(\mathcal{L}_k)$ and define $g_{ij}(\mathcal{L}_k) = f_{ij}(\mathcal{L}_k) \cdot \eta(\mathcal{L}_k)$ and $g_{iB}(\mathcal{L}_k) = f_{iB}(\mathcal{L}_k) \cdot \eta(\mathcal{L}_k)$. For the new nonlinear terms $r_i(\mathcal{L}_k) \cdot \eta(\mathcal{L}_k)$, we define $\mathcal{E}_i(\mathcal{L}_k) = r_i(\mathcal{L}_k) \cdot \eta(\mathcal{L}_k)$. By the new variables $g_{ij}(\mathcal{L}_k)$, $g_{iB}(\mathcal{L}_k)$, and $\mathcal{E}_i(\mathcal{L}_k)$, (3) is reformulated as

$$\sum_{k \in \mathcal{N}}^{k \neq i} g_{ki}(\mathcal{L}_k) + R_i \cdot \eta(\mathcal{L}_k) = \sum_{j \in \mathcal{N}}^{j \neq i} g_{ij}(\mathcal{L}_k) + g_{iB}(\mathcal{L}_k) ,$$

and (5) is rewritten as

$$\mathcal{E}_i(\mathcal{L}_k) = \rho \sum_{k \in \mathcal{N}}^{k \neq i} g_{ki}(\mathcal{L}_k) + \sum_{j \in \mathcal{N}}^{j \neq i} C_{ij} \cdot g_{ij}(\mathcal{L}_k) + C_{iB}(\mathcal{L}_k) \cdot g_{iB}(\mathcal{L}_k) .$$

Second, for p_{vac}, we multiply both sides of (3) and (5) by η_{vac}, and define $g_{ij}(p_{\text{vac}}) = f_{ij}(p_{\text{vac}}) \cdot \eta_{\text{vac}}$, $g_{iB}(p_{\text{vac}}) = f_{iB}(p_{\text{vac}}) \cdot \eta_{\text{vac}}$, and $\mathcal{E}_i(p_{\text{vac}}) = r_i(p_{\text{vac}}) \cdot \eta_{\text{vac}}$. Then (3) and (5) can be reformulated as

$$\sum_{k \in \mathcal{N}}^{k \neq i} g_{ki}(p_{\text{vac}}) + R_i \cdot \eta_{\text{vac}} = \sum_{j \in \mathcal{N}}^{j \neq i} g_{ij}(p_{\text{vac}}) + g_{iB}(p_{\text{vac}})$$

$$\mathcal{E}_i(p_{\text{vac}}) = \rho \sum_{k \in \mathcal{N}}^{k \neq i} g_{ki}(p_{\text{vac}}) + \sum_{j \in \mathcal{N}}^{j \neq i} C_{ij} \cdot g_{ij}(p_{\text{vac}}) + C_{iB}(p_{\text{vac}}) \cdot g_{iB}(p_{\text{vac}})$$

By dividing both sides by τ, constraint (9) can be rewritten as $r_i(p_{\text{vac}}) \cdot \eta_{\text{vac}} + \sum_{k=1}^{K} r_i(\mathcal{L}_k) \cdot \eta(\mathcal{L}_k) \leq \sum_{k=1,...,K}^{D_{iB}(p) \leq D_\delta, p \in \mathcal{A}_k^{u+c}} U_{iB}(\mathcal{L}_k) \cdot \eta(\mathcal{L}_k)$, or equivalently,

$$\mathcal{E}_i(p_{\text{vac}}) + \sum_{m=1}^{K} \mathcal{E}_i(\mathcal{L}_k) \leq \sum_{k=1,...,K}^{D_{iB}(p) \leq D_\delta, p \in \mathcal{A}_k^{u+c}} U_{iB}(\mathcal{L}_k) \cdot \eta(\mathcal{L}_k)$$

Similarly, by dividing both sides by τ, (10) can be rewritten as $r_i(p_{\text{vac}}) \cdot \eta_{\text{vac}} + \sum_{k=1,...,K}^{D_{iB}(p) > D_\delta, p \in \mathcal{A}_k^{u+c}} r_i(\mathcal{L}_k) \cdot \eta(\mathcal{L}_k) \leq \frac{E_{\max} - E_{\min}}{\tau}$, or equivalently,

$$\mathcal{E}_i(p_{\text{vac}}) + \sum_{k=1,...,K}^{D_{iB}(p) > D_\delta, p \in \mathcal{A}_k^{u+c}} \mathcal{E}_i(\mathcal{L}_k) \leq (E_{\max} - E_{\min}) \cdot q .$$

Now the new objective function and new constraints are linear, which makes an LP.

B. PROOF OF LEMMA 3

PROOF. To show that $\psi_{\text{OPT-ub}} = (\tau, \tau_{\text{vac}}, \omega(p), f_{ij}(p), f_{iB}(p), r_i(p))$ is feasible to problem OPT-ub, we need to verify that $\psi_{\text{OPT-ub}}$ satisfies constraints (3), (5), (8), (9), and (10). To do this, we exploit the worst case bounds that are inherited in a logical point representation.

Since $\psi_{\text{OPT-}\mathcal{L}}$ is feasible to problem OPT-$\mathcal{L}$ (based on the K logical points), we know that $\psi_{\text{OPT-}\mathcal{L}}$ satisfies constraints (3), (5), (8), (9), and (10). We now verify each of these constraints for $\psi_{\text{OPT-ub}}$. Since τ and τ_{vac} remain unchanged and $\omega(p_k) = \omega(\mathcal{L}_k)$, where $p_k \in \mathcal{A}_k^{u+c}$, $1 \leq k \leq K$, $\psi_{\text{OPT-ub}}$ satisfies constraint (8). $\psi_{\text{OPT-ub}}$ also satisfies constraints (3) and (5) since $f_{ij}(p_{\text{vac}})$, $f_{iB}(p_{\text{vac}})$ and $r_i(p_{\text{vac}})$ remain unchanged, $f_{ij}(p_k) = f_{ij}(\mathcal{L}_k)$, $f_{iB}(p_k) = f_{iB}(\mathcal{L}_k)$, and $r_i(p_k)$ is determined by (5).

To verify two remaining energy constraints (9) and (10), by (19) and (22), we first have $C_{iB}(p_k) \leq C_{iB}(\mathcal{L}_k), 1 \leq k \leq K$. As a result, $r_i(p_k) = \rho \sum_{k \in \mathcal{N}}^{k \neq i} f_{ki}(p_k) + \sum_{j \in \mathcal{N}}^{j \neq i} C_{ij} \cdot f_{ij}(p_k) + C_{iB}(p_k) \cdot f_{iB}(p_k) \leq \rho \sum_{n \in \mathcal{N}}^{n \neq i} f_{ni}(\mathcal{L}_k) + \sum_{j \in \mathcal{N}}^{j \neq i} C_{ij} \cdot f_{ij}(\mathcal{L}_k) + C_{iB}(\mathcal{L}_k) \cdot f_{iB}(\mathcal{L}_k) = r_i(\mathcal{L}_k)$, where the first equality holds since $\psi_{\text{OPT-ub}}$ satisfies (5), the second equality holds since $f_{ij}(p_k) = f_{ij}(\mathcal{L}_k)$, $f_{iB}(p_k) = f_{iB}(\mathcal{L}_k)$ and $C_{iB}(p_k) \leq C_{iB}(\mathcal{L}_k)$, and the last equality holds since $\psi_{\text{OPT-}\mathcal{L}}$ satisfies (5). By the same token, we have that $r_i(p_{\text{vac}})$ is unchanged since $C_{iB}(p_{\text{vac}})$ is unchanged.

Since $r_i(p_k) \leq r_i(\mathcal{L}_k)$ and $r_i(p_{\text{vac}})$ is unchanged, we have

$$r_i(p_{\text{vac}}) \cdot \tau_{\text{vac}} + \sum_{k=1}^{K} r_i(p_k) \cdot \tau(p_k)$$
$$\leq r_i(p_{\text{vac}}) \cdot \tau_{\text{vac}} + \sum_{k=1}^{K} r_i(\mathcal{L}_k) \cdot \tau(\mathcal{L}_k)$$
$$\leq \sum_{k=1,...,K}^{D_{iB}(p) \leq D_\delta, p \in \mathcal{A}_k^{u+c}} U_{iB}(\mathcal{L}_k) \cdot \tau(\mathcal{L}_k)$$
$$\leq \sum_{k=1,...,K}^{D_{iB}(p_k) \leq D_\delta} U_{iB}(p_k) \cdot \tau(p_k)$$

where the first inequality holds since $r_i(p_{\text{vac}})$ and τ_{vac} are unchanged, $r_i(p_k) \leq r_i(\mathcal{L}_k)$ and $\tau(p_k) = \tau(\mathcal{L}_k)$, the second inequality holds since $\psi_{\text{OPT-}\mathcal{L}}$ meets (9), the third inequality holds by $U_{iB}(\mathcal{L}_k) \leq U_{iB}(p_k)$ (see (15) and (21)) and $\tau(p_k) = \tau(\mathcal{L}_k)$. Thus, constraint (9) holds for $\psi_{\text{OPT-ub}}$. Similarly, we can show that constraint (10) holds for $\psi_{\text{OPT-ub}}$. Therefore, the constructed solution $\psi_{\text{OPT-ub}}$ is feasible to problem OPT-ub. This completes the proof. □

Qute: Quality-of-Monitoring Aware Sensing and Routing Strategy in Wireless Sensor Networks

Shaojie Tang
Department of Computer and Information
Sciences
Temple University
Philadelphia PA 19122
shaojie.tang@temple.edu

Jie Wu
Department of Computer and Information
Sciences
Temple University
Philadelphia PA 19122
jiewu@temple.edu

ABSTRACT

Wireless Sensor Networks (WSNs) are widely used to monitor the physical environment. In a highly redundant sensor network, sensor readings from nearby sensors often have high similarity. In this work, we are interested in how to decide an appropriate sensing rate [1] for each sensor node, in order to maximize the overall Quality-of-Monitoring (QoM), while ensuring that all readings can be transmitted to the sink. Note that a feasible sensing rate allocation should satisfy both energy constraint on each sensor node and flow conservation through the network. In order to capture the statistical correlations among sensor readings, we first introduce the concept of correlation graph. The correlation graph is further decomposed into several *correlation components*, and sensor readings from the same correlation component are highly correlated. For each correlation component, we defined a general utility function to estimate the QoM. The utility function of each correlation component is a non-decreasing submodular function of the total sensing rates allocated to that correlation component. Then we formulate the QoM-aware sensing rate allocation problem as a utility maximization problem under limited power supply on each node. To tackle this problem, we adopted an efficient algorithm, called *Qute*, by jointly considering both the energy constraint on each node and flow conservation through the network. Under some settings, we analytically show that *Qute* can find the optimal QoM-aware sensing rate allocation which achieves the maximum total utility. We conducted extensive testbed verifications of our schemes, and experimental results validate our theoretical results.

Categories and Subject Descriptors

C.2.1 [**Network Architecture and Design**]: Wireless communication, Network topology; G.2.2 [**Graph Theory**]: Network problems, Graph algorithms

---

[1]The number of sensor readings per unit time.

Keywords

Sensing rate allocation, routing design, Quality-of-Monitoring.

1. INTRODUCTION

Wireless Sensor Networks (WSNs) often contain a large amount of sensor nodes which are spatially distributed to monitor physical or environmental conditions, such as temperature, humidity, *etc.*, over a geographic region. Different from traditional networking systems, developing an effective sensor network must take into account both its Quality-of-Monitoring (QoM) [1] and limited energy resource [2]. Sensor nodes periodically gather sensing values from its nearby environment, therefore the QoM crucially depends on the sensing rate of each sensor node. Typically, a high sensing rate is required to achieve high QoM. We note that most sensor nodes used in large scale sensing applications are often resource constrained, with an extremely limited energy budget and wireless communication ability. The massive amount of sensing data posed great challenges on designing efficient data gathering schemes under various network resource constraints. Therefore, there is a great need for developing a QoM-aware sensing rate allocation scheme to decide an appropriate sensing rate for each node, in order to maximize the overall QoM subject to energy constraint on each node. One naive method is to treat each sensor node equally and collect as many readings as possible from the entire network. However, this approach ignores the underlying correlation among sensor nodes. We observed that the sensor readings from nearby sensors are often correlated, resulting in inter-node dependency [3] [4] [5] [6] [7] [8]; This provides us with a good opportunity to design a better sensing rate allocation scheme by avoiding redundant sensor readings.

In this work, we first introduce the correlation graph [4] [9] to capture the statistical correlations among sensor nodes. We further partition the correlation graph into several *correlation components* [10] [11] such that sensor readings from the same component are *highly* correlated. In particular, given the sensor readings gathered from some individual nodes from one correlation component, we can use interpolation to estimate the readings at all other nodes in that component. For each correlation component, we define a general utility function to quantify the Quality-of-Monitoring. The utility function is a non-decreasing submodular function, depending on the total sensing rates from all sensors within that correlation component. The submodularity of the QoM function is due to the correlations among sensor nodes within one component. For example, when using WSNs to monitor the humidity in the forest, we observe that the readings from nearby sensors are often correlated with each other. It implies that allocating an additional sampling to one component results in diminishing improvement to the QoM, as the amount of samplings allocated to the same component grows. The submodularity of QoM for monitoring physical phenomena has been observed in many real-world data sets, including target tracking [12], water quality monitoring [13] and temperature monitoring [14]. In this work, we aim to exploit such correlation to develop a QoM-aware sensing rate allocation scheme, in order to maximize the overall utility, summed over all correlation components. The main contributions of this paper are as follows.

(1) We introduce the concept of correlation component and propose a simple yet general representation, called *utility function* of QoM, which captures the data correlation relationship among sensor nodes in the same correlation component. Particularly, we reveal space-dependence by characterizing the utility function for each correlation component as a non-decreasing submodular function of the total sensing rate from that component.

(2) Based on the proposed utility function, we jointly consider sensing rate allocation and routing design under the energy constraint on each node, and pose those two techniques into a uniform framework, called *Qute*. We theoretically prove that *Qute* can achieve the maximum total utility if the energy consumption for sensing unit data is no less than that for receiving unit data.

(3) We conducted extensive testbed verifications of our protocols, and experimental results show that our protocols perform well in practice, in terms of utility maximization and estimation error minimization.

The remainder of the paper is organized as follows: Section 2 summarizes the related works; Section 3 discusses the motivation of this work and introduces the problem formulation; we adopt an efficient rate allocation protocol in Section 4; some interesting extensions of this work are discussed in Section 5; extensive experimental results are reported in Section 6; we conclude this paper in Section 7.

2. RELATED WORK

Exploiting the correlations among sensor readings in large scale WSNs attracts increasing interest in [1] [5] [15] [16] [17] [18]. In [7] [19], they develop an efficient communication protocol by taking advantage of sensor reading correlations. Our work differs from theirs in the sense that we put our focus on designing a joint sensing rate allocation and routing design, by exploiting spatial correlations among nodes.

In [20] [21], lexicographic maxmin fairness for data collection in traditional WSN and solar powered WSN has been extensively studied. Their objective is to maximize the minimum sensing rate among all sensors, while adhering to energy constraints on each sensor node. In this work, we show that our problem is closely related to fair rate allocation problem, therefore it allows us to adopt a similar rate allocation scheme.

Another category of related works studies the submodularity of sensor readings in various WSN applications such as temperature monitoring [14], water quality monitoring [13], and target tracking [12]. A QoM-aware sensing scheduling scheme is developed in [2]. They propose a simple greedy algorithm for deciding the sensing activity of different sensors. Previous work is extended in [8], they characterize the QoM from both time and space domain as a submodular function. They propose several sensing scheduling schemes under both the centralized and distributed settings. Very recently, Tang *et al.* [22] revisited the distributed sensing scheduling problem, they formulated the problem as a potential game by treating each sensor node as a player, and the convergence of their scheme is guaranteed.

3. MOTIVATIONS AND PROBLEM FORMULATION

3.1 Motivating Application

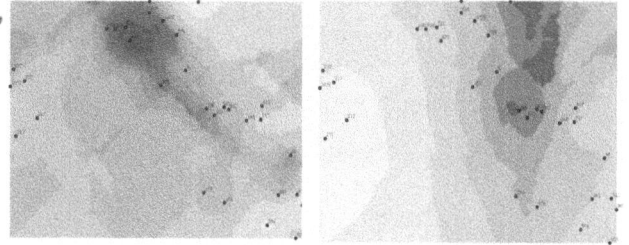

Figure 1: Humidity (left) and temperature (right) distribution over a forest. Dark dots represent the locations of different sensors that are deployed in the forest.

This work was originally motivated by Greenorbs [23] in which thousands of sensors are deployed in the forest for environment monitoring. Each sensor node periodically measures several signal values (such as temperature, humidity, illuminance, and CO_2) from the environment, and continuously transmits them back to the base station through multi-hop relay. To achieve satisfactory quality of monitoring, typical WSN applications require spatially dense sensor deployment [6]. As a result, the readings among neighboring nodes are often spatially correlated. The degree of correlation depends on the internode separation [19]. This kind of spatial redundancy information is referred to as the 'spatial correlation'. For instance, Figure 1 plots the temperature and humidity distribution over a forest, with different colors indicating different senor readings from corresponding areas. We observe that both temperature and humidity readings from nearby sensor nodes tend to be similar to one another; the degree of similarity varies according their locations and spatial distribution of temperature. Those sensors with similar readings naturally form a component or cluster.

Our work aims at QoM-aware sensing rate allocation by exploiting correlation among sensor nodes. We treat all the sensor nodes in one component equally, and define a utility function to quantify the QoM for each component under various sensing rate allocations. The utility function is a submodular function related to total sensing rate from that component. We intend to find the sensing rate for each node, in order to maximize the overall utility, by minimizing the redundant sensor readings from the same component. Note that a feasible rate allocation should take into consideration both the energy budget of each node, and the flow conservation through the network.

3.2 System Models

We first introduce several preliminary concepts which will be frequently used throughout this paper.

3.2.1 Networking Model

Assume a wireless sensor network is composed of n sensor nodes, $\mathcal{V} = \{v_1, v_2, \cdots, v_n\}$, and each sensor has a fixed transmission range. The energy budget of sensor v per unit time is $\mathbf{B}_v$, and the energy consumption for transmitting unit data is δ_t, the energy consumption for receiving unit data is δ_r, the energy consumption for sensing (or generating) unit data is δ_s. We use s_v to represent the sensing rate (the number of generated samplings per unit time) of node v. $S = \{s_{v_1}, s_{v_2}, \cdots, s_{v_n}\}$ is a sensing rate allocation.

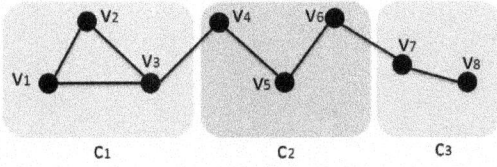

Figure 2: Communication graph and correlation components. Nodes v_1, v_2, v_3 belong to component c_1; nodes v_4, v_5, v_6 belong the component c_2; nodes v_7, v_8 belong to component c_3.

Notice that after each sensor node generates the sensing data, it further transmits the sensing data to the sink node through single- or multi- hop routing. Therefore, a feasible sensing rate allocation should satisfy both energy constraint on each node, and flow conservation through the network.

DEFINITION 1 (COMMUNICATION GRAPH). *Given a sensor network consisting of a set of n sensors, the communication graph over the sensor network is a undirected graph with $\mathcal{V}$ as a set of vertices, and there exists an edge between any two sensors if, and only if, they can communicate with each other.*

3.2.2 Correlation Model

In this work, we use a correlation graph to represent the correlation among sensor nodes. Various graphical models, including Markov random files and Bayesian networks, are widely adopted to represent the statistical correlations among the sensors. Those models [3] [9] are also helpful for designing distributed estimation algorithms. Given a set of n sensors, $\mathcal{V} = \{v_1, v_2, \cdots, v_n\}$, a typical correlation graph of the sensor network is a undirected graph with $\mathcal{V}$ as a set of vertices, and there is a (maybe weighted) edge between any two sensors if, and only if, there exists a conditional dependency or partial correlation between them. The degree of the correlation or similarity between a pair of nodes can be estimated in many ways. One method treats the sensor reading of each node by a variable [3], then the correlation between two sensors is evaluated by the statistical dependence between those two variables. Another method [11] measures the difference between two time series, using the magnitude and the trend of time series. Details of the correlation graph construction is out of the scope of this paper.

According to the correlation graph built in the first phase, we partition all nodes into m disjoint components, $\mathcal{C} = \{c_1, c_2, \cdots, c_m\}$, called *correlation components* by adopting similar approaches used in [3] [4] [5] [10] [11] [24], such that the sensor readings reported by sensor nodes within the same component are highly correlated.

DEFINITION 2 (CORRELATION COMPONENT). *A correlation component is a subset of sensors where the sensor nodes within one component have similar sensing values. Thus, the sensing value reported by any sensor can be approximated or estimated by the readings of any sensor node within the same correlation component.*

Many efficient partitioning algorithms have been proposed to realize the partitioning, based on spatial correlation. In this work, we follow the same clustering techniques as in [11] to partition the correlation graph into a minimum number of cliques, and each clique is treated as a correlation component. As an example, Figure 2 illustrates a typical communication graph and correlation component. Sensor nodes within the same block belong to one correlation

component. There is an edge between two nodes if they can communicate with each other.

3.2.3 Utility Function of QoM

Similar to [1] [8], we define a general submodular function to quantify the Quality-of-Monitoring (QoM) under different sensing rate allocations. Depending on the applications, different sensing rate allocations will provide different levels of QoM. Specifically, for a single component, the contribution of an additional sampling to the QoM crucially depends on the current sensing rate of that component. As pointed out in the literature [1] [8] [25], due to spatial-temporal correlation among sensor readings, the utility function of total sensing rate exhibits diminishing marginal returns. We were motivated by those observations to treat all the nodes in one component equally; we define the utility function, $\mathcal{U}^{c_i}$, for each correlation component c_i by a general, non-decreasing, submodular function $\mathcal{U}()$ in terms of total sensing rate that is allocated to that component: $\mathcal{U}^{c_i} = \mathcal{U}(\sum_{v \in c_i} s_v)$. We say that $\mathcal{U}()$ is submodular if it satisfies a diminishing returns property: the difference from adding an element c to a set a is at least as large as the one from adding the same element to a superset b. The overall utility is defined by summing utilities over all correlation components:

$$\mathbf{U} = \sum_{c_i \in \mathcal{C}} \mathcal{U}^{c_i} = \sum_{c_i \in \mathcal{C}} \mathcal{U}\left(\sum_{v \in c_i} s_v\right)$$

3.3 Problem Formulation

Now we can present the formulation of the problem. Without loss of generality, we assume that there is a set of sensor nodes deployed over a two-dimensional area. In addition, there is one sink node to collect all sensing data from the network. The locations of the sensor nodes are fixed and known a priori, and the correlation components are predefined. We assume that each sensor performs a sensing task, *e.g.*, temperature sampling, at certain rate, and then transmits the sensed data to the sink node through single- or multi-hop routing.

In the remainder of this paper, we use f_{uv} to represent the amount of flow from sensor u to sensor v; s_u to represent the sensing rate of node u; $\mathbf{B}_u$ to denote the energy budget for node u; δ_t to represent the energy consumption for transmitting unit data; δ_r to represent the energy consumption for receiving unit data; δ_s to represent the energy consumption for sensing unit data. Then the **QoM-aware rate allocation** problem, given energy budget on each sensor node, is formulated as:

Problem: *QoM-aware Rate Allocation*
Objective: Maximize $\mathbf{U} = \sum_{c_i \in \mathcal{C}} \mathcal{U}(\sum_{v \in c_i} s_v)$
subject to:

$$\begin{cases} (1) \ s_u + \sum_{v \in N_u} f_{vu} = \sum_{v \in N_u} f_{uv}, \forall u \neq \text{sink} \\ (2) \ \sum_{v \in N_u} f_{vu} \delta_r + \sum_{v \in N_u} f_{vu} \delta_t + s_u \delta_s \leq \mathbf{B}_u, \forall u \neq \text{sink} \\ (3) \ f_{uv} \geq 0, \forall u, v \in \mathcal{V} \end{cases}$$

In the above formulation, N_v represents the set of v's neighbors in the communication graph. Constraint 1 specifies the flow conservation: total amount of inflow plus self-generated data is equal to that of outflow. Constraint 2 specifies that the total energy consumption on each node should not exceed its energy budget. The general objective of this work is to decide an appropriate sensing rate s_v for each node $v \in \mathcal{V}$ and associated flow assignment f_{uv} on each link uv in order to maximize the overall utility while satisfying both the energy constraint and flow conservation.

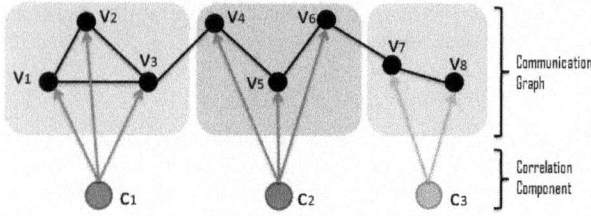

Figure 3: Two-layered Communication-Correlation Graph based on the example in Figure 2.

4. QOM-AWARE SENSING RATE ALLOCATION

We first introduce a two-layered Communication-Correlation Graph (CCG) by integrating the information from both communication graph and correlation component. Based on CCG, we introduce and study a new problem called *fair rate allocation* problem, which shares the same optimal solution with the original QoM-aware rate allocation problem in some scenarios. We adopt an efficient algorithm to find the optimal fair rate allocation, which is also the optimal QoM-aware rate allocation under some settings.

4.1 Communication-Correlation Graph

We first build the Communication-Correlation Graph (CCG) to integrate both the communication graph and the correlation graph. CCG provides a unified framework under which we are able to tackle the QoM-aware rate allocation problem by taking into consideration both the energy/flow constraint of underlying communication graph and the correlation relationship among sensor nodes. The construction of CCG, as a two-layered graph, takes as input the communication graph and the correlation components. The first layer is constituted by the communication graph, as defined in Section 3.2.1. In the second layer, we create a virtual node c_i for each correlation component, and we add a directed edge from virtual node c_i to sensor node v_i in the first layer if, and only if, $v_i \in c_i$, as defined in Section 3.2.2. Please refer to Figure 3 to illustrate: this CCG is built based on the example shown in Figure 2, we add directed edges from c_1 to v_1, v_2 and v_3; from c_2 to v_4, v_5 and v_6; from c_3 to v_7 and v_8.

Next, we re-formulate the QoM-aware rate allocation problem based on CCG. We assume there is no energy constraint on virtual node c_i. The energy consumption for node v to receive unit data from adjacent c_i is δ_s, which is the sensing cost of node v in the original problem. Let s_c represent the total sensing rate from component c: $s_c = \sum_{v \in c} s_v$. We use $S = \{s_{c_1}, s_{c_2}, \cdots, s_{c_m}\}$ to represent a rate allocation where s_{c_i} is the sensing rate assigned to component c_i. It was worth mentioning that in the new problem formulation, we define s_c to represent the total sensing rate of the entire component, instead of defining s_v for each individual node. Therefore, the overall utility function can be rewritten as:

$$\mathbf{U} = \sum_{\mathbf{c} \in \mathcal{C}} \mathcal{U}(\sum_{v \in \mathbf{c}} s_v) = \sum_{\mathbf{c} \in \mathcal{C}} \mathcal{U}(s_{\mathbf{c}})$$

Accordingly, we re-formulate the original problem as follows:

> **Problem:** *QoM-Aware Rate Allocation based on CCG*
> **Objective:** Maximize $\mathbf{U} = \sum_{\mathbf{c} \in \mathcal{C}} \mathcal{U}(s_{\mathbf{c}})$
> subject to:
>
> $$\begin{cases} (1) & s_{\mathbf{c}} = \sum_{v \in N_c} f_{cv}, \forall \mathbf{c} \in \mathcal{C} \\ (2) & \sum_{u \in N_v} f_{uv} + \sum_{\mathbf{c}:N_{\mathbf{c}} \ni v} f_{cv} = \sum_{u \in N_v} f_{vu} \\ (3) & \sum_{\mathbf{c}:N_{\mathbf{c}} \ni v} f_{cv} \delta_s + \sum_{v' \in N_v} f_{vu} \delta_t + \sum_{u \in N_v} f_{vu} \delta_r \leq \mathbf{B}_v \\ (4) & f_{vu} \geq 0, \forall v, u \in \mathcal{V} \end{cases}$$

The above problem formulation is similar to the original one, except for some additional constraints on the virtual node. Constraint 1 ensures that the amount of generated data at each virtual node equals the outflow from that virtual node. Constraint 2 specifies the flow conservation for ordinary sensor nodes. Constraint 3 specifies the energy constraint on each sensor node. In fact, this problem is equivalent to the original problem in the sense that (1) any solution of the above problem can be converted to that of the original problem without utility degradation, and (2) both problems share the same objective function. In particular, after solving the above problem, we immediately obtain a rate allocation for the original problem by setting the sensing rate of v to $s_v = f_{cv}$, and the amount of flow on link uv to f_{uv}. It is easy to verify that this allocation is feasible in satisfying both flow conservation and energy constraint. In the rest of this paper, we will study our problem based on CCG.

4.2 Optimal Fair Rate Allocation

It turns out to be extremely difficult to tackle the original QoM-aware rate allocation problem directly, therefore we start with a new problem, called *fair rate allocation* problem. *Fair rate allocation* problem is also known as the lexicographic maxmin rate allocation problem in [20]. Later, we show that the *fair rate allocation problem* and QoM-aware rate allocation problem share the common optimal solution under some settings. Different from the QoM-aware rate allocation problem, whose objective is to maximize the total utilities, summed over all components, *fair rate allocation* problem seeks a rate allocation which can maximize the minimum sensing rate among all components. Surprisingly, we show that the optimal solutions to those two problems are identical when $\delta_s \geq \delta_r$. In the rest of this section, we first introduce the *fair rate allocation* problem, and then provide an efficient algorithm to tackle this problem. Remember that we use $S = \{s_{c_1}, s_{c_2}, \cdots, s_{c_m}\}$ to represent a rate allocation, where s_{c_i} is the rate assigned to component c_i.

DEFINITION 3 (FAIR RATE ALLOCATION). *Given two feasible sensing rate allocations S_a and S_b, we sort them in non-decreasing order, and obtain two non-decreasing rate vectors Q_a and Q_b. Let Q_a^i and Q_b^i represent the i-th rate in Q_a and Q_b, respectively. We define $S_a = S_b$ if, and only if, $Q_a = Q_b$; $S_a > S_b$ if, and only if, there exists an i such that $Q_a^i > Q_b^i$ and for all $j < i$, $Q_a^j = Q_b^j$. We say a rate allocation S is an optimal fair rate allocation if, and only if, there exists no other rate allocation $S' > S$.*

In order to solve the *fair rate allocation* problem, we adopt a similar approach proposed in [20]. The difference between our problem setting and [20] is that we only care about the sensing rate of each correlation component instead of each individual sensor node. This approach is composed of two parts: (1) Maximum Common Rate Computation: compute a maximum common rate $\bar{s}$ that satisfies all energy constraints and flow conservation; and (2) Maximum Individual Rate Computation: calculate the maximum rate for each component by assuming the sensing rate of all the other components is $\bar{s}$. We compute those two rates iteratively, until the final rate allocation is determined.

I. COMPUTE MAXIMUM COMMON RATE. To compute the maximum common rate, we formulate it as a linear programming problem. In this problem, most constraints are the same as those listed in *QoM Aware Rate Allocation based on CCG*, except that (1) we use the same sensing rate $\bar{s}$ for all correlation components in Constraint 1, and (2) the objective is to find the maximum $\bar{s}$. We can use any linear programming solver, such as simplex methods, to efficiently solve this problem and obtain the maximum common sensing rate $\bar{s}$.

II. COMPUTE MAXIMUM INDIVIDUAL RATE. After the maximum common rate $\bar{s}$ is computed, the second step is to compute the maximum individual sensing rate that can be achieved for each component by assuming all the other components take the same sensing rate $\bar{s}$. For each component $\mathbf{c}$, we formulate the maximum individual rate problem as a linear programming problem. Its formulation is similar to the one defined in the previous phase, except that the first condition is replaced by two constraints (1) $s_{\mathbf{c}} = \sum_{v \in N_{\mathbf{c}}} f_{\mathbf{c}v}$; and (2) $s_{\mathbf{c}'} = \sum_{v \in N_{\mathbf{c}'}} f_{\mathbf{c}'v} = \bar{s}, \forall \mathbf{c}' \in \mathcal{C} \setminus \{\mathbf{c}\}$. Essentially, the objective is to compute the maximum $s_{\mathbf{c}}$ under the constraint that the remaining components have a common sensing rate $\bar{s}$. This problem can still be solved efficiently through linear programming. After obtaining the maximum individual rate for each component, there must exist at least one component, say $\mathbf{c}$, whose maximum individual rate is the same as the maximum common rate. We find all such $\mathbf{c}$, and set their final sensing rate to $s_{\mathbf{c}} = \bar{s}$.

The pseudo codes are listed in Algorithm 1. By solving the previous two problems iteratively, we can determine the rate allocation and associated flow assignment for each correlation component. Theorem 1 shows that the rate allocation returned from Algorithm 1 is an optimal *fair rate allocation*. The general idea of the proof follows that of [20], and the difference is that in our problem, as mentioned earlier, the sensing rate is associated with each correlation component instead of each individual sensor node.

Algorithm 1 Optimal Fair Rate Allocation (FRA)

Input: CCG and associated energy constraint & flow conservation.
Output: Sensing rate for each component and flow assignment on each link.

1: **while** $\mathcal{C} \neq \emptyset$ **do**
2:   Compute the maximum common sensing rate $\bar{s}$ in $\mathcal{C}$;
3:   **for** each component $\mathbf{c}$ in $\mathcal{C}$ **do**
4:     Compute the maximum individual sensing rate $s_{\mathbf{c}}$ by assuming the sensing rate of all other components is $\bar{s}$;
5:     **if** $s_{\mathbf{c}} = \bar{s}$ **then** $\mathcal{C} \longleftarrow \mathcal{C} - \mathbf{c}$
6: **return** the rate allocation.

THEOREM 1. *Algorithm 1 returns the optimal* fair rate allocation.

4.3 QoM-aware Rate Allocation for Utility Maximization

So far, we have demonstrated that Algorithm 1 returns the optimal solution in terms of *fair rate allocation* problem; however, there is still a gap between the optimal *fair rate allocation* and optimal QoM-aware rate allocation. To fill this gap, we show in Theorem 2 that if the per unit data sensing cost is no less than the per unit data receiving cost, then the optimal *fair rate allocation* is also an optimal QoM-aware rate allocation. It implies that, when $\delta_s \geq \delta_r$, Algorithm 1 returns the optimal QoM-aware rate allocation, which achieves the maximum total utility.

We first provide Lemma 1 as a supporting lemma of Theorem 2. Lemma 1 can be easily proved based on the definition of optimal *fair rate allocation*, as in Definition 3. In particular, Lemma 1 reveals an important property of optimal *fair rate allocation*; this property will be used later to establish the equivalence between optimal *fair rate allocation* and optimal QoM-aware rate allocation under some settings.

LEMMA 1. *Given any optimal* fair rate allocation, *in order to increase some correlation component's sensing rate (if possible), we must reduce the sensing rate of some other component who has a lower sensing rate.*

In the following, we prove that Algorithm 1 returns the optimal QoM-aware rate allocation under the assumption that the per unit data sensing cost is no less than the per unit data receiving cost.

THEOREM 2. *When the per unit data sensing cost is no less than the per unit data receiving cost $\delta_s \geq \delta_r$, Algorithm 1 returns the optimal* QoM-aware rate allocation.

PROOF. To better illustrate the big idea used to prove this theorem, we put our focus on a simple case, by assuming that the communication graph is a linear network where the sink is the leftmost node. Later, similar techniques can be extended to the proof for general communication graph.

PROPOSITION 1. *Given any feasible rate allocation S in the linear communication graph with $\delta_s \geq \delta_r$, in order to increase the sensing rate of some component, say $\mathbf{c}^*$, by ϵ, we only need to decrease the total sensing rate of the other components by at most ϵ.*

PROOF. See Figure 4 as an illustration. Given a feasible rate allocation S, we aim to increase the sensing rate of some component, say $\mathbf{c}^*$, by ϵ. For any rate adjustment strategy which can achieve this goal, let $\{v_i : i = 1, 2, \cdots, k\}$ be the set of all nodes (which are not adjacent to $\mathbf{c}^*$) whose sensing rates are decreased. Let ϵ_i represent the decreased rate of node v_i, and $\{v_1, v_2, \cdots, v_k\}$ are ordered in increasing order of their hop distance to the sink, *e.g.*, v_1 is the node closest to the sink.

We study two cases, respectively, in terms of different distributions of ϵ_i, and show that the total amount of decreased rate $\sum_{i=1}^{k} \epsilon_i$ is no larger than ϵ:
▷ If $\sum_{i=2}^{k} \epsilon_i \geq \epsilon$, we can simply set $\epsilon_1 = 0$, without violating the energy constraint of v_1. Notice that, compared to the original rate allocation S, the increased flow from $\mathbf{c}^*$ to v_1 is ϵ. However, since we have $\sum_{i=2}^{k} \epsilon_i \geq \epsilon$, this indicates that the reduced flow from all the other nodes is no less than the increased flow from v_1. In other words, there is no additional flow coming into v_1, compared to S. Because S is a feasible allocation, the energy constraint of v_1 still holds. We perform this operation sequentially to v_1, v_2, $\cdots$ until it meets some node, say v_t, such that $\sum_{i=t}^{k} \epsilon_i > \epsilon$ and $\sum_{i=t+1}^{k} \epsilon_i < \epsilon$, then we apply the following operation.

▷ If $\sum_{i=t}^{k} \epsilon_i > \epsilon$ and $\sum_{i=t+1}^{k} \epsilon_i < \epsilon$, we can set the decreased rate of v_t to $\sum_{i=t}^{k} \epsilon_i - \epsilon$ without violating the energy constraint of v_t. According to similar arguments used in the previous operation, we know that the overall increased flow through v_t is $\epsilon - \sum_{i=t+1}^{k} \epsilon_i$. Because the sensing cost is no less than the receiving cost $\delta_s \geq \delta_r$, therefore decreasing the sensing rate of v_t by at most $\epsilon - \sum_{i=t+1}^{k} \epsilon_i$, the energy constraint is still satisfied. After this stage, we can ensure that the total decreased rate is at most:

$$\epsilon - \sum_{i=t+1}^{k} \epsilon_i + \sum_{i=t+1}^{k} \epsilon_i = \epsilon$$

123

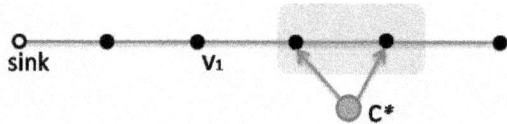

Figure 4: Linear communication graph where sink is the leftmost node. The objective is to increase the sensing rate of component c^* by ϵ. Given any rate adjustment strategy which can achieve this goal, assume v_1 is the leftmost node (which does not belong to c^*) whose sensing rate should be decreased.

This completes the proof. $\square$

LEMMA 2. *For the linear communication graph, any optimal* QoM-aware rate allocation *must also be an optimal* fair rate allocation *if $\delta_s \geq \delta_r$.*

PROOF. We prove this through contradiction. Recall that in Lemma 1, we demonstrate that given any optimal *fair rate allocation*, we cannot increase the sensing rate of a correlation component without reducing the sensing rate of some other correlation component with a lower sensing rate. Given an optimal QoM-aware rate allocation S, assume by contradiction that there exists a component c^* whose sensing rate δ_{c^*} can be increased without decreasing the rate of any other component with a smaller rate. Notice that in Proposition 1, we show that, in order to increase the sensing rate of some component by some amount ϵ, we only need to decrease the sensing rate of the other components by, at most, the same amount ϵ. Therefore, we are able to obtain a new rate allocation S^* by (1) increasing c^*'s sensing rate from δ_{c^*} to $\delta_{c^*} + \epsilon$, and at the same time (2) decreasing the total sensing rate of some other components with a higher rate, by at most ϵ. Recall that the utility function defined for each component is a submodular function with diminishing return property, thus if $s_{c^*} < s_{c_1} < s_{c_2} < \cdots < s_{c_k}$ and $\epsilon = \sum_1^k \epsilon_k$, we have

$$\mathcal{U}(s_{c^*} + \epsilon) + \mathcal{U}(s_{c_1} - \epsilon_1) + \mathcal{U}(s_{c_2} - \epsilon_2) + \cdots + \mathcal{U}(s_{c_k} - \epsilon_k)$$

$$\geq \mathcal{U}(s_{c^*}) + \mathcal{U}(s_{c_1}) + \mathcal{U}(s_{c_2}) + \cdots \mathcal{U}(s_{c_k})$$

It indicates that the total utility of S^* is larger than that of S. This contradicts the assumption that S is an optimal QoM-aware rate allocation. $\square$

A similar approach can be applied to the proof for general graph. In particular, we prove:

PROPOSITION 2. *Given any feasible rate allocation S in general communication graph with $\delta_s \geq \delta_r$, in order to increase the sensing rate of some component, say c^*, by ϵ, we only need to decrease the total sensing rate of the other components by at most ϵ.*

PROOF SKETCH. The technique used to prove this proposition is similar to that used in Proposition 1, except for the selection of v_1. We sort all nodes in topological ordering, then instead of picking the leftmost node as v_1, we pick the node with the smallest order as v_1. The remainder of the proof follows a flow similar to that in Proposition 1. Essentially, we scan all nodes in increasing order of their topological ordering to find all those nodes whose sensing rate can be adjusted. The detailed proof is omitted here to save space.

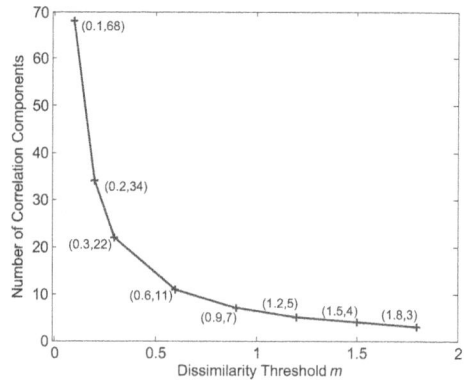

Figure 5: Magnitude dissimilarity threshold versus the number of correlation components.

Based on Proposition 2, and similar proof used in Lemma 2, we are able to show that for general communication graph, any optimal QoM-aware rate allocation is also an optimal *fair rate allocation* if $\delta_s \geq \delta_r$, and vice versa due the uniqueness of the optimal solution.

5. DISCUSSION AND FUTURE WORK

In this work, we study the QoM-aware rate allocation problem. In this section, we briefly discuss the limitations of current work, and further propose some interesting extensions as our future work.

▶ So far we assume that that the per unit data sensing cost is no less than the per unit data receiving cost. This requirement may not always be satisfied in practice, especially for those low energy cost sensors *e.g.*, accelerometer or gyro sensors. One interesting direction for future work is to develop a new rate allocation scheme whose performance can be bounded for the general case.

▶ In this paper, we assume that the utility function $\mathcal{U}()$ is identical to all correlation components; the other possible direction for future research is to generalize existing studies to heterogeneous utility functions, *e.g.*, different components may have different submodular utility functions. In this case, the current approach may fail to achieve the maximum QoM, even under the assumption that the per unit data sensing cost is no less than the per unit data receiving cost.

▶ Another interesting extension is to study the rate allocation problem in multi-application networks. In reality, one node may be involved in many applications [1] [26], such as temperature, humidity, and illuminance monitoring. For different applications, we may partition the network into different correlation components. As a result, one node may belong to multiple components under a multi-application network. Then one interesting problem is to study how to find a rate allocation among sensors and applications, which can achieve the maximum utility across all applications. One possible approach is to add additional virtual nodes for all new added components, and then apply similar approaches as proposed in this work.

6. PERFORMANCE EVALUATION

6.1 Experimental Results

Our outdoor testbed adopts TelosB Mote [27] with a MSP430 processor and CC2420 transceiver. Each mote is equipped with 2 AA batteries. The sensor program is developed based on TinyOS

2.1. The energy consumptions for receiving, transmitting, and sensing unit data are 21.8 mA, 19.5 mA and 22 mA respectively.

6.1.1 Utility Function

We decided to use a multivariate Gaussian, which is widely used in the body of literature [28], as an utility function. We then write our utility function for each component $\mathbf{c}$ in terms of total sensing rate $s_{\mathbf{c}}$ as $\mathcal{U}(s_{\mathbf{c}}) = \log s_{\mathbf{c}} A e^{-\left(\frac{1}{\sigma^2}\right)}$, where A is some constant, and σ is the variance of readings among sensor nodes within one component. It is easy to verify that $\mathcal{U}(s_{\mathbf{c}})$ is indeed a submodular function. The overall utility is defined as $\mathbf{U} = \sum_{\mathbf{c} \in \mathcal{C}} \mathcal{U}(s_{\mathbf{c}}) = \sum_{\mathbf{c} \in \mathcal{C}} \log s_{\mathbf{c}} A e^{-\left(\frac{1}{\sigma^2}\right)}$.

6.1.2 Dissimilarity Score Threshold and Its Impact

In order to exploit the correlation among the data reported by the sensor nodes, and to help reduce the redundant samplings, we dynamically partition the network into a set of disjoint correlation components. The sensor nodes within the same correlation component have strong correlation and, therefore, great similarity in sensing values. At any time instant, only a small fraction of sensor nodes need to be active, serving as the representatives for the whole correlation component. The partition operation is based on the dissimilarity measure, as used in [11]. The dissimilarity measure of time series is computed in a pairwise manner, based on historical observations from individual sensor nodes.

We define two metrics to evaluate the difference of the two time series, *magnitude dissimilarity* and geographic distance. Two time series $\{x_1, x_2, \cdots, x_t\}$ and $\{y_1, y_2, \cdots, y_t\}$ are *magnitude dissimilar* if there is an i $(1 \leq i \leq t)$ such that $|x_i - y_i| > \mathbf{m}$. Here $\mathbf{m}$ denotes the dissimilarity threshold. In our experiments, we put two sensors S_x and S_y into different correlation components if their time series are magnitude dissimilar, or their geographic distance is greater than a threshold value, which is set to be 30 feet.

The goal of this set of experiments is to explore the impact of the dissimilarity threshold value on the number of correlation components. By varying the magnitude dissimilarity threshold value $\mathbf{m}$, we collect a set of performance data, as illustrated in Figure 5. It demonstrates that, with the decrease of dissimilarity threshold value, the number of correlation components increases. This is not surprising, for the reason that a lower dissimilarity threshold value leads to a higher data resolution requirement. If $\mathbf{m} = 0$, each individual sensor node constitutes an independent component itself. On the one hand if $\mathbf{m} = \infty$, we treat the entire network as one component. Intuitively speaking, neither $\mathbf{m} = 0$ nor $\mathbf{m} = \infty$ is a good choice; the first setting completely ignores the potential correlation among sensor nodes, and the second setting somehow overestimates their correlation. As we will discuss later, different selections of $\mathbf{m}$ significantly affects the system performance in terms of estimation accuracy.

6.1.3 Performance in Estimation Error Minimization

An appropriate dissimilarity threshold value is essential to reducing the estimation error under a specific energy budget. The goal of this set of experiments is to find the appropriate dissimilarity threshold value that minimizes the estimation error, given an energy budget. We use the difference distortion measure, which has been broadly used in image compression, to evaluate the accuracy of a reconstructed image against the original image [29]. The difference distortion measure σ^2 is defined by $\sigma^2 = \frac{\sum_{j=1}^{M} \sum_{i=1}^{N} (S_{ij} - R_{ij})^2}{M \times N}$ where S_{ij} is the jth actual sensing value from the ith sensor node, and R_{ij} is the jth restoration value of the ith sensor node at the sink. N denotes the total number of sensor nodes, and M de-

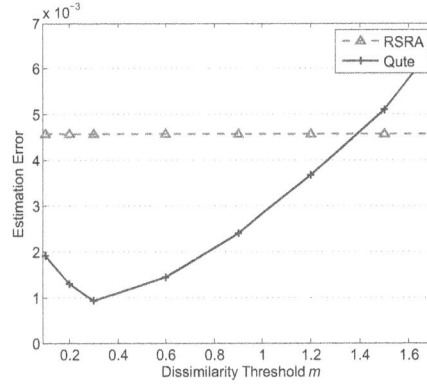

Figure 6: Magnitude dissimilarity threshold versus the estimation error.

notes the total number of samplings from each sensor node. We use a normalized difference distortion measure as the estimation error by normalizing σ^2 by the average variation of samplings. $\sigma_{norm}^2 = \frac{N\sigma^2}{\sum_{i=1}^{N} \text{Var}(S_i)}$ where $\text{Var}(S_i)$ denotes the sensing value variation of the ith sensor node. In our experiment, we set the energy budget for each sensor node as 1 mA per unit time (*i.e.*, minute). We use a random sensing rate allocation (RSRA) protocol as a baseline for performance comparison. Under RSRA protocol, all sensing values are collected through CTP (Collection Tree Protocol) [30]. The sensing rate of each sensor is decided in a distributed manner, where each sensor randomly selects a sensing rate under its current energy budget. The remaining energy is used to relay data for its neighbors. We collect estimation errors under different dissimilarity threshold values and plot the results in Figure 6.

As shown in Figure 6, the estimation error of the RSRA protocol does not change under different dissimilarity threshold values; this is because the rate allocation under RSRA protocol is irrelevant to the dissimilarity threshold. For Qute, the change of dissimilarity threshold values heavily affects the estimation accuracy of the temperature distribution recovered at the sink node. Specifically, a lower dissimilarity threshold value leads to a smaller correlation component. In an extreme case, each individual node stands for an independent component without considering correlation among sensor nodes, which clearly leads to poor estimation accuracy. On the other hand, a higher dissimilarity threshold value leads to misplacement of non-correlated sensor nodes into one correlation component, resulting in failure of collecting a high-resolution temperature distribution at the sink node. We observe from the experimental results that the estimation error is minimized when the dissimilarity threshold value is set to 0.3. Therefore, this value will be used as a default in the following experiments.

6.1.4 Performance in Utility Maximization

Using the utility function defined in Section 6.1.1, we perform a set of experiments to evaluate the performance of *Qute*, in terms of achieved utility under different energy budgets. By varying the energy budget from 0.1 mA per time unit to 8 mA per unit time, we collect a set of performance data, and plot the results in Figure 7. RSRA protocol is used as a baseline for comparison. As shown in the figure, *Qute* outperforms RSRA protocol in all energy budget values. The reason is that *Qute* computes the sensing rate assignment for each sensor node based on their correlation, while

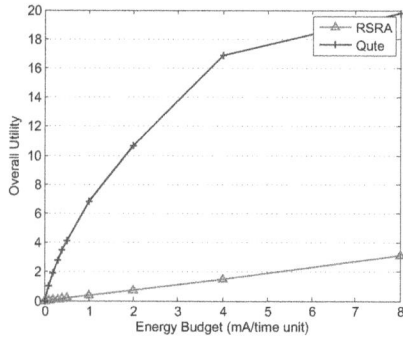

Figure 7: Energy budget versus the overall utility.

RSRA lets the sensor node randomly pick a sensing rate. With the same energy budget, *Qute* therefore achieves significantly higher overall utility by fully exploiting the correlation distribution in the network.

7. CONCLUSION

In this work, we study the QoM-aware rate allocation problem by exploring the correlational relationship among sensor nodes. We adopted an efficient rate allocation strategy, called *Qute*, by jointly optimizing both sensing rate selection and routing. We analytically show that *Qute* can achieve the maximum total utility if the sensing cost is no less than the receiving cost. In the future, we are interested in a more general problem setting, in which different components may have different utility functions.

8. REFERENCES

[1] Y. Xu, A. Saifullah, Y. Chen, C. Lu, and S. Bhattacharya, "Near optimal multi-application allocation in shared sensor networks," in *ACM MobiHoc*. ACM, 2010, pp. 181–190.

[2] S. Tang, X. Li, X. Shen, J. Zhang, G. Dai, and S. Das, "Cool: On coverage with solar-powered sensors," in *ICDCS*. IEEE, 2011.

[3] V. Delouille, R. Neelamani, and R. Baraniuk, "Robust distributed estimation in sensor networks using the embedded polygons algorithm," in *IPSN*. ACM, 2004.

[4] H. Gupta, V. Navda, S. Das, and V. Chowdhary, "Efficient gathering of correlated data in sensor networks," in *ACM MobiHoc*, 2005.

[5] S. Yoon and C. Shahabi, "Exploiting spatial correlation towards an energy efficient clustered aggregation technique (cag)[wireless sensor network applications]," in *ICC*, vol. 5. IEEE, 2005, pp. 3307–3313.

[6] I. Akyildiz, W. Su, Y. Sankarasubramaniam, and E. Cayirci, "Wireless sensor networks: a survey," *Computer Networks*, vol. 38, no. 4, pp. 393–422, 2002.

[7] M. Vuran, O. Akan, and I. Akyildiz, "Spatio-temporal correlation: theory and applications for wireless sensor networks," *Computer Networks*, 2004.

[8] S. Tang and L. Yang, "Morello: A quality-of-monitoring oriented sensing scheduling protocol in sensor networks," in *INFOCOM, 2012*. IEEE, 2012, pp. 2676–2680.

[9] S. Lauritzen, *Graphical models*. Oxford University Press, USA, 1996, vol. 17.

[10] L. Villas, A. Boukerche, D. Guidoni, H. Oliveira, R. Araujo, and A. Loureiro, "Time-space correlation for real-time,

accurate, and energy-aware data reporting in wireless sensor networks," in *ACM MSWIM*, 2011.

[11] C. Liu, K. Wu, and J. Pei, "An energy-efficient data collection framework for wireless sensor networks by exploiting spatiotemporal correlation," *TPDS*, vol. 18, no. 7, pp. 1010–1023, 2007.

[12] V. Isler and R. Bajcsy, "The sensor selection problem for bounded uncertainty sensing models," *Automation Science and Engineering, IEEE Transactions on*, 2006.

[13] A. Krause, J. Leskovec, C. Guestrin, J. VanBriesen, and C. Faloutsos, "Efficient sensor placement optimization for securing large water distribution networks," *Journal of Water Resources Planning and Management*, 2008.

[14] F. Bian, D. Kempe, and R. Govindan, "Utility based sensor selection," in *IPSN*. ACM, 2006, pp. 11–18.

[15] S. Pradhan and K. Ramchandran, "Distributed source coding: Symmetric rates and applications to sensor networks," in *DCC 2000*.

[16] P. Von Rickenbach and R. Wattenhofer, "Gathering correlated data in sensor networks," in *Proceedings of the 2004 joint workshop on Foundations of mobile computing*. ACM, 2004, pp. 60–66.

[17] A. Scaglione and S. Servetto, "On the interdependence of routing and data compression in multi-hop sensor networks," *WINET*, vol. 11, no. 1-2, pp. 149–160, 2005.

[18] C. Wu, Y. Xu, Y. Chen, and C. Lu, "Submodular game for distributed application allocation in shared sensor networks," in *IEEE INFOCOM 2012*.

[19] I. Akyildiz, M. Vuran, and O. Akan, "On exploiting spatial and temporal correlation in wireless sensor networks," in *Intl. Symposium on Modeling and Optimization in Mobile, Ad Hoc, and Wireless Networks 2004*.

[20] S. Chen, Y. Fang, and Y. Xia, "Lexicographic maxmin fairness for data collection in wireless sensor networks," *TMC*, vol. 6, no. 7, pp. 762–776, 2007.

[21] K.-W. Fan, Z. Zheng, and P. Sinha, "Steady and fair rate allocation for rechargeable sensors in perpetual sensor networks," in *ACM SenSys*. ACM, 2008.

[22] S.-J. Tang and J. Yuan, "On distributed sensing scheduling to achieve high quality of monitoring," in *2013 Proceedings IEEE INFOCOM*, Turin, Italy, Apr. 2013.

[23] L. Mo, Y. He, Y. Liu, J. Zhao, S.-J. Tang, X.-Y. Li, and G. Dai, "Canopy closure estimates with greenorbs: sustainable sensing in the forest," in *ACM SenSys*, 2009.

[24] A. Meka and A. Singh, "Distributed spatial clustering in sensor networks," *Advances in Database Technology 2006*.

[25] C. Detweiler, M. Doniec, M. Jiang, M. Schwager, R. Chen, and D. Rus, "Adaptive decentralized control of underwater sensor networks for modeling underwater phenomena," in *ACM SenSys*. ACM, 2010.

[26] A. Tavakoli, A. Kansal, and S. Nath, "On-line sensing task optimization for shared sensors," in *IPSN*, 2010.

[27] J. Polastre, R. Szewczyk, and D. Culler, "Telos: Enabling ultra-low power wireless research," in *IPSN*, 2005.

[28] D. Lynch, D. McGillicuddy *et al.*, "Objective analysis for coastal regimes," *Continental Shelf Research*, 2001.

[29] K. Sayood, *Introduction to data compression*. Morgan Kaufmann, 2005.

[30] O. Gnawali, R. Fonseca, K. Jamieson, D. Moss, and P. Levis, "Collection tree protocol," in *ACM SenSys*. ACM, 2009.

Practical Opportunistic Routing in High-Speed Multi-Rate Wireless Mesh Networks

Wei Hu, Jin Xie, Zhenghao Zhang
Computer Science Department
Florida State University
Tallahassee, FL 32306, USA
{hu,xie,zzhang}@cs.fsu.edu

ABSTRACT

Opportunistic Routing (OR) has been proven effective for wireless mesh networks. However, the existing OR protocols cannot meet all the requirements for high-speed, multi-rate wireless mesh networks, including: running on commodity Wi-Fi interface, supporting TCP, low complexity, supporting multiple link layer data rates, and exploiting partial packets. In this paper, we propose Practical Opportunistic Routing (POR), a new OR protocol that meets all above requirements. The key features of POR include: packet forwarding based on a per-packet feedback mechanism, block-based partial packet recovery, multi-hop link rate adaptation, and a novel path cost calculation which enables good path selection by considering the ability of nodes to select appropriate data rates to match the channel conditions. We implement POR within the Click modular router and our experiments in a 16-node wireless testbed confirm that POR achieves significantly better performance than the compared protocols for both UDP and TCP traffic.

Categories and Subject Descriptors

C.2.2 [**Computer-Communications Networks**]: Network Protocols – *Routing protocols*

General Terms

Algorithms, Design, Performance

Keywords

Opportunistic routing; Multi-rate; Partial packet recovery.

1. INTRODUCTION

Wireless mesh network is an attractive solution to extend network coverage at low cost. In this paper, we consider mesh network that consists of stationary mesh routers communicating with each other via wireless links. The challenge in such networks is to offer high performance over wireless links that are far less reliable than wired links due to noise, interference, and fading. It has

been known that packet overhearing can be exploited to improve the network performance, because the packet forwarding path can be cut short when a packet is opportunistically received by nodes closer to the destination. In the literature, this is often referred to as Opportunistic Routing (OR).

To date, although many OR protocols have been proposed, such as MORE [3], ExOR [5], MIXIT [4], SOAR [6], Crelay [27], none can meet all the following five requirements we identify for wireless mesh networks.

Working with Existing Hardware: For practical reasons, a wireless mesh network prefers existing technologies, such as Wi-Fi. Protocols that require specialized hardware such as MIXIT may have weaker applicability.

Supporting TCP: Many important aspects of TCP, such as congestion control, are optimized under the assumption that the lower layer forwards individual packets. Therefore, protocols that send packets in batches, including ExOR, MORE, and MIXIT, have inherent difficulty in supporting TCP [7]. We believe it is important for a protocol to efficiently support TCP, given the ubiquitous deployment of TCP and the practical infeasibility of modifying the TCP implementations at end devices.

Low Complexity: To support high link layer data rates, the packet processing in an OR protocol should preferably be simple. Protocols such as MORE, MIXIT, and Crelay have high computation complexities in packet processing due to network coding or error correction and cannot support high data rates.

Multi-Rate Support: A wireless link layer usually supports multiple data rates; typically, lower data rates result in longer range and better reception. The actual speed of the network is often largely determined by a good rate selection algorithm. However, existing OR protocols often overlook the rate issue and assume a single operating rate.

Partial Packet Recovery: Partial packets, i.e., packets that contain just a few errors, often exist in wireless transmissions. By default, such packets are retransmitted; however, it is clearly more efficient to repair such packets with partial packet recovery schemes. The existing OR protocols with partial packets recovery are MIXIT and Crelay; however, as mentioned earlier, both suffer high computation complexity.

In this paper, we propose Practical Opportunistic Routing (POR), a new OR protocol which meets all the above requirements. POR achieves this by adopting a software solution that intelligently forwards individual packets with a simple yet efficient per-packet feedback mechanism, augmented by block-based partial packet recovery, efficient rate selection, and a novel path calculation method. We implement POR in the Click modular router [23] and test it in a

	runs on Wi-Fi	TCP friendly	low complexity	multi-rates	partial packets
MORE	√	×	×	×	×
ExOR	√	×	√	×	×
MIXIT	×	×	×	×	√
SOAR	√	√	√	×	×
Crelay	√	√	×	×	√
POR	√	√	√	√	√

Table 1: Protocols on practical requirements of OR.

16-node wireless testbed. Our results show that POR achieves significantly better performance than other compared protocols. We also demonstrate running TCP on top of POR, which, to the best of our knowledge, is the first demonstration of unmodified TCP on top of an OR protocol.

The rest of the paper is organized as follows. Section 2 discusses related work. Section 3 gives an overview of POR. Section 4 describes the packet forwarding protocol. Section 5 describes routing. Section 6 describes rate selection. Section 7 describes our experimental evaluation. Section 8 concludes the paper.

2. RELATED WORK

In recent years, various OR protocols have been proposed and implemented, e.g., MORE [3], ExOR [5], MIXIT [4], SOAR [6], and Crelay [27]. We summarize the core differences between POR and the existing protocols in Table 1. OR has also been studied theoretically, such as in [15] [16]. We note that our focus is the design and implementation of a practical network protocol, which is different from finding theoretical network capacities [16]. The routing algorithm proposed for *anypath* in [15] provides valuable insights; however, the algorithm relies on the assumption that the packet receiving information at a node is known to every node, which can be difficult to implement in practical networks.

Partial packet recovery has attracted much attention in recent years [1, 2, 26]. We note that the existing works are typically developed for one-hop links while POR is developed for multi-hop networks. Rate selection is a classic topic in wireless networks. One of the major components of POR is also handling multiple data rates. However, POR is designed for multi-hop networks, therefore is different from rate selection algorithms for single-hop links such as [10, 11, 12, 13, 14]. Rate selection for multi-hop networks has been studied in [17, 18, 19] with focus on isolating collision loss from channel loss; we note that POR solves many additional problems in addition to rate selection and the solutions in [17, 18, 19] should complement POR. Like OR, network coding is another interesting approach that can improve the performance of wireless mesh networks. Schemes such as COPE [8] have been proposed which combine multiple packets from separate flows into one packet to reduce the number of transmissions. We note that network coding typically depends on the existence of multiple flows intersecting at a common node to create coding opportunities, while OR schemes such as POR can support both single flow and multiple flows because OR is based on packet overhearing. It is possible to combine POR with network coding, which we leave to future works due to the limit of space. It has also been proposed to adopt Time Division Multiple Access (TDMA) in the Medium Access Control (MAC) layer for mesh networks, which has shown notable performance improvements over the Carrier Sensing Multiple Access (CSMA) protocol adopted by Wi-Fi [9]. We note that POR mainly functions in the network layer and can work in collaboration with the new TDMA MAC to achieve even higher performance.

3. OVERVIEW

In POR, a packet is forwarded along a *path*, which is an ordered list of nodes. Any POR path must satisfy the *feedback constraint*; that is, any node on the path must be able to receive from its next hop node to ensure the correct reception of possible feedbacks from its downstream nodes. For simplicity, POR adopts source routing, i.e., the complete path is specified in the header of a packet. Packet in POR is divided into blocks; the checksum of each block is transmitted along with the packet such that a node can determine whether or not a received block is corrupted. A node may attempt to transmit blocks in a packet if none of its downstream nodes has received such blocks; the received blocks are not transmitted. A node announces the receiving status of a packet in a *feedback frame*; to avoid transmitting a packet that has been overheard, a node will not start to transmit a packet until a feedback from its downstream node has been received or until a timeout. With POR, nodes learn the link qualities at different data rates where the link quality is represented by the *Block Receiving Ratio* (BRR) defined as the fraction of correctly received blocks in a packet. A node may send small probe packets at selected rates such that its neighbors may discover the link qualities at such rates and report them back. Based on the link qualities, a node chooses the best data rate to minimize the packet delivery time of a path. The path is selected according to a greedy algorithm.

4. PACKET FORWARDING

The core of POR is its packet forwarding protocol discussed in this section. The goal of the protocol is to achieve high performance by exploiting packet overhearing and partial packets, and the challenge is to achieve this goal at low complexity, low overhead, while being friendly to TCP. To this end, we adopt a simple solution in which nodes send individual packets and determine the best strategy to forward each packet based on the per-packet feedback received from their downstream nodes. We discuss the details of the protocol in the following; for each aspect, we first describe the policy then add remarks to explain the policy. We refer to a source and destination pair as a *flow*.

4.1 Dividing Packets into Blocks

Policy: The source node divides a packet into blocks and calculates the checksums of the blocks which are transmitted in the frame header along with the packet. In our current implementation, each block is 150 bytes; the last block may be less depending on the size of the packet. When a received packet passes the packet level checksum test, there is no need to check the checksum for each block; otherwise, the packet is a partial packet and the node computes the checksums with the received blocks and compares them with those received in frame header to locate the corrupted blocks. In our current implementation, the checksum is 16 bits defined by $x^{16} + x^{15} + x^2 + 1$,

Remark: We note that dividing the packets into blocks allows POR to exploit the partial packets. POR does not use 32-bit checksum due to the better tradeoff between overhead and detection capability of the 16-bit checksum.

4.2 Sending a Packet at a Forwarder

Policy: A forwarder refers to a node on the path that is not the source or the destination. When a forwarder receives a packet with a sufficient amount of correct blocks, set as 50% in our current implementation, it adds the correct blocks of the packet to its queue. The node will not transmit the packet immediately because one of its downstream nodes may have received the packet or some blocks of the packet correctly. Therefore, it will wait for a feedback of the packet for up to a timeout, which is set to be 20 ms in our current implementation, before sending any block of the packet. If the node receives the feedback from the downstream nodes and if the feedback indicates that it has all the blocks currently missing at the downstream nodes, it will schedule a transmission with such blocks; it removes the packet if the feedback indicates that the downstream nodes have no missing blocks. If it does not receive any feedback before the timeout, it will schedule a transmission for the entire packet if it received the packet correctly. The node retransmits a packet or blocks in a packet when a retransmission timer expires if it has not been informed by the downstream nodes that the packet or the blocks have been received correctly; each retransmission is separated by at least 15 ms and a packet is retransmitted up to 5 times, after which it is dropped.

Remark: We note that the key element of POR to exploit overhearing is to hold a packet and wait for the feedback. Although a node may hold a packet for up to 20 ms, in many cases, it may have received a feedback and can send the packet much earlier. In addition, through private communication with *Madcity*, a mesh network service provider, a path in a mesh network typically has no more than 4-5 hops, such that holding a packet for 20 ms will not increase the network delay excessively as the Internet packet delay is often in the order of 100 ms. We also note that a node does not keep packets with too few correct blocks because it will have to send feedbacks for such packets; our current policy is a simple heuristic to reduce the amount of overhead.

4.3 Exploiting Link Layer ACKs

Policy: In POR, when a packet is transmitted, it is encapsulated into a frame with the MAC address of the next hop node as the destination such that the next hop node may reply with a link layer ACK if the packet is received correctly. If a node receives an ACK for a packet, it immediately removes the packet from the queue; otherwise, the node will start the retransmission timer.

Remark: The key advantage of exploiting the link layer ACK is that it is sent immediately after the packet such that the node can determine whether the packet has been received correctly much faster than relying on the feedbacks. We note that encapsulating a packet in a frame with a specific MAC address does not prevent any other downstream nodes from overhearing the packet because the nodes process all overheard packets regardless of the MAC address. One potential problem may arise when a node receives a packet correctly and automatically sends the link layer ACK, but then is forced to drop the packet because the buffer is full. In this case, the sender of the packet has also removed the packet and the packet may be lost in the network. POR solves this problem by making sure that the buffer is rarely full with the congestion control mechanism described in Section 4.6.

4.4 Sending Feedback

Policy: In POR, a node may schedule a feedback when: 1) it has received a partial packet and is the next hop of the sender, or 2) it has received either a complete or a partial packet but is not the next hop of the sender. The feedback of a packet is 8 bytes containing the packet source node and desalination node IDs both in 2 bytes, the sequence number in 2 bytes, and a bitmap of the received blocks in 2 bytes where a bit '1' indicates that the corresponding block has been received correctly. A feedback may be sent immediately or may be combined with other feedbacks belonging to the same flow, in a *feedback frame*, to reduce the overhead. A feedback frame contains only feedbacks and is addressed to the previous hop node. A feedback frame may be sent under the following condi-

tions: 1) 15 ms since a feedback was scheduled, or 2) 8 new feedbacks have been scheduled. A feedback frame may be acknowledged in the link layer; POR does not have other acknowledgment mechanisms for the feedback frames. To improve the reception, a feedback frame is always transmitted twice and may be sent at a lower rate than the data packets; in our current implementation, the rate is highest rate with Packet Receiving Ratio (PRR) no less than 0.8 on the link to the previous hop node. A node also "propagates the good news" when needed; that is, if the node receives a feedback from its downstream node, it will do a bitwise *or* with its own bitmap of the packet and use the new bitmap as the feedback.

Remark: A node does not schedule feedback to its previous hop node for complete packets because it should have sent the link layer ACK. A node waits for up to 15 ms before sending the feedback because its previous hop node holds the packet for up to 20 ms before transmitting the packet.

4.5 Buffer Management

Policy: In POR, a node keeps a single buffer which holds packets of all flows. In our current implementation, the buffer can hold 40 packets. In the buffer, packets belong to different flows are served round-robin. Packets belong to the same flow are served according to the sequence numbers, i.e., the packet with smallest sequence number will be sent first. If a node received a link layer ACK or a feedback for a packet indicating the downstream nodes have obtained all the blocks, it removes the packet from the buffer.

Remark: POR has a shallow buffer because a large buffer may lead to very long delay at the bottleneck links; the best action when congestion occurs should be reducing the sending speed at the source. POR sends packets with smaller sequence numbers first because this helps reducing the out-of-order delivery which may reduce the performance of TCP; we note that packets with larger sequence numbers may be received first due to overhearing.

4.6 Congestion Control

Policy: In POR, if the queue length exceeds a threshold, set to be half of the buffer size in our current implementation, the node will set a "congestion" bit in the feedback frame and the upstream node will cease to transmit any new packet to this node if it is the next hop. The upstream nodes are still allowed to transmit blocks for partial packets. Once the queue length is below the threshold, the node may send a feedback again to allow the upstream node to transmit.

Remark: This simple mechanism basically allows the node to drain its queue before its buffer overflows. The congestion bit will propagate from the bottleneck link back to the source because the upstream nodes will experience the congestion condition in turn when they cannot send packets to their downstream nodes.

5. ROUTING

POR adopts a customized routing module primarily because the path cost metric with POR is different from the existing metrics with its per-packet feedback and multiple data rates. We assume the link quality information of all links in the network at all rates are available to a node when it calculates paths, noting that the link quality information can be propagated by solutions such as that in OSLR [20]. Every node should compute the same path for each source and dentation pair if the nodes have consistent link quality information; as a protection against possible inconsistent link quality information due to packet loss, POR still embeds the entire path calculated by the source node in the header, which will not incur too much overhead as the paths are typically not very long in mesh networks. We note that the routing module involves non-trivial calculations but will not hurt the overall simplicity of the POR protocol because routes are updated infrequently.

5.1 Path Metric

The path metric used in POR is different from existing path metrics such as ETT and ETX with two key distinctions. First, POR considers both the *forward cost* for sending data and *backward cost* for sending feedback, because the feedback cost in POR is not negligible. Second, we note that the links can be of different qualities at different times due to channel fluctuation. In this paper, we refer to each quality as a *state* of the link. The path metric in POR can capture the capability of nodes to adapt to the link qualities by changing rates, while existing metrics typically assume the rates are fixed.

We consider a given path denoted as $(v_1, v_2, ..., v_L)$. The path cost calculation is carried out iteratively, starting from the node closest to the destination. Therefore, when calculating the cost of path $(v_1, v_2, ..., v_L)$, the forward and backward costs of path $(v_i, v_{i+1}, ..., v_L)$ are known for $2 \leq i \leq L$, denoted as C_{v_i} and B_{v_i}, respectively.

5.1.1 Path Cost in Given State

We begin by finding the path cost when the links involving v_1 are in a certain given set of states denoted as τ.

Forward cost: We first consider the forward cost, denoted as $C_{v_1}^\tau$, which is defined as the consumed air time in data transmission in order to deliver a unit size block to the destination when the links are in state τ. The cost of delivering a packet is clearly proportional to the forward cost defined for a block. We assume the feedback is perfect and no node sends duplicate data. We denote the BRR of link $v_1 \rightarrow v_i$ at rate ρ_j as $\mu_i^{\rho_j,\tau}$. We use C_{v_1,ρ_j}^τ to denote the air time consumed to deliver a unit size block to v_L following path $(v_1, v_2, ..., v_L)$, under the condition that v_1 transmits at rate ρ_j. Let ρ^* denote the rate such that $C_{v_1,\rho^*}^\tau \leq C_{v_1,\rho_j}^\tau$ for any j, which is clearly the optimal rate v_1 should use. Assuming ρ^* can be found by the rate selection algorithm, $C_{v_1}^\tau$ is clearly just C_{v_1,ρ^*}^τ. We note that when v_1 sends a block, if v_i receives the block correctly and is the one with the largest index, which occurs with probability $\mu_i^{\rho_j,\tau} \prod_{t=i+1}^L (1 - \mu_t^{\rho_j,\tau})$, v_i will take over the responsibility of forwarding the block to the destination; if none of the downstream nodes of v_1 receives the block, which occurs with probability $\prod_{i=2}^L (1 - \mu_i^{\rho_j,\tau})$, v_1 must repeat the process. Therefore,

$$C_{v_1,\rho_j}^\tau = \sum_{i=2}^L \mu_i^{\rho_j,\tau} \prod_{t=i+1}^L (1 - \mu_t^{\rho_j,\tau})(\frac{1}{\rho_j} + C_{v_i})$$
$$+ \prod_{i=2}^L (1 - \mu_i^{\rho_j,\tau})(\frac{1}{\rho_j} + C_{v_1,\rho_j}^\tau),$$

which can be reduced to

$$C_{v_1,\rho_j}^\tau = \frac{\frac{1}{\rho_j} + \sum_{i=2}^L \mu_i^{\rho_j} \prod_{t=i+1}^L (1 - \mu_t^{\rho_j}) C_{v_i}}{1 - \prod_{i=2}^L (1 - \mu_i^{\rho_j})}. \quad (1)$$

Clearly, ρ^* and $C_{v_1}^\tau$ can be found by a linear search.

Backward cost: We next consider the backward cost at v_1, which is denoted as $B_{v_1}^\tau$ and is defined as the consumed air time in feedback transmission in order to deliver a packet to the destination when the links are in state τ. The backward cost should be considered because the routing algorithm may otherwise select a path with too many nodes that may generate too many feedbacks. There

are two differences between the forward and backward costs calculation. First, for the backward cost, only the complete packets are considered, because the feedback generating mechanism in POR is complicated for partial packets and the partial packet percentage is usually not very large. We note that with this simplification, the PRR is the same as the BRR. Second, unlike the forward cost that considers only the data transmission time, the backward cost considers the transmission time of the entire frame, because the feedbacks are small such that overhead such as frame header becomes significant. Note that when v_1 sends a packet $\mathbf{P}$ at rate ρ^*, if no downstream node receives $\mathbf{P}$, no feedback is sent. If at least one of the downstream nodes receives $\mathbf{P}$, the probability that v_i is the node with the largest index that receives $\mathbf{P}$ is

$$\gamma_i = \frac{\mu_i^{\rho^*,\tau} \prod_{t=i+1}^L (1 - \mu_t^{\rho^*,\tau})}{1 - \prod_{i=2}^L (1 - \mu_i^{\rho^*,\tau})}.$$

If $i = 2$, no feedback frames will be generated because v_2 will send a link layer ACK; otherwise, each of v_3 to v_i will generate a feedback and v_2 will generate a feedback in case it did not receive $\mathbf{P}$ correctly with probability $1 - \mu_2^{\rho^*,\tau}$. Therefore, the backward cost can be calculated according to

$$B_{v_1}^\tau = \gamma_2 B_{v_2} + \sum_{i=3}^L \gamma_i [B_{v_i} + (1 - \mu_2^{\rho^*,\tau})\eta_2 + \sum_{t=3}^i \eta_t] \quad (2)$$

where η_t denotes air time v_t uses to send one feedback. η_t can be calculated according to the simplifying assumption that each feedback frame contains exactly 8 feedbacks, such that η_t is simply one-eighth of the feedback frame duration. We note that the feedback constraint of POR is enforced by the backward cost calculation, because η_t will be very large for some node v_t if v_t cannot find a good rate to send feedback.

5.1.2 Combining Path Cost in Multiple States

As mentioned earlier, a link may be of different qualities at different times and we call each quality a state. Eq. 1 and Eq. 2 find the path cost when the links involving v_1 are in a specific set of states. The actual cost of the path is simply the weighted average of the path costs in all possible sets of the link states, where the weight of a set is simply the probability that the links are in this particular set of states. If the probability that any individual link is in any state is known, the probability that the links are in a set of states can be calculated as the product of the individual link state probabilities by assuming the links are independent.

A practical challenge is to limit the computation complexity, as the number of sets can be very large. Therefore, in our current implementation, we make two simplifications. First, we assume v_1 can only reach nodes v_2, v_3, and v_4 when applying Eq. 1 and Eq. 2. Second, we assume each link has up to 3 states. Therefore, the total number of sets is capped at 27.

We note that the first assumption is a simple heuristics that usually works reasonable well in practice because overhearing at more than 3 hops away is usually rare. We support the second simplification by experimental data. In our experiments, we first measure the link qualities in 1-second intervals and generate for each link a vector where each element represents the BRR of one rate in this interval. In total, we measure 10 intervals, i.e., we have 10 vectors for each link. We note that if a link has only one state, all 10 vectors should be similar; otherwise, the vectors may appear as several clusters where each cluster corresponds to one state. To verify this, we run a clustering algorithm on the vectors. We define the *clustering distance* as the total distance between the vectors and the centers of the clusters they belong to. Fig. 1 shows the clustering

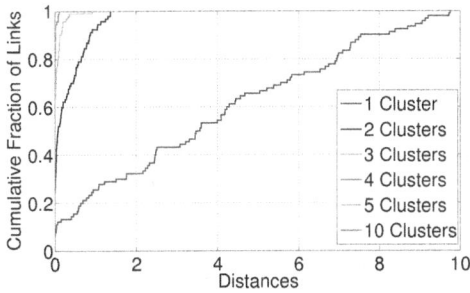

Figure 1: CDF of clustering distance.

distance when we set the maximum number of clusters to be 1 to 10, where 10 has the minimum clustering distance because there are only 10 vectors. We can see that: 1) the clustering distance can be quite large for many links when the number of cluster is 1, which means that many links have more than one state, and 2) the clustering distances are very small if the maximum number of clusters is 3 or above, which means that the links typically have no more than 3 states. The same clustering algorithm is used in POR implementation to classify the link states; we note that the probability that a link is in a certain state can be determined by the number of vectors in the cluster.

5.2 Finding the Path

We define the Optimal Practical Opportunistic Routing (OPOR) problem as finding a path with minimum cost according to the cost defined in Section 5.1. We report our findings about the theoretical aspects of the OPOR problem in [28]. In this paper, due to the limit of space, we describe the simple greedy routing algorithm we adopt, which is inspired by the Dijkstra's algorithm. The algorithm maintains a set π which contains the nodes whose paths to the destination have been determined; the set of nodes not in π is denoted as $\tilde{\pi}$. Each node keeps up to w candidate paths, where w is a constant set as 20 in our current implementation. Initially, π contains only the destination node. The algorithm iterates until all nodes have been added to π. In iteration i:

1. A node denoted as v_i is added to π, if v_i was previously in $\tilde{\pi}$ but has a candidate path with the least cost among all nodes in $\tilde{\pi}$.

2. The candidate paths for all nodes in $\tilde{\pi}$ are updated. That is, for a node in $\tilde{\pi}$ denoted as v_j, the algorithm evaluates its old candidate paths as well as new paths which starts with the link $v_j \rightarrow v_i$ then reach the destination following the candidate paths of v_i, and keeps w candidate paths with lowest costs.

6. RATE SELECTION

POR employs a novel rate selection module to select the data rate a node should use for a given path. The rate selection module considers the link qualities of multiple hops, instead of considering only the link to the next hop node, because a packet may be received by multiple downstream nodes in POR. One challenge, therefore, is to design the protocol to track the link qualities of multiple links. We note that while a node may be able to track the link qualities to its next hop node based on link layer ACKs, it may not be able to track the link qualities to nodes further on the downstream directly. Therefore, in POR, a node sends small dummy packets as probes at selected data rates and the neighboring nodes can measure the

link qualities and report them back in special feedback frames. The key distinction between path calculation and rate selection is that the rate selection adapts to the instantaneous states of the links. A node may carry traffic for multiple flows; however, as each flow follows the same set of rules for rate selection, in the following, we will describe rate selection for one flow. The only additional issue with multiple flows is that the set of probed rates is the union of the probed rates of all flows.

6.1 Candidate Rates

Policy: We refer to the rates that should be probed as the *candidate rates*. The policies for determining the set of candidate rates at any moment are:

1. A rate in *hiatus* is not eligible, where a rate is in hiatus if it has been probed within the last 2 seconds.

2. A rate is eligible if the path cost according to its *estimated BRRs* is lower than the current path cost. For a rate, we define its *lower neighbors* as the rates lower than itself. We define the *immediate lower neighbor* as the highest lower neighbor currently in hiatus. If the immediate lower neighbor exists, the estimated BRRs are set as those of the immediate lower neighbor; otherwise, the estimated BRRs to the next two hops are assumed to be 1 and others 0. To calculate the path cost at any rate, the link BRRs of the target rate are plugged into Eq. 1 and Eq. 2; we note that the calculation does not have to iterate through multiple link states because the path cost for rate selection is the cost under the instantaneous link state.

3. If there are multiple eligible rates, the lowest 3 rates are added to the candidate set.

Remark: We make the following observations about candidate selection.

Observation 1. The hiatus mechanism prevents repeatedly probing a same rate.

Observation 2. If a rate is not in hiatus, its estimated BRRs are likely better than its actual BRRs because the estimated BRRs are either assumed to be the BRRs of the lower rates or simply 1 to nearby nodes. As a result, the path cost of the optimal rate based on its estimated BRRs are lower than the actual cost, which will ensure that it will be probed.

Observation 3. When determining the estimated BRRs for a rate without an immediate lower neighbor, a node assumes ideal BRRs only 2 hops away, because overhearing at more than 2 hops away is very rare if the path has been selected correctly; on the other hand, assuming ideal BRRs at nodes further downstream may be overly optimistic and may result in more unnecessary candidate rates.

Observation 4. No more than 3 rates are probed at any time to limit the overhead in probing. The probed rates will be in hiatus and eventually all eligible rates will be probed.

Observation 5. Adding the lowest 3 rates to the candidate set is a simple heuristic based on the following observation. We note that rate change is needed when channel changes. If the channel condition deteriorates, the optimal rate should be lower than the current rate and giving lower rates higher priority is clearly correct. If the channel condition improves, the optimal rate is higher than the current rate; in this case, the BRRs of the current rate are likely very good such that many lower rates will not appear better than the current rate, therefore not eligible, even when assuming ideal BRRs.

Observation 6. We define the *tracking time* as the time the node switches to the optimal rate when channel changes. We argue that *the tracking time is fast when the channel condition deteriorates and is bounded when the channel condition improves*, which is the desired behavior. We note that

- If the optimal rate is not in hiatus when the channel changes, as explained earlier, it will be probed, and will appear better than the current rate. In POR, the optimal rate is probed after a delay in the order of 100 ms. The node may spend time on rates that appear better than the optimal rate based on measurements taken before the channel change; however, it will soon realize that such rates are worse than the optimal rate. Therefore, the tracking time in this case is in the order of 100 ms.

- If the optimal rate is in hiatus when the channel changes, there are two cases. If the channel condition deteriorates, the path cost based on the old measurements will be better than the true path cost of the optimal rate; therefore, the optimal rate will be set as the current rate even before its measurements expires, and the tracking time should be small. If the channel condition improves, it may happen that the old measurements of the optimal rate do not indicate that it is a good rate; however, it will likely be probed after it comes out of the hiatus, such that the tracking time, although longer, is still bounded.

6.2 Probing

Policy: In our current implementation, a node transmits probe packets which contain one block at each candidate rate every 5 ms; for each probed rate, it sends a total of 20 probe packets each with a unique sequence number, after which it stops probing and will be expecting a special type of feedback called the *link quality feedback*. A node sends a link quality feedback every 100 ms, which contains all the measured BRRs from all links collected in the last 100 ms. A node will update the link quality records after receiving the feedback; it assumes the BRR is 0 if it cannot receive any feedback for a rate to another node 200 ms after it sends the first probe packet.

Remark: The probe packets have sequence numbers and are sent at fixed intervals; therefore, a node knows when the probing finishes and can send the feedback in time, even when it fails to receive the last probe packet in the batch. A node waits for 200 ms for a link quality feedback because by this time, it should have received the feedback unless the other node did not send the feedback or the feedback was lost, in either case an assumption of 0 as the BRR is reasonable. We experimentally evaluate the overhead of POR including probing in Section 7.

6.3 Rate Update

Policy: POR constantly monitors the link qualities at the current rate and updates the current rate if its path cost is higher than another rate currently in hiatus. POR updates the BRR of the current rate on the link to the next hop based on feedbacks and the link layer ACKs. POR updates the BRRs of the current rate on links to nodes further downstream based on the feedbacks from such nodes. The path cost of the current rate is reevaluated every time the BRR entries for the current rate is updated. The set of candidate rates is also refreshed every time the path cost of the current rate is updated.

6.4 Tests in Emulated Wireless Channel

We did a separate set of experiments specifically for the rate selection module because rate selection is difficult to evaluate in real-

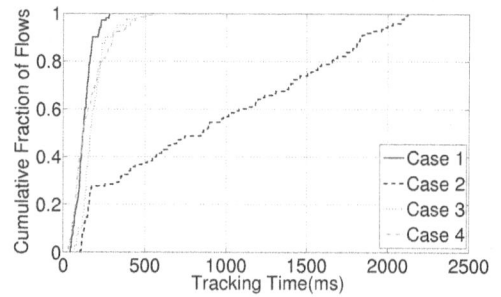

Figure 2: CDF of tracking time in 4 different cases.

world wireless channels. We note that the rate selection module can be verified if it can be shown that the node always converges to the optimal rate in a timely manner. The challenge is that the wireless channels may constantly fluctuate such that it is often not possible learn the actual optimal rates during the experiments. Therefore, we design emulated channel experiments in which the nodes are very closely located such that real packet losses are rare; artificial packet losses and partial packets are introduced by inserting errors to certain packets according to the emulated channel conditions. We deploy a 3-node path and investigate the rate selection at the source node, because packet overhearing at nodes more than 2 hops away are rare such that rate selection is mainly dependent on the link qualities to the next 2 hops. Initially, the emulated channels are set at certain qualities and each node is set to run at the optimal rate under such link qualities; the link qualities are then changed randomly at a random time and stays the same for the rest of experiment.

We show in Fig. 2 the CDFs of tracking time in 4 cases:

- Case 1: The channel becomes better while the optimal rate is not in hiatus at the change. The tracking time is reasonably small, i.e., less than 300 ms, which is because the rate selection module will likely probe the optimal rate shortly after the channel change if the optimal rate is not in hiatus.

- Case 2: The channel becomes better while the optimal rate is in hiatus at the change. The tracking time is much larger and can be as large as around 2000 ms, which is because the optimal rate will be probed after it comes out of hiatus, while it can stay in hiatus for as long as 2000 ms after the channel change.

- Case 3: The channel becomes worse while the optimal rate is not in hiatus at the change. The tracking time is similar to Case 1 because rate selection module will likely probe the optimal rate shortly after the channel changes.

- Case 4: The channel becomes worse while the optimal rate is in hiatus at the change. It is interesting to notice that the track time is much smaller than that in Case 2. This is because at the time of the channel change, the existing record of the optimal rate, which was taken before the channel change, is better than its actual condition; as a result, the optimal rate will appear as a good rate to be attempted and will be subsequently set as the current rate before it comes out of hiatus.

Overall, we can see that the experiments confirm our observation that the rate selection module will converge to the optimal rate after a channel change, while the convergence is much faster if the channel becomes worse.

7. EVALUATION

We implement POR within the Click modular router [23]. In this section, we describe our experimental results under single-flow UDP, multi-flow UDP, and TCP. We compare POR with two other schemes with the implementation from [24]: MORE, which is the benchmark OR protocol, and SPP, which is the traditional shortest path routing protocol without exploiting overheard packets and partial packets.

7.1 POR for A Single UDP Flow

7.1.1 Experiment Setup

Our testbed consists of 16 nodes which are laptop computers with the Cisco Aironet wireless card [22], scattered in one floor of a university building as shown in Fig. 3. We set wireless card in the 802.11a 5 GHz band where is no other traffic. To reduce the communication range and create multi-hop networks, we set the transmission power to be the minimum at 1 dBm and allow link rate at 24 Mbps or higher; we note that the experiments still cover a wide range of data rates from 24 Mbps to 54 Mbps. We connect all nodes with an additional Ethernet network which is a requirement to run MORE and SPP [24]. As MORE does not support multiple data rates, we set all nodes at the same rate of 24 Mbps in MORE experiments, because the results in earlier experiments with similar setup show that 24 Mbps tends to be the best among all rates for single rate networks. For SPP, we enable the auto-rate algorithm of the Madwifi driver [21] to adapt to the best rate of individual link. In each experiment, we send UDP packets of 1500 bytes from a source to a destination, referred to as a *flow*. We collect the results of 105 flows with path length no less than 2 according to POR; the number of flows with path lengths from 2 to 7 are 40, 32, 18, 8, 4, and 3, respectively. Each experiment lasts 65 seconds. Before running an experiment, we collect measurements of link quality, which are fed to SPP, MORE, and POR for routing computations.

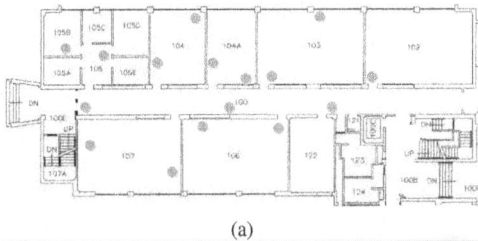

(a)

(b)

Figure 3: (a). The layout of the testbed. (b). Testbed in the hallway.

7.1.2 Throughput Comparison

We run SPP, MORE and POR in turn for the selected flows. Fig. 4 shows the scatter plot of POR v.s. SPP and POR v.s. MORE, where each data point represents a flow with the throughputs of POR and the compared scheme as the y-coordinate and x-coordinate, respectively. We can see that POR is significantly better than SPP or MORE in most flows.

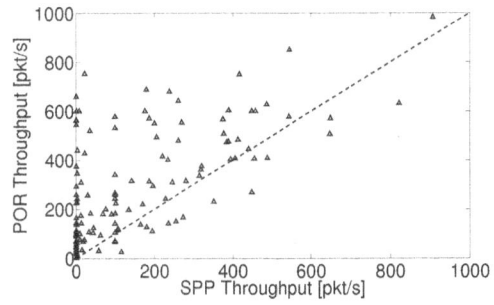

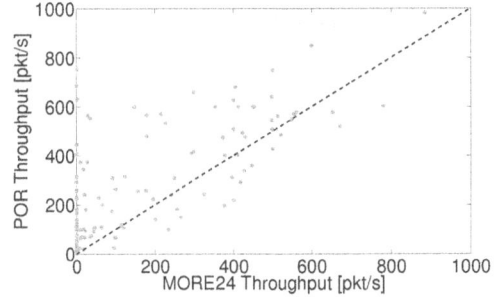

Figure 4: Scatter plot of the flow throughput.

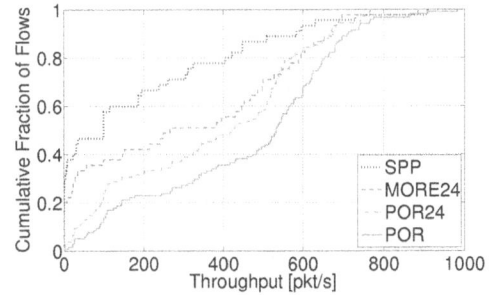

Figure 5: CDF of the flow throughput with POR24.

7.1.3 Throughput Analysis

The throughput gain of POR may be due to many reasons, such as exploiting overheard packets, supporting multiple rates, partial packet recovery, and better path selection. To verify the sources of the gain, we run more experiments with different configurations of POR.

First, we configure POR to run at a fixed rate of 24 Mbps, same as MORE, referred to as POR24. Fig. 5 shows the CDF of the throughputs of SPP, MORE, POR24, and POR. We observe that: 1) POR achieves significant gain due to good rate selection, which is evident from the performance difference between POR and POR24, and 2) POR efficiently exploits overheard packets with its feedback mechanism because POR24 can still achieve notable gains over MORE under the same rate.

Second, we run a modified version of POR referred to as cpPOR, which disables the reception of partial packets, where 'cp' stands for complete packet. Fig. 6 shows the CDF of the throughputs of SPP, MORE, cpPOR, and POR. We observe that partial packets recovery leads to significant gain, which is evident from the performance difference between POR and cpPOR.

Third, we run POR_M, in which the path cost is calculated assuming each link has only one state. Fig. 7 shows the throughputs of SPP, MORE, POR_M, and POR. We can see that when the throughput is high or low, POR_M performs similar as POR; in between, POR achieves higher throughput than POR_M. We believe

133

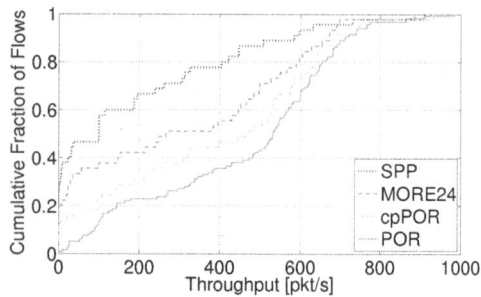

Figure 6: CDF of the flow throughput with cpPOR.

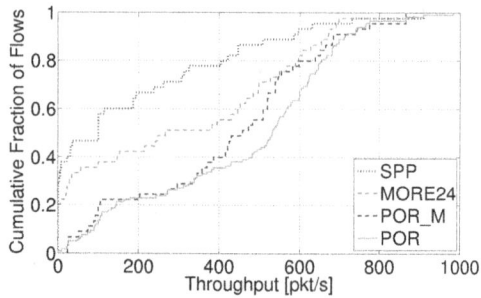

Figure 9: CDF of the duplicate percentage.

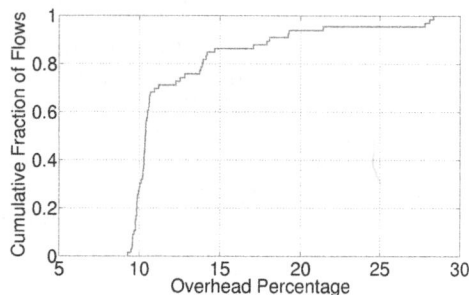

Figure 7: CDF of the flow throughput with POR_M.

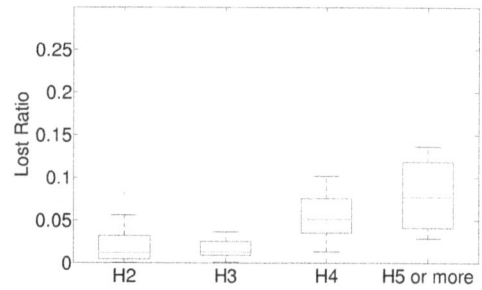

Figure 10: Packet lost ratio.

this is because links have fewer states when they are either very good or very bad; in between, they have more states and POR can better capture the true cost of such links and find better paths.

7.1.4   Various Aspects of POR

We also process the log files to examine various aspects of POR. **Overhead:** POR's overhead include the following. First, the opportunistic routing coordination overhead, which is mainly the feedback frames. Second, the multi-rate probe overhead, which is mainly the probe packets. Third, the partial packet recovery overhead, which includes the headers and checksums in POR data packets. To quantitatively measure the overhead, we define the overhead percentage of a flow as the total air time all nodes transmit the overhead over the total air time all nodes transmit packets. We note that this takes into account the possibility that two nodes may transmit simultaneously due to spatial multiplexing. Fig. 8 shows the CDF of the overhead percentage of POR for each flow, where we can see that the overhead is reasonably low with a median of 10.36%.

Duplicate Percentage: We say the transmission of a data block is a *duplicate* if at least one of the downstream nodes has received the block correctly. To measure the duplicate percentage, in the

experiments, each node increments a counter when it receives a block, and increments another counter when the received block is already in the buffer. To avoid a transmission being counted more than once, a node only increments the counters for transmissions from its previous hop node. In the end, we sum up the two counters of all nodes and use the ratio as an approximation of the fraction of the duplicate. Fig. 9 shows the CDF of the duplicate percentage, where we can see that POR achieves a very small duplicate percentage with a median of 3.29%. This further confirms that the feedback mechanism is effective and allows nodes to avoid unnecessary transmissions.

Packet Loss: Fig. 10 shows the boxplot of the end-to-end packet lost ratio of each flow classified according to path lengths. We can see that POR maintains a low end-to-end packet lost ratio: the median of the lost ratio is less than 1% when the number of hops is no more than 3, and is less than 5% when the number of hops is 4. We note that in practical mesh networks, paths of length more than 4 are rare because such paths cannot sustain high throughput. In addition, we find that the longer paths in our experiments usually contain very bad links and loss is in some sense inevitable. Fig. 11 shows the relation between packet lost ratio and path cost when the hop count of path is 4. The results indicate that packet lost ratio increases when the path cost increases, which is because high cost paths may include some very bad links.

Packet Delay: Fig. 12 shows the boxplot of end-to-end packet delays of POR classified according to path lengths. We can see that the packet delays are typically around 100 ms and the variance is reasonably small for short paths.

Queue Length: Fig. 13 shows the CDF of the average node queue length for nodes on the packet forwarding path in each experiment. We can see that in most cases, the buffers are not full and the queue lengths are around 10 packets, which confirms that the congestion control mechanism works reasonably well.

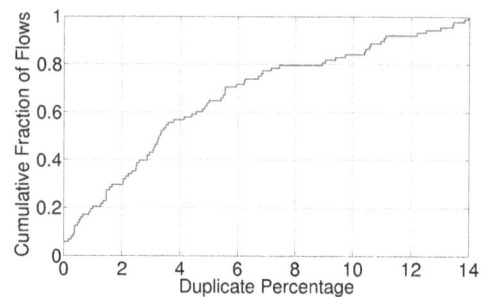

Figure 8: CDF of the overhead percentage.

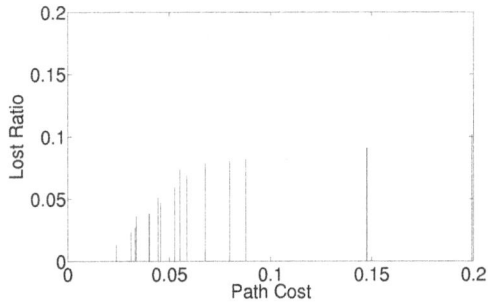

Figure 11: Packet lost ratio and path cost.

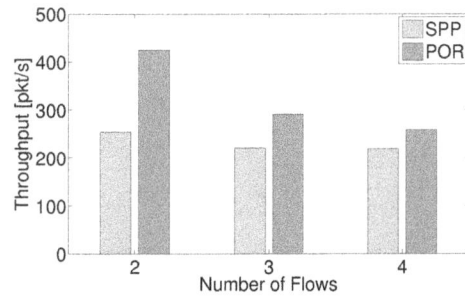

Figure 12: Packet delay.

7.2 POR for Multiple UDP Flows

We also run experiments to demonstrate the capability of POR to support multiple flows in a testbed similar to that for the single-flow UDP experiments. Since the MORE implementation at [24] does not support multiple flows, we compare POR only with SPP. We run 40 experiments by randomly choosing source and destination pairs, varying the number of random active flows in the network from 2 to 4. Fig. 14 show the average per-flow throughput as a function of the number of flows. The throughput of multiple flows depends on many issues out of the scope of this work, such as network-wide load-balancing; still, we can see that POR achieves higher throughput than SPP.

7.3 TCP Performance with POR

We also run TCP experiments on top of POR. To the best of our knowledge, prior to this, no demonstration has been made to run TCP on top of an OR protocol, because many existing OR protocols require packet batching and many will disrupt packet order. In the single UDP flow experiments, we have demonstrated that POR can achieve low end-to-end loss and delay, which should allow POR to support TCP. To verify this, we implement a standard TCP with the

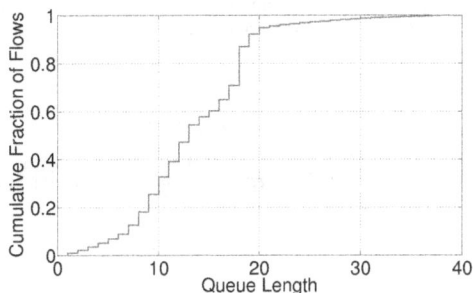

Figure 13: CDF of the average queue length.

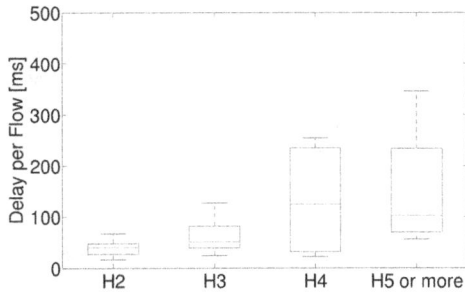

Figure 14: Average flow throughput with multiple flows.

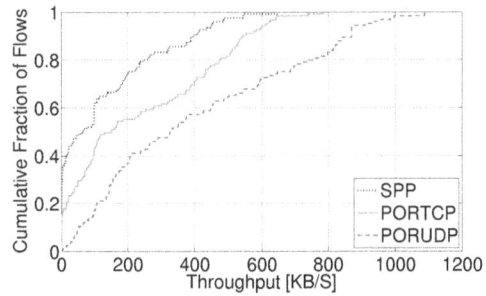

Figure 15: CDF of the TCP throughput on top of SPP and POR and UDP throughput on top of POR.

available code from [25] within the Click modular router [23]. We run TCP on top of POR and SPP in a testbed similar to that for the single-flow UDP experiments for 46 flows. We also run UDP on top of POR during the experiments to find the upper bound of the throughput.

Fig. 15 shows the CDF of the TCP throughput on top of SPP and POR, as well as the CDF of UDP throughput on top of POR. We can see that POR achieves a sizable gain over SPP for TCP. We believe *this is the first real-world demonstration of unmodified TCP over an OR protocol with a performance gain*. The TCP throughput of POR is less than UDP, which is partly because TCP has both the forward flow for data and the backward flow for TCP ACKs, partly because TCP responds to packet loss and packet reordering by cutting down the sending rate.

Fig. 16(a) shows the CDF of the average number of TCP time-outs per second in our experiments. We can see that TCP has less timeouts when running on top of POR than on SPP. This is because in SPP, congestion collapse can also trigger the timeout; on the contrary, POR adopts a congestion control mechanism to prevent congestion in the network. Fig. 16(b) shows the CDF of the average number of TCP fast retransmissions per second in our experiments. We can see that TCP tends to have more fast retransmissions on top of POR than on SPP. In POR, packets within the same TCP flow may travel through different nodes toward the destination, resulting in out-of-order reception at the destination. Such out-of-order reception is undesirable because TCP may interpret an out-of-order reception as a lost packet, and invoke fast retransmission to resend the packet as well as reducing the congestion window size. In an OR protocol, out-of-order receptions are inevitable because overheard packets, by their nature, arrive out-of-order. POR controls out-of-order reception by forwarding packets with smaller sequence number first; we can see that the number of actual fast retransmissions is reasonably small with a median around 5. More importantly, POR compensates for the performance loss due to

135

(a)

(b)

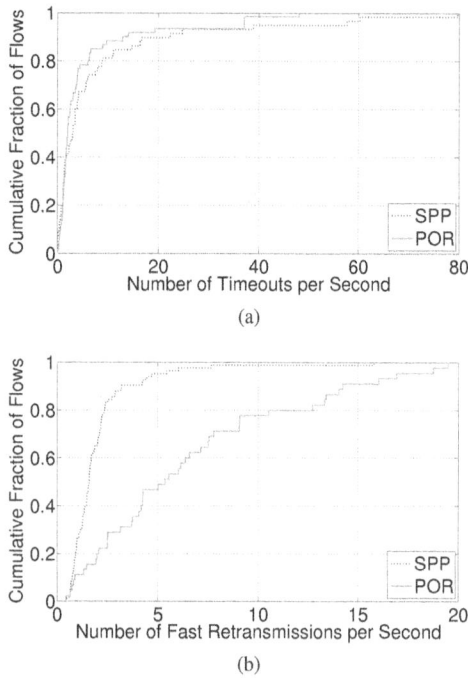

Figure 16: (a). CDF of TCP timeouts per second. (b). CDF of TCP fast retransmissions per second.

out-of-order receptions with other mechanisms and still achieves throughput gain over SPP.

8. CONCLUSIONS

In this paper, we propose POR, a new Opportunistic Routing (OR) protocol for high-speed, multi-rate wireless mesh networks that runs on commodity Wi-Fi interface, supports TCP, has low complexity, supports multiple link layer data rates, and is capable of exploiting partial packets for high efficiency. We believe POR is the first OR protocol that meets all such requirements, and thus significantly advances the practicability of OR protocols. POR adopts a per-packet feedback mechanism in packet forwarding to avoid sending data that has been received by the downstream nodes. POR also incorporates a block-based partial packet recovery scheme, a multi-hop link rate adaptation scheme, and a novel path cost calculation method which enables good path selection. We implement POR within the Click modular router. Our experiments in a 16-node testbed confirm that POR achieves significantly better performance than the compared protocols. We also demonstrate running unmodified TCP on POR, which is the first demonstration of unmodified TCP on an OR protocol with performance gain over traditional routing protocol to the best of our knowledge. Our future work includes extending POR to exploit features in 802.11n and 802.11ac networks.

9. REFERENCES

[1] K. Lin, N. Kushman, and D. Katabi, "ZipTx: Harnessing partial packets in 802.11 networks," in *ACM Mobicom*, 2008.

[2] B. Han, A, Schulman, F. Gringoli, N. Spring, B. Bhattacharjee, L. Nava, L. Ji, S. Lee, and R. Miller, "Maranello: Practical partial packet recovery for 802.11," in *USENIX NSDI, 2010*.

[3] S. Chachulski, M. Jennings, S. Katti and D. Katabi, "Trading structure for randomness in wireless opportunistic routing," in *ACM Sigcomm*, 2007.

[4] S. Katti, D. Katabi, H. Balakrishnan and M. Medard, "Symbol-level network coding for wireless mesh networks," in *ACM Sigcomm*, 2008.

[5] S. Biswas and R. Morris, "Opportunistic routing in multi-hop wireless networks," in *ACM Sigcomm*, 2005.

[6] E. Rozner, J. Seshadri, Y. A. Mehta, and L. Qiu. "SOAR: Simple opportunistic adaptive routing protocol for wireless mesh networks," *IEEE Transactions on Mobile Computing*, vol. 8, no. 12, pp, 1622-1635, 2009.

[7] T. Li, D. Leith, and L. Qiu, "Opportunistic routing for interactive traffic in wireless networks," in *IEEE ICDCS*, 2010.

[8] S. Katti, H. Rahul, W. Hu, D. Katabi, M. Medard, and J. Crowcroft, "XORs in the air: Practical wireless network coding," in *ACM Sigcomm*, 2006.

[9] V. Sevani, B. Raman and P. Joshi, "Implementation based evaluation of a full-fledged multi-hop TDMA-MAC for WiFi mesh networks", *IEEE Transactions on Mobile Computing*, to appear.

[10] D. Halperin, W. Hu, A. Sheth, and D. Wetherall, "Predictable 802.11 packet delivery from wireless channel measurements," in *ACM Sigcomm*, 2010.

[11] G. Judd, X. Wang, and P. Steenkiste, "Efficient channel-aware rate adaptation in dynamic environments," in *ACM MobiSys*, 2008.

[12] S. Sen, N. Santhapuri, R. R. Choudhury, and S. Nelakuditi, "AccuRate: Constellation based rate estimation in wireless networks," in *USENIX NSDI*, 2010.

[13] M. Vutukuru, H. Balakrishnan, and K. Jamieson, "Cross-layer wireless bit rate adaptation," in *ACM Sigcomm*, 2009.

[14] S. H. Y. Wong, H. Yang, S. Lu, and V. Bharghavan, "Robust rate adaptation for 802.11 wireless networks," in *ACM MobiCom*, 2006.

[15] R. Laufer, H. D.-Ferrière, and L. Kleinrock, "Polynomial-time algorithms for multirate anypath routing in wireless multihop metworks," *IEEE/ACM Transactions on Networking*, vol. 20, no. 3, pp. 743-755, Jun. 2012.

[16] K. Zeng, W. Lou, and H. Zhai, "Capacity of opportunistic routing in multi-rate and multi-hop wireless networks," *IEEE Transactions on Wireless Communications*, vol. 7, no. 12, pp. 5118-5128, Dec., 2008.

[17] P. A.K. Acharya, A. Sharma, E. M. Belding, K. C. Almeroth, and K. Papagiannaki, "Rate adaptation in congested wireless networks through real-time measurements," *IEEE Transactions on Mobile Computing*, vol. 9, no. 11, pp. 1535-1550, Nov., 2010.

[18] E. Ancillotti, R. Bruno, and M. Conti, "Experimentation and performance evaluation of rate adaptation algorithms in wireless mesh networks," in *Proceedings of the 5th ACM symposium on performance evaluation of wireless ad hoc, sensor, and ubiquitous networks (PE-WASUN '08)*, pp. 7-14, 2008.

[19] J. C. Park and S. K. Kasera, "Reduced packet probing multirate adaptation for multi-hop Ad Hoc wireless networks," *IEEE Symposium on World of Wireless, Mobile and Multimedia Networks*, June 2009.

[20] Optimized Link State Routing Protocol (OLSR), *http://www.ietf.org/rfc/rfc3626.txt*.

[21] The MadWifi Project, *http://madwifi-project.org/*.

[22] Cisco Aironet 802.11a/b/g wireless cardbus adapter, http://www.cisco.com/.

[23] The Click Modular Router, *http://read.cs.ucla.edu/click/*.

[24] *http://people.csail.mit.edu/szym/more/README.html*.

[25] D. Levin, H, Schioberg, R. Merz and C. Sengul, "TCPSpeaker: Clean and dirty sides of the same slate," *Mobile Computing and Communications Review*, vol. 14, no. 4, pp. 43-45, 2010.

[26] J. Xie, W. Hu, and Z. Zhang, "Revisiting partial packet recovery in 802.11 wireless LANs," in *ACM Mobisys*, 2011.

[27] Z.Zhang, W. Hu, and J. Xie, "Employing coded relay in multi-hop wireless networks," in *IEEE Globecom*, 2012.

[28] W. Hu, J. Xie, and Z. Zhang, "The complexity of the routing problem in POR," *Technical Report TR-130611*, Computer Science Department, Florida State University, 2013.

CSMA over Time-varying Channels:
Optimality, Uniqueness and Limited Backoff Rate

Se-Young Yun[*]
Dept. of Automatic Control
KTH, Sweden
seyoung@kth.se

Jinwoo Shin
Dept. of Electrical Engineering
KAIST, Korea
mijirim@gmail.com

Yung Yi[†]
Dept. of Electrical Engineering
KAIST, Korea
yiyung@kaist.edu

ABSTRACT

Recent studies on MAC scheduling have shown that carrier sense multiple access (CSMA) algorithms can be throughput optimal for arbitrary wireless network topology. However, these results are highly sensitive to the underlying assumption on 'static' or 'fixed' system conditions. For example, if channel conditions are time-varying, it is unclear how each node can adjust its CSMA parameters, so-called backoff and channel holding times, using its local channel information for the desired high performance. In this paper, we study 'channel-aware' CSMA (A-CSMA) algorithms in time-varying channels, where they adjust their parameters as some function of the current channel capacity. First, we show that the achievable rate region of A-CSMA equals to the maximum rate region if and only if the function is exponential. Furthermore, given an exponential function in A-CSMA, we design updating rules for their parameters, which achieve throughput optimality for an arbitrary wireless network topology. They are the first CSMA algorithms in the literature which are proved to be throughput optimal under time-varying channels. Moreover, we also consider the case when back-off rates of A-CSMA are highly restricted compared to the speed of channel variations, and characterize the throughput performance of A-CSMA in terms of the underlying wireless network topology. Our results not only guide a high-performance design on MAC scheduling under highly time-varying scenarios, but also provide new insights on the performance of CSMA algorithms in relation to their backoff rates and underlying network topologies.

[*]This work was done when the first author was with Department of Electrical Engineering, KAIST, Korea and supported by BK21.

[†]This work was supported by the Center for Integrated Smart Sensors funded by the Ministry of Education, Science and Technology as Global Frontier Project (CISS-2012M3A6A6054195).

Categories and Subject Descriptors

C.2.1 [**Computer-Communication Networks**]: Network Architecture and Design

Keywords

CSMA, time-varying channel, Backoff, wireless Ad-hoc network

1. INTRODUCTION

1.1 Motivation

How to access the shared medium is a crucial issue in achieving high performance in many applications, *e.g.*, wireless networks. In spite of a surge of research papers in this area, it's the year 1992 that the seminal work by Tassiulas and Ephremides proposed a throughput optimal medium access algorithm, referred to as Max-Weight [24]. Since then, a huge array of subsequent research has been made to develop distributed medium access algorithms with high performance guarantee and low complexity. However, in many cases the tradeoff between complexity and efficiency has been observed, or even throughput optimal algorithms with polynomial complexity have turned out to require heavy message passing, which becomes a major hurdle to becoming practical medium access schemes, *e.g.*, see [7,26] for surveys.

Recently, there has been exciting progresses that even fully distributed medium access algorithms based on CSMA (Carrier Sense Multiple Access) with no or very little message passing can achieve optimality in both throughput and utility, *e.g.*, see [6,13,16,19]. The main intuition underlying these results is that nodes dynamically adjust their CSMA parameters, *backoff* and *channel holding* times, using local information such as queue-length so that they solve a certain network-wide optimization problem for the desired high performance. We refer the readers to a survey paper [29] for more details.

However, the recent CSMA algorithms crucially rely on the assumption of static channel conditions. It is far from being clear how they perform for time-varying channels, which frequently occurs in practice. Note that it has already been shown that the Max-Weight is throughput optimal for time-varying channels [23] and joint scheduling and congestion control algorithms based on the optimization decomposition, *e.g.*, [2], are utility optimal by selecting the schedules over time, both of which essentially track the channel conditions quickly. However, a similar channel adaptation for CSMA algorithms may not be feasible for the following two reasons. First, each node in a network only knows its local channel information, and cannot track channel conditions of other nodes. Second, there exists a non-trivial coupling between CSMA's performance under time-varying channels and the speed of channel variations. A CSMA schedule at some instant may not have enough time to be close to the desired 'stationary' distribution before the channel changes. In this paper, we formalize and quantify this coupling, and study when and how CSMA algorithms perform depending on the network topologies and the speed of channel variations.

1.2 Our Contribution

In this paper, we model time-varying channels by a Markov process, and study 'channel-aware' CSMA (A-CSMA) algorithms where each link adjusts its CSMA parameters, back-off and channel holding times, as some function of its (local) channel capacity. In what follows, we first summarize our main contributions and then describe more details.

C1 – Achievable rate region of A-CSMA. We show that the achievable rate region of A-CSMA is maximized if and only if the function is exponential. In particular, we prove that A-CSMA can achieve an arbitrary large fraction of the capacity region for exponential functions (see Theorem 3.1), which turns out to be *impossible* for non-exponential functions (see Theorem 3.2).

C2 – Dynamic throughput optimal A-CSMA. We develop two types of throughput optimal A-CSMA algorithms, where links dynamically update their CSMA parameters based on both (a) the exponential function of the channel capacity in *C1* and (b) the empirical local load or the local queue length, without knowledge of the speed of channel variation and the arrival statistics (such as its mean) in advance (see Theorems 4.1 and 4.2).

C3 – Achievable rate region of A-CSMA with limited backoff rates. We provide a lower bound for the achievable rate region of A-CSMA when their backoff rates are highly limited compared to the speed of channel variations (see Theorem 5.1). Our bound depends on a combinatorial property of the underlying interference graph (i.e., its chromatic number), and is independent of backoff rates or the speed of channel variations. Moreover, it is noteworthy that the achievable rate region of A-CSMA includes the achievable rate region of channel-unaware CSMA (U-CSMA) for any limited backoff rate (see Corollary 5.1).

A typical necessary step to analyze and design a CSMA algorithm of high performance in static channels is to characterize the stationary distribution of the Markov chain induced by it [6, 13, 16, 19]. However, this task is much harder for A-CSMA in time-varying channels, since the Markov chain induced by A-CSMA is *non-reversible* (see Theorem 2.1), i.e., it is unlikely that its stationary distribution has a 'clean' formula to analyze, being in sharp contrast to the CSMA analysis for static channels. To overcome this technical issue, we first show that the stationary distribution approximates to a of product-form distribution when backoff rates are sufficiently large. Then, for *C1*, we study the product-form to guarantee high throughput of A-CSMA, where the exponential functions are found. The main novelty lies in establishing the approximation scheme, using the *Markov chain tree theorem* [1], which requires counting the weights of arborescences induced by the non-reversible Markov process to understand its stationary distribution.

For *C2*, we combine *C1* with existing techniques: our first and second throughput optimal algorithms are 'rate-based' and 'queue-based' ones originally studied in static channels by Jiang et al. (cf. [5,6]) and Rajagopalan et al. (cf. [19,21]), respectively. To extend these results to time-varying channels, our specific choice of holding times as exponential functions of the channel capacity plays a key role in establishing the desired throughput optimal performance. To our best knowledge, they are the first CSMA algorithms in the literature which are proved to be throughput optimal under general Markovian time-varying channel models.

C3 is motivated by observing that a CSMA algorithm in fast time-varying channels inevitably has to be of high backoff rates for the desired throughput performance, i.e., high backoff rates are needed for tracking time-varying channel conditions fast enough. However, backoff rates are bounded in practice, which may cause degradation in the CSMA's performance. We note that CSMA algorithms with limited backoff or holding rates have been little analyzed in the literature, despite of their practical importance.[1] *C3* provides a lower bound for A-CSMA throughputs regardless of restrictions on their backoff rates or sensing frequencies. For example, if the interference graph is bipartite (i.e., its chromatic number is two), our bound implies that A-CSMA is guaranteed to have at least 50%-throughput even with arbitrary small backoff rates. Furthermore, one can design a dynamic high-throughput A-CSMA algorithm with limited backoff rates using *C3* (similarly as *C1* is used for *C2*), but in the current paper we do not present further details due to the space limitation.

1.3 Related Work

The research on throughput optimal CSMA has been initiated independently by Jiang et al. (cf. [5,6]) and Rajagopalan et al. (cf. [19,21]), where both consider the continuous time and collision free setting. Under exponential distributions on backoff and holding times, the system is modeled by a continuous time Markov chain, where the backoff rate or channel holding time at each link is adaptively controlled by the local (virtual or actual) queue lengths. Jiang et al. proved that the long-term link throughputs are the solution of an utility maximization problem assuming the infinite backlogged data. Rajagopalan et al. showed that if the CSMA parameters are changing very slowly with respect to the queue length changes, the mixing time is much faster than the queue length changes so that the realized link schedules can provably emulate Max-Weight very well. Although their key intuitions are apparently different, analytic techniques are quite similar, i.e., both require to understand the long-term behavior (i.e. stationarity) of the Markov chains formed by CSMA.

These throughput optimality results motivate further research on design and analysis of CSMA algorithms. The work by Liu et al. [13] proves the utility optimality using a stochastic approximation technique, which has been extended to the multi-channel, multi-radio case with a simpler proof in [17]. The throughput optimality of MIMO networks under SINR model is also shown in [18]. As opposed to the

---

[1]Even in static channels, restrictions on backoff or holding rates may degrade the throughput or delay performances of CSMA algorithms.

continuous-time setting that carrier sensing is perfect and instantaneous (and hence no collision occurs), more practical discrete time settings that carrier sensing is imperfect or delayed (and hence collisions occur) have been also studied. The throughput optimality of CSMA algorithms in discrete time settings with collisions is established in [8], [22] and [9], where the authors in [9] consider imperfect sensing information. In [13], the authors studied the impact of collisions and the tradeoff between short-term fairness and efficiency. The authors in [16] considered a synchronous system consisting of the control phase, which eliminates the chances of data collisions via a simple message passing, and the data phase, which actually enables data transmission based on the discrete-time Glauber dynamics. There also exist several efforts on improving or analyzing delay performance [3, 4, 10, 12, 14, 20], speeding up the convergence [28], and developing a practical protocol based on the CSMA theory with experimental validation [11, 15].

To the best of our knowledge, CSMA under time-varying channels has been studied only in [12] for only complete interference graphs, when the arbitrary backoff rate is allowed, and more seriously, under the time-scale separation assumption, which does not often hold in practice and extremely simplifies the analysis (no mixing time related details are needed).

2. MODEL AND PRELIMINARIES

2.1 Network Model

We consider a network consisting of a collection of n queues (or links) $\{1, \ldots, n\}$ and time is indexed by $t \in \mathbb{R}_+$. Let $Q_i(t) \in \mathbb{R}_+$ denote the amount of work in queue i at time t and let $\boldsymbol{Q}(t) = [Q_i(t)]_{1 \le i \le n}$. The system starts empty, i.e., $Q_i(0) = 0$. We assume work arrives at each queue i as per an exogenous Poisson process with rate $\lambda_i > 0$, where $A_i(s, t) < \infty$ denotes the cumulative arrival to queue i in the time interval $(s, t]$. Each queue i can be serviced at rate $c_i(t) \ge 0$ representing the potential departure rate of work from the queue $Q_i(t)$. We consider finite state Markov time-varying channels [25]: each $\{\boldsymbol{c}(t) = [c_i(t)] : t \ge 0\}$ is a continuous-time, time-homogeneous, irreducible Markov process, where each link has m states channel space such that $c_i(t) \in \mathcal{H} := \{h_1, \ldots, h_m\}$ and $0 < h_1 < \cdots < h_m = 1$. We denoted by $\boldsymbol{\gamma}^{\boldsymbol{u} \to \boldsymbol{v}}$ the 'transition-rates' on the channel state for $\boldsymbol{u} \to \boldsymbol{v}$, $\boldsymbol{u}, \boldsymbol{v} \in \mathcal{H}^n$. For the time-varying channels, we assume that each link i knows the channel state $c_i(t)$ before it transmits.[2] We call $\max_{\boldsymbol{u} \in \mathcal{H}^n} \{\sum_{\boldsymbol{v} \in \mathcal{H}^n : \boldsymbol{v} \ne \boldsymbol{u}} \boldsymbol{\gamma}^{\boldsymbol{u} \to \boldsymbol{v}}\}$ the *channel varying speed*. The inverse of *channel varying speed* indicates the maximum of the expected number of channel transitions during the unit-length time interval. We consider only single-hop sessions (or flows), i.e., once work departs from a queue, it leaves the network.

The queues are offered service as per the constraint imposed by interference. To model this constraint, we adopt a popular graph-based approach, where denote by $G = (V, E)$ the inference graph among n queues, where the vertices $V = \{1, \ldots, n\}$ represent queues and the edges $E \subset V \times V$ represent interferences between queues: $(i, j) \in E$, if queues i and

j interfere with each other. Let $\mathcal{N}(i) = \{j \in V : (i, j) \in E\}$ and $\boldsymbol{\sigma}(t) = [\sigma_i(t)] \in \{0, 1\}^n$ denote the neighbors of node i and a schedule at time t, i.e., whether queues transmit at time t, respectively, where $\sigma_i(t) = 1$ represents transmission of queue i at time t. Then, interference imposes the constraint that for all $t \in \mathbb{R}_+$, $\boldsymbol{\sigma}(t) \in \mathcal{I}(G)$, where

$$\mathcal{I}(G) := \big\{ \boldsymbol{\rho} = [\rho_i] \in \{0, 1\}^n : \rho_i + \rho_j \le 1, \ \forall (i, j) \in E \big\}.$$

The resulting queueing dynamics are described as follows. For $0 \le s < t$ and $1 \le i \le n$,

$$Q_i(t) = Q_i(s) - \int_s^t \sigma_i(r) c_i(r) \mathbf{1}_{\{Q_i(r) > 0\}} \, dr + A_i(s, t),$$

where $\mathbf{1}_{\{\cdot\}}$ denotes the indicator function. Finally, we define the cumulative actual and potential departure processes $\boldsymbol{D}(t) = [D_i(t)]$ and $\widehat{\boldsymbol{D}}(t) = [\widehat{D}_i(t)]$, respectively, where

$$D_i(t) = \int_0^t \sigma_i(r) c_i(r) \mathbf{1}_{\{Q_i(r) > 0\}} dr, \quad \widehat{D}_i(t) = \int_0^t \sigma_i(r) c_i(r) dr.$$

2.2 Scheduling, Rate Region and Metric

The main interest of this paper is to design a scheduling algorithm which decides $\boldsymbol{\sigma}(t) \in \mathcal{I}(G)$ for each time instance $t \in \mathbb{R}_+$. Intuitively, it is expected that a good scheduling algorithm will keep the queues as small as possible. To formally discuss, we define the maximum achievable rate region (also called capacity region) $\boldsymbol{C} \subset [0, 1]^n$ of the network, which is the convex hull of the feasible scheduling set $\mathcal{I}(G)$, i.e.,

$$\boldsymbol{C} = \boldsymbol{C}(\boldsymbol{\gamma}, G) = \Big\{ \sum_{\boldsymbol{c} \in \mathcal{H}^n} \pi_{\boldsymbol{c}} \sum_{\boldsymbol{\rho} \in \mathcal{I}(G)} \alpha_{\boldsymbol{\rho}, \boldsymbol{c}} \boldsymbol{c}^T \cdot \boldsymbol{\rho} : \alpha_{\boldsymbol{\rho}, \boldsymbol{c}} \ge 0 \text{ and}$$
$$\sum_{\boldsymbol{\rho} \in \mathcal{I}(G)} \alpha_{\boldsymbol{\rho}, \boldsymbol{c}} = 1 \text{ for all } \boldsymbol{c} \in \mathcal{H}^n \Big\},$$

where $\boldsymbol{c}^T \cdot \boldsymbol{\rho} = [c_i \rho_i]$ and $\pi_{\boldsymbol{c}}$ denotes the stationary distribution of channel state $\boldsymbol{c}$ under the channel-varying Markov process. The intuition behind this definition comes from the facts: (a) any scheduling algorithm has to choose a schedule from $\mathcal{I}(G)$ at each time and channel state where $\alpha_{\boldsymbol{\rho}, \boldsymbol{c}}$ denotes the fraction of time selecting schedule $\boldsymbol{\rho}$ for given channel state $\boldsymbol{c}$ and (b) for channel state $\boldsymbol{c} \in \mathcal{H}^n$, the fraction in the time domain where $\boldsymbol{c}(t) = [c_i(t)]$ is equal to $\boldsymbol{c}$ is $\pi_{\boldsymbol{c}}$. Hence the time average of the 'service rate' induced by any algorithm must belong to $\boldsymbol{C}$.

We call the arrival rate $\boldsymbol{\lambda}$ *admissible* if $\boldsymbol{\lambda} = [\lambda_i] \in \boldsymbol{\Lambda} = \boldsymbol{\Lambda}(\boldsymbol{\gamma}, G)$, where

$$\boldsymbol{\Lambda}(\boldsymbol{\gamma}, G) := \big\{ \boldsymbol{\lambda} \in \mathbb{R}_+^n : \boldsymbol{\lambda} \le \boldsymbol{\lambda}', \text{ for some } \boldsymbol{\lambda}' \in \boldsymbol{C}(\boldsymbol{\gamma}, G) \big\},$$

where $\boldsymbol{\lambda} \le \boldsymbol{\lambda}'$ corresponds to the component-wise inequality, i.e., if $\boldsymbol{\lambda} \notin \boldsymbol{\Lambda}$, queues should grow linearly over time under any scheduling algorithm. Further, $\boldsymbol{\lambda}$ is called *strictly admissible* if $\boldsymbol{\lambda} \in \boldsymbol{\Lambda}^\circ = \boldsymbol{\Lambda}^\circ(\boldsymbol{\gamma}, G)$ and

$$\boldsymbol{\Lambda}^\circ(\boldsymbol{\gamma}, G) := \big\{ \boldsymbol{\lambda} \in \mathbb{R}_+^n : \boldsymbol{\lambda} < \boldsymbol{\lambda}', \text{ for some } \boldsymbol{\lambda}' \in \boldsymbol{C}(\boldsymbol{\gamma}, G) \big\}.$$

We now define the performance metric.

DEFINITION 2.1. *A scheduling algorithm is called rate-stable for a given arrival rate $\boldsymbol{\lambda}$, if*

$$\lim_{t \to \infty} \frac{1}{t} \boldsymbol{D}(t) = \boldsymbol{\lambda} \qquad \text{(with probability 1)}. \qquad (1)$$

---

[2]The channel information can be achieved using control messages such as RTS and CTS in IEEE 802.11, and links can adapt their transmission parameters to channel transitions for every transmission by changing coding and modulation parameters.

Furthermore, we say a scheduling algorithm has α-throughput if it is rate-stable for any $\boldsymbol{\lambda} \in \alpha \boldsymbol{\Lambda}^o(\gamma, G)$. In particular, when $\alpha = 1$, it is called throughput optimal.

We note that (1) is equivalent to $\lim_{t \to \infty} \frac{1}{t} \boldsymbol{Q}(t) = 0$, since $\lim_{t \to \infty} \frac{A_i(0,t)}{t} = \lambda_i$ (because the arrival process is stationary ergodic). The following lemma implies that the potential departure process suffies to study the rate-stability.

LEMMA 2.1. *A scheduling algorithm is rate-stable if*

$$\lim_{t \to \infty} \frac{1}{t} \widehat{\boldsymbol{D}}(t) > \boldsymbol{\lambda}.$$

We omit the proof due to the space constraint.

2.3 Channel-aware CSMA Algorithm: A-CSMA

The algorithm to decide $\boldsymbol{\sigma}(t)$ utilizing the local carrier sensing information can be classified as CSMA (Carrier Sense Multiple Access) algorithms. In between two transmissions, a queue waits for a random amount of time – also known as *backoff time*. Each queue can sense the medium perfectly and instantly, *i.e.*, knows if any other interfering queue is transmitting at a given time instance. If a queue that finishes waiting senses the medium to be busy, it starts waiting for another random amount of time; else, it starts transmitting for a random amount of time, called *channel holding time*. We assume that queue i's backoff and channel holding times have exponential distributions with mean $1/R_i$ and $1/S_i$, respectively, where $R_i = R_i(t) > 0$ and $S_i = S_i(t) > 0$ may change over time. We define A-CSMA (channel-aware CSMA) to be the class of CSMA algorithms where $R_i(t)$ and $S_i(t)$ are decided by some functions of the current channel capacity, *i.e.*, $R_i(t) = f_i(c_i(t))$ and $S_i(t) = g_i(c_i(t))$ for some functions f_i and g_i. In the special case when $R_i(t)$ and $S_i(t)$ are decided independently of current channel information (*e.g.*, f_i's and g_i's are constant functions), we specially say a CSMA algorithm is U-CSMA (channel-unaware CSMA).

Then, given functions $[f_i]$ and $[g_i]$, it is easy to check that $\{(\boldsymbol{\sigma}(t), \boldsymbol{c}(t)) : t \geq 0\}$ under A-CSMA is a continuous time Markov process, whose kernel (or transition-rates) is given by:

$$
\begin{aligned}
(\boldsymbol{\sigma}, \boldsymbol{u}) &\to (\boldsymbol{\sigma}, \boldsymbol{v}) \quad \text{with rate} \quad \gamma^{\boldsymbol{u} \to \boldsymbol{v}} \\
(\boldsymbol{\sigma}_i^0, \boldsymbol{c}) &\to (\boldsymbol{\sigma}_i^1, \boldsymbol{c}) \quad \text{with rate} \quad f_i(c_i) \cdot \prod_{j:(i,j) \in E} (1 - \sigma_j) \\
(\boldsymbol{\sigma}_i^1, \boldsymbol{c}) &\to (\boldsymbol{\sigma}_i^0, \boldsymbol{c}) \quad \text{with rate} \quad g_i(c_i) \cdot \sigma_i, \quad (2)
\end{aligned}
$$

where $\boldsymbol{\sigma}_i^0$ and $\boldsymbol{\sigma}_i^1$ denote two 'almost' identical schedule vectors except i-th elements which are 0 and 1, respectively. Since $\{\boldsymbol{c}(t)\}$ is a time-homogeneous irreducible Markov process, $\{(\boldsymbol{\sigma}(t), \boldsymbol{c}(t))\}$ is ergodic, *i.e.*, it has the unique stationary distribution $[\pi_{\boldsymbol{\sigma}, \boldsymbol{c}}]$. For example, when functions f_i and g_i are constant (*i.e.*, U-CSMA with fixed $R_i(t) = R_i$ and $S_i(t) = S_i$),

$$
\pi_{\boldsymbol{\sigma}, \boldsymbol{c}} = \pi_{\boldsymbol{c}} \cdot \frac{\exp\left(\sum_i \sigma_i \log \frac{R_i}{S_i}\right)}{\sum_{\boldsymbol{\rho} = [\rho_i] \in \mathcal{I}(G)} \exp\left(\sum_i \rho_i \log \frac{R_i}{S_i}\right)},
$$

and if $\{\boldsymbol{c}(t)\}$ is (time-)reversible, $\{(\boldsymbol{\sigma}(t), \boldsymbol{c}(t))\}$ is as well. In general, $\{(\boldsymbol{\sigma}(t), \boldsymbol{c}(t))\}$ is not reversible unless functions f_i/g_i are constant, as we state in the following theorem.

THEOREM 2.1. *If $\{(\boldsymbol{\sigma}(t), \boldsymbol{c}(t))\}$ is reversible,*

$$\frac{f_i(x)}{g_i(x)} = \frac{f_i(y)}{g_i(y)}, \qquad \text{for all } x, y \in \mathcal{H}, i \in V.$$

PROOF. We prove this by contradiction. Denote by $\boldsymbol{c}_i^u$ and $\boldsymbol{c}_i^v$ two almost identical channel state vectors except i-th elements, which are h_u and h_v, respectively. Suppose that $\{(\boldsymbol{\sigma}(t), \boldsymbol{c}(t))\}$ is reversible and $\frac{f_i(h_u)}{g_i(h_u)} \neq \frac{f_i(h_v)}{g_i(h_v)}$ for some link i. From the reversibility, the transition path $(\boldsymbol{\sigma}_i^0, \boldsymbol{c}_i^u) \to (\boldsymbol{\sigma}_i^0, \boldsymbol{c}_i^v) \to (\boldsymbol{\sigma}_i^1, \boldsymbol{c}_i^v)$ has to satisfy the following balance equations:

$$
\begin{aligned}
\pi_{\boldsymbol{\sigma}_i^0, \boldsymbol{c}_i^u} \gamma^{\boldsymbol{c}_i^u \to \boldsymbol{c}_i^v} &= \pi_{\boldsymbol{\sigma}_i^0, \boldsymbol{c}_i^v} \gamma^{\boldsymbol{c}_i^v \to \boldsymbol{c}_i^u} \\
\pi_{\boldsymbol{\sigma}_i^0, \boldsymbol{c}_i^v} f_i(h_v) &= \pi_{\boldsymbol{\sigma}_i^1, \boldsymbol{c}_i^v} g_i(h_v), \quad (3)
\end{aligned}
$$

Similarly, for the transition path $(\boldsymbol{\sigma}_i^0, \boldsymbol{c}_i^u) \to (\boldsymbol{\sigma}_i^1, \boldsymbol{c}_i^u) \to (\boldsymbol{\sigma}_i^1, \boldsymbol{c}_i^v)$,

$$
\begin{aligned}
\pi_{\boldsymbol{\sigma}_i^0, \boldsymbol{c}_i^u} f_i(h_u) &= \pi_{\boldsymbol{\sigma}_i^1, \boldsymbol{c}_i^u} g_i(h_u), \quad \text{and} \\
\pi_{\boldsymbol{\sigma}_i^1, \boldsymbol{c}_i^u} \gamma^{\boldsymbol{c}_i^u \to \boldsymbol{c}_i^v} &= \pi_{\boldsymbol{\sigma}_i^1, \boldsymbol{c}_i^v} \gamma^{\boldsymbol{c}_i^v \to \boldsymbol{c}_i^u}. \quad (4)
\end{aligned}
$$

From (3) and (4),

$$\frac{\pi_{\boldsymbol{\sigma}_i^0, \boldsymbol{c}_i^u}}{\pi_{\boldsymbol{\sigma}_i^1, \boldsymbol{c}_i^v}} = \frac{\gamma^{\boldsymbol{c}_i^v \to \boldsymbol{c}_i^u} g_i(h_v)}{\gamma^{\boldsymbol{c}_i^u \to \boldsymbol{c}_i^v} f_i(h_v)} = \frac{\gamma^{\boldsymbol{c}_i^v \to \boldsymbol{c}_i^u} g_i(h_u)}{\gamma^{\boldsymbol{c}_i^u \to \boldsymbol{c}_i^v} f_i(h_u)}, \quad (5)$$

which contradicts the assumption $\frac{f_i(h_u)}{g_i(h_u)} \neq \frac{f_i(h_v)}{g_i(h_v)}$. This completes the proof of Theorem 2.1. $\square$

We note that the non-reversible property makes it hard to characterize the stationary distribution $[\pi_{\boldsymbol{\sigma}, \boldsymbol{c}}]$ of the Markov process induced by A-CSMA.

3. ACHIEVABLE RATE REGION OF A-CSMA

In this section, we study the achievable rate region of A-CSMA algorithms given (fixed) functions $[f_i]$ and $[g_i]$. We show that the achievable rate region of A-CSMA is maximized for the following choices of functions:

$$\log \frac{f_i(x)}{g_i(x)} = r_i \cdot x, \quad \text{for } x \in [0, 1], \quad (6)$$

where $r_i \in \mathbb{R}$ is some constant. Namely, the ratio $f_i(x)/g_i(x)$ is an exponential function in terms of x. We let EXP-A-CSMA denote the sub-class of A-CSMA algorithms with functions satisfying (6) for some $[r_i]$. The following theorem justifies the optimality of EXP-A-CSMA in terms of its achievable rate region.

THEOREM 3.1 (OPTIMALITY). *For any arrival rate $\boldsymbol{\lambda} = [\lambda_i] \in \boldsymbol{\Lambda}^o$, interference graph G, and channel transition-rate γ, there exists $[r_i]$, $[f_i]$ and $[g_i]$ satisfying (6) such that the corresponding EXP-A-CSMA algorithm is rate-stable.*

We also establish that Theorem 3.1 is tight in the sense that it does not hold for other A-CSMA algorithms that have different ways of reflecting channel capacity in adjusting CSMA parameters. To state it formally, given a non-negative continuous function $k : [0, 1] \to \mathbb{R}_+$, we define EXP($k$)-A-CSMA as the sub-class of A-CSMA algorithms with the following form of functions:

$$\log \frac{f_i(x)}{g_i(x)} = r_i \cdot k(x), \quad \text{for } x \in [0, 1], \quad (7)$$

where $r_i \in \mathbb{R}$ is some constant. The following theorem states that EXP-A-CSMA is the unique class of A-CSMA maximizing its achievable rate region.

THEOREM 3.2 (UNIQUENESS). *If the conclusion of Theorem 3.1 holds for EXP(k)-A-CSMA, then*

$$EXP(k)\text{-}A\text{-}CSMA = EXP\text{-}A\text{-}CSMA.$$

The proofs of Theorems 3.1 and 3.2 are given in Sections 3.1 and 3.3, respectively. For the proof of Theorem 3.1, Section 3.2 describes the proof of Lemma 3.2, which is a key lemma of this work. In the following proofs (and throughout this paper), we commonly let $[\pi_{\sigma,c}]$, $[\pi_c]$ and $[\pi_{\sigma|c}]$ be the stationary distributions of Markov processes $\{(\sigma(t), c(t))\}$, $\{c(t)\}$ and $\{\sigma(t), c\}$ induced by an A-CSMA algorithm, respectively.

3.1 Proof of Theorem 3.1

To begin with, we recall that the channel varying speed ψ is defined as: $\psi = \max_{u \in \mathcal{H}^n} \{\sum_{v \in \mathcal{H}^n : v \neq u} \gamma^{u \to v}\}$. We first state Lemmas 3.1 and 3.2, which are the key lemmas to the proof of Theorem 3.1.

LEMMA 3.1. *For any $\delta_1 \in (0,1)$, arrival rate $\lambda = [\lambda_i] \in (1-\delta_1)\Lambda^o$, interference graph G and channel transition-rate γ, there exists $[r_i] \in \mathbb{R}^n$ such that*

$$\max_i |r_i| \leq \frac{4n^2 \log |\mathcal{I}(G)|}{\delta_1^2 \min_i \{(\sum_{c \in \mathcal{H}^n} c_i \pi_c)^2\}},$$

and every EXP-A-CSMA algorithm with

$$\log \frac{f_i(h)}{g_i(h)} = r_i \cdot h, \quad \text{for all } i \in V, h \in \mathcal{H}$$

satisfies

$$\lambda_i \leq \sum_{c \in \mathcal{H}^n} c_i \pi_c \sum_{\sigma \in \mathcal{I}(G):\sigma_i=1} \pi_{\sigma|c}, \quad \text{for all } i \in V.$$

LEMMA 3.2. *For any $\delta_2 \in (0,1)$, interference graph G and channel transition-rate γ and A-CSMA algorithm with functions $f = [f_i]$ and $g = [g_i]$ satisfying*

$$\min_{i \in V, h \in \mathcal{H}} \{f_i(h), g_i(h)\} \geq \frac{\psi \cdot m^{2^n m^n (n+1)}}{\delta_2},$$

it follows that

$$\max_{(\sigma,c) \in \mathcal{I}(G) \times \mathcal{H}^n} \left| 1 - \frac{\pi_{\sigma,c}}{\pi_c \pi_{\sigma|c}} \right| < \delta_2.$$

Lemma 3.2 implies that if f_i, g_i are large enough, the stationary distribution $[\pi_{\sigma,c}]$ approximates to a product-form distribution $[\pi_c \pi_{\sigma|c}]$, where under EXP-A-CSMA,

$$\pi_{\sigma|c} \propto \exp\left(\sum_i \sigma_i r_i c_i\right),$$

due to the reversibility of Markov process $\{\sigma(t), c\}$. On the other hand, Lemma 3.1 implies that arrival rate λ is stabilized under the distribution $[\pi_c \pi_{\sigma|c}]$. Therefore, combining two above lemmas will lead to the proof of Theorem 3.1.

We remark that Lemma 3.1 is a non-trivial generalization of Lemma 8 in [5] (for static channels), which corresponds to a special case of Lemma 3.1 with $\pi_c = 1$ for $c = [1]$. The proof of Lemma 3.1 uses a similar strategy with that of Lemma 8 in [5]. Due to the space constraint, we omit the proof which can be found in [27].

Proof of Theorem 3.1. We now complete the proof of Theorem 3.1 using Lemmas 3.1 and 3.2. For a given arrival

rate $\lambda \in \Lambda^o$, there exists $\varepsilon \in (0,1)$ such that $\lambda \in (1-\varepsilon)\Lambda^o$ since $\lambda \in \Lambda^o$. If we apply Lemmas 3.1 and 3.2 with $(1+\varepsilon)\lambda \in (1-\varepsilon^2)\Lambda^o$ (*i.e.*, $\delta_1 = \varepsilon^2$ and $\delta_2 = \frac{\varepsilon}{1+\varepsilon}$), we have that there exists an EXP-A-CSMA algorithm with constant $[r_i]$ and functions $[f_i]$ and $[g_i]$ such that

$$\eta \leq \min_{i \in V, h \in \mathcal{H}} \{f_i(h), g_i(h)\}$$

$$(1+\varepsilon)\lambda_i \leq \sum_{c \in \mathcal{H}^n} c_i \pi_c \sum_{\sigma \in \mathcal{I}(G):\sigma_i=1} \pi_{\sigma|c},$$

where we choose

$$f_i(c_i) = R = \eta \exp(\kappa), \quad g_i(c_i) = R \cdot \exp(-r_i \cdot c_i),$$

$$\kappa = \kappa(\delta_1, G, \gamma) := \frac{4n^2 \log |\mathcal{I}(G)|}{\delta_1^2 \min_i \{(\sum_{c \in \mathcal{H}^n} c_i \pi_c)^2\}},$$

and

$$\eta = \eta(\delta_2, G, \gamma) := \frac{\psi \cdot m^{2^n m^n (n+1)}}{\delta_2}.$$

Therefore, it follows that

$$\begin{aligned}
\lambda_i &\leq \left(1 - \frac{\varepsilon}{1+\varepsilon}\right) \sum_{c \in \mathcal{H}^n} c_i \pi_c \sum_{\sigma \in \mathcal{I}(G):\sigma_i=1} \pi_{\sigma|c} \\
&< \sum_{c \in \mathcal{H}^n} c_i \pi_c \sum_{\sigma \in \mathcal{I}(G):\sigma_i=1} \pi_{\sigma,c} \\
&= \lim_{t \to \infty} \frac{1}{t} \widehat{D}_i(t),
\end{aligned}$$

where the last inequality is from the ergodicity of Markov process $\{(\sigma(t), c(t))\}$. This leads to the rate-stability using Lemma 2.1, and hence completes the proof.

3.2 Proof of Lemma 3.2

Let $\mathcal{G} = (\mathcal{V}, \mathcal{E})$ denote a weighted directed graph induced by Markov process $\{(\sigma(t), c(t))\}$: $\mathcal{V} = \mathcal{I}(G) \times \mathcal{H}^n$ and $((\sigma_1, c_1), (\sigma_2, c_2)) \in \mathcal{E}$ if the transition-rate (which becomes the weight of the edge) from (σ_1, c_1) to (σ_2, c_2) is non-zero in Markov process $\{(\sigma(t), c(t))\}$. Hence, there are two types of edges:

I. $((\sigma_1, c_1), (\sigma_2, c_2)) \in \mathcal{E}$ and $\sigma_1 = \sigma_2$

II. $((\sigma_1, c_1), (\sigma_2, c_2)) \in \mathcal{E}$ and $c_1 = c_2$

A subgraph of $\mathcal{G}$ is called *arborescence* (or spanning tree) with root (σ, c) if for any vertex in $\mathcal{V} \setminus \{(\sigma, c)\}$, there is exactly one directed path from the vertex to root (σ, c) in the subgraph. Let $\mathcal{A}_{\sigma,c}$ and $w(\mathcal{A}_{\sigma,c})$ denote the set of *arborescences* of which root is (σ, c) and the sum of weights of *arborescences* in $\mathcal{A}_{\sigma,c}$, where the weight of an *arborescence* is the product of weight of edges. Then, Markov chain tree theorem [1] implies that

$$\pi_{\sigma,c} = \frac{w(\mathcal{A}_{\sigma,c})}{\sum_{(\rho,d) \in \mathcal{I}(G) \times \mathcal{H}^n} w(\mathcal{A}_{\rho,d})}. \tag{8}$$

Now we further classify the set of *arborescences*. We let $\mathcal{A}_{\sigma,c}^{(i)} \subset \mathcal{A}_{\sigma,c}$ denote the set of *arborescences* consisting of i

edges of type I. Then, we have

$$w(\mathcal{A}_{\boldsymbol{\sigma},\boldsymbol{c}}) = \sum_{i \geq m^n-1} w(\mathcal{A}_{\boldsymbol{\sigma},\boldsymbol{c}}^{(i)}) \overset{(a)}{\leq} w(\mathcal{A}_{\boldsymbol{\sigma},\boldsymbol{c}}^{(m^n-1)}) +$$

$$\sum_{i \geq m^n} \left(\frac{\delta_2}{m^{2^n m^n (n+1)}} \right)^{i+1-m^n} \cdot |\mathcal{A}_{\boldsymbol{\sigma},\boldsymbol{c}}^{(i)}| \cdot w(\mathcal{A}_{\boldsymbol{\sigma},\boldsymbol{c}}^{(m^n-1)})$$

$$\leq w(\mathcal{A}_{\boldsymbol{\sigma},\boldsymbol{c}}^{(m^n-1)}) \cdot \left(1 + \sum_{i \geq m^n} \left(\frac{\delta_2}{m^{2^n m^n (n+1)}} \right)^{i+1-m^n} \cdot |\mathcal{A}_{\boldsymbol{\sigma},\boldsymbol{c}}^{(i)}| \right)$$

$$\leq w(\mathcal{A}_{\boldsymbol{\sigma},\boldsymbol{c}}^{(m^n-1)}) \cdot \left(1 + \frac{\delta_2}{m^{2^n m^n (n+1)}} \cdot |\mathcal{A}_{\boldsymbol{\sigma},\boldsymbol{c}}| \right)$$

$$\overset{(b)}{<} w(\mathcal{A}_{\boldsymbol{\sigma},\boldsymbol{c}}^{(m^n-1)}) \cdot (1 + \delta_2),$$

where (a) is from the condition in Lemma 3.2 and for (b) we use the inequality $|\mathcal{A}_{\boldsymbol{\sigma},\boldsymbol{c}}| < (mn)^{2^n m^n}$. Therefore, using the above inequality, it follows that

$$\frac{\pi_{\boldsymbol{\sigma},\boldsymbol{c}}}{\pi_{\boldsymbol{c}} \pi_{\boldsymbol{\sigma}|\boldsymbol{c}}} = \frac{w(\mathcal{A}_{\boldsymbol{\sigma},\boldsymbol{c}})}{w(\mathcal{A}_{\boldsymbol{\sigma},\boldsymbol{c}}^{(m^n-1)})} \cdot \frac{\sum\limits_{\boldsymbol{d} \in \mathcal{H}^n} \sum\limits_{\boldsymbol{\rho} \in \mathcal{I}(G)} w(\mathcal{A}_{\boldsymbol{\rho},\boldsymbol{d}}^{(m^n-1)})}{\sum\limits_{\boldsymbol{d} \in \mathcal{H}^n} \sum\limits_{\boldsymbol{\rho} \in \mathcal{I}(G)} w(\mathcal{A}_{\boldsymbol{\rho},\boldsymbol{d}})}$$

$$< 1 + \delta_2,$$

where the first equality follows from (8) and

$$\pi_{\boldsymbol{c}} \pi_{\boldsymbol{\sigma}|\boldsymbol{c}} = \frac{w(\mathcal{A}_{\boldsymbol{\sigma},\boldsymbol{c}}^{(m^n-1)})}{\sum\limits_{\boldsymbol{d} \in \mathcal{H}^n} \sum\limits_{\boldsymbol{\rho} \in \mathcal{I}(G)} w(\mathcal{A}_{\boldsymbol{\rho},\boldsymbol{d}}^{(m^n-1)})}.$$

Similarly, one can also show that $\frac{\pi_{\boldsymbol{\sigma},\boldsymbol{c}}}{\pi_{\boldsymbol{c}} \pi_{\boldsymbol{\sigma}|\boldsymbol{c}}} > 1 - \delta_2$. This completes the proof of Lemma 3.2.

3.3 Proof of Theorem 3.2

Consider a star interference graph G, where 1 denotes the center vertex and $\{2, \ldots, n\}$ is the set of other outer vertices. For the time-varying channel model, we set each element of channel by $h_j = \frac{j}{m}$ and assume the channel transition satisfies $\pi_{\boldsymbol{c}} = \frac{1}{m^n}$. For the arrival rate, we choose $\boldsymbol{\lambda} = \arg \max\limits_{\boldsymbol{\lambda} \in (1-\varepsilon)\boldsymbol{C}} \sum_i \lambda_i$, where $\varepsilon \in (0,1)$ will be chosen later.

Under this setup, suppose the conclusion of Theorem 3.1 holds, i.e., there exists a rate-stable EXP(k)-A-CSMA. Then, from the ergodicity of Markov process $\{(\boldsymbol{\sigma}(t), \boldsymbol{c}(t))\}$, we have

$$\lambda_i \leq \sum_{\boldsymbol{c} \in \mathcal{H}^n} \pi_{\boldsymbol{c}} \sum_{\boldsymbol{\sigma} \in \mathcal{I}(G)} c_i \sigma_i \pi_{\boldsymbol{\sigma}|\boldsymbol{c}}, \quad \text{for all } i \in V. \quad (9)$$

Taking the summation over $i \in V$ in both sides of the above inequality and using $\boldsymbol{\lambda} = \arg \max\limits_{\boldsymbol{\lambda} \in (1-\varepsilon)\boldsymbol{C}} \sum_i \lambda_i$, it follows that

$$\sum_i \lambda_i = (1-\varepsilon) \sum_{\boldsymbol{c} \in \mathcal{H}^n} \pi_{\boldsymbol{c}} \max_{\boldsymbol{\rho} \in \mathcal{I}(G)} \sum_i c_i \rho_i$$

$$\leq \sum_{\boldsymbol{c} \in \mathcal{H}^n} \pi_{\boldsymbol{c}} \sum_{\boldsymbol{\sigma} \in \mathcal{I}(G)} \pi_{\boldsymbol{\sigma}|\boldsymbol{c}} \sum_i c_i \sigma_i.$$

By rearranging terms in the above inequality, we have

$$\frac{\varepsilon}{1-\varepsilon} \sum_i \lambda_i \geq \sum_{\boldsymbol{c} \in \mathcal{H}^n} \pi_{\boldsymbol{c}} \left(\max_{\boldsymbol{\rho} \in \mathcal{I}(G)} \sum_i c_i \rho_i - \sum_{\boldsymbol{\sigma} \in \mathcal{I}(G)} \pi_{\boldsymbol{\sigma}|\boldsymbol{c}} \sum_i c_i \sigma_i \right)$$

$$= \sum_{\boldsymbol{c} \in \mathcal{H}^n} \pi_{\boldsymbol{c}} \cdot E \left[\max_{\boldsymbol{\rho} \in \mathcal{I}(G)} \sum_i c_i \rho_i - \sum_i c_i \sigma_i \right],$$

where the expectation is taken with respect to random variable $\boldsymbol{\sigma} = [\sigma_i]$ of which distribution is $[\pi_{\boldsymbol{\sigma}|\boldsymbol{c}}]$. Since we know $\max_{\boldsymbol{\rho} \in \mathcal{I}(G)} \sum_i c_i \rho_i - \sum_i c_i \sigma_i \geq 0$ with probability 1, we further have that for all channel state $\boldsymbol{c}$,

$$E \left[\max_{\boldsymbol{\rho} \in \mathcal{I}(G)} \sum_i c_i \rho_i - \sum_i c_i \sigma_i \right] \leq \frac{\sum_i \lambda_i^{\max} \cdot \frac{\varepsilon}{1-\varepsilon}}{\pi_{\boldsymbol{c}}}$$

$$\leq \frac{n \cdot \frac{\varepsilon}{1-\varepsilon}}{\pi_{\boldsymbol{c}}} = n \cdot m^n \cdot \frac{\varepsilon}{1-\varepsilon}.$$

Markov's inequality implies that

$$\Pr \left[\max_{\boldsymbol{\rho} \in \mathcal{I}(G)} \sum_i c_i \rho_i - \sum_i c_i \sigma_i \geq \frac{1}{m} \right] \leq n \cdot m^{n+1} \cdot \frac{\varepsilon}{1-\varepsilon}.$$

If we choose $\varepsilon = \frac{1}{4n \cdot m^{n+1}}$, then

$$\Pr \left[\max_{\boldsymbol{\rho} \in \mathcal{I}(G)} \sum_i c_i \rho_i - \sum_i c_i \sigma_i \geq \frac{1}{m} \right] < \frac{1}{2}. \quad (10)$$

In the star graph G, $\max_{\boldsymbol{\sigma} \in \mathcal{I}(G)} \sigma_i c_i$ is c_1 or $\sum_{j=2}^n c_j$ and under the channel model, if $c_1 \neq \sum_{j=2}^n c_j$, $\left| c_1 - \sum_{j=2}^n c_j \right| \geq \frac{1}{m}$. If $c_1 > \sum_{i=2}^n c_i$, r_i Thus, from (10), if $c_1 > \sum_{i=2}^n c_i, P[\sigma_1 = 1] > \frac{1}{2}$ and if $c_1 < \sum_{i=2}^n c_i$, $P[\sigma_1 = 1] < \frac{1}{2}$, which implies that $r_1 k(c_1) \geq \sum_{i=2}^n r_i k(c_i)$ and $r_1 k(c_1) \leq \sum_{i=2}^n r_i k(c_i)$, respectively. Therefore, for every channel state $\boldsymbol{c}$ with $0 < \sum_{i=2}^n c_i < 1$,

$$r_1 k \left(\sum_{i=2}^n c_i + \frac{1}{m} \right) > \sum_{i=2}^n r_i k(c_i) > r_1 k \left(\sum_{i=2}^n c_i - \frac{1}{m} \right), \quad (11)$$

which implies that $r_i k(x)$ is a strictly increasing function. In addition, $r_1 k(c_1) > 0$ from (10), because $\sum_{\boldsymbol{\sigma} \in \mathcal{I}(G)} \sigma_1 \pi_{\boldsymbol{\sigma}|\boldsymbol{c}} > \pi_{\boldsymbol{0}|\boldsymbol{c}}$ when $c_1 > \sum_{i=2}^n c_i$. Since $k(x)$ is non-negative and $r_i k(x)$ is strict increasing for all link i, $r_i > 0$. Thus, when we devide both sides of (11) by r_i,

$$k \left(\sum_{i=2}^n c_i + \frac{1}{m} \right) > \sum_{i=2}^n \frac{r_i}{r_1} \cdot k(c_i) > k \left(\sum_{i=2}^n c_i - \frac{1}{m} \right). \quad (12)$$

By choosing $x = c_2 = \cdots = c_n$ and taking $m \to \infty$ in (12), it follows that $\lim_{m \to \infty} \sum_{i=2}^n \frac{r_i}{r_1}$ exists,[3] and for any $0 < x < 1/n$,

$$k((n-1)x) = \lim_{m \to \infty} \sum_{i=2}^n \frac{r_i}{r_1} \cdot k(x),$$

where $\lim_{m \to \infty} \sum_{i=2}^n \frac{r_i}{r_1} > 1$ since $k(x)$ is strictly increasing. Hence, if we take $x \to 0$ in the above inequality, $k(0) = 0$ follows. Similarly, by choosing $x = c_2$, $c_3 = \cdots = c_n = 1/m$ and taking $m \to \infty$ in (12), it follows that that for any $0 < x < 1$,

$$k(x) = \lim_{m \to \infty} \frac{r_2}{r_1} \cdot k(x),$$

where we use $k(0) = 0$ and $\limsup_{m \to \infty} \frac{r_i}{r_1} < \infty$ due to the existence of $\lim_{m \to \infty} \sum_{i=2}^n \frac{r_i}{r_1}$. Thus, $\lim_{m \to \infty} \frac{r_2}{r_1} = 1$, and more generally, $\lim_{m \to \infty} \frac{r_i}{r_1} = 1$ using same arguments. Furthermore, by choosing $x = c_2$, $y = c_3$, $c_4 = \cdots = c_n = 1/m$,

---

[3]Recall that $[r_i]$ is a function of m.

and taking $m \to \infty$ in (12), we have that for any $0 < x+y < 1$,

$$k(x + y) = k(x) + k(y),$$

where we use $\lim_{m \to \infty} \frac{r_2}{r_1} = 1$. This implies that $k(x)$ is a linear function (with $k(0) = 0$), and hence the conclusion of Theorem 3.2 follows.

4. DYNAMIC THROUGHPUT OPTIMAL A-CSMA

In the previous section, it is shown that, for any feasible arrival rate, there exists an EXP-A-CSMA algorithm stabilizing the arrivals. In this section, we describe EXP-A-CSMA algorithms which dynamically update its parameters so as to stabilize the network without knowledge of the arrival statistics. More precisely, the CSMA scheduling algorithm uses $f_i^{(t)}$ and $g_i^{(t)}$ to compute the value of parameters $R_i(t) = f_i^{(t)}(c_i(t))$ and $S_i(t) = g_i^{(t)}(c_i(t))$ at time t, respectively, and update them adaptively over time. We present two algorithms to decide $f_i^{(t)}$ and $g_i^{(t)}$. They are building upon prior algorithms in conjunction with the properties of EXP-A-CSMA established in the previous section, referred to as a rate-based (extension of [5]) and queue-based algorithm (extension of [21]).

4.1 Rate-based Algorithm

The first algorithm, at each queue i, updates $(f_i^{(t)}, g_i^{(t)})$ at time instances $L(j), j \in \mathbb{Z}_+$ with $L(0) = 0$, and $(f_i^{(t)}, g_i^{(t)})$ remains fixed between in the time-interval $[L(j), L(j+1))$ for all $j \in \mathbb{Z}_+$, where we define $T(j) = L(j+1) - L(j)$ for $j \geq 0$. With an abuse of notation, $f_i^{(j)}, g_i^{(j)}$ denotes the value of $f_i^{(t)}, g_i^{(t)}$ for $t \in [L(j), L(j+1))$, respectively. To begin with, the algorithm sets $f_i^{(0)}(x) = g_i^{(0)}(x) = 1$ (i.e, $R_i(0) = S_i(0) = 1$) for all i and all $x \in [0, 1]$.

Now we describe how to choose a varying update interval $T(j)$. We select $T(j) = \exp\left(\sqrt{j}\right)$, for $j \geq 1$, and choose a step-size $\alpha(j)$ of the algorithm as $\alpha(j) = \frac{1}{j}$, for $j \geq 1$. Given this, queue i updates f_i and g_i as follows. Let $\hat{\lambda}_i(j), \hat{s}_i(j)$ be empirical arrival and service observed at queue i in $[L(j), L(j+1))$, i.e.,

$$\hat{\lambda}_i(j) = \frac{1}{T(j)} A_i(L(j), L(j+1)) \quad \text{and}$$

$$\hat{s}_i(t) = \frac{1}{T(j)} \left[\int_{L(j)}^{L(j+1)} \sigma_i(t) c_i(t) \, dt \right].$$

Then, the update rule is defined by, for $x \in [0, 1]$

$$g_i^{(j+1)}(x) = \frac{j+2}{j+1} \cdot g_i^{(j)}(x) \cdot \exp\left(x \cdot \alpha(j) \cdot (\hat{s}_i(j) - \hat{\lambda}_i(j))\right)$$

$$f_i^{(j+1)}(x) = j+2, \tag{13}$$

with initial condition $f_i^{(0)}(x) = g_i^{(0)}(x) = 1$. It is easy to check that the A-CSMA algorithm with functions $[f_i^{(j)}]$ and $[g_i^{(j)}]$ lies in EXP-A-CSMA:

$$\log \frac{f_i^{(j)}(x)}{g_i^{(j)}(x)} = r_i(j) \cdot x,$$

where $r_i(0) = 0$ and

$$r_i(j+1) = r_i(j) + \alpha(j) \cdot (\hat{\lambda}_i(j) - \hat{s}_i(j)).$$

Note that, under this update rule, the algorithm at each queue i uses only its local history. Despite this, we establish that this algorithm is rate-stable, as formally stated as follows:

THEOREM 4.1. *For any given graph G, channel transition-rate γ and $\lambda \in \Lambda^\circ(\gamma, G)$, the A-CSMA algorithm with updating functions as per (13) is rate-stable.*

The proof of Theorem 4.1 uses the same strategy in [5] in conjunction with Lemma 3.1 and Lemma 3.2. Due to the space limitation, we present the proof in [27].

4.2 Queue-based Algorithm

Now we describe the second algorithm which chooses $(f_i^{(t)}, g_i^{(t)})$ as a simple function of queue-sizes as follows.

$$f_i^{(t)}(x) = \left(g_i^{(t)}(x)\right)^2 \quad \text{and}$$
$$g_i^{(t)}(x) = \exp\left(x \cdot \max\left\{w(Q_i(\lfloor t \rfloor)), \sqrt{w(Q_{\max}(\lfloor t \rfloor))}\right\}\right), \tag{14}$$

where $Q_{\max}(\lfloor t \rfloor) = \max_j Q_j(\lfloor t \rfloor)$ and $w(x) = \log\log(x + e)$. One can interpret this as an EXP-A-CSMA algorithm since

$$\log \frac{f_i^{(t)}(x)}{g_i^{(t)}(x)} = r_i(t) \cdot x,$$

where $r_i(t) = \max\left\{w(Q_i(\lfloor t \rfloor)), \sqrt{w(Q_{\max}(\lfloor t \rfloor))}\right\}$. The global information of $Q_{\max}(\lfloor t \rfloor)$ can be replaced by its approximate estimation that can computed through a very simple distributed algorithm (with message-passing) in [19] or a learning mechanism (without message-passing) in [22]. This does not alter the rate-stability of the algorithm that is stated in the following theorem.

THEOREM 4.2. *For any given graph G, channel transition-rate γ and $\lambda \in \Lambda^\circ(\gamma, G)$, the A-CSMA algorithm with functions as per (14) is rate-stable.*

Due to the space constraint, we omit the proof of Theorem 4.2 which can be found in [27].

5. ACHIEVABLE RATE REGION OF A-CSMA WITH LIMITED BACKOFF RATE

In practice, it might be hard to have arbitrary large backoff rate because of physical constraints. From this motivation, in this section, we investigate the achievable rate region of A-CSMA algorithms with limited backoff rate. Note that, in the proof of Theorem 3.1, we choose the backoff rates $[f_i]$ to be proportional to the channel varying speed. Thus, when the backoff rate is limited and the channel varying speed grows up, we cannot guarantee the optimality of EXP-A-CSMA. The main result of this section is that, even with highly limited backoff rate, say at most $\delta > 0$, EXP-A-CSMA is guaranteed to have at least α-throughput, where α is *independent* of the channel varying speed and the maximum backoff rate δ. More formally, we obtain the following result.

THEOREM 5.1. *For any $\phi > 0$, interference graph G, channel transition-rate γ and arrival rate $\lambda \in \alpha \Lambda^\circ$, there exists a* rate-stable EXP-A-CSMA algorithm with functions $[f_i]$ and $[g_i]$ such that

$$\max_{i \in V, x \in [0,1]} f_i(x) \leq \phi,$$

where

$$\alpha = \max \left\{ \min_{i \in V} \sum_{\mathbf{c} \in \mathcal{H}^n} c_i \pi_{\mathbf{c}}, \frac{1}{\chi(G)} \right\}. \quad (15)$$

In above, $\chi(G)$ is the chromatic number of G.

Theorem 5.1 implies that even with arbitrary small back-off rates, A-CSMA is guaranteed to achieve a partial fraction of the capacity region. For example, for a bipartite interference graph, at least 50%-throughput can be achieved since its chromatic number is two. The proof strategy is as follows: (i) We first find the achievable rate region of U-CSMA, and (ii) we then show that for any U-CSMA parameters, there exists a EXP-A-CSMA algorithm satisfying the backoff constraint and achieving ε-close departure rate with that by the U-CSMA (we formally state this in Corollary 5.1).

COROLLARY 5.1. *For any $\phi > 0$, interference graph G, channel transition-rate $\boldsymbol{\gamma}$, and U-CSMA parameters, there exists a EXP-A-CSMA algorithm with functions $[f_i]$ and $[g_i]$ such that $\max_{i \in V, x \in [0,1]} f_i(x) \leq \phi$ and*

$$\limsup_{t \to \infty} \left| 1 - \frac{\widehat{D}_i^A(t)}{\widehat{D}_i^U(t)} \right| < \varepsilon, \text{ for all } i \in V,$$

where $\widehat{D}_i^A$ and $\widehat{D}_i^U$ denote the cumulative potential departure processes of the EXP-A-CSMA and the U-CSMA, respectively.

5.1 Proof of Theorem 5.1

The main strategy for the proof of Theorem 5.1 is that we study U-CSMA (channel-unaware CSMA) to achieve the performance guarantee of A-CSMA. We start by stating the following key lemmas about U-CSMA.

LEMMA 5.1. *Let $P_I(G)$ be the independent-set polytope, i.e.,*

$$P_I(G) = \left\{ \boldsymbol{x} \in [0,1]^n : \boldsymbol{x} = \sum_{\boldsymbol{\rho} \in \mathcal{I}(G)} \alpha_{\boldsymbol{\rho}} \boldsymbol{\rho}, \sum_{\boldsymbol{\rho} \in \mathcal{I}(G)} \alpha_{\boldsymbol{\rho}} = 1, \alpha_{\mathbf{0}} > 0 \right\}. \quad (16)$$

Then, for $\boldsymbol{\lambda} \in P_I(G)$, there exists a U-CSMA algorithm with parameters $\boldsymbol{R} = [R_i]$ and $\boldsymbol{S} = [S_i]$ such that

$$\lim_{t \to \infty} \mathbb{E}[\boldsymbol{\sigma}(t)] > \boldsymbol{\lambda}.$$

PROOF. The proof of Lemma 8 in [5] goes through for the proof of Lemma 5.1 in an identical manner. We omit further details. □

LEMMA 5.2. *For any $\phi > 0$, interference graph G, channel transition-rate $\boldsymbol{\gamma}$ and arrival rate $\boldsymbol{\lambda} \in \alpha \boldsymbol{\Lambda}^\circ$, there exists a rate-stable U-CSMA algorithm with parameters $\boldsymbol{R} = [R_i]$ and $\boldsymbol{S} = [S_i]$ such that*

$$\max_i R_i \leq \phi \quad \text{and} \quad \max_i S_i \leq \phi,$$

where α is defined in (15).

PROOF. It suffices to show that there exists a U-CSMA algorithm stabilizing any arrival rate $\boldsymbol{\lambda}$ such that

$$\boldsymbol{\lambda} \in \frac{1}{\chi(G)} \cdot \boldsymbol{\Lambda}^\circ \quad \text{or} \quad \boldsymbol{\lambda} \in \min_{i \in V} \sum_{\mathbf{c} \in \mathcal{H}^n} c_i \pi_{\mathbf{c}} \cdot \boldsymbol{\Lambda}^\circ.$$

First, consider $\boldsymbol{\lambda} \in \frac{1}{\chi(G)} \cdot \boldsymbol{\Lambda}^\circ$. From Lemma 2.1 and the ergodicity of Markov process $\{(\boldsymbol{\sigma}(t)\}$ and $\{\boldsymbol{c}(t)\}$ under U-CSMA, it suffices to prove that there exists a U-CSMA algorithm satisfying

$$\lim_{t \to \infty} \mathbb{E}[\sigma_i(t) c_i(t)] > \lambda_i \qquad \text{for all } i \in V.$$

Since $\chi(G) \cdot \lambda_i < \lim_{t \to \infty} \mathbb{E}[c_i(t)] = \sum_{\mathbf{c} \in \mathcal{H}^n} c_i \pi_{\mathbf{c}}$ (otherwise, $\chi(G) \boldsymbol{\lambda} \notin \boldsymbol{\Lambda}^\circ$), it is enough to prove that for an appropriately defined $\delta > 0$,

$$\lim_{t \to \infty} \mathbb{E}[\sigma_i(t)] > \frac{1}{\chi(G)} - \delta \qquad \text{for all } i \in V.$$

There exists a U-CSMA algorithm with parameter $\boldsymbol{R} = [R_i]$ and $\boldsymbol{S} = [S_i]$ satisfying the above inequality from Lemma 5.1 and $\left[\frac{1}{\chi(G)} - \delta \right] \in P_I(G)$. Furthermore, we can make R_i and S_i arbitrarily small since $\lim_{t \to \infty} \mathbb{E}[\sigma_i(t)]$ under U-CSMA is invariant as long as ratios R_i/S_i remain same.

Now the second case $\boldsymbol{\lambda} \in \min_{i \in V} \sum_{\mathbf{c} \in \mathcal{H}^n} c_i \pi_{\mathbf{c}} \cdot \boldsymbol{\Lambda}^\circ$ can be proved in an similar manner, where we have to prove that there exists a U-CSMA algorithm satisfying

$$\lim_{t \to \infty} \mathbb{E}[\sigma_i(t)] > \rho_i \qquad \text{for all } i \in V,$$

where we define $\boldsymbol{\rho} = [\rho_i]$ as

$$\boldsymbol{\rho} = \frac{1}{\min_{i \in V} \sum_{\mathbf{c} \in \mathcal{H}^n} c_i \pi_{\mathbf{c}}} \cdot \boldsymbol{\lambda} \in \boldsymbol{\Lambda}^\circ \subset P_I(G).$$

This follows from Lemma 5.1 and $\boldsymbol{\rho} \in P_I(G)$. This completes the proof of Lemma 5.2. □

Lemma 5.2 implies that for any arrival rate $\boldsymbol{\lambda} = [\lambda_i] \in \alpha \boldsymbol{\Lambda}^\circ$, there exist $\varepsilon > 0$ and a rate-stable U-CSMA algorithm with arbitrary small parameters $[R_i]$ and $[S_i]$, which stabilize arrival rate $(1 + \varepsilon)\boldsymbol{\lambda}$, i.e.,

$$(1 + \varepsilon)\lambda_i \leq \sum_{\mathbf{c} \in \mathcal{H}^n} c_i \pi_{\mathbf{c}} \sum_{\boldsymbol{\sigma} \in \mathcal{I}(G): \sigma_i = 1} \pi_{\boldsymbol{\sigma}}^*,$$

where $[\pi_{\boldsymbol{\sigma}}^*]$ is the stationary distribution of Markov process $\{\boldsymbol{\sigma}(t)\}$ induced by the U-CSMA algorithm. In particular, given $\phi > 0$, one can assume $\max_i R_i \leq \phi$. For the choice of $[R_i]$ and $[S_i]$, we consider an EXP-A-CSMA algorithm with functions

$$f_i(x) = R_i \quad \text{and} \quad g_i(x) = R_i \exp(-r_i x),$$

where we choose r_i to satisfy

$$S_i = \sum_{\mathbf{c} \in \mathcal{H}^n} \pi_{\mathbf{c}} R_i \exp(-r_i \cdot c_i).$$

Note that r_i satisfying the above equality always exists for given S_i, and

$$\max_{i \in V, x \in [0,1]} f_i(x) = \max_i R_i \leq \phi.$$

Furthermore, one can observe that the maximum value of $f_i(x)$ and $g_i(x)$ for $x \in [0, 1]$ can be made arbitrarily small due to arbitrarily small R_i, S_i. Using this observation and

144

(a) 5-link complete graph

(b) Random topology

(c) Random: A-CSMA vs U-CSMA and Uniqueness

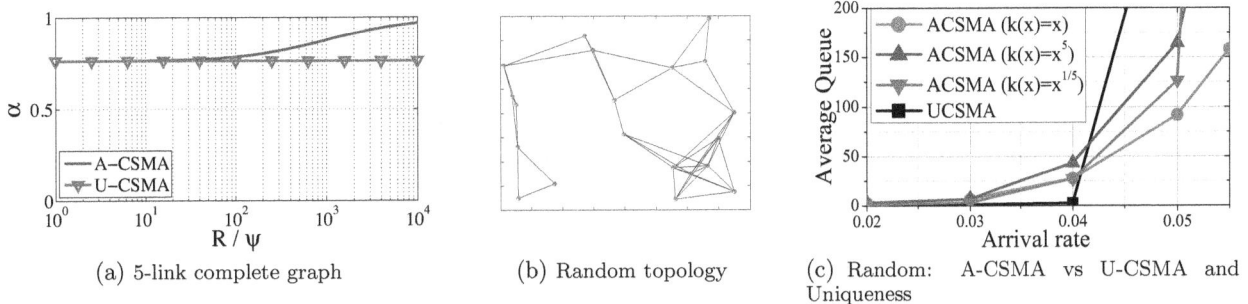

Figure 1: Numerical Results

the Markov chain tree theorem (as we did for the proof of Lemma 3.2), one can show that

$$\max_{(\boldsymbol{\sigma},\mathbf{c})\in\mathcal{I}(G)\times\mathcal{H}^n}\left|1-\frac{\pi_{\boldsymbol{\sigma},\mathbf{c}}}{\pi_{\mathbf{c}}\pi_{\boldsymbol{\sigma}}^*}\right| < \varepsilon,$$

where $[\pi_{\boldsymbol{\sigma},\mathbf{c}}]$ denotes the stationary distribution of Markov process $\{(\boldsymbol{\sigma}(t),\mathbf{c}(t))\}$ by the EXP-A-CSMA algorithm. Therefore, it follows that

$$\begin{aligned}
\lambda_i &\leq \left(1-\frac{\varepsilon}{1+\varepsilon}\right)\sum_{\mathbf{c}\in\mathcal{H}^n} c_i\pi_{\mathbf{c}}\sum_{\boldsymbol{\sigma}\in\mathcal{I}(G):\sigma_i=1}\pi_{\boldsymbol{\sigma}}^* \\
&< \sum_{\mathbf{c}\in\mathcal{H}^n}\sum_{\boldsymbol{\sigma}\in\mathcal{I}(G):\sigma_i=1} c_i\pi_{\boldsymbol{\sigma},\mathbf{c}} \\
&= \lim_{t\to\infty}\frac{1}{t}\widehat{D}_i(t),
\end{aligned}$$

where the last inequality is from the ergodicity of Markov process $\{(\boldsymbol{\sigma}(t),\mathbf{c}(t))\}$. Due to Lemma 2.1, this means that the EXP-A-CSMA algorithm is rate-stable for the arrival rate $\boldsymbol{\lambda}$. This completes the proof of Theorem 5.1.

6. NUMERICAL RESULTS

In this section, we provide several numerical results to demonstrate our analytical findings.

Complete interference graph. We first consider a 5-link complete interference graph, *i.e.*, all 5 links interfere with each other. All queues are homogeneous in terms of time-varying channels, where we assume that the channel space is simply $\{0.5, 1\}$ and the transition-rate $\gamma = \gamma^{0.5\to1} = \gamma^{1\to0.5}$. We compare A-CSMA and U-CSMA, with the following setups:

A-CSMA: $f_i(x) = R$, $g_i(x) = R\cdot 10^{-4x}$
U-CSMA: $f_i(x) = R$, $g_i(x) = R\cdot 10^{-4}$,

so that $\log(f_i/g_i) = 4x$ for A-CSMA and 4 for U-CSMA, respectively. Throughputs of A-CSMA and U-CSMA are evaluated by estimating the average rate in the potential departure process, *i.e.*, $\lim_{t\to\infty}\frac{1}{t}\widehat{D}(t)$. Figure 1(a) shows the results, where in x-axis, we vary the ratio of backoff rate R to channel varying speed ψ (determined by γ) and y-axis represents the fraction of achievable rate region α (note that in a complete interference graph, the rate region is symmetric). We observer that (i) by reflecting the channel capacity in the CSMA parameters as an exponential function, A-CSMA has α-throughput where α approaches 100% (this can be explained by Theorem 3.1), and (ii) U-CSMA

has 76%-throughput. Note that $\alpha \geq 76\%$ even with limited backoff rates (*i.e.*, small R/ψ), and this matches Corollary 5.1 which states that A-CSMA's throughput is at least U-CSMA's throughput.

Random topology. We now study *dynamic* A-CSMA and U-CSMA for a random topology by uniformly locating 20 nodes in a square area and a link between two nodes are established by a given transmission range, as depicted in Figure 1(b). To model interference, we assume the two-hop interference model (*i.e.*, any two links within two hops interfere) as in 802.11. Here, each link has independent and identical channels, where $\mathcal{H} = \{\frac{u}{10} : 1 \leq u \leq 10\}$. For all link i, $\gamma^{u/10\to(u+1)/10} = \gamma^{u/10\to(u-1)/10} = 0.01$, and 0 otherwise.

In Figure 1(c), we increase the arrival rates homogeneously for all links, and plot the average queue lengths to see which arrival rates make the system stable or unstable across the tested algorithms. The average queue length blows up when the algorithm cannot stabilize the given arrival rate. We test *dynamic* A-CSMA and U-CSMA algorithms: the queue-based A-CSMA(x), A-CSMA(x^5), A-CSMA$(x^{1/5})$, and U-CSMA, where for given function $k(x)$, A-CSMA$(k(x))$ denotes the A-CSMA algorithms satisfying (7). Note that if $k(x) = 1$, A-CSMA$(k(x))$ is equal to U-CSMA. The functions $[f_i]$ and $[g_i]$ are defined as stated in Section 4.2 except the channel adaptation function $k(\cdot)$. Figure 1(c) shows that *(a)* A-CSMA(x) stabilize more arrival rates than A-CSMA(x^5) and A-CSMA$(x^{1/5})$, which coincides with our uniqueness result (see Theorem 3.2) and *(b)* dynamic A-CSMA algorithms outperform dynamic U-CSMA when the arrival rate exceeds 0.04, which means that the achievable rate region of A-CSMA includes the achievable rate region of U-CSMA. In the low arrival-rate regime, U-CSMA could be better than A-CSMA in view of delay because the transmission intensity of U-CSMA is always high when the queue is large, while, under A-CSMA algorithm, each link waits until its channel condition being good although the queue length is large.

7. CONCLUSION

Recently, it is shown that CSMA algorithms can achieve throughput (or utility) optimality where 'static' channel is assumed. However, in practice, the channel capacities are typically time-varying. In this paper, we propose A-CSMA which behaves adaptively to channel variations. We first show that the achievable rate region of A-CSMA equals to the maximum rate region under a novel design of the CSMA parameter updating rules as some particular function of instantaneous link capacity. From this result, we also design throughput optimal A-CSMA algorithms. Finally, we also consider a practical case of limited backoff rates. We proved that with any backoff-rate limitation, A-CSMA has the worst-case throughput guarantee without dependence on network topologies, characterized by the achievable throughput of the channel-unaware CSMA which does not adapt to channel variations.

8. REFERENCES

[1] V. Anantharam and P. Tsoucas. A proof of the Markov chain tree theorem. *Statistics & Probability Letters*, 8(2):189 – 192, 1989.

[2] L. Georgiadis, M. J. Neely, and L. Tassiulas. Resource allocation and cross-layer control in wireless networks. *Foundations and Trends in Networking*, 1(1):1–149, 2006.

[3] T. P. Hayes and A. Sinclair. A general lower bound for mixing of single-site dynamics on graphs. *Annals of Applied Probability*, 17(3):931–952, 2007.

[4] L. Jiang, M. Leconte, J. Ni, R. Srikant, and J. Walrand. Fast mixing of parallel Glauber dynamics and low-delay CSMA scheduling. In *Proceedings of Infocom*, 2011.

[5] L. Jiang, D. Shah, J. Shin, and J. Walrand. Distributed random access algorithm: Scheduling and congestion control. *IEEE Transactions on Information Theory*, 56(12):6182 –6207, dec. 2010.

[6] L. Jiang and J. Walrand. A distributed CSMA algorithm for throughput and utility maximization in wireless networks. *IEEE/ACM Transactions on Networking*, 18(3):960 –972, June 2010.

[7] L. Jiang and J. Walrand. *Scheduling and Congestion Control for Wireless and Processing Networks*. Morgan-Claypool, 2010.

[8] L. Jiang and J. Walrand. Approaching throughput-optimality in distributed CSMA scheduling algorithms with collisions. *IEEE/ACM Transactions on Networking*, 19(3):816–829, June 2011.

[9] T. H. Kim, J. Ni, R. Srikant, and N. Vaidya. On the achievable throughput of CSMA under imperfect carrier sensing. In *Proceedings of Infocom*, 2011.

[10] C.-H. Lee, D. Y. Eun, S.-Y. Yun, and Y. Yi. From Glauber dynamics to Metropolis algorithm: Smaller delay in optimal CSMA. In *Proceedings of ISIT*, June 2012.

[11] J. Lee, H. Lee, Y. Yi, S. Chong, B. Nardelli, and M. Chiang. Making 802.11 dcf near-optimal: Design, implementation, and evaluation. In *Proceedings of IEEE Secon*, 2013.

[12] B. Li and A. Eryilmaz. A Fast-CSMA Algorithm for Deadline-Constrained Scheduling over Wireless Fading Channels. *ArXiv e-prints*, Mar. 2012.

[13] J. Liu, Y. Yi, A. Proutiere, M. Chiang, and H. V. Poor. Towards utility-optimal random access without message passing. *Wiley Journal of Wireless Communications and Mobile Computing*, 10(1):115–128, Jan. 2010.

[14] M. Lotfinezhad and P. Marbach. Throughput-optimal random access with order-optimal delay. In *Proceedings of Infocom*, 2011.

[15] B. Nardelli, J. Lee, K. Lee, Y. Yi, S. Chong, E. Knightly, and M. Chiang. Experimental evaluation of optimal CSMA. In *Proc. IEEE INFOCOM*, 2011.

[16] J. Ni, B. Tan, and R. Srikant. Q-CSMA: Queue-length based CSMA/CA algorithms for achieving maximum throughput and low delay in wireless networks. In *Proceedings of Infocom*, 2010.

[17] A. Proutiere, Y. Yi, T. Lan, and M. Chiang. Resource allocation over network dynamics without timescale separation. In *Proceedings of Infocom*, 2010.

[18] D. Qian, D. Zheng, J. Zhang, and N. Shroff. CSMA-based distributed scheduling in multi-hop MIMO networks under SINR model. In *Proceedings of Infocom*, 2010.

[19] S. Rajagopalan, D. Shah, and J. Shin. Network adiabatic theorem: An efficient randomized protocol for contention resolution. In *Proceedings of ACM Sigmetrics*, 2009.

[20] D. Shah and J. Shin. Delay optimal queue-based CSMA. In *Proceedings of ACM Sigmetrics*, 2010.

[21] D. Shah and J. Shin. Randomized scheduling algorithm for queueing networks. *Annals of Applied Probability*, 22:128–171, 2012.

[22] D. Shah, J. Shin, and P. Tetali. Medium access using queues. In *Proceedings of IEEE FOCS*, 2011.

[23] L. Tassiulas. Scheduling and performance limits of networks with constantly changing topology. *IEEE Transactions on Information Theory*, 43(3):1067 –1073, may 1997.

[24] L. Tassiulas and A. Ephremides. Stability properties of constrained queueing systems and scheduling for maximum throughput in multihop radio networks. *IEEE Transactions on Automatic Control*, 37(12):1936–1949, December 1992.

[25] H. S. Wang and N. Moayeri. Finite-state markov channel-a useful model for radio communication channels. *IEEE Transactions on Vehicular Technology*, 44(1):163 –171, feb 1995.

[26] Y. Yi and M. Chiang. *Next-Generation Internet Architectures and Protocols*. Cambridge University Press, 2011. Chapter 9: Stochastic network utility maximization and wireless scheduling.

[27] S.-Y. Yun, J. Shin, and Y. Yi. CSMA over time-varying channels: Optimality, uniqueness and limited backoff rate. *ArXiv e-prints*, 2013.

[28] S.-Y. Yun, J. Shin, and Y. Yi. CSMA using the bethe approximation for utility maximization. In *Proceedings of ISIT*, 2013.

[29] S.-Y. Yun, Y. Yi, J. Shin, and D. Y. Eun. Optimal CSMA: A survey. In *Proceedings of ICCS*, 2012.

ProBeam: A Practical Multicell Beamforming System for OFDMA Small-cell Networks

Jongwon Yoon*
University of Wisconsin,
Madison, WI
yoonj@cs.wisc.edu

Karthikeyan Sundaresan
NEC Labs America Inc.
Princeton, NJ
karthiks@nec-labs.com

Mohammad (Amir)
Khojastepour
NEC Labs America Inc.
Princeton, NJ
amir@nec-labs.com

Sampath Rangarajan
NEC Labs America Inc.
Princeton, NJ
sampath@nec-labs.com

Suman Banerjee
University of Wisconsin,
Madison, WI
suman@cs.wisc.edu

ABSTRACT

Small cells form a critical component of next generation cellular networks, where spatial reuse is the key to higher spectral efficiencies. Interference management in the spatial domain through beamforming allows for increased reuse without having to sacrifice resources in the time or frequency domain. Existing beamforming techniques for spatial reuse, being coupled with client scheduling, face a key limitation in practical realization, especially with OFDMA small cells. In this context, we argue that for a practical spatial reuse system with beamforming, it is important to decouple beamforming from client scheduling. Further, we show that jointly addressing client association with beamforming is critical to maximizing the reuse potential of beamforming.

Towards our goal, we propose *ProBeam* – a practical multi-cell beamforming system for reuse in small cell networks. ProBeam incorporates two key components - a low complexity, highly accurate SINR estimation module that helps determine interference dependencies for beamforming between small cells; and an efficient, low complexity joint client association and beam selection algorithm for the small cells that accounts for scheduling at the small cells without being coupled with it. We have prototyped ProBeam on a WiMAX-based network of four small cells. Our evaluations reveal the accuracy of our SINR estimation module to be within 1 dB, and the reuse gains from joint client association and beamforming to be as high as 115% over baseline approaches.

Categories and Subject Descriptors

C.2.1 [**Network Architecture and Design**]: Wireless Communication

---

*Jongwon Yoon interned in the Mobile Communications and Networking department at NEC Laboratories America Inc, Princeton during this work.

General Terms

Algorithms, Design, Experimentation, Performance

Keywords

Beamforming, OFDMA, WiMAX, Small-cell

1. INTRODUCTION

The proliferation of smartphones and tablet devices has made it necessary for mobile operators to consider new technologies that provide increased network capacity. Small cells (micro and pico cells) provide a promising solution to address this need and are already being deployed for 3G networks, with future rollouts of 4G small cells [1]. With reduced cell sizes and dense deployments, small cells are geared for increased spatial reuse of spectral resources – a valuable and scarce commodity in next generation wireless networks (WiMAX, LTE, LTE-advanced, etc.).

Given the dense deployment of small cells, interference plays a key limiting factor in harnessing their potential. While the sheer scale limits planned deployment of small cells (similar to WiFi), handling interference is a very different problem in small cells compared to WiFi. This can be attributed to their synchronous access mechanism (borrowed from macrocells), coupled with OFDMA (orthogonal frequency division multiplex access) transmissions, wherein multiple users are served in the same frame. Earlier works on interference management in small cell networks [2, 3] employed interference avoidance in the time or frequency domain by allocating orthogonal resources to interfering small cells. In this work, we aim to avoid such sacrifices of spectral resources by exploring interference management for small cells in the spatial domain through beamforming antennas.

Employing beamforming or directional antennas for spatial reuse in a multi-cell set-up has been considered in the context of WiFi [4, 5]. However, such approaches face a key limitation when it comes to practical realization in that a single client is assumed for each AP when computing interference conflicts and determining the spatial reuse schedule. When the client scheduled for an AP changes, the interference conflicts change, requiring a re-computation of the schedule, potentially at the granularity of every packet. This makes it hard to realize such solutions for practically sized WiFi networks and *more so for small-cell networks, where multiple clients are scheduled in each OFDMA frame.* Hence, the goal of this work is to leverage beamforming for spatial reuse across small cells but at the same time decouple it from per-frame scheduling at the small

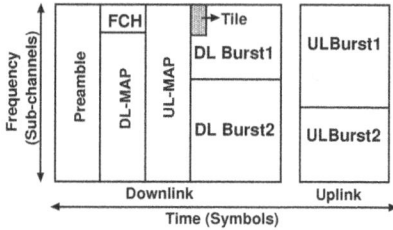

Figure 1: WiMAX frame structure.

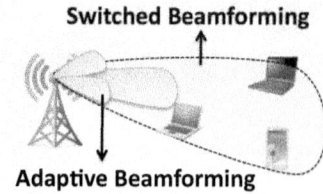

Figure 2: Illustration of beamforming.

cell base station (BS), thereby allowing for beam selections to be computed only at the granularity of seconds (hundreds of frames).

Executing beam selection at coarser time scales compared to client scheduling allows for tangible spatial reuse benefits across cells. However, the beam chosen for a small cell must now deliver good transmission rates to *all* the users that are associated (and hence can be scheduled) with the small cell in order to realize the throughput gains from spatial reuse. Hence, we argue that to realize practical and efficient spatial reuse with small cells, it is important to not just decouple beam selection from scheduling but also integrate beam selection with client association. Towards this goal, we propose *ProBeam* – a practical system that enables joint multi-cell beamforming and client association for increased spatial reuse in small cell networks.

ProBeam incorporates two key components: (i) a *SINR estimation module* – this captures the interference dependencies between small cells in the presence of beamforming. Note that accurate SINR estimation would require measurement *w.r.t* all possible combination of beam choices at small cells, resulting in $O(k^n)$ measurements, where k and n are the number of beam choices and small cells respectively. ProBeam's estimation module indirectly computes SINR from SNR measurements, thereby resulting in only linear number of measurements ($O(kn)$) with an estimation error less than 1 dB for 95% confidence and a maximum error of 1.65 dB. (ii) a *joint beam selection and client association module* – given the hardness of beam selection and client association problems in isolation, their joint problem is significantly challenging to address optimally. ProBeam employs an efficient yet greedy $\frac{1}{2}$-approximation algorithm for client association as a building block to converge to an efficient spatial reuse solution with both beam selection for small cells along with their client associations.

We have implemented ProBeam on a four cell WiMAX-based small cell network. Our experimental evaluations reveal that ProBeam is within 90% of the optimal solution and provides close to 50% throughput gains by addressing the joint problem of client association along with beam selection compared to existing approaches that address only the latter.

Our contributions in this work are multi-fold.

- We propose a low, linear complexity SINR estimation scheme with an error less than 1 dB to generate the interference dependencies needed for computing spatial reuse configurations.
- We establish the hardness of the joint beam selection and client association problem and propose a practical, yet efficient algorithm to address the same.
- We demonstrate the practicality and showcase the benefits of ProBeam by prototyping and evaluating it on a WiMAX-based network of four small cells.

The rest of the paper is organized as follows. Section 2 provides background on OFDMA systems and related work. We motivate the need to couple client association with multi-cell beamforming in Section 3. We describe the algorithm in Section 4 and evaluate the performance of ProBeam using WiMAX testbed in Section 5. Section 6 concludes the paper.

2. BACKGROUND AND RELATED WORK

2.1 WiMAX Preliminaries

OFDMA small cells: Next generation small cell networks for LTE and WiMAX borrow their access mechanism from their macrocell counterparts and are based on OFDMA. Further, they operate on licensed spectrum and follow a synchronous access mechanism (unlike WiFi), wherein frames are transmitted periodically at fixed time intervals (1 ms in LTE, 5 ms in WiMAX). Each OFDMA frame is a two-dimensional template (time and frequency slots) that carries data to *multiple* clients – another key difference compared to WiFi. Transmissions between downlink (DL, BS to client) and uplink (UL, client to BS) are separated either in frequency (FDD) or in time (TDD). Figure 1 shows an example of a WiMAX TDD frame, the underlying structure of which is common to LTE as well. Every frame carries a control and a data part, where the control part (e.g., DL and UL MAPs) provides information to the clients regarding where to pick (place) their respective downlink (uplink) data from the frame and what parameters (MCS - modulation and coding scheme) to use for decoding (encoding) the downlink (uplink) data. Clients use the uplink frame to report their instantaneous CSI (channel state information) to the BS, which in turn is used for diversity scheduling at the BS.

Given the dense deployment of small cells, resource and interference management among small cells happens at the cluster (tens of small cells) granularity, wherein a central entity (SON: self organizing network server [6]) or one of the small cells in the cluster performs centralized resource management for the cells in the cluster and coordination is achieved with the help of a high speed backhaul. While clients use the preamble and control part of the frame to synchronize to the BS, the small cell BSs themselves can synchronize to the macrocell with the help of the SON server or with a GPS antenna module.

Beamforming: Beamforming is one of the core features in next generation networks that is adopted to improve SNR at the intended receivers while decreasing interference at unintended receivers. A beamforming system typically uses multiple antenna elements in an array to form various beam patterns. Beam patterns reinforce transmission energy in desired directions by weighting the signal from the antenna array in both magnitude and phase. Beamforming can be either switched (directional) or adaptive. In switched, a pre-determined set of directional beam patterns covering the azimuth are stored and chosen based on coarse feedback (SNR or RSSI) from the client. In adaptive, fine-grained feedback of channel estimation from the client is used to adapt the beam pattern on the fly to reinforce multipath components and maximize the SINR at the client. By adapting to the instantaneous multipath channel, adaptive provides higher gain (at the cost of increased feedback) compared to switched. However, at the same time, it is more sensitive to channel fluctuations and requires timely feedback to track the channel state - a limiting constraint especially during mobility and in multi-cell resource management.

Both switched and adaptive beamforming co-exist in a complementary manner in cellular systems. Macrocells are sectored in operation (e.g., three $120°$ or six $60°$ directional beams), while adaptive beamforming is enabled to clients within each of the sectors separately. Unlike macrocells, where interference is restricted to cell-edges, thereby allowing for all sectors to operate in tandem, interference is a more pervasive phenomenon in small cells [7]. This requires small cells to select a single sector (switched beam) for operation in a frame (adaptable across frames) so as to avoid interference and maximize reuse among small cells in a dense deployment. Note that adaptive beamforming can still be enabled to clients within the sector of operation at each small cell (see Figure 2 for illustration).

2.2 Related Work

Interference has been shown to be a key performance limiting factor for small cells [7]. This necessitates interference mitigation solutions that incorporate dynamic resource partitioning strategies. There have been studies [8, 9] in this direction but are restricted to theory with several simplifying assumptions that restrict their scope and deployment. Recently [2] and [3] propose centralized and distributed resource management schemes respectively for interference mitigation and demonstrate their efficacy in practice. These solutions allocate orthogonal resources to interfering small cells to avoid interference while reusing resources for the clients that do not incur interference. However, such resource isolation either in time or frequency comes at the cost of sacrificing resources, which in turn can be avoided by addressing interference in the spatial domain through beamforming.

In the space of beamforming, [4, 5] propose to increase the capacity of WLANs through spatial reuse by considering directional antennas only at the APs or at both APs and clients. However, client association is assumed and conflicts and reuse schedule are computed *w.r.t* a single client at each AP. This limits the practical applicability of such solutions (especially for OFDMA systems) since conflicts and reuse schedules have to be recomputed (potentially every packet) every time the client scheduled with any of the AP changes. Several theoretical works [10, 11] have looked at adaptive beamforming in a multi-cell context. However, idealized settings are assumed that require fine grained CSI from all transmitters to all clients be made available to the reuse algorithm at every frame interval. Given the practical feasibility (or lack thereof) of such approaches, experimental works [12] have appropriately focused on adaptive beamforming for SNR improvements within a single cell. Further, none of these works address client association jointly with beamforming.

The focus of our work is to design a *practical* multi-cell spatial reuse system that, decouples client scheduling from beamforming, employs switched beamforming for interference management between small cells, and jointly addresses client association to increase the potential of spatial reuse from beamforming. Being complementary, adaptive beamforming can still be leveraged for SNR improvement within each small cell (although not considered in this work).

3. MOTIVATION

We now motivate the need to couple client association with multi-

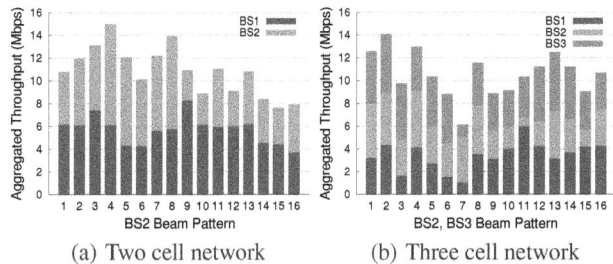

(a) Two cell network   (b) Three cell network

Figure 3: Motivation for coordinated beam selection.

cell beamforming in order to maximize the benefits of spatial reuse. We present results from an experimental WiMAX-based network of four small cells, each equipped with an eight element phased array antenna (details in Section 5) to substantiate our claims.

3.1 Need for Coordinated Beamforming

Beamforming in a multi-cell context has two benefits: (i) increase link capacity through improved SNR, and (ii) increase network capacity through reduced interference (higher SINR) and hence higher spatial reuse. The beam choice of one cell impacts the interference seen by the clients of another cell, thereby requiring a coordinated approach to beam selection across small cells for maximum reuse benefits. However, given the simplicity of un-coordinated, per-cell beamforming (focusing only on SNR), it is important to understand the benefits from coordination and hence the need for it.

We construct a topology with two small cells, each with one scheduled client. First BS1 cycles through all its sixteen beam patterns to determine the one yielding the best rate to its client (C1) in isolation. BS1 is then fixed to use its best beam to C1. Now, in the presence of BS1, BS2 is made to transfer data to its client (C2) on each of its 16 patterns sequentially. We plot the throughput observed at C1 (blue bars) and C2 (grey bars) as a function of the beam pattern used by BS2 in Figure 3(a). Two observations can be made: (i) The interference projected by BS2 on C1 depends tightly on the beam chosen by BS2. C1 achieves its highest throughput (8.3 Mbps) when BS2 employs its 9th pattern and its lowest throughput (3.7 Mbps) when BS2 employs its 16th pattern. (ii) The beam maximizing the throughput of one cell does not necessarily maximize the multi-cell network throughput. While the 9th beam pattern maximizes C1's throughput, it is the 4th pattern that maximizes the aggregate network throughput. A similar behavior is also evident in the three cell experiment presented in Figure 3(b), where the pattern (11th) maximizing throughput for C1 differs from the one (2nd) maximizing the aggregate network throughput. The throughput gain of employing the 2nd pattern over the 11th one is almost 40%.

Thus, *a well-coordinated beamforming algorithm across the small cells is indeed important to maximize the aggregate network throughput.*

3.2 Need for Joint Client Association

Client association has been traditionally employed to load balance clients between multiple cells so as to effectively utilize the capacity of each cell and network as a whole. However, in the context of multi-cell beamforming, client association has a bigger role to play. Note that, unlike in WiFi systems, where a single client is served by a cell at a time, OFDMA systems multiplex multiple clients in the same frame (diversity scheduling). This requires that

(a) SNR based association (C2 is associated with BS1)

(b) Flexible association (C2 is associated with BS2)

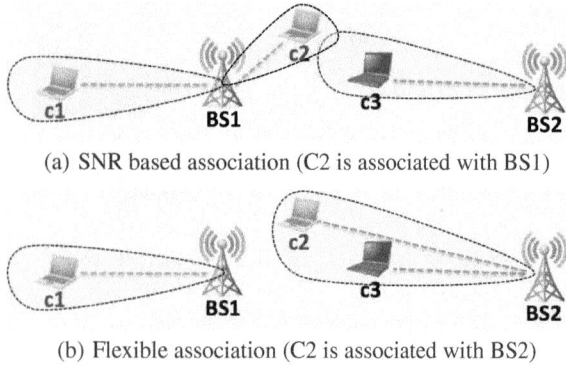

Figure 4: Illustration for flexible client association.

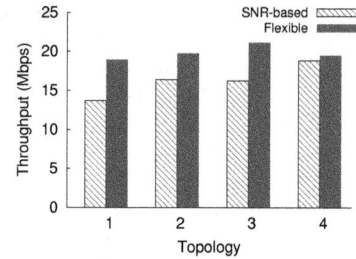

Figure 5: Joint association increases throughput by 40% comparing to decoupled (SNR based) association.

the beam selected for the small cell cater effectively to all its associated and scheduled clients. Further, since the beam choice for a cell impacts the interference and hence performance seen by other cells, this naturally results in client association being closely coupled with multi-cell beamforming.

To see this, consider the following illustration in Figure 4. In conventional association, where SNR is used as a metric for client association, clients C1 and C2 will be associated to BS1, while C3 will be associated to BS2 based on (high) SNR and completely decoupled from beamforming. BSs will then determine the best beams to communicate with their respective clients. Let $b1$ and $b2$ be the only beams on which C2 and C3 can receive good signal strength from their respective BSs. Now, when BS1 is employing beam $b1$ to communicate with C2, this will receive interference from the beam $b2$ used by BS2 to communicate with its client C3. By fixing the client association, depending on the location of associated clients, the ability of beamforming to effectively suppress interference between cells is potentially limited. In contrast, by allowing for flexible association (Figure 4(b)), C2 can be associated with BS2 even though it has a lower SNR to BS2. This would allow BS2 to schedule C2 and C3 jointly on a beam that suffers no interference from that employed by BS1, thereby resulting in a potentially higher SINR for all clients.

To quantify the benefits of coupling client association with beam selection for small cells, we conduct the following experiment with two small cells and three clients, and generate multiple topologies by varying the client locations. We consider two association strategies: *decoupled association*, where the best coordinated beam (for maximum aggregate throughput) for each small cell is selected after client association is done based on SNR; *joint association*, where the client association yielding the highest aggregate throughput is computed among all beam combinations between the two cells. The aggregate throughput results between these two strategies in Figure 5 indicates that joint association can yield gains as high as 40%, with an average gain of about 25%.

This in turn motivates the *need to jointly address client association with beam selection for small cells, whereby client association can be effectively used to maximize the spatial reuse potential of beamforming.*

4. DESIGN

System overview: Small cell networks can be deployed for enterprises as well as outdoors. A central controller (separate entity or one of the small cells) is designated to perform resource and interference management for a cluster (tens) of small cells jointly with a high speed backhaul available for information exchange between them. We expect ProBeam to reside in this central controller (CC). Note that while our primary focus is small cell networks, our system is equally applicable to WiFi networks as well.

ProBeam's spatial reuse solution operates in epochs, which spans several seconds (hundreds of frames). In each epoch, the sequence of operations are as follows. (i) *Interference estimation for beamforming*: The clients measure the average SNR on each of the beams from each of the BSs and forward it to the CC, which then infers their corresponding SINR for various beam combinations at the small cell BSs (details in subsection 4.1). (ii) *Joint beam selection and client association*: Based on the interference information collected, the CC runs its spatial reuse algorithm (for a desired objective) to determine the beam choice for each of the small cells as well as the clients that are associated with it for that epoch (details in subsection 4.2). (iii) Scheduling: Once each small cell BS receives its beam choice and client set, it begins scheduling its clients locally using its own scheduler (proportional fair, max-min fair, etc.) for each frame in the epoch, while applying the beam selected to the frame transmissions (details in subsection 4.3).

4.1 Interference Estimation for Beamforming

Estimating the interference at clients accurately is critical for the efficient operation of ProBeam.

Reducing complexity: Measuring the SINR directly at the clients for various beam configurations (interference) used by small cells is the most accurate approach. However, this would entail that each small cell cycle through each beam pattern, while keeping the beam patterns at other cells fixed and measuring the resulting SINR at all clients. This would however result in a total of $O(k^n)$ measurements, where k in the number of beam patterns and n is the number of small cells. ProBeam measures only the client SNR from each of the small cells in isolation for the various beam choices and then uses this information to estimate the projected client SINR for a given beam configuration at the small cells. By allowing the small cells to operate in isolation during measurements, this significantly reduces the SINR estimation complexity to $O(kn)$. The key question remaining is the accuracy or lack thereof of such an estimation procedure.

Note that SINR can be expressed as $SINR_{ij} = \frac{SNR_{ij}}{\sum_{k \neq i} INR_{kj} + 1}$, where SINR at client j from BS i is related to its SNR and net interference to noise ratio from other BSs ($INR_j = \sum_{k \neq i} INR_{kj}$). Small cells being interference limited, $INR + 1 \approx INR$. In the logarithmic (dB) domain, the relation can be expressed as SINR (dB) = SNR (dB) - INR (dB). Hence, in principle, the SINR at a client can be estimated from its SNR from the desired BS and its aggregate INR from all interfering BSs. For this to be possible, one needs to estimate each INR_{kj}, which can potentially be approximated as the client SNR when associated with the interfering BS in

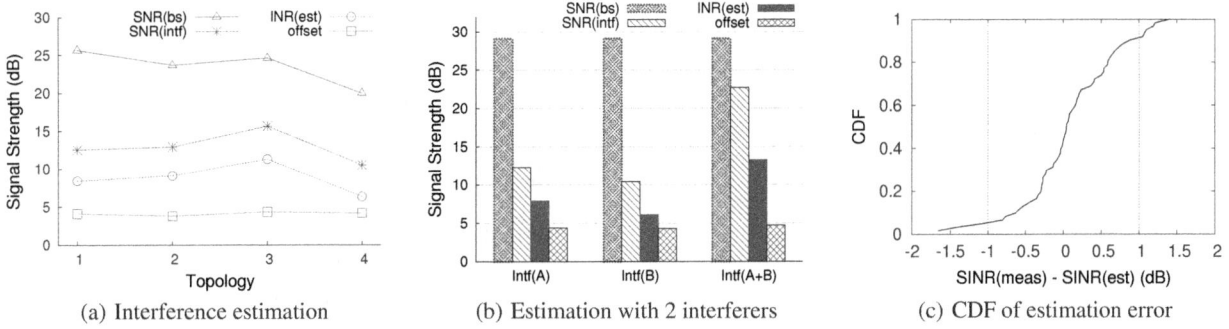

(a) Interference estimation   (b) Estimation with 2 interferers   (c) CDF of estimation error

Figure 6: Accurate estimation of SINR from individual SNR estimates.

isolation (i.e., SNR_{kj}). However, in reality, the accuracy of such an estimation may depend on multiple factors such as quantization, offsets, estimation error, etc.

To verify this approximation, we conducted the following experiment with two small cell BSs, whose beam choices are such that they interfere with the client under consideration. The results are presented in Figure 6(a). First, the client measures the signal strength from the associated BS in the absence (SNR_{BS}) and presence ($SINR_{BS}$) of interference respectively, from which we estimate $INR_{est} = SNR_{BS} - SINR_{BS}$. Then, the client records the signal strength SNR_{intf} after associating with the interfering BS in isolation. Comparing SNR_{intf} with INR_{est} in Figure 6(a), we see that there is a consistent 4 dB offset between the estimated interference and its corresponding signal strength and this remains fixed regardless of the topology and client SNR considered (SNR_{BS} range 19 to 26 dB). We attribute this constant 4 dB difference to the inherent offset β introduced (during client feedback) by the MAC and its quantization of the signal strength value reported from the PHY layer. β being platform dependent, can be calibrated by the client and fed back to the Central Controller for its appropriate estimation of INR. Further, note that when SINR is directly measured, there is only one feedback value from the client. However, when SINR is estimated from SNR and multiple INRs, then each of the SNR feedback (corresponding to INR) introduces an offset that needs to be compensated. When appropriately compensated, the resulting estimation reduces to $SINR_{ij}$ (dB) $= SNR_{ij}$ (dB) $- 10 \log_{10}(\sum_{k \neq i} SNR_{kj}) + \beta$. Note that since interference is aggregated in absolute units, the offset for the aggregate interference remains to be β in dB. This is observed in Figure 6(b), where the offset in the presence of one (A or B) and two (A+B) small cell interferers remains to be the same 4 dB. Thus, with the help of isolated measurements from the small cells, it is indeed possible to estimate SINR, thereby resulting in a linear (in n) complexity of only $O(kn)$.

SINR estimation procedure: ProBeam initiates a measurement phase at the beginning of each epoch, where it operates each small cell BS in the cluster one after another in isolation. When activated, BS i applies its k beam patterns sequentially, each lasting ten frames. All the clients measure the average received SNR from BS i corresponding to beam pattern k. A client j forwards SNR_{ijk}, i.e., measured SNR from BS i with beam pattern k to the CC in ProBeam through its current associated BS. In WiMAX and LTE, clients automatically send Channel State Information (CSI) to BS periodically via dedicated uplink channel resources in every frame. We use such standard feature for obtaining our desired SNR measurements. Once ProBeam gathers SNR measurements from all the clients, then any desired SINR (in dB) for a given beam con-

figuration ($\pi = \{\pi(1)\}$, $\forall i$, beam choices for small cells) can be estimated as,

$$SINR_{ij\pi}(dB) = SNR_{ij\pi(i)}(dB) - 10 \log_{10}(\sum_{k \neq i} SNR_{kj\pi(k)}) + \beta(dB)$$
(1)

Note that SNR measurements can be done within $kn \times 10$ frames. For reasonable values of k (say 10 beams) and n (say 10 cells in a cluster), this would amount to 1 sec in LTE (for 1 ms frames). Also actual data is transmitted during the measurement phase, therefore we do not waste resources for SNR measurements. However, reuse cannot be leveraged, whose overhead (reuse loss) can be amortized as long as the epoch duration is several seconds.

Validation: To validate our estimation procedure, we conduct the following experiments with three small cell BSs and a single client. First, the client measures the SNRs from all three BSs for a given beam configuration in isolation and records them. Then, we make the client associate with one of the BS and measure the SINR in the presence of the other two BSs projecting interference. The beam configuration is chosen so as to project interference to the client under consideration. We repeat the above experiment by changing the beam configuration as well as the topology (i.e., client locations) to obtain confidence in results. Measurements are taken at different client locations to generate plurality of interference scenarios and to also emulate different clients (varying BS deployment is considered in Section 5). We obtain over 100 sets of measurements and present the CDF of the SINR estimation error ($SINR_{meas} - SINR_{est}$) in Figure 6(c). As we can see, 95% of our SINR estimates have less than 1 dB error ($\leq$ 5%), with the highest estimation error being only about 1.65 dB. *Our results clearly indicate the high accuracy of our SINR estimation method, thereby avoiding the complexity of obtaining measurements from all possible combinations of beam patterns at small cells.*

4.2 Joint Client Association and Beam Selection (CABS)

Similar to other resource management problems, we can formulate our problem as a utility maximization problem in every epoch.

$$\text{Maximize} \sum_{j \in \mathcal{K}} U(t_j)$$

where t_j represents the average throughput received by client j in the epoch and $U()$ is a function to capture the corresponding utility. Note that the choice of the utility function determines the fairness policy in the system. We assume utility functions to be concave and non-decreasing. This captures proportional fairness (defined by using the utility function $U(t_j) = \log(t_j)$) that is popular in the standards (WiMAX, LTE). While we need to decouple the time scales of operation for CABS from scheduling, it must be noted that the eventual objective is related to throughput and hence dependent on scheduling. Hence, to allow the decoupling, throughput needs to be modeled as the average throughput received by the client over the epoch for a given scheduling policy. Our problem can be formulated as,

$$(\pi^*, \mathbf{x}^*) = \arg \max_{\pi, \mathbf{x}} \sum_{j \in \mathcal{K}} \sum_{i \in \mathcal{S}} x_{ji} U(t_{ji}^\pi) \text{ s.t. } \sum_{i \in \mathcal{S}} x_{ji} \leq 1, \ \forall j \in \mathcal{K}$$

(2)

where $\mathcal{K}$ and $\mathcal{S}$ represents the set of clients and small cell BSs respectively. Further, $\pi = \{\pi_i, \ \forall i\}$ denotes the beam selection vector for all BSs, while $\mathbf{x} = \{x_{ji}, \ \forall j, i\}$ denotes the association vector for all clients ($x_{ji} = 1$ if client j is associated with BS i and 0 otherwise). t_{ji}^π indicates the client j's average throughput when associated with BS i under beam configuration π and depends on the SINR ($SINR_{ij\pi}$) seen by the client from BS i in the presence of interference from other BSs under the beam configuration π (see Eq. (1)).

4.2.1 A note on fairness

While fairness (starvation) among clients is typically achieved (avoided) over a longer time period, instantaneous per-frame decisions may favor clients with good channel conditions (e.g., proportional fairness). In the case of CABS, decisions are made at the granularity of epochs. Hence, if fairness is ensured over much longer time scales ($\gg$ epoch), then several clients could be subject to starvation in an epoch (several seconds). This would increase the jitter perceived by such clients – a factor critical for real-time media and is hence not desired. Thus, it is more appropriate to ensure fairness within each epoch. This would allow all clients to be scheduled in every epoch. On the other hand, since beam selection decisions are fixed for the entire epoch, accommodating all clients could potentially limit the amount of reuse that can be leveraged in the epoch. Hence, to strike a balance between throughput performance (reuse) and fairness, an alternative is to restrict the utility functions to be non-negative in addition to concave and non-decreasing. This would account for fairness, while at the same time allowing for a small number of clients to be removed from scheduling in an epoch. By weighting the client utility functions inversely proportional to their throughput received (T_j) thus far, one can avoid starvation for all clients across epochs.

In the case of proportional fairness, we can modify the utility function as $U(t_j) = w_j \log(t_j)$; if $t_j > 0$ and 0 otherwise, where $w_j \propto \frac{1}{T_j}$. Further, T_j at current epoch e is updated through an exponentially weighted moving average as $T_j(e) = (1 - \frac{1}{\alpha})T_j(e-1) + (\frac{1}{\alpha})t_j(e)$, where α is the filtering coefficient. Let r_{ji}^π be the average transmission rates (MCS) seen by client j in a slot when associated with BS i under beam configuration π, and N be the total number of time-frequency slots in an OFDMA frame with M frames per epoch. Then, under proportional fairness, it can be easily shown that the number of slots are allocated among all the scheduled clients in the proportion of their weights (equal when $w_j = 1, \ \forall j$). This would in turn result in an average client throughput of $t_{ji}^\pi = \frac{NMw_j r_{ji}^\pi}{\sum_{k \in \mathcal{K}} x_{ki} w_k}$.

4.2.2 Hardness

For a given client association, the problem of beam selection is itself NP-hard [4, 5]. Hence, it comes as no surprise that our joint CABS problem is NP-hard as well. From the perspective of designing algorithms, it helps to understand if beam selection is the

only source of hardness or does client association also contribute to the hardness. In this regard, we have the following result.

THEOREM 1. *For a given beam selection, the CABS problem remains to be NP-hard.*

In the interest of space, we defer the proof to [13].

Algorithm 1 CABS Algorithm

1: INPUT: average SNR ρ_{ji}^b, $\forall i \in \mathcal{S}, j \in \mathcal{K}, b \in \mathcal{B}$
2: OUTPUT: Beam selection $\pi(i)$ and client association $\mathcal{A}_i, \forall i \in \mathcal{S}$
3: Initialization of beam choices, i.e., $\pi(i), \forall i$
4: **for** $i \in [1 : |\mathcal{S}|], b \in [||\mathcal{B}||$ **do**
5:    $\mathcal{L} = \emptyset, u_{ib} = 0$
6:    **while** 1 **do**
7:       $j^* = \arg \max_{j \in \mathcal{K} \setminus \mathcal{L}} \sum_{k \in \mathcal{L} \cup j} U(t_{ki}^b) - u_{ib}$
8:       **if** $j^* = \emptyset$ **then** break
9:       $\mathcal{L} \leftarrow \mathcal{L} \cup j^*; u_{ib} = \sum_{k \in \mathcal{L}} U(t_{ki}^b)$
10:    **end while**
11: **end for**
12: $\pi(i) = \arg \max_b u_{ib}, \forall i$
13:
14: **for** $i \in [1 : |\mathcal{S}|]$ **do**
15:    **for** $b \in [1 : |\mathcal{B}|]$ **do**
16:       % Solve client association by varying only one beam element at a time
17:       $\pi(i) = b, A_i = \emptyset, \forall i$
18:       $(i^*, j^*) = \arg \max_{(i,j) s.t. \ j \notin \cup_i \mathcal{A}_i} \{\sum_{k \in \mathcal{A}_i \cup j} U(t_{ki}^\pi) - \sum_{k \in \mathcal{A}_i} U(t_{ki}^\pi)\}$
19:       $\mathcal{A}_{i^*} \leftarrow \mathcal{A}_{i^*} \cup j^*; u_{ib}^\pi = \sum_i \sum_{j \in \mathcal{A}_i} U(t_{ji}^\pi)$
20:    **end for**
21:    $\pi(i) = \arg \max_b u_{ib}^\pi$
22: **end for**

4.2.3 Algorithm

Since both components of our CABS problem are hard, we must carefully choose the interaction between these components in our solution. Unlike the beam selection problem, the client association problem, although hard, can be solved more efficiently. Hence, ProBeam proposes and employs a simple but efficient client association algorithm as the core building block for solving the CABS problem. At a high level, it solves the client association problem for a given beam configuration and the resulting utility is used to manipulate the beam configuration of small cells in an iterative manner till an efficient CABS solution is attained. The algorithm is given in Algorithm CABS.

The input to the algorithm is the average client SNR (ρ_{ji}^b) for the epoch with respect to its neighboring small cells when they employ different beams ($b \in \mathcal{B}$) in isolation (step 1). Using the approach in Section 4.1, the CC can then determine the average client rates in the presence (r_{ji}^b) and absence (r_{ji}^b) of interference. The CC first determines a bootstrap beam configuration for the small cells as follows (steps 3-12). For each of the small cells, it determines the beam that yields the highest utility in the absence of interference, assuming all active clients can be potentially associated with it, i.e., $\pi(i) = \arg \max_{b \in \mathcal{B}} \{\sum_{j \in \mathcal{K}} x_{ji} U(t_{ji}^b)\}$. Note that t_{ji}^b depends on the scheduling policy and is hence coupled with the set of clients associated with the small cell. For example, in proportional fairness, $t_{ji}^b = \frac{NMw_j r_{ji}^b}{\sum_{k \in \mathcal{K}} x_{ki} w_k}$. Hence, even to determine a beam initialization $\pi(i)$, one needs to determine the set of clients (x_{ji}) that maximize the utility for the given beam in the absence of interference.[1] This can be done optimally (easy to verify) by adding users one by one such that incremental utility is maximized (steps 6-10). Specifically, for proportional fairness, the incremental utility (step 7) would correspond to,

$$j^* = \arg \max_{j \in \mathcal{K} \setminus L} \sum_{k \in \mathcal{L} \cup j} w_k \log(\frac{NMr_{ki}^b}{1 + |\mathcal{L}|}) - \sum_{k \in \mathcal{L}} w_k \log(\frac{NMr_{ki}^b}{|\mathcal{L}|})$$

After the beam initialization, CABS algorithm perturbs the beam choice for each of the small cells, one by one and one beam at a time. For each of the beam choices at a given cell ($\pi(i) = b$), CABS retains the rest of the beam choices for the other cells unchanged and solves the client association problem for all the small cells jointly under the updated beam configuration to determine the new utility (steps 16-19). CABS then fixes the beam choice for the small cell as the one that yields the highest utility among all its choices (step 21). The same process is repeated for updating the beam choice for each of the small cells sequentially (steps 14-22). Note that, although after one complete round of beam updates for each of the small cells (along with joint client re-association), we cannot guarantee convergence to the optimal solution, our evaluations in Section 5 reveal this is sufficient to obtain a performance very close to that of exhaustive search for beam configurations. CABS runs in $O(|\mathcal{K}|^2 |\mathcal{S}|^2 |\mathcal{B}|)$, with a large portion of the complexity coming from the client association module $O(|\mathcal{K}|^2 |\mathcal{S}|)$.

4.2.4 Performance Guarantee

Given the hardness of the joint CABS problem, it is hard to establish an approximation guarantee for the entire algorithm. However, we can establish the following performance guarantee for the core building block in CABS, namely the client association part when the popular proportional fair scheduling policy is considered at the small cells.

THEOREM 2. *CABS is a $\frac{1}{2}$-approximation algorithm under proportional fairness when beam configuration is given.*

We provide some definitions on matroid and sub-modularity that are relevant for the proof.

Partition Matroid: Consider a ground set Ψ and let S be a set of subsets of Ψ. S is a matroid if, (i) $\emptyset \in S$, (ii) If $P \in S$ and $Q \subseteq P$, then $Q \in S$, and (iii) If $P, Q \in S$ and $|P| > |Q|$, there exists an element $x \in P \setminus Q$, such that $Q \cup \{x\} \in S$. A partition matroid is a special case of a matroid, wherein there exists a partition of Ψ into components, $\phi_1, \phi_2, \ldots$ such that $P \in S$ if and only if $|P \cap \phi_i| \leq 1, \forall i$.

Sub-modular function: A function $f(\cdot)$ on S is said to be sub-modular and non-decreasing if $\forall x, P, Q$ such that $P \cup \{x\} \in S$ and $Q \subseteq P$ then,

$$f(P \cup \{x\}) - f(P) \leq f(Q \cup \{x\}) - f(Q)$$
$$f(P \cup \{x\}) - f(P) \geq 0, \quad \text{and } f(\emptyset) = 0$$

PROOF. The sub-optimality of maximizing a sub-modular function over a partition matroid using a greedy algorithm of the form $x = \arg \max_{x \in \phi_i} f(P \cup \{x\}) - f(P)$ in every iteration was shown to be bounded by $\frac{1}{2}$ in [14]. We will now show that CABS is such an algorithm (step 18 being the key step), with our client association objective for a given beam configuration (π) corresponding to a sub-modular function to obtain the desired result.

[1]Note that accommodating all users can hurt the utility due to fixed frame resources but varying client rates.

Consider the ground set to be composed of the following tuples.

$$\Psi = \{(i, j) : i \in [1 : |\mathcal{S}|] \cup \emptyset, j \in [1 : |\mathcal{K}|]\}$$

Now Ψ can be partitioned into $\phi_j = \{(i, j) : i \in [1 : |\mathcal{S}|] \cup \emptyset\}, \forall j$. $i = \emptyset$ allows for the possibility of clients not being scheduled in an epoch. Let R be defined on Ψ as a set of subsets of Ψ such that for all subsets $P \in R$, we have (i) if $Q \subseteq P$, then $Q \in R$; (ii) if element $x \in P \setminus Q$, then $Q \cup \{x\} \in R$; and (iii) $|P \cap \phi_j| \leq 1, \forall j$. This means that R is a partition matroid. Now, it is easy to see that any $P \in R$ will provide a feasible schedule with at most one feasible association to a small cell for each client ($|P \cap \phi_j| \leq 1, \forall j$), thereby allowing the partition matroid R to capture our client association problem. Since each client can associate to only one small cell, our client association objective can be given as,

$$f(P) = \sum_{i \in \mathcal{K}} \mu_i(P)$$
$$\text{where,} \quad \mu_i(P) = \sum_{j:(i,j) \in P} w_j \log(\frac{NMw_j r_{ij}^\pi}{\sum_{k:(i,k) \in P} w_k})$$

It can be seen that if $Q \subseteq P$, then $\mu_i(Q) \leq \mu_i(P)$ since the algorithm picks only elements that result in positive incremental utility. Hence, it only remains to be shown that for an element (i, ℓ) such that $P \cup \{(i, \ell)\}$ forms a valid schedule, then $f(P \cup \{(i, \ell)\}) - f(P) \leq f(Q \cup \{(i, \ell)\}) - f(Q)$. Now, define incremental utility $\Delta_P(i, \ell) = f(P \cup \{(i, \ell)\}) - f(P)$ and similarly define $\Delta_Q(i, \ell)$. Applying the objective function and simplifying, we can show that,

$$\Delta_P(i, \ell) = w_\ell \log(NMw_\ell r_{i\ell}^\pi) - w_\ell \log(w_\ell + \sum_{k:(i,k) \in P} w_k)$$
$$- \sum_{j:(i,j) \in P} w_j \log(\frac{w_\ell + \sum_{k:(i,k) \in P} w_k}{\sum_{k:(i,k) \in P} w_k})$$
$$\Delta_Q(i, \ell) = w_\ell \log(NMw_\ell r_{i\ell}^\pi) - w_\ell \log(w_\ell + \sum_{k:(i,k) \in Q} w_k)$$
$$- \sum_{j:(i,j) \in Q} w_j \log(\frac{w_\ell + \sum_{k:(i,k) \in Q} w_k}{\sum_{k:(i,k) \in Q} w_k})$$

Thus, the difference between $\Delta_P(i, \ell)$ and $\Delta_Q(i, \ell)$ arises in the second (reduction) term, which increases with the number of elements in the allocation thus far. Since $Q \subseteq P$, the reduction term is more for P than for Q, resulting in $\Delta_P(i, \ell) \leq \Delta_Q(i, \ell)$. This establishes that the function $f(P)$ is indeed sub-modular. Further, our client association problem aims to maximize this non-decreasing sub-modular function over a partition matroid. Hence, picking the (client, small cell) pair yielding the highest marginal utility for a given beam configuration in CABS (steps 16-19) would correspond to determining

$$(i^*, j^*) = \arg \max_{(i,j) \in R} \{f(P \cup \{(i, j)\}) - f(P)\}$$

Thus, the sub-optimality of $\frac{1}{2}$ would then follow from the result in [15]. $\square$

4.3 Scheduling

Once the CC determines the beam configuration and client association for the epoch, the appropriate beam and allowable client set are notified to each of the small cell BSs for configuration. Each small cell BS then locally runs its scheduling algorithm (e.g., proportional fair) among the associated clients for each frame in the epoch, while employing the chosen beam for its transmissions. Further, instantaneous channel rate feedback from clients is used in per-frame scheduling for leveraging multi-user diversity.

4.4 Practical Considerations

Mobile clients: While beamforming algorithms work well for static clients, it is important to understand their limitations with respect to mobile clients. Note that, any adaptive beamforming scheme that relies on fine grained channel state information (CSI) will be highly sensitive to lack of timely and accurate CSI, both of which are hard to obtain during mobility. On the other hand, switched beamforming relies only on coarse grained channel feedback (SNR or RSSI) and hence is less sensitive to mobility. As long as the epoch duration is not long enough (several seconds is reasonable), pedestrian to moderate vehicular speeds can be accommodated without warranting a completely new beam to be employed for the client.

Epoch duration: Keeping the epoch duration long is conducive for implementation and overhead. However, it must also be capable of tracking traffic dynamics and client mobility. Allowing for a few seconds of epoch duration strikes a good balance between these objectives.

5. SYSTEM EVALUATION

Testbed and prototype implementation: Our WiMAX testbed consists of four small cells (deployed in an indoor enterprise environment), clients and a central controller as depicted in Figure 7. The small cell BS is a PicoChip [16] WiMAX platform based on IEEE 802.16e standard [17]. The BS is tuned to operate in a 10 MHz bandwidth with the center carrier frequency of 2.59 GHz, for which we have obtained an experimental license to transmit WiMAX signals over the air. In the absence of a macro cell to coordinate with, we use a GPS module to synchronize the WiMAX frame transmissions across the small cells. Each BS has an eight element (analog) phased array antenna [18] connected to its RF port. The antenna array generates sixteen overlapping beam patterns of $45°$ each, spaced $22.5°$ apart to cover the entire azimuth of $360°$. The BS controls the antenna array through a serial port application that we have developed in C. There is a delay of one frame (5 msec) before a particular beam pattern is actually applied by the antenna following the command from the application. This is not an issue given the time scale of epoch or the measurement phase.

ProBeam is standards compatible and works with commercial off-the-shelf clients. We use Windows laptops with a WiMAX interface [19] and omni-directional antennas as our clients. Investigating directionality at the clients is part of our future work. We select 30 locations as marked in Figure 7 for client deployments. The clients are oblivious to beam selection at BS and simply measure the SNR and report them back to the BS for SINR estimations through standard feedback mechanisms. Our experiments have verified that the SNR received on each beam is relatively stable over several seconds for static clients. This gives confidence to the SNR measurements reported by clients in the measurement phase.

All algorithms (CABS and reference schemes) are implemented on the CC and do not require any changes or operational overhead to the BS. All BSs are connected to the CC through an ethernet switch in our set-up.

5.1 Prototype Evaluations

Topologies and rate adaptation: Each data point in our result is averaged over multiple topologies, which are generated by picking

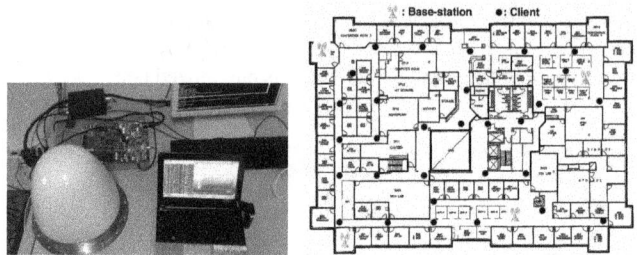

Figure 7: Small cell, beamformer, client and deployment

random subsets of client locations (among 30) for a given number of clients. Further, unless otherwise specified, we consider topologies with four small cells and twenty clients. To remove the influence of rate adaptation algorithms, we consider an ideal PHY rate adaptation by trying out all MCS and record the highest throughput (best MCS) for a client given a network configuration.

Reference schemes: We evaluate the performance of our CABS algorithm in ProBeam against the following benchmark algorithms.

- *Decoupled:* Client association is decoupled and first computed based on SNR, followed by determination of coordinated beams for each BS using the same beam selection component as in CABS.
- *CABS-all:* Allows for joint determination of client association and beam selection as in CABS but requires that all clients be associated and scheduled in every epoch.
- *UB-beam:* Employs the same client association component as in CABS but exhaustively searches over all possible beam combinations at BSs - serves as an upper bound for beam selection in CABS.
- *UB-assoc:* Employs the same beam selection component as in CABS but exhaustively searches over all possible combinations of client association.

Evaluation metrics: We consider the following metrics.

- *Throughput:* Aggregate throughput of all clients in the network.
- *Utility:* Captures both throughput and fairness; aggregate utility of all clients: $\sum_{j \in K} w_j \log(T_j)$ (details in subsection 4.2).
- *Fraction of scheduled clients:* Captures the number of clients not scheduled in an epoch to improve spatial reuse (in CABS and upper bounds).
- *Load balancing factor:* Measures Jain's fairness index among the number of clients associated with each BS.

Throughput: Figure 8(a) presents the throughput results as a function of number of clients in the network. Three observations can be made: (i) CABS' performance is within 96% of that of exhaustive beam search and is not impacted by client density. Given the complexity of the latter, CABS provides a fine balance between performance and complexity. (ii) The increased spatial reuse from jointly addressing client association with beamforming (CABS-all) provides gains as high as 50% (over the decoupled approach). Further, the gains are more pronounced at higher client density, where it becomes harder to isolate interference between small cells without a joint optimization that allows for flexible client association. (iii) Interestingly, by going one step further and allowing some clients from not being scheduled in a given epoch provides CABS with an additional 50% gain over CABS-all, resulting in *a net gain*

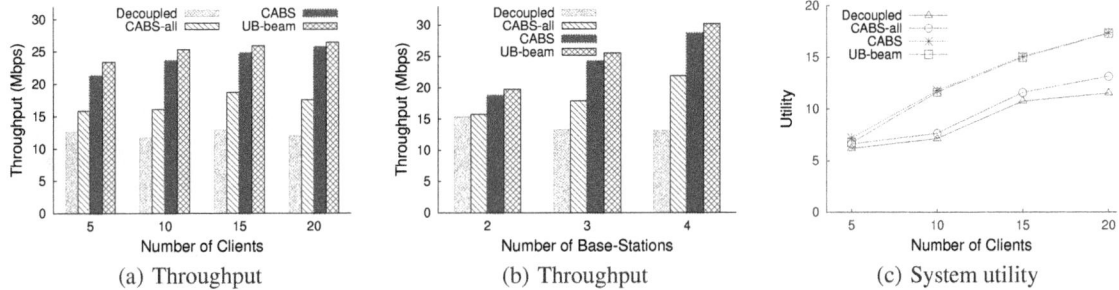

(a) Throughput   (b) Throughput   (c) System utility

Figure 8: Experimental evaluation of ProBeam with 4 small cells.

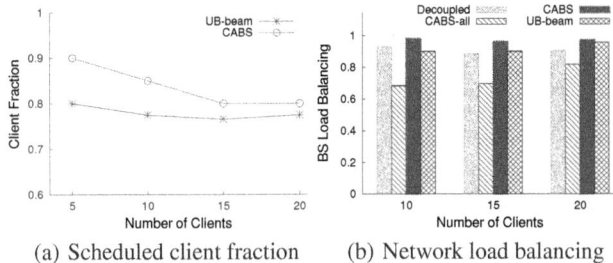

(a) Scheduled client fraction   (b) Network load balancing

Figure 9: Effective client management in ProBeam.

of around 115% over the decoupled approach. Removing even a small fraction of bottleneck clients from scheduling in an epoch can greatly improve the spatial reuse configuration between small cells.

The impact of interference from increased number of BSs is presented in Figure 8(b). The ability to jointly address client association with beam selection helps CABS handle interference effectively, the benefits of which are more pronounced with larger number of interferers.

Fairness: Recall that some of the reuse gains in CABS comes from removing a subset of clients from scheduling in a given epoch. While starvation of such clients is avoided across epochs, it is important to understand if the throughput gains of CABS are not realized at the expense of fairness even within an epoch. The utility measure helps account for fairness within an epoch, whose results are presented in Figure 8(c). It can be clearly seen that CABS' utility is very close to that of its upper bound and outperforms that of the (baseline) decoupled approach. Thus, *adopting a utility based approach to joint CABS, enables ProBeam to bypass some clients from an epoch to maximize reuse gains without compromising on fairness.*

Note that if the number of clients bypassed is large, this would automatically reflect in a reduced system utility. Hence, to further verify this, we present the fraction of scheduled clients in an epoch in Figure 9(a). This clearly shows that only a small fraction of clients (10-20%) are bypassed in CABS. The upper bound is more aggressive in deferring clients to the next scheduling epoch, which in turn contributes to its marginal throughput gains over CABS (Figure 8(a)).

Load balancing: A by-product of *utility maximization in CABS is that it should automatically lead to load balancing.* This is because, given a fixed amount of frame resources, balancing number of users across cells, provides more resources per user and hence better aggregate utility. The load balancing factor, captured thr-

ough Jain's fairness index between number of clients associated with small cells, is presented in Figure 9(b). CABS provides very good load balancing as expected. The decoupled approach does not implicitly account for load balancing, but a uniform distribution of clients automatically provides reasonable load balancing, when SNR-based client association is employed. The interesting observation is that CABS-all's load balancing suffers, especially when the number of clients is not high. Recall that CABS-all's throughput gain (over the decoupled approach) from better interference suppression (and hence reuse) through flexible association, comes at the expense of potential load imbalance across cells, especially when all clients are accommodated.

5.2 Trace-driven Simulations

Our experimental set-up with few tens of clients and three dominant interferers constitutes a realistic set-up for a cluster of small cells. However, to further understand CABS's effectiveness in much denser deployments (10 BSs and 90 clients), we resort to trace based simulations. We collect SNR traces for clients from our experimental network, feed it into a simulator running ProBeam (SINR estimation and CABS) to evaluate the various algorithms. We place our four BSs in various other locations to emulate more small cell BSs and measure SNR traces at the clients from them on all beams. Similarly, we also vary the client locations to emulate a larger set of clients and obtain corresponding SNR traces. Given the traces, we can generate a topology with a specific number of BSs and clients, by sampling BSs and clients randomly from our SNR trace database.

Our simulation results are presented in Figure 10, where throughput is measured as a fraction of that achieved by the upper bound (UB-beam). The trends in these large scale results, including the magnitude of gains possible with CABS, are very similar to those from the experiments, thereby reinforcing our inferences from the prototype evaluation. Hence, in the interest of space, we do not discuss them further. CABS close performance with respect to its upper bound in these results indicates the efficiency of its beam selection component as both the schemes employ the same client association mechanism. Given the hardness of computing a tight upper bound for the joint CABS solution, we now evaluate the efficiency of its client association component as well. We compare it against an upper bound for client association (UB-assoc) that exhaustively searches over all possible client associations, while employing the same beam selection mechanism as in CABS. The results in Figure 11 indicate that, while the sub-optimality of CABS' client association component can at most be within half of the optimal (see Sec.4.2.4) in the worst case, in practice, it yields a performance that is very close to its upper bound. Thus, *the high effi-*

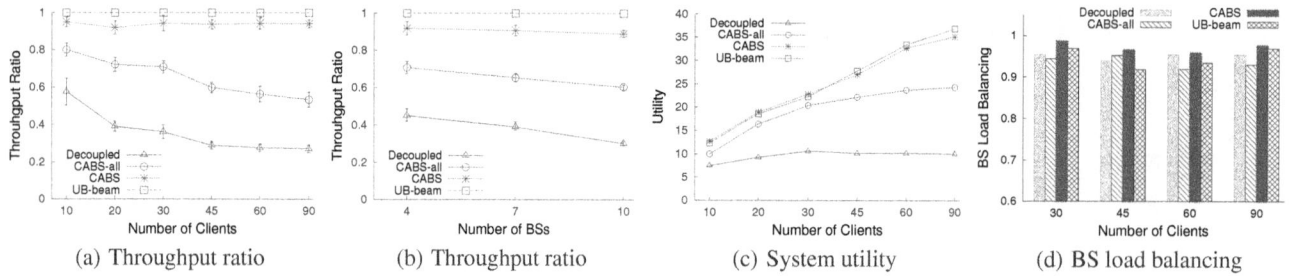

| (a) Throughput ratio | (b) Throughput ratio | (c) System utility | (d) BS load balancing |

Figure 10: Large scale evaluation of ProBeam through trace-driven simulations.

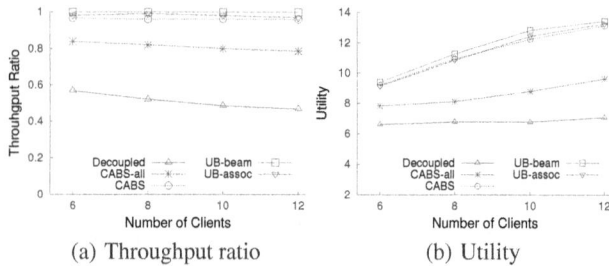

| (a) Throughput ratio | (b) Utility |

Figure 11: Evaluation of client association component in ProBeam.

ciency of the individual components in CABS in turn synergistically contribute to the net gains seen by it.

6. CONCLUSIONS

We design and implement *ProBeam* – a practical system for improving spatial reuse through beamforming in OFDMA based small cell networks. We show that decoupling beamforming from client scheduling is necessary for practical feasibility. Further, we highlight the need to jointly address client association with beamforming to maximize the reuse benefits from the latter. ProBeam incorporates a low complexity, highly accurate SINR estimation module with less than 1 dB error ($\leq 5\%$) to determine interference dependencies between small cells. It also houses an efficient, low complexity joint client association and beam selection algorithm for the small cells that yields close-to-optimal performance. Prototype implementation in a real WiMAX networks of four small cells shows 115% of capacity gain compared to other baseline reuse schemes. We also demonstrate the scalability and efficacy of our system in larger scale settings through simulations. Most of our system components are also applicable to LTE and LTE-A with minor modifications. As part of future work, we plan to investigate synthesis of new beam patterns for beamforming based on client feedback in lieu of a pre-determined set (code-book).

Acknowledgments

We would like to thank the anonymous reviewers for their invaluable feedback. Jongwon Yoon and Suman Banerjee have been supported in part by the following grants of the US National Science Foundation: CNS-1040648, CNS-0916955, CNS-0855201, CNS-0747177, CNS-1064944, and CNS-1059306.

7. REFERENCES

[1] R. Van Nee and R. Prasad, "OFDM for Wireless Multimedia Communications", *Artech House*, 2000.

[2] M. Y. Arslan, J. Yoon, K. Sundaresan, S. V. Krishnamurthy, and S. Banerjee, "FERMI: A FEmtocell Resource Management System for Interference Mitigation in OFDMA Networks", In *ACM MobiCom*, 2011.

[3] J. Yoon, M. Y. Arslan, K. Sundaresan, S. V. Krishnamurthy, and S. Banerjee, "A Distributed Resource Management Framework for Interference Mitigation in OFDMA Femtocell Networks", In *ACM MobiHoc*, 2012.

[4] X. Liu, A. Sheth, M. Kaminsky, K. Papagiannaki, S. Seshan, and P. Steenkiste, "DIRC: Increasing Indoor Wireless Capacity Using Directional Antennas", In *ACM Sigcomm*, 2009.

[5] X. Liu, A. Sheth, M. Kaminsky, K. Papagiannaki, S. Seshan, and P. Steenkiste, "Pushing the Envelope of Indoor Wireless Spatial Reuse using Directional Access Points and Clients", In *ACM MobiCom*, 2010.

[6] Femtocells Core Specification, WMF-T33-118-R016v01.

[7] M. Y. Arslan, J. Yoon, K. Sundaresan, S. Krishnamuthy, and S. Banerjee, "Characterization of Interference in OFDMA Femtocell Networks", In *IEEE Infocom*, 2012.

[8] J. Yun and K. G. Shin, "CTRL: A Self-Organizing Femtocell Management Architecture for Co-Channel Deployment", In *ACM MobiCom*, 2010.

[9] K. Sundaresan and S. Rangarajan, "Efficient Resource Management in OFDMA Femto Cells", In *ACM MobiHoc*, 2009.

[10] S. He, Y. Huang, L. Yang, A. Nallanathan, and P. Liu, "A Multi-Cell Beamforming Design by Uplink-Downlink Max-Min SINR Duality", *IEEE Transactions on Wireless Communications*, vol. 11, no. 8, pp 2858-2867, 2012.

[11] H. Dahrouj, and W. Yu, "Coordinated Beamforming for the Multicell Multi-Antenna Wireless System", *IEEE Transactions on Wireless Communications*, vol. 9, no. 5, pp 1748-1759, 2010.

[12] E. Aryafar, A. Khojastepour, K. Sundaresan, S. Rangarajan, and E. Knightly, "ADAM: An Adaptive Beamforming System for Multicasting in Wireless LANs", In *IEEE Infocom*, 2012.

[13] ProBeam technical report, http://www.cs.wisc.edu/~yoonj/ProBeam-TR.pdf/

[14] L. Fleischer, M. X. Goemans, V. S. Mirrokni, and M. Sviridenko, "Tight Approximation Algorithms for Maximum General Assignment Problems", In *ACM-SIAM symposium on Discrete algorithm*, 2006.

[15] M. Fisher, G. Nemhauser, and G. Wolsey, "An Analysis of Approximations for Maximizing Submodular set Functions-II", *Mathematical Programming Study*, 1978.

[16] PicoChip Femtocell Solutions, http://www.picochip.com/

[17] IEEE 802.16e-2005 Part 16: Air Interface for Fixed and Mobile Broadband Wireless Access Systems, *IEEE 802.16e standard*.

[18] Fidelity Comtech, http://www.fidelity-comtech.com/

[19] Accton Wireless Broadband, http://www.awbnetworks.com/

Connectivity in Obstructed Wireless Networks: From Geometry to Percolation

Marcelo G. Almiron
Dept. of Computer Science
Universidade Federal de
Minas Gerais (UFMG)
Belo Horizonte, Brazil
Computer Science Centre
Université de Genève (UNIGE)
Geneva, Switzerland
malmiron@dcc.ufmg.br

Olga Goussevskaia
Dept. of Computer Science
Universidade Federal de
Minas Gerais (UFMG)
Belo Horizonte, Brazil
olga@dcc.ufmg.br

Antonio A.F. Loureiro
Dept. of Computer Science
Universidade Federal de
Minas Gerais (UFMG)
Belo Horizonte, Brazil
loureiro@dcc.ufmg.br

Jose Rolim
Computer Science Centre
Université de Genève (UNIGE)
Geneva, Switzerland
jose.rolim@unige.ch

ABSTRACT

In this work, we analyze an alternative model for obstructed wireless networks. The model is based on a grid structure of one-dimensional street segments and two-dimensional street intersections. This structure provides a realistic representation of a variety of network scenarios with obstacles and, at the same time, allows a simple enough analysis, which is partly based on percolation theory and partly based on geometric properties. We propose three different ways of modeling the geometric part of the network and derive analytical bounds for the connectivity probability and the critical transmission range for connectivity in the network. Finally, we present extensive simulations that demonstrate that our analytical results provide good approximations, especially for high density scenarios.

Categories and Subject Descriptors

C.2.1 [**Computer-Communication Networks**]: Network Architecture and Design—*Wireless communication*

Keywords

Wireless ad hoc networks, obstructed networks, percolation theory, critical transmission range

1. INTRODUCTION

When modeling and analyzing different problems in wireless communication networks, regardless of their topology

(random [16], regular [10], or arbitrary [7]), it has been typically assumed that communication nodes are deployed in an open space without obstacles. This assumption is quite natural, given that an open space represents the "purest" scenario of wireless communication, in which the communication channel is shared among all communication nodes, a distinguishing and challenging characteristic of wireless technology. Moreover, wireless signal propagation and interference in non-obstructed spaces can be represented by simpler, easier to analyze models [11]. This generates possibilities for more generalized theories and results that can be applied to many network instances of large sizes.

In reality, however, wireless networks are rarely deployed in completely open spaces. Many wireless networks operate in highly obstructed environments, such as dense urban areas and indoor spaces, not to mention networks deployed in constrained spaces like tunnels and subways, or other specialized networks, like smart grid communication networks. The behavior of both wireless signal and interference when obstacles are present is more complex and, therefore, more difficult to model and analyze.

Relatively few attempts have been made to analyze obstructed wireless networks, many of which are quite complex and not easily extended to generic scenarios [13]. One interesting model for obstructed wireless networks is the so-called "line-of-sight networks" [5]. In this model, the network is represented by a grid, and a node is placed on a grid vertex with some constant probability p. Moreover, each node is assumed to have a communication range of ω blocks. If $\omega = 1$, i.e., a node can only see neighboring points, then the model reduces to the well-studied problem of (site) percolation[1] on a lattice [8]. Among other results, Frieze et al. [5] derive asymptotic bounds for k-connectivity of such networks.

In this work, we analyze an alternative model for obstructed wireless networks. We consider a network formed

---

[1] Percolation means existence of a giant connected component of infinite size.

(a) Deployment of nodes over a grid of size $g = 4$.

(b) Discrete percolation model. An **open vertex** is represented by a filled bullet, a **closed vertex** by an empty bullet, an **open edge** by a black line, and a **closed edge** by no line.

(c) An instance of a network with $n = 900$ and $g = 10$. The transmission range corresponds to the CTR. A communication **link** is represented by a black line, and a communication **node** is represented by a black dot.

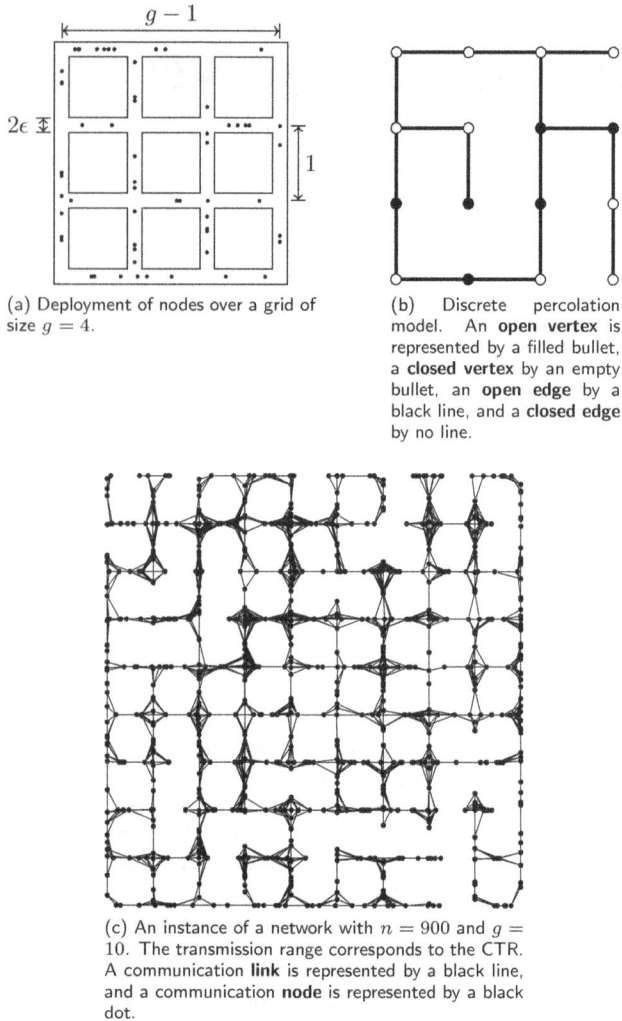

Figure 1: Modeling an obstructed wireless network.

by a grid of $g \times g$ one-dimensional street segments (see Figure 1a). Each vertex of the grid represents a street intersection of two orthogonal streets of a certain width ϵ. The ϵ-parameter is used to model the visibility among nodes located close to an intersection of orthogonal streets. The larger the parameter ϵ, the farther away two nodes can be placed from the intersection and still be able to communicate. We distribute n nodes uniformly at random on this "Manhattan" street lattice, such that, on average, μ nodes are located in each street segment. Figure 1c illustrates an example of a network instance with $g = 10$, $\epsilon = 0.3$, and $n = 900$, which results in $\mu = 5$).

This model has several appealing aspects. From the point of view of applications, it provides a realistic representation of environments with regularly spaced obstacles. As opposed to the model in [5], for instance, where nodes are placed only at street intersections (and there are no nodes along the street segments), placing several nodes along a street might better represent connectivity properties, especially if low-power/short-range radios are used. Therefore, this model can be potentially useful in simulating and analyzing different types of network scenarios, in particular,

those that are comprised of segments of one-dimensional arrays of nodes and regularly distributed arrays of obstacles.

From the point of view of analysis, our model mixes two basic elements. On the one hand, it might be viewed as a percolation model on a lattice (see Figure 1b). On the other hand, it is an intrinsically geometric model on individual edges and vertices of the grid: on the edges, we have a line topology, where connectivity is determined by node density. On the vertices or street intersections, we have a two-dimensional scenario, where connectivity depends on node density and the street width. This division into "percolation" and "geometry" allows us to simplify the model and analyze several important properties of the network, such as connectivity probability and critical transmission range for connectivity.

We propose three different models for connectivity at a street intersection: *MaxNorm*, *Euclidean*, and *Triangular*. The MaxNorm is the simplest of the three models and provides a lower bound on the probability of connectivity at a street intersection, but does not provide a good enough approximation for some scenarios. The Euclidean model is the most realistic among the three and provides an upper bound on the probability of connectivity. We use this model to implement our simulations, however, it is too complex to be useful analytically. The Triangular model represents a compromise between the two other models: it is simple enough to be treated analytically and, at the same time, is a good approximation for the Euclidean model, especially in high-density scenarios. Among other results, we are able to derive a good approximation for the critical transmission range for connectivity in this model.

Our contributions can be summarized as follows:

- We analyze an alternative model for obstructed wireless environments, based on a grid structure of one-dimensional street segments and two-dimensional street intersections;

- We combine elements from percolation theory and geometry to analyze connectivity properties in this model;

- We propose three different geometric models for connectivity at street intersections;

- We derive approximation bounds for probability of connectivity and the critical transmission range for connectivity in the overall network;

- We simulate different network scenarios and demonstrate that our analytical results provide good approximations.

The rest of this paper is organized in the following way. In Section 2, we describe the network model and the three proposed models for connectivity at a street intersection: MaxNorm, Euclidean, and Triangular. In Section 3, we derive a lower bound for the probability of connectivity at street intersections in the Triangular model. In Section 4 we compute the critical transmission range (CTR) for connectivity in the overall network. In Section 5, we validate our analytical results through simulations. In Section 6, we discuss the related literature. Finally, in Section 7, we present the concluding remarks.

2. MODEL

Our goal is to define a model for obstructed wireless networks that captures some essential characteristics of obstructed environments encountered in practice and be simple enough to provide analytical tools for network properties, such as connectivity.

Our solution combines geometric elements that model local interactions between nodes, such as existence of a communication link, and elements from percolation theory [8] that model global properties of the network, such as formation of a giant connected component. To that end, we overlay a regular grid structure, comprised of *vertices* and *edges*, with a random network of n communication *nodes*, deployed uniformly at random, such that a communication *link* is established between a pair of nodes if their distance is less than a certain range r and there is no obstacle between them (see Figure 1a). This combination provides us with enough abstraction to analyze many desirable network properties.

2.1 Percolation "Layer"

The "percolation layer" of our model consists of a lattice square (grid) of size $g \times g$. We refer to the parameter g as the *granularity* of the model. This lattice can be viewed as a Manhattan-style street map. In this way, street segments are represented by edges in the grid, and street intersections are represented by vertices in the grid (see Figure 1b). We assume, without loss of generality (w.l.o.g.), that the length of each street segment has a normalized length of 1, and the width of all streets is 2ϵ, where ϵ is some constant $\epsilon < 1/2$. The street width influences the geometry at street intersections, which determines the communication links between nodes located close to street intersections, as explained later.

We use the (discrete) Mixed Percolation model [8] to analyze global (connectivity) properties of the network[2]. Mixed percolation is defined in a regular grid structure, comprised of a set $\mathbb{V}^2 = \{1, \ldots, g\}^2$ of g^2 vertices. The position of each vertex (x, y) is defined by means of its indices (line and column in the grid). Additionally, the grid has a set of edges $\mathbb{E}^2$, containing all pairs of vertices (u, v) such that $\|x_u - x_v\| + \|y_u - y_v\| \leq 1$.

Each vertex can be *open* ("open" = connected) or *closed* ("closed" = disconnected) with probabilities p_s (*site probability*) and $(1 - p_s)$, respectively. Each edge can be open or closed with probabilities p_b (*bond probability*) and $(1 - p_b)$, respectively.

We define a random graph $\mathbb{G}_g = (\mathbb{V}^2, \mathbb{E}^2)$, and say that $\mathbb{G}_g$ *percolates*[3] if all vertices belong to the same connected component formed by open (connected) vertices and edges. Figure 1b presents an example of a random graph $\mathbb{G}_4$ that does not percolate (filled bullets represent open vertices, and empty bullets represent closed vertices).

2.2 Geometric "Layer"

---

[2]The *Mixed* percolation model combines the so-called *Bond* and *Site* percolation models. In the former, all vertices are open (connected) with probability 1, whereas each edge is open (connected) with a constant probability ≤ 1. In the latter model, edges have probability 1 of being open, whereas each vertex is open with probability ≤ 1.

[3]In the finite random grid, percolation means full connectivity between vertices.

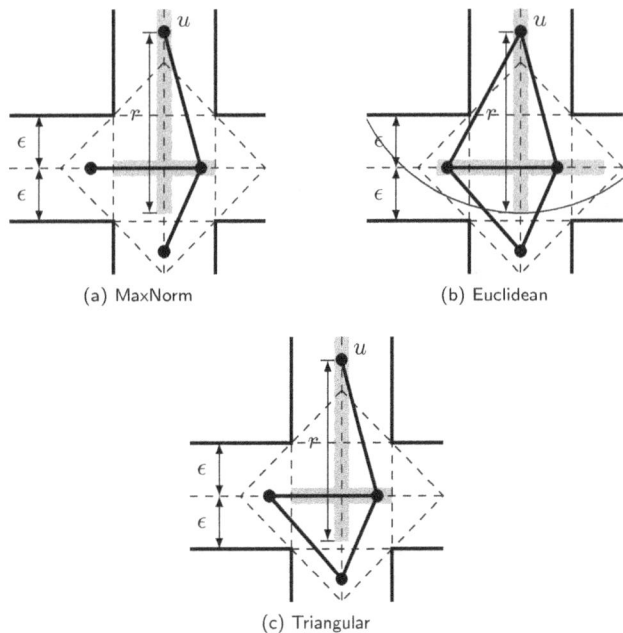

Figure 2: Determining the communication links present at street intersections. The gray region represents the coverage of node u.

In order to represent the actual communication *nodes* and *links*, we overlay the above defined grid structure with a set of n nodes, positioned uniformly at random over the grid. Recall that all streets have width 2ϵ. However, when positioning the nodes, each segment individually is viewed as a one-dimensional space. Thus, all nodes are placed on (imaginary) lines located in the middle of the street segments. Note that the average number of nodes per street segment is $\mu = n/(2g(g-1))$. Additionally, we assume that all nodes in the network have a common transmission range r. As already mentioned, Figure 1c illustrates an example of a network instance with $g = 10$, $\epsilon = 0.3$, and $n = 900$, which results in $\mu = 5$.

We define three different geometric models (see Figure 2) to determine when local communication links exist between two nodes u and v, positioned at (x_u, y_u) and (x_v, y_v), respectively. Note that the existence of such links depends on two criteria: distance and visibility. Also, the most challenging scenario in terms of visibility is when the two nodes are located in perpendicular streets. This happens because if they are in the same street, then only the distance criterion counts, and, if they are in parallel but different streets, they are never visible to each other.

MaxNorm model: u and v are able to communicate with each other if the following two conditions are satisfied: (i) the minimum norm $\min\{\|x_u - x_v\|, \|y_u - y_v\|\}$ does not exceed ϵ, and (ii) the maximum norm $\max\{\|x_u - x_v\|, \|y_u - y_v\|\}$ does not exceed the common transmission range r (see Figure 2a). Informally, this model states that to satisfy the visibility criterion between nodes in perpendicular streets, at least one node of $\{u, v\}$ must be located inside the square of side 2ϵ centered at the street intersection.

Euclidean model: u and v are able to communicate with each other if the following two conditions are satisfied: (i) the Euclidean distance does not exceed the common transmission range r, and (ii) there is a visibility line between u and v (see Figure 2b). Note that this model makes no simplifications when establishing visibility between two nodes in perpendicular streets, i.e., all possible placements of $\{u, v\}$ in perpendicular (and adjacent) street segments must be considered.

Triangular model: u and v are able to communicate with each other if the following two conditions are satisfied: (i) u and v are connected in the MaxNorm model, or (ii) both u and v are at most at a distance 2ϵ far from a shared intersection (see Figure 2c). This model greatly simplifies the definition of visibility between two nodes in perpendicular street segments. It extends the MaxNorm model by stating that if both u and v are located inside the rhombus of diagonal 4ϵ, they are visible to each other. Note that this is a better approximation than the MaxNorm model, since a significantly larger area and, consequently, more links are considered.

A preliminary analysis of the MaxNorm model was presented in [1]. However, as shown in Section 5, it resulted in much higher (worse) values for the critical street width (i.e., for the minimal street width that induces a valid CTR for connectivity) and low-quality approximation to the Euclidean model, which is more realistic but leads to a very complex analysis. The Triangular model is an excellent compromise between the two models above, as we will show in the following sections. It provides much better values for the CTR and better approximations to the Euclidean model. We focus our analysis on the Triangular model in Sections 3 and 4.

2.3 From Geometry to Percolation

We can overlay the two parts of our model by establishing the following relations. The bond probability p_b in the percolation model is equal to the probability of all nodes located in one grid edge (μ, on average) forming a connected topology, to which we refer as $\Pr(S_{con})$ (see Theorem 2). The site percolation p_s is equal to the probability of having a tree connecting all four street segments at the intersection. We refer to this probability as $\Pr(I_{con})$ (see Theorem 1). Note that $\Pr(S_{con})$ is a function of just μ and r, whereas $\Pr(I_{con})$ is a function of μ, r and ϵ.

In order to analyze the connectivity properties of the network, we first compute a critical value p_b^c for bond probability, assuming that $p_s = 1$. We do this because $\Pr(S_{con})$ typically dominates $\Pr(I_{con})$, i.e., to establish connectivity in a street segment is always more difficult than in a street intersection. This allows us to compute the CTR for connectivity in the grid by finding the inverse of the expression for p_b^c. Moreover, p_b^c and p_s determine the critical street width value ϵ_c, such that $\forall \epsilon \geq \epsilon_c$ the CTR derived from p_b^c is valid.

In the following sections we show that when percolation happens in the grid, the overlay network becomes connected with high probability (w.h.p.). Note that a connected component formed by all g^2 vertices of the grid does not necessarily imply that all n nodes will be connected. This happens because there is always the possibility of having isolated nodes or sets of nodes in the middle of a street segment. This situation is a rare event, whose probability goes to zero as the node density increases. Figure 7, discussed in Section 5, will illustrate how the size of the largest connected component in the grid relates to the number of connected components in the network as the communication radius increases.

3. PROBABILITY OF CONNECTIVITY

In this section we compute the probability $\Pr(I_{con})$ of connectivity at street intersections, or vertices of the grid, and present the probability $\Pr(S_{con})$ of connectivity at (one-dimensional) street segments, or edges of the grid, in the Triangular model.

In order to compute connectivity at intersections, we use the following definitions. Let $\mathcal{X}$, $\mathcal{Y}$, $\mathcal{W}$ and $\mathcal{Z}$ be four sets of μ nodes positioned in four adjacent street segments sharing an intersection[4]. This intersection is connected if and only if for every pair of sets, there exists a path between nodes in those sets. This means that there should exist a tree connecting at least one node from each of these four sets. Note that the existence of a tree with four nodes, one per set, is sufficient to grant connectivity at the intersection, and, thus, we only need to consider crossing links from one segment to another.

The first step to calculate the probability of connectivity at street intersections is to compute the probabilities of having links between two perpendicular street segments. The exact probability of existing at least one crossing link between two perpendicular street segments, namely $p_\perp(\mu)$, is given in Lemma 1. Figure 2c can be useful to the reader to follow the proof.

LEMMA 1. *Let $\{X_i\}_{1 \leq i \leq \mu}$, such that $X_i \sim \mathcal{U}(0,1)$, and $\{Y_j\}_{1 \leq j \leq \mu}$, such that $Y_j \sim \mathcal{U}(0,1)$, be two families of independent random variables denoting the position of μ nodes in each one of two perpendicular segments sharing an intersection. Let also r denote the common transmission range of all nodes, where $0 \leq r \leq 1$. Additionally, let 2ϵ be the segments' width. Assuming the Triangular model, the probability $p_\perp(\mu)$ of having at least one link between two nodes in different perpendicular segments is*

$$p_\perp(\mu) = (1 - 2\epsilon)^{2\mu} + 2(1 - \epsilon)^\mu \left((1 - r)^\mu - (1 - 2\epsilon)^\mu\right) + 2\left(1/2 - (1 - r)^\mu\right) \quad (1)$$

whenever $r \geq \sqrt{8}\epsilon$.

PROOF. Let $X_{(1)}, X_{(2)}, \ldots, X_{(\mu)}$ and $Y_{(1)}, Y_{(2)}, \ldots, Y_{(\mu)}$ be the order statistics of $X_1, X_2, \ldots, X_\mu$ and $Y_1, Y_2, \ldots, Y_\mu$, respectively. W.l.o.g., we consider that positions of nodes reflect the Euclidean distances from each node to the intersection point between segments. There exists at least one link between two nodes in different perpendicular segments if and only if there is a link between $X_{(1)}$ and $Y_{(1)}$. Considering $r \geq \sqrt{8}\epsilon$, we can compute $p_\perp(\mu)$ by

$$p_\perp(\mu) = 2 \left(\int_0^{2\epsilon} \int_0^x f_{X_{(1)}Y_{(1)}}(x,y)\, \mathrm{d}y\, \mathrm{d}x + \int_{2\epsilon}^r \int_0^\epsilon f_{X_{(1)}Y_{(1)}}(x,y)\, \mathrm{d}y\, \mathrm{d}x \right), \quad (2)$$

where $f_{X_{(1)}Y_{(1)}}(x,y)$ is the joint distribution function of $X_{(1)}$ and $Y_{(1)}$. It is well known that the k-th order statistic of a

---

[4]Note that in total we consider 4μ nodes.

family of μ independent standard uniform random variables is a Beta random variable $U_{(k)} \sim \mathcal{B}(k, \mu + 1 - k)$. Since all random variables are independent, we have that the joint distribution function of $X_{(1)}$ and $Y_{(1)}$ is $f_{X_{(1)}Y_{(1)}}(x, y) = \mu^2(1 - x)^{\mu-1}(1 - y)^{\mu-1}$. Then, solving (2), we have

$$p_\perp(\mu) = (1 - 2\epsilon)^{2\mu} + 2(1 - \epsilon)^\mu \left((1 - r)^\mu - (1 - 2\epsilon)^\mu \right) + 2\left(1/2 - (1 - r)^\mu \right).$$

$\square$

The next step is to compute the probability of connectivity at street intersections, namely $\Pr(I_{con})$. In Theorem 1, we compute the exact value of $\Pr(I_{con})$ by means of $p_\perp(\mu)$ and a conditional term.

THEOREM 1. *Let* $\{X_i\}_{1 \leq i \leq \mu}$, $\{Y_j\}_{1 \leq j \leq \mu}$, $\{W_k\}_{1 \leq k \leq \mu}$ *and* $\{Z_m\}_{1 \leq m \leq \mu}$ *be four families of independent random variables, such that* $X_i \sim \mathcal{U}(0,1)$, $Y_j \sim \mathcal{U}(0,1)$, $W_k \sim \mathcal{U}(0,1)$ *and* $Z_m \sim \mathcal{U}(0,1)$ *denote the positions of* μ *nodes in four adjacent segments sharing an intersection. Let* r *denote the common transmission range of all nodes, with* $0 \leq r \leq 1$, *and* 2ϵ *be the segments' width. The probability of having a path between at least one node in a segment connecting at least one node in each one of the three other segments, denoted by* $\Pr(I_{con})$, *is*

$$\Pr(I_{con}) = p^2 + 2\delta p(1 - p) + \gamma(1 - p)^2, \quad (3)$$

where $p = p_\perp(\mu)$, $\delta = \Pr(I_{con} \mid e_1 \cap \overline{e_3})$, *and* $\gamma = \Pr(I_{con} \mid \overline{e_1} \cap \overline{e_3})$.

PROOF. W.l.o.g., we consider that positions of nodes reflect the Euclidean distances from each node to the intersection point between segments. Consider the order statistics $X_{(1)}$, $Y_{(1)}$, $W_{(1)}$ and $Z_{(1)}$, from the families $\{X_i\}_{1 \leq i \leq \mu}$, $\{Y_j\}_{1 \leq j \leq \mu}$, $\{W_k\}_{1 \leq k \leq \mu}$ and $\{Z_m\}_{1 \leq m \leq \mu}$, respectively. There exists a path between at least one node in a segment connecting at least one node in each one of the three other segments if and only if there exists a path between $X_{(1)}$, $Y_{(1)}$, $W_{(1)}$ and $Z_{(1)}$.

Let e_1 be the event that indicates the presence of a link between the realizations of $Y_{(1)}$ and $W_{(1)}$ (see Figure 3a). Additionally, let e_2, e_3, e_4, e_5 and e_6 represent the existence of a link between $X_{(1)}$ and $Y_{(1)}$, $X_{(1)}$ and $Z_{(1)}$, $W_{(1)}$ and $Z_{(1)}$, $W_{(1)}$ and $X_{(1)}$ and finally $Y_{(1)}$ and $Z_{(1)}$, respectively. Then, we have $e_1 \perp\!\!\!\perp e_3$ and $e_2 \perp\!\!\!\perp e_4$, where $\perp\!\!\!\perp$ indicates the stochastic independence between events. Consequently, we can compute

$$\Pr(I_{con}) = \Pr(I_{con} \mid e_1 \cap e_3) \Pr(e_1 \cap e_3) + 2 \Pr(I_{con} \mid e_1 \cap \overline{e_3}) \Pr(e_1 \cap \overline{e_3}) + \Pr(I_{con} \mid \overline{e_1} \cap \overline{e_3}) \Pr(\overline{e_1} \cap \overline{e_3}). \quad (4)$$

We are able to compute $\Pr(I_{con} \mid e_1 \cap e_3)$ by observing Figure 3b. Here, we represent e_1 by edges connecting black balls, and e_3 by edges connecting gray balls. Each pair of balls, labeled as (i, i), represents a worst case scenario given that we positioned one of the balls in the vertical line. For instance, Ball 2 in the vertical line, can be connected with a ball in the horizontal line positioned, in the worst case, 2ϵ far from the intersection. We note that for each one of the nine scenarios when both events (e_1 and e_3) happen, it is possible to connect a black ball with a gray ball, which means that

$$\Pr(I_{con} \mid e_1 \cap e_3) = 1. \quad (5)$$

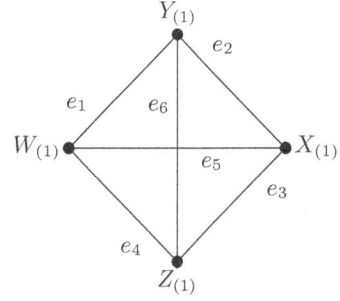

(a) Each edge is an event, and $\Pr(e_1) = \Pr(e_2) = \Pr(e_3) = \Pr(e_4) = p_\perp(\mu)$.

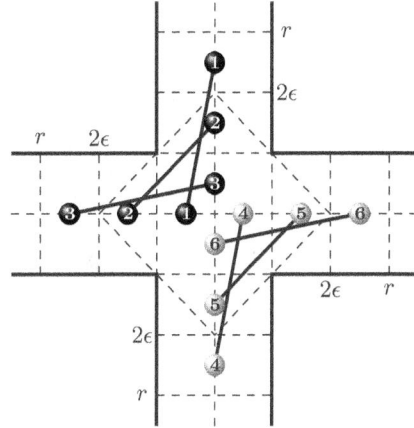

(b) $\Pr(I_{con} \mid e_1 \cap e_3) = 1$.

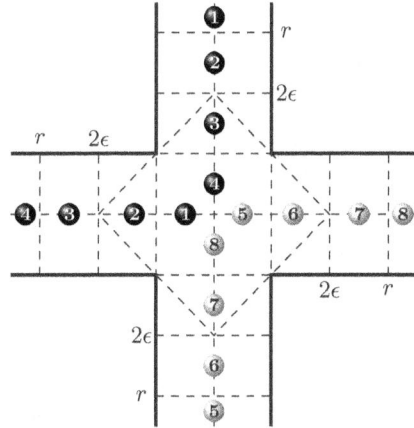

(c) $\Pr(I_{con} \mid \overline{e_1} \cap \overline{e_3})$.

Figure 3: Computing the probability of connectivity at intersections.

Similarly as $\Pr(I_{con} \mid e_1 \cap e_3)$, we compute $\Pr(I_{con} \mid \overline{e_1} \cap \overline{e_3})$ by means of case analysis on all possible scenarios. Figure 3c shows all possible scenarios. Here, we draw a pair of black balls (i, i) to denote the best case scenario for the event $\overline{e_1}$, given that we position a ball in a vertical line. We apply the same criterion for the event $\overline{e_3}$ using gray balls. Observing the figure, we perceive that just in 2 out of 16 scenarios is possible to form a tree. This implies that $\Pr(I_{con} \mid \overline{e_1} \cap \overline{e_3})$ is negligible compared to the other terms in (4).

Since $e_1 \perp\!\!\!\perp e_3$ and $e_1 \perp\!\!\!\perp \overline{e_3}$, we have $\Pr(e_1 \cap e_3) = (p_\perp(\mu))^2$, $\Pr(e_1 \cap \overline{e_3}) = p_\perp(\mu)(1 - p_\perp(\mu))$ and $\Pr(\overline{e_1} \cap \overline{e_3}) = (1 - p_\perp(\mu))^2$.

Then, combining (5) and (4), we conclude that

$$\Pr(I_{con}) = p^2 + 2\Pr(I_{con} \mid e_1 \cap \overline{e_3})p(1-p) +$$
$$\Pr(I_{con} \mid \overline{e_1} \cap \overline{e_3})(1-p)^2,$$

where $p = p_\perp(\mu)$. $\square$

From this last theorem, we perceive that $(p_\perp(\mu))^2$ is a reasonable lower bound for $\Pr(I_{con})$. In Section 5 we validate this through Monte Carlo simulation.

We conclude this section with Theorem 2, from [6], which computes the probability of connectivity at segments.

THEOREM 2. *Let $\{X_i\}_{1 \le i \le \mu}$ be a family of independent random variables, such that $X_i \sim \mathcal{U}(0,1)$, denoting the positions of μ nodes in a segment. Let also r denote the common transmission range of all nodes, where $0 \le r \le 1$. The probability of connectivity between all nodes is*

$$\Pr(S_{con}) = (1 - (1-r)^\mu)^{\mu-1}.$$

PROOF. See [6]. $\square$

Theorems 1 and 2 are used in Section 4 for computing the CTR for connectivity in the overall network.

4. CRITICAL TRANSMISSION RANGE

In this section we proceed to compute the CTR for connectivity in the Triangular model. Formally, the CTR is defined as follows.

DEFINITION 1. *Suppose n nodes are distributed in a region R. The Critical Transmission Range (CTR) for connectivity is the minimum transmission range, r_c, which induces a communication graph with a unique connected component on all n nodes.*

Note that we may have no solution for the CTR under some scenarios. For instance, when ϵ and μ are too small, even using $r = 1$, we will not be able to warrant enough visibility at intersections. Because of this, it is necessary to introduce a minimal value for ϵ that allows enough visibility to warrant connectivity at intersections. We say that this is the *critical value* ϵ_c. For all $\epsilon \ge \epsilon_c$, the CTR problem can be solved, and this solution is given by r_c computed in Theorem 3.

In the following, we present Lemma 2 that computes ϵ_c depending implicitly on μ. We observe that larger values of μ produce smaller values of ϵ_c, as expected.

LEMMA 2. *Let μ be the expected quantity of nodes uniformly deployed per segment, and let also 2ϵ be the segments' width. The critical value ϵ_c is upper bounded by the restriction*

$$q = \left((1 - 2\epsilon_c)^{2\mu}(1 - 2e^{\mu\epsilon_c}) + 1 \right)^2,$$

when $q \approx 1$.

PROOF. We need to characterize the minimum value of ϵ that warrants $\Pr(I_{con}) \approx 1$. A reasonable lower bound for the probability of connectivity at intersections, as we know from Section 3, is $(p_\perp(\mu))^2$. Considering the definition of $p_\perp(\mu)$ for dense scenarios, i.e., for high values of μ we can rewrite $(p_\perp(\mu))^2$ as

$$\left((1-2\epsilon)^\mu((1-2\epsilon)^\mu - 2(1-\epsilon)^\mu) + 1 \right)^2. \qquad (6)$$

Also note that we can rewrite $2(1-\epsilon)^\mu$ as $2e^{\mu\epsilon}(1-2\epsilon)^\mu$. Then, from (6) we obtain

$$q = \left((1-2\epsilon_c)^{2\mu}(1 - 2e^{\mu\epsilon_c}) + 1 \right)^2.$$

$\square$

Having in mind the upper bound for the critical value ϵ_c derived in the lemma above, we present the main result of this section, namely, the CTR for connectivity.

THEOREM 3. *Let $\mathcal{N}$ be a set of nodes deployed uniformly at random in a lattice square of granularity g in the area $[0, g-1]^2$, with segments' width 2ϵ. Let also μ be the expected quantity of nodes per segment. The Critical Transmission Range for Connectivity, denoted by r_c is*

$$r_c = \frac{\ln(g^{a+1/2}) + \ln(\mu - 1)}{\mu}$$

for $a > 0$, whenever $\epsilon \ge \epsilon_c$.

PROOF. Let us abstract the connectivity problem using a bond percolation model within a random grid of $g \times g$ vertices. Here, each vertex corresponds to an intersection in the lattice square and each edge corresponds to a segment. In the percolation model, each edge occurs with probability p_b.

We know, from Theorem 2, the exact connectivity probability at segments, which in the abstraction corresponds to p_b. Additionally, from Theorem 1, we know the probability of connectivity at intersections, which in the abstraction corresponds to having a vertex in the grid, and in the bond model must be one. Here, we relax this assumption and allow the connectivity probability at intersections to be close to one.

Franceschetti and Meester [4] have studied the rate of convergence of p_b (that depends on g), and found that it must scale slightly faster than $\sqrt{g}$ so the random grid becomes fully connected w.h.p., which means that $p_b = 1 - C_g/\sqrt{g}$ guarantees connectivity w.h.p. if and only if the sequence $C_g \to 0$ when $g \to \infty$. In particular, we propose $C_g = 1/g^a$ with $a > 0$ so we can select different convergence's rates. Considering this, from Theorem 2 we have

$$p_b = (1 - (1-r)^\mu)^{\mu-1}$$
$$\frac{\ln\left(1 - \frac{C_g}{\sqrt{g}}\right)}{\mu - 1} = \ln(1 - (1 - r_c)^\mu). \qquad (7)$$

Applying Maclaurin series to both sides of (7), we obtain

$$-\frac{C_g}{\sqrt{g}(\mu-1)} + \frac{c_1}{\mu-1} = -(1-r_c)^\mu + c_2,$$

where c_1 and c_2 are the errors of the Maclaurin approximation taking the first term. Considering that c_1 and c_2 are small constants, let us write

$$-\frac{C_g}{\sqrt{g}(\mu-1)} = -(1-r_c)^\mu.$$

Applying Maclaurin series to the right side of the expression above and replacing C_g by $1/g^a$, we obtain

$$\frac{\ln(g^{a+1/2}) + \ln(\mu - 1)}{\mu} = r_c + c_3, \qquad (8)$$

where c_3 is the error of the Maclaurin approximation.

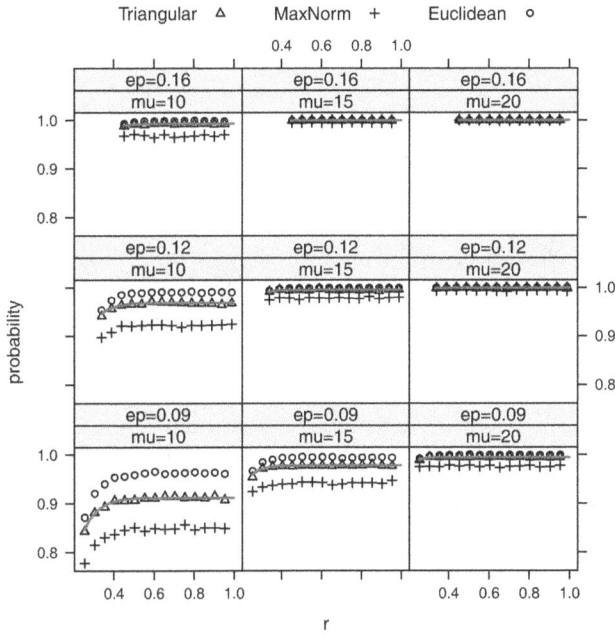

Figure 4: Probability of having at least one crossing link between perpendicular segments. Curves correspond to analytical results.

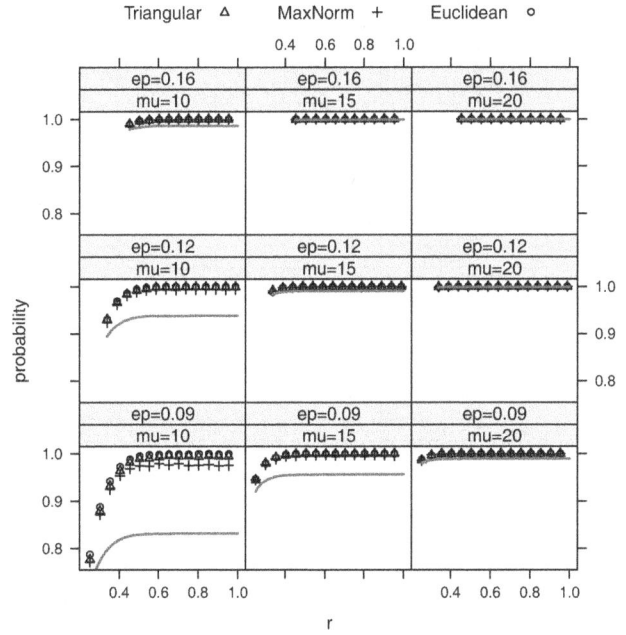

Figure 5: Probability of connectivity at intersections. Curves correspond to the lower bound $p_\perp(\mu)^2$.

Note that Expression (8) corresponds to the CTR whenever connectivity at intersections is guaranteed, i.e., when $\epsilon \geq \epsilon_c$.

□

5. EXPERIMENTAL RESULTS

In this section we present experimental results to validate the analytical results discussed in Sections 3 and 4. Figure 4 shows the results for the probability of having at least one crossing link between perpendicular segments sharing an intersection ($p_\perp(\mu)$, defined in (1)). Figure 5 presents the results for the connectivity probability at intersection ($\Pr(I_{con})$, defined in (3)). In both figures, we observe results for the three models, namely, the MaxNorm, Triangular and Euclidean models. Additionally, we validate the CTR, derived in Section 4, in Figure 6. Finally, we conclude this section by showing in Figure 7 how a unique connected component emerges as we increase the transmission range.

We perform Monte Carlo simulations and, in all cases, we use the R system[5], version 2.13.0. In each figure, curves and lines represent analytical results, and empirical data is represented by bullets.

Each row of Figure 4 corresponds to a particular street width ϵ ("ep"), and each column to a specific network density μ. We consider street width varying according to the critical segments' width ϵ_c for densities $\mu \in \{10, 15, 20\}$. In all cases, we observe that the highest values of $p_\perp(\mu)$ occur for the Euclidean model, and the lowest values appear for the MaxNorm model. The main diagonal of the figure corresponds to cases where $\epsilon = \epsilon_c$, for each value of μ. The

Triangular model represents a compromise between the Euclidean and MaxNorm models, being closer to the Euclidean model. We observe that the lower the density and the lower the street width, the larger the discrepancy between models.

Finally, we highlight that the continuous (non-dotted) lines correspond to the analytical expression (1) of $p_\perp(\mu)$ (for the Triangular model), derived in Theorem 2, fit the empirical data with good precision, validating this analytical result.

Figure 5 presents empirical data for $\Pr(I_{con})$ for all the models considered in this work. The continuous lines in this chart correspond to the analytical lower bound $(p_\perp(\mu))^2$, result that is used in Lemma 2. This implies that, in each case, the difference between circles (empirical data for the Triangular model) and the continuous line represents approximately the value of the second term of (3). The considered scenarios are the same as in Figure 4, that is, we have $\epsilon \in \{0.16, 0.12, 0.09\}$, and $\mu \in \{10, 15, 20\}$. We observe that the smaller the density and smaller the street width ϵ, the larger the probability $\Pr(I_{con} \mid e_1 \cap \overline{e_3})$ (see Equation (3)). This implies that our lower bound $(p_\perp(\mu))^2$ for $\Pr(I_{con})$ improves as μ and ϵ increase. In particular, we observe that for $\epsilon \geq \epsilon_c$ the lower bound $(p_\perp(\mu))^2$ approximate $\Pr(I_{con})$.

Figure 6 presents ECDFs for the CTR for different network sizes and densities, using the critical street width ϵ_c. Here, we represent the theoretical CTR of Theorem 3 by a horizontal line. In particular, we use $a = 1$, which warrants a fast convergence for relatively small grids. We observe that the denser the scenario, the stronger the phase transition.

Finally, Figure 7 presents interesting evidence of how connectivity emerges as we increment the transmission range of nodes. We show two different measures: (i) the normalized quantity of connected components in the network, and (ii) the normalized size of the Giant component, i.e., the

---

[5]Available at http://www.R-project.org/

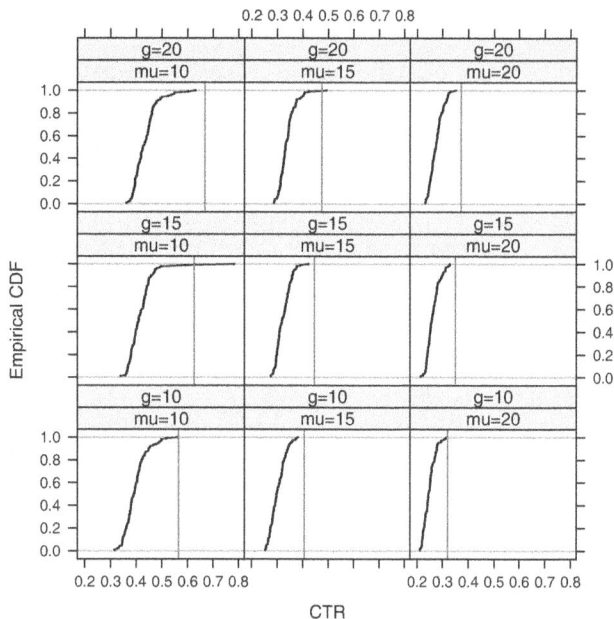

Figure 6: ECDFs and analytical CTRs.

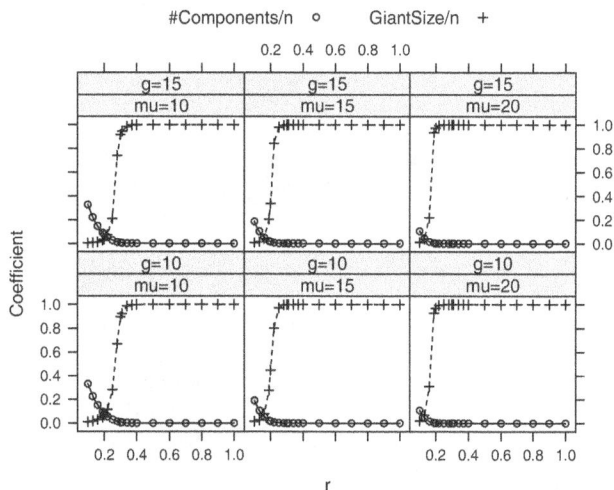

Figure 7: Proportion of connected components and proportional size of the Giant component.

size of the biggest connected component in the network. We observe in this figure that the Giant component grows to a unique connected component very fast, and this phenomenon becomes faster and faster as we increment the density μ. Note that when the Giant component starts to grow, the proportion of connected components diminishes considerably. This implies that it is rare to have isolated nodes and a big Giant component in the same network.

6. RELATED WORK

A popular model when studying wireless (ad hoc) networks is the *Geometric Random Graph* model [4, 14]. In this model, a set of n nodes is deployed randomly in a d-dimensional space, and two nodes can communicate with each other if their distance is less than a certain value r. A variety of analytical results have been derived for this model. Penrose [15] analyzed the influence of parameters n and r on connectivity. In one-dimensional setting, Ghasemi et al. [6] analyzed the exact connectivity probability of connectivity.

Dousse et al. [3] analyzed non-obstructed networks combined with an overlay of a connected backbone of base stations (hybrid networks) in one, two and three dimensions. Among other results, they showed that percolation never occurs (for any node density) in a line or a strip, and that the hybrid structure significantly improves connectivity only when the node distribution is close to one-dimensional.

Shakkottai et al. [17] studied network scenarios with possible failures. They demonstrated that randomness induced by node failure on a regular lattice can lead to similar scaling in terms of critical transmission range as randomness induced by uniform node placement.

Santi and Blough [16] investigated the connectivity in sparse (open-space) networks. They showed that the communication graph in an area $R = [0, \ell]^d$ is connected with high probability if $r^d n = k \ell^d \ln \ell$, for a constant $k > 0$. Also, using results from continuum percolation theory [12], Gupta and Kumar [9] analyzed the critical power level that is necessary for a randomly deployed wireless network to become connected under the assumption that all nodes transmit at the same power level.

Such open-space models, in spite of providing useful analytical tools, make one strong and limiting assumption: any node can freely communicate with all other nodes located within radius r around it. Therefore, these models often do not adequately represent scenarios encountered in practice, like dense urban environments, indoor spaces, or surfaces covered by dense vegetation, in which communication is affected deeply by obstacles.

Studies addressing connectivity properties of obstructed wireless networks are significantly scarce. Frieze et al. [5] proposed a model for wireless environments with regularly distributed obstacles. In this model, nodes are placed on each vertex of an $n \times n$ grid with probability $p(\omega)$, where ω is the communication range of a node. Among other results, they showed that the critical node placement probability is $p(w)^* = O(\ln n/w)$, above which the network becomes k-connected with high probability. Bollobás et al. [2] complemented this result by proving that $\lim_{w \to \infty} wp(w)^* = \log 3/2$.

Nekoui and Pishro-Nik [13] studied how the shape of obstacles in obstructed networks influences the connectivity in a wireless network. Their results, however, can not be applied to more general network topologies.

A simpler version of the model analyzed in this work was first introduced in [1]. A preliminary analysis of the MaxNorm intersection model (see Section 2) was performed, which resulted in low-quality approximations for the connectivity properties of the network. In contrast, in this work, we propose and compare three different ways of modeling connectivity at street intersections and provide a more complete analysis resulting in better approximations, which were confirmed by simulations. In particular, as discussed in Section 5, by using the Triangular model, we obtain a remarkably good compromise between Euclidean and MaxNorm models, which allow us to derive lower values for the critical transmission range for connectivity, as well as lower values for the so-called "critical street width" ϵ_c. As discussed in Section 5, this result improved connectivity for the overall network.

7. CONCLUSIONS

In this work, we proposed and analyzed an alternative model for obstructed wireless networks, based on a grid structure of one-dimensional street segments and two-dimensional street intersections. This model provides a realistic representation of different network scenarios with obstacles and, at the same time, allows us to perform a simple analysis, partly based on percolation theory and partly based on geometry. We analyzed three different ways of modeling the geometric part of the network and derived bounds for the connectivity probability of the critical transmission range for connectivity in this kind of network. Finally, we presented extensive simulations that demonstrated that our analysis provides a remarkably good approximation, especially for high density scenarios.

As a future work, it will be useful to find closed expressions for δ and γ (in Theorem 1) so as to obtain a tight expression for $\Pr(I_{con})$ and, consequently, obtain a tighter bound for ϵ_c. Moreover, this will give us the possibility to perform more rigorous analysis of the scaling rules for site occupation probability in larger grids. Additionally, an interesting extension would be to look at other kinds of regular lattices for the model to replace the underlying grid structure that we used.

8. ACKNOWLEDGEMENTS

This work was partially supported by CNPq and FAPEMIG.

9. REFERENCES

[1] M. G. Almiron, O. Goussevskaia, A. C. Frery, and A. A. Loureiro. Modeling and connectivity analysis in obstructed wireless ad hoc networks. In *Proceedings of the 15th ACM International Conference on Modeling, Analysis and Simulation of Wireless and Mobile Systems (MSWiM)*, pages 195–202, 2012.

[2] B. Bollobás, S. Janson, and O. Riordan. Line-of-sight percolation. *Combinatorics, Probability & Computing*, 18(1–2):83–106, 2009.

[3] O. Dousse, P. Thiran, and M. Hasler. Connectivity in ad-hoc and hybrid networks. In *Proceedings of the 21st Annual Joint Conference of the IEEE Computer and Communications Societies (INFOCOM)*, volume 2, pages 1079–1088, 2002.

[4] M. Franceschetti and R. Meester. *Random Networks for Communication: From Statistical Physics to Information Systems*. Statistical and Probabilistic Mathematics. Cambridge University Press, 2007.

[5] A. M. Frieze, J. M. Kleinberg, R. Ravi, and W. Debany. Line-of-sight networks. In *Proceedings of the 18th Annual ACM-SIAM Symposium on Discrete Algorithms (SODA)*, pages 968–977, 2007.

[6] A. Ghasemi and S. Nader-Esfahani. Exact probability of connectivity in one-dimensional ad hoc wireless networks. *IEEE Communications Letters*, 10(4):251–253, 2006.

[7] O. Goussevskaia, M. M. Halldórsson, R. Wattenhofer, and E. Welzl. Capacity of arbitrary wireless networks. In *Proceedings of the 32nd IEEE International Conference on Computer Communications (INFOCOM)*, pages 1872–1880, 2009.

[8] G. Grimmett. *Percolation*. Springer-Verlag, second edition, 1999.

[9] P. Gupta and P. R. Kumar. Critical power for asymptotic connectivity in wireless networks. *Stochastic Analysis, Control, Optimization and Applications: A Volume in Honor of W. H. Fleming*, pages 547–566, 1998.

[10] P. Gupta and P. R. Kumar. The capacity of wireless networks. *IEEE Transactions on Information Theory*, 46(2):388–404, 2000.

[11] M. Haenggi and R. K. Ganti. Interference in large wireless networks. *Foundations and Trends in Networking*, 3(2):127–248, 2009.

[12] R. Meester and R. Roy. *Continuum percolation*. Cambridge tracts in mathematics. Cambridge University Press, 1996.

[13] M. Nekoui and H. Pishro-Nik. A geometrical analysis of obstructed wireless networks. In *Proceedings of the IEEE Information Theory Workshop (ITW)*, pages 589–593, 2009.

[14] M. Penrose. *Random geometric graphs*. Oxford studies in probability. Oxford University Press, 2003.

[15] M. D. Penrose. On k-connectivity for a geometric random graph. *Random Structures & Algorithms*, 15(2):145–164, Sep 1999.

[16] P. Santi and D. Blough. The critical transmitting range for connectivity in sparse wireless ad hoc networks. *IEEE Transactions on Mobile Computing*, 2(1):25–39, 2003.

[17] S. Shakkottai, R. Srikant, and N. Shroff. Unreliable sensor grids: Coverage, connectivity and diameter. In *Proceedings of the 22nd Annual Joint Conference of the IEEE Computer and Communications Societies (INFOCOM)*, volume 2, pages 1073–1083, 2003.

On the Instantaneous Topology of a Large-Scale Urban Vehicular Network: the Cologne Case

Diala Naboulsi
Université de Lyon, INRIA
INSA-Lyon, CITI-INRIA
F-69621, Villeurbanne, France
diala.naboulsi@insa-lyon.fr

Marco Fiore
CNR – IEIIT
Corso Duca degli Abruzzi 24
10129 Torino, Italy
marco.fiore@ieiit.cnr.it

INRIA
INSA-Lyon, CITI-INRIA
F-69621, Villeurbanne, France
marco.fiore@inria.fr

ABSTRACT

Despite the growing interest in a real-world deployment of vehicle-to-vehicle communication, many topological features of the resulting vehicular network remain largely unknown. We still lack a clear understanding of the level of connectivity achievable in large-scale urban scenarios, of the availability and reliability of connected multi-hop paths, and of the evolution of such features over daytime. In this paper, we investigate how the instantaneous topology of the vehicular network would look like in the case of Cologne, Germany, a typical middle-sized European city. Through a complex network analysis, we unveil the low connectivity, availability, reliability and navigability of the network, and exploit our findings to derive network design and usage guidelines.

Categories and Subject Descriptors

C.2.1 [**Computer-Communication Networks**]: Network Architecture and Design – *network topology, wireless communication*; C.4 [**Performance of Systems**]: *reliability, availability, and serviceability*

General Terms

Performance, reliability

Keywords

Vehicular networking, topology, connectivity, mobility

1. INTRODUCTION

More than a decade after the allocation of dedicated frequency bands in the USA, vehicular communication networks have finally abandoned their long-standing status of fundamental research exercise and have started developing into real-world systems. On-going standardization efforts, including IEEE 802.11p, IEEE 1609, OSI CALM-M5 and ETSI ITS, jointly with the growing interest of the automobile industry, have played a major role in speeding up such a process over the last few years. Thus, it will be soon time to field test the vast amount of vehicular networking solutions proposed by

the scientific community during the last decade, whose scope spans across, e.g., medium access, transmission power control, data rate adaptation, multi-hop routing, content downloading, and data dissemination.

However, the real-world performance of many of such protocols risks to fail the expectations, for the simple reason that they will be confronted by a network they were not designed for. As a matter of fact, dedicated solutions have been – and are being – proposed for vehicular networks whose major topological features remain largely unknown. In urban environments in particular, even basic questions stay unanswered, such as: *is the vehicular network well connected or highly partitioned? Which size can clusters of multi-hop connected vehicles attain? Which is the internal structure of such clusters? How sparse or dense are single-hop communication neighborhoods? How do all these network connectivity features vary in time? How do they depend on the geographical location?*

The responses to these questions directly determine the strengths, weaknesses and overall capabilities of a spontaneous vehicular network, and shall thus be among the main drives to the design of dedicated protocols. Moreover, they are the key to quantifying the *availability* and *reliability* of the network, i.e., the main concerns of car manufacturers when it comes to vehicle-to-vehicle multi-hop communication. Yet, the literature on the topic is surprisingly thin and, despite the interesting studies discussed in Sec. 2, we are still a long way from being able to provide exhaustive answers to the critical questions above.

This paper aims at contributing to the process of better characterizing the major topological features of large-scale urban vehicular networks. Specifically, we focus on the case of Cologne, Germany, a typical middle-sized European city, for which a mobility dataset of unprecedented scale, duration and realism is available, as detailed in Sec. 3. In such a scenario, we are interested in characterizing the *instantaneous topology* of the city-wide vehicular network, and we thus transpose the mobility dataset to a set of instantaneous connectivity graphs, as detailed in Sec. 4. We then study the features and dynamics of the vehicular network topology through a *complex network* approach [1, 2]. The results are analyzed from a communication network perspective, in Sec. 5, 6 and 7. Our conclusions are summarized in Sec. 8.

2. RELATED WORK

The characterization of the instantaneous topology of a vehicular network demands, by its own nature, the knowledge of the exact positions of all the vehicles in a large geographical region at every second or so. Therefore, logs of real-world car movements (e.g., obtained via GPS tracking) would represent the ideal data upon which to carry out a topological analysis. Unfortunately, current real-world vehicular mobility datasets are inadequate, since they only comprise a small percentage of the overall road traffic, limited to specific subsets of vehicles such as buses [3] or taxis [4]. Additionally, the positions of vehicles have a too high update periodicity, in the order of tens of seconds. As a result, studies on the instantaneous connectivity of vehicular networks rely on synthetic traces or analytical tools.

Highway environments are simpler to analyze in terms of connectivity than urban ones, since they result in unidimensional road traffic flows. Studies on the topology of vehicular networks on highways aim at determining which combinations of vehicular density, absolute and relative car speed, technology penetration rate and communication range are required to achieve a full network connectivity. However, the results, be they based on synthetic traces of highway traffic [5,6] or mathematical models [7–9], hardly apply to the urban scenarios we are interested in.

Indeed, urban environments are characterized by complex bidimensional street layouts with heterogeneous restrictions (e.g., speed limits and one-way rules). Highway-oriented analyses do not capture neither those nor the non-trivial mechanisms that regulate urban road intersections, such as traffic lights, roundabouts, stop and yield signs. In fact, these same composite features force analytical studies of the vehicular connectivity in urban areas to resort to strong simplifying assumptions, namely Poisson distribution of cars and regular-grid road layouts [10–13]. Such assumptions make the problem mathematically tractable via, e.g., percolation theory, but dramatically reduce the relevance of the results.

As far as synthetic trace-based approaches are concerned, seminal studies have focused on the impact of microscopic mobility models that describe the acceleration or deceleration behavior of each car with respect to the surrounding driving conditions [14,15]. These works unveiled the high bias that unrealistic random and pseudo-random driver behaviors can induce on the network topology, and the importance of properly modeling road signalization. However, they did not consider the impact of macroscopic traffic parameters, such as the vehicular density, the arrival rates and the routes traveled by drivers.

Therefore, later works focused on the macroscopic dimension, built on top of validated microscopic mobility models. Specifically, the studies in [16,17] aimed at deriving general conclusions on the effect of the diverse macroscopic parameters. To that end, they considered small-scale regular-grid road layouts where the travel demand is random and the traffic density and road junction regulations are controllable system parameter. However, their findings do not necessarily apply to more realistic mobility scenarios with larger scales, real-world street topologies, and actual travel demands and road traffic densities. as an example, our results show that a well-connected vehicular network cannot be formed at all times [16] and that the average vehicular density alone is not sufficient to characterize the overall connectivity [17], due to the high spatial diversity of the network topology.

As a consequence, more recent studies have assessed the vehicular network connectivity in real-world urban areas. In [18], the authors focus on the city of Porto, Portugal. The scope is however limited to the evolution of the average degree of the vehicular network connectivity graph for around 5 minutes. A significantly more complete study is provided in [19], where the authors perform a thorough analysis of the vehicular network topology in Zurich, Switzerland. Our study confirms some of their findings, e.g., the fact that the vehicular network does not exhibit small-world properties, or that cliques of connected cars are easily formed. However, we find completely different results concerning, e.g., the power-law behavior of the vehicular network, the fact that a giant cluster com-

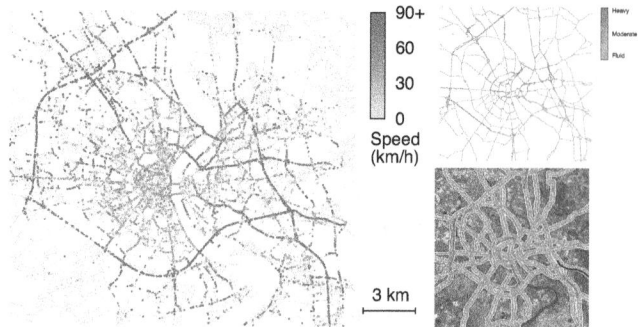

Figure 1: Cologne dataset. Snapshot of the vehicular mobility (left), and road traffic in simulation (top right) and real world (bottom right) at 17:00. This figure is best viewed in colors.

prising more than 50% of the vehicles is present all the time, or that the connectivity within large clusters remains stable over time.

We ascribe these diversities to three aspects. First, the study in [19] is limited to a region of 25 km^2, mapping to downtown Zurich, and to a time span of 3 hours, covering the morning road traffic peak. Our mobility dataset covers a 400-km^2 region, including the center, suburbs and outskirts of Cologne, and a whole day: the unprecedented space and time scale of our dataset allows us to capture spatiotemporal phenomena that are hidden to previous studies. Second, the mobility dataset employed in [19] only includes major road arteries, and the mesoscopic simulator used to generate the vehicle movement yields a rather uniform distribution of cars over the road layout. These factors contribute to a vehicular network significantly biased towards unrealistically high connectivity. The mobility dataset that we adopt features a more complete road layout and a more realistic microscopic mobility description, leading to sounder connectivity results. We refer the reader to [20] for a brief comparison of the two datasets. Third, Zurich and Cologne are two different cities, and some diversities may be inherent to the diverse street layout and road traffic distribution. As we will discuss in our conclusions, this point motivates further studies on urban areas other than Zurich and Cologne.

Finally, we remark that some works in the literature consider metrics, such as roadside infrastructure settings [16] or temporal measures [19,21], that we do not include. As a matter of fact, these are out of the scope of our analysis, which focuses on the instantaneous connectivity of the vehicular network, a complex problem per-se. We acknowledge the relevance of temporal or infrastructure-oriented metrics, and we will consider them in future works.

3. VEHICULAR MOBILITY DATASET

The mobility dataset we employ has been synthetically generated as part of the TAPASCologne initiative of the German Aerospace Center (DLR), that aims at reproducing with the highest level of realism possible the car traffic in the greater urban area of Cologne, Germany. The dataset spans over 24 hours of a typical workday and encompasses 4500 km of roads in an area of 400 km^2, with per-second information on the position and speed of vehicles involved in more than 700000 trips.

This is the largest and most complete synthetic vehicular mobility dataset freely available to date, and is obtained by coupling state-of-art tools dedicated to specific aspects of vehicular traffic modeling. Namely, the street layout of the Cologne urban area is extracted from the OpenStreetMap (OSM) database and fed to the Simulation of Urban Mobility (SUMO) software, the most advanced open-source microscopic vehicular mobility generator available today. Large-scale car flows across the Cologne urban area are determined by coupling a travel demand and a traffic assignment model. The former determines the locations where each vehicle starts and ends its trips and is derived by applying the Travel and Activity PAtterns Simulation (TAPAS) methodology to real-world data collected in the Cologne region [22]. The latter computes the exact path between such locations and is implemented via the relaxation technique proposed by Gawron, that allows to achieve a dynamic user equilibrium [23].

The dataset has been observed to yield similar characteristics to those of real-world road traffic in Cologne. Fig. 1 shows the road topology, as well as the nice match between the road traffic observed at 17:00 in the synthetic trace and in the ViaMichelin live traffic service. We refer the reader to [20] for further information on the dataset.

4. NETWORK MODEL AND METRICS

Our analysis is technology- and protocol-independent, as it targets the physical topology of the vehicular network. To that end, we borrow tools from complex network theory, that have been successfully employed to characterize a number of large-scale real-world networks such as those of Internet routers, World Wide Web pages or interacting social species [1,2]. Next, we provide a description of how we model the instantaneous vehicular network topology into a set of graphs, and we formally define the metrics that will be used in our study.

We sample the vehicular mobility dataset with a fixed frequency[1], and, at each sampled time instant t, we model the vehicular network topology as a graph. We study the *instantaneous* connectivity of the network, by considering each of such graphs separately. We recall that temporal aggregation of the network connectivity graphs or temporal network modeling through tensors are out of the scope of this paper, although we plan to address those in the future.

We name $G(\mathbb{V}(t), \mathbb{E}(t))$ the instantaneous graph at time t. There, $\mathbb{V}(t) = \{v_i\}$ is a set of vertices (or nodes, as the two terms will be used interchangeably in the following) v_i, each representing a vehicle i traveling in the road scenario at time t, and $\mathbb{E}(t) = \{e_{ij}(t) \mid v_i, v_j \in \mathbb{V}, i \neq j\}$ is the set of edges $e_{ij}(t)$, modeling the availability of a communication link between vehicle i and vehicle j at t.

The number of nodes in the network varies with time and is referred to as $\mathcal{N}(t) = \|\mathbb{V}(t)\|$, where $\| \cdot \|$ denotes the cardinality of the included set. The set $\mathbb{E}(t)$ depends instead on the RF signal propagation model adopted, which in this work we will assume to be a simple unit disc model. We acknowledge that, despite being a common practice in the related works presented in Sec. 2, this is a drastic simplification of the reality. However, deterministic propagation models (e.g., ray-tracing ones) do not scale well to our large-scale vehicular scenario, as they require expensive re-computations at each movement of every network node. As for stochastic models, they introduce a random noise that would force us to evaluate a large set of instances for each sampled time instant, again an unbearable task given the size of our scenario. The unit disc model allows to drastically reduce the computational complexity of the analysis, and yet to capture the average behavior of the network. In the following, we denote the unit disc communication range as R. We stress that, as a consequence of the simple propagation model

adopted, only bidirectional links are present in the graph model, i.e., $G(\mathbb{V}(t), \mathbb{E}(t))$ is undirected and $e_{ij}(t) = e_{ji}(t), \forall i, j, t$.

We also denote the shortest multi-hop communication path between any two cars i and j at time t as the ordered sequence of vertices in the shortest path from v_i to v_j, i.e., $\mathbb{p}_{ij}(t) = \{v_i, \ldots, v_j\}$. The length of such a shortest path is denoted as $\mathcal{L}_{ij}(t) = \|\mathbb{p}_{ij}(t)\| - 1$ and maps to the number of transmission hops separating vehicles i and j. If no shortest path exists between i and j at time t, then $\mathbb{p}_{ij}(t) = \emptyset$ and $\mathcal{L}_{ij}(t) = \infty$. If multiple equivalent shortest paths are available between the same node pair, one is randomly picked. Note that, by definition, $\mathbb{p}_{ii}(t) = \{v_i\}$ and $\mathcal{L}_{ii}(t) = 0, \forall i, t$.

The graph model allows us to introduce the metrics below.

Component. Let us associate to each vertex v_i at time t a subset of the vertices $\mathbb{V}_i(t) = v_i \cup \{v_j \mid \mathbb{p}_{ij}(t) \neq \emptyset\}$, and a subset of the edges $\mathbb{E}_i(t) = \{e_{jk}(t) \mid v_j, v_k \in \mathbb{V}_i(t)\}$. We define the subgraph $C_i(t) = G(\mathbb{V}_i(t), \mathbb{E}_i(t))$ as the component within which vertex v_i lies at time t. In other words, the component $C_i(t)$ is the graph representing the portion of the network that vehicle i can reach via multi-hop communication at time t. The size of the component, $\mathcal{S}_i(t) = \|\mathbb{V}_i(t)\|$, is the number of nodes that belong to it. By construction, the subgraph is the same for all vehicles in the same component, i.e., $C_i(t) = C_j(t)$ iff $\mathbb{p}_{ij} \neq \emptyset$, or, from the opposite perspective, no communication path exists between two different components, i.e., $\mathbb{p}_{ij} = \emptyset$ iff $C_i(t) \neq C_j(t)$. Since all node pairs in a same component are connected, given a generic component $C_i(t)$, it holds that $\mathcal{L}_{jk} < \infty, \forall v_j, v_k \in C_i(t)$. We thus define the average shortest path between any two nodes in the component as

$$\mathcal{L}_i(t) = \frac{2}{\mathcal{S}_i(t)\,(\mathcal{S}_i(t) - 1)} \sum_{v_j, v_k \in C_i(t), j < k} \mathcal{L}_{jk}.$$

Finally, we refer to the set of the components present in the network at time t as $\mathbb{C}(t) = \{C_i(t) \mid C_i(t) \cap C_j(t) = \emptyset, \forall i < j\}$. The number of components in the network at time t is then $\mathcal{C}(t) = \|\mathbb{C}(t)\|$.

Giant component. A component $C_i(t)$ is said to be a giant component if $\mathcal{S}_i(t) \geq 0.1 \cdot \mathcal{N}(t)$. That is, a giant component at time t has a size of the same order of magnitude as the whole network at the same time. We also denote the largest component in the network at time t as $C_{max}(t)$, and its size as $\mathcal{S}_{max}(t)$. If the largest component $C_{max}(t)$ is a giant component, we denote its size as $\mathcal{S}_{giant}(t) = \mathcal{S}_{max}(t) \geq 0.1 \cdot \mathcal{N}(t)$. We remark that a giant component is not necessarily present in the network, i.e., it can be that $\mathcal{S}_{max}(t) < 0.1 \cdot \mathcal{N}(t)$.

Degree. Let us consider a subset $\mathbb{V}_i^1(t) = \{v_j \mid \exists e_{ij}(t)\}$ of vertices and a subset $\mathbb{E}_i^1(t) = \{e_{jk}(t) \mid v_j, v_k \in \mathbb{V}_i^1(t)\}$ of edges. We define the subgraph $C_i^1(t) = G(\mathbb{V}_i^1(t), \mathbb{E}_i^1(t))$ as the one-hop communication neighborhood of vertex i at time t. The degree of a vertex v_i is then $\mathcal{D}_i(t) = \|\mathbb{V}_i^1(t)\|$, and corresponds to the number of one-hop communication neighbors of vehicle i.

Betweenness centrality. The betweenness centrality of a vertex v_i at time t is

$$\mathcal{B}_i(t) = \frac{2}{(\mathcal{S}_i(t) - 2)\,(\mathcal{S}_i(t) - 1)} \sum_{\substack{j, k \neq i \\ v_j, v_k \in C_i(t)}} \frac{\|\{\mathbb{p}_{jk}(t) \mid v_i \in \mathbb{p}_{jk}(t)\}\|}{\|\mathbb{p}_{jk}(t)\|}.$$

The betweenness centrality represents how frequently a vehicle i is part of a shortest path between any two other cars, and ranges from zero (no shortest path passes by i) to one (i lies within all the shortest paths of the network).

For the sake of clarity, in the following we will imply that all measures are time dependent and drop the time notation. We will thus denote, for a generic time instant, the number of network nodes as $\mathcal{N}$ and the number of components as $\mathcal{C}$. Also, for the vehicle-

---

[1]In the following, unless stated otherwise, we will employ a sampling periodicity of 10 seconds, that was found to yield results identical to those obtained with higher sampling frequencies, at a lower computational cost.

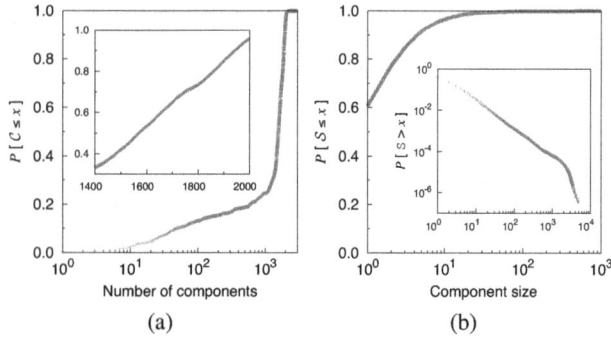

Figure 2: Distributions of (a) the number of components and (b) the component size, when aggregating all the samples of C and S over the whole day.

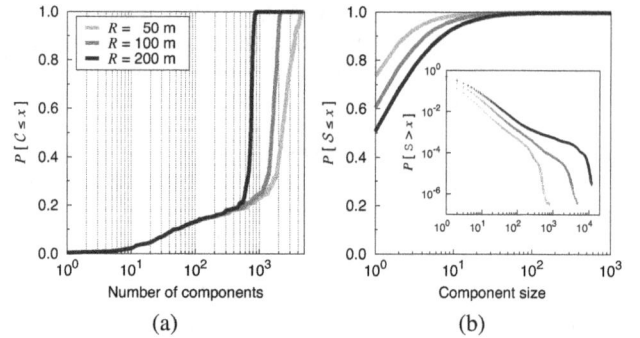

Figure 4: Distributions of (a) the number of components and (b) the component size, when aggregating all the daily samples of C and S, for different R's.

dependent metrics, we will assume that they refer to a generic vehicle i and drop the per-vehicle notation. Therefore, the size of a generic component will be denoted by S, its average path length as $\mathcal{L}$, and the degree and betweenness centrality of a generic node as $\mathcal{D}$ and $\mathcal{B}$.

5. NETWORK-LEVEL ANALYSIS

We start our analysis by considering the vehicular network as a whole and discussing its global properties in terms of connectivity. For the sake of clarity, we initially limit our analysis to one particular value of the communication radius R. To that end, we employ $R = 100$ m, as this is the value referenced by field tests as a typical distance for reliable DSRC vehicle-to-vehicle communication [24, 25]. More precisely, the extensive experimental analysis in [25] shows that a distance of 100 m allows around 80% of the packets to be correctly received in urban environments, under common power levels (15-20 dBm) and with robust modulations (3-Mbps BPSK and 6-Mbps QPSK). We will later generalize our study to $R = 50$ m, experimentally identified as the largest distance at which vehicle-to-vehicle communication has a packet delivery ratio close to one [24, 25], and $R = 200$ m, i.e., the maximum distance granting a reception ratio above 0.5 [25]. We consider wireless links losing more than 50% of the packets hardly exploitable by the network.

Analysis for $R = 100$ m. The level of connectivity of the whole vehicular network is mainly characterized through two metrics: the number of components C, that is an index of the level of network fragmentation, and the component size S, which describes the heterogeneity of the fragmentation.

The Cumulative Distribution Function (CDF) of C, aggregated over all the vehicular network graphs extracted from the whole 24 hours of road traffic, is portrayed in Fig. 2(a). In 80% of cases the vehicular network has more than 1000 disconnected components, and the inset plot shows that most of the probability is concentrated in a linear growth between 1000 and 2000 components. This suggests that the vehicular network is highly partitioned into thousands of separate node groups unable to communicate with each other.

The distributions of the component size S help us clarify whether we face many components of similar size or a heterogeneous network of both large and small components. The CDF in Fig. 2(b) shows that the network is largely made of very small components, with 60% of them being *singletons*, i.e., isolated vehicles, and 95% of the components comprising 10 vehicles or less. However, by looking at the Complementary CDF (CCDF), portrayed in the inset

plot, we can appreciate the heavy tail of the distribution, appearing as linear on a log-log scale. There is thus a non-negligible probability that the aforementioned small components coexist with components that include up to 2000 vehicles connected through direct links or multi-hop routes. After such a component size, the CCDF has an exponential decay, and components as large as 4500 nodes appear with significant lower probability.

Such a general view of a partitioned and heterogeneous network is however aggregated over time and space. In order to unveil the impact of the daytime and the differences between geographical areas, we show in Fig. 3 the instantaneous vehicular network fragmentation measured in the Cologne region at different hours. In each plot, every circle corresponds to one component, its diameter and color mapping to the size of the component it represents: broader circles map to larger components, while the color code is reported on the bars at the right of the plots. The resulting images give a rough, yet intuitive, idea of the behavior of the network connectivity evolution: early in the morning, i.e., before 6:00, the network is very partitioned and only small components of 40 vehicles or less are present. The morning traffic peak, between 7:00 and 8:00, has a very positive impact on the network topology, with the appearance of very large components of thousands of vehicles and a diffuse presence medium-sized components of several tens of cars. That effect disappears later on, and large components do not reappear until the afternoon traffic peak, between 16:00 and 18:00, although a slightly increased connectivity is observed around noon. Moreover, large components mostly appear in the city center, where the traffic is denser. We can infer that both time and space are paramount factors in the characterization of the vehicular network topology, which is generally highly fragmented, with larger components only appearing at specific locations during the traffic peak hours.

Generalization to different R's. We now study how different communication ranges impact the observations above. Fig. 4(a) portrays the CDF of the number of components C when the transmission range R varies from 50 to 200 m. The probability distribution for low C does not vary with R: since we observed in Fig. 3 that the number of components is consistently high throughout the whole day, we can now conjecture that these values of C map to hours when no more than a few hundreds of isolated vehicles are present in the area, i.e., before 5:00 and after 22:00. When C is instead higher than 500, a higher R shifts the probability concentration to the left, or, in other words, a higher communication radius bounds the maximum number of components to a lower value. As a result, we never have more than 1000 components when $R = 200$

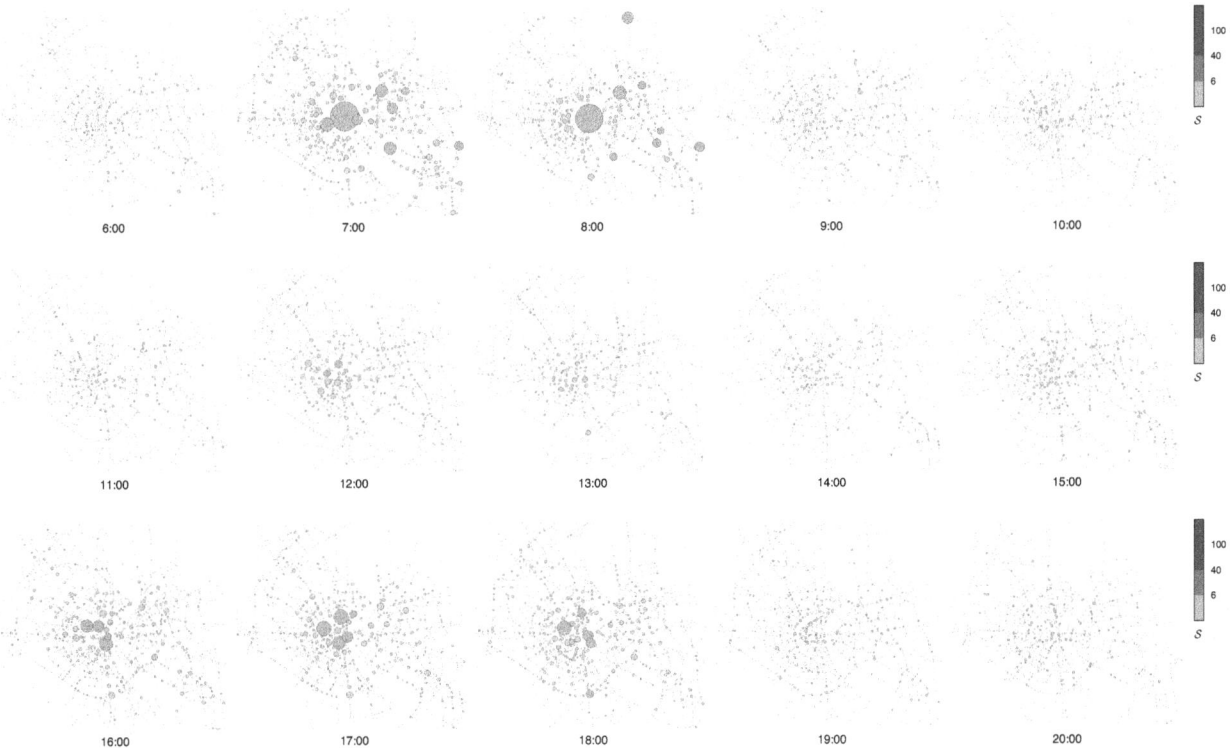

Figure 3: Geographical plots of the network components over the day, when $R = 100$ m. Figure best viewed in colors.

m, while we can have up to 5000 components when $R = 50$ m, a clear symptom that R can significantly enhance or disrupt the network-level connectivity.

A confirmation comes from the component size distributions, in Fig. 4(b). There, higher values of R result in larger components at all levels: both small components, in the CDF, and large ones, in the inset CCDF, are positively affected by R. Thus, when R is reduced to 50 m, no component with more than 1000 vehicles is observed. However, it is interesting to note that the network is still significantly partitioned even when $R = 200$ m, as 90% of the components comprise no more than 10 cars.

A space-time view of the results above is provided by Fig. 5, that displays the geographical distribution of components in sparse (11:00) and dense (7:00) road traffic conditions, when R is set to 50 and 200 m. It is evident that a reduced transmission range yields a completely fragmented network, where multi-hop connectivity is hard to spot even during the traffic peak hours. A higher R leads instead to a network that is diffusely more connected (see the medium- and large-sized components even at 11:00) and allows for very large components, including more than 10000-vehicle components to form in presence of intense road traffic. Once more, we remark the critical impact of both time and space, with the largest components always emerging in downtown Cologne for all R's.

Relationship to $\mathcal{N}$. The strong time dependence of the network outlined by the previous results pushes us to verify if a more rigorous relationship can be found between the individual time instants and the vehicular network topology they yield. Following a complex network common practice, we observe the correlation of the connectivity metrics with the number $\mathcal{N}$ of nodes, i.e., vehicles, present in the network.

Fig. 6 shows the standard deviation of the number of components $\mathcal{C}$ computed among graphs that have a similar $\mathcal{N}$, which in turn varies along the x axis. The deviation is expressed as a percentage of the average $\mathcal{C}$ over the same set of graphs. We observe that the standard deviation is extremely low, typically within 5% from the average value, for any significant $\mathcal{N}$. This behavior, that holds for any R, implies that all the graphs that have a similar number of vertices also have a very similar number of clusters.

Fig. 7 shows the distributions of the component size $\mathcal{S}$ measured at different sample hours, and compares them to distributions obtained by aggregating the data from all the network graphs with a similar number of vertices $\mathcal{N}$. We can observe that for both small (CDF, left) and large (CCDF, right) components, as well as for any value of the transmission range R, the distributions at different daytime with similar $\mathcal{N}$ overlap and follow the aggregate one for that $\mathcal{N}$ range.

We can conclude from these results that it is not the absolute time that drives the vehicular network properties, but the number of vehicles present in the urban region. This is a non-trivial conclusion: it implies that the network behavior in, e.g., the morning and the afternoon is the same, provided that we consider two instants with similar $\mathcal{N}$. Also, it accommunates vehicular networks to many other complex networks, for which the network size $\mathcal{N}$ is the reference parameter driving the evolution of the network [1,2]. In the remainder of the paper, we will study the vehicular topology at the light of the characterizing network parameter $\mathcal{N}$.

Key networking insights. From a networking viewpoint, we can remark that, in the Cologne scenario, *the vehicular network is severely and consistently partitioned*, even when considering idealized RF signal propagation and the maximum reliable vehicle-to-vehicle communication ranges identified by experimental DSRC evaluations. This suggests that the connected multi-hop routing and data dissemination attempted by many protocols in the literature is not feasible through the whole network, and that *carry-and-forward*

(a) 11:00, $R = 50$ m   (b) 11:00, $R = 200$ m

(c) 7:00, $R = 50$ m   (d) 7:00, $R = 200$ m

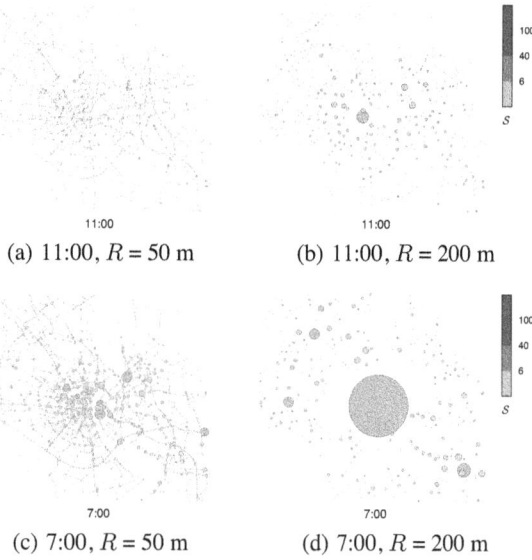

Figure 5: Geographical maps of clusters at 11:00 (top) and 7:00 (bottom), when R is 50 (left) and 200 (right) meters. This figure is best viewed in colors.

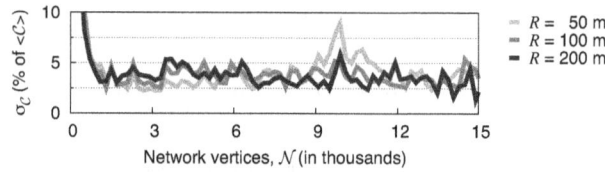

Figure 6: Standard deviation of the component number $\mathcal{C}$, expressed as a percentage of the corresponding average $\mathcal{C}$, versus $\mathcal{N}$ and for different R's.

Figure 7: CDF (left) and CCDF (right) of the component size $\mathcal{S}$, for different values of R. Each plot portrays the distributions at several daytimes, compared with those aggregated over all the snapshots with similar $\mathcal{N}$.

techniques are required to reach all network nodes. Finally, different geographical locations and daytimes imply very diverse connectivity properties: the common practice of *evaluating vehicular network protocols in small-scale scenarios with arbitrary vehicular densities may be harmful.*

6. COMPONENT-LEVEL ANALYSIS

The highly fragmented network topology motivates us to study more deeply the internal structure of the individual components. In particular, our interest lies in the large components we have observed to appear in the network, since it is within them that multi-hop communication can take place and vehicular ad-hoc network protocols can mainly operate.

Component dynamics versus $\mathcal{N}$. Fig. 8 portrays the evolution of the largest network component, C_{max}, in terms of its size $\mathcal{S}_{max}$, versus the network size $\mathcal{N}$ and for different values of the transmission range R. The color of the points refers to the number of components in the network, $\mathcal{C}$, whose landmark values are also pointed out by the arrows.

When R is set to 50 m, the size of C_{max} only slightly increases even for large $\mathcal{N}$'s, implying that large components are never observed in the system. Also, $\mathcal{C}$ monotonically grows with $\mathcal{N}$, suggesting that, in any vehicular traffic condition, new nodes joining the network have high chances of being isolated, forming new singleton components.

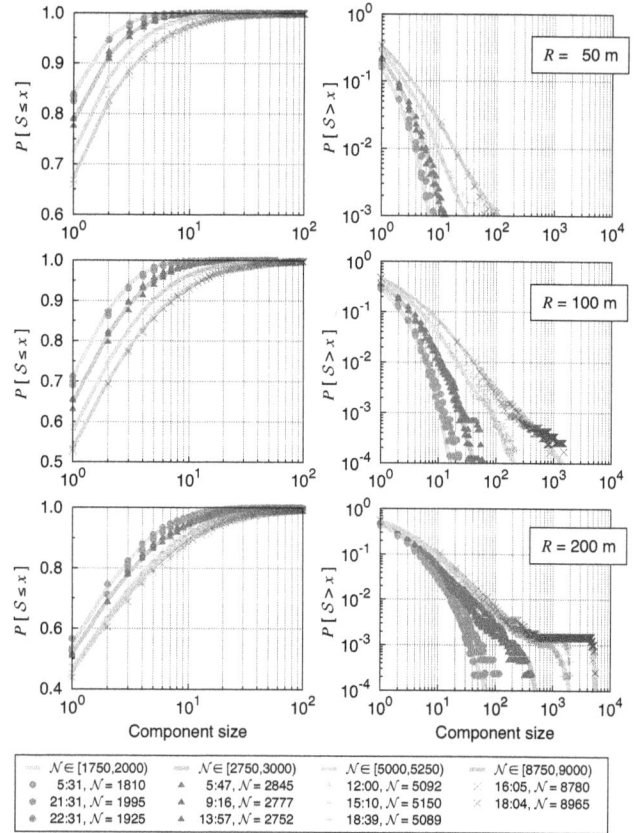

As R is set to 100 m, the behavior significantly differs. The largest component has a negligible size until $\mathcal{N} \sim 6000$, and then grows up to around 4500 nodes. Although there is some variability in the values of $\mathcal{S}_{max}$ for a same $\mathcal{N}$, the positive correlation between the two is evident. Interestingly, the number of components $\mathcal{C}$ grows during the first phase, up to $\mathcal{N} \sim 6000$, and then remains constant at $\mathcal{C} \sim 2000$. What happens is that, once a critical network density is achieved, new nodes joining the network are not isolated anymore, but join existing components: it is not anymore the number $\mathcal{C}$ of components to grow, but their size $\mathcal{S}$, as for $\mathcal{S}_{max}$.

When $R = 200$ m, the behavior changes once more. The critical threshold at which the $\mathcal{S}_{max}$ starts to grow shifts down to $\mathcal{N} \sim 3000$, and is followed by a neat linear relationship between the largest component size and $\mathcal{N}$. Components as large as 12000 vertices can be observed in the network when $\mathcal{N}$ reaches its maximum. The evolution of $\mathcal{C}$ is also notable, as the number of components concurrently present in the network initially grows up to 900 at $\mathcal{N} \sim 5000$, and then starts decreasing. Therefore, when $R = 200$ m, new nodes added to the network after the critical vehicular density has been reached not only join, but even bridge existing components, increasing their size and reducing their number.

Although the dispersion of points at a given $\mathcal{N}$ only allows to individuate rough approximations of the critical density thresholds, the values identified above are qualitative indicators of the network *availability*, i.e., the possibility of leveraging vehicle-to-vehicle multi-hop communication to a meaningful extent. These thresholds are

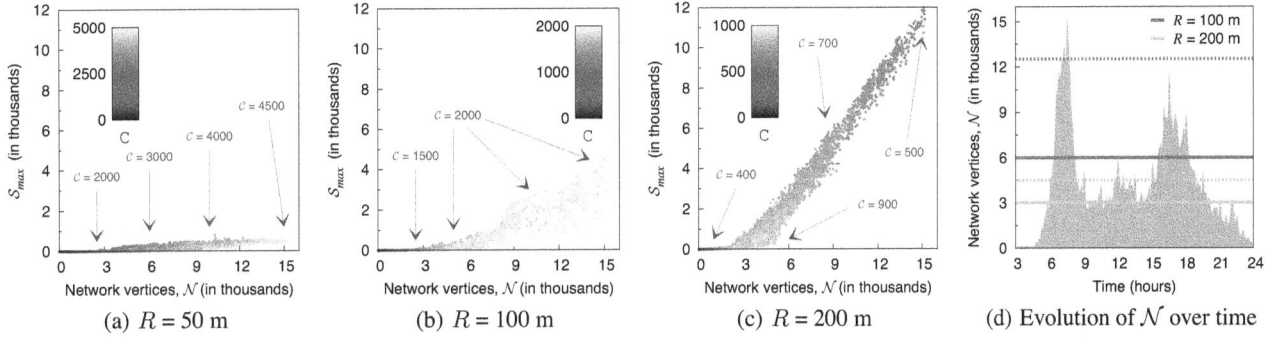

(a) $R = 50$ m   (b) $R = 100$ m   (c) $R = 200$ m   (d) Evolution of $\mathcal{N}$ over time

Figure 8: (a), (b), (c): scatterplots of the largest component size $\mathcal{S}_{max}$ versus the network size $\mathcal{N}$, for different values of communication range R. Colors and arrows indicate the number of components $\mathcal{C}$. (d): evolution of $\mathcal{N}$ over daytime.

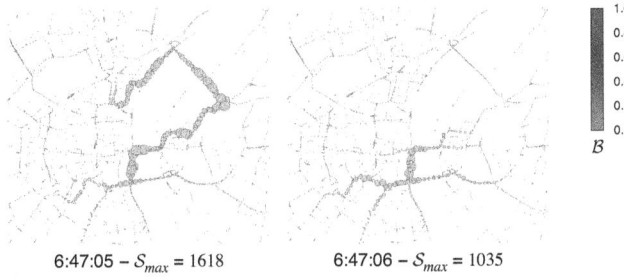

6:47:05 – $\mathcal{S}_{max} = 1618$ 6:47:06 – $\mathcal{S}_{max} = 1035$

Figure 9: Internal structure of $\mathcal{C}_{max}$ at two subsequent time seconds, when $R = 100$ m. Vertices color and size indicate the node's betweenness centrality.

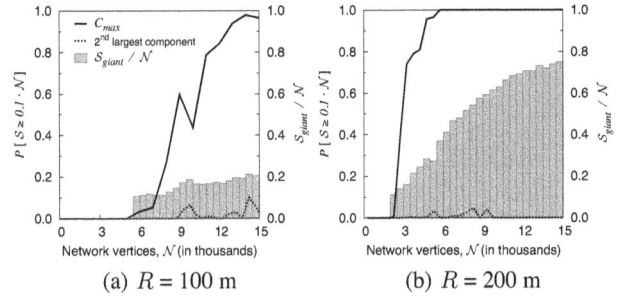

(a) $R = 100$ m   (b) $R = 200$ m

Figure 10: Probability of existence of a first and second giant components and relative size of the first giant component versus $\mathcal{N}$, for different R's.

indicated as solid lines in Fig. 8(d), where they overlap to the evolution of $\mathcal{N}$ over time. Large clusters can be expected to appear in the vehicular network mainly during the morning and afternoon rush hours, with an availability of around 4 hours/day and 8 hours/day when $R = 100$ m and 200 m, respectively. The network is instead never available when $R = 50$ m.

Spatio-temporal dynamics of $\mathcal{C}_{max}$. Not only the availability, but also the *reliability* of large components is a critical factor for the design of vehicular network solutions. In order to analyze this second aspect, we studied the one-second variability of the largest component size $\mathcal{S}_{max}$, and observed that: (i) when $R = 100$ m, $\mathcal{S}_{max}$ undergoes significant variations at each second, its value doubling or halving every few seconds; (ii) when $R = 200$ m, $\mathcal{S}_{max}$ variations still involve thousands of nodes albeit the largest cluster $\mathcal{C}_{max}$ appears more stable due to its larger size, that limits fluctuations to 15% of $\mathcal{S}_{max}$.

The reason behind these major size changes of $\mathcal{C}_{max}$ lies in its internal dynamics. Although a rigorous discussion is not possible due to space limitations, we provide an intuitive example of such dynamics in Fig. 9. The two plots show the vertices in $\mathcal{C}_{max}$ at two subsequent seconds: the internal structure is highlighted in terms of betweenness centrality, the size and color of each node indicating its $\mathcal{B}$ value.

The largest cluster in the left image comprises over 1600 vehicles, and we can note the very high betweenness centrality of nodes in the northeastern region (where a bridge joins the two sides of the Rhine river). This implies that vertices in that area are part of a very large number of shortest paths, or, in other words, they branch together large groups of vehicles. Indeed, at the next time second,

in the right plot, a shift of a few vehicles in that same region is sufficient to break the connectivity over the bridge, disconnecting the whole group of vehicles located in the northern area. As a result, $\mathcal{S}_{max}$ suddenly drops to around 1000 nodes.

These dynamics are very frequent in the vehicular network, where very large components are the result of vehicles traveling between smaller components, branching the latter together: however, such vehicles only build *weak ties* that break as soon as the bridging vertex moves away. We can conclude that, when $R = 100$ m, large components suffer from low reliability, in the sense that they undergo a continuous merge-and-split process that makes long multihop paths appear and disappear at a high frequency. When $R = 200$ m, the sheer size of large clusters reduces the impact of these variations: however, we recall that the unreliability moves to the single-hop level, with wireless links that are significantly more loss-prone. A tradeoff thus appears between unreliabilities at the network and link level, depending on the communication range adopted.

Giant components. The previous results look at the largest component in general, neglecting whether $\mathcal{C}_{max}$ is actually a "large enough" component. We now leverage the rigorous notion of giant component introduced in Sec. 4 to tell apart components that group a meaningful number of nodes, and study their appearance and structure. Fig. 10 displays the probability that the first ($\mathcal{C}_{max}$) and the second largest components in the network are giant components, as $\mathcal{N}$ varies. We focus on R equal to 100 and 200 m only, since no giant component ever appears in the network for $R = 50$ m. In both cases, giant components start to appear at around the approximate critical threshold previously identified, i.e., $\mathcal{N} \sim 6000$ and 3000 respectively; however, they do so with low-to-medium

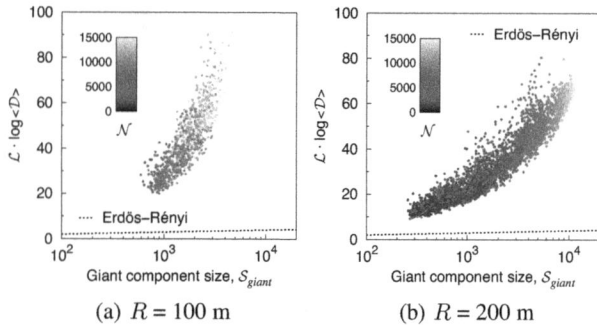

Figure 11: Small world property of the vehicular network versus $\mathcal{N}$ for different R's. The plots also report the result for an Erdös-Rényi random graph.

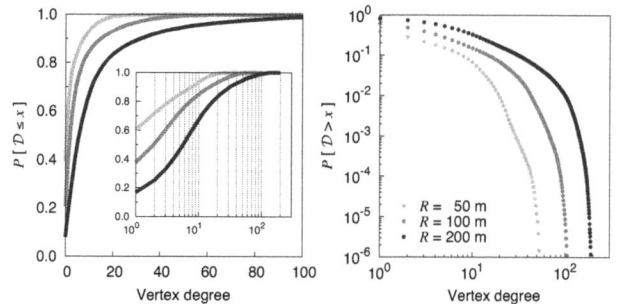

Figure 12: CDF (left) and CCDF (right) of the vertex degree $\mathcal{D}$, when aggregating all the samples over the whole day and for different R's.

probability. In order to observe a giant component with a high probability, e.g., 90%, significantly higher values of $\mathcal{N}$ are to be considered, namely, 12500 for $R = 100$ m and 4500 for $R = 200$ m. These new thresholds, reported as dashed lines in Fig. 8, evidence that, when $R = 100$ m, persistent giant components are basically unavailable. When $R = 200$ m, a persistent giant component is available only during the morning and afternoon rush hours.

Fig. 10 also reports the size of the giant component $\mathcal{S}_{giant}$, expressed as a fraction of the overall network size $\mathcal{N}$. When $R = 100$ m, not only the giant component is less probable to appear, but it never includes more than 20% of the network vertices. Conversely, when $R = 200$ m, $\mathcal{S}_{giant}$ monotonically grows with $\mathcal{N}$, arriving to include 75% of the network nodes during the morning traffic peak.

A second giant component rarely appears in the network, and a third giant component is never observed. Therefore, when it is available, the giant component maps to C_{max}.

We further explore the internal structure of giant components by studying the average length $\mathcal{L}$ of the shortest paths between any two vertices within the component, in Fig. 11. The value of $\mathcal{L}$ is lower for small giant components, i.e., when $\mathcal{S}_{giant}$ is 500 or smaller. However, it then quickly grows as $\mathcal{S}_{giant}$ increases. This behavior holds for both $R = 100$ m and $R = 200$ m, and confirms that larger giant components are not denser and thus better connected, only geographically wider as a result of the (possibly temporary) merge of smaller components. Thus, giant components become more difficult to *navigate* as they grow in size [1,2].

Fig. 11 also compares the vehicular network graph with an Erdös-Rényi random graph, a typical example of *small world* network where each pair of nodes is separated by a low number of hops [1]. Since $\mathcal{L} \approx \frac{log \mathcal{N}}{log \langle \mathcal{D} \rangle}$ in an Erdös-Rényi graph, the latter maps to a line in the plots. In fact, the line stays much lower than all the vehicular network points, a clear indication that vehicular networks are not small worlds.

Key networking insights. Our analysis allows us to observe that very large network components do exist in the Cologne vehicular network, although with much lower probability and smaller size than previously believed [16, 19]. Yet, *multi-hop connected communication among several thousands vehicles is in theory possible*. However, these *very large components are affected by low availability*, only appearing at specific locations during a few hours each day. Moreover, for any communication radius R, *their reliability is poor due to a fundamental tradeoff between network- and link-level reliability*: one has to choose between a merge-and-split network of stable links and a better connected network of loss-prone links.

The latter aspect let us conclude that *carry-and-forward transfer paradigms are compulsory also to route data within large components*. Moreover, the low availability and reliability of the large

components, jointly with their peculiar internal structure, let us argue that *roadside unit (RSU) deployment is critical to achieve large components of connected vehicles that are reliable at both network and link level*. In particular, *RSUs should be deployed where weak ties tend to appear*, so as to act as permanent bridges among small but stable connectivity islands.

Even assuming the presence of bridging RSUs, *large vehicular network components are not easy to navigate*, as they are sparse non-small worlds. Vehicles in these components are in fact connected through long articulated paths, most of which pass through a small set of bridging nodes. We thus deem *geographic information to be indispensable to data routing and dissemination* within complex connected components. Moreover *non-negligible end-to-end latencies are probably unavoidable* in a vehicular network of sensible extension, due to the length of multi-hop paths.

A more controversial conclusion would be that, given the adversity of the vehicular network topology, *vehicles should resort to the cellular infrastructure most of the time*, except for localized transfers (within a few hundreds meters from the transmission origin in the Cologne scenario) or specific situations (e.g., delay-tolerant dissemination during the rush hours in downtown Cologne).

7. NODE-LEVEL ANALYSIS

We conclude our study by descending to the individual node level. Our interest focuses on the characterization of the vertex neighborhood at one and two-hop distance.

One-hop neighborhood. Fig. 12 shows, for each value of R, the CDF (left) and CCDF (right) of the node degree $\mathcal{D}$ in the vehicular network, when aggregating all samples over 24 hours. The plots point out how common small (e.g., $\mathcal{D} < 5$) one-hop communication neighborhoods are for any R, although less probable under larger communication ranges. Higher R's also result in an increased heterogeneity of the one-hop neighborhood: when $R = 200$ m, isolated nodes and vertices with 190 neighbors can coexist in the network.

The degree distribution does not follow a power law for any R, and we can thus conclude that vehicular networks are not *scale-free* in the degree distribution [1,2]. This means that the maximum one-hop neighborhood size is constrained (in our case, by geographical restrictions on the vertices positions) to sizes that are small with respect to the network size $\mathcal{N}$. In other words, although 190 neighbors may seem a lot, they only appear for high $\mathcal{N}$'s and represent a mere 1% of the whole network – again a sign of poor navigability.

Two-hop neighborhood. The characterization of two-hop neighborhoods can be performed by observing the *assortativity* of the network, i.e., the correlation between the degree $\mathcal{D}$ of a node and the average degree of its one-hop neighbors in $\mathbb{V}^1$. Fig. 13 shows that the vehicular network is highly assortative: a strong linear re-

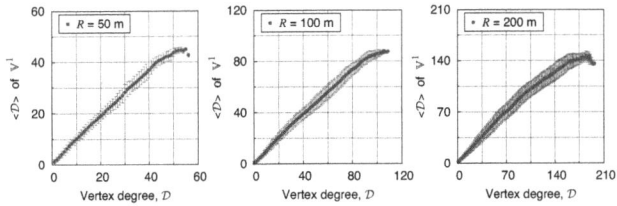

Figure 13: Assortativity for different values of R.

lationship exists between these two measures, with a low standard deviation.

This points out that high-degree vertices are typically connected to high-degree vertices, and low-degree ones to other low-degree vertices. The combination of lack of scale-free property and high assortativity is a proof that, unlike many other real-world networks, vehicular networks do not show a backbone of high-degree *hub nodes* interconnecting low-degree leaf nodes. Rather, the network is structured into weakly tied cliques of cars with similar degree.

Key networking insights. We observed vehicular communication neighborhoods to be heterogeneous and assortative in the Cologne scenario. Apart from validating our previous remarks on the scarce navigability of connected components, these results imply that vehicles can move in a few tens of seconds from quasi-isolation to being part of cliques of hundreds of cars. Then, even for the localized applications the vehicular network topology seems to support the best, *medium access control, data rate adaptation and power control algorithms will play a more important role than expected.* Indeed, the rapidity of such MAC-layer techniques to adapt to the varying network conditions may make the difference between an efficient network and a useless one.

8. DISCUSSION AND CONCLUSIONS

We presented a study of the instantaneous topology of the vehicular network in the urban region of Cologne, Germany. Our large-scale complex network analysis allowed us to draw conclusions on the significant limitations of the topology in terms of connectivity, availability, reliability and navigability, at both network and component levels. We also unveiled the underlying structure of the vehicular network in the Cologne scenario, composed of vehicles gathered into small cliques which are then connected to each other in a weak, intermittent fashion. Overall, the vehicular network topology appears unfit to medium- and long-range delay-bounded data transfers, for which cars should most probably resort to traditional cellular communication. The vehicular network seems instead to best fit delay-tolerant transfers within localized areas: in such cases, our results let us advocate the adoption of carry-and-forward paradigms and/or the deployment of RSUs, and stress the key role of MAC-layer protocols for channel access and rate/power control.

Concerning the limitations of our work, the study is constrained to one specific scenario due to the lack of other mobility datasets that are publicly available and yield a similar level of realism. However, we conjecture that Cologne scenario may be representative of many other urban regions of similar nature and size [26]. Yet, we definitely support the idea of performing studies similar to ours on different urban scenarios, once new vehicular mobility datasets will be available.

We also remark that we focused on the instantaneous vehicular network connectivity, and considered a computationally feasible disc signal propagation model. The characterization of temporal metrics and the impact of more realistic propagation remain open problems that we plan to address in the future.

9. ACKNOWLEDGEMENTS

The authors would like to thank Iago Fraga, Sandesh Uppoor and Nguyen Tien Thanh for their help in processing the dataset.

10. REFERENCES

[1] R. Albert, A. Barabási, "Statistical Mechanics of Complex Networks", *Reviews of Modern Physics*, 74(1):47–97, Jan. 2002.

[2] S Boccaletti, V. Latora, Y. Moreno, M. Chavez, D. Hwang, "Complex Networks: Structure and Dynamics", *Physics Reports*, 424(4-5):175–308, 2006.

[3] M. Doering, T. Pögel, W.-B. Pöttner, L. Wolf, "A new mobility trace for realistic large-scale simulation of bus-based DTNs," *ACM CHANTS*, Chicago, IL, USA, Sep. 2010.

[4] J. Yuan, Y. Zheng, X. Xie, G. Sun, "Driving with knowledge from the physical world," *ACM SIGKDD*, San Diego, CA, USA, Aug. 2011.

[5] M.M. Artimy, W. Robertson, W.J. Phillips, "Connectivity in Inter-vehicle Ad Hoc Networks". *CCECE*, Niagara Falls, NY, USA, May 2004.

[6] H. Füssler, M. Torrent-Moreno, M. Transier, R. Krüger, H. Hartenstein, W. Effelsberg, *Studying Vehicle Movements on Highways and their Impact on Ad-Hoc Connectivity*, ACM MC2R, 10(4):26–27, Oct. 2006.

[7] S. Yousefi, E. Altman, R. El-Azouzi, M. Fathy "Analytical Model for Connectivity in Vehicular Ad Hoc Networks", *IEEE Transactions on Vehicular Technology*, 57(6):3341–3356, Nov. 2008.

[8] M. Khabazian, M.K. Ali, "A Performance Modeling of Connectivity in Vehicular Ad Hoc Networks", *IEEE Transactions on Vehicular Technology*, 57(4):2440–2450, Jul. 2008.

[9] G.H. Mohimani, F. Ashtiani, A. Javanmard, M. Hamdi, "Mobility Modeling, Spatial Traffic Distribution, and Probability of Connectivity for Sparse and Dense Vehicular Ad Hoc Networks", *IEEE Transactions on Vehicular Technology*, 58(4):1998–2006, May 2009.

[10] S. Shioda, J. Harada, Y. Watanabe, T. Goi, H. Okada, K. Mase, "Fundamental Characteristics of Connectivity in Vehicular Ad Hoc Networks", *IEEE PIMRC*, Cannes, France, Sep. 2008.

[11] Y. Zhuang, J. Pan, L. Cai, "A Probabilistic Model for Message Propagation in Two-Dimensional Vehicular Ad-Hoc Networks", *ACM VANET*, Chicago, IL, USA, Sep. 2010.

[12] I. Ho, K.K. Leung, J.W. Polak, "Stochastic Model and Connectivity Dynamics for VANETs in Signalized Road Systems", *IEEE Transactions on Networking*, 19(1):195–208, Feb. 2011.

[13] X. Jin, W. Su, Y. Wei, "Quantitative Analysis of the VANET Connectivity: Theory and Application", *IEEE VTC-Spring*, Budapest, Hungary, May 2011.

[14] I. Ho, K.K. Leung, J.W. Polak, R. Mangharam, "Node Connectivity in Vehicular Ad Hoc Networks with Structured Mobility", *IEE LCN*, Dublin, Ireland, Oct. 2007.

[15] M. Fiore, J. Härri "The Networking Shape of Vehicular Mobility", *ACM MobiHoc*, Hong Kong, PRC, May 2008.

[16] M. Kafsi, P. Papadimitratos, O. Dousse, T. Alpcan, J.-P. Hubaux, "VANET Connectivity Analysis", *IEEE AutoNet*, New Orleans, LA, USA, Dec. 2008.

[17] W. Viriyasitavat, O.K. Tonguz, F. Bai, "Network Connectivity of VANETs in Urban Areas", *IEEE SECON*, Rome, Italy, Jun. 2009.

[18] H. Conceicão, M. Ferreira, João Barros, "On the Urban Connectivity of Vehicular Sensor Networks", *DCOSS*, Santorini, Greece, Jun. 2008.

[19] G. Pallis, D. Katsaros, M. D. Dikaiakos, N. Loulloudes, L. Tassiulas, "On the Structure and Evolution of Vehicular Networks", *IEEE/ACM MASCOTS*, London, UK, Sep. 2009.

[20] S. Uppoor, M. Fiore, "Large-scale Urban Vehicular Mobility for Networking Research", *IEEE VNC*, Amsterdam, Holland, Nov. 2011.

[21] W. Viriyasitavat, F. Bai, O.K. Tonguz, "Dynamics of Network Connectivity in Urban Vehicular Networks", *IEEE Journal on Selected Areas in Communications*, 29(3):515–533, Mar. 2011.

[22] G. Hertkorn, P. Wagner, "The application of microscopic activity based travel demand modelling in large scale simulations", *World Conference on Transport Research*, Istanbul, Turkey, Jul. 2004.

[23] C. Gawron, "An Iterative Algorithm to Determine the Dynamic User Equilibrium in a Traffic Simulation Model", *International Journal of Modern Physics C*, 9(3):393–407, 1998.

[24] D. Hadaller, S. Keshav, T. Brecht, S. Agarwal, "Vehicular Opportunistic Communication Under the Microscope", *ACM MobiSys*, San Juan, Puerto Rico, USA, Jun. 2007.

[25] F. Bai, D.D. Stancil, H. Krishnan, "Toward Understanding Characteristics of Dedicated Short Range Communications (DSRC) From a Perspective of Vehicular Network Engineers", *ACM MobiCom*, Chicago, IL, USA, Sep. 2010.

[26] P. Crucitti, V. Latora, S. Porta, "Centrality measures in spatial networks of urban streets", *Physics Review E*, 73(3), Mar. 2006.

Detecting Epidemics Using Highly Noisy Data

Chris Milling
UT Austin
1 University Station
Austin, TX
cmilling@utexas.edu

Constantine Caramanis
UT Austin
1 University Station
Austin, TX
constantine@utexas.edu

Shie Mannor
The Technion
Haifa, Israel
shie@ee.technion.ac.il

Sanjay Shakkottai
UT Austin
1 University Station
Austin, TX
shakkott@austin.utexas.edu

ABSTRACT

From Cholera, AIDS/HIV, and Malaria, to rumors and viral video, understanding the *causative network* behind an epidemic's spread has repeatedly proven critical for managing the spread (controlling or encouraging, as the case may be). Our current approaches to understand and predict epidemics rely on the scarce, but exact/reliable, expert diagnoses. This paper proposes a different way forward: use more readily available but also more noisy data with *many false negatives and false positives*, to determine the *causative network* of an epidemic. Specifically, we consider an epidemic that spreads according to one of two networks. At some point in time we see a small random subsample (perhaps a vanishingly small fraction) of those infected, along with an *order-wise similar number* of false positives. We derive sufficient conditions for this problem to be detectable, and provide an efficient algorithm that solves the hypothesis testing problem. We apply this model to two settings. In the first setting, we simply want to distinguish between random illness (a complete graph) and an epidemic (spread along a structured graph). In the second, we have a superposition of both of these, and we wish to detect which is the strongest component.

Categories and Subject Descriptors

G.3 [**Probability and Statistics**]: Stochastic Processes

Keywords

epidemic process, network inference

1. INTRODUCTION

The study of epidemic spread over social, communication, and human contact networks, be it a contagion of a hu-

man or computer virus, or a rumor, opinion or trend, begins with two basic questions: do we indeed have a spreading epidemic, and if so, what is the causative network spreading it? Numerous famous examples from the history of epidemiology ([22, 3]) have illustrated the importance and difficulty of determining the causative network. With accurate data collected over time, for example, from high accuracy medical diagnoses of a known illness, the causative network essentially reveals itself. Yet such data are rarely available. More to the point, highly incomplete and noisy data *often are available*. Indeed, the challenge arises in particular, when time lapse data of "true" illness is not available, and when the data we do have is highly noisy with many false positives and negatives. For example, online records (flu-related keywords in social networks [4], or Internet searches such as in Google Flu Trends [11]) provide large but noisy data sources for detecting flu epidemics, but potentially containing many false positives.

A similar fundamental difficulty arises with other epidemics. Consider the adoption of the latest tablet or smartphone. The spread is likely driven *both* by "word-of-mouth" (online social networks via tweets or Facebook posts) advertising, as well as explicit advertising campaigns over television, Internet ads, etc. While both modes likely play a part in driving sales, is it the mass marketing (which is in effect a star network that connects the advertiser to all the television viewers) that serves as the dominant driver of spread, or is it the word-of-mouth viral marketing that is dominant? Surveys may reveal (noisy) data on who owns the new tablet, but pinpointing time-of-acquisition and the causative network is much more difficult.

In a communications setting, carriers need to worry about similar problems. On observing abnormal interactions from some smartphones on their network, they need to decide if this is due to a buggy firmware update, or something more malicious (such as malware/virus spread). Unfortunately however, these carriers rarely have access to these user devices themselves, thus, need to recourse to inferring from the limited and noisy samples (e.g., phone-to-network interactions). While currently small, malwares [15] and viruses that spread via user contact networks [8, 23] are receiving increasing attention.

The key idea in this work, is that different spreading mechanisms have different statistical signatures, in terms of the subset of people infected. This is certainly the case when the causative graphs are very different, and the subset of nodes (people, machines, etc.) the epidemic has reached ("infected nodes") are completely and accurately revealed. As discussed, however, the data available are typically noisy, with many false positives and negatives. Moreover, the larger the fraction of the network the contagion has reached, the more this "network signature" is washed out. This paper explores these tradeoffs. We consider a broad class of graphs: graphs with bounded degree. The degree controls the infection's speed. We give sufficient conditions on when the causative network can be determined when only a vanishing number of infected nodes report, and moreover when a constant fraction of those reporting are false positives. Then, we consider the case most relevant in spread of rumors, technology and ideas: the superposition of two spreading mechanisms. Indeed, in the age of mass advertising and mass media, trends spread friend-to-friend, but also through television, Internet ads, and similar advertising efforts that exhibit a "star-like" contagion network. We provide sufficient conditions for determining which is the dominant effect, again when only a vanishing fraction of infected nodes report, and when no time-lapse data are available.

1.1 Related Work

Analyzing the spread of epidemics under the susceptible-infected (SI) model [9] has been considered in depth for a variety of graphs and circumstances [1, 12]. While there has been much work on what we call the *forward problem*, i.e., predicting what an infection may look like or do, the present work falls under the heading of *backward problems*. Thus, the work in this paper is related to the task of inferring various characteristics of the infection given the infected nodes. We briefly mention several related works. Demiris and O'Neill estimate the transmission rates of the epidemic [5, 6]. Shah and Zaman provide an algorithm to estimate the node most likely to be the source of the infection [20, 21]. Luo and Tai consider a similar problem with multiple infection sources [14]. Myers *et al.* estimate the proportion of infections that occur randomly (from unknown sources external to the network) given the full sequence of infected nodes under a similar mixed infection setting [18].

In even more closely related work, Netrapalli and Sanghavi analyze the problem of estimating the structure of the contagion network given the times each node is infected for several epidemics on the same network [19]. Gomez-Rodriguez *et al.* provide an algorithm to solve a similar problem with a somewhat different infection model [10]. These works use time-lapse information of multiple epidemics, and from those are able to detect the edges of the causative network. This is more general in the sense that very little is known about the underlying spreading mechanics, compared to what amounts to a hypothesis testing problem in our setting. On the other hand, the data-availability regime under which we operate, is much more harsh. We have very limited information, viewing only a very partial and also noisy, subset of the infected nodes of a single contagion *at a single point in time*. We have no time-related information. Under these limited-information conditions, inferring the network structure, as in [19] or [10], would not be possible. The work in [16] and [17] on hypothesis testing for determining the graph corresponding to the epidemic spread, by Milling *et al.*, is also very relevant (we elaborate on the connection in Section 3). However, these papers do not consider

false positives, or superposed spreading (spreading simultaneously via multiple networks), whereas these are precisely the main contributions of this current paper.

1.2 Main Contributions

The fundamental problem we consider is diagnosing the causative network of an epidemic or contagion, *using noisy and highly incomplete* data, and in particular, data without temporal information. The focus and main contribution of this paper is in tempering the effect of this noise in the data, and hence greatly expanding the available data sets that can be used. To the best of our knowledge, this is the first paper that has considered epidemic forensics in this regime.

The first part of the paper attempts to diagnose a contagion as arising from node-to-node contact via a specific contact network, or through a random infection process. The latter can be modeled as a contagion spreading from the center node of a star graph, to the leaves. As discussed above, we assume there is an overwhelming fraction of false negatives (that is, only a vanishing fraction of infected nodes report). Of this vanishing fraction, we assume that a constant fraction are false positives. We note that there is no way to identify false positives or false negatives, but only to provide a system-level diagnosis. Indeed, our goal is to diagnose the spreading network of the contagion.

Next, we consider the superposition of these two contagion processes, and attempt to determine the stronger of the two components. That is, a contagion spreads through a star network as described above, *and also simultaneously through subsequent node-to-node contact*. This is the case with much advertising: television and other mass-media advertisements provide initial "seeds," but subsequent spread may occur through word-of-mouth. For different products, the relative effects of these two mechanisms may differ. Under what circumstances can we determine the stronger component?

We note that robustness is again at the crux: we are attempting to determine system-wide effects; at the local level, it is impossible to say if the contagion reached an infected node through what we have called the star model, or through local interactions (e.g., word of mouth).

Specifically, our contributions are as follows.

- Algorithm Development: We provide an algorithm we call the *Median Ball Algorithm*. This algorithm is simple, efficient to run, and terminates quickly even for very large graphs. In the first case of diagnosing a contagion as a spreading epidemic or a random illness (the star graph), it outputs what it believes is the most likely candidate. Similarly, for the case of a superposition of those two processes, it outputs what it believes is the stronger of the two components.

- Arbitrary (Adversarial) False Positives: We give sufficient conditions in terms of number of total sick nodes, number of sick nodes reporting (i.e., fraction of false negatives) and fraction of false positives, under which the algorithm above correctly diagnoses the causative network, i.e., with the probability of Type I and Type II error going to zero in the size of the graph. In particular, our results show that our algorithm correctly identifies the causative network even when only a vanishingly small fraction of sick nodes report, and moreover, when up to 50% of the reporting nodes are false positives, *even when those false positives are adversarially selected.*

- Random False Positives: We give sufficient conditions for the same problem, when the false positives are randomly (independently and uniformly) distributed in the graph. Here, we show that our algorithm correctly identifies the causative network *for any fraction of false positives up to* 100%.

- Superposed Spreading (Mixed Infection Types): Finally, we consider the setting where the spreading occurs through both random infection (the star spreading graph) representing, for instance, television or Internet advertisement, and subsequently through node-to-node contact (word-of-mouth) and give sufficient conditions for when our algorithm correctly determines which of the two components is the dominant factor in the contagion spread. Again, we require only a vanishingly small fraction of sick nodes to report.

2. MODEL AND ALGORITHM

We have described intuitively the contagion spreading models we attempt to distinguish. In this section, we describe them more precisely and build on the models in [16, 17]. We consider two distinct infection regimes: a contagion spreading through node-to-node contact, versus random spread of the contagion, when each node becomes sick independently of its location in the graph or the status of its neighbors. We can model this, too, as an epidemic spreading over a star graph where initially only the center node is infected. For ease of discussion, we call the node-to-node mode of spreading an *epidemic*, and the star-mode of spreading a *random sickness*. In both cases, we start with a single infected node at time 0.

The Infection Process

Let $G = (V, E)$ denote the graph along which the infection spreads. As discussed above, in the case of an epidemic spreading node-to-node, G is a structured graph (e.g., d-dimensional grid). For the case of a random illness, the graph G is a star graph, with every node connected to a central node assumed to be infected at time zero. We let $n = \text{card}(V)$, the size of the graph. The diameter of the graph is denoted $\text{diam}(G)$.

Given a graph G, the contagion spreads as follows. At time zero, an initial node is selected and called "infected." For the structured graph case, we assume this initial infected node is selected uniformly at random. For the star graph, it is the central node. The infection spreads from that node to its neighbors, across the edges of the graph. The spreading occurs according to a standard susceptible-infected (SI) model [9, 7, 13] for an epidemic. The spreading rate is parameterized by a single number, or rate. To make clear the distinction between the rate for a structured graph or for a star graph, we use η to represent the rate of the structured graph, and γ/n the rate of the star graph. We divide by n in the case of the star graph so that new infections appear at rate γ (ignoring the shrinking number of susceptible nodes). This means the following: for each infected node and for each edge incident to that node, we start an exponential clock, i.e., a clock that expires after an exponentially distributed length of time, of expectation $1/\eta$, i.e., of rate

η for a structured graph, and n/γ, i.e., of rate γ/n, for the star graph. The expiration of a clock indicates that the adjacent node becomes infected (if it is not already infected) and new clocks are started for each edge from this newly infected node. In this way, the infection spreads along the edges of the graph in a node-to-node fashion.

Let S denote the set of infected nodes at a given time. The rate of new infections is (roughly) proportional to the number of uninfected nodes $(V \setminus S)$ incident to an infected node. Thus, for a random sickness (G a star graph) new nodes become infected at a rate $(\text{card}(V \setminus S))\gamma$, and hence the rate of new infections in fact decreases as more nodes become infected. For most graphs, the rate of infections initially increases, before decreasing as more and more nodes become infected. The most challenging regime to pose the problem of diagnosing the causative network of the epidemic, is where the expected number of infected nodes is the same under both models. Thus, for the remainder of this paper, all results are stated under precisely this assumption.

The second half of this paper considers mixtures of both of these types of infections: the star graph infects nodes at rate γ/n, and then these infected nodes infect their neighbors on the structured graph (e.g., the grid) at rate η. Thus, in this superposed process, nodes become infected at random at some rate γ, and the infection then spreads from these nodes as an epidemic at the (different) rate η. In this setting, we consider two different processes: one where the dominant factor is the random infection (the spread from the star graph) and the other where it is the spread along the structured graph that dominates. Thus, in the first setting we have $\gamma \gg \eta$, and the random infection dominates the epidemic, and in the second setting, $\eta \gg \gamma$, and the epidemic spread dominates the infection process.

The Reporting Process

At a given point in time, a subset of the infected nodes is revealed – these nodes are discovered to be infected, or they self-report as infected. Given this snapshot, the task is to determine the spreading process. If we had access to the entire set of infected nodes, then a simple test of the connectivity of the infected nodes would easily distinguish the infection mechanism with overwhelming probability. In most contagion processes, however, only a small – perhaps a vanishingly small – fraction of infected nodes are detected, or self-report. Indeed, most people suffering from flu symptoms do not visit a doctor; no survey or poll reveals more than a minuscule fraction of adopters of a new technology; and likewise, only few virus-infected computers are reported/detected. In short, there may be a large – possibly an overwhelming – fraction of false negatives, i.e., of infected nodes that do not report (or are not detected). As a consequence, reporting infected nodes are likely to be disconnected, possibly with relatively large distance between them. This may be particularly challenging in small-world graphs.

As the theorem statements below make clear, we indeed assume that the fraction of false negatives is overwhelming. Our theorems show that the algorithm we provide can recover the true infection mechanism when only a logarithmic number of infected nodes reports, i.e., when the fraction of reporting infected nodes is *exponentially small*.

Further confounding the task of diagnosing an epidemic versus random illness may be the presence of false positives among the reporting infected nodes. This again is the case with many available data sets. Obtaining accurate diagnoses (i.e., with very low false-positive rate) is difficult. In human illnesses, in many cases (in particular, easily treatable sexually transmitted diseases) public policy prescriptions have focused on tests that are inexpensive, yield results quickly, and have a low false-negative rate. Moreover, data on self-reported illness (i.e., without medical diagnosis) are increasingly available, and should be exploited. Answers to surveys and polls suffer from precisely the same possibility of false positives. Moreover, depending on the setting, it is important to consider the case of correlated (clustered) or even worse, manipulative false positives, that may collude to obscure the infection propagation mechanism, or, simply, may not have an easily describable distribution. On the other hand, one would expect that if false positives are randomly (uniformly, and independently) distributed across the graph, that their effect would be less pernicious. We consider both settings, and indeed, show that this is the case. In the case of adversarially distributed false positives, our algorithm succeeds when up to 50% of the reporting nodes are false positives. When the false positives are randomly distributed, our algorithm can tolerate nearly 100% false positives.

The notation we use is as follows. As above, we let S denote the set of actually infected nodes (revealed or not). We assume that each of these reports its infection with some probability. We note that we do not require the reporting process to be independent, i.e., the set of infected nodes that report may well be correlated. We denote by q the probability that infected nodes report, and we denote the resulting subset by $S_r \subseteq S$ (and hence, card(S_r) has expected value $q$card(S)). We further assume that some fraction of the uninfected nodes, $V \setminus S$, may (falsely) report infection as well. As discussed, we consider both cases where this fraction of falsely reporting nodes is chosen by an adversary and chosen randomly. In the random case, each false positive node is chosen uniformly at random from the entire graph, where repeats are allowed. We allow the "falsely reporting" nodes to be in S_r (so they are not truly false positives) to reduce dependence on S_r. This also ensures that the density of reporting nodes is highest in S_r. We thus denote the set of reporting sick nodes (including both truly infected, and false positives) by $\bar{S}_r \supseteq S_r$. We parameterize the number of false reporting nodes by a constant $f \geq 0$. Let the number of false positives be given by $\lfloor f \cdot \text{card}(S_r) \rfloor$. Thus, $f/(1+f)$ is approximately the fraction of all reporting nodes that are false positives. When $f \to 1$, then fully half the reporting nodes are false positives. As f continues to increase, this fraction approaches 100%.

We consider false positives in the setting where we must determine if the infection spreads as an epidemic (a structured graph) or a random illness (a star graph). We show that our algorithm succeeds against adversarially selected false positives even as $f \to 1$. In the case of randomly selected false positives, our algorithm succeeds for any value of f. In the second half of the paper, we consider the superposition of the two processes. Here we focus only on false negatives. As the proof makes clear, incorporating false positives is a straightforward extension.

2.1 Graphs

Our results apply to a broad family of graphs, with different topologies. The key property we require is that the graph should have bounded degree (where this bound is a constant). From this property, we show that the infection can spread at only limited asymptotic speed, and that the neighborhood sizes are sufficiently small.

That is, there exists a constant $\bar{d}$ greater than or equal to the degree of each node for sufficiently large n. As a result of this property, the infection can only travel at a certain maximum rate through the graph. Define the random variable W as the maximum distance an epidemic has spread from its source. We define the condition *limited epidemic speed* as follows:

Definition 1. A graph has *limited epidemic speed* if there exist finite, positive constants s, λ_1 such that for sufficiently large n and an epidemic starting at any node a and duration t,

$$P(W > st) < e^{-\lambda_1 t}.$$

The speed s mentioned in the above definition is in fact an upper bound on the speed, in that there is no matching lower bound. Nevertheless, we refer to it as the speed for brevity. In addition, we also need a constraint on the neighborhood size.

Definition 2. A graph G has *limited neighborhood size* if diam(G) scales as $\Omega(\log n)$ and there exists a increasing concave function $b(x)$ such that for all $0 < x < 1$, $b(x) > 0$ and all balls of radius no more than $b(x)$ contains less than xn nodes for sufficiently large n with probability tending to 1.

In fact, both of these previous conditions follow from a bounded degree distribution, as stated formally below.

THEOREM 1. *Let G be a graph with maximum degree $\bar{d}$. Then G has both limited epidemic speed and limited neighborhood size.*

PROOF. First, the spread of the epidemic on G can be upper bounded by a tree of degree $\bar{d}$ where nodes are repeated for each path to them. See [17] for details on this bound. Then using a speed upper bound for trees, we find that G has *limited epidemic speed*, where the exponential probability of error follows from a Chernoff bound [2]. Next, using the maximum degree condition, the number of nodes within distance r from an arbitrary node a of G is at most $\bar{d}^{r+1}$. Therefore, for any x, $0 < x < 1$, no ball of radius $\log_{\bar{d}} xn - 1$ contains more than xn nodes. From this, we see that diam(G) $\geq \log_{\bar{d}} n - 1$. Letting $b(x) = \log_{\bar{d}} xn - 1$, we see this satisfies the desired condition for *limited neighborhood size*. This completes the proof. $\square$

Our bounds in the ensuing Theorems depend explicitly on the parameters that define the limited epidemic speed and neighborhood size conditions. We comment that tighter bounds can be derived with additional graph structure. For instance in a grid (lattice), first passage percolation results [13] provide sharper estimates, which in turn, can lead to stronger sufficient conditions in the ensuing Theorems. We refer to [17, 16] for an analogous discussion on graph-specific conditions (however, without false positives or superposed spreading).

Next, the following simple lemma (using a balls-in-bins argument) proves useful in the sequel, so we give it here.

LEMMA 1. *Suppose graph G has limited neighborhood size. Let $0 < x < 1$, $\epsilon > 0$ and R be a collection of nodes with* card(R) $< (1 - \epsilon)xn$. *Let S be a collection of uniformly random nodes with* card(S) $= \omega(\log n)$. *Then the probability that R contains at least x fraction of the random nodes in S decays to 0 as n increases. In particular, there exists a constant $\lambda_2 > 0$ such that*

$$P(\text{card}(R \cap S) \geq x\,\text{card}(S)) < e^{-\lambda_2 \text{card}(S)}.$$

The main way we use this lemma is to show that the probability that a large fraction of randomly selected nodes fall in a ball around a given node, goes to zero.

2.2 Algorithm

We develop a single algorithm we call the Median Ball Algorithm to solve the hypothesis testing problems in this paper – both for the case of detection of epidemic versus random illness (the first part of the paper), as well as the case of determining the dominant factor in the spread of the contagion (the second part of the paper). The Median Ball Algorithm is simple to describe: it searches for the smallest ball that covers a fraction of the reporting infected nodes. Of course, it has no way to tell if a reporting sick node is truly infected or a false positive, and as emphasized above, this is not the goal of this paper. If the resulting radius of this ball is small enough, it declares that there is an epidemic; otherwise, it concludes that the infection process is in fact a random illness. This algorithm is efficient, as even the brute-force implementation runs in time at most $O(|V| \cdot \text{diam}(G))$.

The algorithm takes two parameters α, and r. These parameters are tailored to the problem at hand, including, in the case of r, the size of the graph. As input, it takes a graph G and a set of reporting infected nodes S_r. If the algorithm can cover an α-fraction of the infected nodes in a ball of radius at most r, it declares the infection to be an epidemic; otherwise, it labels the infection a random illness.

Define $Ball_G(a, r)$ as all nodes in G within distance r from node a.

Algorithm 1 Median Ball Algorithm

Input: Graph G; Set of reporting infected nodes S_r;
Output: Epidemic or Random

$c \leftarrow \alpha\left[\text{card}(S_r)\right]$
for all $d \in V$ **do**
  $B \leftarrow Ball_G(d, r)$
  if card($B \cap S_r$) $\geq c$ **then**
    return Epidemic
  end if
end for
return Random

---

3. FALSE POSITIVES

We consider the problem of determining if the spreading mechanism of a contagion is what we have termed an epidemic, or a random illness. We consider the superposition of these two in Section 4.

When all reporting nodes are truly infected (i.e., no false positives) then a special case of our algorithm can solve this problem with asymptotically (as the graph size scales) zero error: taking $\alpha = 1$, our algorithm reduces to the special case considered in [16] where one seeks a small ball containing *all the reporting sick nodes*. However, the algorithm in

[16] fails even with a vanishing fraction of false positives – indeed, even with one single false positive. In contrast, the algorithm we give here succeeds with up to 50% *adversarially placed false positives*, and up to any fixed fraction less than 100% randomly placed outliers. Both of these results are the best possible under our formulation, where the number of false positives is proportional to the number of true infected nodes.

We show that by looking at the α-quantile ball, our algorithm becomes effectively immune to outliers. This happens for the following reason: suppose the true spreading process is an epidemic. We show that there is in fact a small-radius ball that covers all truly infected nodes. Now, the false positives either fall near or within this ball, or far outside it. In the first case, they do not require the ball to be larger and hence do not lead the algorithm to incorrectly pronounce the epidemic a random illness. In the latter case, they are ignored by the quantile ball algorithm, and thus again do not cause the algorithm to produce an error. If the true mechanism is random infection (i.e., the star graph) then the algorithm produces an error if it declares the infection mechanism to be an epidemic. For this to happen, either the true infections must appear clustered, even though they are independently distributed uniformly on the graph, or the false positives must exhibit a sufficient clustering to fool the algorithm. As we see, this is possible under adversarial placement of enough false positives, but probabilistically extremely unlikely otherwise.

Adversarial False Positives: First, consider the case where the false positives are chosen by an adversary with full knowledge of our algorithm and the true type of infection. The adversary can act non-randomly to attempt to confound the algorithm. In particular, the adversary can spread the false positives to look randomly placed when the infection is in fact an epidemic, and when the infection is random, can cluster the false positives to make the random sickness appear like an epidemic. We show that in both of these cases, if $f < 1$, i.e., the reporting nodes are less than 50% false positives, it is possible (for appropriate infection parameters) to distinguish the type of infection with probability of error tending to 0 as the number of nodes, n, tends to infinity.

We consider a graph G, with limited epidemic speed and limited neighborhood size. Let s denote the speed of an epidemic on the graph G (in fact, an upper bound on this, as discussed in Definition 1). Let $b(x)$ denote the limited neighborhood size function as defined previously in Definition 2. Note that from Theorem 1, all graphs with bounded degree have limited speed and neighborhood size.

THEOREM 2. *[**Adversarial False Positives**] Suppose G is as described. Suppose further that $f < 1$ and set $f' = (1-f)/(1+f) > 0$. Suppose t scales such that the number of reporting nodes is $\omega(\log n)$ and $t < b(f'/2)/s$. Then the Median Ball Algorithm with $\alpha = 1/(1+f)$ and $r = st$ correctly determines the type of infection with probability tending to 1 with the number of nodes, n.*

PROOF. First we show that the Type II error probability decays to 0. To this end, suppose the infection is in fact an epidemic. Consider only the true reporting nodes S_r. Note that card(S_r) $\geq \alpha\text{card}(\bar{S}_r)$. By the definition of speed s, the probability the epidemic spreads outside a ball of radius $r = st$ decays to 0, so this ball covers S_r and hence at least

α fraction of the reporting nodes. Therefore it is correctly labeled an epidemic.

Now we show that the Type I error probability also decays to 0. We need to show no ball of radius r can cover $\alpha = 1/(1 + f)$ fraction of the nodes. Since only $f/(1 + f)$ of the nodes are false positives, the ball must contain at least $(1 - f)/(1 + f) = f' > 0$ true reporting nodes. Then it is sufficient that the probability there exists a ball of radius r covering $f' \text{card}(S_r)$ true reporting nodes (which are located randomly) decays to 0.

Since $r < b(f'/2)$, no ball of radius r contains over $f'n/2$ nodes. Consider one of the n balls of radius r (one ball for each possible center node), call it R. Then by Lemma 1, there exists a strictly positive λ_2 such that

$$P(\text{card}(R \cap S_r) \geq f' \text{card}(S_r)) < e^{-\lambda_2 \text{card}(S_r)}.$$

Since $\text{card}(S_r) = \omega(\log n)$, $e^{-\lambda_2 \text{card}(S_r)} = o(1/n^2)$. Therefore, from a union bound, there is some ball of radius r containing over f' fraction of the true reporting nodes with probability at most $o(1/n)$. Hence, no such ball covers α fraction of the nodes in $\bar{S}_r$ with probability tending to 1 so the Type I error probability goes to 0. $\square$

Therefore, as long as the number of false positives is (a fraction) less than the number of true reporting nodes, it is possible to determine whether an infection is due to an epidemic or a random sickness. Given the unlimited adversarial model, it is clear that this is tight. That is, if $f = 1$, it is impossible to distinguish the types of infection in the adversarial setting by any algorithm of any complexity. We state this simple converse result as a theorem.

THEOREM 3. *Suppose $f = 1$ and the random sickness is normalized so that the infection size distribution is equal for both infection processes. Then with adversarial false positives, the probability of error for any algorithm is at least 0.5.*

PROOF. There is a simple adversarial algorithm that guarantees a probability of error of 0.5. Recall the *a priori* probability for each infection process is equal. When the infection is from an epidemic, the adversary chooses nodes randomly exactly as in the random sickness. When the infection is from a random sickness, the adversary chooses nodes exactly as in an epidemic. Therefore, in all cases, exactly half the nodes are due to an epidemic, and half are due to a random sickness. Since the infection size is normalized, each collection of infected nodes is equally likely to be an epidemic as a random sickness. Then the probability of error for every set $\hat{S}_r$ is 0.5 (no matter the algorithm), and hence the overall probability is 0.5. $\square$

Random False Positives: When an adversary places the false positives, the worst case scenario is generally when it places them in a cluster when the infection is in fact a random sickness. Therefore, when the false positives are located randomly over the graph, one would expect that the infection process is distinguishable for a larger range of f. We show that this is in fact the case. It is possible to distinguish an epidemic from a random sickness *for all values of f*. We note, though, that as one would expect, the larger the f, the tighter the constraint on the time of detection, i.e., than total number of infected nodes.

THEOREM 4. *[**Random False Positives**] Let $f > 0$. Suppose t scales such that number of reporting nodes is $\omega(\log n)$ and $t < b\left(\frac{1}{2(1+f)}\right)/s$. Then the Median Ball Algorithm with $\alpha = 1/(1 + f)$ and $r = st$ correctly determines the infection type with prob. tending to 1.*

PROOF. The proof proceeds in a very similar way to Theorem 2. First suppose the infection is an epidemic. We can cover all true reporting nodes with probability scaling to 1 using the speed definition. Since at least an α fraction of the reporting nodes are truly infected, our algorithm correctly reports the infection is an epidemic. Therefore the Type II error probability decays to 0.

Now suppose the infection is a random sickness. Since the false positives are also random, the reporting nodes with the false positives are simply are larger set of random nodes. Note $r = b\left(\frac{1}{2(1+f)}\right)$. Using Lemma 1 in the same way as in Theorem 2, we see that no ball of radius r contains over a $\alpha = 1/(1 + f)$ fraction of the random nodes with probability approaching 1. In this case, our algorithm returns random sickness. Thus the Type I error probability also tends to 0. $\square$

4. MIXED INFECTION TYPES

Now we turn to the problem of mixed infection types. In this case, we deal with infections where the infection is spreading both as an epidemic and a random sickness. We term the nodes that become infected randomly as *seeds*, from which the infection starts spreading as an epidemic. We consider two distinct infection processes. In Process 0, the infection spreads mostly randomly. Let γ_0, η_0 be the infection rates for the random sickness and epidemic respectively and t_0 be the infection time for Process 0. For clarity, we also call Process 0 "Process SR-WE" (Strong random, weak epidemic). In Process 1, the infection is dominated by the epidemic, and let γ_1, η_1, and t_1 be the corresponding parameters as before. We label Process 1 "Process WR-SE" (Weak random, strong epidemic). Note that the infection is the same if the rates are scaled up by the same factor that time is scaled down. Then we can say that the epidemic dominates in Process 1 relative to Process 0 if $\eta_1/\gamma_1 \gg \eta_0/\gamma_0$. Unlike in the previous section, we apply no explicit normalization. Rather, we provide sufficient conditions on the range of the parameters for which the Median Ball Algorithm succeeds.

First we consider Type I errors. Assume the infection spreads by Process SR-WE [Process 0]. We use the Median Ball Algorithm with parameters α and r. Then the following theorem characterizes when the probability of error decays to 0. Let s and $b(x)$ be the speed and neighborhood size function as defined previously.

THEOREM 5. *Consider an infection spreading as in Process 0. Suppose $q\gamma_0 t_0 = \omega(\log n)$. Suppose there exists a constant integer $c_1 \geq 1$ where $\eta_0 t_0 = o\left((\gamma_0 t_0)^{-1/(1+c_1)}\right)$ and for some $\epsilon > 0$, suppose that $r + c_1 < b\left(\frac{\alpha}{\bar{d}^{c_1+1}(1+\epsilon)}\right)$. Then the Type I error probability decays to 0 as n increases.*

PROOF. First we show that no infection (from a single seed) spreads farther than a distance c_1, so each infection contains at most a constant $\bar{d}^{c_1+1}$ nodes (where, recall, $\bar{d}$ is a bound on the maximum degree of the graph). Consider an arbitrary seed a and all paths of length $c_1 + 1$ beginning at

a. There are at most $\bar{d}^{c_1+1}$ such paths. An infection from a must spread over one such path in time t_0 to spread farther than distance c_1. Since the traversal time of an edge has distribution $\mathrm{Exp}(\eta_0)$, the probability the infection can spread over the edge in time t_0 is $1 - e^{-\eta_0 t_0} < \eta_0 t_0$. Then using a union bound, the probability that the infection spreads more than a distance c_1 is less than $(\bar{d}\eta_0 t_0)^{c_1+1}$. Let ϵ_2 satisfy $0 < \epsilon_2 < 1$. By hypothesis, the expected number of seeds is $\omega(\log n)$, so from standard concentration results, the number of seeds is at most $1 + (1 + \epsilon)\gamma_0 t_0$ with probability tending to 1. Let P be the probability the infection spreads farther than distance c_1. Then from a final union bound,

$$
\begin{aligned}
P &< (1 + (1 + \epsilon)\gamma_0 t_0)\,(\bar{d}\eta_0 t_0)^{c_1+1} \\
&= o\left(2\gamma_0 t_0 \bar{d}^{c_1+1}(\gamma_0 t_0)^{-1}\right) \\
&= o(2\bar{d}^{c_1+1}).
\end{aligned}
\tag{1}
$$

Eq. (1) follows from our hypothesis $\eta_0 t_0 = o\left((\gamma_0 t_0)^{-1/(1+c_1)}\right)$. Therefore, $P \to 0$ so the infection travels no more than a distance c_1 with probability tending to 1.

Now we need to show no ball of radius r contains over an α fraction of the reporting nodes. We first consider all infected nodes. Let $\epsilon > 0$ be a constant as specified in the theorem statement. For convenience, let $c_2 = \bar{d}^{c_1+1}$, the maximum number of nodes in a ball of radius c_1. Consider an arbitrary node a, and let $B_{\mathrm{inner}} = Ball(a, r)$, $B_{\mathrm{outer}} = Ball(a, r + c_1)$. Then from the previous result, any seed that has an infection that spreads to a node in B_{inner} must be inside B_{outer} (since it can only travel a distance c_1). By the hypothesis that $r + c_1 < b\left(\frac{\alpha}{c_2(1+\epsilon)}\right)$, $\mathrm{card}(B_{\mathrm{outer}}) < \frac{\alpha n}{c_2(1+\epsilon)}$. Let u be the number of seeds, so $u = \omega(\log n)$, again by hypothesis. Then from Lemma 1, the number of seeds within B_{outer} is less than $\frac{\alpha u}{c_2(1+\epsilon/2)}$ with probability greater than $1 - 1/n^2$. Each of these seeds infects less than c_2 nodes, so the total number of infected nodes within B_{inner} (which must all be from seeds in B_{outer}) is less than $\frac{\alpha u}{1+\epsilon/2}$. Hence, this ball contains less than a $\frac{\alpha}{1+\epsilon/2}$ fraction of the infected nodes.

Finally, we need to show the reporting process does not significantly impact the fraction of infected nodes seen in that ball. We consider an equivalent method of choosing the reporting nodes: first the number of reporting nodes is chosen (with the appropriate distribution), and then these are distributed uniformly over the infected nodes. Let Q be the number of reporting nodes. Then we need to find the probability that αQ reporting nodes are within B_{inner}. Let X be the number of reporting nodes in this region. As we just showed, the probability that any particular reporting node is within that region is at most $\frac{\alpha}{1+\epsilon/2}$. From a standard balls-in-bins argument like in Lemma 1, since $\alpha Q = \omega(\log n)$, $P(X > \alpha Q) < 1/n^2$. That is, the probability that at least αQ of the reporting nodes are in that region is at most $1/n^2$.

Since each ball contains over an α fraction of the reporting nodes with probability no more than $1/n^2$, from a union bound, we find the probability that any of the n possible balls exceeds this bound is at most $1/n$. In this case, our algorithm correctly labels it 'Random'. Therefore, the Type I error probability decays to 0 as desired. □

Next consider the infection spreading by Process WR-SE [Process 1]. Define each of the parameters as before. Then

we can characterize the range for which the Type II error goes to 0 as follows.

THEOREM 6. *Consider an infection spreading as in Process 1. Suppose $r > s\eta_1 t_1$, where s is the speed of the infection when it spreads at rate 1, and $\eta_1 t_1$ scales to infinity. Suppose $\alpha = o((1 + \gamma_1 t_1)^{-1})$, and $\log(1/\alpha) = o(\eta_1 t_1)$. Then the Type II error probability decays to 0 as n increases.*

PROOF. First we show an upper bound on the number of seeds (recall seeds are the nodes randomly infected). The number of seeds is equal to one (the initially infected node) plus a Poisson random variable with mean $\gamma_1 t_1$. Let U be the set of seeds, and let $u = \frac{1}{\alpha}$. Since $\frac{1}{\alpha} = \omega(1 + \gamma_1 t_1)$, from the distribution, $u > \mathrm{card(U)}$ with probability scaling to 1.

From the speed definition, there exists a constant λ_1 such that for each seed a,

$$
P(W > s\eta_1 t_1) < e^{-\lambda_1 \eta_1 t_1},
$$

where W is the radius of the infection starting at a. Now we apply a union bound to see that,

$$
\begin{aligned}
P(\exists a \in U\, s.t.\, W > s\eta_1 t_1) &< \frac{1}{\alpha}e^{-\lambda_1 \eta_1 t_1} \\
&< e^{\lambda_1 \eta_1 t_1/2}e^{-\lambda_1 \eta_1 t_1} \\
&= e^{-\lambda_1 \eta_1 t_1/2} \to 0,
\end{aligned}
\tag{2}
$$

where Equation 2 follows from the fact that $\log(1/\alpha) = o(\eta_1 t_1)$. Therefore, each seed spreads no farther than a distance r with probability tending to 1.

We now show that our algorithm returns 'Epidemic' in this case. Cover the seed with the largest (reporting) infection using a ball of radius r, which we showed covers the entire infection for that seed. Since there are at most $1/\alpha$ seeds total, the fraction of reporting infected nodes covered is at least $1/\frac{1}{\alpha} = \alpha$. Therefore, an α fraction of the reporting infected nodes has been covered a ball of radius r, so the Median Ball Algorithm returns 'Epidemic' as desired. □

The previous theorems establish the set of conditions sufficient for the algorithm to succeed. As the conditions are a little opaque, we summarize them here: *(i)* The total number of nodes that can be covered by a ball of radius $2r$ (where r increases with n) must scale a constant factor less than the total number of nodes times α. *(ii)* In Process SR-WE [Process 0], the expected number of reporting seeds must be order-wise more than $\log n$. *(iii)* In Process SR-WE, the infection spreads no more than a constant distance. *(iv)* For Process WR-SE [Process 1], the threshold r must be set large enough that a ball of radius r covers the largest infection (using the epidemic speed). *(v)* For Process WR-SE, the expected number of seeds must be order-wise less than α^{-1}. *(vi)* For Process WR-SE, α^{-1} must be order-wise less than exponentials in $\eta_1 t_1$.

Finally, recall we can choose the algorithm parameters α and r. Then the question is, when can we choose appropriate algorithm parameters so that the probability of error goes to 0? This is answered by the following theorem.

THEOREM 7. *Suppose there exists c_1 such that $\eta_0 t_0 = o\left((\gamma_0 t_0)^{-1/(c_1+1)}\right)$ and $q\gamma_0 t = \omega(\log n)$. Suppose $\eta_1 t_1 = \omega(\log(\gamma_1 t_1))$, $\gamma_1 t_1 = \omega(1)$, and $s\eta_1 t_1 = o\left(b(\frac{1}{\gamma_1 t_1})\right)$. Then the algorithm parameters can be chosen so that the probability of error tends to 0.*

PROOF. We need to choose r and α so that $s\eta_1 t_1 < r < b\left(\frac{\alpha}{c_2(1+\epsilon)}\right) - c_1$ and $\alpha = o((\gamma_1 t_1)^{-1})$, $\log(1/\alpha) = o(\eta_1 t_1)$, where $c_2 = \bar{d}^{c_1+1}$. First we consider the conditions on α. Define an arbitrary slowly increasing function $g(n) = \theta(1)$, $g(n) = o(\gamma_1 t_1)$. This is possible since $\eta_1 t_1 = \omega(1)$. Choose $\alpha = (\gamma_1 t_1 g(n))^{-1}$. Then we have

$$\log(1/\alpha) = \log(\gamma_1 t_1 g(n))$$
$$< 2\log(\gamma_1 t_1)$$
$$= o(\eta_1 t_1).$$

Thus α satisfies the desired conditions. Now we will show it is possible to choose an appropriate r. By hypothesis, $s\eta_1 t_1 = o(b(\frac{1}{\gamma_1 t_1}))$. From our choice of α, for sufficiently large n, $\frac{\alpha}{c_2(1+\epsilon)} < \frac{1}{\gamma_1 t_1}$. Using the concavity of $b(x)$,

$$b\left(\frac{1}{\gamma_1 t_1}\right) < \frac{\gamma_1 t_1}{\alpha/(c_2(1+\epsilon))} b\left(\frac{\alpha}{c_2(1+\epsilon)}\right)$$
$$= o\left(b\left(\frac{\alpha}{c_2(1+\epsilon)}\right)\right). \qquad (3)$$

Therefore, $s\eta_1 t_1 = o(b(\frac{\alpha}{c_2(1+\epsilon)}))$, with $s\eta_1 t_1 = \omega(1)$ by hypothesis. Thus it is clear r can be chosen with $s\eta_1 t_1 < r < b\left(\frac{\alpha}{c_2(1+\epsilon)}\right) - c_1$, for example by averaging each side. With this choice of parameters, the conditions of Theorem 5 and Theorem 6 are satisfied. Hence, both the Type I and Type II error probabilities will tend to 0. $\square$

5. SIMULATIONS

In the previous sections, we show that the Median Ball Algorithm can distinguish epidemics and random sicknesses in both the cases of false positives, and when the infection process is mixed. In this section we illustrate via simulation the probability of error for reasonable graph sizes and parameters, and how it changes as the parameters are adjusted. While the theory developed so far applies to many types of graphs, here we specifically consider only one structure – grid graphs with wrapping edges – to explore various aspects (false positves, mixed infection) and for different parameter choices, within the page-length space constraints. Naturally, these graphs have bounded degree, and hence have the necessary properties to detect an infection. We use this graph to explore the non-asymptotic behavior of the Median Ball algorithm.

We performed these simulations with false positives and for mixed infections. We evaluate our algorithm by the empirical error probability, the average error probability for both Type I and Type II errors, weighting both equally. The results give insight in how the error probability is affected by graph topology, algorithm parameters, and infection time.

Each simulation was performed as follows. We used a grid graph with $n = 4900$, and infection time $t = 10$. The reporting probability was fixed at $q = 0.25$. The infection was simulated for 10000 trials for each infection processes (a random sickness and an infection), running the Median Ball Algorithm for each set of reporting nodes. We set the ball size parameter (r) to the optimal value as determined empirically. The other parameters were set as described in each caption. The probability of error is mostly plotted against the empirical expected fraction of infected nodes. That is, for each set of parameters, we estimated the expected number of infected nodes from the simulations, which

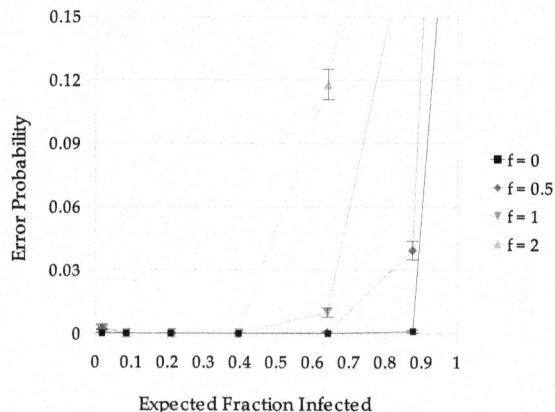

Figure 1: [False Positive Model] This figure shows the overall error probability, the sum of equally weighted Type I and Type II error rates, for a grid graph. The false positives were located randomly on the graph. The x-axis measures the expected fraction of nodes truly infected. As in our results, $\alpha = 1/(1+f)$. The ball radius r was set to the optimal value empirically.

was divided by n to determine the fraction infected. This expected fraction of infected nodes conveys the size of the infection, and hence the difficulty of the problem (since the task is more difficult the larger the infection is). Note that since $q = 0.25$, the expected fraction of reporting nodes is approximately 0.25 times as large. Finally the probability of error was estimated from the frequency at which the Median Ball Algorithm mischaracterized the type of infection.

False Positives: Our first simulation results are on the probability of error for grid graphs for a variety of false positive frequencies. As in our analytical setting, the random sickness infection size was normalized to the same distribution as the epidemic as determined empirically. The results are shown in Figure 1.

The error probability is very low up to a very large number of truly infected nodes. It climbs fairly slowly as the number of false positives increases. Even when two-thirds of the reporting nodes are false positives, the error probability is low even up to an expected 40% of the network infected. Therefore our algorithm works very well in this setting.

Mixed Infection: Next, we present the simulation results for infections with mixed spreading regimes. Unlike for false positives, there is no direct normalization of the infection sizes. Rather, we adjusted the rates so that the infection sizes for both infection processes would be similar. This was done by first choosing the epidemic rate, and then empirically finding the random rate to three significant digits so that expected number of infected nodes hit a target value. This was done so that all the infections (for the various parameters) would be fairly comparable. Process SR-WE [Process 0] used an infection rate of 0.2.

Figure 2 shows the probability of error for various infection sizes. The infection rate for Process WR-SE [Process 1] is given on the x-axis. As expected, the larger the infection, the more difficult it is to use clustering to determine whether an infection is mostly random or mostly an epidemic. When

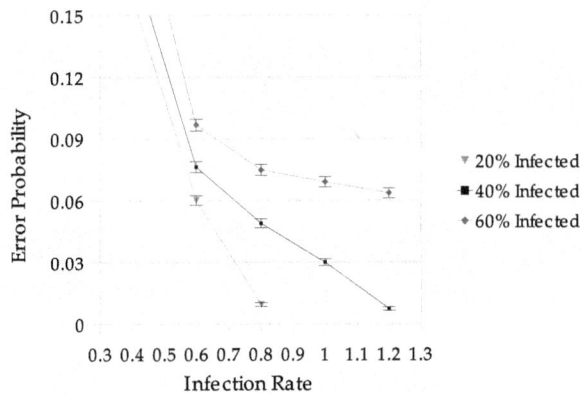

Figure 2: [Mixed Infection Model] This figure shows the overall error probability, the sum of equally weighted Type I and Type II error rates, for various expected fraction infected and Process WR-SE infection rates. The Process SR-WE infection rate is 0.2. The parameter $\alpha = 0.5$. The ball radius r was set to the optimal value empirically.

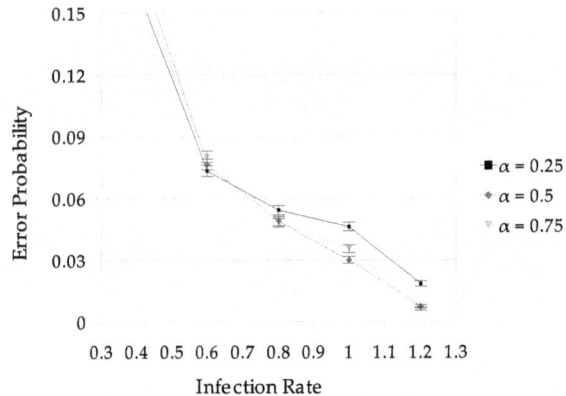

Figure 3: [Mixed Infection Model] This figure shows the overall error probability, the sum of equally weighted Type I and Type II error rates, for various values of α and Process WR-SE infection rates. The Process SR-WE infection rate is 0.2. The expected fraction infected was 40%. The ball radius r was set to the optimal value empirically.

an expected 60% of the nodes in the network are infected, then the probability of error stays high, even for much larger infection rates. Note that there is a maximum infection rate before the target infection size is exceeded regardless of the random sickness rate. We used Process WR-SE infection rates close to that maximum.

Next we determine the effect of α on the probability of error. These results are shown in Figure 3. Surprisingly, changing α has a relatively small effect on the probability of error. The largest effect seen is using too large a value for larger Process WR-SE infection rates (when the probability of error is low). However, that is still relatively small. Then our algorithm seems fairly insensitive to the value of α.

6. ACKNOWLEDGMENTS

This work was partially supported by NSF Grants CNS-1017525, EECS-1056028, DTRA grant HDTRA 1-08-0029 and Army Research Office Grant W911NF-11-1-0265.

7. REFERENCES

[1] F. Ball and P. Neal. Poisson approximation for epidemics with two levels of mixing. *The Annals of Probability*, 32(1B):1168–1200, 2004.

[2] I. Benjamini and Y. Peres. Tree-indexed random walks on groups and first passage percolation. *Probability Theory and Related Fields*, 98:91–112, 1994.

[3] J. Cohen. Making headway under hellacious circumstances. *SCIENCE*, 313:470–473, July 2006.

[4] C. Corley, D. Cook, A. Mikler, and K. Singh. Text and structural data mining of influenza mentions in web and social media. *International Journal of Environmental Research and Public Health*, 7:596–615, 2010.

[5] N. Demiris and P. D. O'Neill. Bayesian inference for epidemics with two levels of mixing. *Scandinavian Journal of Stat.*, 32:265–280, 2005.

[6] N. Demiris and P. D. O'Neill. Bayesian inference for stochastic multitype epidemics in structured populations via random graphs. *Journal of the Royal Stat. Society Series B*, 67(5):731–745, 2005.

[7] R. Durrett. *Random Graph Dynamics*. Cambridge University Press, 2007.

[8] F-Secure. Bluetooth-worm:symbos/cabir, 2012. http://www.f-secure.com/v-descs/cabir.shtml.

[9] A. J. Ganesh, L. Massoulié, and D. F. Towsley. The effect of network topology on the spread of epidemics. In *INFOCOM*, pages 1455–1466, 2005.

[10] M. Gomez-Rodriguez, J. Leskovec, and A. Krause. Inferring networks of diffusion and influence. *ACM Trans. Knowl. Discov. Data*, 5(4):21:1–21:37, Feb. 2012.

[11] Google Flu Trends, http://www.google.org/flutrends/.

[12] A. Gopalan, S. Banerjee, A. Das, and S. Shakkottai. Random mobility and the spread of infection. In *Proc. IEEE Infocom*, 2011.

[13] H. Kesten. On the speed of convergence in first-passage percolation. *The Annals of Applied Probability*, 3(2):296–338, Nov 1993.

[14] W. Luo and W. P. Tay. Identifying infection sources in large tree networks. In *9th Annual IEEE Communications Society Conference on Sensor, Mesh and Ad Hoc Communications and Networks (SECON)*, pages 281–289, June 2012.

[15] New York Times Bits Blog, http://bits.blogs.nytimes.com/2012/12/13/lookout-toll-fraud/.

[16] C. Milling, C. Caramanis, S. Mannor, and S. Shakkottai. Network forensics: random infection vs

spreading epidemic. *SIGMETRICS Perform. Eval. Rev.*, 40(1):223–234, June 2012.

[17] C. Milling, C. Caramanis, S. Mannor, and S. Shakkottai. On identifying the causative network of an epidemic. In *In Proceedings of 50th Annual Allerton Conference on Communication, Control, and Computing*, October 2012.

[18] S. A. Myers, C. Zhu, and J. Leskovec. Information diffusion and external influence in networks. In *Proceedings of the 18th ACM SIGKDD international conference on Knowledge discovery and data mining*, KDD '12, pages 33–41, New York, NY, USA, 2012. ACM.

[19] P. Netrapalli and S. Sanghavi. Learning the graph of epidemic cascades. *SIGMETRICS Perform. Eval. Rev.*, 40(1):211–222, June 2012.

[20] D. Shah and T. Zaman. Detecting sources of computer viruses in networks: Theory and experiment. *SIGMETRICS Perform. Eval. Rev.*, 86:203–214, 2010.

[21] D. Shah and T. Zaman. Rumors in a network: Who's the culprit? *IEEE Transactions on Information Theory*, 57, August 2011.

[22] J. Snow. *On the mode of communication of cholera.* John Churchill, 1855.

[23] Wikipedia. Commwarrior-a — Wikipedia, the free encyclopedia, 2012. [Accessed 30-Sept-2012].

Social Trust and Social Reciprocity Based Cooperative D2D Communications

Xu Chen
School of ECEE
Arizona State University
Tempe, AZ 85287, USA
xchen179@asu.edu

Brian Proulx
School of ECEE
Arizona State University
Tempe, AZ 85287, USA
bbproulx@asu.edu

Xiaowen Gong
School of ECEE
Arizona State University
Tempe, AZ 85287, USA
xgong9@asu.edu

Junshan Zhang
School of ECEE
Arizona State University
Tempe, AZ 85287, USA
junshan.zhang@asu.edu

ABSTRACT

Thanks to the convergence of pervasive mobile communications and fast-growing online social networking, mobile social networking is penetrating into our everyday life. Aiming to develop a systematic understanding of the interplay between social structure and mobile communications, in this paper we exploit social ties in human social networks to enhance cooperative device-to-device communications. Specifically, as hand-held devices are carried by human beings, we leverage two key social phenomena, namely social trust and social reciprocity, to promote efficient cooperation among devices. With this insight, we develop a coalitional game theoretic framework to devise social-tie based cooperation strategies for device-to-device communications. We also develop a network assisted relay selection mechanism to implement the coalitional game solution, and show that the mechanism is immune to group deviations, individually rational, and truthful. We evaluate the performance of the mechanism by using real social data traces. Numerical results show that the proposed mechanism can achieve up-to 122% performance gain over the case without D2D cooperation.

Categories and Subject Descriptors

C.2.1 [**Network Architecture and Design**]: Wireless communication

Keywords

D2D Communication, Cooperative Networking, Mobile Social Networking, Social Trust, Social Reciprocity

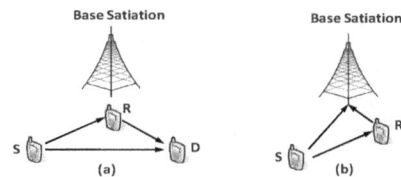

Figure 1: An illustration of cooperative D2D communication for cooperative networking. In sub-figure (a), device R serves as the relay for the D2D communication between devices S and D. In sub-figure (b), device R serves as the relay for the cellular communication between device S and the base station. In both cases, the D2D communication between devices S and R is part of cooperative networking.

1. INTRODUCTION

Mobile data traffic is predicted to grow further by over 100 times in the next ten years [1], which poses a significant challenge for future cellular networks. One promising approach to increase network capacity is to promote direct communications between hand-held devices. Such device-to-device (D2D) communications can offer a variety of advantages over traditional cellular communications, such as higher user throughput, improved spectral efficiency, and extended network coverage [6]. For example, a device can share the video content with neighboring devices who have the similar watching interest, which can help to reduce the traffic rate demand from the network operator.

Cooperative communication is an efficient D2D communication paradigm where devices can serve as relays for each other[1]. As illustrated in Figure 1, cooperative D2D communication can help to 1) improve the quality of D2D communication for direct data offer-loading between devices and 2) enhance the performance of cellular communications between the base station and the devices as well. Hence cooperative D2D communication can be a critical building block

---

[1]There are many approaches for cooperative communications, and for ease of exposition this study assumes cooperative relaying.

for efficient cooperative networking for future wireless networks, wherein individual users cooperate to substantially boost the network capacity and cost-effectively provide rich multimedia services and applications, such as video conferencing and interactive media, anytime, anywhere. Nevertheless, a key challenge here is how to stimulate effective cooperation among devices for cooperative D2D communications. As different devices are usually owned by different individuals and they may pursue different interests, there is no good reason to assume that all devices would cooperate with each other.

Since the hand-held devices are carried by human beings, a natural question to ask is that "is it possible to leverage human social relationship to enhance D2D communications for cooperative networking?". Indeed, with the explosive growth of online social networks such as Facebook and Twitter, more and more people are actively involved in online social interactions, and social relationships among people are hence extensively broadened and significantly enhanced [10]. This has opened up a new avenue for cooperative D2D communication system design – we believe that it has potential to propel significant advances in mobile social networking..

One primary goal of this study is to establish a new D2D cooperation paradigm by leveraging two key social phenomena: social trust and social reciprocity. Social trust can be built up among humans such as kinship, friendship, colleague relationship, and altruistic behaviors are observed in many human activities [8]. For example, when a device user is at home or work, typically family members, neighbors, colleagues, or friends are nearby. The device user can then exploit the social trust from these neighboring users to improve the quality of D2D communication, e.g., by asking the best trustworthy device to serve as the relay. Another key social phenomenon, social reciprocity, is also widely observed in human society [7]. Social reciprocity is a powerful social paradigm to promote cooperation so that a group of individuals without social trust can exchange mutually beneficial actions, making all of them better off. For example, when a device user does not have any trusted friends in the vicinity, he (she) may cooperate with the nearby strangers by providing relay assistance for each other to improve the quality of D2D communications.

As illustrated in Figure 2, cooperative D2D communications based on social trust and social reciprocity can be projected onto two domains: the physical domain and the social domain. In the physical domain, different devices have different feasible relay selection relationships subject to the physical constraints. In the social domain, different devices have different assistance relationships based on social trust among the devices. In this case, each device has two options for relay selection: 1) either seek relay assistance from another feasible device that has social trust towards him (her); 2) or participate in a group formed based on social reciprocity by exchanging mutually beneficial relay assistance. The main thrust of this study is devoted to tackling two key challenges for the social trust and social reciprocity based approach. The first is which option a device should adopt for relay selection: social trust or social reciprocity. The second is how to efficiently form groups among the devices that adopt the social reciprocity based relay selection. We will develop a coalitional game theoretic framework to address these challenges.

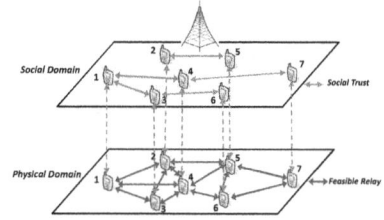

Figure 2: An illustration of the social trust model for cooperative D2D communications. In the physical domain, different devices have different feasible cooperation relationships subject to physical constraints. In the social domain, different devices have different assistance relationships based on social trust among the devices.

1.1 Summary of Main Contributions

The main contributions of this paper are as follows:

- *Social trust and social reciprocity based cooperative D2D communications*: We propose a novel social trust and social reciprocity based framework to promote efficient cooperation among devices for cooperative D2D communications. By projecting D2D communications in a mobile social network onto both physical and social domains, we introduce the physical-social graphs to model the interplay therein while capturing the physical constraints for feasible D2D cooperation and the social relationships among devices for effective cooperation.

- *Coalitional game solutions*: We formulate the relay selection problem for social trust and social reciprocity based cooperative D2D communications as a coalitional game. We show that the coalitional game admits the top-coalition property based on which we devise a core relay selection algorithm for computing the core solution to the game.

- *Network assisted relay selection mechanism*: We develop a network assisted mechanism to implement the coalitional game based solution. We show that the mechanism is immune to group deviations, individually rational, truthful, and computationally efficient. We further evaluate the performance of the mechanism by the real social data trace. Numerical results show that the proposed mechanism can achieve up-to 122% performance gain over the case without D2D cooperation.

A primary goal of this paper is to build a theoretically sound and practically relevant framework to understand social trust and social reciprocity based cooperative D2D communications. This framework highlights the interplay between potential physical network performance gain through efficient D2D cooperation and the exploitation of social relationships among device users to stimulate effective cooperation. Besides the cooperative D2D communication scenario where devices serve as relays for each other, the proposed social trust and social reciprocity based framework can also be applied to many other D2D cooperation scenarios, such as cooperative MIMO communications and mobile cloud computing. We believe that these initial steps presented here open a new avenue for mobile social networking and have great potential to enhance network capacity in future wireless networks.

1.2 Related Work

Much effort has been made in the literature to stimulate, via incentive mechanisms, cooperation in wireless networks. Payment-based mechanisms have been widely considered to incentivize cooperation for wireless ad hoc networks [2]. Another widely adopted approach for cooperation stimulation is reputation-based mechanisms, where a centralized authority or the whole user population collectively keeps records of the cooperative behaviors and punishes non-cooperating users [12]. However, incentive mechanisms typically assume that all users are fully rational and they act in the selfish manner. Such an assumption are not appropriate for D2D communications as hand-held devices are carried by human beings and people typically act with bounded rationality and involve social interactions [8].

The social aspect is now becoming an important dimension for communication system design. Social structures, such as social community which are derived from the user contact patterns, have been exploited to design efficient data forwarding and routing algorithms in delay tolerant networks [5]. The social influence phenomenon has also been utilized to devise effective data dissemination mechanisms for mobile networks [4]. The common assumption among these works, however, is that all users are always willing to help others, e.g., for data forwarding and relaying. In this paper we propose a novel framework to stimulate cooperation among device users while also taking the social aspect into account.

The rest of this paper is organized as follows. We first introduce the system model in Section 2. We then study cooperative D2D communications based on social trust and social reciprocity and develop the network assisted relay selection mechanism in Sections 3 and 4, respectively. We evaluate the performance of the proposed mechanism by simulations in Section 5, and finally conclude in Section 6.

2. SYSTEM MODEL

In this section we present the system model of cooperative D2D communications based on social trust and social reciprocity – a new mobile social networking paradigm. As illustrated in Figure 2, cooperative D2D communications can be projected onto two domains: the physical domain and the social domain. In the physical domain, different devices have different feasible cooperation relationships for cooperative D2D communications subject to the physical constraints. In the social domain, different devices have different assistance relationships based on social relationships among the devices. We next discuss both physical and social domains in detail.

2.1 Physical (Communication) Graph Model

We consider a set of nodes $\mathcal{N} = \{1, 2, ..., N\}$ where N is the total number of nodes. Each node $n \in \mathcal{N}$ is a wireless device that would like to conduct D2D communication to transmit data packets to its corresponding destination d_n. Notice that a destination d_n may also be a transmit node in the set $\mathcal{N}$ of another D2D communication link or the base station. The D2D communication is underlaid beneath a cellular infrastructure wherein there exists a base station controlling the up-link/down-link communications of the cel-

lular devices. To avoid generating severe interference to the incumbent cellular devices, each node $n \in \mathcal{N}$ will first send a D2D communication establishment request message to the base station. The base station then computes the allowable transmission power level p_n for the D2D communication of node n based on the system parameters such as geolocation of the node n and the protection requirement of the neighboring cellular devices. For example, the proper transmission power p_n of the D2D communication can be computed according to the power control algorithm proposed in [15].

We consider a time division multiple access (TDMA) mechanism in which the transmission time is slotted and one node $n \in \mathcal{N}$ is scheduled to carry out its D2D communication in a time slot[2]. At the allotted time slot, node n can choose either to transmit to the destination node d_n directly or to use cooperative communication by asking another node m in its vicinity to serve as a relay.

Due to the physical constraints such as signal attenuation, only a subset of nodes that are close enough can be feasible relay candidates for the node n. To take such physical constraints into account, we introduce the physical graph[3] $\mathcal{G}^P \triangleq \{\mathcal{N}, \mathcal{E}^P\}$ where the set of nodes $\mathcal{N}$ is the vertex set and $\mathcal{E}^P \triangleq \{(n, m) : e_{nm}^P = 1, \forall n, m \in \mathcal{N}\}$ is the edge set where $e_{nm}^P = 1$ if and only if node m is a feasible relay for node n. An illustration of the physical graph is given in Figure 2. We also denote the set of nodes that can serve as a feasible relay of node n as $\mathcal{N}_n^P \triangleq \{m \in \mathcal{N} : e_{nm}^P = 1\}$. A recent work in [16] shows that it is sufficient for a source node to choose the best relay node among multiple candidates to achieve full diversity. For ease of exposition, we hence assume that each node n selects at most one neighboring node $m \in \mathcal{N}_n^P$ as the relay.

For ease of exposition, we consider the full duplex decode-and-forward (DF) relaying scheme [9] for the cooperative D2D communication. Let $r_n \in \mathcal{N}_n^P$ denote the relay node chosen by node $n \in \mathcal{N}$ for cooperative communication. The data rate achieved by node n is then given as [9]

$$Z_{n,r_n}^{DF} = \frac{W}{N} \min\{\log(1 + \mu_{nr_n}), \log(1 + \mu_{nd_n} + \mu_{r_n d_n})\},$$

where W denotes the channel bandwidth and μ_{ij} denotes the signal-to-noise ratio (SNR) at device j when device i transmits a signal to device j. As an alternative, the node n can also choose to transmit directly without any relay assistance and achieve a data rate of $Z_n^{Dir} = \frac{W}{N} \log(1 + \mu_{nd_n})$.

For simplicity, we define the data rate function of node n as $R_n : \mathcal{N}_n^P \cup \{n\} \to \mathbb{R}_+$, which is given by

$$R_n(r_n) = \begin{cases} Z_{n,r_n}^{DF}, & \text{if } r_n \neq n, \\ Z_n^{Dir}, & \text{if } r_n = n. \end{cases} \quad (1)$$

We will use the terminology that node n chooses itself as the relay for the situation in which node n transmits directly to its destination d_n.

2.2 Social Graph Model

We next introduce the social trust model for cooperative D2D communications. The underlying rationale of using social trust is that the hand-held devices are carried by human

---

[2]Our methods are also applicable to other multiple access schemes.

[3]The graphs (e.g., physical graph and social graph) in this paper can be directed.

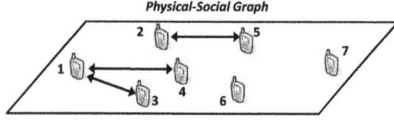

Figure 3: The physical-social graph based on the physical graph and social graph in Figure 2. For example, there exists an edge between nodes 1 and 3 in the physical-social graph since they can serve as the feasible relay for each other and also have social trust towards each other.

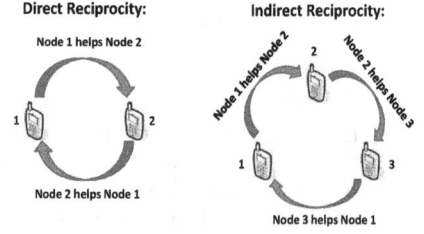

Figure 4: An illustration of direct and indirect reciprocity

beings and the knowledge of human social ties can be utilized to achieve effective and trustworthy relay assistance for cooperative D2D communications.

More specifically, we introduce the social graph $\mathcal{G}^S = \{\mathcal{N}, \mathcal{E}^S\}$ to model the social trust among the nodes. Here the vertex set is the same as the node set $\mathcal{N}$ and the edge set is given as $\mathcal{E}^S = \{(n, m) : e_{nm}^S = 1, \forall n, m \in \mathcal{N}\}$, where $e_{nm}^S = 1$ if and only if nodes n and m have social trust towards each other, which can be kinship, friendship, or colleague relationship between two nodes. We denote the set of nodes that have social trust towards node n as $\mathcal{N}_n^S = \{m : e_{nm}^S = 1, \forall m \in \mathcal{N}\}$, and we assume that the nodes in $\mathcal{N}_n^S$ are willing to serve as the relay of node n for cooperative communication.

Based on the physical graph $\mathcal{G}^P$ and social graph $\mathcal{G}^S$ above, each node $n \in \mathcal{N}$ can classify the set of feasible relay nodes in $\mathcal{N}_n^P$ into two types: nodes with social trust and nodes without social trust. A node n then has two options for relay selection. On the one hand, the node n can choose to seek relay assistance from another feasible device that has social trust towards him (her). On the other hand, the node n can choose to participate in a group formed based on social reciprocity by exchanging mutually beneficial relay assistance. In the following, we will study 1) how to choose between social trust and social reciprocity based relay selections for each node; and 2) how to efficiently form reciprocal groups among the nodes without social trust.

3. SOCIAL TRUST AND SOCIAL RECIPROCITY BASED COOPERATIVE D2D COMMUNICATIONS

In this section, we study the cooperative D2D communications based on social trust and social reciprocity. As mentioned, each node $n \in \mathcal{N}$ has two options for relay selection: social trust based versus social reciprocity. We next address the issues of choosing between social trust and social reciprocity based relay selections for each node and the reciprocal group forming among the nodes without social trust.

3.1 Social Trust Based Relay Selection

We first consider social trust based relay selection for D2D cooperation. The key motivation for using social trust is to utilize the knowledge of human social ties to achieve effective and trustworthy relay assistance among the devices for cooperative D2D communications. For example, when a device user is at home or working place, he (she) typically has family members, neighbors, colleagues, or friends in the vicinity. The device user can then exploit the social trust

from neighboring users to improve the quality of D2D communication by asking the best trustworthy device to serve as the relay.

To take both the physical and social constraints into account, we define the *physical-social graph* $\mathcal{G}^{PS} \triangleq \{\mathcal{N}, \mathcal{E}^{PS}\}$ where the vertex set is the node set $\mathcal{N}$ and the edge set $\mathcal{E}^{PS} = \{(n, m) : e_{nm}^{PS} \triangleq e_{nm}^P \cdot e_{nm}^S = 1, \forall n, m \in \mathcal{N}\}$, where $e_{nm}^{PS} = 1$ if and only if node m is a feasible relay (i.e., $e_{nm}^P = 1$) and has social trust towards node n (i.e., $e_{nm}^S = 1$). An illustration of the physical-social graph is depicted in Figure 3. We also denote the set of nodes that have social trust towards node n and are also feasible relay candidates for node n as $\mathcal{N}_n^{PS} = \{m : e_{nm}^{PS} = 1, \forall m \in \mathcal{N}\}$.

For cooperative D2D communications based on social trust, each node $n \in \mathcal{N}$ can choose the best relay to maximize its data rate subject to both physical and social constraints, i.e., $r_n^S = \arg\max_{r_n \in \mathcal{N}_n^{PS} \cup \{n\}} R_n(r_n)$.

3.2 Social Reciprocity Based Relay Selection

Next, we study the social reciprocity based relay selection. Different from D2D cooperation based on social trust which requires strong social ties among device users, social reciprocity is a powerful mechanism for promoting mutual beneficial cooperation among the nodes in the absence of social trust. For example, when a device user does not have any friends in the vicinity, he (she) may cooperate with the nearby strangers by providing relay assistance for each other to improve the quality of D2D communications. In general, there are two types of social reciprocity: direct reciprocity and indirect reciprocity[4] (see Figure 4 for an illustration). Direct reciprocity is captured in the principle of "you help me, and I will help you". That is, two individuals exchange altruistic actions so that both obtain a net benefit. Indirect reciprocity is essentially the concept of "I help you, and someone else will help me". That is, a group of individuals exchange altruistic actions so that all of them can be better off.

To better describe the possible cooperation relationships among the the set of nodes without social trust, we introduce the *physical-coalitional graph* $\mathcal{G}^{PC} = \{\mathcal{N}, \mathcal{E}^{PC}\}$. Here the vertex set is the node set $\mathcal{N}$ and the edge set $\mathcal{E}^{PC} = \{(n, m) : e_{nm}^{PC} \triangleq e_{nm}^P \cdot (1 - e_{nm}^S) = 1, \forall n, m \in \mathcal{N}\}$, where $e_{nm}^{PC} = 1$ if and only if node m is a feasible relay (i.e., $e_{nm}^P = 1$) and has no social trust towards node n (i.e., $e_{nm}^S = 0$). An illustration of physical-coalitional graph is depicted in Figure 5. We also denote the set of nodes that have no social trust towards user n but are feasible relay candidates of node n as $\mathcal{N}_n^{PC} \triangleq \{m : e_{nm}^{PC} = 1, \forall m \in \mathcal{N}\}$. For social

---

[4]Reciprocity in this study refers to social reciprocity.

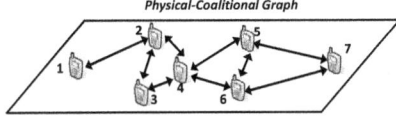

Figure 5: The physical-coalitional graph based on the physical graph and social graph in Figure 6. For example, there exists an edge between nodes 1 and 2 in the physical-coalitional graph since they can serve as the feasible relay for each other and have no social trust towards each other.

reciprocity based relay selection, a key challenge is how to efficiently divide the nodes into multiple groups such that the nodes can significantly improve their data rates by the reciprocal cooperation within the groups. We next develop a coalitional game framework to address this challenge.

3.2.1 Introduction to Coalitional Game

For the sake of completeness, we first give a brief introduction to the coalitional game [13]. Formally, a coalitional game consists of a tuple $\Omega = \{\mathcal{N}, \mathcal{X}_\mathcal{N}, V, (\succ_n)_{n \in \mathcal{N}}\}$, where

- $\mathcal{N}$ is a finite set of players.
- $\mathcal{X}_\mathcal{N}$ is the space of feasible cooperation strategies of all players.
- V is a characteristic function that maps from every nonempty subset of players $\mathcal{S} \subseteq \mathcal{N}$ (a coalition) to a subset of feasible cooperation strategies $V(\mathcal{S}) \subseteq \mathcal{X}_\mathcal{N}$. This represents the possible cooperation strategies among the players in the coalition $\mathcal{S}$, given that other players out of the coalition $\mathcal{S}$ do not participate in any cooperation.
- $\succ_n$ is a strict preference order (reflexive, complete, and transitive binary relation) on $\mathcal{X}_\mathcal{N}$ for each player $n \in \mathcal{N}$. This captures the idea that different players may have different preferences over different cooperation strategies.

In the same spirit as Nash equilibrium in a non-cooperative game, the "core" plays a critical role in the coalitional game.

DEFINITION 1. *The core is the set of $\boldsymbol{x} \in V(\mathcal{N})$ for which there does not exist a coalition $\mathcal{S}$ and $\boldsymbol{y} \in V(\mathcal{S})$ such that $\boldsymbol{y} \succ_n \boldsymbol{x}$ for all $n \in \mathcal{S}$.*

Intuitively, the core is a set of cooperation strategies such that no coalition can deviate and improve for all its members by cooperation within the coalition [13].

3.2.2 Coalitional Game Formulation

We then cast the social reciprocity based relay selection problem as a coalitional game $\Omega = \{\mathcal{N}, \mathcal{X}_\mathcal{N}, V, (\succ_n)_{n \in \mathcal{N}}\}$ as follows:

- the set of players $\mathcal{N}$ is the set of nodes.
- the set of cooperation strategies $\mathcal{X}_\mathcal{N} = \{(r_n)_{n \in \mathcal{N}} : r_n \in \mathcal{N}_n^{PC} \cup \{n\}, \forall n \in \mathcal{N}\}$, which describes the set of possible relay selections for all nodes based on the physical-coalitional graph $\mathcal{G}^{PC}$.
- the characteristic function $V(\mathcal{S}) = \{(r_n)_{n \in \mathcal{N}} \in \mathcal{X}_\mathcal{N} : \{r_n\}_{n \in \mathcal{S}} = \{k\}_{k \in \mathcal{S}}$ and $r_m = m, \forall m \in \mathcal{N} \backslash \mathcal{S}\}$ for each coalition $\mathcal{S} \subseteq \mathcal{N}$. Here the condition "$\{r_n\}_{n \in \mathcal{S}} =$

$\{k\}_{k \in \mathcal{S}}$" represents the possible relay assistance exchange among the nodes in the coalition $\mathcal{S}$. The condition "$r_m = m, \forall m \in \mathcal{N} \backslash \mathcal{S}$" states that the nodes out of the coalition $\mathcal{S}$ would not participate in any cooperation and choose to transmit directly. For example, in Figure 4, the coalition $\mathcal{S} = \{1, 2\}$ in the direct reciprocity case adopts the cooperation strategy $r_1 = 2$ and $r_2 = 1$ and the coalition $\mathcal{S} = \{1, 2, 3\}$ in the indirect reciprocity case adopts the cooperation strategy $r_1 = 3$, $r_2 = 1$ and $r_3 = 2$.

- the preference order $\succ_n$ is defined as $(r_m)_{m \in \mathcal{N}} \succ_n (r'_m)_{m \in \mathcal{N}}$ if and only if $r_n \succ_n r'_n$. That is, node n prefers the relay selection $(r_m)_{m \in \mathcal{N}}$ to another selection $(r'_m)_{m \in \mathcal{N}}$ if and only if its assigned relay r_n in the former selection $(r_m)_{m \in \mathcal{N}}$ is better than the assigned relay r'_n in the latter selection $(r'_m)_{m \in \mathcal{N}}$. In the following, we define that $r_n \succ_n r'_n$ when $R_n(r_n) > R_n(r'_n)$, and if $R_n(r_n) = R_n(r'_n)$ then ties are broken arbitrarily.

The core of this coalitional game is a set of $(r_n^*)_{n \in \mathcal{N}} \in V(\mathcal{N})$ for which there does not exist a coalition $\mathcal{S}$ and $(r_n)_{n \in \mathcal{N}} \in V(\mathcal{S})$ such that $(r_n)_{n \in \mathcal{N}} \succ_n (r_n^*)_{n \in \mathcal{N}}$ for all $n \in \mathcal{S}$. In other words, no coalition of nodes can deviate and improve their relay selection by cooperation in the coalition. We will refer the solution $(r_n^*)_{n \in \mathcal{N}}$ as the core relay selection in the sequel.

3.2.3 Core Relay Selection

We now study the existence of the core relay selection. To proceed, we first introduce the following key concepts of coalitional game.

DEFINITION 2. *Given a coalitional game $\Omega = \{\mathcal{N}, \mathcal{X}_\mathcal{N}, V, (\succ_n)_{n \in \mathcal{N}}\}$, we call a coalitional game $\Phi = \{\mathcal{M}, \mathcal{X}_\mathcal{M}, V, (\succ_m)_{m \in \mathcal{M}}\}$ a coalitional sub-game of the game Ω if and only if $\mathcal{M} \subseteq \mathcal{N}$ and $\mathcal{M} \neq \varnothing$.*

In other words, a coalitional sub-game Φ is a coalitional game defined on a subset of the players of the original coalitional game Ω.

DEFINITION 3. *Given a coalitional sub-game $\Phi = \{\mathcal{M}, \mathcal{X}_\mathcal{M}, V, (\succ_m)_{m \in \mathcal{M}}\}$, a non-empty subset $\mathcal{S} \subseteq \mathcal{M}$ is a top-coalition of the game Φ if and only if there exists a cooperation strategy $(r_m)_{m \in \mathcal{M}} \in V(\mathcal{S})$ such that for any $\mathcal{K} \subseteq \mathcal{M}$ and any cooperation strategy $(r'_m)_{m \in \mathcal{M}} \in V(\mathcal{K})$ satisfying $r_m \neq r'_m$ for any $m \in \mathcal{S}$, we have $r_m \succ_m r'_m$ for any $m \in \mathcal{S}$.*

That is, by adopting the cooperation strategy $(r_m)_{m \in \mathcal{S}}$, the coalition $\mathcal{S}$ is a group that is mutually the best for all its members [3].

DEFINITION 4. *A coalitional game $\Omega = \{\mathcal{N}, \mathcal{X}_\mathcal{N}, V, (\succ_n)_{n \in \mathcal{N}}\}$ satisfies the top-coalition property if and only if there exists a top-coalition for any its coalitional sub-game Φ.*

We then show that the proposed coalitional game for social reciprocity based relay selection satisfies the top-coalition property. For simplicity, we first denote $\tilde{\mathcal{N}}_n^{PC} \triangleq \mathcal{N}_n^{PC} \cup \{n\}$. For a coalitional sub-game $\Phi = \{\mathcal{M}, \mathcal{X}_\mathcal{M}, V, (\succ_m)_{m \in \mathcal{M}}\}$, we denote the mapping $\gamma(n, \mathcal{M})$ as the most preferable relay of node $n \in \mathcal{M}$ in the set of nodes $\mathcal{M} \cap \tilde{\mathcal{N}}_n^{PC}$, i.e., $\gamma(n, \mathcal{M}) \succ_n i$ for any $i \neq \gamma(n, \mathcal{M})$ and $i \in \mathcal{M} \cap \tilde{\mathcal{N}}_n^{P}$. Based on the mapping γ, we can define the concept of reciprocal relay selection cycle as follows.

DEFINITION 5. *Given a coalitional sub-game* $\Phi = \{\mathcal{M}, \mathcal{X}_{\mathcal{M}},$ $V, (\succ_m)_{m \in \mathcal{M}}\}$, *a node sequence* $(n_1, ..., n_L)$ *is called a reciprocal relay selection cycle of length* L *if and only if* $\gamma(n_l, \mathcal{M}) = n_{l+1}$ *for* $l = 1, ..., L - 1$ *and* $\gamma(n_L, \mathcal{M}) = n_1$.

Notice that when $L = 1$ (i.e., $\gamma(n, \mathcal{M}) = n$), the most preferable choice of node n is to choose to transmit directly; when $L = 2$, this corresponds to the direct reciprocity case; when $L \geq 3$, this corresponds to the indirect reciprocity case. Since the number of nodes (i.e., $|\mathcal{M}|$) is finite, there hence must exist at least one reciprocal relay selection cycle for the coalitional sub-game Φ. This leads to the following result.

LEMMA 1. *Given a coalitional sub-game* Φ, *there exists at least one reciprocal relay selection cycle. Any reciprocal relay selection cycle is a top-coalition of the coalitional sub-game* Φ.

According to Lemma 1, we have the following result.

LEMMA 2. *The coalitional game* Ω *for cooperative D2D communications satisfies the top-coalition property.*

Based on the top-coalition property, we can construct the core relay selection in an iterative manner. Let $\mathcal{M}_t$ denote the set of nodes of the coalitional sub-game $\Phi_t = \{\mathcal{M}_t, \mathcal{X}_{\mathcal{M}_t}, V, (\succ_m)_{m \in \mathcal{M}_t}\}$ in the t-th iteration. Based on the mapping γ and the given set of nodes $\mathcal{M}_t$, we can then find all the reciprocal relay selection cycles as $\mathcal{C}_1^t, ..., \mathcal{C}_{Z_t}^t$ where each cycle $\mathcal{C}_z^t = (n_1^t, ..., n_{|\mathcal{C}_z^t|}^t)$ is a node sequence and Z_t denotes the number of cycles at the t-th iteration. Abusing notation, we will also use $\mathcal{C}_z^t$ to denote the set of nodes in the cycle $\mathcal{C}_z^t$. We can then construct the core relay selection as follows. For the first iteration $t = 1$, we set $\mathcal{M}_1 = \mathcal{N}$ and find the reciprocal relay selection cycles as $\mathcal{C}_1^1, ..., \mathcal{C}_{Z_1}^1$ based on the set of nodes $\mathcal{M}_1$. For the second iteration $t = 2$, we can then set that $\mathcal{M}_2 = \mathcal{M}_1 \backslash \cup_{i=1}^{Z_1} \mathcal{C}_i^1$ (i.e., remove the nodes in the cycles in the previous iteration) and find the new reciprocal relay selection cycles as $\mathcal{C}_1^2, ..., \mathcal{C}_{Z_2}^2$ based on the set of nodes $\mathcal{M}_2$. This procedure repeats until the set of nodes $\mathcal{M}_t = \varnothing$ (i.e., no operation can be further carried out). We summarize the above procedure for constructing the core relay selection in Algorithm 1.

Suppose that the algorithm takes T iterations to converge. We can obtain the set of reciprocal relay selection cycles in all T iterations as $\{\mathcal{C}_i^t : \forall i = 1, ..., Z_t \text{ and } t = 1, ..., T\}$. Since the mapping $\gamma(n, \mathcal{M}_t)$ is unique for each node $n \in \mathcal{M}_t$, we must have that $\cup_{i=1,...,Z_t}^{t=1,...,T} \mathcal{C}_i^t = \mathcal{N}$ (i.e., all the nodes are in the cycles) and $\mathcal{C}_i^t \cap \mathcal{C}_j^{t'} = \varnothing$ for any $i \neq j$ and $t, t' = 1, ..., T$ (i.e., there do not exist any intersecting cycles). For each cycle $\mathcal{C}_i^t = (n_1^t, ..., n_{|\mathcal{C}_i^t|}^t)$, we can then define the relay selection as $r_{n_l^t}^* = n_{l+1}^t$ for any $l = 1, 2..., |\mathcal{C}_i^t| - 1$ and $r_{n_{|\mathcal{C}_i^t|}^t}^* = n_1^t$. We show that $(r_n^*)_{n \in \mathcal{N}}$ is a core relay selection of the coalitional game Ω for the social reciprocity based relay selection.

THEOREM 1. *The relay selection* $(r_n^*)_{n \in \mathcal{N}}$ *is a core solution to the coalitional game* Ω *for the social reciprocity based relay selection.*

PROOF. We prove the result by contradiction. We assume that there exists a nonempty coalition $\mathcal{S} \subseteq \mathcal{N}$ with another relay selection $(r_m)_{m \in \mathcal{N}} \in V(\mathcal{S})$ satisfying $(r_m)_{m \in \mathcal{N}} \succ_n (r_m^*)_{m \in \mathcal{N}}$ for any $n \in \mathcal{S}$. Let $\mathcal{C}^t = \cup_{i=1}^{Z_t} \mathcal{C}_i^t$ be the set of nodes in the reciprocal relay selection cycles obtained in the t-th iteration. According to Lemma 1, we know that each cycle

---

Algorithm 1 Core Relay Selection Algorithm

1: **initialization:**
2:     **set** initial set of nodes $\mathcal{M}_1 = \mathcal{N}$.
3:     **set** iteration index $t = 1$.
4: **end initialization**

5: **loop** until $\mathcal{M}_t = \varnothing$:
6:     **find** all the reciprocal relay selection cycles $\mathcal{C}_1^t, ..., \mathcal{C}_{Z_t}^t$.
7:     **remove** the set of nodes in the cycles from the current set of nodes $\mathcal{M}_t$, i.e., $\mathcal{M}_{t+1} = \mathcal{M}_t \backslash \cup_{i=1}^{Z_t} \mathcal{C}_i^t$.
8:     **set** $t = t + 1$.
9: **end loop**

$\mathcal{C}_i^1$ is a top-coalition given the set of nodes $\mathcal{M}_1 = \mathcal{N}$. By the definition of top-coalition, we must have that $\mathcal{S} \cap \mathcal{C}^1 = \varnothing$. In this case, we have that $\mathcal{S} \subseteq \mathcal{M}_2 \triangleq \mathcal{M}_1 \backslash \mathcal{C}^1$. Similarly, each cycle $\mathcal{C}_i^2$ is a top-coalition given the set of nodes $\mathcal{M}_2$. We thus also have that $\mathcal{S} \cap \mathcal{C}^2 = \varnothing$. Repeating this argument, we can find that $\mathcal{S} \cap \mathcal{C}^t = \varnothing$ for any $t = 1, ..., T$. Since $\mathcal{N} = \cup_{t=1}^T \mathcal{C}^t$, we must have that $\mathcal{S} \cap \mathcal{N} = \varnothing$, which contradicts with the hypothesis that $\mathcal{S} \subseteq \mathcal{N}$ and $\mathcal{S} \neq \varnothing$. This completes the proof. $\square$

3.3 Social Trust and Social Reciprocity Based Relay Selection

According to the principles of social trust and social reciprocity above, each node $n \in \mathcal{M}$ has two options for relay selection. The first option is that node n can choose the best relay $r_n^S = \arg\max_{r_n \in \mathcal{N}_n^{PS} \cup \{n\}} R_n(r_n)$ from the set of nodes with social trust $\mathcal{N}_n^{PS}$. Alternatively, node n can choose a relay $r_n \in \mathcal{N}_n^{PC}$ from the set of nodes without social trust by participating in a directly or indirectly reciprocal cooperation group.

We next address the issue of choosing between social trust and social reciprocity based relay selections for each node, by generalizing the core relay selection $(r_n^*)_{n \in \mathcal{N}}$ in Section 3.2.3. The key idea is to adopt the social trust based relay selection r_n^S as the benchmark for participating in the social reciprocity based relay selection. That is, a node n prefers social reciprocity based relay selection to social trust based relay selection if the social reciprocity based relay selection offers better performance. More specifically, we define that $r_n \succ_n n$ if and only if $r_n \succ_n r_n^S$ and the selection "$r_n = n$" represents that node n will select the relay r_n^S based on social trust. Based on this, we can then compute the core relay selection $(r_n^*)_{n \in \mathcal{N}}$ according to Algorithm 1. In this case, if we have $r_m^* = m$ in the core relay selection $(r_n^*)_{n \in \mathcal{N}}$, then node m will select the relay r_n^S based on social trust. If we have $r_m^* \neq m$ in the core relay selection $(r_n^*)_{n \in \mathcal{N}}$, then node m will select the relay based on social reciprocity.

4. NETWORK ASSISTED RELAY SELECTION MECHANISM

In this section, we turn our attention to the implementation of the core relay selection for social trust and social reciprocity based cooperative D2D communications. A key issue here is how to find the reciprocal relay selection cycles in the proposed core relay selection algorithm (see Algorithm 1). In the following, we will first propose an algorithm for finding the reciprocal relay selection cycles, and then develop a network assisted mechanism to implement the core relay selection solution in practical D2D communication systems.

4.1 Reciprocal Relay Selection Cycle

We first consider the issue of finding the reciprocal relay selection cycles in the core relay selection algorithm. We introduce a graphical approach to address this issue. More specifically, given the set of nodes $\mathcal{M}_t$ and the mapping γ, we can construct a graph $\mathcal{G}^{\mathcal{M}_t} = \{\mathcal{M}_t, \mathcal{E}^{\mathcal{M}_t}\}$. Here the set of vertices is $\mathcal{M}_t$ and the set of edges $\mathcal{E}^{\mathcal{M}_t} = \{(nm) : e_{nm}^{\mathcal{M}_t} = 1, \forall n, m \in \mathcal{M}_t\}$ where there is an edge directed from node n to m (i.e., $e_{nm}^{\mathcal{M}_t} = 1$) if and only if $\gamma(n, \mathcal{M}_t) = m$.

We next introduce the concept of path in graph theory. A path of length I on a graph is a sequence of nodes $(n_1, n_2, ..., n_I)$ where there is an edge directed from node n_i to n_{i+1} on the graph for any $i = 1, ..., I - 1$. A cycle of the graph is a path in which the first and last nodes are identical. A reciprocal relay selection cycle of the coalitional game then corresponds to a cycle of the graph $\mathcal{G}^{\mathcal{M}_t}$. When $\gamma(n, \mathcal{M}_t) = n$, the cycle degenerates to a self-loop of node n. In the following, we say a path $(n_1, n_2, ..., n_I)$ induces a cycle if there exists a path beginning from node n_I that is a cycle. If two cycles are a cyclic permutation of each other, we will regard them as one cycle.

LEMMA 3. *Any sufficiently long path beginning from any node on the graph $\mathcal{G}^{\mathcal{M}_t}$ induces one and only one cycle.*

Based on Lemma 3, we propose an algorithm to find the reciprocal relay selection cycles in Algorithm 2. The key idea of the algorithm is to explore the paths beginning from each node. More specifically, if a path beginning from a node induces an unfound cycle, then we find a new cycle. We will set the nodes in both the path and cycle as visited nodes since any path beginning from these nodes would induce the same cycle. If a path beginning from a node leads to a visited node, the path would induce a cycle which has already been found if we continue to construct the path on the visited nodes. We will also set the nodes in the path as visited nodes. Since each node will be visited once in the algorithm, the computational complexity of the reciprocal relay selection cycles finding algorithm is $\mathcal{O}(|\mathcal{M}_t|)$.

4.2 NARS mechanism

We now propose a network assisted relay selection (NARS) mechanism to implement the core relay selection, which works as follows.

- Each node $n \in \mathcal{N}$ first determines its preference list $\mathcal{L}_n^P$ for the set of feasible relay selections $\tilde{\mathcal{N}}_n^P \triangleq \mathcal{N}_n^P \cup \{n\}$ based on the physical graph $\mathcal{G}^P$. Here $\mathcal{L}_n = (r_n^1, ..., r_n^{|\tilde{\mathcal{N}}_n^P|})$ is a permutation of all the feasible relays in $\tilde{\mathcal{N}}_n^P$ satisfying that $r_n^i \succ_n r_n^{i+1}$ for any $i = 1, ..., |\tilde{\mathcal{N}}_n^P| - 1$. This step can be done through the channel probing procedure to measure the achieved data rate resulting from choosing with different relays.

- Each node $n \in \mathcal{N}$ then computes the best social trust based relay selection $r_n^S = \arg\max_{r_n \in \mathcal{N}_n^{PS} \cup \{n\}} R_n(r_n)$ based on the physical-social graph $\mathcal{G}^{PS}$ and the preference list $\mathcal{L}_n^P$.

- Each node $n \in \mathcal{N}$ next determines its preference list $\mathcal{L}_n^{PC}$ for the set of relay selections $\mathcal{N}_n^{PC} \cup \{n\}$ based on the physical-coalitional graph $\mathcal{G}^{PC}$. Notice that we have that $r_n \succ_n n$ in the preference list $\mathcal{L}_n^{PC}$ if and only if $r_n \succ_n r_n^S$ in the preference list $\mathcal{L}_n^P$.

Algorithm 2 Algorithm For Finding Reciprocal Relay Selection Cycles

```
1: initialization:
2:     construct the graph G^Mt based on the set of nodes M_t
       and the mappings {γ(n, M_t)}_{n∈M_t}.
3:     set the set of visited nodes V = ∅ and the set of unvisited
       nodes U = M_t\V.
4:     set the set of identified cycles △ = ∅.
5: end initialization

6: loop until U = ∅:
7:     select one node n_a ∈ U randomly.
8:     set the set of visited nodes in the current path H = {n_a}.
9:     set the flag F = 0.
10:    loop until F = 1:
11:        generate the next node n_b = γ(n_a, M_t).
12:        if n_b ∈ V then
13:            set V = V ∪ H and U = M_t\V.
14:            set F = 1.
15:        else if n_b ∈ H then
16:            set the identified cycle as C = (n_1 = n_b, ..., n_i =
               γ(n_{i-1}, M_t), ..., n_I = n_a).
17:            set the set of identified cycles △ = △ ∪ {C}.
18:            set V = V ∪ H and U = M_t\V.
19:            set F = 1.
20:        else
21:            set H = H ∪ {n_b}.
22:            set n_a = n_b.
23:        end if
24:    end loop
25: end loop
```

- Each node $n \in \mathcal{N}$ then reports its preference list $\mathcal{L}_n^{PC}$ to the base-station.

- Based on the preference lists $\mathcal{L}_n^{PC}$ of all nodes, the base-station computes the core relay selection $(r_n^*)_{n \in \mathcal{N}}$ according to Algorithms 1 and 2 and broadcasts the relay selection $(r_n^*)_{n \in \mathcal{N}}$ to all nodes.

As mentioned in Section 3.3, if $r_m^* = m$ in the core relay selection $(r_n^*)_{n \in \mathcal{N}}$, then node m will select the relay r_n^S based on social trust. If $r_m^* \neq m$ in the core relay selection $(r_n^*)_{n \in \mathcal{N}}$, then node m will select the relay based on social reciprocity.

We now use an example to illustrate how the NARS mechanism works. We consider the network of $N = 7$ nodes based on the physical graph $\mathcal{G}^P$ and the social graph $\mathcal{G}^S$ in Figure 2. According to NARS mechanism, each node n first determines its preference list $\mathcal{L}_n$ for the set of feasible relay selections $\mathcal{N}_n^P \cup \{n\}$. We will use the preference lists $\mathcal{L}_n^P$ in Table 1. For example, in the table the feasible relays for node 7 on the physical graph $\mathcal{G}^P$ are $\{5, 6, 7\}$. The preference list $(5, 6, 7)$ represents that $5 \succ_7 6 \succ_7 7$, i.e., node 7 prefers choosing node 5 as the relay to choosing node 6 and transmitting directly offers the worst performance. Then based on the physical-social graph $\mathcal{G}^{PS}$ in Figure 3 and the preference list $\mathcal{L}_n^P$, each node n computes the best social trust based relay selection r_n^S. For example, node 4's best social trust based relay selection $r_n^S = 1$ (i.e., node 1). Each node n next determines the preference list $\mathcal{L}_n^{PC}$ based on the physical-social graph $\mathcal{G}^{PS}$ in Figure 5.

All the nodes then report the preference lists $\mathcal{L}_n^{PC}$ to the base-station. Based on the preference lists, the base-station will compute the core relay selection $(r_n^*)_{n \in \mathcal{N}}$ according to the core relay selection algorithm in Algorithm 1. We illustrate the iterative procedure of the core relay selection

Table 1: The preference lists of $N = 7$ nodes based on the physical graph $\mathcal{G}^P$ and social graph $\mathcal{G}^S$ in Figure 2.

Node n	Preference List $\mathcal{L}_n^P$	Relay r_n^S	Preference List $\mathcal{L}_n^{PC}$
1	(1,2,3,4)	1	(1,2)
2	(1,3,2,4,5)	2	(1,3,2,4)
3	(2,3,4,1)	3	(2,3,4)
4	(2,1,4,3,5,6)	1	(2,4,3,5,6)
5	(4,6,7,5,2)	5	(4,6,7,5)
6	(7,5,4,6)	6	(7,5,4,6)
7	(5,6,7)	7	(5,6,7)

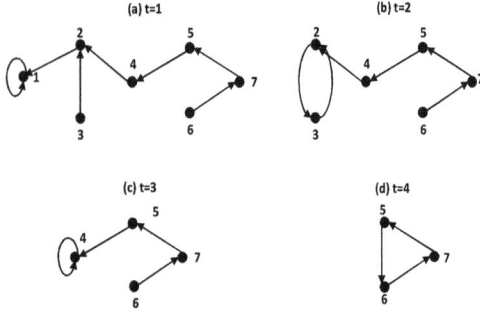

Figure 6: An illustration of the resulting graphs $\mathcal{G}^{\mathcal{M}_t}$ at each iteration t of the core relay selection algorithm.

algorithm in Figure 6 by adopting the graphical representation $\mathcal{G}^{\mathcal{M}_t}$ introduced in Section 4.1. Recall that there is an edge directed from node n to node m on graph $\mathcal{G}^{\mathcal{M}_t}$ if node m is the most preferable relay of node n given the set of nodes $\mathcal{M}_t$. At iteration $t = 1$, given that $\mathcal{M}_1 = \mathcal{N}$, the base-station identifies one cycle, i.e., a self-loop formed by node 1. At iteration $t = 2$, given that $\mathcal{M}_2 = \mathcal{M}_1 \backslash \{1\}$, the base-station then identifies one cycle formed by nodes 2 and 3. Notice that graph $\mathcal{G}^{\mathcal{M}_2}$ can be derived from graph $\mathcal{G}^{\mathcal{M}_1}$ by removing node 1 and any edges directed to node 1. For each node (e.g., node 2) from which there is a removed edge directed to node 1, we add a new edge directed from the node to its most preferable node among the set of nodes $\mathcal{M}_2$ (e.g., the edge $2 \to 3$). We continue in this manner until all the nodes have been removed from the graph. Figure 7 shows all the reciprocal relay selection cycles identified by the core relay selection algorithm in Figure 6. In this case, the core relay selection is: (a) since $r_1^S = 1$, node 1 transmits directly; (b) nodes 2 and 3 serves as the relay of each other (i.e., direct reciprocity based relay selection); (c) since $r_4^S = 1$, node 4 seeks relay assistance from node 1 (i.e., social trust based relay selection); (d) node 5 serves as the relay of node 7, which in turn serves as the relay of node 6 and node 6 in turn is the relay of node 5 (i.e., indirect reciprocity based relay selection).

4.3 Properties of NARS mechanism

We next study the properties of the proposed NARS mechanism. First of all, according to the definition of the core solution of coalitional game, we know that

LEMMA 4. *The core relay selection $(r_n^*)_{n \in \mathcal{N}}$ by NARS mechanism is immune to group deviations, i.e., no group of nodes can deviate and improve by cooperation within the group.*

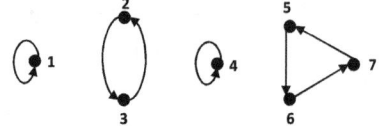

Figure 7: The reciprocal relay selection cycles identified by the core relay selection algorithm in Figure 6

We can then show that the mechanism guarantees individual rationality, which means that each participating node will not achieve a lower data rate than that when the node does not participate (i.e., in this case the node will transmit directly).

LEMMA 5. *The core relay selection $(r_n^*)_{n \in \mathcal{N}}$ by NARS mechanism is individually rational, i.e., each node $n \in \mathcal{N}$ will be assigned a relay r_n^* which satisfies either $r_n^* \succ_n n$ or $r_n^* = n$.*

PROOF. If the assigned relay $r_n^* \prec_n n$ for some node $n \in \mathcal{N}$, then the node n can deviate from the current coalition and improve its data rate by transmitting directly (i.e., $r_n^* = n$). This contradicts with the fact that $(r_n^*)_{n \in \mathcal{N}}$ is a core relay selection. □

We next explore the truthfulness of NARS mechanism. A mechanism is truthful if no node can improve by reporting a preference list different from its true preference list, given that other nodes report truthfully.

LEMMA 6. *NARS mechanism is truthful.*

PROOF. Let $\mathcal{C}^t$ be the set of nodes in the reciprocal relay selection cycles obtained in the t-th iteration of core relay selection algorithm. Suppose that the node m reports another preference list that is different from its true preference list. Let τ be the index such that $m \in \mathcal{C}^\tau$. Given that the nodes in the set $\cup_{t=1}^{\tau-1} \mathcal{C}^t$ truthfully report, they will be assigned the relays in the core relay selection regardless of what the nodes out of the set $\cup_{t=1}^{\tau-1} \mathcal{C}^t$ report. In this case, given the set of remaining nodes $\mathcal{M}_\tau = \mathcal{N} \backslash \cup_{t=1}^{\tau-1} \mathcal{C}^t$, the most preferable relay of node m is the relay r_m^* in the core relay selection. This is exactly what the node m achieves by reporting truthfully. Thus, the node m can not improve by reporting another preference list. □

We finally consider the computational complexity of NARS mechanism. We say the mechanism is computationally efficient if the solution can be computed in polynomial time.

LEMMA 7. *NARS mechanism is computationally efficient.*

PROOF. Recall that the reciprocal relay selection cycle finding algorithm in Algorithm 2 has a complexity of $\mathcal{O}(|\mathcal{M}_t|)$. Since the reciprocal relay selection cycle finding algorithm is the dominating step in each iteration, the core relay selection algorithm hence has a complexity of $\mathcal{O}(\sum_{t=1}^T |\mathcal{M}_t|)$. As $\sum_{t=1}^T |\mathcal{M}_t| = N + \sum_{t=2}^T (N - \sum_{\tau=1}^{t-1} |\mathcal{C}^\tau|)$ and $\sum_{t=1}^T |\mathcal{C}^t| = N$, by setting $|\mathcal{C}^\tau| = 1$ for $\tau = 1, ..., T$, we have the worst case that $\sum_{t=1}^T |\mathcal{M}_t| = \sum_{i=1}^N i = \frac{N(N+1)}{2}$. Thus, the mechanism has a complexity of at most $\mathcal{O}(N^2)$. □

The above four Lemmas together prove the following theorem.

THEOREM 2. *NARS mechanism is immune to group deviations, individually rational, truthful, and computationally efficient.*

5. SIMULATIONS

In this section we evaluate the performance of the proposed social trust and social reciprocity based relay selection for cooperative D2D communications through simulations.

We consider that multiple nodes are randomly scattered across a square area with a side length of 1000 m. Two nodes are randomly matched into a source-destination D2D communication link. We compute the SNR value μ_{ij} according to the physical interference model, i.e., $\mu_{ij} = \frac{p_i}{\omega_0 \cdot ||i,j||^\alpha}$ with the transmission power $p_i = 1$ Watt, the background noise $\omega_0 = 10^{-10}$ Watts, and the path loss factor $\alpha = 4$. Based on the SNR μ_{ij}, we set the bandwidth $W = 10$ Mhz and then compute the data rate achieved by using different relays according to Equation (1). We construct the physical graph $\mathcal{G}^P$ by setting $e^P_{nm} = 1$ (i.e., node m is a feasible relay of node n) if and only if the distance between nodes n and m is not greater than a threshold $\delta = 500$ m (i.e., $||n,m|| \leq \delta$). For the social trust model, we will consider two types of social graphs: Erdos-Renyi social graph and real data trace based social graph.

5.1 Erdos-Renyi Social Graph

We first consider $N = 100$ nodes with the social graph $\mathcal{G}^S$ represented by the Erdos-Renyi (ER) graph model [14] where a social link exists between any two nodes with a probability of P_L. To evaluate the impact of social link density of the social graph, we implement the simulations with different social link probabilities $P_L = 0, 0.05, 0.1, ..., 1.0$, respectively. For each given P_L, we average over 1000 runs. As the benchmark, we also implement the solution that each node transmits directly, the solution that each node selects the relay based social trust only (i.e., $r_n = r^S_n$), and the solution that each node selects the relay based on social reciprocity only by assuming that there is no social trust among the nodes. Furthermore, we also compute the throughput upper bound by letting each node select the best relay $\bar{r}_n = \arg\max_{r_n \in \mathcal{N}^P_n \cup \{n\}} R_n(r_n)$ among all its feasible relays. Notice that the throughput upper bound can only be achieved when all the nodes are willing to help each other (i.e., all the nodes are cooperative).

We show the average system throughput in Figure 8. We see that the performance of the social trust and social reciprocity based relay selection dominates that of social trust only based relay selection and social reciprocity only based relay selection. When the social link probability P_L is small, the social trust and social reciprocity based relay selection achieves up to 64.5% performance gain over the social trust only based relay selection. When the social link probability P_L is large, the social trust and social reciprocity based relay selection achieves up to 24% performance gain over the social reciprocity only based relay selection. We also observe that the social trust and social reciprocity based relay selection achieves up-to 100.4% performance gain over the case that all the nodes transmit directly. Compared with the throughput upper bound, the performance loss of the social trust and social reciprocity based relay selection is at most 24%. As the social link probability P_L increases, the social trust and social reciprocity based relay selection improves and approaches the throughput upper bound. This is due to the fact that when the social link probability P_L is large, each node will have a high probability of having social trust from any other node and hence each node is likely to have social trust from its best relay node. This can be illustrated by Figure 9 that the average size of the reciprocal relay selection cycles in the social trust and social reciprocity based relay selection decreases as the social link probability P_L increases.

5.2 Real Trace Based Social Graph

We then evaluate the proposed social trust and social reciprocity based relay selection with the social graphs generated according to the friendship network of the real data trace Brightkite [11]. We implement simulations with the number of nodes $N = 250, 500, ..., 1500$, respectively. The total number of social links among these nodes of the social graphs is shown in Figure 10.

We show the average system throughput in Figure 11. We see that the system throughput of the social trust and social reciprocity based relay selection increases as the number of users N increases. This is because that more cooperation opportunities among the nodes are present when the number of users N increases. Moreover, the social trust and social reciprocity based relay selection achieves up-to 122% performance gain over the solution that all users transmit directly. Compared with the throughput upper bound, the performance loss by the social trust and social reciprocity based relay selection is at most 21%. We also show the computational complexity of the NARS mechanism for computing the social trust and social reciprocity based relay selection solution in Figure 12. We see that the average number of iterations of the mechanism grows linearly as the number of nodes N increases. This demonstrates that the proposed NARS mechanism is computationally efficient (i.e., has a polynomial convergence time).

6. CONCLUSION

In this paper we studied cooperative D2D communications based on social trust and social reciprocity. We introduced the physical-social graphs to capture the physical constraints for feasible D2D cooperation and the social relationships among devices for effective cooperation. We proposed a coalitional game theoretic approach to find the efficient D2D cooperation strategy and developed a network assisted relay selection mechanism for implementing the coalitional game solution. We showed that the devised mechanism is immune to group deviations, individually rational, truthful, and computationally efficient. We further evaluated the performance of the mechanism based on Erdos-Renyi social graphs and real data trace based social graphs. Numerical results show that the proposed mechanism can achieve up-to 122% performance gain over the case without D2D cooperation.

We are currently generalizing the notion of social trust from the current one-hop setting (e.g., friends) to the multi-hop setting (e.g., friend's friends). Intuitively, as the number of social hops between two nodes increases, the strength of social trust decreases. Mathematically, we can introduce a weighted social graph to model such features by defining the weight as the strength of social trust. It is of great interest to design efficient stimulation mechanisms for D2D cooperation by taking both generalized social trust and social reciprocity into account.

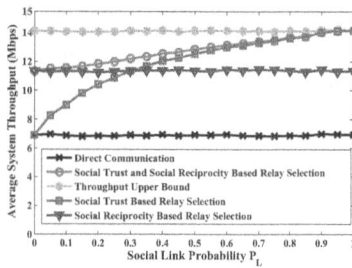

Figure 8: System throughput with the number of nodes $N = 100$ and different social network density.

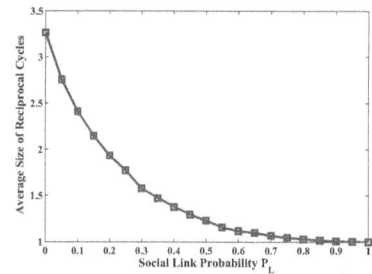

Figure 9: Average size of the reciprocal relay selection cycles in the social trust and social reciprocity based relay selection with $N = 100$ and different social network density.

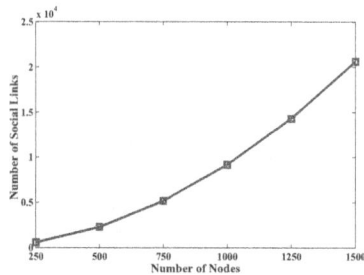

Figure 10: The number of social links of the social graphs based on real trace Brightkite.

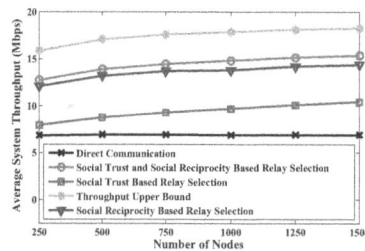

Figure 11: Average system throughput with different number of nodes.

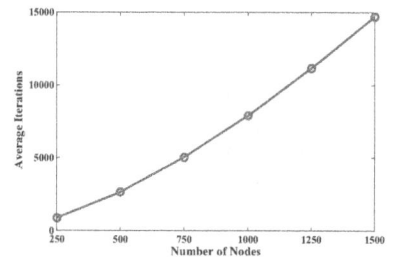

Figure 12: Average number of iterations of the NARS mechanism.

7. ACKNOWLEDGMENTS

This research was supported in part by the U.S. National Science Foundation under Grants CNS-1117462, CNS-1218484, and DoD MURI project No. FA9550-09-1-0643.

8. REFERENCES

[1] Global mobile data traffic data forecast update 2011-2016. In *Cisco white paper*, 2012.

[2] L. Anderegg and S. Eidenbenz. Ad hoc-vcg: a truthful and cost-efficient routing protocol for mobile ad hoc networks with selfish agents. In *ACM MobiCom*, pages 245–259. ACM, 2003.

[3] S. Banerjee, H. Konishi, and T. Sönmez. Core in a simple coalition formation game. *Social Choice and Welfare*, 18(1):135–153, 2001.

[4] C. Boldrini, M. Conti, and A. Passarella. Contentplace: social-aware data dissemination in opportunistic networks. In *ACM MSWiM*, 2008.

[5] P. Costa, C. Mascolo, M. Musolesi, and G. Picco. Socially-aware routing for publish-subscribe in delay-tolerant mobile ad hoc networks. *IEEE JSAC*, 26(5):748–760, 2008.

[6] G. Fodor, E. Dahlman, G. Mildh, S. Parkvall, N. Reider, G. Miklós, and Z. Turányi. Design aspects of network assisted device-to-device communications. *IEEE Communications Magazine*, 50(3):170–177, 2012.

[7] H. Gintis. Strong reciprocity and human sociality. *Journal of Theoretical Biology*, 206(2):169–179, 2000.

[8] T. Govier. *Social trust and human communities*. McGill-Queen's University Press, 1997.

[9] A. Host-Madsen and J. Zhang. Capacity bounds and power allocation for wireless relay channels. *IEEE TIT*, 51(6):2020–2040, 2005.

[10] N. Kayastha, D. Niyato, P. Wang, and E. Hossain. Applications, architectures, and protocol design issues for mobile social networks: A survey. *Proceedings of the IEEE*, 99(12):2130–2158, 2011.

[11] J. Leskovec. Brightkite dataset. Stanford University. Available at http://snap.stanford.edu/data /loc-brightkite.html, 2012.

[12] R. Molva and P. Michiardi. Core: a collaborative reputation mechanism to enforce node cooperation in mobile ad hoc networks. *Institute EurecomResearch Report RR-02-062*, 2001.

[13] R. Myerson. *Game theory: analysis of conflict*. Harvard University Press, 1997.

[14] M. Newman, D. Watts, and S. Strogatz. Random graph models of social networks. *PANS*, 99(1):2566–2572, 2002.

[15] C.-H. Yu, O. Tirkkonen, K. Doppler, and C. Ribeiro. On the performance of device-to-device underlay communication with simple power control. In *IEEE VTC-Spring*, pages 1–5. IEEE, 2009.

[16] Y. Zhao, R. Adve, and T. Lim. Improving amplify-and-forward relay networks: optimal power allocation versus selection. In *IEEE ISIT*, 2006.

Realtime Streaming with Guaranteed QoS over Wireless D2D Networks*

Navid Abedini
Qualcomm Flarion
Technologies
Bridgewater, NJ
navida@qti.qualcomm.com

Swetha Sampath
IBM
Systems & Technology Group
Houston, TX
swesam@gmail.com

Rajarshi Bhattacharyya
Dept. of ECE
Texas A&M University
College Station, TX
rajarshibh@tamu.edu

Suman Paul
Dept. of ECE
Texas A&M University
College Station, TX
sumanpaul@tamu.edu

Srinivas Shakkottai
Dept. of ECE
Texas A&M University
College Station, TX
sshakkot@tamu.edu

ABSTRACT

We consider a group of co-located wireless peer devices that desire to synchronously receive a live content stream. The devices are each equipped with an expensive unicast base-station-to-device (B2D) interface, as well as a broadcast device-to-device (D2D) interface over a shared medium. The stream is divided into blocks, which must be played out soon after their initial creation. If a block is not received within a specific time after its creation, it is rendered useless and dropped. The blocks in turn are divided into random linear coded chunks to facilitate sharing across the devices. We transform the problem into the two questions of (i) deciding which peer should broadcast a chunk on the D2D channel at each time, and (ii) how long B2D transmissions should take place for each block. We analytically develop a provably-minimum-cost algorithm that can ensure that QoS targets can be met for each device. We study its performance via simulations, and present an overview of our implementation on Android phones using the algorithm as a basis.

Categories and Subject Descriptors

C.2.1 [**Computer-Communication Networks**]: Network Architecture and Design—*wireless communication*

General Terms

Theory and Algorithms

*Research was funded in part by NSF grants CNS-0904520, CNS-0963818, CNS-1149458, DTRA grant HDTRA1-13-1-0030, AFOSR grant FA9550-13-1-0008 and the Google Faculty Research Awards program. The first two authors are former students of Texas A&M University

Keywords

Live streaming, quality of service, linear network coding, multiple wireless interfaces, queueing

1. INTRODUCTION

There has recently been a sharp increase in the use of smart, handheld devices for content consumption. These devices, such as smart phones and tablets, are equipped with multiple orthogonal wireless communication interfaces. Interfaces include expensive (both dollar-cost and energy) long-range base-station-to-device (B2D) interfaces (*e.g.* 3G or 4G), as well as low-cost short-range interfaces like WiFi. More recently, the use of short-range interfaces such as WiFi-Direct and FlashLinQ for device-to-device (D2D) communication is starting to make an appearance. Simultaneously, there has been an explosion in available content, and it is expected that streaming of live events will play a big part in future demand [3].

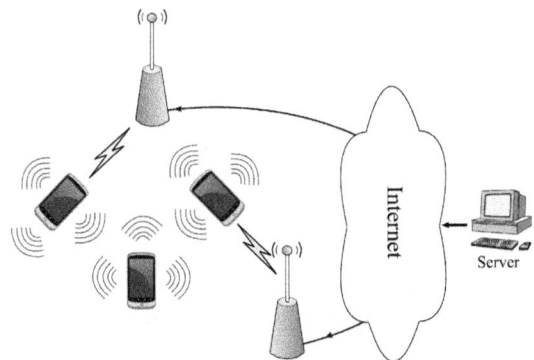

Figure 1: Each device can simultaneously utilize unicast base-station-to-device (B2D) as well as broadcast device-to-device (D2D) communication.

In this paper, we focus on live-streaming of content to multiple co-located devices, as shown in Figure 1. Here, data is generated in realtime by a server, and must be delivered to all the devices quickly in order to maintain the "live" aspect of the event. Each device is simultaneously capable of unicast communication via a B2D interface, and broadcast D2D communication with peer devices. Devices desire to minimize the usage of their B2D interfaces to reduce cost, while maintaining synchronous reception and playout of content. While it might be possible for a cellular base station to broadcast live events to multiple handsets, such content would be restricted to a few selected channels, and only available to subscribers of a single provider. Utilizing both interfaces enables users to pick any event of interest, and "stitch together" their B2D capacities regardless of provider support. Apart from content consumption in a social setting, such a scheme would also be valuable in an emergency response or battlefield setting.

In our model, content is generated in realtime in the form of *blocks*, with one block corresponding to a playout time called a *frame*. In other words, each device must receive one block of data within one frame of time for smooth playout. The blocks are divided into *chunks* for sharing via different interfaces. Coordination of chunk identities across all devices and all base-stations is near impossible, so a coding solution is needed for transmitting chunks. Hence, the server performs random linear coding [5,8] across each block, and unicasts the resulting coded chunks to the devices via the Internet and through the B2D connections. Thus, each coded chunk can be thought of as an element in a vector space with the scalars in a finite field. The information at each device can then be represented by a matrix that contains all the vectors that it has received thus far, and the block of information can be decoded when this matrix is of full rank. Both the Internet and B2D links are lossy, but since the chunks are coded, there is no need for feedback and the server simply generates a new chunk for each transmission. The devices receive coded chunks from the server, and in turn transmit these chunks to each other over a potentially lossy D2D broadcast channel, again without feedback. If each device receives enough chunks to decode the block, it can play it, else it has to skip that block.

We illustrate the timing sequence in Figure 2. Here, we divide time into frames, which are subdivided into T *slots*. Devices must synchronously play out block k in frame k. Since our application is that of *live* streaming, which requires short delays between the creation of a block and its playout, we assume that block k is only created at the beginning of frame $k-2$. The server immediately divides the block into N chunks and performs random linear coding over these chunks. The server then transmits some number of these coded chunks to the devices via their individual B2D interfaces. This number is to be kept small to reduce B2D usage. Next, in frame $k-1$, the devices use the broadcast D2D network to disseminate these chunks amongst themselves. At the end of frame $k-1$ the devices attempt to decode block k. If a device i has received enough coded chunks to decode the block, it plays out the block during frame k. Otherwise, i will be idle during this frame. Note that in our framework, the B2D and D2D interfaces can simultaneously transmit chunks corresponding to different blocks.

We use a recent model of service quality for realtime wireless applications in which each station's QoS is parametrized by a *delivery-ratio*, which is the average ratio of blocks desired to the blocks generated [6]. For example, a delivery ratio of 90% would mean that 10% of the blocks can be skipped without noticeable impairment (which depends on the video coding used). We desire an algorithm that min-

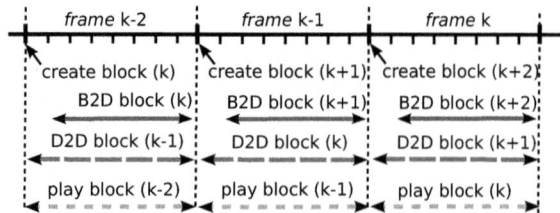

Figure 2: Frames are divided into slots, and one B2D and one D2D transmission can simultaneously take place in each slot. A block must be delivered within two frames of its creation time or it is lost.

imizes the usage of the expensive B2D connections, while maintaining an acceptable quality of service for each device. We note that the scenario is different from traditional peer-to-peer sharing in a wired network, wherein streaming with full chunk coordination using tree construction is the norm. In our situation, each device is independently connected to a base-station (perhaps using different service providers), and there is no coordination across base stations. The broadcast D2D medium is shared across the devices, and interference across the devices is a limiting factor.

The objectives of this work are three-fold. First, we would like to systematically design provably-cost-optimal chunk exchange algorithms that would minimize the usage of expensive B2D transmissions, while ensuring that quality constraints for live streaming can be met. Second, we would like to conduct performance analysis of our schemes using simulations to understand how different parameters affect the algorithms. Third, we desire to implement the developed algorithms on Android smart phones in order to observe real-world behavior. In what follows, we will present our results in attaining these objectives.

Related Work

The problem of exchanging random linear coded information while using the minimum number of transmissions over a broadcast D2D network (starting from some initial condition) was studied in [4,10]. There is also recent work on quick file transfer to a set of peer devices using a hybrid network, while minimizing the amount of B2D usage [2]. All consider a reliable model for the D2D links, and their common objective is to be able to decode a single block of information at the end. Unlike these papers, our objective and QoS metric—minimum cost timely synchronization of a live stream—is quite different.

There is a rich body of literature in the area of P2P streaming. For example, Lava [11] is one of the first streaming systems built based on the idea of network coding. While these systems only consider unicast (wired) transmissions among peers, our D2D framework is one of the few that allows having broadcast transmissions over the D2D network.

Closest to our work is [9], which investigates the problem of cooperatively managing multiple interfaces for content sharing. However, unlike our block-by-block delay sensitive model for realtime streaming, they maximize a utility function of the average information flow rate achieved by the peers, which is relevant to elastic traffic such as file transfer or streaming of stored content.

Organization and Main Results

In Section 2, we present the live streaming model and the QoS metric. In our model, we assume that time is slotted, and in each time slot there can be one unicast B2D transmission to each device, and one broadcast D2D transmission over all devices. The probabilities of success of each are different, and these are assumed to be known. There is no feedback in the system, consistent with the idea of using UDP-based communications. Our frame timing structure allows us to study the B2D and the D2D communications as two sub-problems, namely, (i) B2D stopping time: how long should each device use its B2D channel for each block? and (ii) D2D broadcast scheduling: which device should broadcast at each time?

We consider the problem of D2D scheduling in Section 3. Since the B2D communications result in a random initialization of coded chunks with each peer device at the beginning of each frame, our first objective is to find a set of necessary and sufficient conditions for achievability of a given QoS. We then use ideas from queueing theory and the Foster-Lyapunov stability criterion to find an optimal D2D scheme for sufficiently large field sizes over which we perform coding. We next study the special case of fully reliable D2D broadcast in Section 4. The optimal D2D scheduling algorithm in this case has a simple and intuitive form, which allows for easy implementation, as well as development of good heuristics.

Our next problem is to answer the question of how to determine the stopping time for B2D, which we tackle in Section 5. This problem is hard to solve explicitly, given the form of the necessary conditions for QoS achivability. However, it has to be solved only once for any particular set of users, and we provide a general framework to find the optimal B2D usage times for any given cost structure.

We assumed in the above sections that coding is performed using fields of sufficiently large size as to ignore the performance degradation effects of receiving linearly dependent (and hence useless) chunks. In Section 6, we show that the degradation decreases inversely proportional to the field size, and is independent of other parameters like the number of devices or the number of chunks in a block, which is a useful result as the system scales.

We validate the proposed algorithms through simulations in Section 7, comparing with some intuitive heuristic algorithms. Finally, we discuss implementation ideas for music streaming in Section 8. Our system is more like an emulation for B2D, as we did not yet have 3G service on the phones, and hence initialized each device with coded chunks periodically. However, the D2D part was achieved accurately through WiFi broadcast. Finally, we conclude in Section 9.

2. SYSTEM MODEL

We have M co-located peer devices in our system, denoted by $i \in \{1, \dots, M\}$, all interested in synchronously receiving the same stream of data. The data source generates the stream in the form of a sequence of blocks. Each block is further divided into N chunks for transmission. We use random linear network coding over the chunks of each block, which implies that each coded chunk (a degree of freedom) is now a random linear combination (with coefficients in finite field F_q of size q) of the N original chunks in the corresponding block, and can be represented by a vector in F_q^N.

Time is divided into slots, and at each time slot τ, each device can receive up to one chunk using its B2D interface, and one using its D2D interface. We assume that the two interfaces are orthogonal. Also, both B2D unicasts and D2D broadcasts are made in a connectionless fashion with no feedback. The probabilities of success when using each interface are different, and we can use these probabilities to reflect the relative throughputs of the two interfaces.

Thus, each device has an expensive but unreliable B2D unicast channel to a base-station, whose success probability depends on the loss probability over the Internet as well as the condition of the B2D channel. For each device i, we model the chunk reception event for the B2D interface by a Bernoulli random variable with parameter β_i, independent of the other devices. Each device also has a low-cost D2D broadcast interface, and only one device (denoted by $u[\tau] \in \{1, \dots, M\}$) can broadcast over the D2D network at each time τ. Losses over the D2D channel could happen due to collisions that occur as an overhead of distributed scheduling. Hence, we assume that a D2D broadcast is either received by all devices with a probability equal to δ, or lost[1]. Wireless devices could apply channel estimation techniques to determine their success probabilities β_i and δ, and we will develop algorithms assuming that these values are known. We will also study a system (similar to the ones investigated in [4,10]) in which the D2D broadcast is assumed to be completely reliable.

Since each D2D broadcast is either received by all devices or none, there is no need to rebroadcast any information received via D2D. Following this observation, it is straightforward to verify that the order of D2D transmissions has no impact on the performance. Thus, we only need to keep track of the number of chunks transmitted and received over the interfaces during a frame in order to determine the state of the system. We denote the total number of coded chunks of block k delivered to device i via the B2D network during frame $k-2$ using $b_i^{(k)}$. Also, $t_i^{(k)}$ and $r_i^{(k)}$ are used to denote, respectively, the total number of transmitted and received chunks of block k by device i via D2D during frame $k-1$ (i.e., before frame k, which is the block k's play out time). Note that according to the model

$$\sum_i t_i^{(k)} \leq T, \qquad (1)$$

i.e., only one device can broadcast over the D2D network at each slot, and hence at most T transmissions can occur during frame $k-1$.

Let $\hat{n}_i^{(k)}$ denote the total number of coded chunks of block k possessed by device i at the beginning of frame k,

$$\hat{n}_i^{(k)} = b_i^{(k)} + r_i^{(k)}.$$

Hence, device i has a set of $\hat{n}_i^{(k)}$ vectors of coefficients corresponding to the coded chunks that it has received. We denote by $n_i^{(k)}$ (called the kth rank of device i) the dimension of the subspace spanned by these vectors. In order to decode the original N chunks of block k at the beginning of frame k, device i must have $n_i^{(k)} = N$. Otherwise, i skips the block and will be idle during this frame. Let $R_i[k] \in \{0, 1\}$ denote the success or failure of the kth block

---

[1]It is straightforward to extend our results and framework to a more general D2D model in which only a subset of devices receive each broadcast chunk, but at the cost of greater algorithmic complexity.

w.r.t i, *i.e.*, $R_i[k] = 0$ means device i is idle during frame k, and $R_i[k] = 1$ denotes the event of successful decoding of block k. Clearly, $R_i[k] = 0$, if $n_i^{(k)} < N$.

Each device i has a delivery ratio $\eta_i \in (0, 1]$, which is the minimum acceptable long-run average number of frames device i must playout. Hence, we require

$$\eta_i \leq \lim_{K \to \infty} \frac{1}{K} \sum_{k=1}^{K} \mathbb{E}[R_i[k]]. \qquad (2)$$

The objective of this paper is to find a scheme to coordinate both interfaces that would satisfy the delivery ratio requirement of all devices at the lowest cost of using B2D transmissions. Since the B2D and D2D transmissions corresponding to a particular block k occur during the disjoint frames $k-2$ and $k-1$, respectively, we will explore schemes that consist of two parts, namely:

1. **B2D Stopping Time:** In order to keep the communication overhead with the server as small as possible, we use a fixed number of transmissions on the B2D channel for each block and for each device. Hence, for each block k, device i receives some random number $\mathbf{b}_i^{(k)}$ of coded chunks via B2D at the beginning of frame $k-1$. By assumption, the values of $\mathbf{b}_i^{(k)}$ are independent over devices. Also, since we fix the B2D usage in each frame, $\mathbf{b}_i^{(k)}$ are independent and identically distributed over blocks k for each device i. We need to determine the minimum stopping time such that delivery requirements can be met.

2. **D2D Scheduling Algorithm:** Given the B2D stopping time, each device receives chunks at the beginning of each frame in an *i.i.d* fashion according to an arrival process denoted by $\mathbf{b}_i$. We must determine a scheduling algorithm that can achieve the delivery requirements if at all it is possible to do so, given the D2D broadcast channel reliability.

In what follows, we will first solve the problem of D2D scheduling, and will then show how to select the B2D stopping time. We will do so under different assumptions on channel reliability and coding schemes.

3. D2D SCHEDULING ALGORITIIM

Under the assumption of *i.i.d* B2D arrivals, we first need to determine whether the desired QoS metric $(\eta_1, ..., \eta_M)$ is achievable for the given arrival process $(\mathbf{b}_1, ..., \mathbf{b}_M)$. For large field sizes, distinct randomly coded chunks are linearly independent with high probability. In what follows, we assume that the field size q is large enough that we can ignore the effect of its finiteness on the performance of the linear coding. We will consider the impact of finite field size on performance in Section 6.

DEFINITION 1. *We say the QoS $(\eta_1, ..., \eta_M)$ is achievable, if there exists a feasible policy to coordinate the D2D transmissions such that, on average each device i successfully receives η_i fraction of the blocks before their deadlines.*

Since the B2D arrivals are assumed to be *i.i.d.* over frames, the achievability of the QoS metric can be evalu-

ated based on the existence of a randomized stationary D2D policy as follows:[2]

LEMMA 1. *The QoS $(\eta_1, ..., \eta_M)$ is (strictly) achievable if and only if there exists a D2D policy $\mathbf{P}^*$ such that, for each block k, given the B2D arrivals $b^{(k)} = (b_1^{(k)}, ..., b_M^{(k)})$, $\mathbf{P}^*$ chooses a feasible D2D schedule $t^{(k)} = (t_1^{(k)}, ..., t_M^{(k)})$ (satisfying (1)) with probability $\mathbb{P}\left(t^{(k)}|b^{(k)}\right)$, such that for each device i,*

$$\mathbb{E}_{\mathbf{b}}\left[\sum_{t^{(k)}} \mathbb{E}\left[R_i[k] \mid t^{(k)}, b^{(k)}\right] \mathbb{P}\left(t^{(k)}|b^{(k)}\right)\right] > \eta_i \qquad (3)$$

where $\mathbb{E}_{\mathbf{b}}[.]$ is expectation with respect to the B2D arrival processes.

Optimal D2D Scheme

In order to keep track of quality of service, each device maintains a so-called deficit queue. The length of this queue $d_i[k]$, for device i at frame k, follows the dynamic below

$$d_i[k] = d_i[k-1] + \eta_i - R_i[k]. \qquad (4)$$

Recall that $R_i[k] = 0$ means device i is not successful in receiving the block k before its deadline, and hence is idle in frame k. In this case, the deficit value of this device increases by an amount equal to its delivery ratio η_i. Otherwise, $R_i[k] = 1$ and the deficit value decreases by $1 - \eta_i$. Therefore, the deficit queue can be thought of as a counter that captures the accumulated "unhappiness" of a device experienced thus far,

$$d_i[k] = k\eta_i - \sum_{l=1}^{k} R_i[l].$$

The evolution of these deficit queues can be understood by using a Markov chain $\mathcal{D}$, whose state at each step k is $([d_1[k]]^+, ..., [d_M[k]]^+)$, where $[a]^+ = \max\{a, 0\}$. If our D2D algorithm is such that $\mathcal{D}$ is stable (positive recurrent), then $\mathbb{E}[[d_i[k]]^+] < \infty$ for all k, and we will have

$$\lim_{K \to \infty} \frac{1}{K} \mathbb{E}[d_i[k]] \leq \lim_{K \to \infty} \frac{1}{K} \mathbb{E}[[d_i[k]]^+] = 0.$$

Hence, $\eta_i \leq \lim_{K \to \infty} \frac{1}{K} \sum_{k=1}^{K} \mathbb{F}[R_i[k]]$, which implies that the QoS requirement of device i is satisfied. We next define a D2D scheme in Algorithm 1, whose optimality is shown in Theorem 2.

THEOREM 2. *The D2D scheme in Algorithm 1 is throughput optimal, i.e., it can satisfy all achievable QoS metrics $(\eta_1, ..., \eta_M)$.*

PROOF. In this proof, we will use the Lyapunov criterion to show the stability of Markov chain $\mathcal{D}$. Consider the candidate Lyapunov function $V[k] = \frac{1}{2} \sum_i ([d_i[k]]^+)^2$. We will show that for any achievable QoS, the proposed D2D algorithm results in an expected drift $\Delta V[k] =$

$$\mathbb{E}\left[V[k] - V[k-1] \mid \text{state of the system at frame } k-1\right],$$

---

[2]It is easy to verify this characterization of achievability, which is used in much recent work. For instance, a detailed proof can be found in [7].

Algorithm 1 Optimal D2D scheme (unreliable broadcast)
___

At the beginning of each frame $k - 1$: Given the B2D arrivals $(b_1^{(k)}, ..., b_M^{(k)})$ and the deficit values $([d_1[k - 1]]^+, ..., [d_M[k - 1]]^+) = (d_1, ..., d_M)$, solve the following maximization problem to find the optimal number of transmissions $(t^{(k)})^* = ((t^{(k)})_1^*, ..., (t^{(k)})_M^*)$:

$$\max \sum_i d_i \mathbb{P}(\sum_{j \neq i} \min(\hat{t}_j^{(k)}, b_j^{(k)}) \geq N - b_i^{(k)})$$
$$\text{s.t.}$$
$$\hat{t}_j^{(k)} = Bin(\delta, t_j^{(k)}), \quad \forall j \qquad (5)$$
$$\sum_j t_j^{(k)} \leq T$$

Devices take turns (in any arbitrary order) to broadcast over the D2D network: Device i first broadcasts $\min(b_i^{(k)}, (t^{(k)})_i^*)$ of its initial coded chunks received from the B2D network. For the remaining $\max((t^{(k)})_i^* - b_i^{(k)}, 0)$ transmissions, device i randomly combines the initial $b_i^{(k)}$ B2D chunks.

Note that above, $Bin(\delta, t)$ refers to a Binomial random variable with probability of success δ and t trials.
___

which is negative except in a finite subset of the state space. Hence, the Lyapunov Theorem implies that the Markov chain $\mathcal{D}$ is stable. Now, we present the details of the proof.

$$\Delta V[k] = \mathbb{E}\left[V[k] - V[k-1] \Big| [d_i[k-1]]^+ = d_i : \forall i \right]$$
$$= \frac{1}{2} \mathbb{E}\left[\sum_i \left([d_i[k-1] + \eta_i - R_i[k]]^+ \right)^2 \Big| [d_i[k-1]]^+ \right]$$
$$- \frac{1}{2} \sum_i (d_i)^2 \overset{(a)}{\leq} \frac{1}{2} \mathbb{E}\left[\sum_i (d_i + \eta_i - R_i[k])^2 - (d_i)^2 \right]$$
$$= \mathbb{E}\left[\sum_i d_i (\eta_i - R_i[k]) \right] + \frac{1}{2} \mathbb{E}\left[\sum_i (\eta_i - R_i[k])^2 \right]$$
$$\overset{(b)}{\leq} M/2 + \sum_i d_i \eta_i - \mathbb{E}\left[\sum_i d_i R_i[k] \right],$$

where (a) follows from $([X + Y]^+)^2 \leq ([X]^+ + Y)^2$, and (b) holds since $(\eta_i - R_i[k])^2 \leq \max((\eta_i)^2, (1 - \eta_i)^2) \leq 1$.

In order to get a negative drift (except in a finite subset), we minimize the above upperbound. Hence, at the beginning of each frame $k - 1$, for given deficit values $([d_1[k - 1]]^+, ..., [d_M[k - 1]]^+) = (d_1, ..., d_M)$ and any realization of the B2D arrivals $(b_1^{(k)}, ..., b_M^{(k)})$, we solve the following

$$\max_{t^{(k)}} \sum_i d_i \mathbb{E}\left[R_i[k] \mid t^{(k)}, b^{(k)} \right], \qquad (6)$$

to find the optimal schedule $(t^{(k)})^*$. Note that policy $\mathbf{P}^*$ (in Lemma 1) randomly chooses a feasible $t^{(k)}$ according to some distribution $\mathbb{P}\left(t^{(k)} | b^{(k)}\right)$, hence one can easily verify that

$$\sum_i d_i \mathbb{E}\left[R_i[k] \mid (t^{(k)})^*, b^{(k)} \right] \geq$$
$$\sum_{t^{(k)}} \sum_i d_i \mathbb{E}\left[R_i[k] \mid t^{(k)}, b^{(k)} \right] \mathbb{P}\left(t^{(k)} | b^{(k)}\right). \qquad (7)$$

Taking expectation on both sides of the above inequality with respect to the arrival processes results in

$$\sum_i d_i \mathbb{E}\left[R_i[k] \mid (t^{(k)})^* \right]$$
$$\geq \sum_i d_i \mathbb{E}\left[\sum_{t^{(k)}} R_i[k] \mid t^{(k)}, b^{(k)} \right] \mathbb{P}\left(t^{(k)} | b^{(k)}\right) \qquad (8)$$
$$\overset{(c)}{>} \sum_i d_i \eta_i \geq \sum_i d_i (\eta_i + \epsilon)$$

for some small enough $\epsilon > 0$, where (c) follows from (3). By considering (8) in inequality (b), we conclude that using schedule $(t^{(k)})^*$ will result in

$$\Delta V[k] \leq \frac{M}{2} - \epsilon \sum_i d_i,$$

which is negative for large enough d_i values, and hence can stabilize the deficit queues.

We have just shown that we can satisfy any achievable QoS metric by using the schedule obtained by solving (6) for each frame. In what follows, we will show (6) is equivalent to the optimization problem (5) that appears in Algorithm 1.

Recall that $R_i[k] = 1_{\{n_i^{(k)} = N\}}$, where the indicator variable $1_{\{A\}}$ is equal to 1 if A is true, and 0 otherwise. Hence, $\mathbb{E}[R_i[k]] = \mathbb{P}\left(n_i^{(k)} = N\right)$. Suppose device j broadcasts $t_j^{(k)}$ chunks generated from random linear combinations of its initial $b_j^{(k)}$ chunks. Since each transmission is successful with probability δ, the number of successful transmissions by device j is distributed as a Binomial random variable $\hat{t}_j^{(k)} = Bin(\delta, t_j^{(k)})$ with parameters δ and $t_j^{(k)}$.

Now, we have assumed that the random linear coding is performed over a field of sufficiently large size q such that N distinct randomly coded chunks are linearly independent. Consequently, each successful transmission increases the rank of the receiving devices by one. Thus, the transmissions by device j introduces $\min(\hat{t}_j^{(k)}, b_j^{(k)})$ new degrees of freedom (DoF) at the other devices. Also, device i is full-rank (i.e., $n_i^{(k)} = N$) if and only if it has received at least $N - b_i^{(k)}$ new DoFs from other devices, that is $b_i^{(k)} + \sum_{j \neq i} \min(\hat{t}_j^{(k)}, b_j^{(k)}) \geq N$. Therefore, we have

$$\mathbb{E}[R_i[k]] = \mathbb{P}\left(n_i^{(k)} = N\right)$$
$$= \mathbb{P}\left(b_i^{(k)} + \sum_{j \neq i} \min(\hat{t}_j^{(k)}, b_j^{(k)}) \geq N\right). \qquad (9)$$

Substituting (9) in (6) results in the optimization (5). $\square$

We have just shown the throughput optimality of Algorithm 1 for the unreliable D2D case. However, we note that in practice we need a central entity with significant computational capabilities to implement this optimal algorithm, since it takes a combinatorial form. In the following section, we will consider the special case when D2D broadcast is completely reliable. We will show that the optimal algorithm under this case possesses an intuitively appealing form that lends itself to easy implementation. We also will develop a heuristic modification of this algorithm when the D2D broadcast is not reliable, which does not have the complexity of the optimal algorithm.

4. D2D ALGORITHM UNDER RELIABLE BROADCAST

In this section, we assume that D2D broadcasts are always successfully received by all intended recipients. Of course, the constraint that only one device can broadcast at a time still applies. As before, we first model the B2D arrivals as *i.i.d.* random variables $\mathbf{b}_i$ for each device i, and focus on scheduling the D2D transmissions. Since broadcasts are deterministically successful, $\sum_{j \neq i} t_j^{(k)} = r_i^{(k)}$. In other words, the total number of D2D chunks received by device i is equal to the number of transmissions performed by all other devices. Also, each device i can transmit at most $b_i^{(k)}$ times during frame $k-1$, since any further transmissions by i will not add to other devices' information, i.e.,

$$t_i^{(k)} \le b_i^{(k)}. \tag{10}$$

Consequently, it is easy to see that $n_i^{(k)} = \min\{N, \hat{n}_i^{(k)}\}$ holds for large field size q, i.e., device i will obtain full-rank if it receives at least a total of N coded chunks from B2D and D2D interfaces.

Further, since the chunks received from the B2D interfaces are initially randomly coded, combining them further with random coefficients cannot improve performance. Hence, at time τ, the device chosen to transmit, $u[\tau]$, can simply broadcast any of the chunks received via its B2D interface, which it has not transmitted yet. We are primarily interested in the case where $N > T$, because otherwise there are enough number of time slots in each frame for devices to broadcast all N degrees of freedom and hence the optimal D2D scheme becomes trivial.

4.1 Achievability of QoS Metric

In Section 3, Lemma 1 implicitly characterizes the achievability conditions of a given QoS requirement $(\eta_1, ..., \eta_M)$, and can be applied for the reliable D2D case as well. However, we will see below that the QoS achievability condition can be determined explicitly based on a set of necessary and sufficient conditions for this case.

Devices have T slots in each frame to exchange the received B2D chunks. Hence, each device i can potentially recover block k, only if (i) it has received enough B2D chunks initially (i.e, $b_i^{(k)} \ge N - T$) and (ii) the whole system is full-rank at the beginning of the frame (i.e, $\sum_j b_j^{(k)} \ge N$). Therefore, for each device i we have,

$$R_i[k] \le 1_{\{b_i^{(k)} \ge N-T,\ \sum_j b_j^{(k)} \ge N\}}. \tag{11}$$

Since $(b_1^{(k)}, ..., b_M^{(k)})$ is assumed to be identically and independently distributed across frames according to $(\mathbf{b}_1, ..., \mathbf{b}_M)$, from (2) we get the following necessary condition on the achievability of η_i :

$$\eta_i \le \mathbb{P}\left(\mathbf{b}_i \ge N-T,\ \sum_j \mathbf{b}_j \ge N\right). \tag{12}$$

Let $N_s(b_1^{(k)}, ..., b_M^{(k)}) = \sum_i R_i[k]$ be the number of devices that successfully receive the whole block k, given the B2D arrivals $(b_1^{(k)}, ..., b_M^{(k)})$. The following lemma provides an upper bound on $N_s(b_1^{(k)}, ..., b_M^{(k)})$.

LEMMA 3. *Given the B2D arrivals* $(b_1^{(k)}, ..., b_M^{(k)})$, *we have*

$$\begin{aligned} &N_s(b_1^{(k)}, ..., b_M^{(k)}) \\ &\le N_s^*(b_1^{(k)}, ..., b_M^{(k)}) \\ &\equiv \frac{\min\left(|\mathcal{S}|(N-T), \left[\sum_i b_i^{(k)}-T\right]^+\right)}{N-T}, \end{aligned} \tag{13}$$

where $\mathcal{S} = \{i \in \{1, ..., M\} : N - b_i^{(k)} \le T,\ b_i^{(k)} + \sum_{j \ne i} b_j^{(k)} \ge N\}$.

PROOF. Since $n_i^{(k)} \le \hat{n}_i^{(k)}$, we have

$$R_i[k] = 1_{\{n_i^{(k)} \ge N\}} \le 1_{\{\hat{n}_i^{(k)} \ge N\}} = 1_{\{b_i^{(k)} - t_i^{(k)} + \sum_j t_j^{(k)} \ge N\}}.$$

Therefore, we can solve the following maximization problem in order to find an upper bound on $N_s(b_1^{(k)}, ..., b_M^{(k)})$,

$$\begin{aligned} \max\ & \sum_i 1_{\{b_i^{(k)} - t_i^{(k)} + \sum_j t_j^{(k)} \ge N\}} \\ \text{s.t.}\ & t_i^{(k)} \le b_i^{(k)}\quad \forall i \\ & \sum_j t_j^{(k)} \le T \end{aligned} \tag{14}$$

The first constraint implies $\sum_j t_j^{(k)} \le \sum_j b_j^{(k)}$, and to achieve the maximum objective we let $\sum_j t_j^{(k)} = \min(\sum_j b_j^{(k)}, T)$.

We partition the set of devices $\{1, ..., M\}$ into sets $\mathcal{S}$ and $\mathcal{S}^c = \{1, ..., M\} \backslash \mathcal{S}$. Here, $\mathcal{S}^c$ is the set of devices which, either individually or collectively, have not received enough number of B2D arrivals to possibly recover the block, and $R_i[k] = 0$ for $i \in \mathcal{S}^c$.

Suppose $\sum_j b_j^{(k)} < N$. Then we have $N_s(b_1^{(k)}, ..., b_M^{(k)}) = |\mathcal{S}| = 0$. Otherwise $\sum_j b_j^{(k)} \ge N$, and with our assumption that $T < N$, the optimization in (14) can be rewritten as

$$\begin{aligned} \max\ & \sum_{i \in \mathcal{S}} 1_{\{b_i^{(k)} - t_i^{(k)} \ge N-T\}} \\ \text{s.t.}\ & t_i^{(k)} \le b_i^{(k)}\quad\quad\quad\quad\quad \forall i \\ & \sum_j t_j^{(k)} = \min(\sum_j b_j^{(k)},\ T) = T. \end{aligned} \tag{15}$$

The maximum value above can be shown to be

$$\left\lfloor \frac{\min\left(|\mathcal{S}|(N-T), \left[\sum_i b_i^{(k)}-T\right]^+\right)}{N-T} \right\rfloor.$$

Consequently, we obtain

$$N_s(b_1^{(k)}, ..., b_M^{(k)}) \le \frac{\min\left(|\mathcal{S}|(N-T), \left[\sum_i b_i^{(k)}-T\right]^+\right)}{N-T}.$$

$\square$

Now, from Lemma 3 and (2), since $(b_1^{(k)}, ..., b_M^{(k)})$ is *i.i.d* over frames, the following necessary condition on $\sum_i \eta_i$ can be obtained:

$$\sum_i \eta_i \le \mathbb{E}\left[\lfloor N_s^*(\mathbf{b}_1, ..., \mathbf{b}_M) \rfloor\right]. \tag{16}$$

The following theorem summarizes our results.

THEOREM 4. *The QoS metric* $(\eta_1, ..., \eta_M)$ *is achievable with respect to i.i.d B2D arrivals* $(\mathbf{b}_1, ..., \mathbf{b}_M)$ *if and only if the following conditions are satisfied*

$$\begin{aligned} (C1)\ \forall i :\ & \eta_i \le \mathbb{P}\left(\mathbf{b}_i \ge N-T,\ \sum_j \mathbf{b}_j \ge N\right) \\ (C2)\ & \sum_i \eta_i \le \mathbb{E}\left[\lfloor N_s^*(\mathbf{b}_1, ..., \mathbf{b}_M) \rfloor\right]. \end{aligned} \tag{17}$$

Further, we can show that for the symmetric case where $\eta_i = \eta$, *and* $\mathbf{b}_i$ *are identically distributed for all devices* i, *the necessary and sufficient condition reduces to*

$$(C2')\quad \eta \le \frac{1}{M}\mathbb{E}\left[\lfloor N_s^*(\mathbf{b}_1, ..., \mathbf{b}_M) \rfloor\right]. \tag{18}$$

PROOF. The necessity part was shown in (12) and (16). To prove the sufficiency of these conditions, we will propose an algorithm in the next subsection which can fulfill any QoS constraints satisfying (C1) and (C2). The proof of (C2') is not hard, and is omitted here due to space constraints. $\square$

4.2 Optimal Reliable D2D Scheme

In this subsection, we propose a simple algorithm that can achieve any QoS metric satisfying the conditions in (17). As a result, $(C1)$ and $(C2)$ are sufficient conditions on achievability of a QoS metric (Theorem 4). Also, the proposed algorithm is throughput optimal.

We follow the same approach as in Section 3 to study the D2D system using deficit queues $d_i[k]$. Algorithm 2 describes an optimal D2D scheme that can stabilize these deficit queues for any achievable QoS metrics.

Algorithm 2 Optimal D2D scheme (reliable broadcast)

At the beginning of each frame $k-1$, given the arrivals $(b_1^{(k)}, ..., b_M^{(k)})$:
Partition the devices into sets $\mathcal{S} = \{i \in \{1, ..., M\} : N - b_i^{(k)} \leq T, b_i^{(k)} + \sum_{j \neq i} b_j^{(k)} \geq N\}$ and $\mathcal{S}^c$.
If $\mathcal{S} = \emptyset$, nobody can get full-rank. Otherwise,
Phase 1) Let all the devices in $\mathcal{S}^c$ transmit all they have initially received for the next $T_1 = \min\{\sum_{i \in \mathcal{S}^c} b_i^{(k)}, T\}$ slots.

If there exist time and a need for more transmissions,
Phase 2) Let each device $i \in \mathcal{S}$ transmit up to $b_i^{(k)} + T - N$ of its initial chunks.
Phase 3) While there exist time and a need for more transmissions, let devices in $\mathcal{S}$ transmit their remaining chunks in an increasing order of their deficit values.

THEOREM 5. *The D2D scheme in Algorithm 2 is throughput optimal when the D2D transmissions are reliable.*

PROOF. As in the proof of Theorem 2, we use the Lyapunov criterion. Equivalent to the optimization in (6), we need to solve the following

$$\max \sum_i d_i R_i[k] = \sum_i d_i \mathbf{1}_{\{b_i^{(k)} + \sum_{j \neq i} t_j^{(k)} \geq N\}}$$
$$\text{s.t.} \quad t_i^{(k)} \leq b_i^{(k)} \quad \forall i \qquad (19)$$
$$\sum_i t_i^{(k)} \leq T.$$

The optimization in (19) is similar to the one in (14). Hence, we can apply a similar argument and verify that the following optimization problem is equivalent to (19),

$$\max \sum_i d_i z_i$$
$$\text{s.t.} \quad z_i \leq \mathbf{1}_{\{b_i^{(k)} \geq N-T, \sum_j b_j^{(k)} \geq N\}} \quad \forall i \qquad (20)$$
$$\sum_i z_i \leq N_s^*(b_1^{(k)}, ..., b_M^{(k)}),$$

where $N_s^*(b_1^{(k)}, ..., b_M^{(k)})$ is defined in (13).
Using the solution to the above maximization in the drift formula $\Delta V[k]$ will result in

$$\Delta V[k] \leq M/2 + \sum_i d_i \eta_i - \mathbb{E}\left[\max \sum_i d_i z_i\right]$$
$$\leq M/2 + \sum_i d_i \eta_i - \max \sum_i d_i \mathbb{E}[z_i]. \qquad (21)$$

From (20), we notice that $\mathbb{E}[z_i]$ must satisfy

$$\mathbb{E}[z_i] \leq \mathbb{P}\left(b_i^{(k)} \geq N - T, \sum_j b_j^{(k)} \geq N\right)$$
$$\sum_i \mathbb{E}[z_i] \leq \mathbb{E}\left[N_s^*(b_1^{(k)}, ..., b_M^{(k)})\right]. \qquad (22)$$

In Subsection 4.1, we have shown that an achievable QoS metric $(\eta_1, ..., \eta_M)$ needs to satisfy the above conditions. This suggests that for a strictly achievable QoS metric that is

prametrized by $(\eta_1, ..., \eta_M)$, for which these conditions hold with strict inequalities, there exists some $\epsilon > 0$ such that

$$\max \sum_i d_i \mathbb{E}[z_i] \geq \sum_i d_i \eta_i (1 + \epsilon). \qquad (23)$$

Consequently, the drift in (21) reduces to

$$\Delta V[k] \leq M/2 - \epsilon \sum_i d_i \eta_i. \qquad (24)$$

Thus, for large enough deficit values d_i, the drift is negative. This means the D2D scheme implied by solving (19) at each frame can satisfy any achievable QoS metric. Furthermore, it proves the sufficiency of the conditions in Theorem 4 on the achievability of a QoS metric.

The argument showing that the D2D scheme in Algorithm 2 actually solves the optimization problem in (19) is straightforward, and follows from dividing devices into those that cannot complete (and so might as well transmit all that they have), then considering those that can complete (and hence can transmit a limited number of chunks), and finally considering those with the largest deficit (and so should stop transmitting after phase two). The full argument is skipped for brevity. □

4.3 Simple Suboptimal Scheme for Unreliable D2D Network

The optimal D2D scheme, presented in Section 3 for the unreliable D2D system is hard to implement in practice due to high complexity. Further, this algorithm requires estimates on the channel quality. Here, we propose a modified version of Algorithm 2, which inherits its simple 3-phase form and tries to account for the unreliability of the broadcast network by letting devices retransmit a number of times.

We incorporate the following modifications in Algorithm 2:

1. In the second phase, we choose devices in an increasing order of their deficit values to transmit upto the same threshold that Algorithm 2 suggests.

2. Each device chosen to transmit by Algorithm 2 holds onto the channel for ρ time slots. For each transmission, it generates a new chunk as a random combination of its initial B2D chunks.

The parameter ρ can be chosen in accordance with the quality of broadcast channel. The heuristic algorithm attempts to improve the performance of Algorithm 2 when loss rates are not too high by accounting for deficits during Phase 2, as well as improving reliability by retransmissions. We will investigate the performance of this algorithm in Section 7.

5. SELECTION OF B2D STOPPING TIME

We will now determine the number of time slots that the B2D interface should be used by each device over each frame. We consider this as an offline stopping time problem in order to ensure low communication overhead with the server. Thus, we want to find $0 \leq T_B(i) \leq T$, which is the number of times in each frame that device i attempts to receive a coded chunk from the server using the B2D interface.

The B2D usage times should ensure that the QoS target can be met with lowest total cost $C(T_B(1), ..., T_B(M))$, where $C(.)$ is a general cost function. Note that the function could be linear $(\sum_i T_B(i))$ if usage-based costs are considered. Furthermore, it makes intuitive sense that devices that desire a higher QoS η_i, should contribute more so as to maintain a level of fairness. This requirement can also be captured in the cost function by simply choosing weights for each device that depend on its desired QoS value.

Recall that the B2D interface of each device i is assumed to be Bernoulli with success probability β_i. This means that the B2D arrivals $b_i^{(k)}$ to each device i are independently and identically distributed over frames k as a Binomial random variable $Bin(\beta_i, T_B(i))$ with parameters β_i and $T_B(i)$:

$$\mathbb{P}(b_i^{(k)} = a) = 1_{\{0 \leq a \leq T_B(i)\}} \binom{T_B(i)}{a} \beta_i^a (1 - \beta_i)^{T_B(i)-a}.$$

For the fully reliable D2D scenario, we can determine the achievability of a QoS requirement based on conditions $(C1)$ and $(C2)$ in Theorem 4. In order to achieve a given QoS metric $(\eta_1, ..., \eta_M)$, $T_B(i)$ values must be large enough such that these conditions are satisfied. Hence, the optimal values of $T_B^*(i)$ can be obtained by solving the following problem:

$$\min C(T_B(1), ..., T_B(M))$$
$$\text{s.t.}$$
$$0 \leq T_B(i) \leq T \qquad \text{for all } i \qquad (25)$$
$$\mathbf{b}_i = Bin(\beta_i, T_B(i)) \quad \text{for all } i$$
$$(C1) \text{ and } (C2) \text{ in } (17) \text{ are satisfied.}$$

Note that the minimization problem in (25) does not have a simple form that can be solved efficiently. However, since the region for feasible $T_B(i)$ values is finite (i.e., $\{0, 1, ..., T\}^M$), and the values need to be determined only once for the given set of system parameters, we could conduct an exhaustive search to find the optimal $T_B^*(i)$ values. We can also improve the search algorithm by applying more efficient search methods like *branch-and-bound*.

In the case of the unreliable D2D channel, there is only an implicit achievability condition as given by Lemma 1. We therefore cannot analytically solve for the optimal B2D usage times. Since $T_B(i)$ values are to be calculated in an offline manner, we can run stochastic simulations of the optimal D2D scheme presented in Algorithm 1 in order to find the optimal values numerically. We present results of this nature in Table 1 of Section 7.

6. FINITE FIELD CASE

So far we assumed that the field size q is large enough that all randomly coded chunks are linearly independent almost surely. Under this assumption, it was sufficient for each device to receive N distinct coded chunks in order to recover the original block. In this section, we turn our attention to the case where field size $q < \infty$, and there is a non-zero probability that randomly coded chunks are linearly dependent. More specifically, we are interested in evaluating the performance of our proposed B2D and D2D schemes in the case of finite field sizes. We require the following useful Lemma, whose proof is omitted for brevity.

LEMMA 6. *A matrix of dimension $R \times N$, whose elements are drawn uniformly at random from a finite field F_q, is full-rank with probability at least $1 - \frac{1}{q^{|N-R|}(q-1)}$, where q is the field size.*

We now have the following result.

THEOREM 7. *Suppose that the coefficients for coding the chunks are drawn uniformly at random from a field of size*

$q \geq 2$. *If we apply Algorithm 1 (or Algorithm 2 for the reliable D2D case) to coordinate the D2D broadcasts and choose the B2D usage times as discussed in Section 5, then for each device i we have*

$$\lim_{K \to \infty} \frac{1}{K} \sum_{k=1}^{K} \mathbb{E}[R_i[k]] \geq \eta_i - \frac{1}{q-1}. \qquad (26)$$

PROOF. We define $\hat{R}_i[k] = 1_{\{b_i^{(k)} + \sum_{j \neq i} \min(\hat{t}_j^{(k)}, b_j^{(k)}) \geq N\}}$. For the case of infinite field size, we had $n_i^{(k)} = \min\left(N, b_i^{(k)} + \sum_{j \neq i} \min(\hat{t}_j^{(k)}, b_j^{(k)})\right)$ and $\hat{R}_i[k] = R_i[k]$.

In the proof of throughput optimality of Algorithms 1 and 2, we showed that

$$\lim_{K \to \infty} \frac{1}{K} \sum_{k=1}^{K} \mathbb{E}[\hat{R}_i[k]] \geq \eta_i \qquad (27)$$

holds true, when we employ the proposed D2D schemes. For $R_i[k]$, we have

$$\begin{aligned} \mathbb{E}[R_i[k]] &= \mathbb{E}[R_i[k]|\hat{R}_i[k] = 1]\mathbb{P}(\hat{R}_i[k] = 1) \\ &+ \mathbb{E}[R_i[k]|\hat{R}_i[k] = 0]\mathbb{P}(\hat{R}_i[k] = 0) \\ &\stackrel{(a)}{=} \mathbb{E}[R_i[k]|\hat{R}_i[k] = 1]\mathbb{P}(\hat{R}_i[k] = 1) \\ &= \mathbb{E}[R_i[k]|\hat{R}_i[k] = 1](1 - 1 + \mathbb{E}[\hat{R}_i[k]]) \\ &\stackrel{(b)}{\geq} \mathbb{E}[R_i[k]|\hat{R}_i[k] = 1] + \mathbb{E}[\hat{R}_i[k]] - 1 \stackrel{(c)}{\geq} \mathbb{E}[\hat{R}_i[k]] - \frac{1}{q-1}, \end{aligned} \qquad (28)$$

where (a) holds because if $\hat{R}_i[k] = 0$, then i did not receive enough distinct coded chunks and cannot get full-rank (i.e., $R_i[k] = 0$). (b) follows from $-1 + \mathbb{E}[\hat{R}_i[k]] \leq 0$ and $\mathbb{E}[R_i[k]|\hat{R}_i[k] = 1] \leq 1$, and (c) holds by Lemma 6, since $\mathbb{E}[R_i[k]|\hat{R}_i[k] = 1]$ is equal to the probability that i who has received at least N distinct randomly coded chunks is full-rank. We then sum both sides of (28) from $k = 1$ to K, divide the result by K, and let $K \to \infty$. Then the desired bound in (26) follows from (27). $\square$

Note that the effect of finiteness of the field size is limited by the value $\frac{1}{q-1}$, for example for field size $q = 32$, there is only around 3% reduction in the quality of service. In the simulation results presented in Figure 3, we see the actual reduction in the QoS is even less than this value. Note that the achieved delivery ratio is sometimes higher than desired since T_B is chosen as the smallest integer value that can support η.

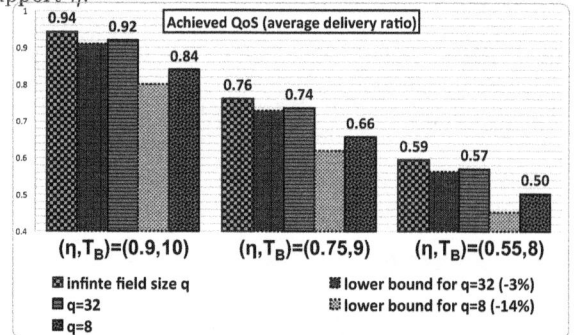

Figure 3: Achievable delivery ratios with finite field sizes ($N = 20$, $T = 15$, $M = 4$, $\beta_i = 0.9$ in a reliable D2D network)

7. SIMULATION RESULTS

We now study the performance of our different algorithms through Matlab simulations. We chose the chunks per block $N = 20$, frame length $T = 15$, B2D success probability $\beta_i = 0.9$ for all i, and varied the other parameters.

B2D Stopping Time. In Section 5, we saw that the B2D stopping time problem needs a numerical solution. We present the optimal stopping time T_B^* for a system with M devices with QoS requirement η, probability of success of the D2D channel δ, and a linear cost criterion in Table 1. We use values from this table as the optimal B2D stopping times T_B^* in the other simulations presented in this section.

Table 1: Optimal B2D stopping time T_B^*

η \ δ		1	0.95	0.9	0.85	0.8
$M = 4$	0.8	9	11	12	13	14
	0.9	10	12	13	14	15
$M = 5$	0.8	8	10	11	12	13
	0.9	9	11	12	13	14

Heuristic algorithm for unreliable D2D: In Section 4.3, we proposed a simple heuristic scheme based on the optimal algorithm derived for the reliable D2D case. Table 2 depicts the achieved QoS by this algorithm for number of per-chunk transmissions $\rho = 2$, in a network of $M = 4$ devices and a target QoS of $\eta = 0.8$. We observe that if we

Table 2: QoS achieved by the heuristic algorithm for $T_B \geq T_B^*$

δ \ T_B	11	12	13	14	15
0.8				73.88%	74.78%
0.85			72.54%	74.45%	74.93%
0.9		68.92%	73.54%	74.77%	74.96%
0.95	61.79%	70.78%	74.11%	74.85%	74.97%

set $T_B = T_B^*$, the degradation in the QoS when using the heuristic algorithm is around $7\% - 22\%$ for different values of δ. We can improve the achieved QoS by increasing the B2D usage time, e.g., the degradation reduces to $6\% - 11\%$ if we let $T_B = T_B^* + 1$. The results suggest that the simple heuristic algorithm can be used successfully if the D2D channel quality is reasonably good, or if the frame length T is long enough to permit retransmissions.

Other intuitive D2D schemes: In Figure 4, we compare the optimal scheme, Algorithm 2, with other algorithms in the case of reliable D2D: (i) Round Robin: each device broadcasts in turn, (ii) Min-Deficit-First: the device with smallest deficit broadcasts, and (iii) Max-Rank-First: the device with highest current rank broadcasts. We see that the optimal algorithm outperforms the others.

Playout smoothness: Since our QoS metric is in terms of the average delivery ratio, we do not know if a device loses a few chunks with a small periodicity, or a large number of chunks with a large periodicity, or indeed if it is periodic at all. The first case is preferred, since by using an outer code over the blocks, the device will be able to recover the dropped blocks to some extent. We define a *smooth play out interval* as the time between two consecutive block drops. Figure 5 demonstrates the distribution of these smooth play out intervals. We observe that the actual QoS implied by our

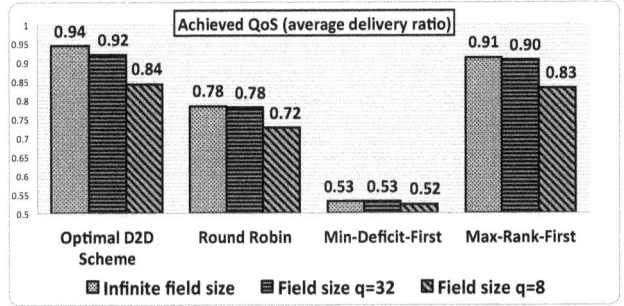

Figure 4: Comparison of Algorithm 2 with others for $M = 4$ and $\eta = 0.9$

algorithm is acceptable, since the smooth play out interval has a small variance (i.e., almost periodic reception) and a small average (no loss of big portions).

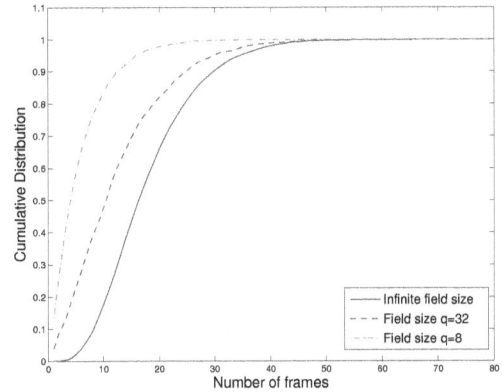

Figure 5: Smooth playout time distribution ($\eta = 0.9$).

8. ANDROID EXPERIMENTS

We now describe experiments on an Android smart phone testbed. A full scale system with peer discovery, multimedia coding and so on is beyond the scope of this paper, and our objective is primarily to observe real-world performance of the D2D algorithms. As discussed earlier, we emulate B2D transmissions by initializing each device periodically with chunks representing their B2D arrivals to test the D2D algorithms. Since Android does not support ad-hoc UDP broadcast (D2D) over 802.11 (WiFi), we rooted the phones to enable this service. To allow experiments with more (unrooted) phones, we also used the approximation of UDP broadcast via a WiFi access point (AP). Here, the phone unicasts to the AP, which broadcasts to the rest.

As in the analytical model, we divide the file into blocks, which are subdivided into chunks. We chose the chunk size as 1450 Bytes to roughly correspond to the usual packet size in 802.11. Chunks are generated using an open source random linear coding library [1] with field size 256, and the degrees of freedom per block is 10, i.e., a block is decodable with high probability if 10 chunks are received successfully. Each chunk has a header that contains the block number that it corresponds to and the device's current deficit.

Unlike the analytical model, we do not have fine grain control of when exactly transmissions take place. We transmit a pilot chunk (whose header contains the frame number) periodically to indicate the beginning of a new frame, but do not have time slots within the frame. Recall that our D2D scheduling in Algorithm 2 proceeds in three phases. We approximate these phases by backing off by different amounts of time based on an estimate of which phase that device is in and send the chunk to the UDP socket. Note that after doing this, 802.11 also follows a back off procedure.

Now, the three backoff times are chosen as follows. Devices that cannot complete (*i.e.,* Phase 1 devices) should be aggressive and transmit all their chunks. Hence, we set them to randomly backoff between 1 and 5 ms before transmission. Devices that can afford to transmit some number of chunks should be less aggressive, and transmit chunks upto the limit as given by Phase 2 of Algorithm 2 by backing off between 1 and 15 ms. Finally, once each device completes Phase 2, it enters Phase 3, and modulates its aggressiveness based on deficits. Each device normalizes its deficit based on the maximum deficit that it sees from other transmissions, and backs off proportional to this deficit within the interval of 10 to 20 ms. To allow for the transmission of 10 or more D2D chunks, we set the frame duration as 450 ms.

Devices ensure that they are all mutually synchronized by observing the frame ID present in the header of each chunk. If a device is not able to recover a block by the time a new frame pilot chunk or a chunk bearing a new frame ID is received, it reports that block as lost and updates its deficit.

We conducted experiments involving 3 to 5 phones, using a standard MP3 audio file as the source, with playout using the built in player on Android. An example trajectory of the (smoothed) deficit queue for a run with 3 phones, with each phone being initialized with 4 chunks in each frame, and desired delivery ratio 0.95 is shown in Figure 6. The deficit queue exhibits periodic decrease, and clearly does not increase to infinity, showing stabilization.

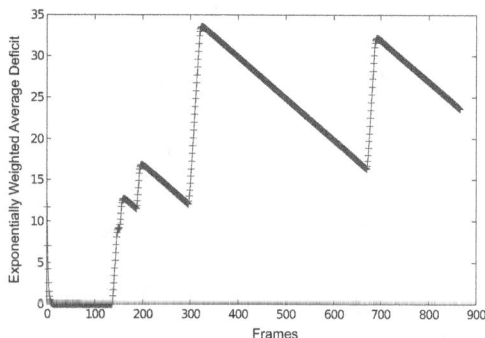

Figure 6: Sample trajectory of deficit. It is clear that the deficit queue is stable

We also conducted experiments to determine the stable delivery ratio achieved using D2D for different B2D initializations per frame. We present these results for 5 phones (labeled Ph $1-5$), with delivery ratio indicated out of 100% in Table 3. It is clear we can have significant savings in B2D usage. For instance, with a target delivery ratio of 94%, we save about 60% of the B2D costs for each phone.

9. CONCLUSION

We studied the problem of realtime streaming applications for wireless devices that can receive coded data chunks via both expensive long-range unicast and an inexpensive short-range broadcast wireless interfaces. QoS was defined in terms of the average fraction of blocks recovered. We utilized a Lyapunov stability argument to propose a minimum cost D2D transmission algorithm that can attain the required QoS, in the case of unreliable and reliable D2D transmissions. We also showed how to calculate the minimum cost B2D usage per device. We showed how the infinite field size assumption under which the algorithms are derived does not have a significant impact on scaling and performance, and then showed how performance changes with parameters using simulations. Finally, we described our design decisions for an Android implementation. Future work includes a full scale implementation, as well as multihop D2D transmission.

Table 3: Delivery ratios for different initializations.

B2D	Ph 1	Ph 2	Ph 3	Ph 4	Ph 5	Avg
2	56.38	38.44	56.37	41.7	43.1	47.198
3	86.9	70.85	81.8	75.3	83.6	79.69
4	85.75	94.6	94.3	98.31	98.45	94.282
5	98.34	98.03	97.3	99.48	100	98.63

10. REFERENCES

[1] Network coding utilities. Library available at `http://arni.epfl.ch/software`.

[2] N. Abedini, M. Manjrekar, S. Shakkottai, and L. Jiang. Harnessing multiple wireless interfaces for guaranteed QoS in proximate P2P networks. In *Proc. Intl. Conference on Communications in China*, 2012.

[3] Cisco. *Cisco Visual Networking Index: Global Mobile Data Traffic Forecast Update, 2010-2015*. Cisco, February 2011.

[4] T. Courtade and R. Wesel. Coded cooperative data exchange in multihop networks. *Arxiv preprint arXiv:1203.3445v1*, 2012.

[5] S. Deb, M. Médard, and C. Choute. Algebraic gossip: A network coding approach to optimal multiple rumor mongering. *IEEE Trans. on Information Theory*, 52(6):2486–2507, 2006.

[6] I. Hou, V. Borkar, and P. Kumar. A theory of QoS for wireless. In *IEEE INFOCOM 2009*, Rio de Janeiro, Brazil, April 2009.

[7] M. Neely. Energy optimal control for time varying wireless networks. *IEEE Trans. Information Theory*, 52(2):2915–2934, July 2006.

[8] A. ParandehGheibi, M. Medard, A. Ozdaglar, and S. Shakkottai. Avoiding interruptions-a QoE reliability function for streaming media applications. *IEEE Journal on Selected Areas in Communications*, 29(5), May 2011.

[9] H. Seferoglu, L. Keller, B. Cici, A. Le, and A. Markopoulou. Cooperative video streaming on smartphones. In *Allerton*, pages 220–227. IEEE, 2011.

[10] A. Sprintson, P. Sadeghi, G. Booker, and S. El Rouayheb. A randomized algorithm and performance bounds for coded cooperative data exchange. In *Proc. of IEEE ISIT*, pages 1888–1892, 2010.

[11] M. Wang and B. Li. Lava: A reality check of network coding in peer-to-peer live streaming. In *INFOCOM 2007*, pages 1082–1090.

On Bridging The Gap Between Homogeneous and Heterogeneous Rendezvous Schemes for Cognitive Radios

Ching-Chan Wu, Shan-Hung Wu
Dept. of Computer Science, National Tsing Hua University
Hsinchu, Taiwan, ROC
ccwu@netdb.cs.nthu.edu.tw, shwu@cs.nthu.edu.tw

ABSTRACT

Cognitive radio allows radio devices to access the idle spectrum opportunistically, thus alleviates the huge demand for spectrum. Rendezvous, where two radios complete handshaking in an idle channel, is a key step for cognitive radios to start communication. Radios may have the same (homogeneous) or different (heterogeneous) spectrum sensing capabilities. Currently, there is a "gap" between the rendezvous algorithms for homogeneous and heterogeneous cognitive radios—existing homogeneous algorithms incur high delay when applied to heterogeneous radios; while heterogeneous algorithms incur high congestion when applied to homogeneous radios. Since mixtures of these two types of radios appear commonly in practice, it is crucial to bridge the gap between the respective rendezvous algorithms. In this paper, we propose a new rendezvous algorithm, named the ICH scheme, for arbitrary mixtures of radios with homogeneous or heterogeneous spectrum sensing capabilities. Rigorous analysis and extensive simulations are conducted and show that ICH is the first rendezvous scheme that guarantees rendezvous for arbitrary mixtures of homogeneous and heterogeneous radios without incurring large delay and congestion.

Categories and Subject Descriptors

C.2.1 [**Network Architecture and Design**]: Wireless communication; C.2.4 [**Distributed Systems**]: Distributed applications

Keywords

cognitive radio; blind rendezvous; homogeneous; heterogeneous

1. INTRODUCTION

The tremendous demand for the radio spectrum continues growing, as more and more wireless devices have spread around people in the past few years. Due to the fixed and uneven spectrum allocation, the spectrum resources are inadequate for applications in some places. However, the spectrum is rarely used cross channels, time, and space continuously [11, 1]. There are usually spectrum holes, which consist of idle channels, at some time and space. To alleviate the spectrum demand, cognitive radio is proposed as a means of DSA [1] to utilize those spectrum holes by allowing the *Secondary Users* (SUs, or simply *radios*) to sense the spectrum ranging within their own device capability and to dynamically tune into different idle channels not currently used by *Primary Users* (PUs) to communicate with each other opportunistically.

Rendezvous is a key step for cognitive radios to start communication. Two radios are said to rendezvous with each other if they complete handshaking (for the purposes of neighbor discovery, data transmission, etc.) in an idle channel. One popular technique to guide a pair of radios to rendezvous is to use the *Channel Hopping* (CH) *schemes* [5, 9, 10, 13, 14, 17, 18]. A CH scheme, programmed in each radio in a network, divides the time of a radio evenly into *time slots*, and requires the radio to hop to a sequence of channels in some predefined order at consecutive slots. This sequence is called the *CH sequence* for that radio. The CH scheme ensures that, by following their CH sequences, two radios can rendezvous with each other within a finite delay, called *Time To Rendezvous* (TTR). In contrast to other centralized techniques [6, 7], CH schemes allow radios to obtain their own CH sequences in a distributed manner, thereby walking around a single point of failure. CH schemes also help avoid congestion, since at each time slot, different radios may hop to different channels. Recent CH schemes [14, 21, 4, 16] give another advantage that rendezvous can be guaranteed without assuming timer synchronization between radios. Since timer synchronization is hard to achieve in practice (especially before rendezvous), these schemes have broader applicability. In this paper, we focus on CH schemes for asynchronous radios.

Existing CH schemes focus on either homogeneous or heterogeneous cognitive radios. Let V_i be the spectrum sensing capability of a radio i (that is, a set of channels with which the radio i is capable of sensing), and P_i be a set channels in V_i that are detected to be occupied by PUs. Homogeneous CH schemes assume radios to have homogeneous sensing capability, i.e., $V_i = V_j = V$, and guarantees rendezvous if $(V \setminus P_i) \cap (V \setminus P_j) \neq \emptyset$ within a worst-case delay, called *Maximum Time To Rendezvous* (MTTR), of $O(|V|^2)$ slots [14, 21, 4]. Heterogeneous CH schemes, on the other hand, assume $V_i \neq V_j$ and guarantees rendezvous within $O(|V_i||V_j|)$ MTTR [18, 19] if $(V_i \backslash P_i) \cap (V_j \backslash P_j) \neq \emptyset$.

We observe a "gap" between the homogeneous and heterogeneous CH schemes. Applying homogeneous schemes to heterogeneous radios results in either loss of the guarantee (if we let the CH scheme generate a CH sequence for radio i using V_i directly) or $O(|U|^2)$ MTTR (if we let $V_i = U$ for all i and regard channels in $U \backslash V_i$ as occupied, where U, $|U| \gg |V_i|$, is the set of universal channels) which is too high to make the schemes feasible. Similarly, applying heterogeneous schemes to homogeneous radios either loses the rendezvous guarantee [18] or incurs serious congestion [19]. In real networks, mixtures of homogeneous and heterogeneous radios are common. For example, there may be radios from different troop/organizations in a network, and radios from the same troops/organization are likely to have the same spectrum sensing capabilities. It is crucial to have a new rendezvous technique that bridges the gap between homogeneous and heterogeneous CH schemes.

In this paper, we propose a new CH scheme, named the *Interlocking Channel Hopping* (ICH) scheme that guarantees rendezvous for arbitrary mixtures of homogeneous and heterogeneous radios. In addition, the ICH scheme is carefully designed to achieve two goals—minimizing the MTTR for heterogeneous radios and minimizing the level of congestion (called *load*, to be explained later) for homogeneous radios—that are currently conflicting due to the aforementioned gap.

To the best of our knowledge, the ICH scheme is the first CH scheme that guarantees rendezvous for both homogeneous and heterogeneous radios without incurring large delay and congestion. This study largely increases the practicability of CH schemes to real networks. Following summarizes our contributions:

- We identify a gap between the homogeneous and heterogeneous CH schemes and propose the ICH scheme that guarantees rendezvous between radios i and j as long as $(V_i \backslash P_i) \cap (V_j \backslash P_j) \neq \emptyset$, no matter $V_i = V_j$ or $V_i \neq V_j$.

- The ICH scheme ensures $O(|V|^2)$ MTTR when $V_i = V_j = V$, which is the same as the shortest MTTR achieved by existing homogeneous schemes [14, 21, 4]. When $V_i \neq V_j$, the ICH scheme ensures $O(|V_i||V_j|)$ MTTR, which is again as short as the best MTTR achieved by current heterogeneous scheme [18, 19].

- We study the degrees of congestion (denoted by *load*) for cognitive radios, and carefully design our scheme without incurring congestion. The simulation results show that the load of ICH is very close to the optimal load, $E[load]_{opt} = \frac{1}{|D|}$, as $|V_i|$ is usually not small.

- The ICH scheme takes into account the clock shift between radios, therefore supports both synchronous and asynchronous environments.

- Extensive simulations are conducted and the results show that under various combination of radios, our scheme is either 10 times faster than extensions of existing homogeneous CH schemes, or incurs 50% lighter load than existing heterogeneous CH schemes.

The rest of this paper is organized as follows. In section 2, we formally define the problem and review a CH scheme

Variable	Description
c_x	The channel numbered x
U	The set of universal channels
V_i	Device capability of radio i
P_i	The set of PU occupied channels that radio i detects
$start_i$	The starting channel of V_i
$t_i^{[x]}$	The xth time slot of radio i
$S_i^{[x]}$	CH sequence of radio i in the xth round
$s_i^{[x,0]}$	The yth element in $S_i^{[x]}$ the xth round
F_i	The fixed subsequence of S_i
R_i	The rotating subsequence of S_i
N_i	The insurance subsequence of S_i
k_i	Rotating amount of R_i
M_i	The rotating subsequence of N_i
B_i	The insurance subsequence of N_i
a_i	Rotating amount of M_i
b_i	The insurance channel of B_i
D	The set of all radios.

Table 1: Notation.

that is relevant to our study. We then explain why minimizing the MTTR for heterogeneous radios and minimizing the load for homogeneous radios are conflicting goals in existing CH schemes, and propose the ICH scheme for these two goals in Section 3. Section 4 evaluates the performance of our proposals. In Section 5, we review existing works on rendezvous for cognitive radios. Finally, Section 6 concludes the paper.

2. PRELIMINARIES

In this section, we formally define the rendezvous problem. We also review state-of-the-art CH schemes. Table 1 lists the notations used throughout this paper.

2.1 Problem Definition

Assume that the universal spectrum can be divided into a set $U = \{c_0, c_1, \cdots, c_{|U|-1}\}$ of channels. Each radio i can sense a range of spectrum consisting of a set $V_i = \{c_x, c_{x+1}, \cdots, c_{x+|V_i|-1}\}$ of continuous channels starting from c_x [12, 2, 8, 11, 1]. We denote c_x as $start_i$. Each channel in V_i is either occupied by nearby Primary Users (PUs) or available for opportunistic usage, and we let P_i be the set of PU occupied channels that radio i detects. Two radios i and j are said to have *capability-overlap* if they can sense common channels, i.e., $V_i \cap V_j \neq \emptyset$. The time of each radio i is divided evenly into *time slots*, denoted as $t_i^{[0]}, t_i^{[1]}, \cdots$. We do *not* assume any timer synchronization between radios. So given an index x, slots $t_i^{[x]}$ and $t_j^{[x]}$ of two radios i and j may have arbitrary shift in time. We say that two slots $t_i^{[x]}$ and $t_j^{[y]}$ have *time-overlap* if they overlap for an interval longer than half of a slot, as shown in Fig. 1.

We adopt the channel hopping scheme, where each radio hops to a channel at each time slot and waits for rendezvous with other radios. Specifically, given a Channel Hopping (CH) sequence $S_i = [s_i^{[0]}, s_i^{[1]}, \cdots]$, where $s_i^{[x]} \in V_i$, the radio i hops to channel $s_i^{[0]}$ at slot $t_i^{[0]}$, and $s_i^{[1]}$ at slot $t_i^{[1]}$, and so on.

Definition 2.1 (Rendezvous). Given a pair of capability-overlapping radios i and j in a network, the radios i and j

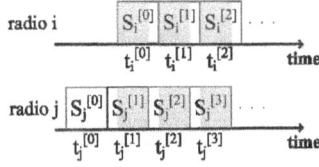

Figure 1: Despite of the asynchronous timers, a slot of radio i must overlap with one slot of radio j over an interval (shaded) longer than half of a slot. For example, the slot $t_i^{[1]}$ is *time-overlapping* with $t_j^{[2]}$, but not with $t_j^{[3]}$.

rendezvous if $s_i^{[x]} = s_j^{[y]} = c$ for some x, y, and c, where $t_i^{[x]}$ time-overlaps with $t_j^{[y]}$ and c is in both $V_i \backslash P_i$ and $V_j \backslash P_j$.

Two radios are said to *rendezvous* if they hop to some common available channel at a pair of time-overlapping slots. We assume that the duration of a time slot is set long enough such that the handshaking (for, say, neighbor discovery or data transmission) can be done within half of a slot at which rendezvous takes place [5, 18].

We formally define our problem as follows:

Problem 2.2. Design a CH scheme such that a) given any pair of capability-overlapping radios i and j in a network, the scheme is able to return two CH sequences S_i and S_j and guarantee that by following S_i and S_j respectively, the radios i and j will rendezvous within finite delay (called *Time to Rendezvous*) as long as $(V_i \backslash P_i) \cap (V_j \backslash P_j) \neq \emptyset$; and b) at any time slot, the number of radios which rendezvous on a particular channel should be minimized to avoid congestion.

Note that it is impossible for two radios to rendezvous if they have no available channels in common, e.g., $V_i \cap V_j = \emptyset$ or $(V_i \backslash P_i) \cap (V_j \backslash P_j) = \emptyset$.

To simplify the delay analysis, one common metric is the *Maximum Time to Rendezvous* (MTTR), which measures the maximum time (in number of slots) required for two radios to rendezvous. The shorter the MTTR the better.

2.2 State of the Arts

Many CH schemes are proposed for the rendezvous problem, and can be generally classified into the homogeneous schemes [14, 21, 4] (which assume $V_i = V_j$ for all radios i and j) and heterogeneous [16, 18, 19] schemes ($V_i \neq V_j$). Next, we briefly summarize the HH scheme [19] as it provides some lemmas that are useful to our study.

To start, we need to extend the notation for a CH sequence first. A CH sequence $S_i = [s_i^{[0]}, s_i^{[1]}, \cdots]$ can be partitioned evenly into *rounds* $S_i^{[x]} = [s_i^{[x,0]}, s_i^{[x,1]}, \cdots, s_i^{[x,|S_i^{[x]}|]}]$, where $s_i^{[x,y]}$ denotes the yth element in the xth round, as shown in Fig. 2. Note that $s_i^{[x,y]} = s_i^{[x \cdot |S_i^{[x]}|+y]}$.

In the HH scheme, we partition S_i into rounds of length 3:

$$s_i^{[x,y]} = \begin{cases} f_i^{[x]}, & y = 0, \\ r_i^{[x]}, & y = 1, \\ n_i^{[x]}, & y = 2, \end{cases}$$

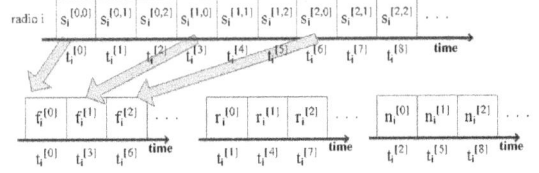

Figure 2: An example CH sequence. S_i is divide into the *fixed* F_i, *rotating* R_i, and *insurance* N_i subsequences.

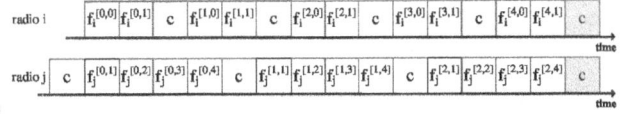

Figure 3: Rounded fixed sequences F_i and F_j with a common available channel c, where $|F_i^{[x]}| = 3$, $|F_j^{[x]}| = 5$. The MTTR is bounded by $O(|F_i^{[x]}||F_j^{[x]}|)$ time slots.

and denote the three elements in each round x, $f_i^{[x]}$, $r_i^{[x]}$, $n_i^{[x]}$, respectively. This effectively divide S_i into three subsequences, namely the *fixed sequence* $F_i = [f_i^{[0]}, f_i^{[1]}, \cdots]$, *rotating sequence* $R_i = [r_i^{[0]}, r_i^{[1]}, \cdots]$, and *insurance sequence* $N_i = [n_i^{[0]}, n_i^{[1]}, \cdots]$ (see Fig. 2).

The fixed sequence F_i is further partitioned into rounds $F_i^{[x]} = [f_i^{[x,0]}, f_i^{[x,1]}, \cdots, f_i^{[x,|F_i^{[x]}|-1]}]$. Let $|F_i^{[x]}|$ be the least prime number larger than $|V_i|$. The HH scheme assigns channels to F_i by

$$f_i^{[x,y]} = \begin{cases} v_i^{(y)}, & x = 0 \text{ and } y < |V_i|, \\ \text{an arbitrary element of } V_i, & x = 0 \text{ and } y \geq |V_i|, \\ f_i^{[x-1,y]}, & \text{otherwise,} \end{cases}$$

where $v_i^{(y)}$ is the yth element in V_i (indexed from 0). An example is shown in Fig. 3. Notice that if $|V_i|$ is a prime already, then $|F_i^{[x]}|$ needs to be the next prime number.

The rotating sequence R_i is also partitioned into rounds $R_i^{[x]} = [r_i^{[x,0]}, r_i^{[x,1]}, \cdots, r_i^{[x,|R_i^{[x]}|-1]}]$. Let $|R_i^{[x]}| = |F_i^{[x]}|$, the least prime larger than $|V_i|$. The HH scheme assigns channels to R_i by

$$r_i^{[x,y]} = f_i^{[(-x \cdot k_i + y) \bmod |R_i^{[x]}|]},$$

where $k_i = (start_i \bmod (|R_i^{[x]}| - 1)) + 1$. Basically, elements in $R_i^{[x]}$ are rotated k_i slots forward to produce the next round $R_i^{[x+1]}$. An example is shown in Fig. 4. Notice that $r_i^{[x,y]}$ and $r_i^{[x+1,y]}$ must be different since $1 \leq k_i \leq |R_i^{[x]}| - 1$. Finally, all slots of the insurance sequence N_i are filled in the starting channel $start_i$.

The authors of the HH scheme give the following lemmas:

Lemma 2.3. *Let p be a prime and m be an integer coprime with p. Then for any d, the integers $d, d+m, d+2m, \cdots, d+(p-1)m$ are all distinct under modulo-p arithmetic.*

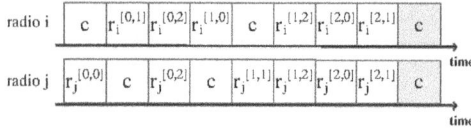

Figure 4: Rounded rotating sequences R_i and R_j with a common available channel c, where $|R_i^{[x]}| = |R_j^{[x]}| = 3$, $k_i = 1$, and $k_j = 2$. The MTTR is bounded by $O(|R_i^{[x]}||R_j^{[x]}|)$ time slots.

Consider two radios i and j, $(V_i \backslash P_i) \cap (V_j \backslash P_j) \neq \emptyset$, and two CH sequences S_i and S_j adopted by i and j respectively having the same round length.

Lemma 2.4. *Given that S_i and S_j have the fixed sequences F_i and F_j respectively and slots in F_i and F_j are time-overlapping. The MTTR between radios i and j is bounded by $O(|V_i||V_j|)$ if $|F_i^{[x]}| \neq |F_j^{[x]}|$.*

Lemma 2.5. *Given that S_i and S_j have the rotating sequences R_i and R_j respectively and slots in R_i and R_j are time-overlapping. The MTTR between i and j is bounded by $O(|V_i||V_j|)$ if $|R_i^{[x]}| = |R_j^{[x]}|$ and $k_i \neq k_j$.*

Based on the above lemmas, the author further show that the HH scheme guarantees rendezvous for i and j despite of their clock shift, and the MTTR is always bound by $O(|V_i||V_j|)$. Interested reader may refer to [19] for the proofs and detailed discussions. It is important to note that Lemmas 2.4 and 2.5 are applicable to CH sequences generated by any other scheme.

3. RENDEZVOUS FOR HOMO AND HETERO RADIOS

In this section, we demonstrate the gap between existing homogeneous and heterogeneous CH schemes and then propose a new rendezvous algorithm, named the Interlocking Channel Hopping (ICH) scheme.

3.1 The Gap

Existing homogeneous CH schemes incur high delay when applied to heterogeneous radios. At the same time, heterogeneous CH schemes result in severe congestion when applied to homogeneous radios.

To see this, consider the homogeneous CH schemes first, which assume $V_i = V_j = V$ and give $O(|V|^2)$ MTTR. When $V_i \neq V_j$, homogeneous schemes lose guarantee for rendezvous. One extension to these schemes to ensure rendezvous is to let $V_i = U$ for all i, and those channels in $U \backslash V_i$ as PU occupied (in P_i). However, this leads to $O(|U|^2)$ MTTR, which is unlikely to be acceptable for most of applications as $|U| \gg |V_i|$.

To see the problems of heterogeneous schemes, we need to measure the degree of congestion, called *load*, incurred by a CH scheme first.

Definition 3.1 (Channel Load). The *channel load* of a channel c is defined as $load_{ch}(c) = \max_t \sum_{i \in D} \delta(i,t)/|D|$, where D is the set of all radios and $\delta(i,t)$ is an indicating function that equals to 1 if a radio i hops to c at t, otherwise 0.

The $load_{ch}(c)$ is the proportion of maximum number of radios that hop to c at the same time to the total number of radios, which indicates the degree of congestion of a channel. A CH scheme should result in a low channel load for all c. However, $load_{ch}(c)$ is dependent with both device capabilities and CH sequences. To distinguish the load incurred by capabilities and by CH scheme, we give the following definitions:

Definition 3.2 (Capability Load). The *capability load* of a set of radios D is defined as $load_{cap} = \max_c \sum_{V_i} \lambda(i,c)/|D|$, where V_i is the capability of radio i, and $\lambda(i,c)$ is an indicating function that equals to 1 if $c \in V_i$, otherwise 0.

The $load_{cap}$ is the proportion of maximum number of radios capable of sensing the same channel to the total number of radios, which indicates the degree of congestion in the worst. Note that $load_{ch}(c) \leq load_{cap}$ for all c, and $load_{ch}(c*) = load_{cap}$ when $c*$ is sensible to the most radios and all these radios hop to $c*$ at the same time.

Definition 3.3 (Load). The *load* incurred by a CH scheme is defined as $load = \max_c load_{ch}(c)/load_{cap}$.

The load is the proportion of the maximum channel load to the capability load, which measures the degree of congestion incurred by a CH scheme. Note that a high load does not always imply a high channel load. For example, given a set of heterogeneous radios where the capability-overlaps between these radios are evenly distributed among the radio spectrum, we have a low $load_{cap}$. Clearly, congestion does not occur even when $load$ is high, as $load_{ch}(c) \leq load_{cap}$ for all c. However, the load is good a measurement of congestion for the networks having high $load_{cap}$ (e.g., network consisting of homogeneous radios mostly, where $load_{cap}$ is close to the highest 1). In this case, a large number of radios may crowd into a channel, and a high load implies a high channel load (congestion). A CH scheme should keep a low $load$ when $load_{cap}$ is high.

In the following, we focus on homogeneous environments where $load_{cap} = 1$. Most homogeneous CH schemes proposed recently [21, 4] give a balanced load across channels. For example, the JS scheme [14] spreads out the rendezvous opportunities uniformly over the device capability V and time, thus it has an optimal expected load, $E[load]_{opt} = E[\max_c load_{ch}(c)] = E[load_{ch}(c)] = [\sum_{k=0}^{|D|} \binom{|D|}{k} \cdot k \cdot (\frac{1}{|V|})^k (1 - \frac{1}{|V|})^{|D|-k}]/|D| = \frac{1}{|V|}$, where k is the number of radios hop to channel c in a slot and $\frac{1}{|V|}$ is the probability that a radio hop to channel c in a slot.

When applied to homogeneous environments, some heterogeneous CH schemes [18, 16] lose guarantee for rendezvous. The HH scheme [19], although guaranteeing rendezvous between homogeneous radios, incurs a very high load. This is because that with HH homogeneous radios rely on their insurance sequences to rendezvous with each other, and the starting channel in V (i.e., $start_i$, which is the same for all i here) is the only channel in these insurance sequences (see Section 2.2). Specifically, the HH scheme has the expected load $E[load]_{HH} = E[\max_c load_{ch}(c)] = E[load_{ch}(start_i)] = \frac{2}{3} \cdot \frac{1}{|V|} + \frac{1}{3} \cdot 1 = \frac{|V|+2}{3|V|}$, as a) $start_i$ is the channel with the highest channel load; b) each CH sequence has the round length 3, implying that at a time slot, there are two-third of the radios that hop to channels in the fixed and rotating radio i: | b_i | b_i | b_i | $m_i^{[0,0]}$ | $m_i^{[0,1]}$ | b_i | $m_i^{[0,2]}$ | b_i | b_i | b_i | $m_i^{[1,0]}$ | $m_i^{[1,1]}$ | b_i | $m_i^{[1,2]}$ | → time

radio j: | b_j | b_j | b_j | $m_j^{[0,0]}$ | $m_j^{[0,1]}$ | b_j | $m_j^{[0,2]}$ | b_j | b_j | b_j | $m_j^{[1,0]}$ | $m_j^{[1,1]}$ | b_j | $m_j^{[1,2]}$ | → time

Figure 5: Example insurance sequences N_i and N_j of the ICH scheme, where $|N_i^{[x]}| = |N_j^{[x]}| = 7$, $Q_i = Q_j = \{0, 1, 2, 5\}$, and $|M_i| = |M_j| = 3$. B_i and B_j will have time-overlaps every round.

sequences, and one-third of the radios to $start_i$ in the insurance sequences; c) the fixed and rotating sequences have an optimal load $\frac{1}{|V|}$ as in JS [14], while the in the insurance sequence has the worst load 1. As we can see, since $\frac{|V|+2}{3|V|} > \frac{1}{3}$, this heterogeneous scheme leads more than $\frac{|D|}{3}$ radios to hop to the starting channel at the same time, results in serious congestion when $|D|$ is large.

It turns out that minimizing the MTTR for heterogeneous radios and minimizing the load for homogeneous radios are two conflicting goals in state of the arts. This severely limits the practicability of CH schemes, as mixtures of homogeneous and heterogeneous radios are common in real networks.

3.2 Interlocking Channel Hopping Scheme

In this subsection, we propose a new scheme, called *Interlocking Channel Hopping* (ICH) scheme, that minimizes MTTR and congestion for both homogeneous and heterogeneous radios.

The ICH is defined as follows. We partition S_i into rounds of length 5:

$$s_i^{[x,y]} = \begin{cases} f_i^{[x]}, & y = 0, \\ f_i^{[x]}, & y = 1, \\ f_i^{[x]}, & y = 2, \\ r_i^{[x]}, & y = 3, \\ n_i^{[x]}, & y = 4, \end{cases}$$

By repeating the channel in the first three slots, we divide S_i into three subsequences, namely the *fixed sequence* $F_i = [f_i^{[0]}, f_i^{[1]}, \cdots]$, *rotating sequence* $R_i = [r_i^{[0]}, r_i^{[1]}, \cdots]$, and *insurance sequence* $N_i = [n_i^{[0]}, n_i^{[1]}, \cdots]$. The F_i and R_i have the same settings as in the HH scheme. However, N_i is further partitioned into rounds $N_i^{[x]} = [n_i^{[x,0]}, n_i^{[x,1]}, \cdots, n_i^{[x,|N_i^{[x]}|-1]}]$. Let $U = \{0, 1, ..., |N_i^{[x]}| - 1\}$ be the set of slots in a round. We employ the optimal cyclic quorum algorithm [15] to construct a cyclic quorum Q_i, $Q_i \subseteq U$ and $Q_i \neq \emptyset$, for radio i.

Definition 3.4 (Coterie). Let X be a set of nonempty subsets of U. We call X an coterie iff for all $Q, Q' \in X$, $Q \cap Q' \neq \emptyset$.

Definition 3.5 (Cyclic Set). Given an integer l, where $0 \leq l \leq |N_i^{[x]}| - 1$. Let Q be a subset of U. We call $C_l(Q)$ an l-cyclic set of Q iff $C_l(Q) = \{(q + l) \bmod |N_i^{[x]}| : \forall q \in Q\}$.

For convenience, we denote a group of cyclic set as $C(Q) = \{C_l(Q) : \forall l\}$.

Definition 3.6 (Cyclic Quorum System). Let $X = \{Q_0, Q_1, \cdots\}$ be a set of nonempty subsets of U. We call X an cyclic quorum system iff the set of sets $C(Q_0) \cup C(Q_1) \cup ...$ is a coterie.

We call elements of X the cyclic quorums. By Definition 3.6, we have the following theorem.

Theorem 3.7. *Given two insurance sequences N_i and N_j, and some cyclic quorums Q_j and Q_j defined for $N_i^{[x]}$ and $N_j^{[x]}$ respectively. Despite of clock shift between radios i and j, in each round $N_i^{[x]}$ (or $N_j^{[x]}$) there must exists a pair of slots from Q_j and Q_j that are time-overlapping with each other if $|N_i^{[x]}| = |N_j^{[x]}|$.*

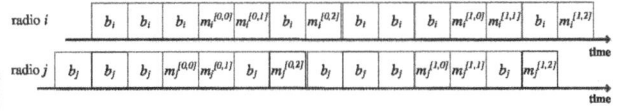

In the ICH scheme, N_i is divided into two subsequences, namely *sub-rotating sequence* M_i and *sub-insurance sequence* B_i. Let b_i be an arbitrary channel in V_i chosen by radio i, called the *insurance channel*. We define N_i as follows:

$$N_i^{[x,y]} = \begin{cases} m_i, & y \notin Q_i, \\ b_i, & y \in Q_i, \end{cases}$$

where $M_i = [m_i^{[0]}, m_i^{[1]}, \cdots]$, and $B_i = [b_i, b_i, \cdots]$. The sub-rotating sequence M_i is further partitioned into rounds $M_i^{[x]} = [m_i^{[x,0]}, m_i^{[x,1]}, \cdots, m_i^{[x,|M_i^{[x]}|]}]$. Let $|M_i^{[x]}|$ be $|N_i^{[x]}| - |Q_i|$. The ICH scheme assigns channels to M_i by

$$m_i^{[x,y]} = \begin{cases} v_i^{(y)}, & x = 0 \wedge y < |V_i|, \\ m_i^{[x-1,(y+|R_i^{[x]}|-a_i) \bmod |R_i^{[x]}|]}, & x \neq 0 \wedge y < |R_i^{[x]}|, \\ v_i^{(x \bmod |V_i|)}, & \text{otherwise}, \end{cases}$$

where $v_i^{(y)}$ is the yth element in V_i (indexed from 0) and a_i is the rotating amount of $M_i^{[x]}$. Basically, we fill B_i with b_i, and $M_i^{[x]}$ is similar to the rotating sequence $R_i^{[x]}$, except that the rotating amount a_i of is determined by the insurance channel b_i, i.e., $a_i = b_i \pmod{(|M_i^{[x]}| - 1)} + 1$. An example is shown in Fig. 5. We let $|N_i^{[x]}|$ be a prime number such that $|N_i^{[x]}| - |Q_i| \geq |R_i^{[x]}|$, and employ the optimal cyclic quorum algorithm [15] to construct $N_i^{[x]}$, that we put B_i in the quorum positions Q_i. Due to the space limitation, we do not discuss the construction of cyclic quorum systems here. Interested reader may refer to [15].

Theorem 3.7 guarantees that there are time-overlapping slots between B_i and B_j. With insurance sequences N_i and N_j, two radios will rendezvous at either the time-overlapping slot pair (denoted by (B_i, B_j)) if they choose the same insurance channel, or otherwise at the (M_i, M_j), (M_i, B_j), or (B_i, M_j) pairs using Lemma 2.5.

However, the problem is how to choose the insurance channel b_i. Considering two radios i and j with $|F_i^{[x]}| = |F_j^{[x]}|$ and $start_i = start_j$, as V_i and V_j may still be different, V_i needs to pick b_i such that $b_i \in V_i \cap V_j$ without knowing V_j in advance, and vise versa for V_j to pick s_j. We let $p = |F_i^{[x]}| = |F_j^{[x]}|$ and q_i (q_j) be the largest prime number that smaller than $|V_i|$ ($|V_j|$). As we know that $|F_i^{[x]}|$ is the least prime number that larger than $|V_i|$, we have $q_i = q_j = q$ and $q < |V_i|, |V_j| < p$ given $|F_i^{[x]}| = |F_j^{[x]}|$. Hence, V_i can be sure that $[start_i, start_i + q_i - 1] \subseteq V_i \cap V_j$ and can pick any element in $[start_i, start_i + q_i - 1]$ as b_i. Similarly V_j can pick b_j from $[start_j, start_j + q_j - 1]$.

Finally, the ICH scheme assigns the insurance channel to the sub-insurance sequence. Note that, b_i cannot be selected from occupied channels by PUs, i.e., $b_i \in (V_i \setminus P_i)$. Accord-

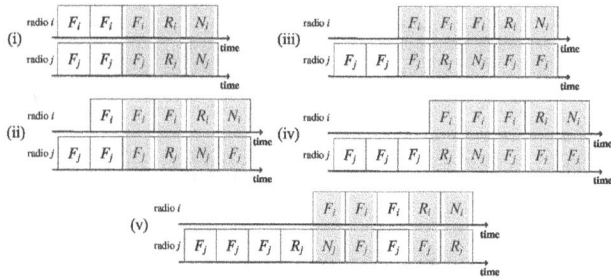

Figure 6: Each round of a CH sequence S_i of the ICH scheme is partitioned into three consecutive F_is followed by a R_i and a N_i. In terms of time-overlapping slots, the shifts between two CH sequences range from 0 to 4 slots.

ingly, we allow radios to use different insurance channels even when $start_i = start_j$, thereby avoiding homogeneous radios crowding into the the same channel.

3.3 Minimizing Delay

Next, we verify that the MTTR of ICH is $O(|V_i||V_j|)$ for heterogeneous radios and $O(|V|^2)$ for homogeneous radios. Firstly, let's discuss the MTTR using the time-overlapping slots between N_i and N_j. Considering two radios i and j with $|F_i^{[x]}| = |F_j^{[x]}|$ and $start_i = start_j$, we discuss two cases: 1) $b_i = b_j$, 2) $b_i \neq b_j$.

- Case 1: By Theorem 3.7, when $|N^{[x]}| = |N_i^{[x]}| = |N_j^{[x]}|$, there is at least one time-overlapping in a round ($|N^{[x]}|$ slots) between B_i and B_j, and i and j rendezvous as $b_i = b_j$. By [15], we have $|N_i^{[x]}| \leq 2|R_i^{[x]}| = O(|V_i|)$, that is, the MTTR of this case is $O(|V_i|)$.

- Case 2: $b_i \neq b_j$ implies $a_i \neq a_j$. If M_i time-overlaps M_j entirely, by Lemma 2.5, the MTTR is $O(|M_i^{[x]}| |M_j^{[x]}|) = O(|V_i||V_j|)$. Otherwise, considering the time-overlapping pair (M_i, B_j), by Lemmas 2.3 and 2.5, b_j will meet all different channels in M_i and $b_j \in V_i$, thus i and j will rendezvous in $O(|M_i^{[x]}||N_j^{[x]}|) = O(|V_i||V_j|)$ slots, vise versa for time-overlapping pair (B_i, M_j).

Theorem 3.8. *Given that two radios i and j with common available channels that adopt two CH sequences generated by the ICH scheme with N_i, N_j time-overlapping. The MTTR between i and j is bounded by $O(|V_i||V_j|)$ as long as a) $(V_i \backslash P_i) \cap (V_j \backslash P_j) \neq \emptyset$ and b) $|F_i^{[x]}| = |F_j^{[x]}|$ and $start_i = start_j$.*

Then, as shown in Fig. 6, we verify the MTTR of this ICH scheme by considering three time-overlapping cases: (i) F_i with F_j, R_i with R_j, and N_i with N_j, (ii) F_i with F_j, F_i with R_j, R_i with N_j, and N_i with F_j, (iii) F_i with F_j, F_i with R_j, F_i with N_j, R_i with F_j and N_i with F_j. Notice that the case (iv) in Fig. 6 is covered by case (iii), and case (v) is covered by the (ii). Without loss of generality, each of the above cases can be further classified into three subcases in terms of capabilities of radios i and j: (a) $|V_i| \neq |V_j|$, (b) $|V_i| = |V_j| \wedge start_i \neq start_j$, and (c) $|V_i| = |V_j| \wedge start_i = start_j$. Since $|F_i^{[x]}|$ depends on $|V_i|$, we rewrite the cases as: (a) $|F_i^{[x]}| \neq |F_j^{[x]}|$, (b) $|F_i^{[x]}| = |F_j^{[x]}| \wedge start_i \neq start_j$

(c) $|F_i^{[x]}| = |F_j^{[x]}| \wedge start_i = start_j$. We assume $(V_i \backslash P_i) \cap (V_j \backslash P_j) \neq \emptyset$ and verify the MTTR of the ICH scheme for each of these subcases.

- Case (i-a) & (ii-a) & (iii-a): Since F_i time-overlaps F_j and $|F_i^{[x]}| \neq |F_j^{[x]}|$, by Lemma 2.4, we have $O(|V_i||V_j|)$ MTTR.

- Case (i-b): Since R_i time-overlaps R_j and $|R_i^{[x]}| = |R_j^{[x]}| \wedge start_i \neq start_j$, which implies $k_i \neq k_j$, by Lemma 2.5, we have $O(|V_i||V_j|)$ MTTR.

- Case (i-c): Since N_i time-overlaps N_j and $|F_i^{[x]}| = |F_j^{[x]}| \wedge start_i = start_j$, by Theorem 3.8, we have $O(|V_i||V_j|)$ MTTR.

- Case (ii-b) & (iii-b): Since F_i time-overlaps R_j and $|F_i^{[x]}| = |R_j^{[x]}| \wedge start_i \neq start_j$, which implies $k_i' \neq k_j$ by viewing the rotating amount of F_i as $k_i' = 0$ and $k_j \in [1, |R_j^{[x]}| - 1]$, by Lemma 2.5, we have $O(|V_i||V_j|)$ MTTR.

- Case (ii-c) & case (iii-c): N_i time-overlaps F_j and $|F_i^{[x]}| = |F_j^{[x]}| \wedge start_i = start_j$, which implies $|N_i^{[x]}|$, $|F_j^{[x]}|$ coprime and $s_i \in V_j$ (i.e., $\{s_i\} \cap V_j \neq \emptyset$). Also, the positions of s_i are fixed in $N_i^{[x]}$, by Lemma 2.4, we have $O(|V_i||V_j|)$ MTTR.

In summary of all the above cases, the MTTR of ICH is bounded by $O(|V_i||V_j|)$ for heterogeneous radios ($V_i \neq V_j$) and $O(|V|^2)$ for homogeneous radios ($V_i = V_j = V$). Note that, let $V_i' = V_i \backslash P_i$, the CH sequence, which is constructed on V_i' by setting $start_i$ and $last_i$ be the first and last unoccupied channels respectively and $|V_i'| = last_i - start_i + 1$, still has the same rendezvous guarantee with shorter TTR, due to $|V_i'| \leq |V_i|$.

3.4 Minimizing Congestion

To demonstrate ICH minimizes the congestion, we conduct a series of simulations and compare ICH with HH and JS in terms of the degree of congestion, namely load. First, we show that the load is minimized for mixtures of heterogeneous radios in Fig. 14(b), where ICH has the load as low as the JS's load, which is the optimal load. Also we examine the worst case for load. Fig. 16 (b) exhibits the load induced by a set of homogeneous radios versus the number of radios. ICH has 25% to 40% reduction from HH in load as increasing the number of radios, and we can expect that the reduction will be higher when there are more radios. Stand by the simulation results, ICH is shown to minimize the congestion and have the load close to the optimal load.

4. EXPERIMENTAL RESULTS

In this section, we evaluate the performance of Interlocking Channel Hopping (ICH) scheme. We compare ICH with the state of the arts in homogeneous and heterogeneous environments, respectively Jump-Stay (JS) [14] and the HH scheme [19]. According to the assumption mentioned in Sec. 3.1 that $V_i = V_j = U$, we let JS generates CH sequences based on U. Note that we do not compare our study with other works due to loss of guarantee and their limitations stated in Sec. 5. We investigate the basic behaviors and proper functioning of compared schemes by pairwise radios, and study the performance for mixtures of homogeneous and heterogeneous radios with TTR and load. We also verify the reduction on load compared with the HH scheme in the homogeneous environment.

Since timer synchronization is hard to achieve in practice, we focus on the asynchronous environment. We implement these works based on the Network Simulator 3 (NS3). WiFi MAC is modified to support the channel hopping function, and the IEEE 802.11b is adopted as the MAC layer protocol. Each radio is capable to switch amongst channels in its device capability, and is either receiving or transmitting at a time. Also, radios can detect whether a channel is idle or not. Considering the large range of currently available spectrum which is up to 3 GHz or higher, we assume 5MHz per channel and set the number of universal channels $|U|$ to 600 by default. The period of a time slot is set to 10 ms. The ratio of non-idle channels to the universal channels is set to 0.1, where those non-idle channels are randomly selected from U. We set the default average capability $|V| = 25$, and control the overlapping ratio, which is the proportion of channels that can be operated by more than two radios to the union of capability $\bigcup_i V_i$. We set the overlapping ratio to 0.1. We focus on the affects of capabilities of radios, to avoid unnecessary complexity, we let each radio's transmission range cover other radios. Note that, a possible rendezvous is treated as a failed rendezvous if it spends more than 600 seconds, and it is *not* counted into average TTR.

4.1 Pairwise Radios

We verify the rendezvous guarantee of ICH and observe its fundamental behaviors in this series of simulation. Two radios i and j are used in a simulation run and each data point is averaged from 120 runs. We generate two different range of capabilities that $|V|+x$, $|V|-x$, where x is randomly selected from $[1, |V|/2]$. First, we vary the overlapping ratio from 0.5 to 0.1. As the lower overlapping ratio indicating less common operable channels, intuitively, the TTR may get larger. In Fig. 7(a), the average TTR of JS substantially increases from 10 to 17 seconds, on the contrary, the average TTR of ICH and HH have no obvious growth. We can see the advantage that construct CH sequence by capability instead of universal channels. Fig. 7(b) shows that all these schemes can guarantee rendezvous. Since JS construct CH sequences based on U, it incurs large TTR but does not lose guarantee. Fig. 8 shows that how many ratio of rendezvous can be achieved in 60 seconds. While overlapping ratio is 0.1, JS has most TTRs larger than ICH's. Moreover, JS has more excessively large TTRs, showing that the infeasibility and instability from U affects the average TTR greatly. The HH scheme has much more rendezvous than others in the first five second. One of reasons is that, while overlapping ratio is not high, the insurance sequence containing only one channel has much higher probability to rendezvous. This advantage of the HH scheme becomes not obvious while overlapping ratio is 0.5.

Then, we change non-idle channel ratio from 0.1 to 0.5. The results are shown in Fig. 9. The effect of non-idle channel ratio is similar to overlapping ratio, that the number of common operable channels decreases while non-overlapping ratio increases. As we can see, the trends of average TTR and success rate are similar to the ones in varying overlapping ratio, and we believe that the reasons are the same. Although the one-channel insurance sequence of the HH

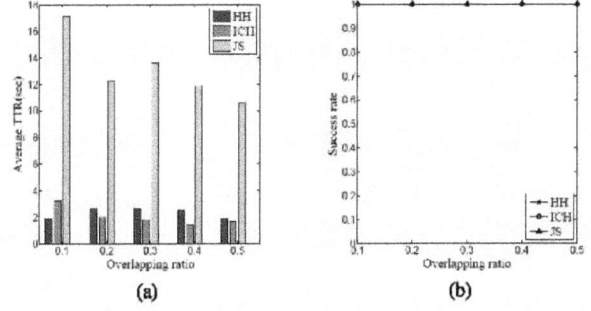

Figure 7: (a) average TTR vs. overlapping ratio (b) success rate vs. overlapping ratio

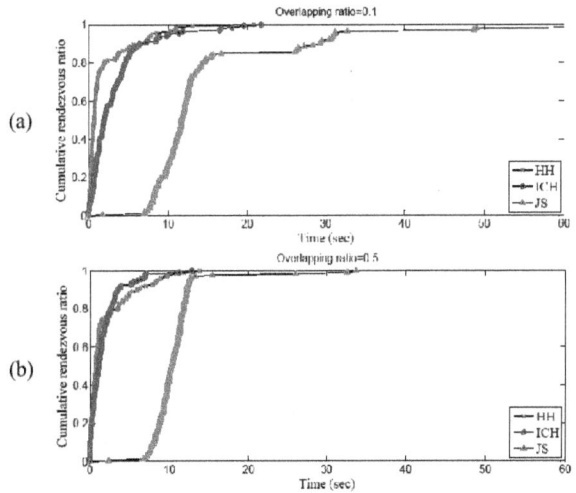

Figure 8: Cumulative rendezvous ratio vs. time while (a) overlapping ratio = 0.1 and (b) 0.5 respectively

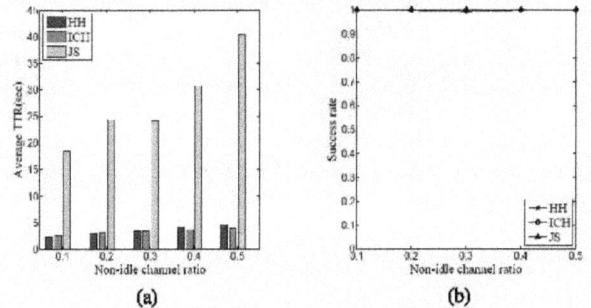

Figure 9: (a) average TTR vs. non-idle channel ratio (b) success rate vs. non-idle channel ratio

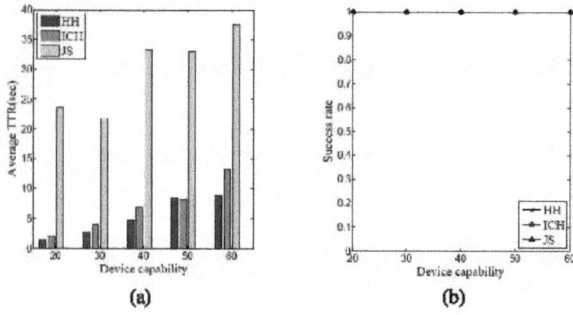

Figure 10: (a) average TTR vs. device capability (b) success rate vs. device capability

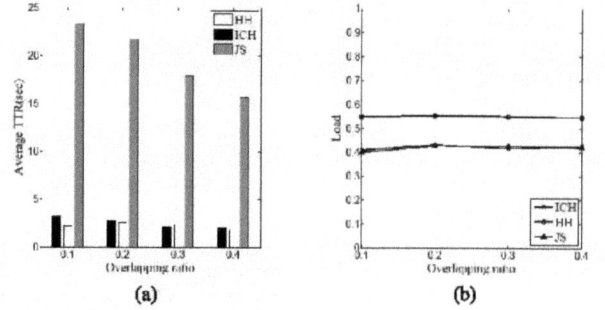

Figure 12: Mixtures of homo & hetero radios: (a) average TTR vs. overlapping ratio (b) load vs. overlapping ratio

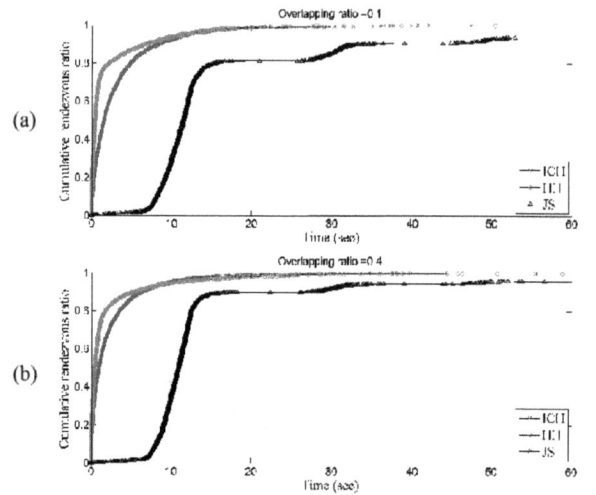

Figure 11: Cumulative rendezvous ratio vs. time while (a) device capability = 20 and (b) 60 respectively

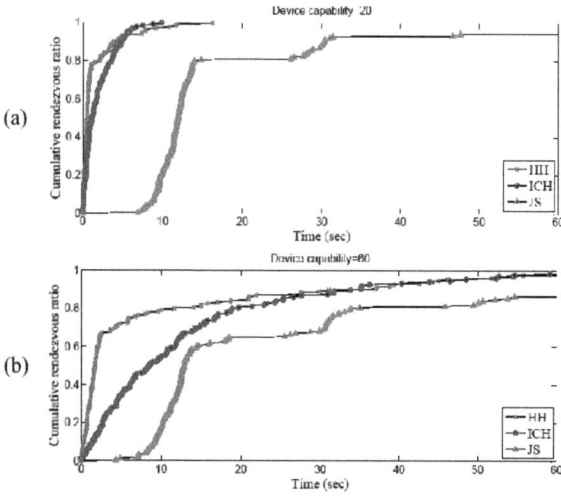

Figure 13: Mixtures of homo & hetero radios: cumulative rendezvous ratio vs. time while (a) overlapping ratio = 0.1 and (b) 0.4 respectively

scheme may speed up the TTR, it takes a risk that one-third of the time will be wasted while the starting channel is non-idle. In the Fig. 9(b), we find that the success rate of the HH is not 1 while non-idle ratio is 0.3. We believe that is because non-idle insurance channel delays TTR to be larger than 000 seconds. Another reason may be most of common operable channels happen to be non-idle in that case.

Next, we change average capability $|V|$ from 20 to 60. We may think that the performance of JS may be better while $|V|$ is closer to $|U|$, where the environment seems more homogeneous. The results are shown in Fig. 10. The TTR of ICH and HH increase as we expected, but TTR of JS also increases due to the heterogeneity of radios. However, ICH and HH still have much lower TTR. Again, the advantage of one-channel insurance sequence of the HH scheme appears in Fig. 11, that reduces much of its average TTR. Overall, the ICH scheme have better performance while $V_i \neq V_j$.

4.2 Mixtures of Homogeneous and Heterogeneous Radios

In this set of simulation, we generate three kinds of heterogeneous radios and five radios for each kind. Similar to the trends in the above pairwise cases, in Fig. 12(a) and

14(a), TTR of ICH and the HH are stable to overlapping ratio and non-idle ratio while TTR of JS changes largely. Shown in Fig. 13 and 15, ICH can complete all possible rendezvous within one minute while JS cannot. ICH even can achieve 90% rendezvous in 10 seconds. In Fig. 12(b) and 14(b), we show that ICH reduces the load by 25% from HH for heterogeneous radios, and the load of ICH is low as JS. Note that, mentioned in Sec. 3.1, JS is load-balanced and has optimal load. On the whole, ICH has low TTR and low load for mixtures of heterogeneous and homogeneous radios.

4.2.1 Worst Case for Load

As mentioned in Sec. 3.1, the load will be worst while the mixtures consisting of homogeneous radios only. We verify the load reduction of ICH compared to the HH in the homogeneous environment. We vary the number of homogeneous radios from 10 to 40, as exhibited in Fig. 16, while the average TTR of ICH is competitive with HH, the ICH largely reduces the load by 35% to 50% from the HH. This is because that the congestion gets severer quickly due to the high load of the HH scheme. The load non-intuitively decreases while number of radios increases, however, the number of radios,

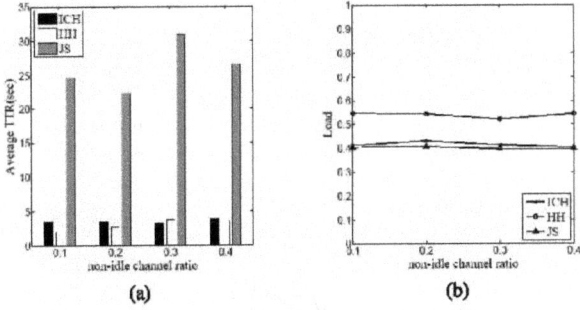

Figure 14: Mixtures of homo & hetero radios: (a) average TTR vs. non-idle channel ratio (b) load vs. non-idle channel ratio

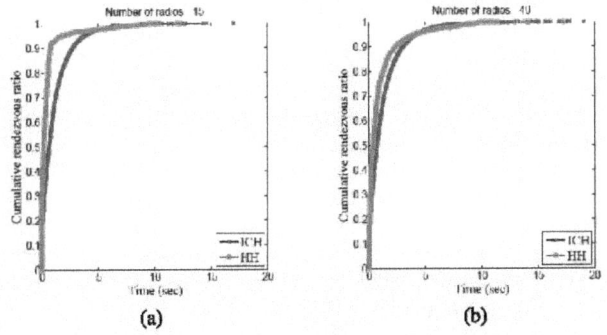

Figure 15: Mixtures of homo & hetero radios: cumulative rendezvous ratio vs. time while (a) non-idle channel ratio = 0.1 and (b) 0.4 respectively

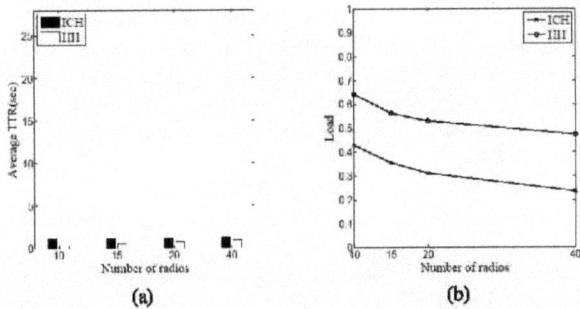

Figure 16: Multiple homogeneous radios: (a) average TTR vs. number of radios (b) load vs. number of radios

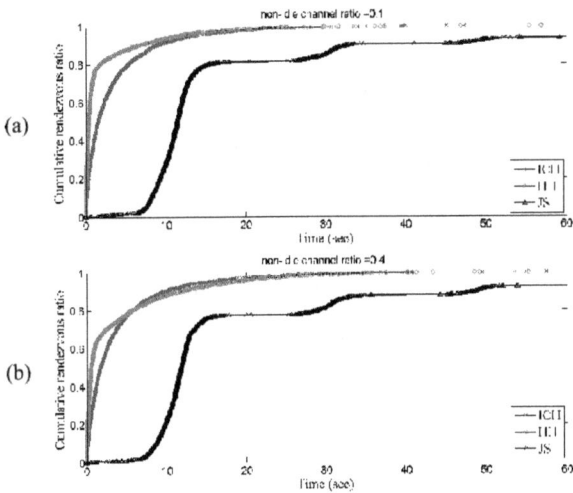

Figure 17: Multiple homogeneous radios: cumulative rendezvous ratio vs. time while number of radios = 15 and 40 respectively

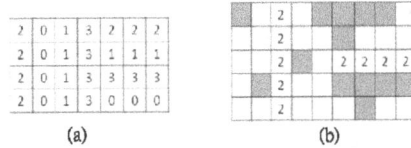

Figure 18: (a) The mapping grid of $V_i = \{0, 1, 2, 3\}$ (b) Let $|V_j| = 5$ and $V_i \cap V_j = \{2\}$, the mapping grid of V_j only shows assignments of channel 2, and marks the corresponding slots V_i hops to channel 2. It cannot guarantee rendezvous in $O(|V_j|^2)$ (or $O(|V_i||V_j|)$) slots.

which hop to the same channel at the same slot, increases. Therefore the congestion becomes more serious when using the HH scheme. Also, the HH has impact of congestion on its immediate slowing down of TTR, where the up-left angle becomes smooth as seen in Fig. 17. This set of simulations show that the ICH is robust to congestion.

5. RELATED WORK

The typical blind rendezvous technique is Channel Hopping (CH). Traditionally, CH schemes for homogeneous radios can be divided into synchronous and asynchronous depend on their environments. In the synchronous environment, the timer is synchronized among all the radios, so that all the radios can start and hop to a channel simultaneously [13, 20, 17, 3]. However, timer synchronization may not be easily achieved in practice. Asynchronous homogeneous CH schemes are proposed. Bian et al. [5, 4] proposed A-MOCH, and Sym-ACH. A-MOCH requires senders and receivers are known in advance which are not likely known before rendezvous. Sym-ACH assumes each node has an unique ID, and the MAC address seems the only choice, causing the $2^{48}O(|V|^2)$ MTTR. DaSilva et al. [10] proposed another CH scheme under the condition that all radios have the same available channels. Zhang et al. [21] proposed the Asyn-ETCH that needs to pre-construct possible schedules with guaranteed rendezvous, but it is not likely to pre-construct overlapping schedules while $V_i \neq V_j$ and V_j is unknown by i. Lin et al. [14] proposed the JS that uses jump and stay patterns, with length of $2|V|$ and $|V|$ respectively, to ensure rendezvous even when timers are not synchronized.

For asynchronous heterogeneous environments, Theis et al. proposed the MC [18] based on the number theory. The MC scheme and its modified version cannot guarantee rendezvous if two radios accidentally make the same decision to construct their CH sequences either on the "rate" parameter or on the augmented capability. Romaszko et al. proposed the MtQS-DSrdv [16] that verifies the rendezvous when $|V_i|$, $|V_j| \leq 8$, but does not provide a theoretical guarantee for all cases under heterogeneous environments. Since the search method for difference set is not detailed in [16], we simplify the MtQS-DSrdv as follows: each radios i i) creates a slot to channel mapping grid of size $|V_i| \cdot (2|V_i| - 1)$, where each row r has slots from $t_i^{[0+r(2|V_i|-1)]}$ to $t_i^{[2|V_i|-2+r(2|V_i|-1)]}$, ii) assigns each channel in V_i to a randomly selected column of slots from the first $|V_i|$ columns, iii) assigns each channel in V_i to a randomly selected rows of the rest empty slots. An example is depicted in the Fig. 18 (a). While $V_i \neq V_j$, the mapping grids need to be pre-constructed to ensure rendezvous, however, it is not sure to find the mapping grids with guarantee. We show a failed example in the Fig. 18 (b).

6. CONCLUSION

In this paper, we study the channel hopping schemes for homogeneous and heterogeneous cognitive radios. We proposed the ICH scheme, an efficient channel hopping scheme which minimizes MTTR and congestion for homogeneous and heterogeneous radios. Extensive simulations showed that our proposal achieves 10 times faster TTR than the state-of-the-art homogeneous scheme and 50% reduction of congestion from the state-of-the-art heterogeneous scheme.

7. REFERENCES

[1] I.F. Akyildiz, W.Y. Lee, M.C. Vuran, and S. Mohanty. Next generation/dynamic spectrum access/cognitive radio wireless networks: a survey. *Computer Networks*, 50(13):2127–2159, 2006.

[2] R. Bagheri, A. Mirzaei, S. Chehrazi, M. Heidari, M. Lee, M. Mikhemar, M. Tang, and A. Abidi. An 800mhz to 5ghz software-defined radio receiver in 90nm cmos. In *IEEE Int'l Solid-State Circuits Conference (ISSCC)*, pages 1932–1941, 2006.

[3] P. Bahl, R. Chandra, and J. Dunagan. Ssch: slotted seeded channel hopping for capacity improvement in ieee 802.11 ad-hoc wireless networks. In *Proc. of ACM Int'l Conf. on Mobile computing and networking (MobiCom)*, pages 216–230, 2004.

[4] K. Bian and J. Park. Maximizing rendezvous diversity in rendezvous protocols for decentralized cognitive radio networks. *IEEE Transactions on Mobile Computing*, 2012.

[5] K. Bian, J.M. Park, and R. Chen. Control channel establishment in cognitive radio networks using channel hopping. *IEEE Journal on Selected Areas in Communications*, 29(4):689–703, 2011.

[6] V. Brik, E. Rozner, S. Banerjee, and P. Bahl. Dsap: a protocol for coordinated spectrum access. In *IEEE Int'l Symp. on New Frontiers in Dynamic Spectrum Access Networks (DySPAN)*, pages 611–614, 2005.

[7] M.M. Buddhikot, P. Kolodzy, S. Miller, K. Ryan, and J. Evans. Dimsumnet: new directions in wireless networking using coordinated dynamic spectrum. In

[8] D. Cabric, S.M. Mishra, and R.W. Brodersen. Implementation issues in spectrum sensing for cognitive radios. In *Proc. of Asilomar Conf. on Signals Systems and Computers*, pages 772–776, 2004.

[9] C. Cormio and K.R. Chowdhury. Common control channel design for cognitive radio wireless ad hoc networks using adaptive frequency hopping. *Ad Hoc Networks*, 8(4):430–438, 2010.

[10] L.A. DaSilva and I. Guerreiro. Sequence-based rendezvous for dynamic spectrum access. In *IEEE Int'l Symp. on New Frontiers in Dynamic Spectrum Access Networks (DySPAN)*, pages 1–7, 2008.

[11] A. Ghasemi and E.S. Sousa. Spectrum sensing in cognitive radio networks: requirements, challenges and design trade-offs. *IEEE Communications Magazine*, 46(4):32–39, 2008.

[12] M. Ingels, V. Giannini, J. Borremans, G. Mandal, B. Debaillie, P. Van Wesemael, T. Sano, T. Yamamoto, D. Hauspie, J. Van Driessche, et al. A 5 mm^2 40 nm lp cmos transceiver for a software-defined radio platform. *IEEE Journal of Solid-State Circuits*, 45(12):2794–2806, 2010.

[13] Y.R. Kondareddy and P. Agrawal. Synchronized mac protocol for multi-hop cognitive radio networks. In *Proc. of IEEE Int'l Conf. on Communications (ICC)*, pages 3198–3202, 2008.

[14] Z. Lin, H. Liu, X. Chu, and Y.W. Leung. Jump-stay based channel-hopping algorithm with guaranteed rendezvous for cognitive radio networks. In *Proc. of IEEE Int'l Conf. on Computer Communications (INFOCOM)*, pages 2444–2452, 2011.

[15] W.S. Luk and T.T. Wong. Two new quorum based algorithms for distributed mutual exclusion. In *Proc. of IEEE Int'l Conf. on Distributed Computing Systems (ICDCS)*, pages 100–106, 1997.

[16] S. Romaszko and P. Mahonen. A rendezvous protocol with the heterogeneous spectrum availability analysis for cognitive radio ad hoc networks. *Journal of Electrical and Computer Engineering*, 2012.

[17] C.F. Shih, T.Y. Wu, and W. Liao. Dh-mac: A dynamic channel hopping mac protocol for cognitive radio networks. In *Proc. of IEEE Int'l Conf. on Communications (ICC)*, pages 1–5, 2010.

[18] N.C. Theis, R.W. Thomas, and L.A. DaSilva. Rendezvous for cognitive radios. *IEEE Transactions on Mobile Computing*, 10(2):216–227, 2011.

[19] S.H. Wu, C.C. Wu, and W.K. Han. Rendezvous for heterogeneous cognitive radios. In *http://www.cs.nthu.edu.tw/ shwu/reports/shwu-tr12-hh.pdf*.

[20] C. Xin, M. Song, L. Ma, and C.C. Shen. An approximately optimal rendezvous scheme for dynamic spectrum access networks. In *Proc. of IEEE Int'l Conf. on Gloal Telecommunications (GLOBECOM)*, pages 1–5, 2011.

[21] Y. Zhang, Q. Li, G. Yu, and B. Wang. Etch: Efficient channel hopping for communication rendezvous in dynamic spectrum access networks. In *Proc. of IEEE Int'l Conf. on Computer Communications (INFOCOM)*, pages 2471–2479, 2011.

[7] IEEE Int'l Symp. on World of Wireless Mobile and Multimedia Networks (WoWMoM), pages 78–85, 2005.

Model-Driven Energy-Aware Rate Adaptation

Muhammad Owais Khan, Vacha Dave, Yi-Chao Chen, Oliver Jensen
Lili Qiu, Apurv Bhartia and Swati Rallapalli
The University of Texas at Austin
{owais,vacha,yichao,ojensen,lili,apurvb,swati}@cs.utexas.edu

ABSTRACT

Rate adaptation in WiFi networks has received significant attention recently. However, most existing work focuses on selecting the rate to maximize throughput. How to select a data rate to minimize energy consumption is an important yet under-explored topic. This problem is becoming increasingly important with the rapidly increasing popularity of MIMO deployment, because MIMO offers diverse rate choices (*e.g.*, the number of antennas, the number of streams, modulation, and FEC coding) and selecting the appropriate rate has significant impact on power consumption.

In this paper, we first use extensive measurement to develop a simple yet accurate energy model for 802.11n wireless cards. Then we use the models to drive the design of an energy-aware rate adaptation scheme. A major benefit of a model-based rate adaptation is that applying a model allows us to eliminate frequent probes in many existing rate adaptation schemes so that it can quickly converge to the appropriate data rate. We demonstrate the effectiveness of our approach using trace-driven simulation and real implementation in a wireless testbed.

Categories and Subject Descriptors

C.2.1 [**Computer-Communication Networks**]: Network Architecture and Design—*Wireless communication*

General Terms

Experimentation, Performance

Keywords

IEEE 802.11, MIMO, Rate Adaptation, Energy.

1. INTRODUCTION

Motivation: Multiple Input Multiple Output (MIMO) is an exciting breakthrough that offers large capacity increase for wireless networks. For example, the current IEEE 802.11n standard [1] supports up to 4 antennas and data rates of up to 600Mbps. The upcoming IEEE 802.11ac standard plans to increase the number of antennas up to 8 to achieve 10Gbps.

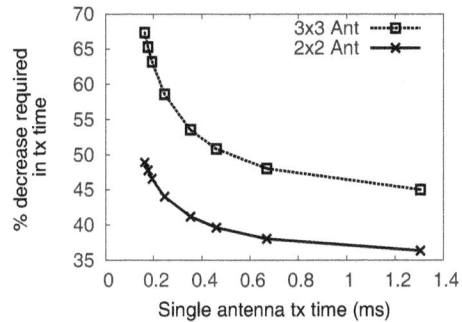

Figure 1: % reduction in transmission time for MIMO needed over SISO for energy improvement.

While MIMO provides a large capacity gain, using multiple antennas can consume significantly more energy, which is undesirable for mobile devices [11]. For a fixed number of antennas, reducing the transmission time always results in a decrease in energy consumption. But for the same transmission time, the energy consumed by multiple antennas is much higher than a single antenna. This is because MIMO transmission requires additional hardware and RF chains for MIMO processing, which increases energy consumption. On the other hand, using multiple antennas reduces transmission time by allowing multiple data streams to transmit simultaneously. Hence, there is a trade-off between minimizing the transmission time using multiple antennas and the additional energy cost associated with using multiple antennas.

Figure 1 compares transmission time of a single antenna with that of using two and three antennas. The plot is based on the transmitter energy model for Intel 5300 WiFi card, which is presented in Section 3. The x-axis shows transmission time of a single antenna transmission. The y-axis shows the percentage of transmission time that two and three antenna MIMO transmissions must reduce in order for them to have the same energy as the single antenna transmission. From the figure, we can see that for a single antenna transmission time of $0.2ms$, using 3 antennas is only beneficial if the transmission time can be reduced by more than 68%. In comparison, for transmission time of $1.3ms$, the number reduces to 50%. So in the best case scenario where the three antenna MIMO transmission uses the same modulation and coding rate as the single antenna transmission but transmits three streams, the transmission time will decrease by 66% and exceed the minimum required 50% reduction in transmission time, therefore leading to energy saving.

The above examples indicate that there is no single setting that minimizes energy in all cases and a single antenna does not always lead to minimum energy. The exact rate and antenna configuration

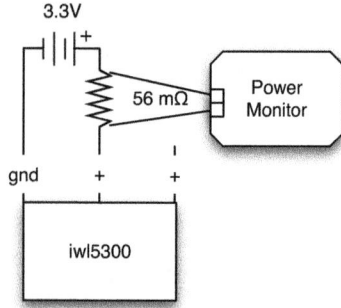

Figure 2: Circuit diagram of measurement setup for Intel card

	Intel	Atheros	Phone
A	$0.24 \times n_{tx} + 0.425 \times MIMO + 1.02$	$0.38 \times n_{tx} + 0.108$	1.53
B	$0.045 \times n_{tx} + 0.108$	$0.040 \times n_{tx} + 0.062$	0.036
C	$0.30 \times n_{rx} + 0.61$	$0.142 \times n_{rx} + 0.30$	1.23
D	$0.064 \times n_{rx} + 0.167$	$0.048 \times n_{rx} + 0.106$	0.002

Table 1: Parameters in the energy models.

that minimize energy depends on a number of factors, such as channel condition, wireless card energy profile, and frame size. These factors explored in detail in section 5. Therefore it is essential to have a comprehensive understanding about how energy consumption relates to these factors and design a rate adaptation scheme that automatically selects the rate to minimize energy according to the current network condition and wireless device.

Our approach: In this paper, we first conduct extensive measurements using different wireless cards to understand the relationship between the data rate and resulting energy consumption. Our main observation is that for a fixed number of antennas, the energy consumed in transmitting or receiving a frame is proportional to the expected transmission time (ETT) [8] (*i.e.*, the total amount of time required to successfully deliver a frame to the receiver), and the slope of the energy consumption versus ETT depends on the number of antennas being used. Based on these insights, we develop a simple yet accurate model to predict the energy consumption when a specified rate is used. We then develop a model-driven rate adaptation scheme on top of the model to select the rate that optimizes energy consumption. In addition, we also design a simple variant that can effectively trade off between energy and throughput. We evaluate our approach using trace-driven simulation and real implementation. Our results show that our approach yields 14-35% energy savings compared with the existing approaches.

Paper outline: The remainder of this paper is organized as follows. We describe our measurement methodology in Section 2. We present our energy model in Section 3, and develop a model-driven rate adaptation in Section 4. We evaluate our approach using trace-driven simulation in Section 5 and using testbed implementation in Section 6. We overview related work in Section 7. We conclude in Section 8.

2. MEASUREMENT METHODOLOGY

To derive power models, we conduct fine-grained power measurements for the following wireless cards: (i) Intel 5300 N series wireless adapter [16], (ii) Atheros 802.11n wireless adapter, and (iii) embedded IEEE 802.11b/a WiFi device on a Windows Mobile smartphone with a single antenna. The first two are commonly used in laptops and can transmit or receive using up to three antennas. The third one is used to verify if the energy model carries over to the embedded WiFi device on a phone. Since multi-antenna WiFi devices for smartphones were not available in the market at the time of our study, we use a single antenna device.

To measure the power consumption of the wireless adapter cards, we use a desktop computer equipped with a PEX1-MINI PCI Express X1 Bus to PCI MINI Bus adapter [24]. It allows us to bypass the PCI bus power supply, and powers the wireless cards using an external source as shown in Figure 2. We supply the power

to the wireless card using a Monsoon power monitor [26], which measures the current using a 56 milli-Ohm resistor. The power monitor samples instantaneous power at the rate of one reading per microsecond and returns a maximum power value for every $200\mu s$ period. We measure energy consumption of the embedded wireless adapter in a mobile phone by bypassing the battery and ground connector and supplying power to the phone as a whole using the same power monitor.

To control the frames involved in transmissions and to avoid unexpected frames, we use UDP packets, set retransmission threshold to zero, and turn off RTS/CTS. We vary data rate and antenna configuration by modifying device drivers of the Intel and Atheros cards. To force the phone into a particular data rate, we use HostAP daemon [15] as our access point and let it advertise only the required data rate in beacons.

3. MEASUREMENT-BASED MODEL

We collect and analyze power measurements from a variety of transmission and reception configurations. We vary the frame size from 250 to 1500 bytes. For Intel iwl5300 card, we collect power measurements for all high throughput (HT) 11n data rates using one, two, and three antennas supported by the card. The same process is repeated for the Atheros card and the phone. Figures 3 and 4 plot the energy consumption versus the expected transmission time (ETT) [8], which is defined as the expected time required to successfully transmit the frame from the source to the destination. ETT can be computed as

$$ETT = \frac{s}{r} \frac{1}{1-p},$$

where p denotes the frame loss rate, r denotes the data rate, and s denotes the frame size. As we can see, in all the figures, the energy consumption is proportional to the expected transmission time (ETT) [8]. The slope of the line depends on the number of transmitting and receiving antennas being used. This holds for all three cards we use.

Based on these observations, we develop simple energy models by performing least-square fitting to find the coefficients that best match the energy consumption of the different cards. The energy models are as follow:

$$E_{tx} = A \times ETT + B \qquad (1)$$
$$E_{rx} = C \times ETT + D \qquad (2)$$

where the parameters in the models A, B, C, D vary across different wireless cards and are shown in Table 1.

We make several observations. First, the energy consumption is a linear function of ETT, as mentioned earlier. The slope depends on the number of transmitting or receiving antennas. This is intuitive since using more antennas consumes more energy and the amount of extra energy that is consumed relates to how long the antennas are used. The y-intercept of the linear function reflects a constant processing cost for each frame regardless of their duration. Second, the exact parameters across different cards are similar but not identical. For example, the Intel transmitter requires an additional parameter $MIMO$, which indicates whether MIMO mode is enabled. This is a well documented anomaly of the Intel card,

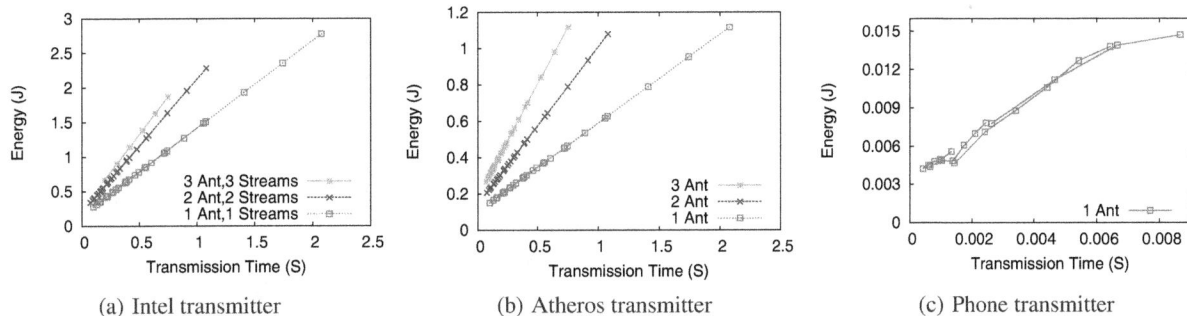

(a) Intel transmitter   (b) Atheros transmitter   (c) Phone transmitter

Figure 3: Measured energy consumption under different transmission configurations as a function of ETT.

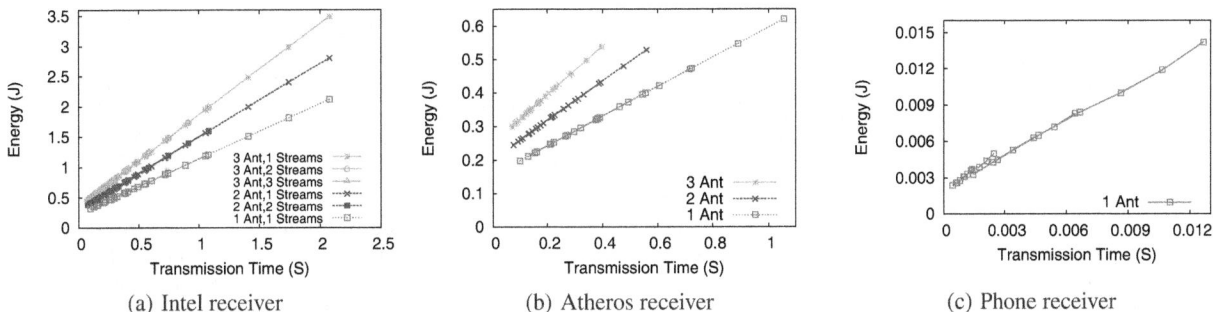

(a) Intel receiver   (b) Atheros receiver   (c) Phone receiver

Figure 4: Measured energy consumption under different reception configurations as a function of ETT.

where two antennas turn on almost all the hardware required for three antennas, with only 5% energy difference between two and three antennae configurations. This is also reported in [11]. The model for the phone is similar in spirit to the other cards. But since we do not have a smartphone with an embedded MIMO enabled Wi-Fi card, we cannot separate which parts in A and B are from n_{tx} and n_{rx}. The values for the phone are higher than those of the other two cards under 1 antenna because the measured energy from the phone includes everything, such as display, CPU, as well as wireless cards. Third, the energy consumption depends on the number of antennas, but not the number of streams. For example, as shown in Figure 4(a), the energy consumptions under 3 antennas using 1, 2, and 3 streams are identical and overlap; similarly for 2 antennas using 1 and 2 streams. Finally, we note that our receiver energy model is conservative (*i.e.*, it may sometimes over-estimate the energy consumption). This is because depending on where the reception fails (*e.g.*, if preamble detection fails, the receiver will stop further processing the signals and the energy consumption is likely to be lower than that of a successful reception). We conservatively assume every transmissions (regardless failures or success) consumes the same amount of receiving energy. Since preambles are quite reliable compared to data symbols, which may be sent at a higher data rate, the approximation error is likely to be small.

Table 2 shows mean absolute percentage error (MAPE) of our energy models versus the measurement data, defined as

$$MAPE = mean(|\frac{x - x'}{x}|),$$

where x and x' are the actual and estimated energy consumption, respectively. As we can see, the error is consistently below 5%, indicating a close match.

4. ENERGY-AWARE RATE ADAPTATION

In this section, we develop an energy aware rate adaptation protocol based on the energy models. Our goal is to select the data rate

Card	transmission	reception
Atheros	3.4%	1.3%
Intel	0.65%	1.4%
Phone	4.9%	3.6%

Table 2: Mean absolute percentage error of energy models.

for the next transmission in order to minimize the energy consumption. In IEEE 802.11n, the data rate is defined as Modulation and Coding Scheme (MCS), which specifies the modulation, FEC coding, and antenna configuration. To achieve this goal, the protocol first obtains Channel State Information (CSI) seen by the receiver, then computes the delivery ratio and energy consumption under different MCS, and selects the MCS that yields the lowest estimated energy. Below we describe each step in detail.

Channel State Information (CSI): IEEE 802.11n standard specifies how to calculate and report CSI. The CSI values are a collection of $M \times N$ matrices H_s, each of which specifies amplitude and phase between pairs of N transmit and M receive antennas on subcarrier s. SNR and amplitude A have the following relationship: $SNR = 10log_{10}(A^2/N)$, where N denotes the average power of white noise. For example, Intel Wi-Fi Link 5300 (iwl5300) IEEE a/b/g/n wireless network adapters collects the CSI of each frame preamble across all subcarriers for up to three antennas.

Using the CSI values, we calculate the post-processed SNR (pp-SNR) values for each subcarrier under every supported transmission configuration. The post-processed SNR is the SNR value obtained after MIMO decoding. In MIMO, since a transmitted symbol is received on multiple antennas, the final SNR experienced by the symbol is the combination of the multiple receptions and the combined SNR dictates whether it will be decoded correctly. For spatial multiplexing modes, we use a Minimum Mean Squared Error (MMSE) equalizer to calculate the post-processed SNR. The SNR value for the mth stream on subcarrier s after MMSE equal-

modulation	BER
BPSK	$Q(\sqrt{2snr})$
QPSK	$Q(\sqrt{snr})$
QAM-16	$\frac{3}{4}Q(\sqrt{snr/5})$
QAM-64	$\frac{7}{12}Q(\sqrt{snr/21})$

Table 3: BER for different modulations as a function of SNR

ization can be written as:

$$SNR_m^{MMSE} = \frac{E_s}{N_t N_0} \frac{1}{\left[H^H H + \left(\frac{E_s}{N_t N_0} \right)^{-1} I \right]_{m,m}^{-1}} \quad (3)$$

where E_s is the total transmission energy across all transmit antennas, N_t is the number of transmit antennas, N_0 is the noise power, H is the channel matrix for subcarrier s (H_{ij} is the channel coefficient of the j-th transmitting antenna to i-th receiving antenna), I is an identity matrix, and H^H is the Hermitian transpose of H matrix. The pp-SNR expression in equation 3 is applicable for all cases, including when the number of spatial streams is equal to the number of transmit antennas ($N_{ss} = N_t$) and when the number of transmit antennas is less than or equal to the number of receive antennas ($N_t \leq N_r$). Hence, equation 3 is used for all receive diversity cases since $N_t < N_r$ is for receive diversity.

The calculation of pp-SNR for transmit diversity modes depends on the mechanism used to achieve diversity. The two supported mechanisms in IEEE 802.11n are Space Time Block Coding (STBC) and Cyclic Delay Diversity (CDD). For STBC, which provides full diversity, the pp-SNR can be calculated as:

$$SNR^{STBC} = \frac{E_s}{N_t N_0} \sum_{i=1}^{N_r} \sum_{j=1}^{N_t} |h_{ij}|^2 \quad (4)$$

where h_{ij} is the channel coefficient of the j-th transmitting antenna to i receiving antenna, N_t and N_r are the numbers of transmit and receive antennas, respectively.

For CDD modes, the SNR can be estimated by [5]:

$$SNR_s^{CDD} = \frac{E_s}{N_t N_0} \sum_{i=1}^{N_r} \left| \sum_{k=1}^{N_t} h_{ik} e^{-j\frac{2\pi s}{N_{fft}}\delta_{cy(k)}} \right|^2 \quad (5)$$

where $\delta_{cy(k)}$ is the delay defined by the IEEE 802.11n standard for cyclic delay transmission for transmit antenna k. N_{fft} is the FFT size, and s is the subcarrier index. It should be noted that Equation 5 depends on the subcarrier index.

Computing loss rate: To compute the loss rate, we first map the pp-SNR of each subcarrier to the uncoded BER using the well-known relationship between SNR and BER as shown in Table 3. Then to take into account the frequency diversity (*i.e.*, SNR varies across different subcarriers), as [12] suggests, we compute average BER across all the subcarriers. Next we derive the BER after FEC coding using the error-probability upper bound defined for the Viterbi decoder to map the uncoded BER to coded BER. The Viterbi decoder's probability of bit error is upper bounded as follows according to [27]:

$$BER_{coded}(\rho) = \sum_{d=d_{free}}^{\infty} a_d . P_d(\rho) \quad (6)$$

$$P_d(\rho) = \begin{cases} \sum_{k=(d+1)/2}^{d} \binom{d}{k} . \rho^k . (1-\rho)^{d-k}, \text{if d is odd} \\ \frac{1}{2} . \binom{d}{d/2} . \rho^{d/2} . (1-\rho)^{d/2} + \\ \sum_{k=(d+1)/2}^{d} \binom{d}{k} . \rho^k . (1-\rho)^{d-k}, \text{if d is even} \end{cases} \quad (7)$$

where ρ is the uncoded BER, d_{free} is the minimal hamming distance between two coded sequences, and a_d is the number of incorrect paths of hamming distance d that diverge from the correct path and then re-merge sometime later [10]. The coded BER value can then be used to approximate the frame error rate (FER) as $1 - (1 - BER_{coded})^L$ assuming independent bit error rate, where L is the frame size.

To further enhance performance, Partial Packet Recovery (PPR) [17] is proposed to let a receiver extract correct bits from a partially corrupted frame. When PPR is used, our goal is to maximize the expected number of delivered bits, which can be computed as $(1 - HeaderLoss)(1 - BER_{uncoded}) \times L'$, where $HeaderLoss$ is the loss rate of the frame header, L' is the payload size, and $BER_{uncoded}$ is uncoded BER. $BER_{uncoded}$ is used since the FEC is no longer useful for a corrupted frame.

Estimating energy consumption: To accurately estimate the energy consumption, an AP or a back-end server should keep a table of the energy models for commonly used Wi-Fi cards. Whenever a new client arrives, it checks the make and model of the wireless card based on either explicit feedback or passive detection of 802.11 wireless drivers [9] or fingerprinting techniques [22] using 802.11 protocol fields. For example, "more fragments", "retry", or "power management" bits in the protocol field reveals the wireless card information. Then it computes ETT based on frame loss rate and applies the corresponding energy model to derive the energy consumption for the next transmission under different MCS. When a client's wireless card has unknown energy profile, it is possible to infer the energy model based on data transmissions. For example, the AP can let the client report the energy consumption at a few data rates under different numbers of antennas to estimate the slope in the energy model. The model is then inserted to the table and can be updated as more measurements become available. As part of our future work, we plan to investigate how quickly we can infer the energy model using such online measurement.

MCS and Antenna Selection: Based on the frame error rate calculated for all MCS, we identify the MCS that have a reasonable delivery rate (*e.g.*, 90% or above). Among these MCS, we select the MCS that yields the minimum energy. Note that we can easily incorporate different objectives in this process, such as minimizing energy or minimizing energy subject to throughput constraint (*e.g.*, throughput is within $X\%$ from the optimal throughput, where X is a configurable knob), or other combinations of throughput and energy. In our evaluation, we also consider several variants that jointly optimize energy and throughput.

5. TRACE DRIVEN SIMULATION

We first evaluate various rate adaptation schemes using trace-driven simulation. We quantify the performance of different schemes in terms of their energy consumption and throughput.

5.1 Simulation Methodology

We develop a simulator in python using the CSI traces. For each frame, the data rate is selected according to different rate adaption schemes. Then we determine if the frame is successfully received using pp-SNR and taking into account FEC. The simulator also supports Partial Packet Recovery (PPR), which uses uncoded BER

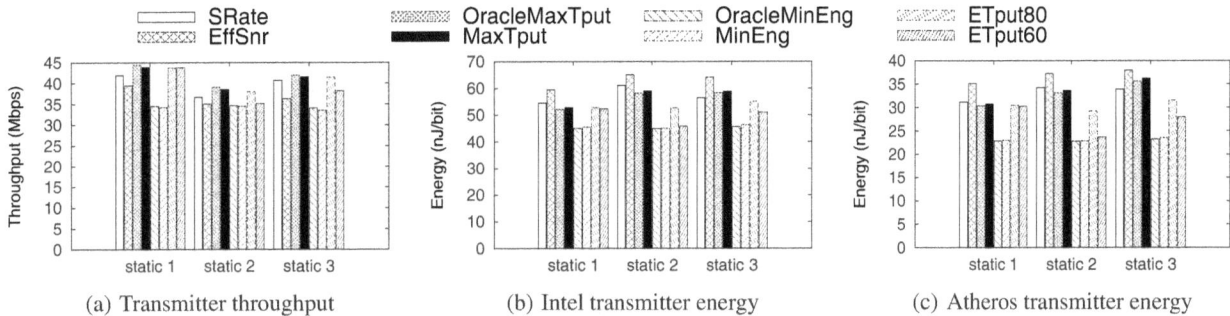

(a) Transmitter throughput   (b) Intel transmitter energy   (c) Atheros transmitter energy

Figure 5: Transmitter Energy comparison in static networks.

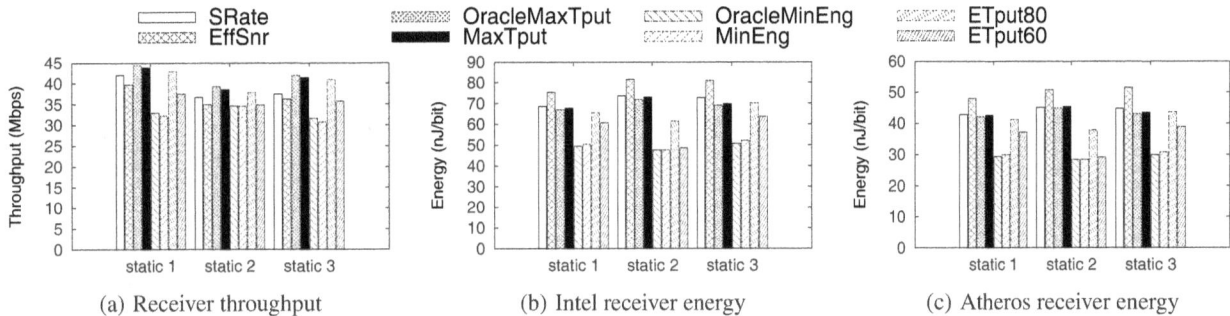

(a) Receiver throughput   (b) Intel receiver energy   (c) Atheros receiver energy

Figure 6: Receiver Energy comparison in static networks.

to determine the number of bits correctly received. We compare the following rate adaptation schemes:

- **Sample Rate (SRate):** Sample Rate [3] is a widely used rate adaptation scheme. It probes the network at a random rate every 10 frames and selects the rate that minimizes transmission time including retransmission time. Its goal is to maximize throughput without considering energy consumption. We implement an extended version of Sample Rate which supports MIMO transmission modes. The original Sample Rate starts at the highest rate and reduces the rate based on channel conditions. We extend this idea and start at the highest rate using all antennas and then reduce or increase the MCS or the number of antennas based on throughput of the previous transmissions.

- **Effective SNR (EffSNR):** [12] proposes selecting the data rate based on effective SNR derived from the CSI values. It computes the post-processed SNR for each subcarrier and maps it to BER. Then it calculates the average BER across all subcarriers and converts the average BER to effective SNR with the same BER. Effective SNR also aims to maximize throughput and does not consider energy consumption.

- **Maximum Throughput (MaxTput):** Maximum Throughput rate adaptation uses the rate selection scheme in Section 4. Unlike energy minimization scheme, it picks the MCS that maximizes throughput.

- **Minimum Energy (MinEng):** Minimum Energy is our proposed rate adaptation scheme from Section 4. It picks the MCS that minimizes the energy consumption while ensuring the frame delivery rate is above 90%.

- **Minimum Energy with Throughput Constraint (ETput$X$):** This scheme aims to select the MCS that minimizes the energy provided the throughput is no less than X% of the maximum throughput. We vary X to yield different variants. For example,

ETput80 means minimizing energy while ensuring throughput is at least 80% of the maximum throughput.

The energy consumption is derived using the energy models for Intel and Atheros as described in Section 3. We collect three channel traces from static environments, and another three traces from mobile environments with human walking speed. The three mobile traces are collected in an office environment using 1 moving receiver and 3 static senders. The three static senders are 7m away from each other. Each trace corresponds to one of the three senders transmitting while the receiver is moved at a walking speed.

We use Intel Wi-Fi Link 5300 (iwl5300) IEEE a/b/g/n wireless network adapters to collect the CSI of each frame preamble across all subcarriers. These NICs have three antennas. We enable all three antennas at both the sender and receiver. The modified driver [13] reports the channel matrices for 30 subcarrier groups, which is about one group for every two subcarriers in a 20 MHz channel according to the standard [1] (i.e., 4 groups have one subcarrier each, and the other 26 groups have two subcarriers each). We use 1000-byte packets and MCS-16, with a transmission power of 15 dBm. MCS-16 has 3 streams, so the NICs report CSI in the form of 3×3 matrices for each frame.

5.2 Simulation Results

Static networks: First, we evaluate the performance in static networks using three traces collected in a static environment. Each trace contains 2000 CSI samples. Figure 5 plots the throughput and energy consumption for the transmitter. As we can see, compared to the scheme that maximizes throughput, the energy-aware rate adaptation scheme consumes 14-24% less energy for the Intel card and 25-35% less energy for the Atheros card. The throughput loss for both cards is 10-22%. Compared with Effective SNR and Sample Rate, minimum energy reduces transmitter energy by 17-31% for the Intel card and 26-39% for the Atheros card while the throughput loss is 1-19%. The energy saving is higher and throughput reduction is lower in the latter cases because Effec-

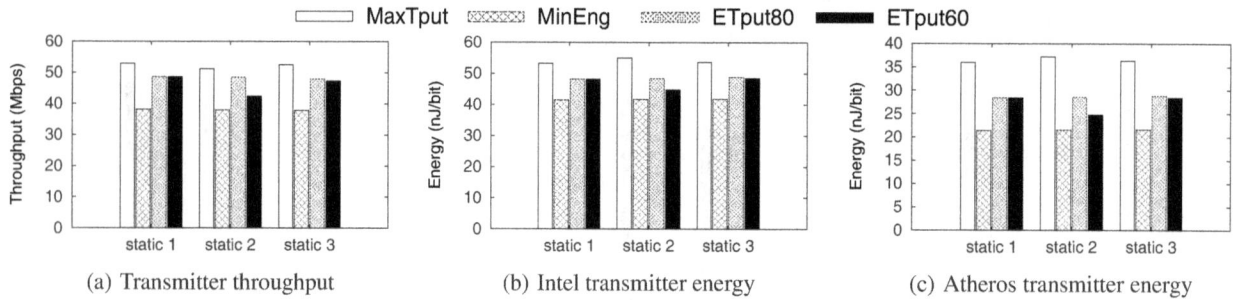

Figure 7: Transmitter Energy comparison in static networks using PPR.

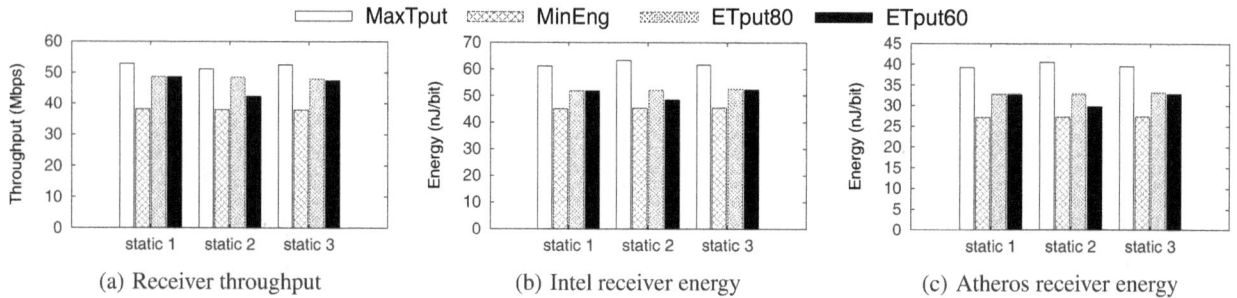

Figure 8: Receiver Energy comparison in static networks using PPR.

tive SNR or Sample Rate are not optimal for either throughput or energy. ETput$X$ balances the throughput and energy. For example, compared with the maximum throughput scheme, ETput80, which minimizes energy while ensuring at least 80% of the maximum throughput, saves energy of up to 10% and 13% for the Intel and Atheros transmitters, respectively, while reducing throughput within 1%. Moreover, OracleMinEng and OracleMaxTput know the exact CSI of the next frame and eliminate the performance degradation caused by prediction error. As we can see, the CSI prediction error causes only 1-2% more energy consumption and 1-2% throughput reduction, indicating the impact of prediction error is small.

Figure 6 shows the performance results for the receiver. Compared with the scheme that maximizes throughput, the energy-aware rate adaptation scheme reduces the receiver's energy by 25-35% for the Intel card and 30-37% for the Atheros card at the cost of 10-26% throughput reduction. Compared with Effective SNR and Sample Rate, minimum energy reduces receiver energy by 26-42% for the Intel card and 30-44% for the Atheros card while the throughput loss is 1-23%. As before, ETput$X$ balances energy and throughput: ETput80 reduces energy by 10% and 13% for the Intel and Atheros receivers, respectively, with almost no throughput loss. In addition, compared with OracleMinEng and OracleMaxTput, MinEng incurs only 1-4% more energy and 1-5% throughput loss.

Figure 11 shows the number of antennas used by each scheme. We can see that the energy-aware rate adaptation tends to use one antenna to minimize energy consumption. Meanwhile, it also uses two antennas in some cases whenever the reduced transmission time can offset the additional energy required by an extra antenna. The maximum throughput scheme, on the other hand, does not care about the energy consumption and uses as many antennas as possible to achieve better throughput. ETput$X$ schemes try to balance MinEng and MaxTput schemes and the number of antennas they use is between those used by the two schemes.

We also ran simulations using a Partial Packet Recovery(PPR). As shown in Figure 7, in this case the energy-aware rate adaptation reduces the transmission energy by 22-24% for the Intel card and

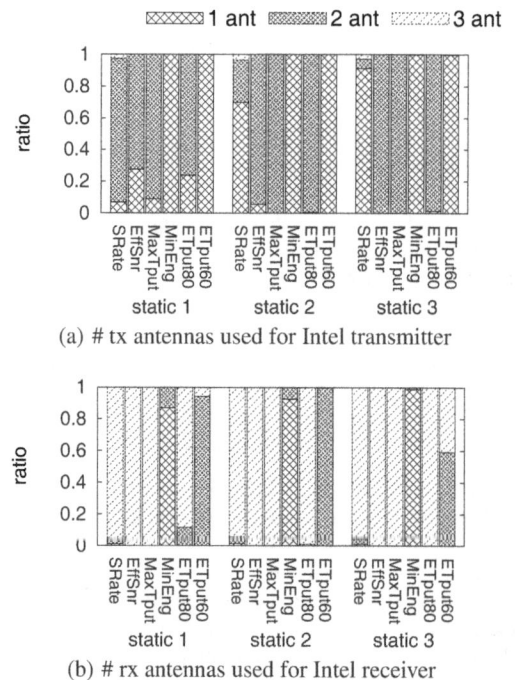

(a) # tx antennas used for Intel transmitter

(b) # rx antennas used for Intel receiver

Figure 11: Number of antenna used in static networks.

by 40-42% for the Atheros card. These energy savings are achieved at the cost of 26-28% throughput reduction for both cards.

As shown in Figure 8, the energy savings for the PPR receiver are 26-28% and 31-33% for the Intel and Atheros cards, respectively. The throughput loss for these cards is 26-28%. To trade off between throughput and energy savings, ETput80 saves energy by 9% and 21% for Intel and Atheros, respectively. The throughput reduction is within 9%. Moreover, comparing PPR energy saving with non PPR energy savings, we see PPR based scheme improves

(a) Transmitter throughput   (b) Intel transmitter energy   (c) Atheros transmitter energy

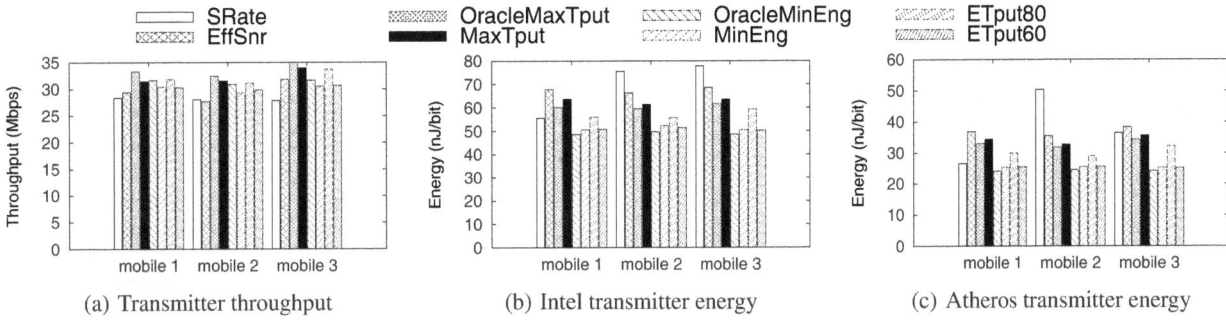

Figure 9: Transmitter Energy comparison in mobile networks.

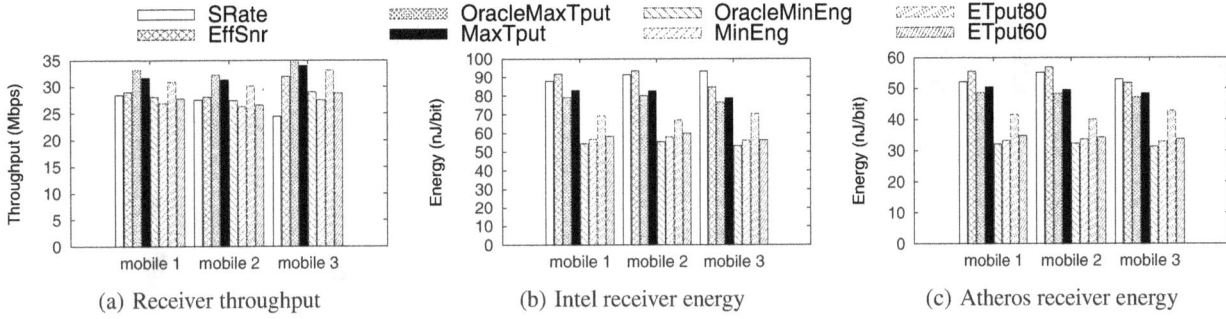

(a) Receiver throughput   (b) Intel receiver energy   (c) Atheros receiver energy

Figure 10: Receiver Energy comparison in mobile networks.

the energy by 6-23% by extracting correct symbols from partially corrupted frames.

Mobile networks: Next we evaluate the different schemes using the three mobile traces. Figure 9 and 10 summarize the results. Compared with the scheme that maximizes throughput, minimum energy reduces transmitter energy by 15-21% for the Intel card and 22-29% for the Atheros card. For both Intel and Atheros, the throughput loss is 3-10%. Compared with Effective SNR and Sample Rate scheme, minimum energy reduces transmitter energy by 9-35% for the Intel card and 5-49% for the Atheros card. The throughput of minimum energy is higher than Effective SNR and Sample Rate in some mobile traces since the latter two are not optimal for throughput.

For the receiver, minimum energy reduces energy by 29-31% for the Intel card and 32-34% for the Atheros card while reducing the throughput by 15-19% compared to maximum throughput scheme. Compared with Effective SNR and Sample Rate scheme, minimum energy reduces receiver energy by 34-40% for the Intel card and 36-41% for the Atheros card. To trade off between throughput and energy savings, ETput80 scheme reduces the throughput by 2% compared to maximum throughput scheme while providing energy savings of 16% and 18% for Intel and Atheros receivers, respectively. Compared with OracleMinEng and OracleMaxTput, the CSI prediction error causes only 2-6% more energy consumption and 3-6% throughput reduction. The degradation in mobile traces is slightly larger than that in static traces as expected since the channel variation in mobile traces increases the CSI prediction error. Nevertheless, the degradation in this case is still small. As in the static networks, the energy-aware rate adaptation uses one antenna in most cases, and uses more antennas to reduce transmission time if possible. The maximum throughput scheme uses as many antennas as the channel condition allows.

Figure 12 and 13 further show the performance of various PPR versions of rate adaptation schemes. In this case, the minimum energy scheme reduces Intel transmitter energy by 26-28% and

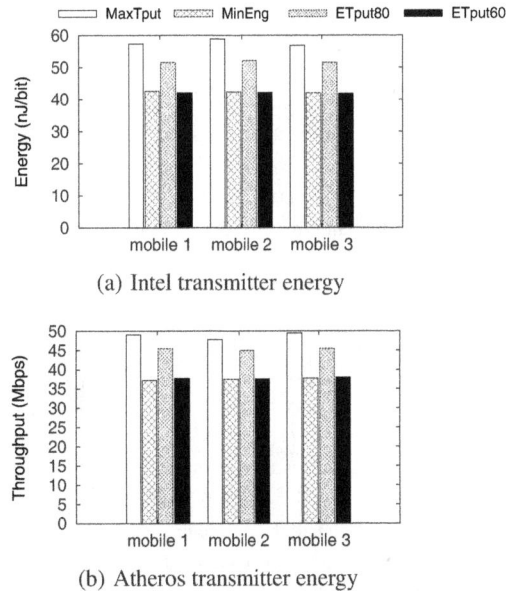

(a) Intel transmitter energy

(b) Atheros transmitter energy

Figure 12: Transmitter Energy comparison in mobile networks using PPR.

Atheros energy by 43-45%. The throughput loss is 22-24%. For receiver, the energy savings for Intel are 30-32% and for Atheros are 34-36%. The throughput loss is 22-24%. Compared with non-PPR counterparts, the PPR versions lead to 13-20% energy savings. To balance the throughput and energy savings, ETput80 scheme reduces the throughput by 8% while providing energy savings of 10% and 22% for Intel and Atheros transmitters, respectively.

Impact of frame sizes: In order to take full advantage of the high data rates offered by IEEE 802.11n, using large frames is strongly recommended. Therefore, we further evaluate the impact of frame sizes. Figure 14 shows the number of antennas selected by MinEng

223

(a) Intel receiver energy

(b) Atheros receiver energy

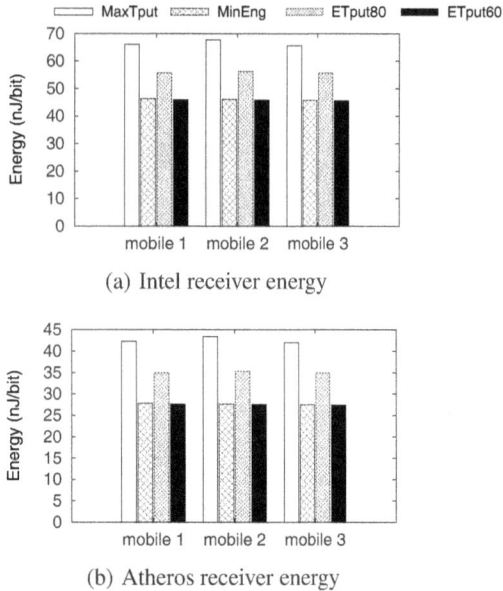

Figure 13: Receiver Energy comparison in mobile networks using PPR.

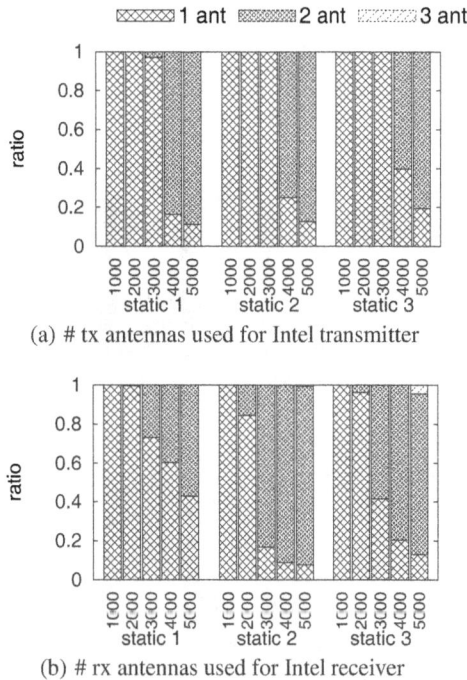

(a) # tx antennas used for Intel transmitter

(b) # rx antennas used for Intel receiver

Figure 14: Number of antennas used for different frame sizes.

for the Intel card as we vary the frame sizes from 1000 bytes to 5000 bytes. As we can see, MinEng always selects the one antenna rate for 1000-byte frames in our traces. However, as the frame size increases, we see more transmissions use multiple antennas. For 5000-byte frames almost all transmissions use two antennas. This indicates as frame size increases, it becomes more advantageous to use multiple antenna rates to minimize energy.

Multiple antennas provide energy saving for larger frames because for small frames the preamble transmission time dominates the total transmission time. Hence, using multiple antennas only results in small reduction in ETT, which does not offset the additional energy required to power up multiple antennas. As the frame size increases, using multiple antennas leads to larger reduction in ETT,

which more than offsets the additional energy required to power up more antennas.

Other energy objectives: Our scheme is general and can easily support other energy objectives. To give another example, here we consider minimizing the total energy consumption from both sender and receiver, which is especially interesting in ad hoc networks where the sender and receiver are both mobile nodes with limited energy. Figure 15 shows the performance of MaxTput and MinEng scheme with different objectives in static traces. The performance of mobile traces is similar and omitted for brevity. As it shows, MinEng leads to 19-30% total energy saving with 10-26% throughput reduction. ETput80 balances the total energy consumption and throughput, and reduces energy by 1-13% at a 1-2% throughput loss. ETput60 reduces the total energy by 2-28% with a 5-9% throughput reduction.

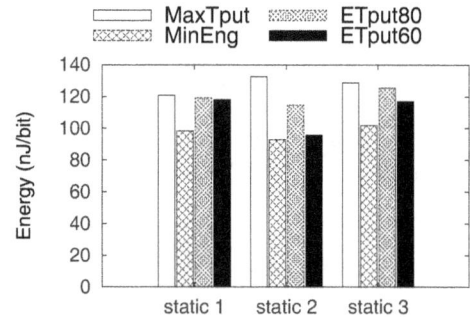

Figure 15: Comparing total energy consumption in static networks.

6. TESTBED EVALUATION

Testbed implementation: We implement different rate adaptation schemes in the Intel Wi-Fi link 5300 driver. We use the tool in [12] to extract CSI from the Intel card at the receiver. The receiver uses the extracted CSI information to calculate the throughput and energy consumption for each MCS. The receiver then uses these calculated values to select the appropriate MCS and informs the transmitter to use the selected MCS.

We conduct testbed experiments using two desktop machines. For each experiment, we send 200 UDP packets with 1000-byte payload. The experiments are conducted in static and mobile scenarios. For mobile experiments, initially the machines are placed close to one another and then the receiver is moved away from the transmitter at a walking speed. For each configuration, we report the average throughput and energy consumption across 10 runs for static experiments and across 5 runs for mobile experiments.

Testbed results: Figure 16 shows the throughput and energy consumption for static experiments. As we can see, MinEng reduces the energy consumption by 19% for the transmitter and by 28% for the receiver. The throughput reduction is 24% for the transmitter and 22% for the receiver. ETputX smoothly trades-off between the two objectives. For example, ETput80 reduces energy by 6% at a 11% throughput loss for the transmitter. For the receiver, ETput80 reduces the energy by 16% with a throughput reduction of 2%. Figure 17 shows the number of transmit and receive antennas used during the experiment. Due to the static channel, the schemes use the same MCS for most transmissions which is expected. MinEng uses a single antenna at both the transmitter and receiver to reduce energy. In comparison, MaxTput utilizes two and three antennas to achieve higher throughput at the cost of additional energy.

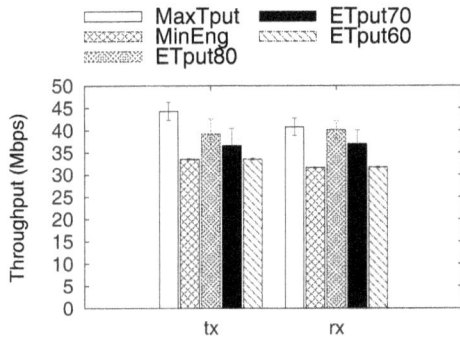

(a) Throughput of the static trace in testbed

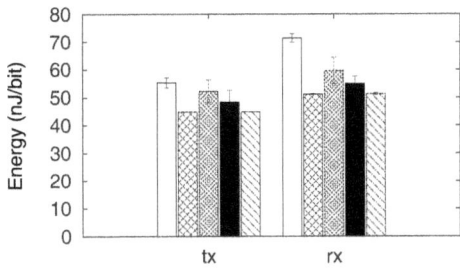

(b) Energy of the static trace in testbed

Figure 16: Comparison of performance of the static trace in the testbed.

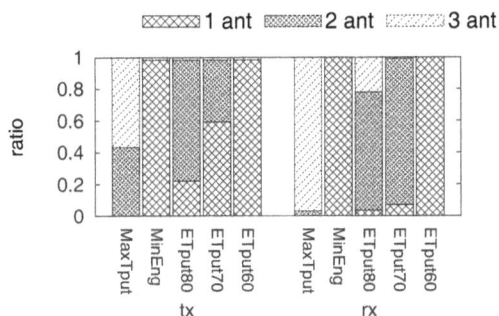

Figure 17: Number of antenna used in the static testbed.

Figure 18 shows how MCS changes over an mobile experiment for MaxTput and MinEng. MCS 0 to 7 use 1 antenna, MCS 8 to 15 use 2 antennas, and MCS 16 to 23 use 3 antennas. In each case, the number of spatial streams is equal to the number of antennas. In region 1, when the channel is good, MaxTput transmits using all 3 antennas at MCS 22. Since MinEng tries to minimize energy, it uses MCS 6, the highest 1-antenna rate that can be supported by the current channel. MinEng saves 16.9% energy over MaxTput in this region. As the receiver moves away from the transmitter, the channel condition degrades and forces MaxTput to drop to MCS 14, while MinEng continues to use MCS 6. The energy improvement reduces to 11.9% because MCS 14 used by MaxTput consumes less energy than its previous MCS 22 due to a fewer number of antennas used. In region 3, MaxTput drops from MCS 14 to MCS 12. Since MCS 12 still uses 2 antennas but takes longer to transmit than MCS 14, MCS 12 consumes 15.5% more energy than MCS 14. In comparison, MinEng continues to use MCS 6 and its energy saving jumps to 21%. In region 4, the MinEng drops to MCS 5, resulting in longer transmission time. Since MaxTput still uses MCS 12, the energy saving of MinEng reduces to 20.06%. It is interesting to note that even though the channel degrades continuously, the energy savings do not follow the trend. In fact, region 2 has the least

gap between MaxTput and MinEng while region 3 has the highest. In all cases, MinEng yields significant energy savings.

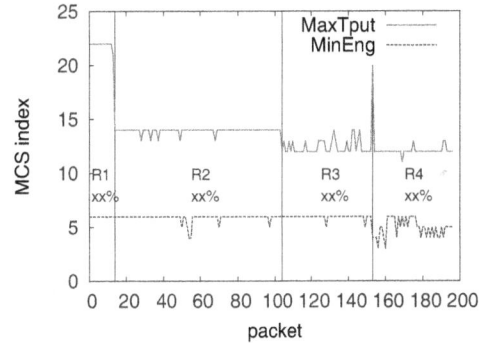

Figure 18: The evolution of MCS over time for MinEng and MaxTput in a mobile experiment.

7. RELATED WORK

We classify related work into the following areas: (i) energy measurement and models, (ii) power saving, (iii) rate adaptation.

Energy measurement and models: Carvalho *et al.* [4] present a simple model for power consumption in 802.11 ad-hoc networks as a function of the number of bytes and a constant radio overhead for all antenna configurations. They also augment it to account for channel contention costs. Balasubramanian *et al.* [2] present an empirical study of energy consumption on mobile phones for 3G, GSM, and WiFi energy consumption, and formulate an energy model for WiFi based on the transfer energy cost (per transfer size) and the maintenance cost of WiFi. Neither model considers the effects of multiple antennas, data rate, and transmit power. Sesame [7] is a system in which a mobile device creates its own energy model by using the battery interface with high accuracy. The scheme does not specifically model the energy consumption of the WiFi Adapter. Halperin *et al.* [11] study power consumption of the iwl5300 under different transmit power levels, card mode (*e.g.*, sleep, idle, transmit, receive), the number of active antennas and spatial streams, channel width and data rate. While their empirical observations are insightful, they do not develop an energy model.

Power saving: Motivated by the power-hungry nature of network interfaces, several works try to minimize time in idle listening mode. Rozner *et al.* [28] use virtualization techniques and energy-aware scheduling algorithm to reduce background traffic and allow 802.11 cards to enter Power Saving Mode (PSM) to save energy by 70%. Jang *et al.* [18] propose an energy management technique for 802.11n by configuring a client's sleep duration and antenna configuration. *Sleepwell* [21] is a system that achieves energy efficiency by evading network contention among multiple APs in the vicinity of a mobile client. E-mili [31] is a scheme that reduces power consumed in idle listening by down-clocking radio. Catnap [6] allows a device to sleep by combining small gaps between packets into meaningful intervals, while [23] detects mobile phone bugs that prevent the phone from sleeping. DozyAp [14] allows power-efficient WiFi Tethering. All these works are complimentary to our work, which focuses on optimizing MIMO transmissions to save energy.

Rate Adaptation: Many rate adaptation algorithms have been proposed for SISO systems, including commonly used SampleRate [3] and RRAA [30]. [12] shows effective SNR is a good metric for rate adaptation to maximize throughput. More recently, the success of IEEE 802.11n has motivated researchers to develop rate adaptation for IEEE 802.11n. Since IEEE 802.11n offers a wide range of rate configurations, rate adaptation becomes more challenging. [25] proposes an interesting ZigZag search to find the rate to optimize throughput. Turborate [29] is another MIMO rate adaptation algorithm. All the above works, however, focus on maximizing throughput and do not consider energy consumption. [19] is one of the few that considers energy in rate adaptation. It formulates the MIMO-OFDM minimum energy link adaptation problem as a geometric programming (GP) problem with an augmented parameter set under the control of the link adaptation protocol, but they do not empirically measure or derive energy models for wireless adapters. [20] also studies rate adaptation to reduce energy consumption. But unlike our work, which optimizes power based on the energy model, [20] uses probes to search for the rate that reduces energy. In general, it takes a longer time for a probing-based scheme to converge to a desirable rate than a model-based approach, which directly computes the rate that minimizes the energy. Moreover, the data rate used by the probes may not be appropriate (*e.g.*, it may incur losses or consume higher energy), which limits its effectiveness.

8. CONCLUSION

In this paper, we collect and analyze power measurement from different wireless cards and derive simple energy models for transmission and reception. Based on the models, we develop a model-driven energy-aware rate adaptation scheme. Our simulation and experiments show our approach reduces energy by 14-37% over the existing approaches. The PPR version is even more effective: it leads to 22-45% energy reduction over the PPR extension of the existing rate adaptation schemes and 6-23% energy reduction over the non-PPR version of MinEng. As part of our future work, we plan to explore energy minimization under more extensive scenarios, such as under multiple clients and more diverse traffic patterns.

Acknowledgements: This work is supported in part by NSF Grants CNS-0916106 and CNS-1017549.

9. REFERENCES

[1] LAN/MAN Standards Commmittee of the IEEE Computer Society. Part 11: Wireless LAN Medium Access Control and Physical Layer (PHY) Specifications. *IEEE Standard 802.11*. http://standards.ieee.org/getieee802/download/802.11n-2009.pdf.

[2] N. Balasubramanian, A. Balasubramanian, and A. Venkataramani. Energy consumption in mobile phones: A measurement study and implications for network applications. In *Proc. of IMC*, Nov. 2009.

[3] J. Bicket. Bit-rate selection in wireless networks. In *MIT Master's Thesis*, 2005.

[4] M. M. Carvalho, C. B. Margi, K. Obraczka, and J. J. Garcia-Luna-Aceves. Modeling energy consumption in single-hop IEEE 802.11 ad hoc networks. In *Proc. of ICCCN*, Oct. 2004.

[5] A. Dammann and R. Raulefs. Comparison of space-time block coding and cyclic delay diversity for a broadband mobile radio air interface. In *Interface, 6th Int. Symp. Wireless Personal Multimedia Communications (WPMC 2003)*, pages 411–415, 2003.

[6] F. R. Dogar, P. Steenkiste, and K. Papagiannaki. Catnap: exploiting high bandwidth wireless interfaces to save energy for mobile devices. In *Proc. of ACM MobiSys*, 2010.

[7] M. Dong and L. Zhong. Self-constructive high-rate system energy modeling for battery-powered mobile systems. In *Proc. of ACM MobiSys*, 2011.

[8] R. Draves, J. Padhye, and B. Zill. Routing in multi-radio, multi-hop wireless mesh networks. In *Proc. of ACM MobiCom*, Sept. - Oct. 2004.

[9] J. Franklin, D. McCoy, P. Tabriz, V. Neagoe, J. Van Randwyk, and D. Sicker. Passive data link layer 802.11 wireless device driver fingerprinting. In *Proc. of USENIX Security*, 2006.

[10] D. Haccoun and G. Begin. High-rate punctured convolutional codes for Viterbi and sequential decoding. *IEEE Transactions on Communications*, 37(11), 1989.

[11] D. Halperin, B. Greensteiny, A. Shethy, and D. Wetherall. Demystifying 802.11n power consumption. In *Proc. of HOTPOWER*, 2010.

[12] D. Halperin, W. Hu, A. Sheth, and D. Wetherall. Predictable 802.11 packet delivery from wireless channel measurements. In *Proc. of ACM SIGCOMM*, 2010.

[13] D. Halperin, W. Hu, A. Sheth, and D. Wetherall. Tool release: Gathering 802.11n traces with channel state information. *ACM SIGCOMM CCR*, 41(1):53, Jan. 2011.

[14] H. Han, Y. Liu, G. Shen, Y. Zhang, and Q. Li. Dozyap: power-efficient wi-fi tethering. In *Proc. of ACM MobiSys*, 2012.

[15] hostapd: IEEE 802.11 AP, IEEE 802.1X/WPA/WPA2/EAPRADIUS authenticator. http://hostap.epitest.fi/hostapd/.

[16] Intel Ultimate N WiFi Link 5300 and Intel WiFi Link 5100 products. http://www.intel.com/products/wireless/adapters/5000/index.htm.

[17] K. Jamieson and H. Balakrishnan. PPR: Partial packet recovery for wireless networks. In *Proc. of ACM SIGCOMM*, 2007.

[18] K.-Y. Jang, S. Hao, A. Sheth, and R. Govindan. Snooze: Energy Management in 802.11n WLANs. In *Proc. of ACM CoNEXT*, 2011.

[19] H. S. Kim and B. Daneshrad. Energy-constrained link adaptation for MIMO OFDM wireless communication systems. *IEEE Transactions on Wireless Communications*, 2010.

[20] C.-Y. Li, C. Peng, S. Lu, and X. Wang. Energy-based rate adaptation for 802.11n. In *Proc. of ACM MobiCom*, 2012.

[21] J. Manweiler and R. Roy Choudhury. Avoiding the rush hours: WiFi energy management via traffic isolation. In *Proc. of ACM MobiSys*, 2011.

[22] J. Pang, B. Greenstein, R. Gummadi, S. Seshan, and D. Wetherall. 802.11 user fingerprinting. In *Proc. of MobiCom*, 2006.

[23] A. Pathak, A. Jindal, Y. C. Hu, and S. P. Midkiff. What is keeping my phone awake?: characterizing and detecting no-sleep energy bugs in smartphone apps. In *Proc. of ACM MobiSys*, 2012.

[24] PCI EXPRESS X1 to PCI Express Mini interface adapter. http://www.adexelec.com/pciexp.htm.

[25] I. Pefkianakis, Y. Hu, S. H. Wong, H. Yang, and S. Lu. MIMO rate adaptation in 802.11n wireless networks. In *Proc. of ACM MobiCom*, 2010.

[26] Monsoon solutions power monitor. http://www.msoon.com/LabEquipment/PowerMonitor.

[27] D. Qiao, S. Choi, and K. Shin. Goodput analysis and link adaptation for IEEE 802.11a wireless LANs. *IEEE TMC*, Oct. 2002.

[28] E. Rozner, V. Navda, R. Ramjee, and S. Rayanchu. NAPman: Network-assisted power management for WiFi devices. In *Proc. of ACM MobiSys*, June 2010.

[29] W.-L. Shen, Y.-C. Tung, K.-C. Lee, K. C.-J. Lin, S. Gollakota, D. Katabi, and M.-S. Chen. Rate adaptation for 802.11 multiuser MIMO networks. In *Proc. of ACM MobiCom*, 2012.

[30] S. H. Y. Wong, H. Yang, S. Lu, and V. Bharghavan. Robust rate adaptation for 802.11 wireless networks. In *Proc. of ACM MobiCom*, 2006.

[31] X. Zhang and K. G. Shin. E-mili: energy-minimizing idle listening in wireless networks. In *Proc. of ACM MobiCom*, 2011.

Quantize-Map-Forward (QMF) Relaying: An Experimental Study

Melissa Duarte
École Polytechinque Fédérale
de Lausanne (EPFL)
CH-1015 Lausanne,
Switzerland
melissa.duartegelvez@epfl.ch

Ayan Sengupta
École Polytechinque Fédérale
de Lausanne (EPFL)
CH-1015 Lausanne,
Switzerland
ayan.sengupta@epfl.ch

Siddhartha Brahma
École Polytechinque Fédérale
de Lausanne (EPFL)
CH-1015 Lausanne,
Switzerland
siddhartha.brahma@epfl.ch

Christina Fragouli
École Polytechinque Fédérale
de Lausanne (EPFL)
CH-1015 Lausanne,
Switzerland
christina.fragouli@epfl.ch

Suhas Diggavi
University of California Los
Angeles (UCLA)
Los Angeles, CA 90095
USA
suhas@ee.ucla.edu

ABSTRACT

We present the design and experimental evaluation of a wireless system that exploits relaying in the context of WiFi. We opt for WiFi given its popularity and wide spread use for a number of applications, such as smart homes. Our testbed consists of three nodes, a source, a relay and a destination, that operate using the physical layer procedures of IEEE802.11. We deploy three main competing strategies that have been proposed for relaying, Decode-and-Forward (DF), Amplify-and-Forward (AF) and Quantize-Map-Forward (QMF). QMF is the most recently introduced of the three, and although it was shown in theory to approximately achieve the capacity of arbitrary wireless networks, its performance in practice had not been evaluated. We present in this work experimental results—to the best of our knowledge, the first ones—that compare QMF, AF and DF in a realistic indoor setting. We find that QMF is a competitive scheme to the other two, offering in some cases up to 12% throughput benefits and up to 60% improvement in frame error-rates over the next best scheme.

Categories and Subject Descriptors

C.2.1 [**Computer-Communication Networks**]: Network Architecture and Design—*wireless communication*; E.4 [**Coding and Information Theory**]: Error control codes.

Keywords

Cooperative Communication; Relaying; 802.11; Quantize-Map-Forward; Software Defined Radio; WARP.

1. INTRODUCTION

In theory, it is well established that relaying can extend the range of a wireless network and increase its bandwidth efficiency. In practice, although relaying is increasingly becoming part of standards, it is not clear which of the many proposed relaying strategies performs best, and when. We present in this paper an experimental study of state-of-the-art relaying strategies, and compare their performance for several indoor 802.11 (WiFi) topologies.

WiFi is what most homes already have: if we can enhance its performance, we pave the road for a number of interesting applications. Ease of installation, low-cost, interoperability and ubiquity have made WiFi the most popular WLAN choice with homeowners. Smart homes seem a tangible possibility thanks to this wide penetration of WiFi, as many appliance manufacturers are looking into WiFi-enabled machines that communicate with each other. Green home solutions as well rely on WiFi connectivity for controlling energy efficiency. High rate wireless data, home support for impaired people, entertainment, the list of potential applications is long. Yet, the reality is that the indoor WiFi links are not always reliable and cannot support high enough rates, both because of physical obstruction (*e.g.*, an elevator) and interference (*e.g.*, a cordless phone)—leading to a disappointing performance for a number of interesting applications. WiFi is thus a good example where relaying can have high impact.

We focus in this paper on the most basic topology: a source sends data to a destination with the help of a relay, as depicted in Fig. 1. In our home, the source could be a WiFi router that streams video to a tablet with the help of a WiFi enabled appliance that boosts the network performance by acting as a relay. We select this setup for several reasons: the simpler the experiment the easier to interpret results; it can serve as a building block towards more complex configurations; this is commonly accepted as the topology that would be used in practice, at least in the immediate future, thus making it more meaningful for performance comparisons; and finally, surprisingly, even for this small topology, few experimental results have been published.

In information theory, relaying has been a topic of research for the last forty years, and many approaches have been proposed and analyzed. Perhaps the two most studied are Amplify and Forward (AF) and Decode and Forward (DF). DF in particular has been shown to be within one bit from the capacity of the single relay network we are focusing on. Recently, a third strategy, Quantize-Map-Forward (QMF), was also shown to achieve the capacity of the single relay network within one bit, and moreover, to achieve within a constant gap the capacity of *arbitrary* wireless networks.

QMF has in theory a number of attractive attributes. It dictates a simple relay operation: the relay captures the signal, quantizes it, performs an appropriate mapping and forwards it to the destination, which is much simpler than decoding for instance. It only requires receive channel information at each relay. It is scalable, in the sense that adding or removing relays does not require to change what the operating relays do. Most importantly, it is proved that it approximately achieves the capacity of *arbitrary* wireless networks.

However, it is not clear how many of these advantages translate into practical systems. The information-theoretic analysis assumes infinite length coding and no complexity constraints on the transceivers; even then there is a possible gap to information-theoretic capacity; and of course the analysis assumes perfect estimates of the noise and the channel coefficients for decoding. It is well possible that for networks at moderate SNR these advantages disappear – this is especially so, if we want to operate with backwards compatible coding schemes at the encoder, such as those that 802.11 supports. Moreover, for small networks, such as the one we examine, DF is also (approximately) optimal and may perform better in practice. As far as we know, there has been no experimental evaluation of QMF relaying (before our work).

We present in this paper the first system design, deployment and experimental comparison of QMF, AF and DF relaying. We compare these results to a baseline direct transmission (DT) which does not employ a relay. Our system design emulates the 802.11 physical layer procedures, such as the frame structure, use of Orthogonal Frequency Division Multiplexing (OFDM) and use of standardized LDPC encoders. We implement all schemes and deploy them simultaneously on the same network to offer a fair comparison. We find that relaying schemes significantly outperform the baseline DT. We also find that QMF is a competitive scheme to DF and AF, offering in some cases up to 12% throughput benefits and up to 60% improvement in frame error-rates over the next best scheme. However, we also see that in a single-relay network, there are scenarios where DF does better than QMF.

Our main contributions are:

- We design and deploy the first QMF relaying system, as well as *coded* AF and DF relaying systems.

- We report a number of experiments that compare the three schemes, and evaluate the utility of relaying in the context of WiFi.

The rest of the paper is organized as follows. Section 2 presents our physical layer signaling as well as the network operation; Section 3 describes the relaying schemes and our designs for WiFi; Section 4 details the implementation of

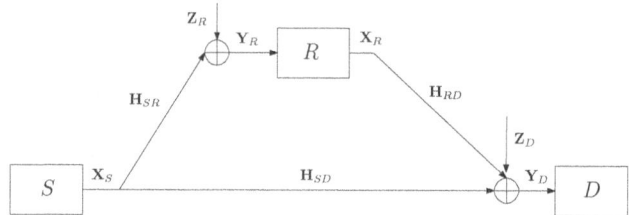

Figure 1: The 3-node network considered in this paper: the source S communicates with the destination D with the help of the relay R.

our system; and finally Section 5 provides and discusses our experimental results.

2. BACKGROUND AND MODEL

We consider a network that consists of three wireless nodes: a source S communicates with a destination D with the help of a relay R (see Fig. 1). The relay R is half-duplex in nature, i.e., it cannot simultaneously listen and transmit over the same frequencies. Our focus is on physical layer communication schemes: we are interested in comparing the performance of QMF with DF and AF, and more generally, evaluate how much physical layer cooperation through relaying can improve the performance as compared to routing. Our performance metrics are frame error rate and throughput, and are described in Section 5.

At a high level, the communication over our network depends on three choices: (i) the physical layer signaling; we emulate the physical layer operation prescribed by 802.11. We highlight the parts we need to describe our relaying schemes in Section 2.1 and give specific details in Section 4. (ii) the network operation; that is, whether we use the relay and in what network topology. For example, we may use the relay only when direct transmissions fail, or we may use it to form a two-hop path that connects the source to the destination. We describe our choices in Section 2.2. (iii) the relaying scheme; that is, if we do use the relay, which relaying strategy we deploy? We discuss several strategies and how we optimize their system implementation in Section 3.

2.1 Signal Model

The communication happens in frames. We work with coded systems, i.e., to improve the error resiliency, the source encodes the information bits using the 802.11 standard LDPC codes (we give more details on the codes we implement in Section 5). After encoding, the codeword bits are first mapped to constellation symbols (we use QAM constellations supported in WiFi), and then modulated using OFDM. We employ OFDM modulation as specified in the WiFi physical layer to combat channel frequency selectivity. We denote by $\mathbf{X} = [X_1, X_2, \ldots, X_m]^T$ an input vector to an OFDM modulator and by $\mathbf{Y} = [Y_1, Y_2, \ldots, Y_m]^T$ the output vector of the corresponding OFDM demodulator, where the subscript denotes the inputs and outputs at subcarrier $l \in \{1, \ldots, m\}$, with m being the total number of subcarriers (we use $m = 64$). The inputs X_l are complex numbers corresponding to the signal constellation used. With OFDM, the point-to-point signal model [24] for each subcarrier is:

$$Y_l = H_l X_l + Z_l, \tag{1}$$

where Z_l is the additive Gaussian noise and H_l is the channel for subcarrier $l \in \{1, \ldots, m\}$.

In our network, both the source and the relay may transmit at the same time. From the superposition property of wireless channels, if the two transmitters S, R each emit an OFDM symbol $\mathbf{X}_i = [X_{i1}, X_{i2}, \ldots, X_{im}]^T$, where $i \in \{S, R\}$, then the demodulated OFDM symbol at node D, denoted by $\mathbf{Y}_D = [Y_{D1}, Y_{D2}, \ldots, Y_{Dm}]^T$, is given by [24]

$$Y_{Dl} = H_{SD,l}X_{Sl} + H_{RD,l}X_{Rl} + Z_{Dl}, \qquad (2)$$

where $H_{SD,l}$ is the channel from S to destination D and $H_{RD,l}$ is the channel from R to destination D.

2.2 Network Operation

Even in a small network with a single relay, there are several possibilities for the *network operation, i.e.,* which links of the network we use and when we use them. We explore the following four choices in our experimentation.

1. Direct Transmission (DT): In the DT mode, we do not use the relay, i.e., the source S and destination D communicate directly using only the S–D direct link. The failure recovery is simply through retransmission of the same frame. We deploy the DT mode in our experiments to assess how useful the relay and physical layer cooperation are in our scenario.

2. Direct Transmission at Half Rate (DTHR): This scheme also uses the S–D direct link only, but incorporates a failure recovery mechanism through rate adaptation as advocated in the WiFi standard [1]. More specifically, we first attempt a direct transmission using a 16 QAM constellation. If this fails, we retransmit the same frame at half the rate. This is implemented by using a constellation of half the rate for the second transmission, i.e., 4 QAM[1]. As a result, the time duration of the second transmission doubles. Note that we need to account for the change in the rate when comparing the throughput performance with the other schemes. DTHR offers a recovery mechanism well suited to a direct link that is in general strong, but may infrequently experience deep fades.

3. Link Switching (LS): In LS we employ all three links: in a first attempt we employ broadcasting from the source, i.e., the S–D and the S–R links; in a second attempt we employ only the R–D link. First, the source broadcasts a frame that is intended for the destination. The relay R also overhears the source transmission. If the destination successfully decodes the frame through the direct link, we declare success and proceed with the next frame. However, if the first attempt fails and the destination cannot decode, then in a second attempt, the relay retransmits the frame it has overheard from the source, while the source remains silent. The operation for LS is summarized in Table 1, and the corresponding signal model is:

$$\begin{aligned} \mathbf{Y}_R^{(1)} &= \mathbf{H}_{SR}^{(1)}\mathbf{X}_S^{(1)} + \mathbf{Z}_R^{(1)} \\ \mathbf{Y}_D^{(1)} &= \mathbf{H}_{SD}^{(1)}\mathbf{X}_S^{(1)} + \mathbf{Z}_D^{(1)} \\ \mathbf{Y}_D^{(2)} &= \mathbf{H}_{RD}^{(2)}\mathbf{X}_R^{(2)} + \mathbf{Z}_D^{(2)}, \end{aligned} \qquad (3)$$

where the superscript (k), ($k \in \{1, 2\}$), denotes the attempt, and $\mathbf{H}_{ij}^{(k)} = \mathrm{diag}(H_{ij,1}^{(k)}, \ldots, H_{ij,m}^{(k)})$ denotes the (sub-carrier)

[1]Clearly, adaptation to different rates is possible; we use this to get an indicative performance benchmark and because it ties well with 802.11 rate adaptation mechanisms.

channels from node i to node j at attempt k. Note that since the transmissions from the source and the relay are orthogonalized in time, LS does not require synchronized transmissions.

Attempt	Source	Relay	Destination
T1	Transmit	Receive	Receive
T2	Silent	Transmit	Receive

Table 1: Schedule for Link Switching

LS can be interpreted as trying to utilize two paths: first the direct S–D link and then, if this fails, the S–R–D path. When the relay uses DF (Decode-Forward is described in Section 3) the LS mode of failure recovery is the single-relay analog of *routing* over larger networks.

4. Link Cooperation (LC): In LC as well, we employ all three links. Similar to LS, in a first attempt we employ broadcasting from the source, i.e., the S–D and the S–R links; however, if we fail, we then simultaneously employ both the S–D and R–D links. More specifically, if the destination fails to decode the first direct transmission of the source, in the second attempt both the source and relay cooperatively transmit. The schedule for LC is summarized in Table 2, and the corresponding signal model is:

$$\begin{aligned} \mathbf{Y}_R^{(1)} &= \mathbf{H}_{SR}^{(1)}\mathbf{X}_S^{(1)} + \mathbf{Z}_R^{(1)} \\ \mathbf{Y}_D^{(1)} &= \mathbf{H}_{SD}^{(1)}\mathbf{X}_S^{(1)} + \mathbf{Z}_D^{(1)} \\ \mathbf{Y}_D^{(2)} &= \mathbf{H}_{RD}^{(2)}\mathbf{X}_R^{(2)} + \mathbf{H}_{SD}^{(2)}\mathbf{X}_S^{(2)} + \mathbf{Z}_D^{(2)}, \end{aligned} \qquad (4)$$

where the notation is similar to (3).

Attempt	Source	Relay	Destination
T1	Transmit	Receive	Receive
T2	Transmit	Transmit	Receive

Table 2: Schedule for Link Cooperation

LC is the mode of operation that is the most promising: it puts in use all the network resources–both broadcasting and transmit cooperation to forward the information and thus, if we use a "good" relaying scheme, it has the potential to offer the best performance no matter which of the network links are stronger. The question is whether one of the relaying schemes (that we describe in the next Section) makes this possible.

Discussion. The focus of our work is in evaluating the potential of PHY-layer cooperation; we leave open what a well-matched MAC-layer protocol would be. However, we do not think that heavy MAC-layer redesign would be required; there exist schemes in the literature that already implement the simple functionalities we would need. For instance, the modes that use a relay (LS and LC) operate *on-demand*, i.e., the relay is invoked only when the first attempt to communicate with the destination has failed. MAC-layer protocols for activating on-demand relaying in 802.11-like networks have been discussed in [26, 11, 9].

3. RELAYING SCHEMES

In this section we describe the three relaying schemes we will compare: Decode-Forward (DF), Amplify-Forward (AF) and Quantize-Map-Forward (QMF). We first give the high level operation and then describe our system design choices. The relaying schemes are only relevant for the LS and LC network operations (since in DT and DTHR we do not employ the relay at all).

3.1 Relay Operation

Decode Forward (DF): Once the relay receives a frame, it attempts to decode it and retrieve the information bits. If decoding is successful (i.e., the CRC passes *after* running the LDPC decoder), the relay can re-encode the decoded information, remodulate and forward the frame to the destination. If correct decoding is possible, it is an optimal relay operation, as it removes the S–R link noise. If decoding at the relay fails however, the relay remains silent and cannot cooperate for that frame.

Amplify Forward (AF): The relay does not attempt to decode; rather it simply amplifies its received signal to the maximum transmit power of its radio (by multiplying with an appropriate amplification factor A) and retransmits it. This is advantageous in cases where the relay cannot decode, but the destination can do so with the help of both the source and relay transmissions.

Quantize-Map-Forward (QMF): The relay quantizes the received symbols, collects a sequence of quantized values, and operates on the entire sequence to produce a transmit sequence. This is distinct from DF since the relay does not decode and is distinct from AF since it operates on a *sequence* of quantized symbols.

3.2 System Design

3.2.1 Cooperative Transmissions

In the LC mode, effectively we have the source and relay cooperating like a *distributed* 2×1 transmit antenna system to the destination. For point-to-point links with 2 transmit and 1 receive antennas, the WiFi standard recommends an Alamouti space-time code [3, 1], as it gives the best rate-reliability tradeoff for the 2×1 MISO channel, asymptotically in SNR (*i.e.*, it is diversity-multiplexing-tradeoff optimal). Accordingly, to optimize the performance of AF and DF, we implement a *distributed* Alamouti code in the LC mode. QMF cannot take advantage of Alamouti coding, because of the random mapping at the relay. We describe in the following, the transmissions for all schemes.

Decode Forward (DF): In the LC mode, if the relay successfully decodes the source frame, then we can exactly create a 2×1 multiple transmit antenna situation. That is, we can implement a distributed Alamouti scheme by using the decoded symbols at the relay in cooperation with the source. If S_1 and S_2 are two QAM symbols transmitted on a particular subcarrier l over two OFDM symbols, then the transmission scheme for the distributed Alamouti scheme is given in Table 3. Using the OFDM signal model of (2), the demodulated symbols at the destination D for that subcarrier l are

$$Y_{Dl}^{(2,1)} = H_{RD,l}^{(2)}S_1 + H_{SD,l}^{(2)}S_2 + Z_{Dl}^{(2,1)} \quad (5)$$

$$Y_{Dl}^{(2,2)} = -H_{RD,l}^{(2)}S_2^* + H_{SD,l}^{(2)}S_1^* + Z_{Dl}^{(2,2)},$$

where $Y_{Dl}^{(2,k)}$ is the demodulated signal on subcarrier l at D across two OFDM symbols, $k = 1, 2$ and $H_{RD,l}^{(2)}$, $H_{SD,l}^{(2)}$ denote the channels on subcarrier l in attempt 2. Standard Alamouti combining at D [3] results in the following effective

point-to-point channels per subcarrier:

$$\tilde{Y}_{Dl}^{(2,1)} = S_1\sqrt{|H_{RD,l}^{(2)}|^2 + |H_{SD,l}^{(2)}|^2} + \tilde{Z}_{Dl}^{(2,1)} \quad (6)$$

$$\tilde{Y}_{Dl}^{(2,2)} = S_2\sqrt{|H_{RD,l}^{(2)}|^2 + |H_{SD,l}^{(2)}|^2} + \tilde{Z}_{Dl}^{(2,2)},$$

where $\tilde{Z}_{Dl}^{(2,k)}$, $k = 1, 2$ is Gaussian noise of the same variance as the noise in (5).

	OFDM Symbol 1	OFDM Symbol 2
Relay	S_1	$-S_2^*$
Source	S_2	S_1^*

Table 3: Transmitted signals from source and relay per subcarrier for DFLC

Amplify Forward (AF): For AF, since the relay does not decode, the distributed Alamouti scheme of Table 3 is modified by using the amplified received signal at subcarrier l, instead of the decoded symbols as in DF (see Table 4).

	OFDM Symbol 1	OFDM Symbol 2
Relay	$AY_{Rl}^{(1,1)}$	$AY_{Rl}^{(1,2)}$
Source	S_2	S_1^*

Table 4: Transmitted signals from source and relay per subcarrier for AFLC

where

$$Y_{Rl}^{(1,1)} = H_{SR,l}^{(1)}S_1 + Z_R^{(1,1)}$$

$$Y_{Rl}^{(1,2)} = -H_{SR,l}^{(1)}S_2^* + Z_R^{(1,2)}$$

The received signals are modified from (5) as:

$$Y_{Dl}^{(2,1)} = H_{RD,l}^{(2)\prime}S_1 + H_{SD,l}^{(2)}S_2 + Z_{Dl}^{(2,1)\prime} \quad (7)$$

$$Y_{Dl}^{(2,2)} = -H_{RD,l}^{(2)\prime}S_2^* + H_{SD,l}^{(2)}S_1^* + Z_{Dl}^{(2,2)\prime}$$

where $H_{RD,l}^{(2)\prime} = AH_{RD,l}^{(2)}H_{SR,l}^{(1)}$. Note that the noise-variances corresponding to $Z_{Dl}^{(2,k)\prime}$ are larger than $Z_{Dl}^{(2,k)}$ in (5) since we are forwarding noise in AF. Using the same Alamouti combining at D as in (6), we get

$$\breve{Y}_{Dl}^{(2,1)} = S_1\sqrt{|H_{RD,l}^{(2)\prime}|^2 + |H_{SD,l}^{(2)}|^2} + \breve{Z}_{Dl}^{(2,1)} \quad (8)$$

$$\breve{Y}_{Dl}^{(2,2)} = S_2\sqrt{|H_{RD,l}^{(2)\prime}|^2 + |H_{SD,l}^{(2)}|^2} + \breve{Z}_{Dl}^{(2,2)},$$

where $\breve{Z}_{Dl}^{(2,k)}$, $k = 1, 2$ is still Gaussian noise with same (larger) variance as $Z_{Dl}^{(2,k)\prime}$ in (7).

Quantize-Map-Forward (QMF): We implement it as follows: the relays employ a scalar quantizer (symbol-by-symbol on a QAM symbol level) to quantize the received signals from the source (as proposed in [22, 4]). Instead of the random mapping operation proposed in [5], we use a permutation mapping (randomly chosen bit-interleaver) on the quantized bits corresponding to each codeword in the frame. This mapping was shown to perform well in [22], and it also facilitates a simpler decoder operation. After mapping, the relay re-modulates the frame as per the 802.11 specifications, and if the direct transmission in the first time slot fails, forwards it to the destination.

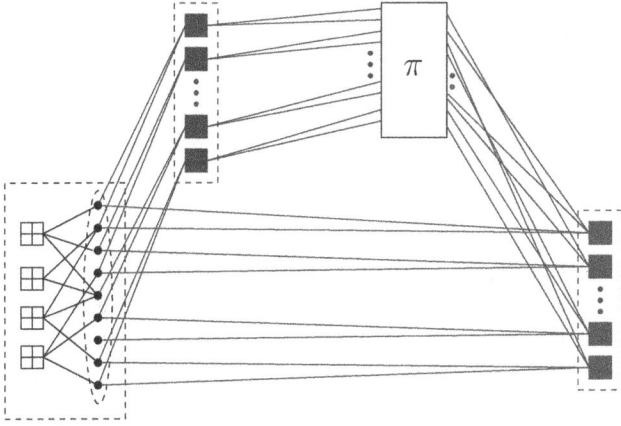

Figure 2: LDPC-based joint decoder for QMF

3.2.2 Decoding at the Receiver

In attempt 2 of the LC mode, as we already described in (2), the destination receives a superposition of the source and relay transmissions

$$Y_{Dl} = H_{SD,l}X_{Sl} + H_{RD,l}X_{Rl} + Z_{Dl}.$$

For AF and DF, we have already seen that effectively we have point-to-point channels with parameters that take into account the operations at the relay. Thus, standard point-to-point LDPC decoders are sufficient for decoding AF and DF frames.

QMF, on the other hand, uses non-linear operations (quantization) at the relay and hence a point-to-point decoder does not suffice for decoding. As an example, Fig. 2 outlines the graphical structure of the decoder for one codeword with QMF LC. This is *not* a standard structure for iterative (LDPC) codes due to two reasons: (i) the quantization at the relay which takes the received signal Y_{Rl} and quantizes it to $\hat{Y}_{Rl}$. (ii) the superposition of the two streams X_{Sl}, X_{Rl}.

For the QMF iterative decoder, we introduce two novel features: (i) We use the stochastic quantizer check nodes (red boxes in Fig. 2) for the functions $p(\hat{Y}_{Rl}|X_{Sl})$ that capture the quantization operation at the relay. We use the multiple-access check nodes (blue boxes in Fig. 2) to capture the signal superposition at the destination.(ii) We use a graphical structure *across* the nodes of the network and run our decoding iterations over the graph. That is, the iterative decoder runs across the graphical structure of Fig. 2, with the bit level reliabilities (represented by log-likelihood ratios, LLRs) flowing along the edges in the graph.

In addition, to reduce the occurrence of low-weight error events, we use a further *enhancement layer* for the LDPC decoders of all three schemes, QMF, AF and DF, that treats LLR magnitudes above a certain pre-determined threshold as correct, and others as *erasures* (lost bits) after a fixed number of iterations. Then another erasure correction round of iterations is performed to clean up the residual errors.

4. SYSTEM IMPLEMENTATION

Overall system. The source, relay, and destination nodes were implemented using the WARP SDR platform [2]. We used the WARPLab framework, which allows interaction with the WARP hardware via a host PC running MATLAB

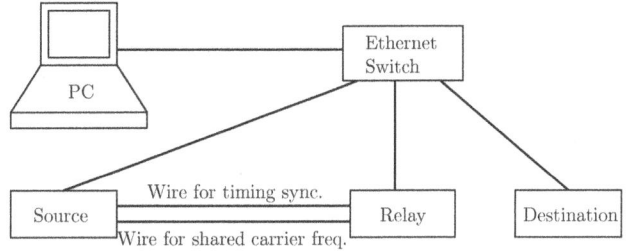

Figure 3: Node and host PC configuration

(see Fig. 3). The host PC was used to control real-time over-the-air transmission and reception.

We implemented the four network operation methods described in Section 2.2 coupled with the three relaying strategies defined in Section 3.1. Table 5 summarizes the implemented schemes when the relay is active; additionally, we implemented Direct Transmission (DT) and Direct Transmission Half Rate (DTHR).

4.1 Frame Structure and Operation

We use a frame structure specified in the 802.11 standard [1], which stipulates that we first transmit training for Automatic Gain Control (TAGC), followed by training for timing synchronization (TSYNC), training for channel estimation (TCHE) and then the payload. We use the signal that the 802.11 standard defines as long training symbols for timing synchronization and channel estimation. Our experiments correspond to a 20 MHz bandwidth system with 64 subcarriers. During the payload transmission, for each OFDM symbol transmitted, 48 subcarriers are data subcarriers, 4 subcarriers carry pilots, and the rest of the subcarriers are unused, as per the 802.11 payload structure. The signals transmitted over-the-air were centered at a frequency of 2.4 GHz in a 20MHz bandwidth. Next, we give more specific details of the frame structures for each scheme for the attempts over the two time-slots.

Time-slot operation: To make a fair comparison between schemes with and without cooperation, we allow a maximum of two attempts for every source information frame to be successful, as described in Section 2.2. We send each attempt over a *time slot*, which is one frame in length. If the communication is not successful after two attempts then the frame is declared in error and the source moves on to the first attempt for the next information frame.

First Time Slot: For all schemes, the first attempt is always only via the direct link from source to destination. The source transmits using the standard frame structure we previously described; the frame structure is shown in Fig. 4. If the destination is able to decode successfully, we proceed with the first attempt of a new frame. If the transmission attempt fails, then we move to the next attempt in the second time-slot described below.

Second Time Slot: The source re-transmits at lower rate in DTHR mode. In LS mode, only the relay transmits, using a standard frame structure. In the LC mode, both the source and the relay transmit. The different relaying operations for each of the modes are described in Section 3.1. The frame structures for the second attempt are shown in Fig. 4. As noted in Fig. 4, in AFLC/AFLS and QMFLC/QMFLS, the relay sends one more OFDM symbol than in DFLC/DFLS. This extra OFDM symbol is only sent by the relay and it is

	Amplify-Forward (AF)	Decode-Forward (DF)	Quantize-Map-Forward (QMF)
Link Switching (LS)	AFLS	DFLS	QMFLS
Link Cooperation (LC)	AFLC	DFLC	QMFLC

Table 5: Implemented schemes when the relay is active. The network modes (rows) are described in Section 2.2 and the relaying strategies (columns) in Section 3.1.

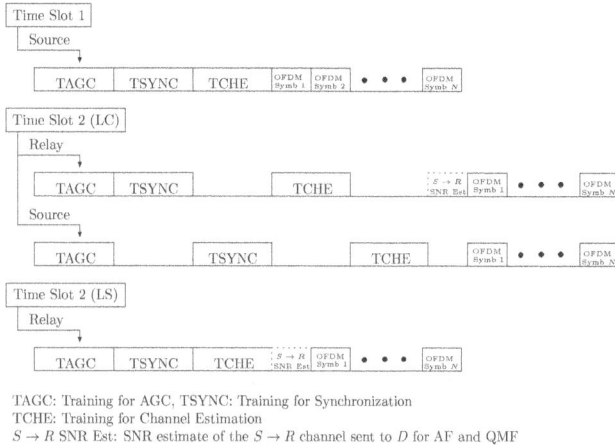

TAGC: Training for AGC, TSYNC: Training for Synchronization
TCHE: Training for Channel Estimation
$S \rightarrow R$ SNR Est: SNR estimate of the $S \rightarrow R$ channel sent to D for AF and QMF

Figure 4: Time-slot frame structure for LS and LC.

used to forward an estimate of the source to relay SNR. This estimate is used by the iterative decoder at the destination. The SNR estimate is not forwarded by DFLC/DFLS because the bits are decoded at the relay before forwarding them.

4.2 Timing and Carrier Synchronization

The WARPLab framework allows a coarse timing synchronization of the nodes involved in communication. However, for 20MHz OFDM time scales this coarse synchronization is not accurate enough; more elaborate mechanisms that enable the required timing and carrier frequency synchronization are required both at the receivers and distributed transmitters.

Operation at a receiver: A node in receiver mode can solve timing synchronization by exploiting the autocorrelation properties of the training sent for timing synchronization. The signal used for channel estimation, labeled as TCHE in Fig. 4, is also used for time-domain estimation and correction of the carrier frequency offset (CFO). The time domain correction is applied to the training for channel estimation and to the OFDM symbols before performing the FFT for conversion to the frequency domain. To correct for residual phase noise present, another level of estimation and correction of CFO and a phase noise correction is applied in the frequency domain. The frequency domain CFO and phase noise are estimated based on the four known pilot subcarriers that are included in each OFDM symbol. Since DT/DTHR/LS modes effectively create single-transmitter, single-receiver (point-to-point) channels, we distinguish its operation from the LC modes.

Link Cooperation (LC): In order for the destination to be able to set the AGC correctly for the reception of the sum signal, the source and relay send training for AGC simultaneously. The waveform for AGC sent by the source is similar to the one sent by the relay, with the only difference being that it is cyclically shifted in order to avoid accidental

nulling. The source and relay both send training for timing synchronization and channel estimation and these training signals are orthogonal in time. This orthogonality ensures that the two node diversity is also present in the timing synchronization phase, since the destination can solve timing synchronization from any of the two copies it receives. The orthogonal training for channel estimation is needed in order to compute clean channel estimates from each of the links; these two channel estimates are needed for decoding in the LC mode. To implement the orthogonality of training/timing in link cooperation mode, the four pilot subcarriers are split and alternated between the source and the relay in the following way. In odd OFDM symbols the relay transmits pilots 1 and 3 and the source transmits pilots 2 and 4. In even OFDM symbols the relay transmits pilots 2 and 4 and the source transmits pilots 1 and 3. This is the same method for pilot assignment proposed in [17, 21].

The mechanism described above for CFO correction enables a receiver to solve for only one CFO. However, in LC mode, at the destination there would be two CFOs to correct for–one due to the R–D link and another due to the S–D link. In [18, 21] this issue of two CFOs is solved by having the relay lock on to the carrier of the source by estimating its CFO with respect to the source and applying a time domain correction so that the destination observes only one CFO. Although feasible to implement, this was not the focus of our study. Consequently, for our implementation, we used a wire to share the carrier frequency clock of the source between the source and the relay, as shown in Fig. 3. **Distributed synchronization:** In LC, in the second time slot, the relay and the source transmit synchronously to ensure required orthogonality in the training phase and avoid intersymbol interference when receiving the payload. Thus, the synchronization mechanism must be accurate enough to ensure that at the destination, the time of arrival of the signals from the source and the relay is within the cyclic prefix of the OFDM symbols. In our implementation, the duration of cyclic prefix is 0.8 μs, which is the exact same duration of the cyclic prefix specified in 802.11 for a 20 MHz bandwidth system. Recent work [18, 21, 6] has demonstrated that one can design protocols that achieve accurate timing synchronization (between 20 ns and 100 ns accuracy) for distributed OFDM communication enabled by implementing a large part of the distributed timing synchronization mechanisms in real time in the FPGA. Incorporating this fast turnaround time to the WARPLab framework, although feasible, was not the focus of our study. Consequently, we synchronized the time of transmission of distributed transmitters via a wired connection between source and relay, as shown in Fig. 3, for all schemes.

4.3 Estimation of Effective Noise

To decode, all our schemes require an estimate of the effective noise variance in the digital (sampled) domain. This estimate is used to compute the log likelihood probabili-

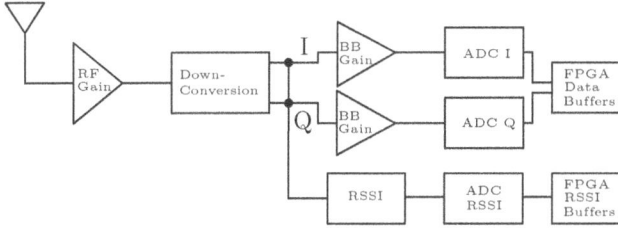

Figure 5: Radio board receiver path

ties that serve as the input to the decoders, and thus an accurate estimation is important for the iterative (LDPC) decoder performance. However, accurate estimation of the noise variance can be a challenging and complex task, since there are many sources of noise in a hardware implementation. For example, there is additive noise, quantization noise at ADC and DAC, IQ imbalance, channel estimation errors, carrier frequency offset errors and timing synchronization inaccuracy. Also, the noise added by the radios is a function of the gains set by the AGC and the received signal power. To balance the need for an accurate estimate and the system need for low complexity, we developed the following simple algorithm that builds on four components: (i) the (analog) RSSI measurement provided by the radios (ii) the approximate (calibrated) noise level of the radios we use (iii) the channel estimates per subcarrier (iv) the model for the received signal. We describe how we put these together in the following two steps.

• *Step 1 - Computation of the Analog SNR:* We compute the SNR in the analog domain as

$$\text{SNR-analog } (dB) = \text{PayloadRSSI } (dBm) - N_0 \ (dBm) \quad (9)$$

The analog thermal noise value N_0 is initialized with the calibrated value for the WARP boards. The receiver radio path is shown in Fig. 5. The RSSI reading is made after downconversion to baseband and the radios provide the value of the receiver RF Gain and the mapping of RSSI readings to RSSI at the receiver antenna. Hence, from the RSSI reading we actually get the RSSI of the signal at the receiver antenna. This is the PayloadRSSI that we use to compute the SNR-analog in (9). The analog RSSI values are reported every $0.1\mu sec$ by the radios or 40 RSSI readings per OFDM symbol. Hence, for a payload of L OFDM symbols we have $40L$ RSSI readings that are averaged to get the payload RSSI represented in (9).

• *Step 2 - Solve for the effective digital noise variance:* In the digital domain, we model the received signal as shown in (1). Since we have m subcarriers, the SNR in the digital domain can be estimated as

$$\text{SNR-digital} = \frac{\sum_{l=1}^{m} |H_l|^2}{m\sigma^2}, \quad (10)$$

where σ^2 is effective (digital) noise power per subcarrier (the signal power is normalized to 1). We estimate that the SNR computed in the digital domain is the same as the SNR computed in the analog domain minus an estimate Δ of effects like imperfect channel estimation and imperfect CFO correction.

$$\text{SNR-digital } (dB) = \text{SNR-analog } (dB) - \Delta \ (dB) \quad (11)$$

From (9), (10) and (11) we solve for the effective noise variance σ^2, which is used in the iterative decoder. Guided by the WARP radio calibrations, we use values of $N_0 = -95 \ dBm$ and $\Delta = 5 \ dB$.

As for any estimator, the method for σ^2 estimation described in the two steps above can have errors that affect the decoder performance. To the best of our knowledge, there is currently no work on the comparison of DF, AF and QMF schemes under errors in σ^2. We leave the development and implementation of more robust estimators for future work.

For AF and QMF, the decoding at the destination needs the SNR-digital computed for the source to relay link. The relay computes this information and forwards it to the destination using one extra OFDM symbol, as shown in Fig. 4. To do so, we first quantize the SNR estimate to one out of 40 possible values ranging from -10 to 30 dB (in steps of 1 dB). We can describe these 40 values using 6 bits. We repeat these 6 bits 8 times, modulate them with BPSK and allocate them to the 48 data subcarriers in the OFDM symbol used to forward the SNR information.

5. EXPERIMENTATION

Setup. Our testbed covers a rectangular area of 175 m^2, and spans across three rooms. We report the results for three scenarios, which we describe in terms of the average Received Signal Strength Indicator (RSSI) in dBm. We note that the RSSI values not only depend on the distance between devices, but also on the multipath effects in our indoor environment. Moreover, in addition to the relative positioning, we also needed to adjust the transmit power of the radios in some cases, in order to achieve the RSSI values for our scenarios that are typical for 802.11 systems and recommended for the modulation we employ. The RSSI values for all the scenarios is given in Fig. 7.

- **Scenario 1** captures the case where the source and the relay are close: we deployed the relay at a distance of 1.5m from the source as depicted in Fig. 6(a). As a result, the S–R link is much stronger than the S–D and R–D links, which is reflected in the average RSSI values we measure at our receivers.

- **Scenario 2** captures the case where all three links have approximately equal strength. The node placement is shown in Fig. 6(b).

- **Scenario 3** captures the case where the S–R and R–D links are much stronger than the S–D direct link. The nodal arrangement is depicted in Fig. 6(c).

In all our experiments we used 16 QAM modulation, an LDPC code of rate 3/4 with codeword length of 1944 bits and the parity matrix defined in the 802.11 standard. Each frame encapsulated four codewords, leading to 5832 information bits (payload) per frame. For each experiment we transmitted at least 1600 frames, which corresponds to at least 9331200 information bits transmitted per experiment. **Metrics:** We use two performance metrics:
(i) *Frame Error Rate (FER)* quantifies the error probability our frame transmissions have with each scheme.
(iii) *Throughput* quantifies the amount of successfully decoded information bits that reach the destination, normalized by the the total over-the-air transmission time and

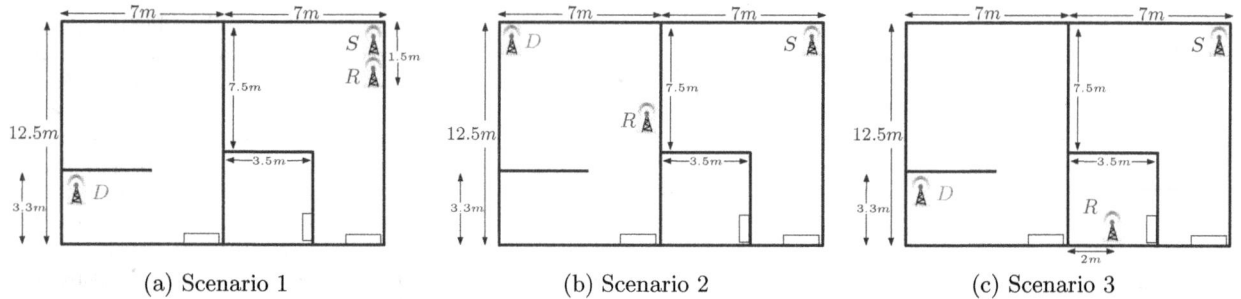

(a) Scenario 1   (b) Scenario 2   (c) Scenario 3

Figure 6: Node placement in our testbeds

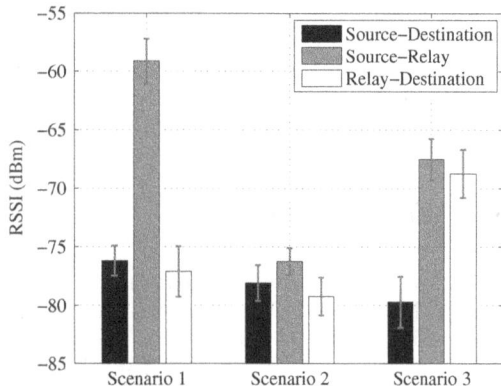

Figure 7: Average measured RSSI values per scenario.

bandwidth used. Thus, we count the amount of decoded information bits per channel use in bps/Hz.

Discussion of results. The following observations are based on the FER results shown in Fig. 8 and the throughput results shown in Fig. 9.

(1) *Relaying helps over DT:* In all three scenarios, LC schemes significantly outperform DT. The FER gains of relay transmission ranged from a factor of 10 in Scenarios 1 and 2, to nearly a factor of 100 in Scenario 3, where the direct path was very weak. The throughput gains were 20-40% in Scenarios 1 & 2, and was nearly a factor of 15 in Scenario 3. DTHR[2] can improve the throughput performance with respect to DT, for example in Scenarios 2 and 3. However, in all the scenarios the performance of DTHR was always worse than all of the LC schemes. This is because the reliability benefit of transmitting at a lower rate does not offset the diversity gain obtained with link cooperation.

(2) *Link Cooperation (LC) helps:* In Scenarios 1 and 2, link cooperation gives a factor of 10–40 gain in FER and a 40–70% gain in throughput over link switching. In Scenario 3, link switching performs marginally better than link cooperation. There are two main reasons why in Scenario 3 LC does not improve performance over LS. First, at an average RSSI of −80 dBm, the *S–D* link in Scenario 3 is 13 dB weaker than the *R–D* link. Hence, the contribution from the S-D link is minimal. Second, in LC mode, the pilots for frequency domain CFO and phase noise corrections are

split between the source and the relay as was explained in Section 4.2. However, since the *S–D* link is very weak, the pilots assigned to the source are received with very weak power and hence do not help towards CFO or phase noise correction. Consequently, in Scenario 3, the last hop in LC is effectively a single *R–D* link, reminiscent of LS, with the added disadvantage of having only half the pilots as compared to LS.

(3) *Universality of QMF:* QMFLC shows the most competitive performance across all the scenarios. QMF outperforms AF because the frames it sends in the second time-slot are less noisy since the quantization removes some of the noise. In Scenarios 1 and 3 there is very little difference between the performance of QMF and DF. In these two scenarios, since the *S–R* link is at a medium-to-high RSSI (−60 dBm in Scenario 1 and −67 dBm in Scenario 3), DF can decode most of the time and hence exhibits good performance. In Scenario 2, the *S–R* link is at a low RSSI (−78 dBm); hence decoding at the relay fails more often, affecting the performance of DF negatively. As a result, the gains from QMF are most pronounced in Scenario 2, as the relay always transmits a reasonable quantized signal that the destination can exploit in decoding. Scenario 2 shows a 12% throughput benefit for QMFLC and up to 60% improvement in FER over the next best scheme (DFLC).

Note: Although we observe that cooperation helps and QMFLC outperforms the other schemes in Scenarios 1 and 2, we neither believe nor expect that this is true for all possible scenarios; for example, if the direct link S–D is much stronger than the rest, we expect DT to do best; and if this link essentially does not exist (effectively creating a line network as seen in Scenario 3) DFLS (*i.e.*, routing) does better than all others schemes. In this paper we focused on scenarios where the outcome of the comparison is not obvious, and thus more interesting.

Why frames are lost: We grouped frame failures in categories: failed due to timing, or a certain number of bit errors between [a, b] remained in the decoded frame (and thus the CRC check failed). We plot an example of this analysis for Scenario 2 in Fig. 10. We observe that for DF and QMF, frame errors primarily occur when a *large* fraction of the bits are in error, indicating that relay processing reduces low-weight error patterns. In contrast, in AF we see frames failing even with a small number of bit errors: a plausible explanation being that since AF does not process its received signal at the relay (DF decodes, QMF quantizes and remaps), the amplified noisy signals propagate to the destination resulting in a (relatively) less sharp threshold behavior in the LDPC decoder.

---

[2]For DTHR, we only compared throughput since the transmission rate was lower than the other schemes. This implied that the FER was not directly comparable.

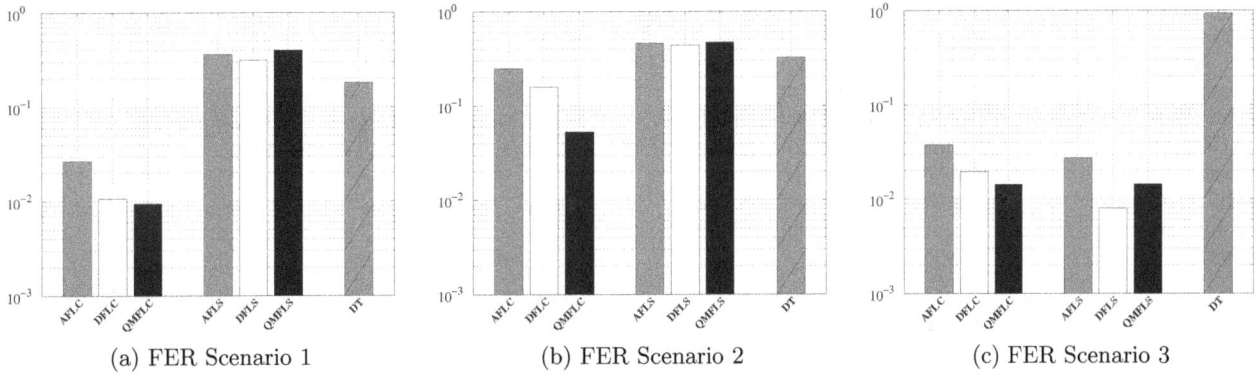

(a) FER Scenario 1   (b) FER Scenario 2   (c) FER Scenario 3

Figure 8: Frame error rates.

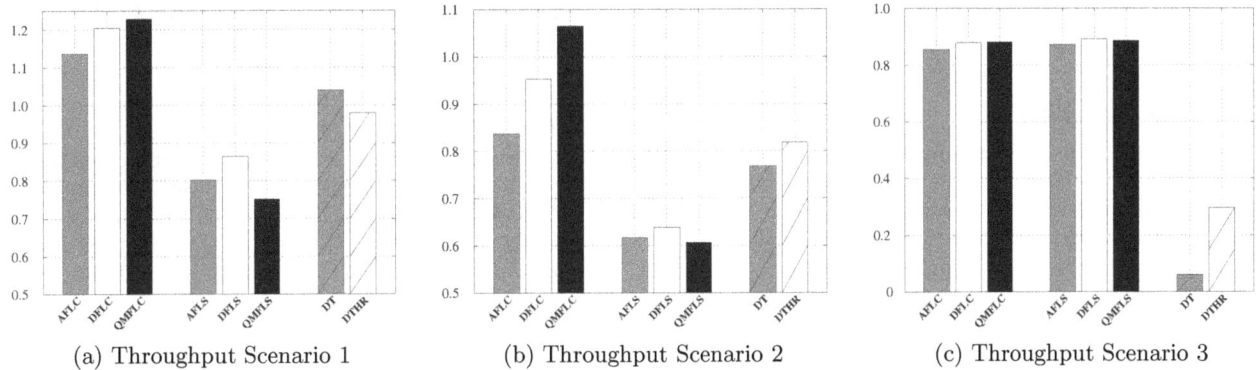

(a) Throughput Scenario 1   (b) Throughput Scenario 2   (c) Throughput Scenario 3

Figure 9: Throughput in bps/Hz.

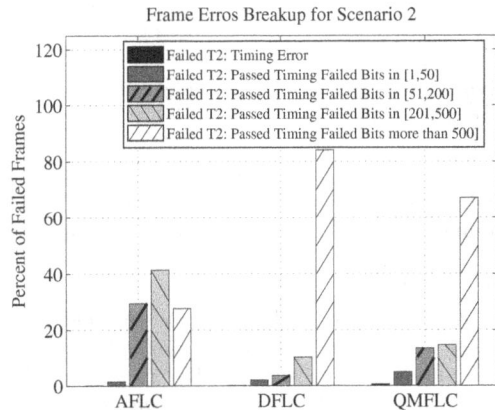

Figure 10: Distribution of percentage of frames that fail in $T2$.

6. RELATED WORK

Surveys covering the field of testbed implementations of physical layer cooperative communications are found in [13], [8]. In [7], a testbed implementation of PHY layer cooperation using uncoded DF on a single relay network is investigated. A WARP radio testbed implementation of cooperative relaying is presented in [12] where again DF based schemes are implemented. Both [7, 12] assume orthogonal transmissions from the source and relay. In [25] uncoded

DF protocols with one and two relays on the Universal Software Radio Peripheral (USRP) platform are studied. In all these studies advanced error correction or broadband OFDM modulation are not implemented as done in our paper. The work [18] is perhaps closest to our paper. They implement both (uncoded) AF and DF relaying strategies along with distributed Alamouti-based transmission. It is important to note however, that this implementation also did not use error correcting codes at any point in the network. Other testbed implementations that specifically focus on tackling the synchronization problem for multiple simultaneous transmissions are found in [21] and [26].

Testbed implementations focussing on the benefits of MAC-level cooperative communication over WiFi was presented in [15], using (DF-based) multi-hop transmission. Other approaches to increase the throughput of WiFi include intelligently selecting access points based on channel utilization and adaptive switching [23], and recommendations to improve the way in which rate adaptation is done in WiFi ([16] and references therein).

Many of the fundamental information theoretic ideas of relaying were laid in the seminal work by Cover and El-Gamal [10], where the DF strategy was formalized. They also proposed Compress-Forward (CF). which is closely related to posed QMF proposed in [5]. The distinction is that the CF needs complete network state at the relays and therefore does not allow distributed operation. QMF also extends seamlessly for multicast, which is not possible with CF [5]. It was shown that QMF approximately achieves network capacity (within a constant gap, independent of the channel gains and SNR). Similar schemes using lattice-based codes were developed in [20] and QMF was extended to discrete memoryless networks in [14]. Practical coding schemes for QMF relaying with LDPC codes were proposed for an orthogonalized version of the half-duplex single relay network in [19] and for the full-duplex diamond network in [22], where interleavers as relay maps were also proposed. The use of demodulation instead of quantization was used in [22] and [4]. None of these papers had an experimental evaluation of performance.

7. ACKNOWLEDGMENTS

The authors would like to thank the anonymous reviewers for their comments that improved the presentation of the work. We also acknowledge funding from the European Research Council grant NOWIRE ERC-2009-StG-240317, the EU project CONECT FP7-ICT-2009-257616, the NSF award 1136174 and MURI award AFOSR FA9550-09-064.

8. REFERENCES

[1] Local and metropolitan area networks–specific requirements part 11: Wireless LAN medium access control (MAC) and physical layer (PHY) specifications. *IEEE Std 802.11-2012*.

[2] WARP Project, http://warpproject.org.

[3] ALAMOUTI, S. A simple transmit diversity technique for wireless communications. *IEEE Journal on Select Areas in Communications 16*, 8 (October 1998), 1451–1458.

[4] ATSAN, E., KNOPP, R., DIGGAVI, S., AND FRAGOULI, C. Towards integrating quantize-map-forward relaying into LTE. In *Proceedings of the IEEE Information Theory Workshop* (September 2012), pp. 212–216.

[5] AVESTIMEHR, A. S., DIGGAVI, S. N., AND TSE, D. N. C. Wireless network information flow: A deterministic approach. *IEEE Transactions on Information Theory 57*, 4 (April 2011), 1872–1905.

[6] BALAN, H. V., ROGALIN, R., MICHALOLIAKOS, A., PSOUNIS, K., AND CAIRE, G. AirSync: Enabling distributed multiuser MIMO with full spatial multiplexing. *IEEE/ACM Transactions on Networking*, 99 (2013).

[7] BRADFORD, G., AND LANEMAN, J. N. An experimental framework for the evaluation of cooperative diversity. In *Proceedings of the IEEE CISS* (March 2009), pp. 641–645.

[8] BRADFORD, G., AND LANEMAN, J. N. A survey of implementation efforts and experimental design for cooperative communications. In *Proceedings of the IEEE ICASSP* (2010), pp. 5602–5605.

[9] CHANG, K. ET AL. Relay operation in 802.11ad. In *IEEE 802.11ad TGad 1-/0494r1* (2010).

[10] COVER, T. M., AND GAMAL, A. E. Capacity theorems for the relay channel. *IEEE Transactions on Information Theory 25*, 5 (September 1979), 572–584.

[11] HUNTER, C., MURPHY, P., AND SABHARWAL, A. Real-time testbed implementation of a distributed cooperative MAC and PHY. In *Proceedings of the IEEE CISS* (March 2010), pp. 1–6.

[12] KNOX, M., AND ERKIP, E. Implementation of cooperative communications using software defined radios. In *Proceedings of the IEEE ICASSP* (March 2010), pp. 5618–5621.

[13] KORAKIS, T., KNOX, M., ERKIP, E., AND PANWAR, S. Cooperative network implementation using open-source platforms. *IEEE Communications Magazine 47*, 2 (2009), 134–141.

[14] LIM, S. H., KIM, Y.-H., GAMAL, A. E., AND CHUNG, S.-Y. Noisy network coding. *IEEE Transactions on Information Theory 57*, 5 (May 2011), 3132–3152.

[15] LIU, P., TAO, Z., NARAYANAN, S., KORAKIS, T., AND PANWAR, S. CoopMAC: A cooperative MAC for wireless LANs. *IEEE Journal on Selected Areas in Communications 25*, 2 (February 2007), 340–354.

[16] LOIACONO, M., ROSCA, J., AND TRAPPE, W. The snowball effect: Detailing performance anomalies of 802.11 rate adaptation. In *Proceedings of the IEEE GLOBECOM* (November 2007), pp. 5117–5122.

[17] MURPHY, P. *Design, Implementation, and Characterization of a Cooperative Communications System*. PhD thesis, Rice University, 2010.

[18] MURPHY, P., AND SABHARWAL, A. Design, implementation, and characterization of a cooperative communications system. *IEEE Transactions on Vehicular Technology 60*, 6 (July 2011), 2534–2544.

[19] NAGPAL, V., WANG, I.-H., JORGOVANOVIC, M., TSE, D., AND NIKOLIĆ, B. Coding and system design for quantize-map-and-forward relaying. *IEEE Journal on Selected Areas in Communications* (August 2013). See also arXiv:1209.4679 [cs.IT].

[20] ÖZGÜR, A., AND DIGGAVI, S. Approximately achieving Gaussian relay network capacity with lattice codes. In *Proceedings of the IEEE International Symposium on Information Theory* (June 2010), pp. 669–673. See also arXiv:1005.1284 [cs.IT].

[21] RAHUL, H., HASSANIEH, H., AND KATABI, D. SourceSync: a distributed wireless architecture for exploiting sender diversity. In *Proceedings of the ACM SIGCOMM* (August 2010), pp. 171–182.

[22] SENGUPTA, A., BRAHMA, S., OZGUR, A., FRAGOULI, C., AND DIGGAVI, S. Graph-based codes for quantize-map-and-forward relaying. In *Proceedings of the IEEE Information Theory Workshop* (October 2011), pp. 140–144.

[23] SUN, T., ZHANG, Y., AND TRAPPE, W. Improving access point association protocols through channel utilization and adaptive switching. In *Proceedings of the 8th IEEE International Conference on Mobile Ad-Hoc and Sensor Systems* (October 2011), pp. 155–157.

[24] TSE, D., AND VISWANATH, P. *Fundamentals of Wireless Communication*. Cambridge University Press, May 2005.

[25] ZHANG, J., JIA, J., ZHANG, Q., AND LO, E. M. K. Implementation and evaluation of cooperative communication schemes in software-defined radio testbed. In *Proceedings of the IEEE INFOCOM* (March 2010), pp. 1307–1315.

[26] ZHANG, X., AND SHIN, K. G. DAC: Distributed asynchronous cooperation for wireless relay networks. In *Proceedings of the IEEE INFOCOM* (March 2010), pp. 1064–1072.

Near-Optimal Truthful Spectrum Auction Mechanisms With Spatial and Temporal Reuse in Wireless Networks

He Huang[1]   Yu-e Sun[2*]   Xiang-Yang Li[3,4]   Zhili Chen[5]   Wei Yang[5]   Hongli Xu[5]

[1]School of Computer Science and Technology, Soochow Univ., Suzhou, China
[2]School of Urban Rail Transportation, Soochow Univ., Suzhou, China
[3]Dept. of Computer Science, Illinois Institute of Technology, USA
[4]Tsinghua National Laboratory for Information Science and Technology (TNLIST), Tsinghua Univ., China
[5]Dept. of Computer Science and Technology, Univ. of Science and Technology of China, China
{huangh, sunye12}@suda.edu.cn, xli@cs.iit.edu, {zlchen3,qubit,xuhongli}@ustc.edu.cn

ABSTRACT

In this work, we study spectrum auction problem where each spectrum usage request has spatial, temporal, and spectral features. After receiving bid requests from secondary users, and possibly reserve price from primary users, our goal is to design truthful mechanisms that will either optimize the social efficiency or optimize the revenue of the primary user. As computing an optimal conflict-free spectrum allocation is an NP-hard problem, in this work, we design near optimal spectrum allocation mechanisms separately based on the techniques: derandomized allocation from integer programming formulation, and its linear programming (LP) relaxation. We theoretically prove that 1) our derandomized allocation methods are monotone, thus, implying truthful auction mechanisms; 2) our derandomized allocation methods can achieve a social efficiency or a revenue that is at least $-\frac{1}{e}$ times of the optimal respectively; Our extensive simulation results corroborate our theoretical analysis.

Categories and Subject Descriptors

C.2.1 [**Computer Systems Organization**]: Computer Communication Networks; D.2.8 [**Management of Computing and Information Systems**]: Installation Management—*Pricing and resource allocation*

Keywords

Spectrum auction; Truthful; Approximation mechanism; Social efficiency; revenue

1. INTRODUCTION

The growing demand for limited spectrum resource poses a great challenge in spectrum allocation and usage. One of the most promising methods is spectrum auction, which provides sufficient incentive for primary user (*a.k.a seller*) to sublease spectrum to secondary users (*a.k.a buyers*). The design of spectrum auction mechanisms are facing two major challenges. First, spectrum channels can be reused in spatial, temporal, and spectral domain if the buyers are conflict-free with each other, and thus, allocating the requests of buyers in channels optimally is often an NP-hard problem. Second, truthfulness is regarded as one of the most critical properties, however, it's difficult to ensure truthfulness in a spectrum auction mechanism with performance guarantee. Recent years, many mechanisms were proposed to address some of the auction challenges [1–3,7–11]. However, to the best of our knowledge, there is no truthful spectrum auction mechanism with performance guarantee, in which spectrums can be reused both in spatial and temporal domains.

Maximization of the *social efficiency*, *i.e.* allocating a channel to buyers who **value** it most, and maximization of the *expected revenue*, *i.e.* allocating a channel to buyers who **pay** it most, both are the natural goals for spectrum auctions. Thus, we design a framework for spectrum auction which can maximize the social efficiency or the expected revenue. Since channels can be reused in both spatial and temporal domain, the problem of allocating requests of buyers in channels optimally to maximize the social efficiency or the expected revenue is NP-hard. To tackle this challenge, we first relax the integer programming formulation of the channel allocation problem into a linear program (LP) problem, which is solvable in polynomial time. A fractional solution for channel allocation can be obtained by solving this LP optimally. Then, we transform this fractional solution into a feasible integer solution of the original channel allocation problem by using a carefully designed randomized rounding procedure that ensures the feasibility of the solution and good approximation to the objective functions. We prove that the **expected** weight of the feasible integer solution is at least $- /e$ times of the weight of the optimal solution. To complete our allocation mechanism, we also propose a derandomization algorithm to get a feasible solution whose weight is always guaranteed to be at least $- /e$ times of the weight of the optimal solution. Then, we propose a revised derandomization algorithm and prove that this new allocation method does satisfy the bid-monotone property, thus, implying a truthful mechanism. We point out that our allocation mechanisms can either approximate the social efficiency or the expected revenue, but not both simultaneously.

2. PRELIMINARIES

Auctions in our model are executed periodically. In each round, the primary user subleases the access right of m channels in the fixed areas during time interval $, T$, and n buyers request the usage of channels in fixed time intervals and geographical location-

*Yu-e Sun is the corresponding author.

s/areas. Our goal is to allocate these requests of buyers in channels, such that we maximize either the social efficiency or the expected revenue. We consider two models of the requests of buyers. The first one is the *Point model*, in which each buyer requests the usage of channels in a particular geographical location and during a fixed time interval. The second one is *Area model*, in which each buyer requests the usage of channels in a particular geographical area and also during a fixed time interval.

We use $\mathcal{S}$ to denote the set of channels, and define each channel $s_j \in \mathcal{S}$ as $s_j = \langle R_j, A_j \rangle$, where A_j is its license area, and R_j is the interference radius of a transmission when a user transmits in channel s_j. Let $\mathcal{B}$ be the set of buyers, in which each buyer $i \in \mathcal{B}$ is assumed to have a request r_i. Let $\mathcal{R}$ be the set of requests of buyers. Each request $r_i \in \mathcal{R}$ is defined as $r_i = \langle L_i, b_i, v_i, a_i, t_i, d_i \rangle$, where L_i is i's geographical location in *Point model* or the area where i wants to access the channel in the *Area model*, b_i denotes its bidding price, v_i stands for its true valuation, and a_i, d_i, and t_i denote the arrival time, deadline, and duration time (or time length), respectively. In this paper, we only consider the case of $d_i - a_i = t_i$, which means that the time request from the buyer is a fixed time interval. We leave the case of $d_i - a_i > t_i$ as the future work.

We say that two requests r_i and r_k conflict with each other, if they satisfy the following constrains: (1) The distance between L_i and L_k is smaller than twice of the interference radius in the *Point model*, or $L_i \bigcap L_k \neq \emptyset$ in the *Area model*; and (2) The required time intervals from r_i and r_k overlap with each other. We denote the conflict relationships among requests by a conflict graph $\mathcal{G} = \langle \mathcal{V}, \mathcal{E} \rangle$, where $\mathcal{V}$ is the set of requests of buyers, and edge $\langle r_i, r_k \rangle \in \mathcal{E}$ if requests r_i and r_k conflict with each other. Note that, for the same requests r_i and r_k, different interference radii of channels will lead to a different conflict relationship. We use a matrix $Y = \{y_{i,k,j}\}_{n \times n \times m}$ to represent the conflict relationships in graph $\mathcal{G}$, in which if requests r_i and r_k conflict with each other in channel s_j, $y_{i,k,j} = 1$; otherwise, $y_{i,k,j} = 0$. Since the spectrum is a local resource, we also need to define a location matrix $C = \{c_{i,j}\}_{n \times m}$ to represent whether L_i is in the license regions of channel s_j. $c_{i,j} = 1$ if L_i is in the license regions of channel c_j; otherwise, $c_{i,j} = 0$. Therefore, two requests r_i and r_k can share channel s_j only if $y_{i,k,j} = 0$, and $c_{i,j} = 1$, $c_{k,j} = 1$.

The objective of our work is to design a mechanism satisfying *truthfulness* constraint, while maximizing the *social efficiency* or *revenue*. In our problem definition, an auction is said to be truthful if revealing true valuation is the *dominant strategy* for each bidder, regardless of other bidders' bids. An auction mechanism is truthful if its allocation algorithm is monotonic and it always charges critical values from its buyers [5]. The *critical value* for a buyer is the minimum bid value, with which the buyer will win the auction.

Social Efficiency Maximization: Social efficiency for an auction mechanism is defined as $\max \sum_{r_i \in \mathcal{R}} v_i x_i$, where $x_i = 1$ if buyer i wins in the auction; otherwise, $x_i = 0$.

Revenue Maximization: The revenue of an auction is the total payment of buyers. An auction maximizing the revenue for the auctioneer is known as an optimal auction in economic theory [4]. In the optimal auction, Myerson introduces the notion of virtual valuation $\phi_i(b_i)$ as

$$\phi_i(b_i) = b_i - \frac{1 - F_i(b_i)}{f_i(b_i)} \quad (1)$$

where $F_i(b_i)$ is the probability distribution function of true valuations of buyer i, and $f_i(b_i) = \frac{dF_i(b_i)}{db_i}$. According to the theory of optimal auction [4], maximizing the expected revenue is equivalent to finding the optimal solution of $\max \sum_{r_i \in \mathcal{R}} \phi_i(b_i) x_i$.

3. A STRATEGYPROOF SPECTRUM AUCTION FRAMEWORK

In this section, we propose a general truthful spectrum auction framework with the goal of maximizing social efficiency or revenue, as shown in *Algorithm 1*. In our framework, we can flexibly choose different optimization targets according to the practical requirements of auction problems. To ensure the worst case profit, the primary user will set a virtual reservation price η^θ, which is the minimum virtual price for spectrums per unit time. Let $\{\xi\}_\Psi$ denote a set whose elements satisfy property Ψ. The details are depicted as follows.

Algorithm 1 Our truthful spectrum auction framework

Input: conflict graph $\mathcal{G}$, location matrix C, set of channels $\mathcal{S}$, set of requests $\mathcal{R}$, monotone allocation and payment mechanism $\mathcal{A}$;
Output: channel assignment X, payment P;
  1: $\mathcal{R}' = \mathcal{R}$;
  2: **for** each $r_i \in \mathcal{R}$ **do**
  3:   **if** the target is maximization of social efficiency **then**
  4:     $\phi_i(b_i) = b_i$;
  5:   **else**
  6:     $\phi_i(b_i) = b_i - \frac{1 - F_i(b_i)}{f_i(b_i)}$;
  7:   **if** $\phi_i(b_i) < \eta^\theta t_i$ **then**
  8:     $\mathcal{R}' = \mathcal{R}' \backslash \{r_i\}$; // Delete r_i from set $\mathcal{R}'$
  9: Run $\mathcal{A}$ using the set of virtual bids $\{\phi_i(b_i)\}_{r_i \in \mathcal{R}'}$;
 10: Let $X = \{x_i\}_{r_i \in \mathcal{R}'}$ be the channel allocation and $\tilde{P} = \{\tilde{p}_i\}_{r_i \in \mathcal{R}'}$ be the corresponding payment returned by $\mathcal{A}$;
 11: **for** each x_i **do**
 12:   **if** the target is maximization of social efficiency **then**
 13:     $p_i = \tilde{p}_i$;
 14:   **else**
 15:     $p_i = \phi_i^{-1}(\tilde{p}_i)$;
 16: **return** X, P;

---

4. ALLOCATION MECHANISM WITH APPROXIMATION RATIO (1-1/e)

4.1 The Optimal Channel Allocation

In the channel allocation problem, we need to match the requests and channels optimally under their constraints. In order to simplify the matching model between requests and channels, we would like to segment the available time of each channel into many time slices. Recall that the available time of each channel is [0,T] in each auction period. Then, we use the arrival time a_i and deadline d_i of each request r_i to partition the time interval [0,T]. Each arrival time/deadline of requests divides the time axis of one channel into two parts. Suppose there are n requests, the time interval [0,T] is divided into no more than $2n$ time slices.

After the introduction of segmentation process, we give the detailed description of the channel allocation problem. First, for each partitioned time slice derived from channel s_j, it can only be allocated to the requests within the license area of channel s_j. Let $x_{j,i}^l$ represent whether the l-th time slice of channel s_j is allocated to the request r_i, then we get a constraint $x_{j,i}^l \leq c_{i,j}$. Second, each time slice can only be allocated to requests conflict-free with each other. Thus, we get another constraint $\sum_{k \neq i} x_{j,k}^l y_{i,k,j} = x_{j,i}^l \leq 1$. Let t_j^l be the length of l-th time slice in channel s_j. Modify a_i to be the first time slice r_i wants to use, and d_i to be the last time slice r_i wants to use. Moreover, if we allocate request r_i in channel s_j,

238

the time assigned to request r_i from channel s_j should be equal to the required time of request r_i, so we get $\sum_{l=a_i}^{d_i} x_{j,i}^l t_j^l \ge t_i x_{i,j}$. From the analysis above, the allocation problem can be formulated as follows.

$$\max \sum_{s_j \in \mathcal{S}} \sum_{r_i \in \mathcal{R}'} \phi_i b_i x_{i,j}, \text{ subject to,} \qquad \text{(IP (1))}$$

$$
\begin{cases}
\sum_{s_j \in \mathcal{S}} x_{i,j} \le 1, \forall r_i \in \mathcal{R}' \\
x_{j,i}^l \le c_{i,j}, \forall s_j \in \mathcal{S}, \forall r_i \in \mathcal{R}', \forall l \\
\sum_{k \ne i} x_{j,k}^l y_{i,k,j} + x_{j,i}^l \le 1, \forall s_j \in \mathcal{S}, \forall r_i \in \mathcal{R}', \forall l \\
\sum_{l=a_i}^{d_i} x_{j,i}^l t_j^l \ge t_i x_{i,j}, \forall s_j \in \mathcal{S}, \forall r_i \in \mathcal{R}' \\
x_{i,j} \in \{0, 1\}, \forall s_j \in \mathcal{S}, \forall r_i \in \mathcal{R}' \\
x_{j,i}^l \in \{0, 1\}, \forall s_j \in \mathcal{S}, \forall r_i \in \mathcal{R}', \forall l
\end{cases}
$$

where $x_{i,j}$ stands for whether channel s_j is allocated to request r_i or not, $y_{i,k,j}$ represents whether request r_i conflicts with request r_k or not in channel s_j.

4.2 (1-1/e)-Approximation methods

LP relaxation technique can be introduced to solve NP-hard problems, and it often leads to a good approximation algorithm. We release IP(1) to linear programming LP(2) by replacing $x_{i,j} \in \{0, 1\}$ with $0 \le x_{i,j} \le 1$, and replacing $x_{j,i}^l \in \{0, 1\}$ with $0 \le x_{j,i}^l \le 1$.

Recall that the number of time slices is no more than $2n$ for each channel, so LP(2) has a polynomial number of variables and constraints, and can be solved optimally in polynomial time.

4.2.1 Randomized Rounding

Suppose O_{LP2} is the optimal solution of LP(2), we apply a standard randomized rounding on it to obtain an integral feasible solution f_{IP1} to IP (1). The rounding procedure is presented as follows.

1. Randomly choose a channel s_j, randomly choose a request r_i with $x_{i,j} > 0$, and set $x_{i,j} = 1$;
2. If $x_{i,j} = 1$, set $x_{k,j} = 0$ for all requests r_k with $y_{i,k,j} = 1$;
3. If $x_{i,j} = 1$, set $x_{i,k} = 0$ for all channels with $k \ne j$.
4. Repeat steps 1 to 3 until all requests have been processed.

Through the randomized rounding procedure above, the optimal solution of LP(2) is converted into a feasible solution of IP(1). Let $w_{O_{LP2}}$ be the weight of O_{LP2}, and let $E[w_{f_{IP1}}]$ be the expected weight of f_{IP1}. We show in Theorem 1 that $E[w_{f_{IP1}}] \ge 1 - 1/e \cdot w_{O_{LP2}}$.

THEOREM 1. *The expected weight of the rounded solution is at least $1 - 1/e$ times of the weight of the optimal solution to LP (2).*

PROOF. Due to the page limit, all the detailed proofs can be referred to [6]. □

We have shown that the expected weight of feasible solution f_{IP1} of IP (1) obtained by our randomized rounding is larger than $1 - 1/e$ times of the weight of the optimal solution of LP (2). Obviously, the weight of the optimal solution of LP (2), which is denoted by $w_{O_{LP2}}$, is larger than the optimal solution of IP (1), which is denoted by $w_{O_{IP1}}$. Therefore, we can get that

THEOREM 2. *The expected weight of the rounded solution is at least $1 - 1/e$ times of the weight of the optimal solution to IP (1).*

4.2.2 Deterministic Methods

The rounding procedure only makes sure that the expected weight of f_{IP1} is larger than $1 - 1/e$ times of the weight of O_{LP2}. What we need is to find a feasible solution of IP(1) whose weight is exactly larger than $1 - 1/e$ times of the $w_{O_{LP2}}$. In the following, we show that the rounding procedure can be derandomized and how the method of conditional probabilities can be used in our setting.

Algorithm 2 DCA: Derandomized Channel Allocation Based on Linear Programming

---

Input: Conflict graph $\mathcal{G}$, location matrix C, set of channels $\mathcal{S}$, set $\mathcal{R}'$ sorted in increasing order according to a_i;
Output: channel assignment X^*;
1: Solve LP(2) optimally;
2: $E[w_{f_{IP1}}] = \sum_{s_j \in \mathcal{S}} \phi_i b_i (1 - \prod_{s_j \in \mathcal{S}}(1 - x_{i,j}))$;
3: **for** $i = 1$ to n **do**
4:    **if** $x_i > 0$ **then**
5:       **for** $j = 1$ to m **do**
6:          **if** $E[w_{f_{IP1}}] \le E[w_{f_{IP1}}|i, j]$ **then**
7:             set $x_{i,j} = 1, x_i = 1$;
8:             set all $x_{i,k} = 0$ and $x_{i,k}^l = 0$ if $k \ne j$;
9:             set all $x_{k,j} = 0$ and $x_{k,j}^l = 0$ if $k \ne i$ and $y_{i,k,j} = 1$;
10:             Break
11:    **if** $x_i \ne 1$ **then**
12:       $x_i = 0$;
13: **return** X^*;

Let $E[w_{f_{IP1}}|r_i \to s_j]$ be the expected weight when request r_i is allocated in channel s_j, and let $E[w_{f_{IP1}}|\tilde{i}]$ be the expected weight when request r_i will not be allocated in any channel. Next, we will show how our derandomization algorithm works. We first sort all the requests by their arrival time a_i in the ascending order. Let $x_i = \sum_{j \in \mathcal{S}} x_{i,j}$, and then scan all the requests one by one to decide which request can be allocated in channels. When request r_i is considered, we scan all of the channels that are available for r_i to check if r_i can be allocated in one of them. If $E[w_{f_{IP1}}|r_i \to s_j] < E[w_{f_{IP1}}]$, set $x_{i,j} = 0$; otherwise, allocate r_i in channel s_j, and set $x_{i,j} = 1, x_i = 1, x_{i,k} = 0$ if $k \ne j$. Meanwhile, if r_i is allocated in channel s_j, we set $x_{k,j}^l = 0$ if $y_{i,k,j} = 1$.

Suppose r_i is the first request that satisfies $x_i > 0$ in the ordered requests. Let $q_{i,j}$ denote the probability that request r_i is allocated in channel s_j and let $q_{\tilde{i}}$ denote the probability that r_i is not allocated in any channel. By the formula for conditional probabilities, we have

$$E[w_{f_{IP1}}] = \sum_{r_j \in \mathcal{S}} E[w_{f_{IP1}}|r_i \to s_j] q_{i,j} + E[w_{f_{IP1}}|\tilde{i}] q_{\tilde{i}} \quad (2)$$

In particular, there exists at least one conditional expectation in $E[w_{f_{IP1}}|r_i \to s_1], \cdots, E[w_{f_{IP1}}|r_i \to s_m], E[w_{f_{IP1}}|\tilde{i}]$, which is larger than $E[w_{f_{IP1}}]$. If it is $E[w_{f_{IP1}}|r_i \to s_j] \ge E[w_{f_{IP1}}]$, we allocate request r_i in channel s_j; otherwise, $E[w_{f_{IP1}}|\tilde{i}] \ge E[w_{f_{IP1}}]$ holds, reject request r_i, and set $x_{i,j} = 0$ for each $s_j \in \mathcal{S}$. This can be done since $E[w_{f_{IP1}}] = \sum_{r_i \in \mathcal{R}'} \phi_i b_i q_i$, and q_i can be computed precisely by

$$q_i = 1 - \prod_{s_j \in \mathcal{S}}(1 - x_{i,j}) \quad (3)$$

Let $q_{r_i \to s_j, k}$ stand for the probability that request r_k is allocated in a channel when request r_i is allocated in s_j. Then $q_{r_i \to s_j, k}$ can be calculated by

$$q_{r_i \to s_j, k} = \begin{cases} 1 - \prod_{o \ne j}(1 - x_{k,o}), y_{i,k,j} = 1 \\ q_k, otherwise \end{cases} \quad (4)$$

For each request r_i, we can compute $E[w_{f_{IP1}}|r_i \to s_j]$ precisely as the follows

$$E[w_{f_{IP1}}|r_i \to s_j] = \phi_i b_i + \sum_{k \neq i} \phi_k b_k q_{r_i \to s_j,k} \quad (5)$$

Given the selections in the prior requests, we can continue deterministically to allocate other requests and do the same thing while maintaining the invariant that the conditional expectation $E[w_{f_{IP1}}]$, never deceases. After allocating all of the requests, we can get a feasible solution of IP(1) whose weight is as good as $E[w_{f_{IP1}}]$, i.e. at least $(1 - 1/e)w_{O_{LP2}}$.

Recall that to ensure the truthfulness of our auction mechanism, the allocation algorithm must be bid-monotone. However, we cannot prove or disprove the bid-monotone property of the allocation method DCA (presented in Algorithm 2). Thus, it is unknown whether we can design a truthful mechanism based on this method. In the rest of the section, we revise this method and show that the revised method does satisfy the bid-monotone property.

Since that there exists at least one of the conditional expectations between $max_{s_j \in \mathcal{S}} E[w_{f_{IP1}}|r_i \to s_j]$ and $E[w_{f_{IP1}}|\tilde{i}]$, which is larger than $E[w_{f_{IP1}}]$. Thus, if we allocate r_i in the channel with the maximal conditional expectation as long as $max_{s_j \in \mathcal{S}} E[w_{f_{IP1}}|r_i \to s_j] \geq E[w_{f_{IP1}}|\tilde{i}]$, and do not allocate r_i in any channel otherwise, we can also get a feasible solution of IP(1), whose weight is as good as $E[w_{f_{IP1}}]$.

This can be done since we can compute $E[w_{f_{IP1}}|i,j]$ and $E[w_{f_{IP1}}|\tilde{i}]$ precisely as follows:

$$E[w_{f_{IP1}}|r_i \to s_j] = \phi_i b_i + E_{k \neq i}[w_{f'_{IP1}}|r_i \to s_j] \quad (6)$$

where $E_{k \neq i}[w_{f'_{IP1}}|r_i \to s_j]$ is the expected weight of all other requests when request r_i has been allocated in channel s_j. We can get it by allocating r_i in channel s_j first, and then solve LP(2) optimally with other requests.

$$E[w_{f_{IP1}}|\tilde{i}] = E_{\mathcal{R}'/r_i}[w_{f'_{IP1}}] \quad (7)$$

where $E_{\mathcal{R}/r_i}[w_{f'_{IP1}}]$ is the expected weight of all other requests when request r_i is not allocated in any channel. We can get it by solving LP(2) optimally with requests except r_i.

Based on the observation above, we give a revised version (called MDCA) of Algorithm DCA as follows.

Algorithm 3 MDCA: Monotone Derandomized Channel Allocation Based on Linear Programming

---

Input: Conflict graph $\mathcal{G}$, location matrix C, set of channels $\mathcal{S}$, set of $\mathcal{R}'$ sorted in increasing order according to a_i;
Output: channel assignment X^*;
1: Solve LP(2) optimally;
2: **for** $i = 1$ to n **do**
3:    **for** $j = 1$ to m **do**
4:      **if** $x_{i,j} > 0$ **then**
5:        $E[w_{f_{IP1}}|r_i \to s_k] = max_{s_j \in \mathcal{S}} E[w_{f_{IP1}}|r_i \to s_j]$
6:        **if** $E[w_{f_{IP1}}|r_i \to s_k] \geq E[w_{f_{IP1}}|\tilde{i}]$ **then**
7:          set $x_{i,j} = 1, x_i = 1$;
8:          set all $x_{i,k} = 0$ and $x^l_{i,k} = 0$ if $k \neq j$;
9:          set all $x_{k,j} = 0$ and $x^l_{k,j} = 0$ if $k \neq i$ and $y_{i,k,j} = 1$;
10:        Break
11:    **if** $x_i \neq 1$ **then**
12:      $x_i = 0$;
13: **return** X^*;

In MDCA, we first sort all of the requests by their arrival times in the ascending order, and then we scan all requests one by one to decide which request can be allocated in channels. When request r_i is considered, we compute $E[w_{f_{IP1}}|r_i \to s_j]$ for all channels $s_j \in \mathcal{S}$ that no request conflicting with it has been allocated in. We allocate r_i in channel s_k when $E[w_{f_{IP1}}|r_i \to s_k] = max_{s_j \in \mathcal{S}} E[w_{f_{IP1}}|r_i \to s_j] \geq E[w_{f_{IP1}}|\tilde{i}]$, and reject it otherwise. After the last request was considered in MDCA, we get a feasible solution of IP (1), whose weight is as good as $E[w_{f_{IP1}}]$.

THEOREM 3. *MDCA (see Algorithm 3) is bid monotone.*

Since the revised Algorithm MDCA is bid-monotone, there exists a critical value for each winner. Thus, we can design a truthful auction mechanism through charging each winner its critical value.

5. CONCLUSION

In this paper, we have studied the case that spectrum can be reused spatial domain, temporal domain. We have designed a general truthful spectrum auction framework which can maximize the social efficiency or revenue. As allocating channels optimally is NP-hard in our model, we have also proposed a set of near-optimal channel allocation mechanisms with an approximation factor of $(1 - 1/e)$.

Acknowledgement

The research of authors is partially supported by the National Grand Fundamental Research 973 Program of China (No.2011CB302905, No.2011CB302705), National Natural Science Foundation of China (NSFC) under Grant No. 61202028, No. 61170216, No. 612282 02, and NSF CNS-0832120, NSF CNS-1035894, NSF ECCS-1247 944. Specialized Research Fund for the Doctoral Program of Higher Education (SRFDP) under Grant No. 20123201120010.

6. REFERENCES
[1] L. Deek, X. Zhou, K. Almeroth, and H. Zheng. To preempt or not: Tackling bid and time-based cheating in online spectrum auctions. In *IEEE INFOCOM 2011*, pages 2219–2227, 2011.
[2] M. Dong, G. Sun, X. Wang, and Q. Zhang. Combinatorial auction with time-frequency flexibility in cognitive radio networks. In *IEEE INFOCOM 2012*, pages 2282–2290, 2012.
[3] A. Gopinathan, Z. Li, and C. Wu. Strategyproof auctions for balancing social welfare and fairness in secondary spectrum markets. In *IEEE INFOCOM 2012*, pages 2813–2821, 2011.
[4] R.B. Myerson. Optimal auction design. *Mathematics of operations research*, 6(1):58–73, 1981.
[5] N. Nisan. *Algorithmic game theory*. Cambridge Univ Pr, 2007.
[6] Yu-e Sun, He Huang, Xiang-Yang Li, Zhili Chen, Wei Yang, Hongli Xu, and Liusheng Huang. Near-optimal truthful auction mechanisms in secondary spectrum markets. *arXiv preprint arXiv:1305.6390*, 2013.
[7] H. Xu, J. Jin, and B. Li. A secondary market for spectrum. In *IEEE INFOCOM 2010*, pages 1–5, 2010.
[8] P. Xu and X.Y. Li. TOFU: semi-truthful online frequency allocation mechanism for wireless network. *IEEE/ACM Transactions on Networking (TON)*, 19(2):433–446, 2011.
[9] P. Xu, X.Y. Li, and S. Tang. Efficient and strategyproof spectrum allocations in multichannel wireless networks. *IEEE Transactions on Computers*, 60(4):580–593, 2011.
[10] P. Xu, S.G. Wang, and X.Y. Li. SALSA: Strategyproof online spectrum admissions for wireless networks. *IEEE Transactions on Computers*, 59(12):1691–1702, 2010.
[11] X. Zhou, S. Gandhi, S. Suri, and H. Zheng. ebay in the sky: strategy-proof wireless spectrum auctions. In *ACM Mobicom 2008*, pages 2–13, 2008.

Characteristic Function of Connectivity with Obstacles

Yun Wu, Xinbing Wang
Department of Electronic Engineering
Shanghai Jiao Tong University, China
{the_sun_shines, xwang8}@sjtu.edu.cn

ABSTRACT

Previous works usually simplify the network area as a disk, a square or a torus without obstacles. However, geometry in practical environment is much more intricate. Rectilinear communication between nodes is often blocked by external surroundings, such as buildings and mountains. Complicated shapes of the obstacles may cause the computation intractable. In this paper, we propose a characteristic function of connectivity to capture the geometric features of an ad-hoc network. The function can be expressed as power series, of which the orders and coefficients is based on the characteristics of the network. We formulate the shapes with fundamental factors and investigate their algebraic impact on connectivity. The critical transmission range is also calculated to mitigate the influence of obstacles.

Categories and Subject Descriptors

C.2.1 [**Computer-Communication Networks**]: Network Architecture and Design—*Wireless Communications*

General Terms

Theory and Algorithms

Keywords

Connectivity, Obstacle

1. INTRODUCTION

In wireless networks, connectivity is of paramount significance for the success of information interaction. The transmitting range in ad-hoc networks needs to be carefully designed to satisfy the requirement of connectivity and energy consumption. In [1], Gupta and Kumar proved that in flat static network, with range $r(n) = \sqrt{\frac{\log n + \kappa(n)}{\pi n}}$, overall connectivity can be established with probability one as $n \to \infty$

if and only if $\kappa(n) \to \infty$. In modern society, radio frequency network designers often has to take obstacles into consideration. The complication of neighborhood may harm the connectivity by blocking the rectilinear communication. With the widespread of wireless radio technology, obstacles become an inevitable challenge for designers that will impair the quality of communication.

There are several previous works concerning the relationship between connectivity and obstacles. In [3], Tan, etc. provided connectivity-guaranteed and obstacle-adaptive deployment schemes to achieve maximum coverage for mobile sensor network. In [5], the author deployed the nodes by robots in partitioned region separated by the distance from obstacle and boundaries. [6] explored the robotic problem of preserving desired links in the duration of movement. P. D. Holmes and E. Jungert [4] offered 2 alternative methods for routing planning: heuristic, symbolic processing and the other employing geometric calculation.

However, in other circumstances, the distribution of nodes is random. The number of nodes may be too substantial for human beings to locate them. The sensors may be carried by moving things like animals or vehicles and their location cannot be fixed. Or the target area cannot be accessed easily, like the battlefield controlled by the enemy. Therefore, it is essential to determine the preset critical transmission range in given environment to guarantee connectivity. In these situations, the influence of different geometric feature is difficult to investigate. Our work provides an algebraic method to investigate the geometry of the network. Based on our result, the critical transmission range of sensors can be computed directly. The influence of the shapes and boundaries can be studied by investigating the coefficients of a characteristic function.

In this paper, we concentrate on connectivity in ad-hoc networks with obstacles. We develop a characteristic function of connectivity to capture the geometric features of an ad-hoc network for the first time. One can easily interpret the geometry to algebraic functions based on the methods we provided. The impact of different geometric features can be studied by investigating the coefficients in the power series. Specifically, we explore the crucial features in 2-D network. We formulate the shapes with three fundamental factors, i.e., area, perimeter and angle of intersection. Then we extend the result to high dimension space and provide some general properties of characteristic function. Meanwhile, the critical transmitting radius is calculated to mitigate the influence from barriers.

2. NETWORK MODEL

In this section, we explain our network model with obstacles. We consider the following problem: Let $\mathcal{D}$ be a torus in $\mathbb{R}^2$ normalized to have unit area. Let $\mathcal{G}(n, r(n))$ be the network formed when n nodes are placed randomly in $\mathcal{D}$ with a uniform probability distribution, and independently of each other. All nodes in the network share same transmitting radius $r(n)$. Suppose that $\mathbf{x}_k$ is the location of node k. Nodes i and j can communicate in one hop if $\|\mathbf{x}_i - \mathbf{x}_j\| \leq r(n)$ (The norm used here is the Euclidean norm), provided that no obstacles lie on the connection of 2 nodes. These obstacles are considerable and signal bypassing by diffraction cannot be stably detected and correctly interpreted by current technology. Then the problem is to determine $r(n)$ that guarantees $\mathcal{G}(n, r(n))$ to be asymptotically connected with probability one. That is, the probability that $\mathcal{G}(n, r(n))$ is connected, denoted by $\mathbb{P}_c(n, r(n))$, goes to one as $n \to \infty$.

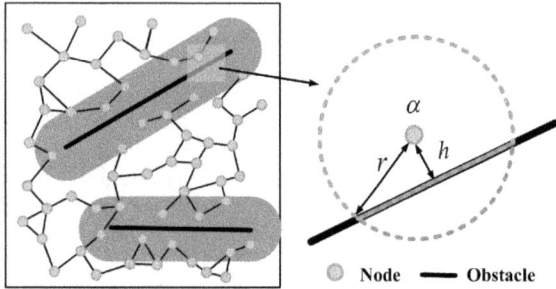

Figure 1: nodes affected by 1 obstacle.

2.1 Definition

Before we state our main results, we first give a rigorous definition of critical transmission range(CTR) in wireless sensor network. Note that these factors are very important indicators for connectivity and power consumption issue in the wireless network.

DEFINITION 1. *For wireless sensor networks, $r(n)$ is the critical transmission range if*

$$\lim_{n \to \infty} \mathbb{P}_c(\mathcal{G}) = 1, \; if \; r \geq cr(n) \; for \; any \; c > 1;$$

$$\lim_{n \to \infty} \mathbb{P}_c(\mathcal{G}) < 1, \; if \; r \leq \hat{c}r(n) \; for \; any \; 0 < \hat{c} < 1.$$

3. PROPERTIES OF CHARACTERISTIC FUNCTIONS

In this work, we focus on the characteristic function of connectivity. The variable in the function is the transmission range of nodes in the network. The function can be expressed as power series and the coefficients deliver the message about the geometry of network.

Suppose that the critical transmission range can be regarded as the root of equation $f(x) = 0$, where $x \in (0, 1)$. Let

$$f(x) = \sum_i a_i x^i. \tag{1}$$

Then we can find the following properties to help determine $f(x)$.

- In k-D network with obstacles, $a_i = 0 (i > 2k)$.

- The constant term in the function is determined by the density of the network. The sum of the other terms shows the expected transmission range of nodes in the network area.

- Given the geometric feature, the order it determines depends on its influential area and the transmission scope of the nodes. The value of coefficient it determines depends on the expected blocked range of nodes in its influential area.

- The existence of obstacle in network always harms the connectivity and increase the critical transmission range. Such impair on connectivity results from the edge of the obstacles. Higher terms in the function will mitigate the impair but can never offset it completely.

4. GEOMETRIC ANALYSIS OF THE IMPACT OF OBSTACLES

In this section, we will analyze the influence of different geometric features in 2-D space. According to our work[2], the expected transmission range is essential to the calculation of critical transmission range. In other words, we can investigate the impact of obstacles from the perspective of reachable transmitting scope.

4.1 Area

Suppose that the area of the obstacles can be denoted as S_o. Intuitively, we can find that nodes are not allowed to locate in/on the obstacle on a planar. With the same network area of $\mathcal{D}$ and number of nodes n, the density of nodes actually increases to $n' = n/(1 - S_o)$. Therefore, the influence of the area is embodied by the increase of the density, despite the specific shape of the obstacles.

4.2 Perimeter

Firstly, we will consider the situation where only 1 linear obstacle with length l locates in $\mathcal{D}$. Stable communication cannot set up if the obstacle lies between two nodes' connection. We define effective transmitting scope of node i as the area where other nodes are able to connect to node i. Considering the influence of the obstacle, such area may not be a circle as previous works suggested ([1]).

Without the loss of generality, we arbitrarily choose node i. There are two situations of its position. Firstly, the distance between node i and the obstacle is larger than its transmission distance, r. (Here we use $r = r(n)$ to avoid redundancy.) In this situation, no obstacles appear in the transmitting area. Nodes within i's effective transmitting area $S = \pi r^2$ are able to maintain communication with i.

In the second case, the linear obstacle lies in the communication scope of wireless node. In this situation, nodes behind the obstacle cannot transmit message to i even though it located in the r-neighborhood of i. The effective transmission area can be calculated as the union of a triangle and a sector.

$$S = r^2 \arccos \frac{h}{r} + h\sqrt{r^2 - h^2}.$$

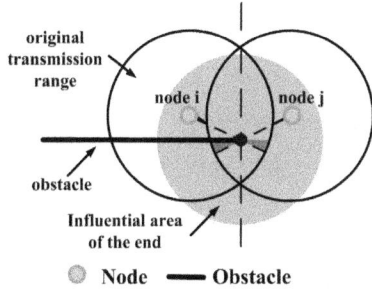

Figure 2: nodes affected by the end of the obstacle.

Here, the transmission range of each node is r. The length of linear obstacle is l. h denotes the distance between the wireless node i and the obstacle.

Therefore, the relationship between the effective transmission scope and the location of the node can be expressed as

$$S = \begin{cases} \pi r^2 & h > r \\ (\pi - \arccos \frac{h}{r})r^2 + h\sqrt{r^2 - h^2} & h < r. \end{cases}$$

Then we use integral to find the expected transmission area of a certain node.

$$\begin{aligned} E &= (1 - 2lr)\pi r^2 + \\ & \quad 2lr \cdot \frac{1}{r} \int_0^r (\pi - \arccos \frac{h}{r})r^2 + h\sqrt{r^2 - h^2} dh \\ &= (1 - 2lr)\pi r^2 + 2lr(\pi r^2 - \frac{2}{3}r^2) \\ &= \pi r^2 - \frac{4}{3}lr^3. \end{aligned} \quad (2)$$

Note that we regard the area influenced by the obstacle as a rectangle. We can prove the equivalence by Figure. 2. Suppose that the bypassing signal is too weak to be detected. Now we consider the situation where the node locates near the end of the obstacle. As Figure 2 shows, the influential area of the end is a circle with radius r (the light blue area). The effective transmission range of nodes within the circle cannot be calculated as the previous conditions. The dark blue area on the right is the additional transmitting area of node i in the second situation. And the dark blue area on the left side is the blocked area of node j in the first situation. Given the position on the left side of the influential area, one can always find its symmetrical point on the right. The extra transmission range on the right equates to the blocked area of the corresponding node on the left. Therefore, when we compute the expected transmission range, we can classify the nodes on the right into the first situation and left into the second case.

In real environment, obstacles may not stand in one line. There may be multiple barriers and their geometric relationships vary. As we mentioned before, nodes in different regions suffer from different impact and every obstacle has its functioning region. Sometimes, they are far away from each other such that they do not interact with each other. However, as they become closer, the functioning regions overlap. Will such change make difference on the polynomial equation? Our answer is no until the intersection of obstacles. We have proved that the coefficient of r^3 is determined by total perimeter. The influence of intersection can be found in higher order.

4.3 Intersection

The problem will become more complicated with intersection. We begin with 2 obstacles and the result can be easily generalized.

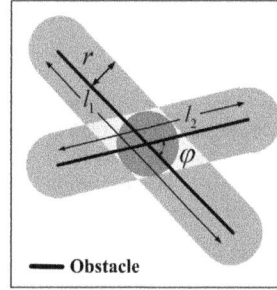

Figure 3: Network with intersection

As Figure 3 shows, there are four kinds of regions in network with intersected obstacles. We will calculate the expected transmission range respectively. In the white region, the node is not affected by obstacles and the effective transmission scope is πr^2. For nodes blocked by 1 obstacle, the expected transmission range is $\pi r^2 - \frac{2}{3}r^2$ as proved in Section 4.2.

We can easily obtain the area where the nodes are influenced by 2 obstacles as $4r^2/\sin\varphi$ and φ is the angle of intersection. There are 2 kinds of nodes with in such area, i.e. intersection happens in or out of the node's transmission range. In the first situation, area blocked by 2 obstacles does not overlap and the effective transmission range of the node is the linear combination of 2 individual obstacle. In the second situation(Figure 4), the overlap will decrease the area blocked by barriers. In this section, we denote the arch

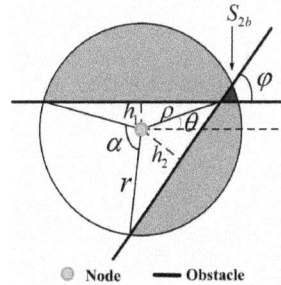

Figure 4: Transmission range with intersected obstacles

area blocked by one obstacle as S_{1b} and two obstacles as S_{2b}.

Then the expected blocked area is

$$\begin{aligned} S_b &= \int_{1 - 2lr + \frac{4r^2}{\sin\varphi}} 0 dS + \int_{2lr - 2 \cdot \frac{4r^2}{\sin\varphi}} S_{1b} dS \\ & \quad + \int_{\frac{4r^2}{\sin\varphi} - \pi r^2} 2S_{1b} dS + \int_{\pi r^2} 2S_{1b} - S_{2b} dS \\ &= \int_{2lr} S_{1b} dS - \int_{\pi r^2} S_{2b} dS \\ &= 2lr \cdot \frac{2}{3}r^2 - \int_{\pi r^2} S_{2b} dS \end{aligned}$$

243

We use polar coordination to simplify our computation and we can find that

$$\int_{\pi r^2} S_{2b}\mathrm{d}S = 2\int_0^r \int_0^\pi S_{2b}(\rho,\theta)\rho\mathrm{d}\theta\mathrm{d}\rho$$
$$= r^4\left(\frac{\pi^2}{2} + \frac{5}{6} - \frac{1}{2}\cos 2\varphi + \pi\varphi - \varphi^2\right)$$

With increase of φ, the influence of the intersection becomes stronger and reaches the peak when $\varphi = \frac{\pi}{2}$. The fact that derivative of $\int_{\pi r^2} S_{2b}\mathrm{d}S$ is always above zeros when $\varphi \in (0, \pi/2)$ also verifies the monotonous increase of the function.

Therefore, the expected transmission range with intersected obstacles can be calculated as

$$E = \pi r^2 - S_b$$
$$= \pi r^2 - 2lr\cdot\frac{2}{3}r^2 + r^4\left(\frac{\pi^2}{2} + \frac{5}{6} - \frac{1}{2}\cos 2\varphi + \pi\varphi - \varphi^2\right)$$
$$= \pi r^2 + f_l(l)r^3 + f_\varphi(\varphi)r^4$$

where l is total length of 2 obstacles, $f_l(l) = -\frac{4}{3}l$, and $f_\varphi(\varphi) = \frac{\pi^2}{2} + \frac{5}{6} - \frac{1}{2}\cos 2\varphi + \pi\varphi - \varphi^2$.

5. ALGEBRAIC ANALYSIS OF THE IMPACT OF OBSTACLES

According to previous analysis, we can suppose that a characteristic function in 2-D network can be expressed as

$$f(x) = a_0 + a_1 x + a_2 x^2 + a_3 x^3 + a_4 x^4.$$

Then the critical transmission range can be regarded as the root of $f(x) = 0$, $x \in (0, 1)$. Based on [2], we are able to provide the value of every coefficient and discuss its influence on connectivity.

- $a_0 = -S_n \frac{\log \lambda}{\lambda}$, where λ is density of the nodes and S_n is the network area. a_0 decreases with the increase of λ. Meanwhile, we can see that $a_0 < 0$ and $a_0 \to 0$ when $\lambda \to \infty$.

- $a_1 = 0$ when the nodes are static.

- $a_2 = (1 - S_o)\pi$, where S_o is the total area of obstacles. Since $0 \le S_o \le 1$, $a_2 \in (0, \pi]$.

- $a_3 = kl$, where l is the perimeter of obstacles and k is a constant. When obstacles are regarded as lines, $k = -\frac{4}{3}$. Otherwise, nodes can only locate on one side of obstacles and $k = -\frac{2}{3}$. Here, the perimeter of one linear obstacle must be less than $\pi/2$ and the perimeter of a polygon must be less than π. In either case, $a_3 \le 0$. If $a_3 = 0$, then $a_4 = 0$ and no obstacles exist in the network area.

- $a_4 = f_\varphi(\varphi)$. Here, f_φ is a function of intersection angles. Despite the complexity, we find that $a_4 \ge 0$ always holds. Thus, the existence of interception will improve connectivity by reducing CTR. However, such improvement will alleviate, but never offset the negative impact of obstacles. If $a_4 = 0$, there is no interception between obstacles.

Then we will prove the existence of the root of the critical transmission range. According to the discussion above, for the network with sufficiently large density,

$$f(1) = a_0 + a_2 + a_3 + a_4 > 0, n \to \infty.$$

Meanwhile, suppose that r_0 is the CTR without obstacles. Then $a_0 + a_2 r_0^2 = 0$

$$f(r_0) = a_0 + a_2 r_0^2 + a_3 r_0^3 + a_4 r_0^4 < 0, n \to \infty$$

Concerning the continuity of quartic function, there will always exist a root of $f(x) = 0$ in $(r_0, 1)$ as the critical transmission range of the network.

6. CONCLUSION

In this paper, we provide characteristic function to study connectivity in a practical environment. The influence from external surroundings like obstacles has been taken into consideration. Our contributions are twofold. First of all, the characteristic function can help us analyze different geometry in ad-hoc network. One can also infer the realistic network according to the characteristic function. In 2-D network, the constant term in the function is determined by the density of the ad-hoc network. The coefficient of x^2, x^3, and x^4 relate to the area, linear boundaries and intersection in network respectively. In addition, The requisite transmission range can be easily computed to mitigate the blockage of obstacles. Therefore, our work provides a novel insight on the understanding of connectivity in ad-hoc networks.

7. REFERENCES

[1] P. Gupta and P. R. Kumar, "Critical Power for Asymptotic Connectivity in Wireless Networks," in *Stochastic Analysis, Control, Optimization and Applications:* A Volume in Honor of W. H. Fleming, W. M. McEneaney, G. Yin, and Q. Zhang, Boston, MA: Birkhauser, pp. 547-566, 1998.

[2] Y. Wu and X. Wang, "Characteristic Function of Connectivity with Obstables." technical report, 2013, available at http://iwct.sjtu.edu.cn/Personal/xwang8/paper/MOBIHOC2013_Tech.pdf.

[3] G. Tan, S. A. Jarvis, and A. -M. Kermarrec, "Connectivity-Guaranteed and Obstacle-Adaptive Deployment Schemes for Mobile Sensor Networks," in *IEEE TMC*, vol. 8, no. 6, pp. 836-848, June 2009.

[4] P. D. Holmes, and E. R. A. Jungert, "Symbolic and geometric connectivity graph methods for route planning in digitized maps," in *IEEE Trans.on Pattern Analysis and Machine Intelligence*, vol. 144, no. 5, pp. 549-565, May 1992.

[5] Y-C. Wang, C-C. Hu, and Y-C. Tseng, "Efficient deployment algorithms for ensuring coverage and connectivity of wireless sensor networks," in *Proc. First International Conference on Wireless Internet*, pp. 114-121, July 2005.

[6] J. M. Esposito, and T. W. Dunbar, "Maintaining wireless connectivity constraints for swarms in the presence of obstacles," in *IEEE Proc. Robotics and Automation*, pp. 946-951, May 2006.

Maximizing Network Throughput with Minimal Remote Data Transfer Cost in Unreliable Wireless Sensor Networks

Xu Xu[†] Weifa Liang[†] Xiaohua Jia[‡] Wenzheng Xu[†§]

[†]Australian National University
[‡]City University of Hong Kong
[§]Sun Yat-Sen University, China

ABSTRACT

In this paper we consider the use of a link-unreliable wireless sensor network for remote monitoring, where the monitoring center is geographically located far away from the region of the deployed sensor network. The sensing data is transferred to the monitoring center by the third party communication service, which incurs service cost. We first formulate a novel optimization problem of maximizing the network throughput with minimal service cost, which is shown to be NP-hard. We then develop approximation algorithms. We finally evaluate the performance of the proposed algorithms by simulations. Experimental results demonstrate that the solutions delivered by proposed algorithms are fractional to the optimum.

Categories and Subject Descriptors

C.2.2 [**Computer-Communication Network**]: Network Protocols; G.1.6 [**Numerical Analysis**]: Optimization

Keywords

Unreliable data transmission, load-balanced forest, combinatorial optimization problem

1. INTRODUCTION

In this paper we consider a renewable wireless sensor network (WSNs) [1] with unreliable wireless links deployed in a remote region to monitor phenomena of interest, where the monitoring center is geographically located far away from the monitored region [2]. The transfer of sensing data from the deployed WSN to the monitoring center is carried by the third party communication service, such as 3G/4G networks or satellite telecommunication. The accumulative volume of data received by the center within a specified period is referred to as the *network throughput* and the cost incurred by the use of the third party service is referred to as the *service cost*. Our objective is to maximize the network throughput

of the deployed unreliable sensor network, while minimizing the service cost of transferring data to the monitoring center.

Deploying wireless sensor networks for environmental monitoring has been extensively studied in the past, and most existing studies in literature focused on optimizing network performance by assuming (i) the base station and the sensors are located in the same region; and (ii) wireless communication in the sensor network is reliable. In contrast, we here deal with an essentially different application scenario where (i) the data monitoring center is geographically located far away from the region of the sensor network, and the sensing data needs to be forwarded to the center via a third party network which does incur service cost, and (ii) wireless communication in the sensor network are not reliable, which means that data loss is unavoidable during its transfer. Load-balancing adopted in this paper was extensively studied in the past [3, 4]. However, the load-balancing studied here is different from previous studies in that the loads at tree roots are allowed to exceed their capacities, and the load of each tree root is determined by not only the number of its descendants but also the end-to-end reliability of each routing path between a node and the root. Thus, traditional solutions are not applicable, and new algorithms need to be developed.

2. PRELIMINARIES

2.1 System model

We consider a heterogeneous, unreliable wireless sensor network $G = (V \cup GW, E)$, where V is the set of sensors, GW is the set of gateways equipped with 3G/4G radios, and E is the set of links. There is a link between two sensors or a sensor and a gateway if they are within the transmission range of each other. $n = |V|$ and $K = |GW|$. We assume that both sensors and gateways are stationary and their locations are known a *priori*, where gateways are deployed in some strategic locations in the monitoring region. The data generation rate of all sensors is identical, denoted by r. We consider a long-term periodic environmental monitoring application scenario, in which sensors have low data generation rates and the generated data is transmitted to the gateways through multi-hop relays. Thus, data burst and bandwidth capacity constraint are not the major issues for such an application. Denote by R the transmission range of each sensor. As wireless communication is unreliable, denote by p_e the *reliability* of link $e \in E$ with $0 \le p_e \le 1$. We assume that the sensing data generated from each sensor will be sent to one of the gateways through the routing tree rooted at the gateway. The collected data by the gateways will then be forwarded to the remote monitoring center, through a long-distance communication service provided by a third party company.

The accumulative volume of data received by the K gateways within a specified monitoring period τ is defined as the *network throughput*. Let T_i be the tree rooted at gateway $g_i \in GW$ and $V(T_i)$ the set of sensor nodes in T_i, $1 \leq i \leq K$. All sensors are to be included in the K trees, i.e., $\bigcup_{1 \leq i \leq K} V(T_i) = V$. Let $e_1, e_2, \ldots, e_l$ be the link sequence of a routing path $P(v, g_i)$ in T_i from a sensor node $v \in V(T_i)$ to gateway g_i. Denote by p_{v,g_i} *the end-to-end reliability* of $P(v, g_i)$ between v and g_i, then $p_{v,g_i} = \prod_{i=1}^{l} p_{e_i}$. Let $L(g_i)$ be the amount of data received by gateway g_i through tree T_i within the period of τ, referred to as *the load* of gateway g_i, then $L(g_i) = \sum_{v \in V(T_i)} (p_{v,g_i} \cdot \tau \cdot r)$. The network throughput thus is

$$D^{(\tau)} = \sum_{i=1}^{K} L(g_i). \qquad (1)$$

The cost of transferring the collected data from the gateways to the remote monitoring center for a monitoring period of τ is referred to as *the service cost*. In this paper, we adopt a popular data plan provided by telecommunication companies, where a fixed cost C_f for a data quota Q at each gateway is charged within the period of τ, and the extra charge will be applied with a *penalty* rate c_p for every MB of exceeding data over the data quota, assuming that the penalty rate is much higher than the fixed cost rate c_f for data quota Q, i.e., $c_p > c_f = \frac{C_f}{Q}$. Let C_p be the penalty cost of a data quota Q, i.e., $C_p = c_p \cdot Q$, then $C_p > C_f$. The service cost on each individual gateway depends on not only the volume of collected data but also the chosen data plan. Denote by C the service cost for the period of τ,

$$C = K \cdot C_f + \sum_{i=1}^{K} \max\{0, (L(g_i) - Q) \cdot c_p\}. \qquad (2)$$

2.2 Problem definition

Given an unreliable sensor network $G(V \cup GW, E)$, a specified monitoring period of τ, and a data plan that has a fixed cost C_f for a data quota Q and a penalty rate c_p, the *maximizing network throughput with minimal service cost* (MTMC) problem is to find a forest $\mathcal{F} = \{T_i \mid T_i$ is a routing tree rooted at gateway $g_i \in GW, 1 \leq i \leq K\}$ in G spanning all sensor nodes such that the accumulative volume of data received by the K gateways $D^{(\tau)}$ in Eq. (1) is maximized while the service cost of transferring $D^{(\tau)}$ to the remote monitoring center is minimized.

THEOREM 1. *The decision version of MTMC problem is NP-complete.*

PROOF. This claim is proved by reducing a NP-complete problem, the subset sum problem, to the MTMC problem. Details are omitted, due to space limitation. $\square$

3. ALGORITHM FOR UNIFORM LINK RELIABILITY

In this section we consider the MTMC problem in a sensor network where each link has a uniform link reliability p with

$0 < p \leq 1$, for which we devise a novel approximation algorithm that can achieve the maximum network throughput while the service cost is bounded.

Given the network $G(V \cup GW, E)$, we construct another network $G' = (V \cup GW \cup \{s'\}, E')$ as follows. A virtual sink s' and an edge between s' and each gateway node $g \in GW$ are added to G', i.e., $E' = E \cup \{(g_i, s') \mid g_i \in GW\}$. Let T^{BFS} be a Breadth-First search tree in G' rooted at the virtual sink s' with the depth h, where the virtual sink s' is in layer 0 and all gateway nodes are in layer 1 of T^{BFS}. Let V_l be the set of nodes of T^{BFS} in layer l for all l with $0 \leq l \leq h$. Then, the nodes in $V \cup GW$ is partitioned into h disjoint subsets $V_1, V_2, \ldots, V_h$ such that $V_1 = GW$, $\cup_{l=2}^{h} V_l = V$, and $V_i \cap V_j = \emptyset$ if $i \neq j$ and $1 \leq i, j \leq h$. Note that $l - 1$ is the minimum number of hops in G from $v \in V_l$ to its nearest gateway in GW, with $2 \leq l \leq h$. For each node $v \in V_l$, its contribution to the network throughput for a period of τ is identical and at its maximum, which is $d_l = d_{max}^v = p^{l-1} \cdot (\tau \cdot r)$. The maximum network throughput thus is $D_{max} = \sum_{l=2}^{h} \sum_{v \in V_l} p^{l-1} \cdot (\tau \cdot r) = \sum_{l=2}^{h} (|V_l| \cdot p^{l-1} \cdot \tau \cdot r)$.

3.1 Load-balanced forest

We first iteratively construct a forest $\mathcal{F}$ consisting of K routing trees rooted at K gateways such that the load among the gateways is well balanced while the maximum network throughput is maintained. Within each iteration, a level expansion of each tree is conducted. Let $\mathcal{F}_l$ be the forest spanning the nodes in the first l layers. Initially, $\mathcal{F}_1$ spans all gateways $g_1, g_2, \ldots, g_K \in GW$ and the load of each gateway $L(g_i)$ is 0, for all i with $1 \leq i \leq K$. Assuming that forest $\mathcal{F}_l$ has been built, we now expand the forest to $\mathcal{F}_{l+1}$ by including the nodes in V_{l+1} with the objective that the maximum load among the K gateways in $\mathcal{F}_{l+1}$ is minimized. To this end, we adopt a similar strategy as the one in [4]. The algorithm of constructing a load-balanced forest is referred to as Load_Balanced_Forest, or LBF for short.

3.2 Dynamic load readjustment

Having the forest of load-balanced routing trees, we then readjust the load among the gateways dynamically to further reduce the service cost. To balance the load among the K gateways, we consider the following cases. If the loads of all gateways are no greater than or no less than the data quota Q, we do nothing because the service cost is already the minimum one. Otherwise, the cost can be improved through readjusting the load among the K gateways through a series of *edge swapping* that replaces a tree edge by a non-tree edge while keeping network throughput unchanged. To this end, we only consider edge swapping in the same neighboring layer to reduce the service cost.

Given the load-balanced trees in forest $\mathcal{F}$, the dynamic load readjustment algorithm examines the tree edges in the forest from the lower layer to the higher layer. Let $E_{l,l+1} = (V_l \times V_{l+1}) \cap E$ be the edge set between two neighboring layers l and $l+1$ for all l with $1 \leq l \leq h-1$. Let $e_1 = (u, v) \in E_{l,l+1}$ be a tree edge considered at this moment and nodes v and u are in layers l and $l+1$, respectively. Let $L(u)$ be the volume of data collected at node u in T_i from all its descendant nodes in the subtree rooted at u, and the volume of data received at the gateway derived from node u is $p^l \cdot L(u)$. We remove this tree edge and add another non-tree edge $e_2 = (u, v') \in E_{l,l+1}$ to form a new forest if this leads to a less service cost. Assume that nodes v and v' are in trees

T_i and T_j rooted at gateways g_i and g_j. Perform swapping only if the following conditions are met: (i) T_i and T_j are not the same tree, (ii) $L(g_i) > Q$ while $L(g_j) < Q$, and (iii) $L(g_i) - p^l \cdot L(u) > L(g_j)$. This procedure continues until all tree edges have been examined. We refer to this dynamic load readjustment procedure as algorithm `Refine_Cost`.

3.3 Algorithm

Algorithm `Uniform_Link` for the uniform link reliability case is described as follows.

Algorithm 1 `Uniform_Link`

Input: $G(V \cup GW, E)$, monitoring period of τ, the data plan consisting Q, C_f and c_p, and link reliability p
Output: The network throughput $D^{(\tau)}$ and the cost C
1: $\mathcal{F} \leftarrow \emptyset$;
2: Partition the nodes in $V \cup GW$ into h disjoint subsets $V_1, V_2, \ldots, V_h$;
3: Construct $\mathcal{F}_h$ by calling **algorithm LBF**, the network throughput $D^{(\tau)}$ is then obtained;
4: Readjust the load among the gateways in $\mathcal{F}_h$ by calling **algorithm Refine_Cost**. Let C be the cost;
5: **return** $D^{(\tau)}$ and C.

THEOREM 2. *Given an unreliable sensor network $G(V \cup GW, E)$ with uniform link reliability p with $0 < p < 1$, there is an approximation algorithm* `Uniform_Link` *which can achieve the maximum network throughput with at most $(1 + \frac{C_p}{C_f})$ times of the minimal service cost. The time complexity of the proposed algorithm is $O(|V||E|^2)$, assuming that $|GW| << |V|$.*

PROOF. Omitted, due to space limitation. $\square$

4. ALGORITHM FOR NON-UNIFORM LINK RELIABILITY

In this section we deal with the MTMC problem in wireless sensor networks with non-uniform link reliabilities.

4.1 Simple heuristic

Recall that $T_1, T_2, \ldots, T_K$ are the K routing trees rooted at the K gateway nodes. We now show that how to construct the K routing trees such that the network throughput is maximized. Initially, each tree T_i contains only gateway $g_i \in GW$ for each i with $1 \leq i \leq K$. Let $V' \subseteq V$ be the set of nodes that are not included in these trees but one hop neighbors of the nodes in $\cup_{i=1}^{K} V(T_i)$, i.e., $V' = \{v \mid (u, v) \in E, u \in \cup_{i=1}^{K} V(T_i), v \notin \cup_{i=1}^{K} V(T_i)\}$. The proposed algorithm proceeds iteratively. Within each iteration, only one node in V' is added to one of the K trees. A node $v' \in V'$ is added if the routing path between v' and the gateway of the tree that it joined is the most reliable. Node v' is then removed from V'. This procedure continues until $V' = \emptyset$. The service cost then can be calculated by Eq. (2). We refer to this iterative algorithm as `Simple_Alg`.

THEOREM 3. *Given an unreliable sensor network $G(V \cup GW, E)$ with link reliability p_e for each link $e \in E$, algorithm* `Simple_Alg` *delivers a solution that has the maximum network throughput but the service cost is not optimized.*

PROOF. Every node in V is added to forest $\mathcal{F}$ via the most reliable path, thus the accumulative volume of data collected by all gateways is the maximum one. Notice that the service cost is not taken into account in algorithm `Simple_Alg`, thus it is not optimized. $\square$

4.2 Approximation algorithm

We now take into account the service cost in the design of an improved algorithm. For the sake of convenience, we assume that the reliability of each link in $G(V \cup GW, E)$ is within the range of $[p, (1+\delta)p]$, where $p > 0$ and $(1+\delta)p \leq 1$. The improved algorithm proceeds as follows.

We first call algorithm `LBF` with uniform link reliability of p to construct a forest consisting of load-balanced routing trees $\mathcal{F} = \{T_1, T_2, \ldots, T_K\}$, clearly $\mathcal{F}$ is a feasible solution to the problem. We then perform dynamic load readjustment on the trees in $\mathcal{F}$ to further improve the network throughput while optimizing the service cost too. Similar to the discussions in the previous section, we only consider the edge swapping in the same neighboring layers. Consider swapping a pair of edges: a tree edge $e_1 = (u, v)$ and a non-tree edge $e_2 = (u, v')$, where node u is in layer $l + 1$ and nodes v and v' are in layer l of trees T_i and T_j rooted at gateways g_i and g_j respectively. We distinguish into five cases. Case one: when $T_i = T_j$ and $L(g_i) > Q$, perform swapping only if $\Delta d = (p_{u,v'} \cdot p_{v',g_i} - p_{u,v} \cdot p_{v,g_i}) \cdot L(u) > 0$. Case two: when $T_i = T_j$ and $L(g_i) < Q$, perform swapping only if $\Delta d = (p_{u,v'} \cdot p_{v',g_i} - p_{u,v} \cdot p_{v,g_i}) \cdot L(u) > 0$. Case three: when $T_i \neq T_j$, $L(g_i) > Q$ and $L(g_j) < Q$, perform swapping only if (i) $\Delta d = (p_{u,v'} \cdot p_{v',g_i} - p_{u,v} \cdot p_{v,g_i}) \cdot L(u) > 0$, and (ii) either $L'(g_i) = L(g_i) - p_{u,v} \cdot p_{v,g_i} \cdot L(u) \geq Q$ or $Q - L(g_j) > Q - L'(g_i)$. Case four: when $T_i \neq T_j$, $L(g_i) < Q$ and $L(g_j) < Q$, perform swapping only if (i) $\Delta d = (p_{u,v'} \cdot p_{v',g_j} - p_{u,v} \cdot p_{v,g_j}) \cdot L(u) > 0$, and (ii) either $L'(g_j) \leq Q$ or $Q - L(g_j) \geq p_{u,v} \cdot p_{v,g_j} \cdot L(u)$. Case five: when $T_i \neq T_j$, $L(g_i) > Q$ and $L(g_j) > Q$, perform swapping only if (i) $\Delta d = (p_{u,v'} \cdot p_{v',g_j} - p_{u,v} \cdot p_{v,g_j}) \cdot L(u) > 0$, and (ii) $L'(g_i) = L(g_i) - p_{u,v} \cdot p_{v,g_i} \cdot L(u) \geq Q$.

Having performed a series of edge swapping, the network throughput is increased while the service cost is further optimized. We refer to this improved algorithm as `Impro_Alg`.

THEOREM 4. *Given an unreliable wireless sensor network $G(V \cup GW, E)$ with link reliability in $[p, (1 + \delta)p]$ and $0 \leq \delta \leq \frac{1}{p} - 1$, there is an approximation algorithm* `Impro_Alg` *for the MTMC problem, with delivered a throughput no less than $\frac{1}{(1+\delta)^h}$ times of the maximum one and the service cost is no more than $(1 + \frac{C_p}{C_f})$ times of the minimum one, where h is the maximum number of hops from any sensor to its nearest gateway. The algorithm takes $O(|V| \cdot |E|^2)$ time.*

PROOF. Omitted, due to limited space. $\square$

5. PERFORMANCE EVALUATION

In this section we evaluate the performance of the proposed algorithms in terms of network throughput and the service cost. We consider a sensor network consisting of 100 to 300 sensors randomly deployed in a $1000m \times 1000m$ square region. The transmission range of sensors is 120 meters, and the data generation rate is $r = 100bytes/s$. The number of gateways K varies from 4 to 10, and they are deployed as follows. The monitoring region is divided into roughly equal-size K sub-regions with each containing one gateway randomly deployed. In our experiments, the following three different data plans provided by Vodafone [5] for one month monitoring period (i.e. $\tau = 30days \times 24hours \times 3,600seconds$) will be examined: (I) $Q = 2GB$ and $C_f = \$19$; (II) $Q = 4GB$ and $C_f = \$29$; and (III) $Q = 10GB$ and $C_f = \$39$. Each of these three data plans has the same penalty rate $c_p = \$0.02/MB$. Each value in the figures is the mean of the results by applying the mentioned algorithm to 20 different network topologies of the same size.

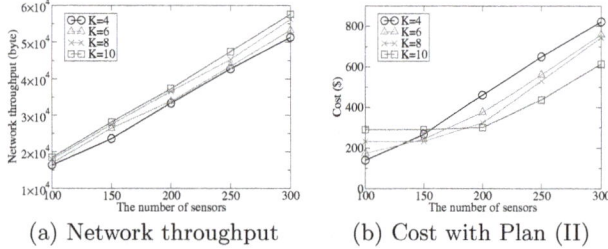

(a) Network throughput   (b) Cost with Plan (II)

Figure 1: Impact of K on the network performance when $p = 0.8$ and Plan (II) is adopted.

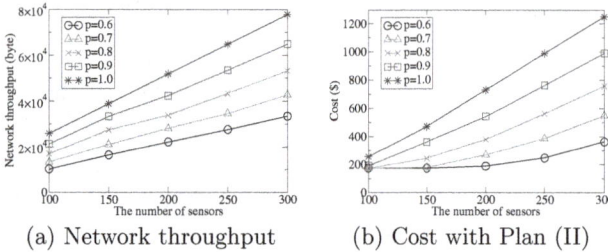

(a) Network throughput   (b) Cost with Plan (II)

Figure 2: Impact of link reliability p on the network performance when $K = 6$ and Plan (II) is adopted.

We first study the performance of algorithm `Uniform_Link`. Fig. 1 shows the larger number of gateways K, the higher network throughput the algorithm delivers and so is the service cost. Fig. 2 indicates that the higher the link reliability, the larger the network throughput, and the higher the service cost.

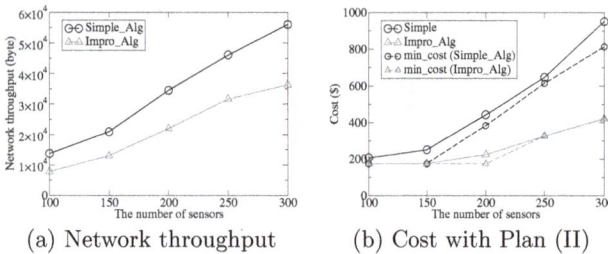

(a) Network throughput   (b) Cost with Plan (II)

Figure 3: Performance of different algorithms when $K = 6$, $p_e \in [0.1, 1]$ for each link $e \in E$, and Plan (II) is adopted.

We next evaluate the performance of algorithms `Simple_Alg` and `Impro_Alg`. Fig. 3 plots that the network throughput delivered by algorithm `Impro_Alg` is no less than 78% of the maximum throughput, while its service cost is no more than 103% of the minimal cost. The dashed curve in Fig. 3(b) is

a lower bound on the minimum service cost, which is calculated by assigning data relay workload equally to gateways. The costs delivered by algorithms `Simple_Alg` and `Impro_Alg` are respectively 116% and 103% of their minimum costs. Fig. 4 illustrates that a larger K will result in a higher network throughput but does not necessarily incur a higher service cost. Fig. 5 shows that with the growth in the range of link reliability, the network throughput increases, so does the service cost.

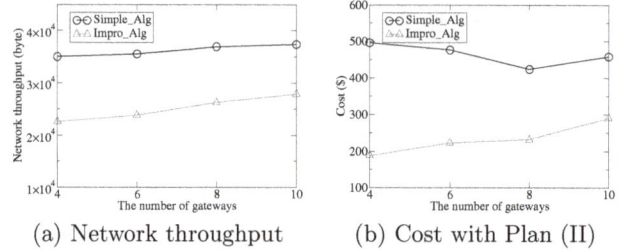

(a) Network throughput   (b) Cost with Plan (II)

Figure 4: Impact of K on the network performance when $n = 200$, $p_e \in [0.1, 1]$ for each link $e \in E$, and Plan (II) is adopted.

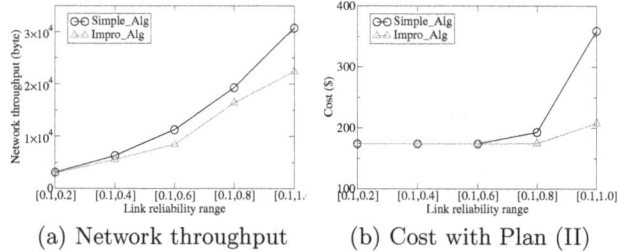

(a) Network throughput   (b) Cost with Plan (II)

Figure 5: Impact of the range of link reliability on the network performance when $K = 6$, $n = 200$, and Plan (II) is adopted.

6. CONCLUSION

In this paper we first formulated the MTMC problem and showed its NP-completeness. We then proposed an approximation algorithm with guaranteed approximation ratio. We finally conducted experiments by simulations to evaluate the performance of the proposed algorithms. Experimental results demonstrated that the proposed improved algorithm is very promising, and the solution is fractional of the optimum.

7. REFERENCES

[1] X. Jiang, J. Polastre, and D. Culler. Perpetual environmentally powered sensor networks. *Proc. of IPSN*, ACM, 2005.

[2] X. Xu, W. Liang, and Z. Xu. Minimizing remote monitoring cost of wireless sensor networks. *Proc. of WCNC*, IEEE, 2013.

[3] W. Liang, P. Schweitzer, and Z. Xu. Approximation Algorithms for Capacitated Minimum Forest Problems in Wireless Sensor Networks with a Mobile Sink *IEEE Trans. on Computers*, PrePrints, 2012.

[4] F. Shan, W. Liang, J. Luo, and X. Shen. Network Lifetime maximization for time-sensitive data gathering in wireless sensor networks. *Computer Networks*, Vol. 57, pp. 1063-1077, 2013.

[5] Vodafone. *http://www.vodafone.com.au*.

Route-Switching Games in Cognitive Radio Networks

Qingkai Liang, Xinbing Wang, Xiaohua Tian
Department of Electronic Engineering
Shanghai Jiao Tong University, Shanghai, China
victor856, xwang8, xtian@sjtu.edu.cn

Qian Zhang
Department of Computer Science and Engineering
Hong Kong University of Science and Technology
qianzh@cse.ust.hk

ABSTRACT

In Cognitive Radio Networks (CRNs), Secondary Users (SUs) are provided with the flexibility of accessing Primary Users' (PUs') idle spectrum bands but the availability of spectra is dynamic due to PUs' uncertain activities of channel reclamation. In the multi-hop CRNs consisting of SUs as relays, such spectrum mobility will cause the invalidity of pre-determined routes of data flows since some of the channels pre-assigned to the routes become unavailable. In this paper, we investigate the spectrum-mobility-incurred route-switching problem in both spatial and frequency domains for CRNs, where the spatial switching determines which relays and links should be re-selected and the frequency switching decides which channels ought to be re-assigned to the spatial routes. We further formulate the route-switching problem as the *Route-Switching Game* which is shown to be a *potential game* and has a pure Nash Equilibrium (NE). Accordingly, an efficient algorithm for finding the NE is proposed. The proposed route-switching scheme not only avoids conflicts with PUs but also mitigates spectrum congestion. Meanwhile, tradeoffs between *routing costs* and *channel switching costs* are achieved.

Categories and Subject Descriptors

C.2.1 [[**Computer-Communication Networks**]: Network Architecture and Design—*Wireless Communications*

General Terms

Algorithms, Design

Keywords

Cognitive Radio Networks, Game Theory, Routing, Spectrum Mobility

1. INTRODUCTION

Cognitive Radio has been proposed as a promising paradigm for relieving the spectrum shortage. In Cognitive Radio Networks (CRNs), Secondary Users (SUs) who require the usage of spectra can access Primary Users' (PUs') idle channels. Such dynamic spectrum access provides SUs with high flexibility in selecting spectra but brings new challenges to the design of CRNs at the same time, one of which is the *spectrum mobility*.

In CRNs, PUs can reclaim their licensed channels at any time due to their high priority of channel occupation, and SUs must cease their transmission[1] on those spectrum bands. Hence, from SUs' perspective, the availability of spectra is dynamic due to PUs' uncertain activities of channel reclamation, which causes the spectrum mobility in CRNs.

In the context of multi-hop CRNs where multiple SUs act as potential relays, one of the major influences of spectrum mobility is the break of routes of incoming data flows since the unavailability of PUs' channels disables the transmission over some links on the pre-determined routes. To avoid conflicts with PUs and resume routing, each flow initiator can either inform intermediate SU relays to switch their accessing channels or re-select a new spatial route[2] where channels are not reclaimed. However, the following tradeoff implies that the **two-dimensional route switching** (i.e., the combination of both channel switching and spatial route re-selection) is a better choice.

On one hand, by maintaining previous (and possibly the best) spatial routes and choosing less congested (and idle) channels as the switching destination, channel switching can efficiently avoid conflicts with PUs and reduce *routing costs* incurred by relaying data, including the transmission delay, energy consumption, etc. Unfortunately, frequent channel switching could also cause significant *switching costs* such as the additional power consumption and switch delay. On the other hand, re-selecting a new spatial route can yield fewer *switching costs* but may lead to additional *routing costs* at the same time. Consequently, there's a *tradeoff* between the two costs, which must be achieved by switching routes in both spatial and frequency domains.

In this paper, we propose Route-Switching Games to address the above *spectrum-mobility-incurred route-switching* problem in CRNs. The contributions of this paper include the following aspects.

• To our best knowledge, this paper is the first to investigate the spectrum-mobility-incurred route-switching problem in CRNs. Our scheme not only avoids conflicts with PUs

---

[1] In this paper, we only consider the spectrum overlay mode.
[2] In the following, we will refer the selection of intermediate nodes and edges as *spatial routes* and the choice of channels exploited on the spatial routes as *frequency routes*.

but also mitigates the spectrum congestion and achieves the tradeoff between routing costs and channel switching costs.

• We formulate the proposed problem as Route-Switching Games which is proved be a *potential game*. An efficient algorithm for finding the Nash Equilibrium (NE) is provided.

The remainder of this paper is organized as the following. We will first introduce the system model in section 2. Next, Route-Switching Games and simulation results will be demonstrated in sections 3 and 4 respectively. Finally, we will give future works in section 5.

2. NETWORK MODEL

2.1 Architecture of Multi-hop CRNs

We consider a multi-hop CRN where multiple SUs act as routers for the incoming data flows, and there're C orthogonal channels (with the same bandwidth) that can be accessed by SUs when they are not occupied by PUs (each channel is denoted by $j \in \mathcal{C} = \{1, 2, \cdots, C\}$). For the simplicity of analysis, we assume the entire secondary network lies in the same "collision domain" with PUs[3], i.e., the perceived channel states (either busy or idle) at each SU are identical in the considered network. This assumption is valid for many geographically-centric secondary networks coexisting with powerful PU transceivers, like PU base stations in cellular networks, as is shown in Figure 1.

Formally, the entire secondary network can be characterized by a topological graph $G = (V, E)$. Here, V is the set of nodes (SUs) and E is the set of edges in the topological graph. Note that there's an edge $E_{u,v}$ between a pair of nodes (u, v) iff they're within the transmission range of each other, so an edge corresponds with a data link. However, for a link to be able to transmit data, it must be allocated a traffic channel. As our focus is the *route-switching* problem, we suppose there have been pre-determined channel allocations on each link (but these pre-assigned channels may be reclaimed by PUs and become unavailable now). Here, we denote matrix $\mathbf{A}$ the indication of pre-assigned channels on different links. Specifically, its element $A_{e,j} = 1$ implies that channel j was pre-assigned to link e ($A_{e,j} = 0$ otherwise). Hence, matrix $\mathbf{A}$ represents a part of the "routing history".

2.2 Flow Model

Suppose there're M concurrent and constant[4] data flow inputs (denoted by F_k, $k \in \mathcal{M} = \{1, 2, \cdots, M\}$) which need routing among the secondary network, and denote the source and destination of data flow F_k by a pair (S_k, D_k). For the efficiency and reliability of flow transmission, F_k ($k \in \mathcal{M}$) segments its data by packets, each with size μ_k. Additionally, we denote the flow rate of F_k by r_k and assume that those data flows are from different initiators, each hoping to minimize its own costs incurred in the routing process.

2.3 Spectrum Mobility and Route Switching

When high-priority PUs reclaim their licensed channels, SUs must cease their transmission on those spectrum bands, which causes *spectrum mobility*. Here, we denote Γ the set of all currently *unavailable* channels due to PUs' reclamation.

---

[3]Note that our scheme can also be modified to incorporate the spatial diversity of PUs' spectra in secondary networks.
[4]We assume those data flows can last for a period of time like minutes or hours, which is particularly suitable for characterizing multimedia streaming, P2P downloading, etc.

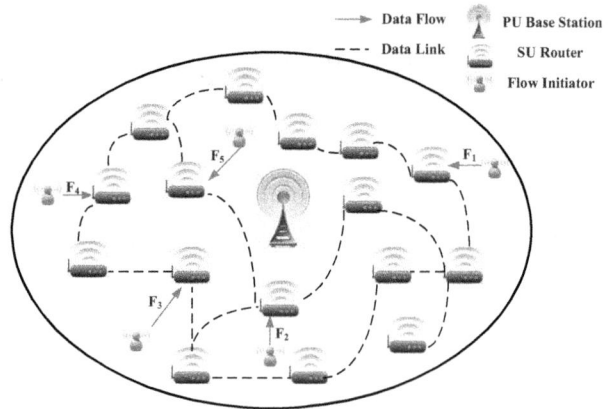

Figure 1: An example of the multi-hop CRN. Note that the entire secondary network is within the transmission range of the same PU base station, so the spectrum opportunities perceived at each SU are identical.

Unlike many previous works proposing statistical models to characterize PUs' reclaiming activities [2, 3], we do not predict PUs' behaviors, i.e., our scheme is *reactive*, since the precision of predictions still remains a major problem. Besides, route-switching schemes should provide routing reliability as much as possible, instead of probabilistic results, because the focus of the proposed problem is exactly to handle the negatives effects of PUs' spectrum uncertainty, which is the other reason why we don't choose proactive models.

In face of spectrum mobility, routes must be switched in both spatial and frequency domains so as to avoid conflicts with PUs. Here, we use a 3-dimensional matrix $\mathbf{X}$ to characterize the *newly selected* spatial and frequency routes in CRNs, which is also the strategy variable in the considered problem. Specifically, its element $X_{e,j}^k = 1$ when link e is included in the new spatial route of flow F_k and channel j is re-selected for this link ($X_{e,j}^k = 0$ otherwise).

2.4 Interference Model

We use the protocol interference model [1], where the transmission in channel j over link e succeeds if all potential interferers in the *interference neighborhood* of link e remain silent in channel j for the transmission duration. Here, the *interference neighborhood* of link e, i.e., $I(e)$, is the set of links whose end nodes have interference links or data links incident on the end nodes of e. Further, when channel j is perceived idle over link e, the contention window is activated and link e will contend for the transmission opportunities with all interferers in $I(e)$ (the transmitter on link e execute the contention). This model resembles CSMA/CA in IEEE 802.11, based on an RTS-CTS-Data-ACK sequence.

2.5 Cost Model

We mainly consider two types of costs in our model: routing costs and switching costs, which will be further discussed in the rest of this subsection. Additionally, it should be mentioned that the following costs are derived under the premise that channel $j \in \mathcal{C}$ is sensed to be *idle*, i.e., $j \notin \Gamma$.

2.5.1 Routing Cost

Routing costs characterize the potential expense incurred by relaying packets on established routes, including the routing delay, power consumption, etc. Here, we concentrate on the following two major routing costs.

• *Delay Cost:* Under the interference model mentioned above, significant transmission delay will be incurred if a channel is congested, since SUs must contend and wait for transmission opportunities. By comparison, other minor delay (e.g., propagation delay) is neglected in this paper.

As is typical of many random access protocols [7, 8], we make the following assumption: in any contention window and any channel $j \in \mathcal{C}$, a certain link $e \in E$ and all its interfering links have the same probability of wining the access to this channel. Following the methods provided in [9], we can calculate the *expected* transmission delay under our interference model, which characterizes the expected waiting time before a certain link e wins the opportunity to transmit *one packet* in channel j:

$$\lambda_{e,j} = \sum_{e' \in I(e)} \sum_{k \in \mathcal{M}} X^k_{e',j} \omega_k, \qquad (1)$$

where $\omega_k := \frac{\mu_k}{r_k}$ is the transmission time required by flow F_k for one packet. The derivation of (1) is beyond the scope of this paper, so we omit it for brevity. The intuition behind (1) is explained as the following. $\sum_{k \in \mathcal{M}} X^k_{e',j} \omega_k$ in (1) represents the traffic demands (for transmission time) in channel j over link e' imposed by all passing data flows, and thus (1) is the aggregate traffic demands in channel j from the interference neighborhood of link e. Generally speaking, $\lambda_{e,j}$ reflects the *congestion level* of channel j perceived over link e, and delay costs can also be interpreted as the *congestion costs* in our model.

Further, the total expected delay incurred on flow F_k's two-dimensional route is:

$$DC_k = \sum_{e \in E} \sum_{j \in \mathcal{C}} X^k_{e,j} \lambda_{e,j}. \qquad (2)$$

In the rest of this paper, we will use DC_k to characterize the delay costs for flow F_k.

• *Energy Cost:* We mainly consider the energy used for data transmission. Under our interference model, when one SU transmits in channel j over link e, other SUs within $I(e)$ must remain silent in channel j, so the SINR perceived at each SU receiver is merely dependent on the intrinsic channel quality and geographical conditions, such as the AWGN, path loss, etc. Further, according to Shannon Formula, energy costs incurred by transmitting per packet of F_k's data in channel j over link e are only related to the flow rate, packet size (which influences the transmission duration), channel quality and geographical conditions. We model such energy costs as a general form $\varphi_{e,j}(r_k, \mu_k)$ and $\varphi^k_{e,j}$ for short. Then, the total energy costs on F_k's route are

$$EC_k = \sum_{e \in E} \sum_{j \in \mathcal{C}} X^k_{e,j} \varphi^k_{e,j}. \qquad (3)$$

2.5.2 Switching Cost

The other kind of costs are caused by channel switching. Each time switching happens at intermediate SU routers, switching costs are incurred, including additional energy consumption used for establishing new connections, additional costs of sensing, switching delay, etc. Here, we use γ to indicate the overall costs per time of channel switching. Note that γ includes the costs of both ON→OFF switching (tearing down the old channel connection) and OFF→ON

switching (establishing the new channel connection), so the two types of switching costs can be seen as being incurred altogether in *either* switching scenario. In this paper, we assume the overall switching costs γ are incurred only in the OFF→ON transition. Therefore, the total channel switching costs caused by F_k's re-selection strategy are

$$SC_k = \sum_{e \in E} \sum_{j \in \mathcal{C}} X^k_{e,j}(1 - A_{e,j})\gamma. \qquad (4)$$

Corresponding to the above discussion, (4) implies that only when $A_{e,j} = 0$ and $X^k_{e,j} = 1$ (i.e., link e did not use channel j previously, but now this channel is allocated to e according to F_k's strategy), switching costs are incurred.

As is stated in the introduction, switching costs need to be balanced with routing costs, so we model flow F_k's *total costs* as:

$$TC_k = DC_k + EC_k + SC_k. \qquad (5)$$

3. ROUTE-SWITCHING GAMES

Equation (1) indicates that a certain flow's delay costs are also dependent on other flow's route-switching strategies (i.e., the selection of $X^k_{e',j}$, $\forall k \in \mathcal{M}$), so we formulate the above problem as Routing-Switching Games, where players (i.e., flow initiators) distributively and selfishly *switch* their two-dimensional routes in face of spectrum mobility, aiming at minimizing their *total costs*.

In this short paper, we only discuss the *complete-information* game, where each player's (flow's) information (i.e., data rate r_k and packet size μ_k, $\forall k \in \mathcal{M}$) is known to others. Results for the *incomplete-information* scenario can be found in the full version of this paper [10].

3.1 Game Formulation

We can use a tuple $\mathcal{G} = \{G, \mathbf{A}, \Gamma, r, \mu, TC, S\}$ to denote the Route-Switching Game. Here, $r = \{r_1, \cdots, r_M\}$ and $\mu = \{\mu_1, \cdots, \mu_M\}$ are publicly-known flow parameter. TC is the set of players' cost functions, shown in (5). $S = \{(e,j) | e \in E, j \in \mathcal{C}\}$ is the two-dimensional strategy space. In this paper, we consider the *symmetric game* where all players have the same strategy space. Further, we denote $s = \{s_1, \cdots, s_M\}$ the strategy profile, where $s_k = \{(e,j) \in S | X^k_{e,j} = 1\}$ is F_k's route-repick strategy. Since F_k's costs are relevant to the strategy profile s, we denote them by $DC_k(s)$, $EC_k(s)$, $SC_k(s)$, $TC_k(s)$ and $\lambda^k_{e,j}(s)$, respectively.

In addition, it's worthy mentioning that the above formulation does not impose any constraints on the connectivity of switched routes but such an omission won't influence any of the following analytical results. Instead, we guarantee the connectivity through our algorithm implementation (see Theorem 3 in section 3.2).

3.2 Existence and Computation of the NE

In this subsection, we will first prove that the Route-Switching Game is essentially a *potential game* [4, 5, 6]. Then an algorithm for computing the NE will be provided. However, due to space limitation, we present the introduction of potential games and the proofs to the following theorems in our online full-version paper [10].

Theorem 1: Any Route-Switching Game is a finite potential game which has the potential function:

$$\Phi(s) = \sum_{k \in \mathcal{M}} \omega_k [DC_k(s) + 2EC_k(s) + 2SC_k(s)]. \qquad (6)$$

Theorem 2: There's a unique Nash Equilibrium in the Route-Switching Game.

Next, we will design an algorithm for finding the NE in the Route-Switching Game, shown in **Algorithm 1**. This algorithm is actually an iterative algorithm following the Finite Improvement Property of potential games (see [10] for details). Due to the space limitation, we won't provide the detailed demonstration of this algorithm, and readers can refer to our full-version paper [10] for further explanations. The correctness of Algorithm 1 is shown in Theorem 3.

Algorithm 1 Find the Nash Equilibrium s^* in the Route-Switching Game with *complete information*

1: Initialize $X_{e,j}^k = 1$, $\forall k \in \mathcal{M}, e \in E, j \in \mathcal{C}$;
    $\Phi_0 = +\infty$, $n = 0$, $k = 0$, $m = 0$;
2: Extend edge e ($\forall e \in E$) to C parallel edges
    (each extended edge is denoted by (e, j), $\forall j \in \mathcal{C}$);
3: **repeat**
4:    $n = n + 1$, $k = (k \mod M) + 1$;
5:    **for** each $e \in E, j \in \mathcal{C}$ **do**
6:        Update edge weight $W_{(e,j)}$;
7:    **end for**
8:    Set $W_{(e,j)} = +\infty$, $\forall j \in \Gamma, e \in E$;
9:    Set $W_{(e,j)} = +\infty$, $\forall (e, j) \in \Omega$;
10:    Set $W_{(e,j)} = +\infty$, $\forall v \in \Lambda, e \in E(v), j \in \mathcal{C}$;
11:    Call *Dijsktra Algorithm* to find the shortest path for F_k in
        the **extended graph** with weight $W_{(e,j)}$ ($\forall e \in E, j \in \mathcal{C}$),
        and set (Source, Destination)$\leftarrow (S_k, D_k)$;
12:    Compute Φ_n according the shortest path;
13:    **if** $\Phi_n < \Phi_{n-1}$ **then**
14:        Update F_k's strategy according to the shortest path;
15:        $m = 0$;
16:    **else**
17:        $m = m + 1$, $\Phi_n = \Phi_{n-1}$;
18:    **end if**
19: **until** $m = M$
20: $s_k^* = \{(e, j) | X_{e,j}^k = 1, e \in E, j \in \mathcal{C}\}$, $\forall k \in \mathcal{M}$;
21: END.

Theorem 3: Each improvement step in Algorithm 1 can reduce the potential function to the maximum extent and guarantee the route connectivity in polynomial time with time complexity $O(|E||\mathcal{C}| + |V|^2)$.

4. SIMULATION

In this paper, we utilize MATLAB as the simulation tool, and the detailed simulation settings can be found in our full-version paper [10].

We first simulate the Finite Improvement Property (see [10]) of Route-Switching Games, shown in Figure 2. Initially, each player randomly picks a two-dimensional route, which incurs a large value of the potential function. After each improvement step, the potential is gradually reduced and finally reaches the minimum, where the NE is reached. It also shows games with more players require more improvement steps but the minimum can still be fast reached. Figure 3 shows the variation of total channel switching times in the CRN with M and $|V|$. From this figure, we can observe that channel switching times increase with M almost linearly since more players imply more congestion and they are more likely to execute channel switching to avoid mutual interference. Besides, larger network size also leads to higher switching frequency since the length of both frequency and spatial routes increases and the chances that players get influenced by spectrum dynamics are greatly raised.

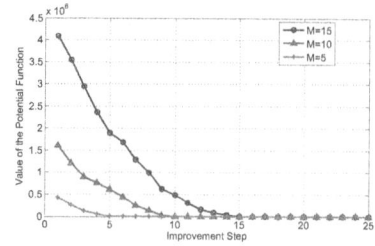

Figure 2: Finite Improvement Property of the Route-Switching Game.

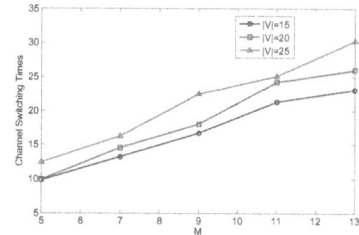

Figure 3: Channel switching times vary with M and $|V|$.

5. FUTURE WORKS

In this paper, we investigate the spectrum-mobility-incurred route-switching problem in spatial and frequency domains for CRNs. However, due to the page limits, we can only present the core analysis of this problem and leave more works for the future, which mainly includes the following aspects: (1) development of a fast algorithm for the approximate NE, (2) extension to the incomplete-information scenario, (3) measurement of the social efficiency (analysis of *price of anarchy*), and (4) improvement of the social efficiency. So far, we have gained good results about the first three aspects, which is available in the full-version paper [10].

6. REFERENCES

[1] P. Gupta and P. R. Kumar. The Capacity of Wireless Networks. *IEEE Transactions on Information Theory*, 46(2):388-404, 2000.

[2] R. Southwell, J. Huang and X. Liu. Spectrum Mobility Games. *IEEE INFOCOM*, 2012.

[3] Q. Liang, S. Han, F. Yang, G. Sun and X. Wang. A Distributed Centralized Scheme for Short and Long Term Spectrum Sharing with a Random Leader in Cognitive Radio Networks. *IEEE Journal on Sel. Areas in Comm.*, 30(11):2274-2284, 2012.

[4] X. Chen and J. Huang. Spatial spectrum access game: nash equilibria and distributed learning. *ACM MobiHoc*, 2012.

[5] Y. Song, H. Y. Wong and K. W. Lee. Optimal Gateway Selection in Multi-domain Wireless Networks: A Potential Game Perspective. *ACM MobiCom*, 2011.

[6] D. Monderer and L. S. Shapley. Potential Games. *Games and Economic Behavior*, 14(44):124-143, 1996.

[7] N. Abramso. The Aloha System: another alternative for computer communications. *ACM AFIPS*, 1970.

[8] W. Stallings. *Wireless Communications and Networks*, 2nd edition. Prentice Hall, 2005.

[9] T. Rappaport. *Wireless communications: Principles and Practice*. Prentice Hall, 2002.

[10] Q. Liang, X. Wang, X. Tian and Q. Zhang. Route-Switching Games in Cognitive Radio Networks. [Online Full Version]. Available online at http://iwct.sjtu.edu.cn/Personal/xwang8/paper/route.pdf

MINT: Maximizing Information Propagation in Predictable Delay-Tolerant Network

Shaojie Tang
Department of Computer and
Information Sciences
Temple University
Philadelphia PA 19122
shaojie.tang@temple.edu

Jing Yuan
Department of Computer
Science
University of Illinois at Chicago
Chicago IL
csyuanjing@gmail.com

Xiang-Yang Li
Department of Computer
Science
Illinois Institute of Technology
Chicago IL
xli@cs.iit.edu

Yu Wang
Department of Computer
Science
UNC at Charlotte
Charlotte NC 28223
yu.wang@uncc.edu

Cheng Wang
Department of Computer
Science
Tongji University
Shanghai China
3chengwang@gmail.com

Xuefeng Liu
Department of Computing
Hong Kong Polytechnic
University
Hong Kong China
csxfliu@gmail.com

ABSTRACT

Information propagation in delay tolerant networks (DTN) is difficult due to the lack of continues connectivity. Most of previous work put their focus on the information propagation in static network. In this work, we examine two closely related problems on information propagation in predicable DTN. In particular, we assume that during a certain time period, the interacting process among nodes is known a priori or can be predicted. The first problem is to select a set of initial source nodes, subject to budget constraint, in order to maximize the total weight of nodes that receive the information at the final stage. This problem is well-known influence maximization problem which has been extensively studied for static networks. The second problem we want to study is minimum cost initial set problem, in this problem, we aim to select a set of source nodes with minimum cost such that all the other nodes can receive the information *with high probability*. We conduct extensive experiments using $10,000$ users from real contact trace.

Categories and Subject Descriptors

C.2.1 [**Network Architecture and Design**]: Wireless communication, Network topology; G.2.2 [**Graph Theory**]: Network problems, Graph algorithms

Keywords

DTN, propagation, temporal, submodular.

1. INTRODUCTION

Information propagation in delay tolerant networks (DTN) is difficult due to the lack of continues connectivity and unstable net-

MobiHoc'13, July 29–August 1, 2013, Bangalore, India.
Copyright 2013 ACM 978-1-4503-2193-8/13/07 ...$15.00.

work structure. Such dynamics are often ignored in most of previous works *e.g.,* most researches put their focus on the information propagation in static network [1] [2]. One simple way to model the time dependent DTN is to represent it using a static graph, nodes in the graph represent individuals, where we add an edge between any two nodes if and only if they have been interacted or met during one period. We may further assign certain weight to each edge to reflect their interacting frequency, *e.g.,* if two nodes meet each other m times among n time slots, we will add an edge between them with a probability (weight) m/n. Information propagation processes in such static networks have already been well studied [1]. However, this kind of static view fails to capture the network dynamics, that is shown to be very critical to the information propagation problem. For example, assume that there are three nodes v_a, v_b and v_c in DTN. Node v_a interacts with node v_b at the first timestep, and v_b meets v_c at the next timestep. When taking a static view of this dynamic network, we may conclude that v_a and v_c has the same importance since they meet v_b with the same frequency. It turns out that the final results will be totally different by choosing different one as the initial source node. In particular, if we select v_a as the source node, both v_b and v_c have chance to receive the information. In contrast, if we pick v_c as the source node, neither v_a nor v_b has chance to get the information.

In this work, we study two closely related problems on information propagation in predictable delay tolerant network (DTN). The first problem is weighted influence maximization problem, we assume that we are given a fixed set of nodes, and the interactions that happen among nodes over a period of time T can be known a priori or can be predicted. The objective is to select a subset of nodes as initial source nodes such that the total weight of nodes that can successfully receive the information within a period is maximized. As shown in [1] [3], find an optimal solution is NP-Hard in most existing models. We also study the minimum cost initial set problem. Given a predictable DTN, we aim to select a set of initial source nodes with minimum total cost such that *all* the nodes can receive the information with high probability.

The rest of the paper is organized as follows. Section 2 briefly review the related results. System model and problem formulation are presented in Section 3. We study the weighted influence maximization problem in Section 4 and the minimum cost initial set problem in Section 5. We report extensive experimental results in Section 6 and conclude the paper in Section 7.

2. RELATED WORK

In many applications, data propagation in DTN is needed. Propagation protocols (including multicast and broadcast protocols) allow sending data packets from a source to multiple receivers. They are more effective, for data dissemination and multi-party communications, than unicast routing protocols. Chaintreau *et al.* [4] and Chierichetti *et al.* [5] both studied simple gossip-based protocols (including geographic and social gossip) to propagate information in mobile social networks. Zhou *et al.* [6] proposed new semantic multicast models for DTN multicast and developed several multicast routing algorithms with different routing strategies. Recently, Gao *et al.* [7] studied multicast in DTNs from the social network perspective. Unweighted influence maximization in dynamic networks is studied in [8]. Similar problem is also studied in [9].

3. SYSTEM MODEL

3.1 Network Model

In this work, we represent the time-evolving DTN by a sequence of static networks. Each static network represents the contact information among users at that timeslot. See Fig. 1 (a) (b) for illustration. Here we highlight two important observations which can be found in many kinds of social based DTNs (*e.g.*, cell phone network, bus-based disruption-tolerant network) [10] [11]: 1) nodes usually interact with each other in a regular way instead of moving randomly; 2) the interaction occurs among nodes can be predicted with sufficient history information. We call these networks predictable DTN, and we mainly study the information propagation problem on predictable DTN in this paper. Throughout this paper, we assume that the contact information of the dynamic network within a certain period can be pre-obtained through certain existed prediction method or statistical analysis.

3.2 Information Propagation Model

We adopt *Independent Cascade Model* [1] [8] [12] [13] in this work. *Independent Cascade Model* describes a spreading process comprising of two sets of individuals, source node and non-source node. In a predictable DTN, a source node v tries to send the information to each of its current neighbors with success probability $p_{(vw)}$. If v succeeds in propagating information to w at timestep t, then w will act as new source node in the following timesteps. If v fails to transmit information to w at timestep t, v will try again in the subsequent timesteps as long as they meet each other. This process continues until timestep T, let $\mathrm{ProgSet}(S)$ denote the set of nodes which received the information until T, corresponding to the set of initial source nodes S. Let v^t denote node v at timestep t, and let V^t denote all nodes at timestep t.

3.3 Discussion on Network Modeling

Remember that in our problem formulation, we assume that the temporal contact pattern is known deterministically in advance. One may challenge the practicability of this assumption *e.g.* the current and future network topologies may not be predicted *accurately*. Actually, it will not affect the generality of our results even we take the "uncertainty" associated with the contact pattern into consideration. One way to resolve this issue is to regard the "uncertainty" as part of the success probability. For example, assume that the probability that nodes v_1 and v_2 contact each other during t and $t+1$ is **p**, and the success probability in the original problem formula-

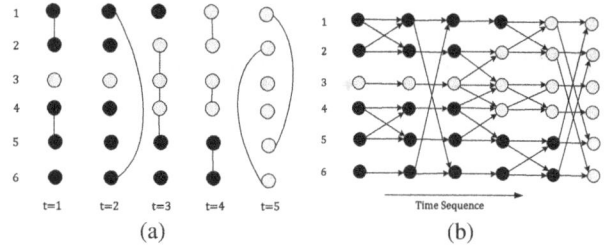

Figure 1: (a) illustrates the information propagation happens at each timestep by selecting node 3 as the initial source node; (b) describes the equivalent processing under the new constructed graph G. As shown in (a), if we choose node 3 as the initial source node, all of the nodes will receive the information (become light) at last. Equivalently, all nodes in V^T can be reached by node 3 in G.

tion is $p_{(v_1^t, v_2^{t+1})}$. Then we can define a new success probability as $\mathbf{p} \times p_{(v_1^t, v_2^{t+1})}$. By introducing the new success probability, we can ensure the feasibility of our results even under uncertain contact pattern. We may also formulate it as a robust optimization or stochastic optimization formulation which is left as possible future works.

4. WEIGHTED INFLUENCE MAXIMIZATION PROBLEM

We first study the weighted influence maximization problem. We assume that each node is associated with certain cost at which it can be selected as initial source node. We are interested at finding a set of initial source nodes, A, under the budget constraint such that the total weight of the propagated set is maximized.

Let $w(S)$ denote the the expected weight of the final propagated set under the initial source nodes S, we can prove that $w(S)$ is submodular monotone and non-negative. Our problem can be tackled by a simple greedy algorithm. We first compute two candidate sets for A: The first candidate set $A_{[1]}$ contains a single node which can maximize the expected total weight; the second candidate set $A_{[2]}$ is computed by Hill Climbing Algorithm in which we always add the node v that can maximize the expected incremental marginal gain: $w(A_{[2]} \cup \{v\}) - w(A_{[2]})/c(v)$ until the budget constraint is violated. Then we choose the better one as A. Based on the submodularity of $w(S)$ and similar proof in [14], we can prove that this algorithm achieves $\frac{1}{2} \cdot (1 - \frac{1}{e} - \epsilon)$ approximation.

5. MINIMUM COST INITIAL SET PROBLEM

In the previous sections, we mainly study how to select a set of initial source nodes under limited budget such that the final propagation range is maximized. We next study a symmetric problem: Minimum Cost Initial Set Problem. In particular, given a predictable DTN, each node is associated with certain cost, we aim to select a set of initial source nodes with minimum total cost such that *all* the nodes can receive the information at last. We study this problem under both deterministic model and probabilistic model. Note that when we study the minimum cost initial set problem under the probabilistic model, we are interested at finding the set of initial source nodes in order to let all the nodes receive the information *with high probability*.

Deterministic Model: Under the deterministic model, we assume that the success probability on each link is 1. Therefore, each node is associated with a determined propagated set. Naturally, our problem can be converted to a standard weighted set cover problem where the ground set is composed of the nodes in V_T, and each node in V_0 acts as a subset which can cover the nodes in its corresponding propagated set, and the classic greedy algorithm is a $\ln n$-approximate algorithm.

Probabilistic Model: We also studied this problem under the probabilistic model. In the probabilistic model, the success probability on each link is some value in [0,1], we aim to select the minimum cost initial source nodes such that the information can be propagated to the whole network with high probability (*e.g.*, the probability is lower bounded by certain value). We propose several heuristics to tackle this problem. In some scenarios, the hill climbing greedy algorithm can achieve constant approximation ratio.

6. EXPERIMENT RESULTS

In this section, we conduct extensive experiments to evaluate the performance of our algorithm.

The Network Data: A typical predictable DTN can be found in university campuses, *e.g.*, the National University of Singapore (NUS) student contact trace model [15]. We download the schedules of the 4,885 classes and enrollment of 22,341 students for each week of 77 class hours.

Information Propagation Model: We conduct extensive experiments under three basic information propagation models. In the first model, which is called *uniformly setting*, we assigned a uniform success probability p to each edge of the graph, and for simplicity we choose $p = 10\%$ in results reported here. In the second model, which is called *randomly setting* we assign p to each edge randomly at uniform, choosing from 0% to 100%. In the third model, which is called *deterministic setting*, we assigned the propagation probability as 1 to each edge of the graph.

The Algorithms and Implementation: For the weighted influence maximization (WIM) problem, we implement a greedy algorithm (HC Selection) that selects the initial source nodes one by one such that the expected incremental marginal gain is maximized, until the budget is violated. For the Minimum Cost Initial Set (MCIS) problem, we modify the greedy algorithm to select the nodes one by one until the influential probability for each node is met. We compare our greedy algorithm with the following heuristics: 1). *Randomized Selection:* for the WIM problem, we randomly select the initial source nodes one by one until the budget constraint is violated; for the MCIS problem, we randomly select the nodes one by one until the influential probability for each node is met. We run the experiment 1000 times and choose the average weight as final results; 2). *Top-k Selection:* We first order all the nodes in the decreasing order of $|\operatorname{ProgSet}(v)|$. For the WIM problem, we pick the nodes one by one in order until the budget is violated; and for the MCIS case, we select the nodes in order until the influential probability constraint is met. 3). *Minimum Cost Selection:* This method is evaluated only for the WIM problem. We select the nodes one by one in the increasing order of their cost until the budget is violated.

The Results: First we present the results of our algorithms under the WIM problem settings. In this set of experiments, we compare all above algorithms under weighted probabilistic model with budget constraint. The network size is fixed to 10,000 if not other specified.

Fig. 2(a) and Fig. 2(b) demonstrate the comparison results under the weighted probabilistic model, with $p = 10\%$ and random $p \in [0,1]$ respectively. As shown in Fig. 2(a), when the budget is relatively small, *e.g.* less than 100, there is no huge difference on the performance among these four algorithms. However, as budget goes to larger, our greedy algorithm performs much better than the other three methods, *e.g.* with a budget over 700, the expected weight of propagation set exceeds 10,000 using greedy algorithm, the greedy algorithm outperforms the other heuristics by 30%. Fig. 2(b) tell us similar results, there is no huge difference among different algorithms when the budget is small, when the budget becomes larger and larger, the greedy algorithm outperforms the other heuristics by up to 30%. Remember that in the Top-k Selection, we choose the nodes which can perform best individually. The experimental results show that the performance of this algorithm is almost the worst. This basically tells us that significantly better marketing results can be obtained by explicitly considering the dynamics of information in a network, rather than relying solely on some properties of single node.

Furthermore, we present the results of our algorithms under the MCIS problem settings. We compare three algorithms: greedy algorithm, Top-k Selection and Random Selection. For random algorithm, the result is averaged over 1,000 experiments. Fig. 2(c) shows the results under the weighted deterministic model where propagation probability of each link is set to be 1. We plot the cost of the initial node set such that all the nodes receive the information with at least a certain probability (influential probability). As shown in the figure, when the influential probability is low, there is no significant difference observed among the three algorithms. However, as the influential probability increases, the greedy algorithm outperforms the other two algorithms. Moreover, we evaluate the performance of the algorithms under the weighted probabilistic model. The results in Fig. 2(d) and Fig. 2(e) compare the algorithms with $p = 10\%$. Fig. 2(d) shows that the cost of initial node set increases as the influential probability increases, and that the greedy algorithm achieves the lowest initial cost. Then we fix the influential probability to 90% and vary the number of nodes and the results are shown in Fig. 2(e). Again, the greedy algorithm performs the best. At last, we fix the influential probability to 90% and evaluate the performance of the algorithms under varied propagation probability p. As shown in Fig. 2(f) the greedy algorithm achieves the lowest initial cost and outperforms the other two algorithms in all cases.

7. CONCLUSION

In this work, we studied several information propagation problems in DTN by exploring the temporal information of the intermittent connections. For both problems, we develop several efficient algorithms. There are a number of interesting problems left for future research. One problem is to design practically efficient methods when we know that the DTN network is some random network, instead of being an arbitrary network. For example, the network structure at every time slot could be a graph with power-law degree distribution, while the graphs at different time slots are independent. The other direction is to study both problems under new information propagation models such as linear threshold model.

8. ACKNOWLEDGEMENT

The research of authors is partially supported by NSF CNS-0832120, NSF CNS-1035894, NSF ECCS-1247944, NSF CNS-0915331, NSF CNS-1050398, National Natural Science Foundation of China under Grant No. 61170216, No. 61228202, China 973 Program under Grant No.2011CB302705, No. 61202383, the Program for New Century Excellent Talents in University under grant No. NCET-12-0414, the Natural Science Foundation of

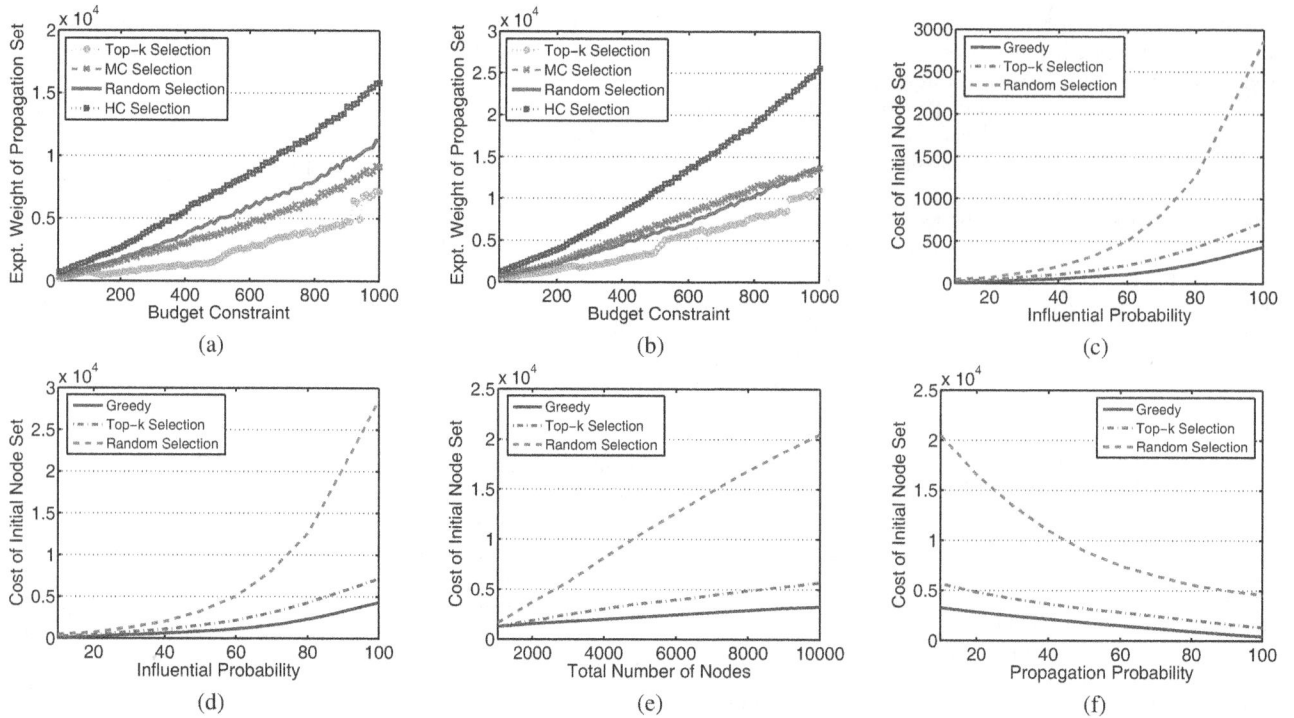

Figure 2: Performance evaluation under weighted probabilistic model and deterministic model.

Shanghai under grant No. 12ZR1451200, the National Science Foundation for Postdoctoral Scientists of China under grant No. 2012M510118, the Research Fund for the Doctoral Program of Higher Education of China under grant No. 20120072120075, and the Special Foundation of China Postdoctoral Science under grant No. 2013T60463.

9. REFERENCES

[1] D. Kempe, J. Kleinberg, and É. Tardos, "Maximizing the spread of influence through a social network," in *KDD*. ACM, 2003, pp. 137–146.

[2] S. Tang, J. Yuan, X. Mao, X.-Y. Li, W. Chen, and G. Dai, "Relationship classification in large scale online social networks and its impact on information propagation," in *INFOCOM 2011*. IEEE, 2011, pp. 2291–2299.

[3] P. Domingos and M. Richardson, "Mining the network value of customers," in *KDD*. ACM, 2001, pp. 57–66.

[4] A. Chaintreau, P. Fraigniaud, and E. Lebhar, "Opportunistic spatial gossip over mobile social networks," in *Proceedings of the first workshop on Online social networks*. ACM, 2008, pp. 73–78.

[5] F. Chierichetti, S. Lattanzi, and A. Panconesi, "Gossiping (via mobile?) in social networks," in *Proceedings of the fifth international workshop on Foundations of mobile computing*. ACM, 2008, pp. 27–28.

[6] W. Zhao, M. Ammar, and E. Zegura, "Multicasting in delay tolerant networks: semantic models and routing algorithms," in *Proceedings of the 2005 ACM SIGCOMM workshop on Delay-tolerant networking*. ACM, 2005, pp. 268–275.

[7] W. Gao, Q. Li, B. Zhao, and G. Cao, "Multicasting in delay tolerant networks: a social network perspective," in *Proceedings of the tenth ACM international symposium on*

Mobile ad hoc networking and computing. ACM, 2009, pp. 299–308.

[8] T. Y. Berger-Wolf, "Maximizing the extent of spread in a dynamic network," 2007.

[9] B. Han, P. Hui, V. A. Kumar, M. V. Marathe, J. Shao, and A. Srinivasan, "Mobile data offloading through opportunistic communications and social participation," *TMC*, vol. 11, no. 5, pp. 821–834, 2012.

[10] Q. Yuan, I. Cardei, and J. Wu, "Predict and relay: an efficient routing in disruption-tolerant networks," in *Proceedings of the tenth ACM international symposium on Mobile ad hoc networking and computing*. ACM, 2009, pp. 95–104.

[11] X. Zhang, J. Kurose, B. N. Levine, D. Towsley, and H. Zhang, "Study of a bus-based disruption-tolerant network: mobility modeling and impact on routing," in *Proceedings of the 13th annual ACM international conference on Mobile computing and networking*. ACM, 2007, pp. 195–206.

[12] J. Goldenberg, B. Libai, and E. Muller, "Talk of the network: A complex systems look at the underlying process of word-of-mouth," *Marketing letters*, vol. 12, no. 3, pp. 211–223, 2001.

[13] G. etal., "Using complex systems analysis to advance marketing theory development: Modeling heterogeneity effects on new product growth through stochastic cellular automata," *Academy of Marketing Science Review*, vol. 9, no. 3, pp. 1–18, 2001.

[14] S. Khullera, A. Mossb, and J. Naor, "The budgeted maximum coverage problem," *Information Processing Letters*, vol. 70, pp. 39–45, 1999.

[15] V. Srinivasan, M. Motani, and W. T. Ooi, "Analysis and implications of student contact patterns derived from campus schedules," in *MobiCom*. ACM, 2006, pp. 86–97.

Community Based Cooperative Content Caching in Social Wireless Networks

Mahmoud Taghizadeh
Google Inc.
Mountain View, CA 94043
mtaghim@google.com

Subir Biswas
Michigan State University
Lansing, Michigan, 48823
sbiswas@egr.msu.edu

ABSTRACT

This paper presents a conceptual framework of social community based cooperative caching for minimizing electronic content provisioning cost in Mobile Social Wireless Networks (MSWNET). Drawing motivation from Amazon's Kindle electronic book delivery model, this paper develops practical network, service, and pricing models which are then used for creating an optimal cooperative caching strategy based on social community abstraction in wireless networks.

Categories and Subject Descriptors

C.2.1 [Network Architecture and Design]: Wireless communication

Keywords

Mobile Social Wireless Networks, Cooperative Caching, Community based Mobility, Content Provisioning

1. INTRODUCTION

Recent emergence of mobile devices and wireless-enabled data applications have fostered new content dissemination models in today's mobile ecosystem. A list of such devices includes Apple's iPhone, iPad, Google's Android, Amazon's Kindle and electronic book readers from other vendors. The contents related to these devices include phone Apps, electronic books, mp3 music, video and news clips and etc. Conventionally, a user downloads content directly from a Content Provider's (CP) server over a Communication Service Provider's (CSP) network. Downloading content through communication service provider's network involves a cost which must be paid either by end users or by the content provider. For instance, in Amazon Kindle electronic book delivery business model, CP (Amazon) pays to CSP (Sprint) the cost of network usage due to downloaded e-books by Kindle users.

When users carrying mobile devices physically gather in public places, Mobile Social Wireless Networks (*MSWNETs*) are formed using ad hoc wireless connections. Within an *MSWNET*, an alternative content access approach would be to first search the local *MSWNET* for the requested content before downloading it from the CP's server. This can lower content provisioning cost significantly since the download cost to CSP would be avoided when the content is found within the *MSWNET*. This mechanism is termed as *cooperative content caching*. In order to entice end-consumers (in spite of the storage and energy costs of caching) to

cache downloaded content and to share it with others, a peer-to-peer rebate mechanism is proposed. This rebate is factored in the content provider's overall cost.

A key question for cooperative content caching is how to store contents in devices such that the network wide provisioning cost is minimized. In addition to finding such optimal policies we also show that those policies can depend heavily on user interaction patterns and their resulting social community formation.

2. SERVICE AND PRICING MODELS

After an object request is originated, a mobile device first searches its local cache. If the search fails, it searches the object within its local *MSWNET*. Depending on the user tolerable delay (UTD), a device can opt for searching the content in nodes not only within its current physical partition, but also in nodes within its future partitions. This may lead to a larger *Temporal Partition Size* (TPS), thus improving its odds of finding the content. If the search in *MSWNET* fails, the object is downloaded directly from the CP's server using CSP's network.

We follow the Amazon Kindle[TM] pricing model in which Content Provider (e.g. Amazon) pays a download cost C_d to a Communication Service Provider (CSP) when an End-Consumer (EC) downloads an object from the CP's server through CSP's cellular network. Also, as shown in Fig. 1, Whenever an EC provides a locally cached object to another EC within its *MSWNET*, the provider EC is paid a rebate C_r by the CP.

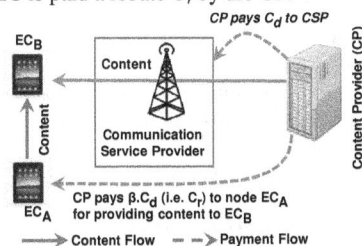

Fig. 1: Content and cost flow model

Note that these cost items, namely, C_d and C_r, do not represent the selling price of an object (e.g. e-book). The selling price is directly paid to the content provider (e.g. Amazon) by an end consumer (e.g. a Kindle user) through an out-of-band secure payment system.

3. COMMUNITY BASED CACHING

Human mobility is often abstracted by underlying social community structures, which can and should be leveraged for cooperative caching. Node contact probabilities in community based mobility can have various forms of locality which should be considered while designing caching policies. Our aim here is to develop a hierarchical split-cache strategy for leveraging such locality structures in inter-contact times. The key point here is to limit the cooperation only among nodes belongs to the same community. In next section we briefly introduce the community detection algorithms evaluated in this paper.

3.1 Community Detection Algorithms

As a first step, the contact pattern of mobile users is mapped to a weighted graph in which the vertices represent mobile nodes and edges represent connections between nodes. The weight of an edge indicates the strength of connection between two nodes. For example, if two nodes always move together (i.e. they are always connected), the weight of the edge between them will be 1. In the other extreme, two nodes that never meet should have an edge with weight zero. Note that the weight of an edge between any pair of nodes must be computed based on the preset user tolerable delay value. Such weight is defined as the probability that those nodes meet within that delay duration.

Girvan-Newman [1,2,3] is one of the popular hierarchically divisible community detection algorithms that works based on finding and removing network edges with high *betweenness*. Edge *betweenness* is defined as the number of shortest paths between pairs of vertices that pass through it. The procedure of finding and removing edges stops when the cumulative *modularity* of all communities in the resulting network is maximized. Modularity of a community is defined as the actual number of edges within the community minus the expected number of edges in an equivalent network (with the same community divisions) when edges placed at random. In this paper we term this algorithm as *fastcommunity*.

Blondel et al. [4] propose a two-phase heuristic algorithm, named as *fastfold*, which is based on a local optimization of Girvan-Newman modularity. As a first step, each network node is assigned to a different community and then the algorithm attempts to increase the cumulative network modularity by combining nodes and their neighbors. A node is grouped with other nodes that result in a network with increased modularity. The process of grouping stops when a local maximum of the network modularity is attained, i.e. when no individual grouping move can improve the modularity any further.

Hierarchical clustering algorithms [5] group *similar* nodes within communities. The similarity between two nodes can be defined based on criteria such as Euclidean distance, Pearson correlation and cosine similarity. To find similar nodes (finding *cliques*) we used Pearson correlation which is computed as:

$$x_{ij} = \frac{\frac{1}{n}\sum_k(A_{ik} - \mu_i)(A_{jk} - \mu_j)}{\sigma_i \sigma_j}$$

where $\mu_i = \frac{1}{n}\sum_j A_{ij}$, $\sigma_i^2 = \frac{1}{n}\sum_j(A_{ij} - \mu_i)^2$.

With Pearson correlation measure, nodes with more of common neighbors have higher similarity. A stopping criterion is used in which no clique bigger than 3 nodes remains in the entire network.

3.2 Hierarchical Split Caching

From each node's standpoint, any other node in a mobile network can be in-community or out-of-community (i.e. *stranger*). Nodes in the same community meet more frequently an often for longer duration. Frequencies of contacts between strangers are usually low and last shorter These properties are experimentally found in a number of real-life human mobility traces [6,7] and also mobility models [8,1,9].

Cache Space (C)		
Duplicate	Unique in community	Unique in Network
$\lambda_i C$	$\mu_i C$	$(1-\lambda_i-\mu_i)C$

Fig. 2: Hierarchical partitioning for supporting communities

These observations lead to the following Hierarchical Split Cache or HSC approach in which the level of cache collaboration between two nodes depends on whether they are in-community or out-of-community. In HSC the cache space in each node is divided into three separate areas (see Fig. 2). Certain number of very popular objects are cached in the *first* area in order to secure their access locally. Objects stored in this area can be duplicated at nodes across and outside communities. The *second* area is reserved for cooperation among in-community nodes. The objects stored in this *second* cache area of a node cannot be duplicated across its in-community member nodes, but duplication of such objects can be allowed in its stranger nodes. In other words, these are unique within a community. Finally, the *third* area in the cache is used for implementing global cooperation among all nodes by maintaining global uniqueness. This *third* area of cache helps nodes in a small community to take advantage of cooperation with stranger nodes. This is especially useful in situations in which one meets many strangers in public places such as train stations and airports. One's device in such situations can retrieve content from the *third* cache area of those strangers' devices. The best split factors between the three areas need to be determined based on the social community properties of the network.

Cost Computation: We first define function $f(k)$ to be the probability of finding an arbitrary object within a device's cache that is filled with the k most popular objects. This function can be expressed as $\sum_{j=1}^{k} p_j$ where p_j is the popularity of object j-th.

The average local hit rate for nodes in the ith community can be computed as:

$$L_i = f(\lambda_i C) + \frac{f((m_i\mu_i + \lambda_i)C) - f(\lambda_i C)}{m_i} + \frac{C(1 - \lambda_i - \mu_i)}{N_u} \times H_u$$

where m_i is community size, N_u is the total number of unique objects stored in the *third* cache area of all network nodes, and H_u is the corresponding hit rates of these N_u objects.. N_u can be computed using the following equation:

$$N_u = \sum_{\forall i} m_i(1 - \lambda_i - \mu_i)C$$

Hit rate of unique objects can be written as:

$$H_u = f(O_{max} + N_u) - f(O_{max})$$

where the parameter O_{max} is the least popular among all objects that have been stored in the *second* cache area. It can be computed as:

$$O_{max} = \max_i\big((\mu_i - \lambda_i)m_i + \lambda_i\big)C$$

Let ϕ be the contact probability between in-community nodes with a specified user tolerable delay. Then the percentage of requests that can be found in remote caches within the ith community can be computed as:

$$R_{ic} = (m_i - 1) \times \phi \times$$
$$\left(\frac{\left(f((m_i\mu_i + \lambda_i)C) - f(\lambda_i C) \right)}{m_i} + \frac{C(1 - \lambda_i - \mu_i)}{N_u} \times H_u \right) \quad (1)$$

The first term refers to hit rate of objects stored in the *second* cache area of nodes in the ith community, and the second term indicates hit rate of objects stored in the *third* cache area.

A node in i^{th} community may also meet strangers from other communities. The average remote hit rate of objects found in the stranger nodes can be computed as:

$$R_{is} \approx \frac{\Gamma_i}{M} H_u \quad (2)$$

where Γ_i represents the average number of stranger nodes that an ith community node is expected to see during a specified user tolerable delay. The average object provisioning cost for an ith community node can be calculated as [10]:

$$Cost_i = (1 - (1 - \beta)R_i - L_i)C_d$$

where $R_i = R_{ic} + R_{is}$ is the total remote hit rate for nodes in community-i and the β shows rebate to download cost ratio. The overall cost can be computed as:

$$Cost = \frac{\sum_{\forall i} m_i Cost_i}{M} \quad (3)$$

where M is the total number of nodes in the entire network. To minimize the overall cost, we need to differentiate the above function with respect to all unknown parameters (i.e. λ_i and μ_i) and equate it to zero.

Observe that when the communities are completely independent (i.e. when nodes never meet strangers from other communities), the overall cost can be minimized by simply minimizing the cost in each community. In this case, the problem reduces to the stationary partition case [10].

4. EVALUATION

4.1 Simulation Setup and Mobility Traces

The performance of proposed community based hierarchical split-caching has been evaluated with two human mobility traces (i.e. Infocom '06 trace [6] and UCSD trace [11]) and a synthetic human mobility generator HCMM [8].

In [11] project 275 freshman were equipped with WiFi enabled PDA devices that periodically scanned and logged the visible access points for a period of 77 days. The contacts between nodes, determined by access point sharing, were extracted from the log file. The average inter-contact time in this trace was about 19 hours. Fig. 3(a) reports the number of associated users per hour across all access points in the campus during the first 14 days of the experiment.

The Infocom '06 trace includes records of Bluetooth sightings by a group of 78 users attending Infocom 2006. Each of the 78 users was carrying a small iMote device during the conference. Fig.3(b) reports the number of contacts between nodes per hour during 4 days of experiment. Bluethooth discovery is done with a frequency of once every 4 minutes.

For all trace files, a proximity matrix is generated based on the number of recorded contacts between users. To compute proximity matrix, the trace is first divided to a series of UTD long intervals. The proximity between node i and j is then computed based the number of intervals that these nodes had a contact. The proximity

matrix indicates the strength of connections among all participating nodes. The set of pairs that have proximity higher than a specific threshold are fed into the community detection algorithm.

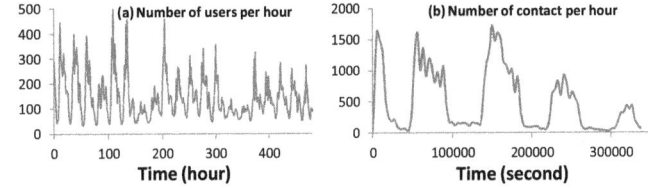

Fig 3: (a) Number of associated users per hour in the UCSD trace; (b) Number of contacts per hour in Infocom '06 trace

4.2 Mobile Social Wireless Networks

Fig. 4(a) reports the measured object provisioning cost for a mobile network of 50 nodes generated using **HCMM** model [8]. With increasing user tolerable delay, each node is able to meet more nodes and therefore, the object provisioning cost reduces. This is especially pronounced when the rebate to download cost ratio (i.e. β) is small. In extreme case, when $\beta=0$, the provisioning cost can be as low as 6 per object with UTD of 1200 seconds. It goes up to 7.8 when the UTD reduces to 10 seconds. The reasons for lower cost with higher UTDs are as follows: 1) with increasing UTD the percentage of isolated nodes reduces due to increasing community size. Isolated nodes are nodes that do not belong to any community, thus unable to leverage cooperative caching, leading to higher provisioning costs. Thus by reducing the number of isolated nodes it is possible to reduce the average network-wide cost. 2) With increasing UTD we get higher community cohesions. Cohesion factor is defined as the ratio of number of edges between nodes in a community to the number of edges in a fully connected community with the same size. This metric is defined to measure the strength of connections in a community.

For the **Infocom '06** trace there were very few contacts between nodes during the night time part of it. Due to this reason, we have first removed the night time part from the trace in order to avoid any skew in the results. The proximity matrix is then computed based on the reduced trace. From the proximity matrix, all pairs with proximity higher than a threshold are extracted and fed into the *fastcommunity* detection algorithm.

Fig. 4(b) reports the provisioning cost for a mobile network derived from the reduced Infocom 06 trace. It can be seen that with increasing β the provisioning cost also increases. This is expected as higher β implies higher rebate which, in turn, reduces the benefit of cooperative caching. The second observation is that when users opt to wait longer (higher user tolerable delay), they have a higher chance to meet more number of nodes and therefore the overall probability of finding the requested object within the network increases. This in turn results in lower cost.

Average provisioning cost from the UCSD trace is reported in Fig. 4(c). Similar to the Infocom 06 scenario, as expected, the provisioning cost shows a decreasing trend when the UTD goes up, and the cost reduces with decreasing β. However, for the UCSD trace cost reduction starts at a higher UTD (i.e. 8 hours) compared to that for the Infocom trace (i.e. 10 minutes). The main reason for this is higher inter-contact times for the UCSD case in comparison to the Infocom case. An interesting observation in Fig. 5(c) is that the minimum provisioning cost for the UCSD trace with 273 users

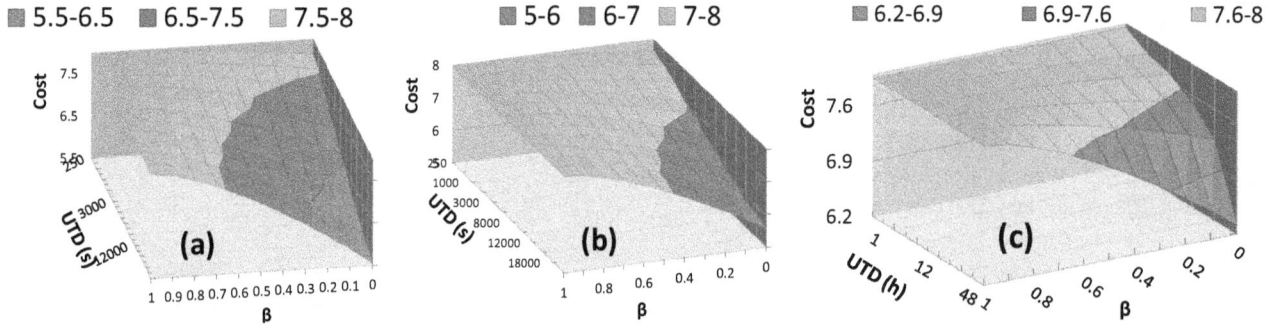

Fig. 4: Object provisioning cost for (a) HCMM trace; (b) Infocom 06 trace and (c) UCSD trace

is 6.3 (when β is *0* and UTD is *160* hours). This number for the Infocom trace with 98 users is about 5.5 (when β is 0 and UTD is 20000 seconds). This result shows that many of the users in the UCSD trace never meet each other (even when we increase the UTD to 160 hours). To confirm this we measured the percentage of isolated nodes for Infocom and UCSD trace and observed that at least 20 percent of nodes in UCSD trace are isolated. This number of Infocom trace was close to zero.

4.3 Compare Algorithms

In this section we compare the impacts of different community detection algorithms on object provisioning cost. The UCSD trace with UTD of 48 hours has been used for these comparisons presented in Fig. 5.

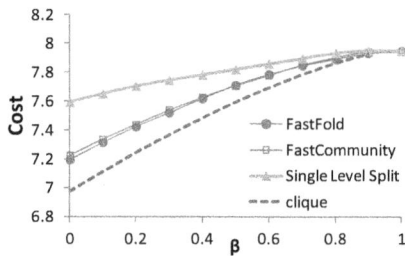

Fig. 5: Comparison of different community detection algorithms

The object provisioning costs for *fastcommunity* and *fastfold* community detection algorithms are very similar and both of them provides better cost compared to the single level split caching. Both of these algorithms are heuristic in nature, and they work based on maximizing the network modularity. The community detection algorithm *clique* provides lower cost solutions when compared to the other two. The *clique* algorithm is also more computationally expensive compared to the other two, which are heuristics based.

With *clique* all nodes in a community will have a strong connection and therefore the cohesion factor in community found by *clique* is very high which results in lower provisioning cost for clique as it is shown in Fig. 5.

4.4 Comparison with Simple Caching

In this section we compare the HSC strategy with non-hierarchical split caching (i.e. single level split caching that was proven to be optimal for stationary networks [10]). This single split does not leverage information about community abstraction.

The minimum provisioning cost for single level split is numerically found by measuring the performance of Split caching over all possible values of the split factor λ. The percentage improvement of hierarchical split cache with respect to the best result obtained by single level caching is reported in the following graphs.

Fig. 6(a) reports the percentage improvement of hierarchical split caching on single level split caching for the HCMM model. Initially when UTD is small, community based hierarchical split-caching caching shows no improvement over the single level split caching. With small UTDs, the number of contacts between nodes is also very low which eliminates any chance of cooperation with in-community peers. In this case, the object provisioning cost can be reduced only by local caching and therefore performance of hierarchical based caching and single level split caching are the same. By increasing UTD however nodes have higher change to meet other in-community nodes. Intuitively the number of contacts among nodes in the same community is much higher than the number of contacts between nodes in different communities. Considering this fact in its design, community based hierarchical caching is able to take advantage of cooperation between nodes in the same community. The single level split caching, however, lacks this feature, which leads to higher provisioning costs. The percentage improvement generally reduces as we increase the rebate to download cost ratio (i.e. β) because by increasing β the cooperation loses its role in cost reduction.

A very similar trend in Fig. 6(b) can be observed for the Infocom 06 trace. Unlike for the HCMM case, however, the percentage improvement of hierarchical based caching in the Infocom 06 trace starts falling at relatively higher values of UTD. The reason is that by increasing UTD, the communities merge together. In the extreme case, when UTD is large enough, all nodes in the network will be in the same community. In this case, the performance of split caching and community based caching would be the same. For the Infocom '06 trace, only two communities are detected for large UTDs. By increasing UTD even more, one community starts growing and the other one shrinks. Note that performance improvement of hierarchical based caching is maximized when the sizes of all communities are the same. Performance improvement for the UCSD trace is shown in Fig. 6(c) Similar to the two other traces, the community based caching shows higher improvement for lower βs and higher UTDs.

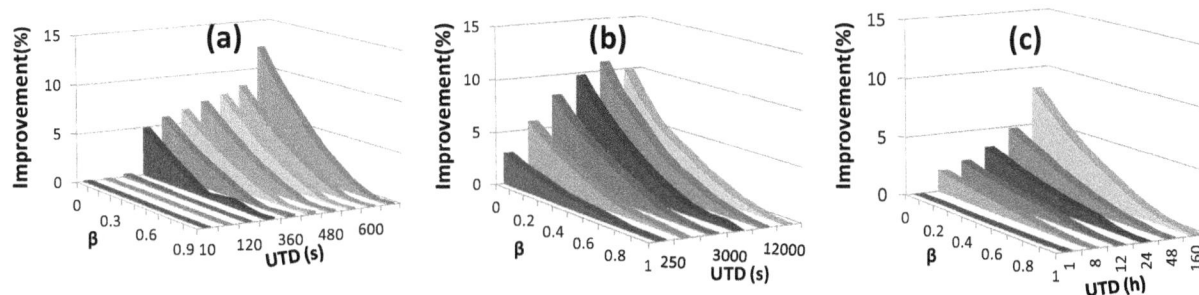

Fig. 6: Performance improvement of HSC with respect to Split Caching for (a) HCMM (b) Infocom '06 and (c) UCSD trace

5. RELATED WORK

There is a rich body of existing literature on several aspects of cooperative caching including object replacements, reducing cooperation overhead, and cooperation performance in traditional wired networks. The Social Wireless Networks (MSWNET) explored in this paper, which are often formed using mobile ad hoc network protocols, are different in the caching context due to their additional constraints such as topological insatiability and limited resources. As a result, most of the available cooperative caching solutions for traditional static networks are not directly applicable for the MSWNETs.

Three schemes for MANET have been presented in [12]. In the first one, CacheData, a forwarding node checks the passing-by objects and caches the ones deemed useful according to some pre-defined criteria. The second approach, CachePath, is different in that the intermediate nodes do not save the objects; instead they only record paths to the closest node where the objects can be found. The last approach in [12] is the HybridCache in which either CacheData or CachePath is used based on the properties of the passing-by objects through an intermediate node.

Recently, caching solutions have been proposed to improve the performance of data access in social based networks. Authors in [13] propose to intentionally cache data at a set of network central locations which can be easily accessed by other nodes. Zhuo et al. [14] propose a social-based caching that considers the effects of contact duration. The community based caching proposed in this paper chooses a different strategy. Instead of finding the most central nodes, our community based caching attempts to manage the cache space based on structure of community in the network.

In summary, in most of the existing collaborative caching work, there is a focus on maximizing the cache hit rates without considering its effects on the overall cost which depends heavily on the content service and pricing models. The pricing model in our work which is based on the practical Amazon Kindle business model is substantially unique and practical compared to those used in [12-15].

6. SUMMARY AND CONCLUSIONS

Drawing motivation from Amazon's Kindle electronic book delivery model, this paper develops a set of practical network, service, and pricing models which are used for creating a cost-optimal cooperative caching strategy in Mobile Social Wireless Networks. It was shown that the proposed strategy can reduce content provisioning cost under a range of various user mobility scenarios with and without social community abstractions. We also construct analytical models and simulation results for evaluating

the performance of the proposed strategy with prevalent mobility models and real human mobility traces from the literature.

7. ACKNOWLEDGEMENTS

This work was partially supported by a grant (NeTS 1017477) from National Science Foundation.

8. REFERENCES

[1]   M. Girvan, M. E. J. Newman. Community structure in social and biological networks. Proc. Natl. Acad. Sci. 99(2002).

[2]   M. Newman, "Modularity and Community Structure in Networks," Proc. Nat'l Academy of Sciences, pp. 8577-8582, 2005.

[3]   M.E.J. Newman, "Fast Algorithm for Detecting Community Structure in Networks," Physical Rev. E, vol. 69, no. 066133, 2004

[4]   V. Blondel, J. Guillaume, R. Lambiotte, E. Lefebvre, "Fast unfolding of communities in large networks", in Journal of Statistical Mechanics: Theory and Experiment 2008 .

[5]   Hastie, Trevor; Tibshirani, Robert; Friedman, Jerome (2009). Hierarchical clustering The Elements of Statistical Learning (2nd ed.). New York: Springer. pp. 520–528. SBN0-387-84857-6.

[6]   J. Scott, R. Gass, J. Crowcroft, P Hui, C. Diot, and A. Chaintreau. "CRAWDAD data set cambridge/haggle". http://crawdad.cs.dartmouth.edu/cambridge/, Sep 2006.

[7]   M. McNett and G. Voelker, "Access and mobility of Wireless PDA users,"ACM SIGMOBILE Mobile Computing and Communications Review, vol. 9, no. 2, pp. 40–55, 2005.

[8]   C. Boldrini, M. Conti, and A. Passarella. "Users mobility models for opportunistic networks: the role of physical locations". In Proc. of IEEE WRECOM, 2007.

[9]   M. Gonzalez, C. Hidalgo, A. Barabasi Understanding individual human mobility patterns Nature, 453 (7196) (2008), pp. 779–782

[10] M. Taghizadeh, A. Plummer, A. Aqel and S. Biswas, "Towards Optimal Cooperative Caching in Social Wireless Networks", in Proc. of IEEE GLOBECOME 2010.

[11] M. McNett and G. Voelker, "Access and mobility of Wireless PDA users,"ACM SIGMOBILE Mobile Computing and Communications Review, 2005

[12] L. Yin and G. Cao, "Supporting cooperative caching in ad hoc fnetworks," IEEE Transactions on Mobile Computing, 2006

[13] W. Gao, G. Cao, A. Iyengar, and M. Srivatsa, "Supporting Cooperative Caching in Disruption Tolerent Networks," in Proc. of IEEE ICDCS, 2011.

[14] X. Zhuo, Q. Li, G. Cao, Y. Dai, B. Szymanski, and T. L. Porta, "Social-Based Cooperative Caching in DTNs: A Contact Duration Aware Approach", IEEE MASS, 2011.

[15] B. Tang, H. Gupta, and S. Das, "Benefit-based Data Caching in Ad Hoc Networks," IEEE TMC, vol. 7, no. 3, pp. 289–304, 2008.

Mobile Social Networking through Friend-to-Friend Opportunistic Content Dissemination

Kanchana Thilakarathna
NICTA & UNSW, Australia
kanchana.thilakarathna
@nicta.com.au

Aline Carneiro Viana
INRIA, France
aline.viana@inria.fr

Aruna Seneviratne,
Henrik Petander
NICTA & UNSW, Australia
[aruna.seneviratne;
henrik.petander]
@nicta.com.au

ABSTRACT

We focus on dissemination of content for delay tolerant applications, (i.e. content sharing, advertisement propagation, etc.) where users are geographically clustered into communities. We propose a novel architecture that addresses the issues of lack of trust, delivery latency, loss of user control, and privacy-aware distributed mobile social networking by combining the advantages of decentralized storage and opportunistic communications. The content is to be replicated on friends' devices who are likely to consume the content. The fundamental challenge is to minimize the number of replicas whilst ensuring high and timely availability. We propose a greedy heuristic algorithm for computationally hard content replication problem to replicate content in well-selected users, to maximize the content dissemination with limited number of replication. Using both real world and synthetic traces, we show the viability of the proposed scheme.

Categories and Subject Descriptors

C.2.1 [**Computer-Communication Networks**]: Network Architecture and Design—*Store and forward networks*

Keywords

Opportunistic Content Dissemination; User Generated Content Sharing; Mobile Social Networking; Dynamic Centrality Metrics

1. INTRODUCTION

The past few years have seen the rapid growth in the use of free social networking applications such as Facebook, Google+ and other applications that support the distribution of user generated content (UGC) such as YouTube and Flickr. Today, more and more UGC is being generated by mobile devices and all predictions are that mobile users will be the dominant generators and consumers of content in the near future. This will impact the user in two ways. Firstly, it will exacerbate the problem associated with the exponential increase in mobile data traffic [4]. Secondly, it will make the

problems associated with privacy concerns and data ownership even more acute, as the service providers will have full control of the user data.

There have been numerous proposals for dealing with the ever increasing mobile data traffic by taking advantage of ubiquitous availability of mobile devices and/or access to heterogeneous networks [10, 6]. As a separate body of work, there have been a number of proposals aimed at addressing the issues associated with privacy and the user losing control of their data. This work has lead to the development of *distributed decentralized social networking* architectures [7, 9], which allow an individual user or a community of users to host their own data.

Despite offering many advantages for mobile users, opportunistic communication solutions are not adopted for social networking services such as Facebook, or Google+ for two primary reasons. Firstly, there is an inherent reluctance by the users to interact with strangers or third parties, due to security and privacy concerns of social networking. Although, this has been partially addressed by the use of secure communication and storage methods which use encryption, it is still a drawback. Secondly, the high delivery latency, i.e. when a user wants to share information, none of the couriers of information may be in the vicinity. In addition, the opportunistic solutions do not address the issues of loss of control of data nor the loss of privacy. Finally, distributed decentralized social networking architectures, in contrast, address the issue of loss of control of data and the loss of privacy. However in a mobile system, they increase mobile data traffic because of content/profile replication on distributed servers to increase the content availability [10]. With the current trend of mobile operators moving to *capped* data plans, the communication costs limits the applicability of these service in mobile networks.

In this paper, we propose a new architecture that addresses the issues of lack of trust, delivery latency, loss of user control, and user privacy by combining the advantages of decentralized storage and opportunistic communications. Although such approaches are not new, the proposed hybrid method of combining the two ideas is novel. In particular, we propose to exploit content replication to bridge the gap between the user and the mobile social networking *friends* who are not in the vicinity of the user, thereby reducing the content delivery latency and increasing the delivery success rate. Since the proposed scheme relies on devices that are used to replicate the content, the devices are only selected among the users who are previously identified as friends as it improves the privacy of the users. We propose dynamic

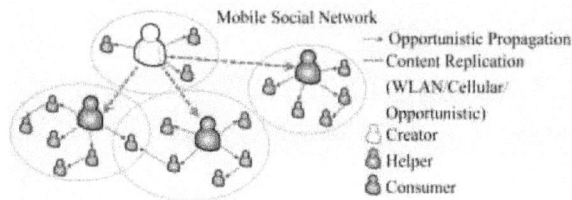

Figure 1: Overview of the system architecture

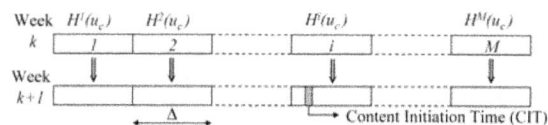

Figure 2: Periodic weekly helper selection

centrality metrics to rank users for community based greedy helper selection algorithm. Through extensive trace driven simulations with both real world and synthetic trace data sets, we show that approximately 60% of content can be delivered in less than one day latency with less than 10% of replication and the dynamic centrality measures are highly dependent on the surrounding environment.

The remainder of the paper is organized as follows: Section 2 presents the overview of the proposed system followed by the content replication algorithm in Section 3. Section 4 evaluates the content delivery performance. Finally, Section 5 concludes the paper.

2. THE SYSTEM ARCHITECTURE

Consider a scenario where a user, a *creator*, wants to share content with a set of users who have previously been identified as *friends* through a social networking service. However, each and every friend may not be interested in the shared content even though it is pushed to the device [12]. Assume that potential *consumers* among the friends can be predicted through the social networking service based on the history of content access patterns. To this end, the aim is to propagate the content only to the consumers and leave the option to other friends to fetch the content from the creator only if they are interested. If this predictive pushing (pre-fetching) can be scheduled to use low-cost networks, uncongested networks or opportunistic networks, it is possible to minimize the communication costs, energy usage and storage requirements of mobile devices. Although it is difficult to predict the content consumption of a user with a high degree of accuracy, even with low prediction accuracies, it will be possible to have significant impact.

If opportunistic networks are to be used, the challenge is the timeliness of the delivery due to limited number of consumers in the vicinity of the creator. It has been suggested that this can be addressed by replicating the content to some carefully selected consumers, namely *helpers*, and then trust these helpers to propagate the content to other consumers through opportunistic communication, as shown in Figure 1. The novelty of the proposed system stems from the selection of helpers, and minimization of replication.

There is a central entity performing the helper selection based on the connectivity information as described in Section 3. Since helpers are only selected from the consuming friends, the privacy of the users is better preserved. As shown in Figure 1, initial content replication is carried out by either a pre-existent network such as WLANs and cellular networks, or short-range opportunistic communication, if the helpers are in the vicinity of the creator. At the same time, the creator initiates content dissemination to the consumers in the vicinity via opportunistic communication.

To perform content dissemination among the users, a peer-to-peer (P2P) protocol, a modified version of BitTorrent, will be used. In particular, a consumer becomes a *propagator* or a seeder only after the consumer has downloaded all pieces of the content. This enables the prioritization of the energy consumption of mobile devices, e.g. firstly for its own purposes (i.e. to finish downloading the full content for its own use) and secondly to help others in content propagation. Due to space limitations, the P2P mechanism is not described further in this paper.

3. CONTENT REPLICATION

To investigate the effectiveness of opportunistic communication for content dissemination using only interactions among the creator and the consumers, in [11] we analyzed the contact traces generated from Dartmouth data set [1]. The results show that the number of consumers is the limiting factor for opportunistic content delivery success rate. Even in the large systems like Facebook, the median friend size is approximately 100 [12], i.e. the number of consumers are limited. Hence, opportunistic communication is not effective for content dissemination in mobile social networks. To this end, it is possible to use content replication to improve the performance of opportunistic content delivery. However, since content replication to helpers consumes more network resources due to the use of infrastructure based communication, there is an obvious trade-off between delivery performance and content replication overhead. Thus, the content replication challenge is to *maximize the content delivery rate with limited content replication.*

Opportunistic encounters are highly dependent on user mobility patterns, which essentially demonstrates social behavior of users. These encounter patterns have been extensively analyzed in the areas of context-aware services and mobile social networking [3]. Usually, social behavior of the majority of users have weekly routines. Furthermore, there is higher probability that a user meets the same people at the same time in every week. We take advantage of this predictive regularity of encounter patterns for the purpose of content replication.

To facilitate instant content dissemination, helpers have to be selected in advance such that the helpers for week $k+1$ during the week k as shown in Figure 2. Let C be a set of consumers and a creator u_c wants to share a content via a mobile social networking application/service with consumers $u \in C$. The week (Δ_w) is divided into Δ time slots, where the Δ is the content delivery deadline. Since we do not know when creators are going to generate content during the week $k+1$, we select several sets of helpers, $H_{k+1}(u_c) = \left\{ H_k^j(u_c) : u \in C \text{ and } j \in [1 : \Delta_w/\Delta] \right\}$ during week k. At the end of every week, a central entity performs helper selection and inform all creators the respective sets of helpers.

3.1 Dynamic Centrality Metrics

The helper selection problem is NP-Hard as shown in [11]. However the objective can be approximated efficiently using heuristics. Thus, we present a community based greedy algorithm COMMUNITY-GREEDY for content replication. The most influential user in the network is the one who has contacts with (covers) the maximum number of consumers, which can be intuitively used as a greedy choice property.

As the first step, we aggregate every contact for a single graph without loosing any temporal information. Let an aggregated weighted graph $G = (C, E)$ consists of all edges in $G_t(C, E_t)$, $\forall t \in (1, 2, \cdots, n)$ such that $G = G_1 \cup G_2 \cup \cdots \cup G_n$. There exists an edge $e \in E_t$ between two consumers if they are in the communication range of each other at time t. Let $\alpha_{u,v}^t$ be the edge weights which represents the contact duration between consumers u and v at time t. Then, we focus on centrality metrics in G having an impression similar to expected number of covered consumers. Hereafter, we propose two kinds of centrality metrics: local and global.

3.1.1 Local metrics

One of the simplest centrality metric that implies the capability of neighbourhood coverage is the degree centrality $C_{LD}(u) = |N(u)|$ where $N(u)$ is the set of neighbours of u in the aggregated graph G. Degree centrality identifies popular nodes in the network and thus has higher influence on content propagation.

However simple degree centrality does not guarantee that all counted encounters are practically realizable due to the lack of temporal information. Further, the contacts that happen early are important in propagation than those that happen later. Hence a centrality metric which captures temporal information could be more realistic to be considered in dynamic networks. We define the initial contact time as $I(u, v) = min\{t\} : \alpha_{u,v}^t > 0$ for all $t \leq \Delta$ and the total contact duration $D(u, v) = \sum_{t=1}^{\Delta} \alpha_{u,v}^t$ for an edge (u, v). We calculate the weight $w_{u,v} = I(u, v) + (1/D(u, v))$ for all $(u, v) \in E$. $w_{u,v}$ has the meaning of earliness and solidity of the contact (u, v). In practice, each mobile device can calculate w locally for all other devices it encounters for a given period. To this end, we define an improved dynamic degree centrality metric: $C_{LID}(u) = |N(u)| + \frac{|N(u)|}{\sum_{v \in N(u)} w_{u,v}}$ $N(u)$ is the set of neighbours of u. C_{LID} has the impression of how early and how independently the user makes other users into content propagators. We aim to use C_{LID} in the greedy choice for helper selection.

3.1.2 Global metrics

Here, we define two path-based centrality metrics for node ranking. We first define a naive simple metric $C_{GP}(u) = \sum_{v \in C} p(u, v)$ for all $t \leq \Delta$ where $p(u, v) = 1$ if there is a path between u and v and $p(u, v) = 0$ otherwise. This can be viewed as an extended degree for node u giving heuristics about the popularity and the availability of the node. Simplicity of the metric is the main advantage, which requires only information about existence of a path.

On the other hand, simplicity does not provide accurate heuristics. Therefore, we construct a directed aggregated graph $G' = (C', E')$ by directing all edges in G for both directions with same weights. Next, we prune all unrealizable edges in the network, i.e. if a content is to be propagated via a node, its outgoing contact has to take place after at least its first incoming contact. Then, at each node there is a guarantee that content will be propagated to other nodes if the content has arrived at the node. We calculate shortest-path $sp(u, v)$ for all node pairs $(u, v) \in C'$ in terms of edge weights $w_{u,v} = I(u, v) + (1/D(u, v))$. We define a path-based dynamic centrality metric $C_{GIP}(u) = \sum_{v \in C'} p(u, v) + \frac{\sum_{v \in C'} p(u, v)}{\sum_{v \in C'} sp(u, v)}$. Similar to C_{LID}, C_{GIP} implies how early and how independently the user makes other content propagators.

3.2 Community based Greedy Algorithm

In this section, we present our content replication algorithm which combines social sub-structural properties such as communities with previously defined dynamic centrality metrics. We find communities using *k-clique* community detection algorithm. Then, we distribute helpers among communities based on their ranking given by the proposed dynamic centrality metrics as in Algorithm 1. First, the consumer with highest centrality value is selected as a helper and rely on that helper to propagate the content within the community. Then, the next highest consumer from a different community is selected, i.e. line 9 of the Algorithm 1. If the threshold of replication (λ) is lower than the number of communities, initial content propagators will not be selected from the creator's community, assuming that the creator is capable to propagate the content within its community. There can be a scenario where the majority of the consumers do not belong to communities. Then, the selection is purely based on the centrality value of consumers.

Algorithm 1 COMMUNITY-GREEDY$(G, G', \Delta, \lambda, u_c)$

---

1. $D \leftarrow H(u_c) \leftarrow \emptyset$
2. **for** all $u \in C$ **do**
3.   Find centrality metric
    $C_{LD}(u), C_{LID}(u), C_{GP}(u), C_{GIP}(u)$
4. communities $\leftarrow$ k-clique-algorithm$(G', 3)$
5. Let a community $com(u)$ be the u's community
6. $D \leftarrow com(u_c)$
7. $H(u_c) \leftarrow u_c$
8. **while** $|H(u_c)| \leq \lambda(u_c)$ or $D \neq C$ **do**
9.   Let $u \in (C \setminus (D \cup H(u_c)))$
    maximizing $C_{LD}(u), C_{LID}(u), C_{GP}(u), C_{GIP}(u)$
10. $H(u_c) \leftarrow u$
11. $D \leftarrow D \cup com(u)$
12. **return** $H(u_c)$

Furthermore, it is expected that the selected helpers will become quasi static over the time due to the regular behavior of people [5]. Hence the helper calculation will be carried out only for users who have changed their behavioral pattern. To this end, we believe that the helper selection algorithm will scale up with the increasing number of users. In the next section, we evaluate the performance of this approach with respect to different centrality metrics.

4. PERFORMANCE EVALUATION

Two real-world data sets were used in the evaluation: Dartmouth data set [1] contains wireless connectivity patterns of 1138 devices for 8 weeks and USC [1] data set, which also contains wireless connectivity patterns of 1846 users in a campus environment. Moreover, we use the synthetic trace SWIM-500 that is generated using the SWIM simulator [8] by extending the Cambridge data set [1] of 36 Bluetooth enabled iMotes. We consider that two users are in contact when they are connected to the same WiFi access point for both Dartmouth and USC and when two users are within the Bluetooth communication range of each other for SWIM.

The trace data sets were divided into weeks as shown in Figure 2. We select the set of helpers according to the proposed algorithm during week k and use them for content replication during week $k+1$. The content delivery deadline Δ is considered as 3 days and the creator and consumers are selected randomly. In content propagation, we only use opportunistic contacts among the consumers. We consider that the number of consumers for a creator is 100 [12]. We select

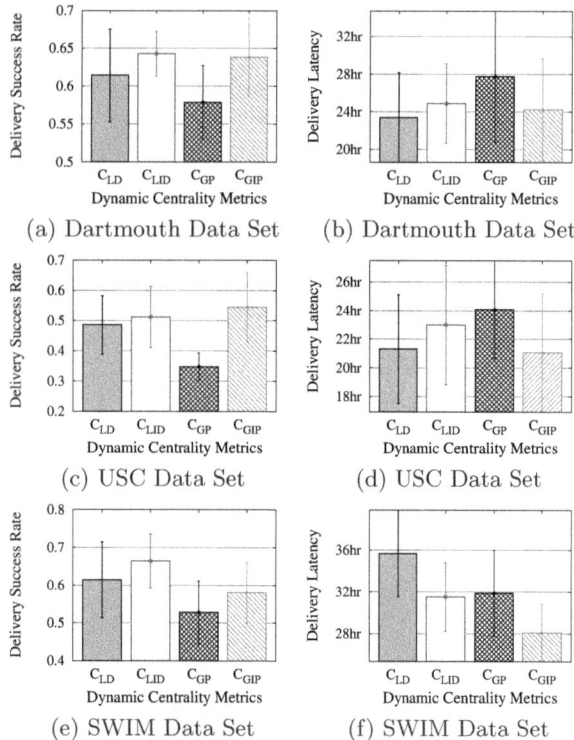

Figure 3: Comparison of different dynamic centrality metrics. $\lambda = 10\%$

consumers randomly from the users in the trace to evaluate the worst case scenario where the connectivity patterns of the users within the group has minimum level of correlation. Each creator will generate a content of size 8.4MB, which is the median content size in YouTube [2] and the transfer rate among consumers are considered as uniform and 2Mbps. Hence a consumer has to have aggregated contact duration of 33.6 seconds with the creator or any of the helpers or the propagators to completely download the content. All simulations are carried out varying the monitoring and evaluation periods through out the data sets.

4.1 Evaluation of Dynamic Centrality Metrics

Figure 3 shows the content delivery success rate and the mean content delivery latency when the helpers are selected based on different centrality metrics according to the Algorithm 1. Approximately 60% of content can be delivered in less than one day latency. When we compare the two local centrality metrics, C_{LID} has a slightly better delivery success rate with lower standard deviation than C_{LD} in all three data sets. Similarly, the improved global centrality metric C_{GIP} has better performance in terms of both delivery rate and latency compared to the naive C_{GP}. This is due to the fact that each improved metric, C_{LID} and C_{GIP}, considers the time dependency in connectivity patterns.

However in some cases there is no significant difference between the performance of local and global metrics. All these general similarities are related to dynamics of the contact patterns among consumers in these environments. For instance, when the degree of the majority of the consumers are similar, e.g. skewed degree distribution for SWIM [11], the performance of a simple degree based local centrality metric becomes significant as depicted in Figure 3e for SWIM. In contrast, when the degree distribution is not skewed, the intelligent path based selection will perform better, similar

to C_{GIP} performance for Dartmouth and USC. In particular, USC has the highest increment of approximately 20% in coverage for C_{GIP} compared to C_{GP} because USC has the most distributed degree distribution. On the other hand, due to the large number of isolated consumers in USC [11], it does not have much gain in delivery latency. In SWIM, C_{GIP} has much lower delivery latency because it has the lowest number of isolated users. Hence the selection of the appropriate centrality metric to identify the most influential users in content dissemination is highly environment dependent and the appropriate centrality metric can be identified by analysing the user behavior of the particular community.

5. CONCLUSION

We proposed a novel opportunistic content dissemination architecture that addresses the issues of lack of trust, timeliness of delivery, loss of user control. We proposed dynamic centrality metrics to identify most influential users to replicate content within a community. Then, a community based greedy algorithm was presented for efficient content replication by taking advantage of routine behavioral patterns of users. Using both real world and synthetic traces, we compare the delivery performance when selecting helpers based on different dynamic centrality metrics. The results show that approximately 60% of content can be delivered in less than one day latency and the dynamic centrality measures are highly dependent on the user behaviour in the community. In future, we aim to further investigate the performance of social-aware content propagation and compare with alternative strategies.

6. REFERENCES

[1] Crawdad:. In *http://crawdad.cs.dartmouth.edu*.
[2] A. Abhari and M. Soraya. Workload generation for youtube. *Multimedia Tools and Applications*, 46(1):91–118, 2010.
[3] M. Barbera, J. Stefa, A. Viana, M. de Amorim, and M. Boc. Vip delegation: Enabling vips to offload data in wireless social mobile networks. In *DCOSS*, 2011.
[4] Cisco. Cisco visual networking index: Global mobile data traffic forecast update, 2011–2016.
[5] H. Falaki, R. Mahajan, S. Kandula, D. Lymberopoulos, R. Govindan, and D. Estrin. Diversity in smartphone usage. In *MobiSys '10*, pages 179–194, San Francisco, California, USA, 2010.
[6] B. Han, P. Hui, V. Kumar, M. Marathe, J. Shao, and A. Srinivasan. Mobile data offloading through opportunistic communications and social participation. *Mobile Computing, IEEE Transactions on*, (99):1–1, 2011.
[7] http://joindiaspora.org.
[8] A. Mei and J. Stefa. Swim: A simple model to generate small mobile worlds. In *INFOCOM 2009, IEEE*, pages 2106–2113. IEEE, 2009.
[9] R. Sharma and A. Datta. Supernova: Super-peers based architecture for decentralized online social networks. In *COMSNETS'12*, pages 1–10. IEEE, 2012.
[10] K. Thilakarathna, H. Petander, J. Mestre, and A. Seneviratne. Enabling Mobile Distributed Social Networking on Smartphones. In *Proc. of ACM MSWiM'12*, pages 357–366, oct 2012.
[11] K. Thilakarathna, A. C. Viana, A. Seneviratne, and H. Petander. The Power of Hood Friendship for Opportunistic Content Dissemination in Mobile Social Networks. Technical report, INRIA, France, 2012.
[12] J. Ugander, B. Karrer, L. Backstrom, and C. Marlow. The anatomy of the facebook social graph. *Arxiv preprint arXiv:1111.4503*, 2011.

Scheduling of Access Points for Multiple Live Video Streams

I-Hong Hou
CESG and Department of ECE
Texas A&M University
College Station, TX 77843, USA
ihou@tamu.edu

Rahul Singh
CESG and Department of ECE
Texas A&M University
College Station, TX 77843, USA
rsingh1@tamu.edu

ABSTRACT

This paper studies the problem of serving multiple live video streams to several different clients from a single access point over unreliable wireless links, which is expected to be major a consumer of future wireless capacity. This problem involves two characteristics. On the streaming side, different video streams may generate variable-bit-rate traffic with different traffic patterns. On the network side, the wireless transmissions are unreliable, and the link qualities differ from client to client. In order to alleviate the above stochastic aspects of both video streams and link unreliability, each client typically buffers incoming packets before playing the video. The quality of the video playback subscribed to by each flow depends, among other factors, on both the delay of packets as well as their throughput.

In this paper we address how to schedule packets at the access point to satisfy the joint per-packet-delay-throughput performance measure. We test the designed policy on the traces of three movies. From our tests, it appears to outperform other policies by a large margin.

Categories and Subject Descriptors

C.2.1 [**COMPUTER-COMMUNICATION NETWORKS**]: Network Architecture and Design —*Wireless communication*

Keywords

Wireless networks; video streaming; delays; deadlines

1. INTRODUCTION

Multimedia applications for video streaming have been predicted to become a dominant portion of future wireless traffic [1]. In order to provide smooth playback to end users with multimedia applications, their packets need to be delivered in a timely manner and with a sufficient throughput. Providing

This material is based upon work partially supported by NSF under Contract Nos. CNS-1302182 , CPS-1239116, CPS-1232601, CPS- 1232602 and ECCS-1150944, and AFOSR under Contract No. FA 9550-13-1-0008.

the per-packet delay guarantees is particularly challenging for two reasons. First, on the application side, the video streams generate variable-bit-rate (VBR) traffic that congests the network when multiple streams generate a burst of packets at the same time. Second, on the network side, wireless transmissions are inherently unreliable due to shadowing, fading, and interference.

In order to mitigate the effect of the uncertainties in traffic bit rate as well as channel reliability, packets are typically buffered at the receivers. Each receiver waits a specified amount of time buffering incoming packets before it starts playing the video. On the positive side, this approach has the benefit of greatly reducing the impact of network congestion and channel unreliability. However, buffering also increases delay as the receiver cannot start playing the video immediately. For live video streaming, such as sports events and instant news, it is important to provide smooth playback while waiting only a reasonably small amount of time before playing the video.

In this paper, we analyze a model that addresses the three characteristics of stochastic packet arrivals, unreliable wireless channel, and buffering delay. We determine a simple online scheduling policy, the Earliest Positive-Debt Deadline First (EPDF) policy, for serving the live video streams that achieves the delay-throughput capacity. This policy does not require knowledge of the exact traffic pattern of each video stream, enhancing its suitability for implementation.

We test the proposed EPDF policy on three movie traces. We compare its performance with three other well known policies in ns-2. The EPDF policy outperforms the other policies by a large margin on these three video traces. We also propose a modification to enhance short-term performance, and investigate the tradeoff between long-term and short-term performance guarantees.

2. RELATED WORK

There have been several studies on providing delay guarantees for flows that require delay guarantees over wireless networks. Liu, Wang, and Giannakis [12] have proposed a scheduler that updates link priorities dynamically to provide QoS. Dua and Bambos [2] have proposed a heuristic that jointly considers wireless channel quality and packet deadlines. Raghunathan et al [13] and Shakkottai and Srikant [15] have proposed scheduling policies that minimize the number of packets that exceed their delay bounds.

Hou el al [4] have proposed a model that jointly considers the per-packet delay bound and the per-flow throughput requirement. This model has been extended to consider variable-bit-rate traffic [5], fading wireless channels [6] [9], the mixture of real-time and non-real-time traffic [8], and

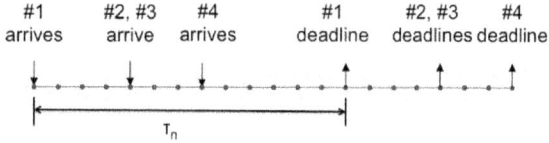

Figure 1: Packet arrivals and deadlines of a stream, with sequence numbers indicated.

multi-hop wireless transmissions [10]. However, these studies assume that all flows are synchronized and generate packets at the same time, and the results depend critically on this assumption. Moreover, they assume that all flows start playback immediately without buffering any packets, which is a critical feature of the playback process. Dutta el al [3] have studied serving video streams when receivers may buffer packets before playing them. They assume, however, that all packets are available at the server when the system starts, which is applicable to on-demand videos, but not to live videos.

3. SYSTEM MODEL

Consider a system with N wireless clients, $\{1, 2, \ldots, N\}$, and one access point (AP). Each client subscribes to a live video stream through the AP.

Video streams generate variable-bit-rate traffic, and the exact number of packets generated in a time slot is influenced by not only the video encoding mechanism, but also the context of the current frame. We suppose that the number of packets generated by each flow in a time slot is described by some irreducible finite-state Markov chain, and is bounded. However, we do not assume that either the AP or the clients know this Markov chain model. We only suppose that each client n knows the long-term average number of packets generated by its subscribed video stream, which we denote by r_n. In each time slot the access point can transmit one such packet, including all overhead.

To smooth the bursty arrival process as well as channel unreliability, each client buffers incoming packets before it commences playing the video. To be more specific, a video frame that is generated at time slot t for client n will be played by the client at time slot $t + \tau_n$, with the AP informed of the value of τ_n. This is equivalent to specifying a hard per-packet delay bound of τ_n time slots for the video stream of client n. Fig. 1 illustrates an example of packet arrivals and their respective deadlines. If a packet cannot be delivered within its delay bound, it is dropped from the system. Packet drops result in glitches in the live videos, and hence need to be limited.

We model the channel reliability between the AP and client n by a probability p_n. When the AP schedules a transmission for client n, the transmission is successful with probability p_n. The AP obtains immediate feedback on whether a transmission is successfully received by ACKs. A transmission is considered to be successful if both the packet and ACK are correctly received. When a transmission fails, the AP may retransmit the packet, if the packet has not already expired, i.e., exceeded its delay bound, or it may schedule a different packet in the next time slot.

For enforcing a minimum quality for each video, each client n requires that its throughput is at least q_n packets per time slot; that is, it requires that $\liminf_{T \to \infty} \frac{\sum_{t=1}^{T} e_n(t)}{T} \geq q_n$, where $e_n(t)$ is the indicator function that a packet for client n is delivered in time slot t.

4. EARLIEST POSITIVE-DEBT DEADLINE POLICY AND ITS OPTIMALITY

In this section we establish the optimality of a simple online scheduling policy called the Earliest Positive-Debt Deadline First (EPDF). We will show that it can support every vector of throughputs that is strictly in the capacity region.

Denote by $w_n := \frac{q_n}{p_n}$ the implied workload of client n. As noted in [4], the throughput of client n is at least q_n packets per time slot if and only if the AP, on average, schedules client n for w_n times per time slot. The EPDF policy is then based on the concept of truncated time debt, which differs from the debt in [4]. The choice of M below is related to the usage of a multistep negative drift of the Lyapunov function.

DEFINITION 1. The truncated time debt of client n at time t, denoted by $d_n(t)$, is defined recursively by:

$$d_n(t+1) = [d_n(t) + M w_n 1(t+1 \equiv 1 \pmod{M}) - 1(\text{the AP schedules } n \text{ in time } t+1)]^+,$$

where $1(\cdot)$ is the indicator function.

Note that it increases by $M w_n$ every M time slots, where M is an adjustable parameter, and decreases by 1 in each time slot that the AP schedules client n for transmission. Finally, we only retain the positive part. We will hereafter say that the M time slots form a frame. The choice of M will be discussed in the sequel. It will be the single parameter that will be tuned by the policy to adapt to all the above uncertainties.

We now present the scheduling policy, which we call the Earliest Positive-Debt Deadline First (EPDF) policy: In each time slot, the AP schedules the packet with the earliest deadline from those whose associated clients have strictly positive truncated time debts, that is, $d_n(t) > 0$. Ties are broken arbitrarily. If the associated client of every packet in the system has $d_n(t) = 0$, the AP schedules the packet with the earliest deadline. The AP only idles in a time slot when there are no packets to be transmitted. We note that this policy differs from the EDF policy in that it restricts the AP from serving clients with zero truncated time debts, so as to prevent the AP from providing too much service to some clients while starving others.

The EPDF policy only needs to be tuned with an appropriate choice of M. It does not need any explicit information on the traffic pattern of each client besides the knowledge of the implied workload and the choice of M. It is a simple online scheduling policy that is readily implementable. We now show that this policy supports every vector of throughputs that is strictly in the capacity region through an appropriate choice of M.

The proof employs the Lyapunov function

$$L(k) := \sum_{n=1}^{N} d_n(kM).$$

In our context the Foster-Lyapunov Theorem is used as follows:

THEOREM 1. If, under some scheduling policy η, there exists a positive number δ, and a finite subset $\mathcal{D}_0$ of $\mathbb{R}^N$ such that:

$$E\{L(k+1) - L(k)|[d_n(kM)]\} \leq -\delta, \text{ if } [d_n(kM)] \notin \mathcal{D}_0,$$
$$E\{L(k+1) - L(k)|[d_n(kM)]\} < \infty, \text{ if } [d_n(kM)] \in \mathcal{D}_0,$$

then the throughput of each client n is at least q_n.

THEOREM 2. The EPDF policy supports every vector $[q_n]$ that is strictly in the capacity region, with properly chosen M.

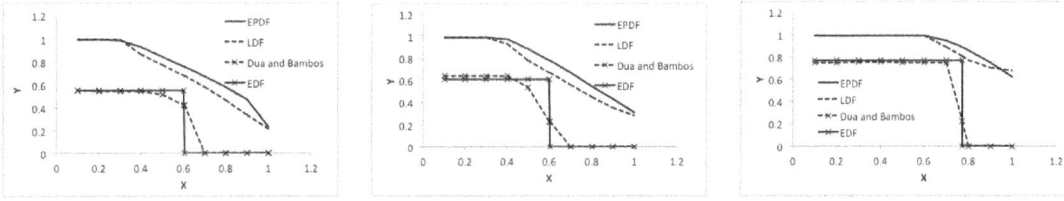

(a) Harry Potter   (b) Finding Neverland   (c) Transporter 2

Figure 2: Achieved throughput regions under different policies

PROOF. Due to space limitation, we only provide a sketch of the proof. Interested readers can find the complete proof in [7].

Consider a vector of throughputs $[q_n]$ that is strictly in the capacity region. There exists some $\alpha > 0$ such that the vector $[q_n + \alpha]$ is still in the capacity region. Therefore, we can assume that there exists a scheduling policy η, under which the throughput of each client n is at least $q_n + \alpha$. We then show that there exists a large enough M such that both conditions in Theorem 1 are satisfied when using η.

Next, we demonstrate that the value of $E\{L(k+1) - L(k)\}$ under the EPDF policy is smaller than that under η. Hence, the condition in Theorem 1 is also satisfied under the EPDF, and the throughput of each client n is at least q_n under the EPDF policy. □

In the proof of Theorem 2, the number of time slots in a frame, M, can be large, which means the truncated time debt of each client is updated infrequently. While the EPDF policy guarantees that the long-term average throughput of each client is at least as large as its requirement, a large M may lead to undesirable short-term performance. To improve short-term performance, we may want to set M to be a small number. However, the following example suggests that the EPDF policy may fail to support a vector of throughputs that is strictly in the capacity region when M is too small.

EXAMPLE 1. *Consider a system with three clients. Client 1 generates a packet in every time slot, has channel reliability $p_1 = 1.0$, and requires that $\tau_1 = 1$, $q_1 = 0.5$, and hence $w_1 = 0.5$. Client 2 generates a packet in every time slot, has channel reliability $p_2 = 1.0$, and requires that $\tau_2 = 1$ and $q_2 = 0$. Finally, client 3 generates a packet in time slots of the form $4m+2$, $m = 0, 1, 2, \ldots$, i.e., in time slots $\{2, 6, 10, 14, \ldots\}$. Client 3 has channel reliability $p_3 = 0.5$, and requires that $\tau_3 = 2$, $q_3 = 3/16$, and hence $w_3 = 3/8 = 0.375$.*

Suppose we set $M = 4$. Then the EPDF policy schedules client 1 in time slots of the form $4m + 1$ and $4m + 2$, schedules client 3 in time slots of the form $4m+3$, and schedules client 3 in time slots of the form $4m + 4$ if the transmission in time slot $4m + 3$ fails, for all $m = 0, 1, 2, \ldots$. It is easy to check that the EPDF policy supports the required throughput to every client.

On the other hand, suppose we set $M = 2$. At the beginning of time slot 1, we have $d_1(1) = 1, d_2(1) = 0, d_3(1) = 0.75$. The EPDF policy may schedule client 1 in time slot 1, and client 2 in time slot 2. Note that the packet of client 3 has not been generated in time slot 2, and hence cannot be scheduled in the time slot. At the beginning of time slot 3, we have $d_1(3) = 1 - 1 + 1 = 1$, $d_2(3) = 0$, and $d_3(3) = 0.75 + 0.75 = 1.5$. The EPDF policy then schedules client 1 in time slot 3, as its deadline is earlier than that of client 3, and schedules client 3 in

time slot 4. The same schedule then repeats for all the following time slots. Hence, under the EPDF policy, client 3 is scheduled once every 4 time slots, and has a throughput of $1/8$, which is less than its requirement. □

5. SIMULATIONS ON THREE MOVIES

We now present simulation results based on real traces of video streaming. We use the traces provided by the Video Trace Library of the Arizona State University [14,16]. We conduct our simulations on three different HD movies, namely, "Harry Potter," "Finding Neverland," and "Transporter 2."

We have implemented the EPDF policy, as well as three other policies for performance comparison, in ns-2. The three other policies are the EDF policy, which is proposed in [11], the Largest Debt First (LDF) policy proposed in [4], and a policy from Dua and Bambos [2].

We consider IEEE 802.11a, which supports up to 54Mbps data rate, as the MAC protocol. We assume that the system has 30 clients arranged in a 6×5 grid, where adjacent clients are separated by 170 meters, and one AP that is placed at the center of the grid. We use the Shadowing module in ns-2 to model the unreliable wireless links. Each simulation lasts 100 seconds. The delay bounds of clients, τ_n, are assumed to be evenly distributed between $20000 - 30000$ time slots.

To better illustrate the simulation results, we assume that half of the clients require a portion X of their packets to be delivered on time, while the other half of the clients require a portion Y of their packets to be delivered on time. The throughput requirements of clients are then computed accordingly. We define the *achieved throughput region* of a policy as the region composed of all (X, Y) that can be achieved by the policy. A pair (X, Y) is considered to be achieved if the resulting throughput of each client is at least 95% of its required throughput.

We first present the performance of all the four scheduling policies. Simulation results are shown in Fig. 2. It can be seen that the EPDF policy outperforms all the other three policies in all three movies. For a fixed X, the difference of achieved Y between the EPDF policy and the LDF policy can be as large as 0.13. Further, both the EDF policy and the policy in [2] result in poor performance.

Next, we investigate the influence of frame sizes on system performance. We simulate the performance of the EPDF policy under two extreme cases: $M = 1$ and $M = 1000$. Simulation results are shown in Fig. 3. Surprisingly, the achieved throughput regions for the two cases are virtually indistinguishable. For any movie and any fixed X, the difference between the achievable Y is less than 0.03 under the two cases.

We further investigate the short-term performance under both cases of $M = 1$ and $M = 1000$. We assume that all

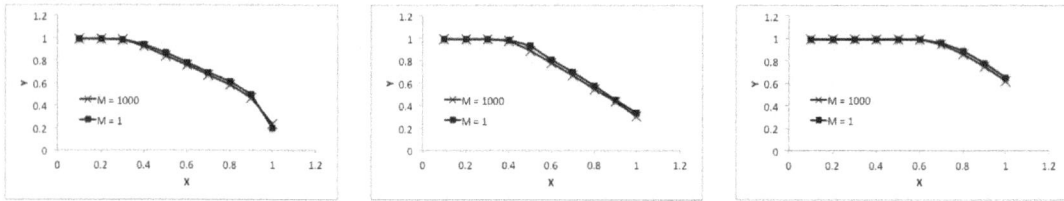

(a) Harry Potter (b) Finding Neverland (c) Transporter 2

Figure 3: Achieved throughput regions for different frame sizes.

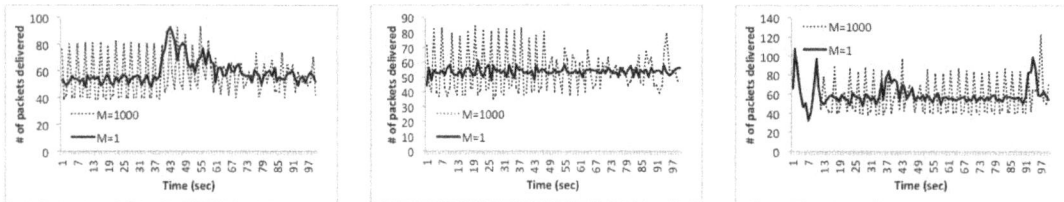

(a) Harry Potter (b) Finding Neverland (c) Transporter 2

Figure 4: Short-term throughputs for different frame sizes.

clients require a portion X of their packets to be delivered on time. The value of X is set to be 0.65 for Harry Potter, 0.7 for Finding Neverland, and 0.8 for Transporter 2. For each movie, we find the client that generates the most packets during the simulation, and then plot the number of packets that this client receives in every second of the simulations.

Simulation results are shown in Fig. 4. The number of packets delivered to the client in each second varies much from second to second when $M = 1000$. On the other hand, when $M = 1$, the number of packets delivered to the client does not vary as much. Although both $M = 1000$ and $M = 1$ result in similar long-term throughputs for each client, smaller M can lead to much better short-term performance.

6. CONCLUSIONS

We have studied the problem of serving multiple live video streams to wireless clients with various delay and throughput requirements, as well as heterogeneous link reliabilities. We have proposed a simple scheduling policy, the EPDF policy, and proved that it is able to support every vector of throughputs that is strictly in the capacity region. The utility of the EPDF policy has been demonstrated through trace-based simulations, where it is seen that it appears to outperform other policies by a large margin.

7. REFERENCES

[1] CISCO. Cisco visual networking index: Global mobile data traffic forecast update, 2011-2016.

[2] DUA, A., AND BAMBOS, N. Downlink wireless packet scheduling with deadlines. *IEEE Transactions on Mobile Computing 6*, 12 (Dec. 2007).

[3] DUTTA, P., SEETHARAM, A., ARYA, V., CHETLUR, M., KALYANARAMAN, S., AND KUROSE, J. On managing quality of experience of multiple video streams in wireless networks. In *Proc. of IEEE INFOCOM* (Mar. 2012).

[4] HOU, I.-H., BORKAR, V., AND KUMAR, P. A theory of QoS in wireless. In *Proc. of IEEE INFOCOM* (2009).

[5] HOU, I.-H., AND KUMAR, P. Admission control and scheduling for QoS guarantees for variable-bit-rate applications on wireless channels. In *Proc. of ACM MobiHoc* (2009).

[6] HOU, I.-H., AND KUMAR, P. Scheduling heterogeneous real-time traffic over fading wireless channels. In *Proc. of IEEE INFOCOM* (2010).

[7] HOU, I.-H., AND SINGH, R. Capacity and scheduling of access points for multiple live video streams. arXiv:1306.2360.

[8] JARAMILLO, J., AND SRIKANT, R. Optimal scheduling for fair resource allocation in ad hoc networks with elastic and inelastic traffic. In *Proc. of IEEE INFOCOM* (Mar. 2010).

[9] JARAMILLO, J., SRIKANT, R., AND YING, L. Scheduling for optimal rate allocation in ad hoc networks with heterogeneous delay constraints. *IEEE Journal on Selected Areas in Communications 29*, 5 (May 2011).

[10] LI, R., AND ERYILMAZ, A. Scheduling for end-to-end deadline-constrained traffic with reliability requirements in multi-hop networks. In *Proc. of IEEE INFOCOM* (Apr. 2011).

[11] LIU, C. L., AND LAYLAND, J. W. Scheduling algorithms for multiprogramming in a hard-real-time environment. *J. ACM 20*, 1 (1973).

[12] LIU, Q., WANG, X., AND GIANNAKIS, G. A cross-layer scheduling algorithm with qos support in wireless networks. *IEEE Transactions on Vehicular Technology 55*, 3 (May 2006).

[13] RAGHUNATHAN, V., BORKAR, V., CAO, M., AND KUMAR, P. Index policies for real-time multicast scheduling for wireless broadcast systems. In *Proc. of IEEE INFOCOM* (2008).

[14] SEELING, P., AND REISSLEIN, M. Video transport evaluation with H.264 video traces. *IEEE Communications Surveys and Tutorials, in print* (2012). Traces available at trace.eas.asu.edu.

[15] SHAKKOTTAI, S., AND SRIKANT, R. Scheduling real-time traffic with deadlines over a wireless channel. *Wireless Networks 8*, 1 (2002).

[16] VAN DER AUWERA, G., DAVID, P., AND REISSLEIN, M. Traffic and quality characterization of single-layer video streams encoded with the h.264/mpeg-4 advanced video coding standard and scalable video coding extension. *IEEE Transactions on Broadcasting 54*, 3 (Sep. 2008).

A Simple Approximate Analysis of Floating Content for Context-Aware Applications

Shahzad Ali
Institute IMDEA Networks,
Madrid, Spain
Universidad Carlos III de
Madrid, Spain
shahzad.ali@imdea.org

Gianluca Rizzo [*]
HES-SO Valais, Sierre,
Switzerland
gianluca.rizzo@hevs.ch

Balaji Rengarajan
Institute IMDEA Networks,
Madrid, Spain
balaji.rengarajan@imdea.org

Marco Ajmone Marsan
Institute IMDEA Networks
Madrid, Spain
Politecnico di Torino, Italy
marco.ajmone@imdea.org

ABSTRACT

Context-awareness is a peculiar characteristic of an expanding set of applications that make use of a combination of restricted spatio-temporal locality and mobile communications, to deliver a variety of services. Opportunistic communications satisfy well the communication requirements of these applications, because they naturally incorporate context. Recently, an opportunistic communication paradigm called "Floating Content" (FC) was proposed, to support infrastructure-less, distributed content sharing. But how good is floating content in supporting context-aware applications? In this work, we present a simple approximate analytical model for the performance analysis of context-aware applications that use floating content. We estimate the "success probability" for a representative category of context-aware applications, and show how the system can be configured to achieve the application's target QoS. We validate our model using extensive simulations under different settings and mobility patterns, showing that our model-based predictions are highly accurate under a wide range of conditions.

Categories and Subject Descriptors

I.6.4 [**Computing Methodologies**]: SIMULATION AND MODELING—*Model Validation and Analysis*

Keywords

Context-Aware Applications, floating content, anchor zone.

*Gianluca Rizzo was with Institute IMDEA Networks, Madrid, Spain during this work

1 Introduction

The growth of mobile computing, and the pervasiveness of smart user devices is progressively driving applications towards context-awareness, i.e., towards the exploitation of "any information that can be used to characterize the situation of an entity" [7], in order to implement new services that better suit the needs of users. One example of context information is given by spatio-temporal coordinates. For a parking finding application, the information about a vacant parking spot is characterized by the position of the spot, and it is valid for a limited time (until the space is filled). Many more examples of context-aware applications are emerging, that make use of spatio-temporal context and wireless communications to deliver a variety of services. For most context-aware applications, the scope of generated content is limited in time and space. The local relevance of the content generated and exchanged by context-aware applications makes their communication requirements significantly different from those of ordinary applications. Indeed, this locality feature makes traditional centralized, infrastructure-based content distribution services unfit, as they imply a heavily suboptimal use of storage and connectivity resources when supporting context-aware applications. Opportunistic communications can play a special role when coupled with context-awareness. Indeed, they naturally incorporate context and locality as spatial proximity is closely associated with connectivity. In our work we consider a specific opportunistic communication paradigm, known as *floating content* (FC) [3], conceived to support server-less distributed content sharing. It aims at ensuring the availability of some content within a certain geographic area called *anchor zone* (AZ), and for a given time period. Initially, only one node generates the content and defines the AZ for that content. Within the AZ, any time a user who is unaware of the content enters the transmission range of another user possessing it, the content is shared. By being replicated and made available to nodes within the AZ, even after that the original node which generated it leaves the AZ, the content 'floats' within the AZ. Users who traverse the AZ while content is floating have an opportunity to learn the content, provided they meet a node with content prior to leaving the AZ. Users delete content if they move outside AZ. Due to stochastic fluctuations of the number of nodes with content within the AZ, and to their traversal patterns, the content ultimately disappears from the AZ. In the previous modeling studies of FC [3, 5] the focus was on understanding the asymptotic properties of the floating lifetime, i.e. the duration of time for which content floats in the AZ, and asymptotic conditions were derived (called criticality conditions) for the expected floating lifetime to be large under some large population assumption. Instead, our objective is to characterize the performance of context-aware applications relying on the FC service. Indeed, from an application perspective, it is not sufficient that the content asymptotically floats: what matters is the probability for a node traversing the AZ to get the content which is floating. In this paper we address this issue, investigating the effect of the system design parameters on the performance of applications using FC. We restrict ourselves to a regime where floating lifetime is expected to be large, and we focus on the *success probability*, which captures the likelihood for a user to receive the content when traversing the AZ. A key characteristic of our modeling approach is that success probability is computed from few primitive system parameters, most notably the probability density function of the length of the path followed by users within the AZ. This allows our analysis to be generalized to a variety of AZ shapes, user mobility patterns, user speed distributions, and applications. Our main contributions are: i) we develop an approximate analytical model for success probability, with key parameters the AZ radius, the node density, and the node transmission range; ii) we apply our model to a representative category of context-aware applications, and derive expressions for their success probability; iii) we demonstrate how the model predictions can be used to tune key system parameters to achieve the desired application performance; and iv) we validate our analysis with extensive simulations, showing that the predicted success probability is very accurate. The rest of the paper is organized as follows. In Section 2, we present the system model and we define the performance metrics. We present our analysis in Section 3, and in Section 4 we validate it via simulations. Finally, we present our conclusions in Section 5.

2   System Model

We consider mobile nodes in $\mathbb{R}^2$ distributed at any time t according to a homogeneous Poisson point process with intensity λ. This assumption is justified by the considered mobility model.

Mobility Model: We assume that nodes move according to the Random Direction (RD) mobility model [2], in which nodes independently travel along a straight line, with an angle of movement uniformly distributed between 0 and 2π, and at a constant velocity v. In [4] it is shown that under this mobility model, the spatial node distribution remains uniform at all time instants. The reason for choosing this mobility model is its simplicity and analytical tractability.

Content Replication: We assume that at time $t = 0$ we have a "seeder" node with a given information item, which has to be made available to (possibly all) nodes within its *anchor zone* (AZ). AZ is a circular area with radius R, centered at the position of the seeder at $t = 0$. Then for $t > 0$ every time two nodes *within the AZ* come in transmission range of each other (we call this a contact), if only one of

the two nodes possesses the item, it transfers it to the other. We assume there is no supporting infrastructure available, so that nodes must rely exclusively on ad-hoc communication. All nodes are assumed to have the same transmission range r, with $R \gg r$, since these are the cases where content floating has practical utility. When nodes move outside of the AZ, they delete their own copy of the item. We assume every node knows its exact position in space at any time.

Performance Metric: In this paper, we focus on scenarios where node density and anchor zone radius are such that the criticality condition derived in [3] holds, so that the expected lifetime of content floating is infinite under the fluid limit approximation of [3]. In practice, this implies large average floating lifetimes. The performance metric that we use is the *success probability*:

DEFINITION 1   (SUCCESS PROBABILITY). *Consider an information item floating in its anchor zone. The success probability $P_s(\tau)$ of the item is the probability that a node receives it within a time τ after entering the AZ (if the node is still in the AZ), or by the time it leaves the AZ (if it leaves it before time τ), averaged over the duration of the floating lifetime.*

3   Analysis

In this section we derive an expression for approximating the success probability for the general floating content model described in Section 2. Then, we extend our analysis to a family of applications, and derive approximate expressions for success probability for these applications.

General Floating Content: We assume in our derivation that the system is in an equilibrium state (as assumed in [3]), in which the content floats indefinitely, and in which the average rate at which nodes get the information item inside its AZ equals the average rate at which nodes having the item leave the AZ. In such an equilibrium, the average number of nodes with and without the item within the AZ remains constant.

RESULT 1. *Consider an AZ with radius R, node density λ, and nodes with transmission range r and speed v. Let Q denote the probability that two nodes successfully transfer the content while they are in contact. Then $P_s(\tau)$ for $\tau \leq 2R/v$ can be approximated as*

$$P_s(\tau) = \int_0^{2R} \frac{l^2}{\pi R^2 \sqrt{4R^2 - l^2}} \cdot$$

$$\cdot \sum_{k=1}^{\infty} \left[1 - \left(1 - \frac{Q\overline{n}}{(\overline{m} + \overline{n})} \right)^k \right] \frac{(2r\lambda(l \wedge v\tau))^k e^{-2r\lambda(l \wedge v\tau)}}{k!} dl$$

$$(1)$$

where $\overline{m} = \min(\frac{v}{Q\nu R}, \lambda \pi R^2)$, $\overline{n} = \lambda \pi R^2 - \overline{m}$, and $\nu = \frac{2rv^2}{(\pi R^2)}$.

Here, $a \wedge b$ stands for $\min(a, b)$. $\overline{n}$ and $\overline{m}$ are respectively the average number of nodes with and without content within the anchor zone. For the derivation of Result 1, please see the Appendix.

Note that in the expression above, the integral is over l which is the length of the AZ chord traversed by a node. The expression calculates the probability that a node meets k other nodes during its traversal as the product of the pdf of the chord length and the conditional pdf of the number of contacts given the chord length. The term in square brackets is the probability that at least one out of the k nodes met has the content, and that the transfer is successful. In deriving the above result, we assume that the distribution of nodes with content in the AZ is uniform, and that the odds of meeting nodes possessing the information are uncorrelated. In reality, both assumptions are not satisfied, since there is some spatial clustering of nodes with content, and also a higher density of such nodes near the center of the AZ. However, as we show through simulations in the sequel, these are second-order effects and the result in (1) captures the success probability well. In ideal conditions, when node density is sufficiently high and Q is close to 1, we see that $P_s(\tau) \approx Q$, as in those conditions the main limit to the performance of the FC service is due to the effectiveness of the information transfer process.

For some applications of floating content, an important performance parameter is the probability P_s that a node obtains the content before leaving the AZ. An expression for it can be derived from Result 1, by letting $\tau = 2R/v$. In Result 1, all factors influencing the probability of successful content transfer between two nodes in range of each other are captured by the parameter Q. This allows detailed models for information transfer at physical and/or MAC layer to be easily incorporated in the analysis, without changing the structure of the formula. In the following theorem we provide a simple expression for the probability of successful content transfer, which accounts for finite bandwidth availability and transmission errors, assuming that nodes continually retry on failure as long as they stay within transmission range.

THEOREM 1. *If X' is the minimum time necessary for the transfer of the content, the probability of successful transfer is given by*

$$Q(X') = \sum_{k=1}^{\infty} \int_{kX'}^{(k+1)X'} [1 - (1-S)^k] f_\tau(t) dt \qquad (2)$$

where $f_\tau(t)$ is the pdf of the contact duration under the RD mobility model, given by

$$f_\tau(t) = \int_0^{min(2v, \frac{2r}{t})} \frac{2\omega^3 t^2}{\pi^2 r^2 \sqrt{4r^2 - \omega^2 t^2}\sqrt{4v^2 - \omega^2}} d\omega \qquad (3)$$

and S is the probability of no transmission failures (errors, collisions, etc) for each content transfer attempt.

For the proof of Theorem 1, please refer to [1].

Applications performance: The anchor zone defined so far plays a double function. On one side, it is the area within which the content should be available to users, and it is used as a reference for the computation of the success probability. On the other, it is the area within which the content replication mechanism, which makes the content float, is active. To accomplish the first function, in some applications the AZ could be so small to make it impossible for the content to float. In order to better tune system (rather, application) performance, we separate these two functions, and we define the Range Of Interest (ROI) as the zone (circular, and centered at the same point as the AZ) within which the content should be available to users. Necessarily, the ROI is contained in the AZ, as otherwise the content would not be available in some portions of the ROI. Here we focus on a family of applications for which the content must be delivered to users by the time they leave an area, because the content is expected to trigger some specific actions once they leave that area. One example of such application can be targeted advertising, when users visiting a given site should be notified about some offer/discount (e.g. people in vicinity of a shopping mall notified about some ongoing special offers). The expression for success probability for this kind of applications is derived in Result 2 using the following definition:

DEFINITION 2. *The success probability for getting content before leaving the ROI (P_{SBL}) is the probability that a node entering the ROI gets the content before exiting the ROI, averaged over the duration of the floating lifetime.*

RESULT 2. *For an AZ with radius R_2 and a ROI with radius $R_1 \le R_2$, P_{SBL} can be approximated as*

$$P_{SBL} = \sum_{k=1}^{\infty} \left(\int_{\sqrt{R_2^2 - R_1^2}}^{R_1 + R_2} \left(f_{L_1}(\ell) \frac{(2r\ell\lambda)^k e^{-2r\ell\lambda}}{k!} \right) d\ell \right)$$
$$\left[1 - \left(1 - \frac{Q\overline{n}}{(\overline{m} + \overline{n})} \right)^k \right] \qquad (4)$$

with $f_{L_1}(l) = \frac{2\sqrt{R_2^2 - (g^{-1}(l))^2}\sqrt{R_2^2 - (g^{-1}(l))}}{(\int_0^{R_1} g(y)dy)g^{-1}(l)}$ and $g(y) = \sqrt{R_2^2 - y^2} + \sqrt{R_1^2 - y^2}$.

For the derivation of Result 2, please refer to [1].

4 Simulations and Results

In this section, we present and discuss numerical results obtained from our analysis, and from simulations. The goal here is, on one hand, to validate the approximate expressions derived for success probability, showing that they are accurate under varied conditions. On the other hand, to show the effectiveness of our analysis in selecting the floating content parameters for a family of applications and under different scenarios. For all simulation experiments, OMNeT++ based framework called INET [6] is used. Confidence intervals at 95% confidence level were evaluated for all cases. The transmission range is 50 meters and nodes move with a constant speed of 10 meter per second. We measure success probability in each instance by averaging over the floating lifetime or until a maximum of 50000 seconds have elapsed. Fig. 1 shows both the analytical predictions and the empirically determined values of success probability as a function of the AZ radius. In addition to the random direction mobility model (RDMM), we evaluate success probability under the generalized Manhattan Grid Mobility Model (MGMM) with block sizes of 100 m × 150 m, and probability of turning left or right at an intersection set to be 0.25 each. To model success probability for MGMM, we empirically measure the average number of nodes met by a node while traversing the AZ as well as the overall rate of contacts. We assume, as we did in the analysis of the RDMM, that node contacts are uniformly distributed in space. In order to compute the predicted success probability, we use an analog of Result 1 with the number of contacts during a node traversal modeled by a Poisson random variable parametrized with the empirical mean.

It can be seen that the model predictions match very well with the simulated results for both RDMM and MGMM, suggesting that the model indeed captures successfully the first order-effects on success probability. The curves also show, as expected, that an increase in both node density and AZ radius improves the success probability, as they both increase the chances of meeting a node having content. For

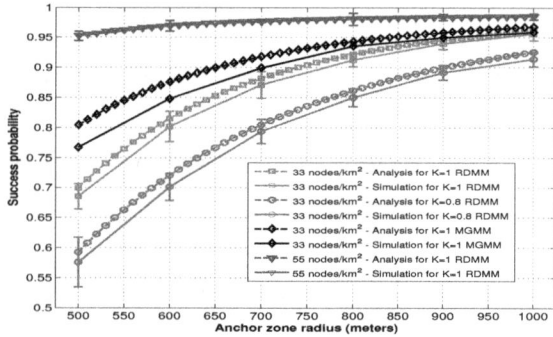

Figure 1: Success probability vs. AZ radius.

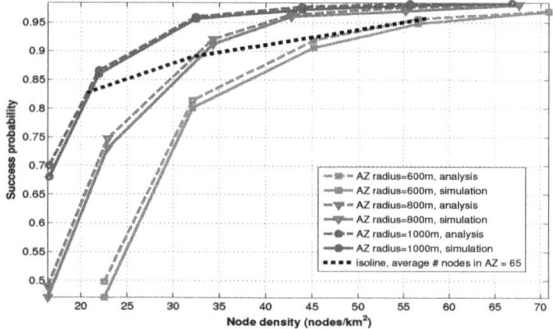

Figure 2: Success probability vs. node density.

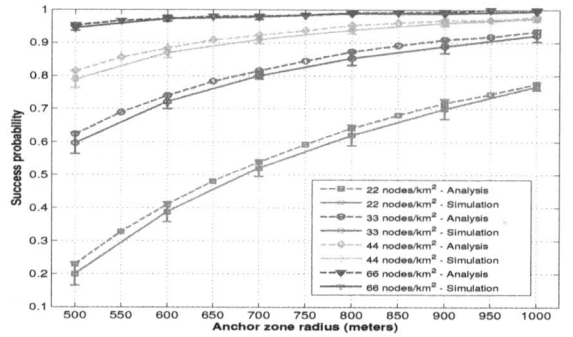

Figure 3: Application success probability with $ROI = 200 \; meters$.

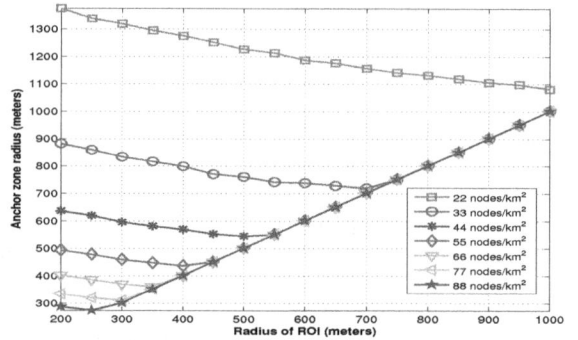

Figure 4: AZ radius to achieve 90% application success probability.

identical node density (33 nodes per km^2), under MGMM the success probability is higher than RDMM. The major reason behind this is the higher contact rate under MGMM compared to RDMM. This increases the population of nodes with content in the AZ, and also the odds of a node meeting a node with content under MGMM. Fig. 1 also shows the impact of transmission errors and finite bandwidth for RDMM. It can be seen that under a finite bandwidth model with a data rate of 11 Mbps, for transmission error probability of 0.2 with a file size of 2MB, the success probability decreases. As the transmission errors and limited contact times reduce the rate of contacts where communication is successful, this reduces the fraction of nodes in the AZ with content as well as the overall success probability. Fig. 2 shows success probability versus node density for different choices of AZ radius. As node density increases, a large increase in success probability can be observed. The analytical model captures this effect, and is a very good predictor for success probability. Further, Fig. 2 shows that as AZs grow larger, a given success probability threshold can be achieved at lower node densities. This is due to node paths through the AZ getting larger with AZ size, resulting in more opportunities to obtain content. However, if node densities and AZ size are varied jointly such that the average number of nodes in the AZ stays unchanged (see the isoline corresponding to an average of 65 nodes in the AZ), larger AZs result in lower success probabilities, since nodes have less chances to come in range of each other and exchange content. Fig. 3 shows curves of success probability versus the AZ radius, for different node densities, for the considered family of applications, when ROI radius is 200 meters. Even in this case, increases in either AZ radius or node density result in in-

creased success probability for applications. An interesting feature shown by the curves is that increasing AZ radius has diminishing returns. Indeed, increasing the radius of the AZ affects the density of users with content in the AZ, which is known to be quite flat away from borders [3]. The figure also shows that our analytical results are very close to simulations results. The proposed analytical model can be used to tune AZ radius to achieve a target application success probability. In Fig. 4 we assumed the threshold to be 90%, and used our models to compute the minimum required AZ radius at different ROIs and node densities. Interestingly, for a given node density, as the radius of ROI increases, the required AZ radius decreases, until the condition is reached where replicating the content within the ROI is sufficient to achieve the desired success probability.

5 Conclusions

In this paper, we focus on floating content as a communication service supporting context-aware applications. We defined success probability as the primary performance indicator, and developed a simple, approximate analytical modeling framework, which can be adapted to several different settings. Our models can be adapted to several different categories of context-aware applications, and the model predictions can be used in order to tune key parameters of the system to achieve the required performance with minimum overhead. We show by simulation that our approximation computes very accurate success probability values for a wide range of anchor zone radii and node densities demonstrating the viability of floating content as an enabler for context-aware applications.

References

[1] S. Ali, G. Rizzo, B. Rengarajan, and M. Ajmone Marsan. A simple approximate analysis of floating content for context-aware applications. Technical report, 2013, Available at: http://eprints.networks.imdea.org/511/.

[2] C. Bettstetter. Mobility modeling in wireless networks: categorization, smooth movement, and border effects. *SIGMOBILE Mob. Comput. Commun. Rev.*, 5(3):55–66, July 2001.

[3] E. Hyytiä, J. Virtamo, P. Lassila, J. Kangasharju, and J. Ott. When does content float? characterizing availability of anchored information in opportunistic content sharing. In *INFOCOM*, pages 3123–3131, Shanghai, China, Apr. 2011.

[4] P. Nain, D. Towsley, B. Liu, and Z. Liu. Properties of random direction models. In *INFOCOM '05*, volume 3, pages 1897 – 1907, March 2005.

[5] J. Ott, E. Hyytia, P. Lassila, T. Vaegs, and J. Kangasharju. Floating content: Information sharing in urban areas. In *PerCom 2011*, pages 136 –146, March 2011.

[6] A. Varga. The omnet++ discrete event simulation system. *ESM 2001*, June 2001.

[7] A. Zimmermann, A. Lorenz, and R. Oppermann. An operational definition of context. In *CONTEXT'07*, pages 558–571, 2007.

Appendix

Proof of Result 1: In order to prove Result 1, we need to introduce the following lemmas.

LEMMA 1. *Under the RD mobility model with node density λ, when nodes have a transmission radius of r, and velocity equal to v, the number of contacts made by a node in a time interval τ is Poisson distributed with mean $\mu_C = 2rv\tau\lambda$.*

PROOF. For the proof, please refer to [1]. □

LEMMA 2. *In equilibrium state, for a node traversing an AZ under the Random Direction mobility model, the probability of meeting k nodes is given by*

$$\int_0^{2R} \frac{\ell^2}{\pi R^2 \sqrt{4R^2 - \ell^2}} \frac{(2r\ell\lambda)^k e^{-2r\ell\lambda}}{k!} d\ell \qquad (5)$$

PROOF. Due to the properties of the RD mobility model, node distribution is uniform in space at any point in time [4], and as the nodes move in a straight line therefore the pdf of chord length L inside AZ is given by

$$f_L(\ell) = \frac{\ell^2}{R^2\pi\sqrt{4R^2 - \ell^2}} \qquad (6)$$

with $\ell \in [0, 2R]$.

Using Lemma 1 with $l = v\tau$, the probability of meeting k nodes along this trajectory can be expressed as a Poisson distribution, with intensity $2r\ell\lambda$:

$$P(meet\ k\ nodes|\ell) = \frac{(2r\ell\lambda)^k e^{-2r\ell\lambda}}{k!} \qquad (7)$$

The probability that a node meets k other nodes during its traversal is expressed as the product of the pdf of the chord length and the conditional pdf of the number of contacts given the chord length ℓ. Therefore, the probability of meeting k nodes is given by (5). □

LEMMA 3. *Consider an AZ in equilibrium state, with an average number of nodes equal to $\overline{N}$, and let $\overline{n}$ and $\overline{m}$ denote the average numbers of nodes with and without content respectively. Then $\overline{m} = \min(\frac{v}{Q\nu R}, \lambda\pi R^2)$ and $\overline{n} = \lambda\pi R^2 - \overline{m}$, with ν given by $\frac{2rv^2}{(\pi R^2)}$.*

PROOF. For a detailed proof, please refer to [1]. We denote with $n(t)$ and $m(t)$ the number of nodes with and without content, respectively, at a given time t, and define $N(t) = n(t) + m(t)$. All these quantities vary over time, as nodes move in and out of the anchor zone, and as content is exchanged. We build a set of differential equation which describe how these quantities vary over time. $n(t)$ varies over time: due to (1) nodes with content exiting the AZ and (2) nodes without content in the AZ getting the content. After computing both of these quantities we get

$$\frac{dn(t)}{dt} = \nu n(t)(N(t) - n(t))Q - \frac{n(t)v}{R} \qquad (8)$$

The number of nodes inside the AZ without content varies in time due to: (1) nodes without content exiting the AZ, (2) new nodes entering the AZ, and (3) nodes without content in the AZ getting the content, by meeting other node(s) with content. After computing these quantities we get

$$\frac{dm(t)}{dt} = \lambda v\pi R - \nu n(t)(N(t) - n(t))Q - \frac{m(t)v}{R} \qquad (9)$$

Since we assume that the system is in equilibrium, the time averages of $m(t)$ and $n(t)$, indicated respectively as $\overline{n}$ and $\overline{m}$, remain constant over time. We can write for $\overline{n}$ and $\overline{m}$ differential equations very similar to those derived above, and we can set both $\frac{d\overline{m}}{dt}$ and $\frac{d\overline{n}}{dt}$ equal to zero. Solving for $\overline{n}$ and $\overline{m}$ we get the expressions in the lemma. □

LEMMA 4. *With the same assumptions as before, the probability that a node gets content given that it meets k nodes is given by*

$$1 - \left(1 - \frac{Q\overline{n}}{(\overline{m} + \overline{n})}\right)^k \qquad (10)$$

PROOF. The probability for a node to successfully get the content upon meeting another node, can be computed by the product of the probability that the encountered node has the content, (equal to the average fraction of nodes having content in the AZ, and given by $\frac{\overline{n}}{\overline{n}+\overline{m}}$), and the probability of successful information transfer Q. The probability in the lemma is then derived as the probability that at least one out of k encounters with other nodes results into a successful transfer of content. □

PROOF. (Result 1) Consider a node traversing the AZ. This node gets content if, during its traversal: (1) at least one of the encountered nodes has the content, and (2) the information is transferred successfully between the two nodes. In steady state, the success probability P_s can be written as

$$P_s = \sum_{k=1}^{\infty} P(meet\ k\ nodes)P(get\ content|\ meet\ k\ nodes) \qquad (11)$$

The probability of meeting k nodes is given by Lemma 2, while the probability to successfully get the content upon meeting k nodes is given by Lemma 4. Substituting, we get (1). □

Randomized Skip Lists-Based Private Authentication for Large-Scale RFID Systems

Kazuya Sakai[1]   Min-Te Sun[2]   Wei-Shinn Ku[3]   Ten H. Lai[1]

[1]Department of Computer Science and Engineering, The Ohio State University, OH 43210, USA
[2]Department of Computer Science and Information Engineering, National Central University, Taiwan
[3]Department of Computer Science and Software Engineering, Auburn University, AL 36849, USA
{sakai,lai}@cse.ohio-state.edu, msun@csie.ncu.edu.tw, weishinn@auburn.edu

ABSTRACT

The performance of key authentication and the degree of privacy in large-scale RFID systems are considered by many researchers as tradeoffs. Based on how keys are managed in the system, the privacy preserving tag authentications proposed in the past can be categorized into tree-based and group-based approaches. While a tree-based approach achieves high performance in key authentication, it suffers from the issue of low privacy should a fraction of tags be compromised. On the contrary, while group-based key authentication is relatively invulnerable to compromise attacks, it is not scalable to the large number of tags. In this paper, we propose a new private tag authentication protocol based on skip lists, named Randomized Skip Lists-based Authentication. Without sacrificing the authentication performance, our scheme provides a strong privacy preserving mechanism.

Categories and Subject Descriptors

H.2.7 [**Database Administration**]: Security, integrity, and protection

General Terms

Security

Keywords

RFID, privacy, authentication

1. INTRODUCTION

Radio Frequency Identification (RFID) is widely used to smooth the way of various applications, such as library managements [7], transportation payment, natural habitat monitoring, indoor localization [5, 11, 12], and so on. In these systems, the administrator manages and monitors a large number of objects by reading passive RF tags attached to the objects with an RF reader. To protect the tag's content [9], low-cost cryptographic operations [4] are conducted during singulation process. Hence, on receiving the tag's reply, the reader must scan all keys to find the corresponding key in order to decrypt the content. When it comes to a large-scale RFID system, the authentication process can take a long time.

To accommodate this issue, a number of private tag authentication protocols with structured key management have been proposed. In these approaches, a unique key and a set of group keys are assigned to each tag. The group keys are shared among several tags and are used to confine the search space of the unique key corresponding to a tag's reply. Based on how group keys are managed, they are categorized into two types: tree-based [2, 6, 7] and group-based protocols [1, 3]. In a tree-based protocol, tags are mapped to leaf nodes in the tree and keys are assigned to internal nodes. Each tag has its unique key and a set of shared keys associated with the nodes from the leaf to the root. By traveling the tree, the reader can securely singulate tags. This results in high authentication efficiency, but discloses a large amount of information once tags in the system are compromised. On the contrary, in a group-based protocol, each tag has two kinds of keys: a unique key and a group key. With this approach, even if one of the group members is compromised, tags in other groups are intact. However, the authentication efficiency of this approach is low.

Therefore, for large-scale RFID systems, the performance and privacy/security of key authentication are commonly seen as tradeoffs. In this research, we propose a scheme that provides both good performance and a high level of privacy/security for a large-scale RFID system. Since both tree-based and group-based structures have pros and cons, we take a different approach based on *skip lists* [8], a data structure with which operations are performed in a logarithmic order like a balanced tree. We propose a new private tag authentication protocol, named Randomized Skip Lists-based Authentication (RSLA), which provides strong privacy protection and high performance of authentication like the tree-based approach. In our proposed scheme, an interrogator authenticates a tag by traveling skip lists from top to bottom with a random rotation at each level.

The rest of this paper is organized as follows. In Section 2, we propose RSLA. Section 3 concludes this paper.

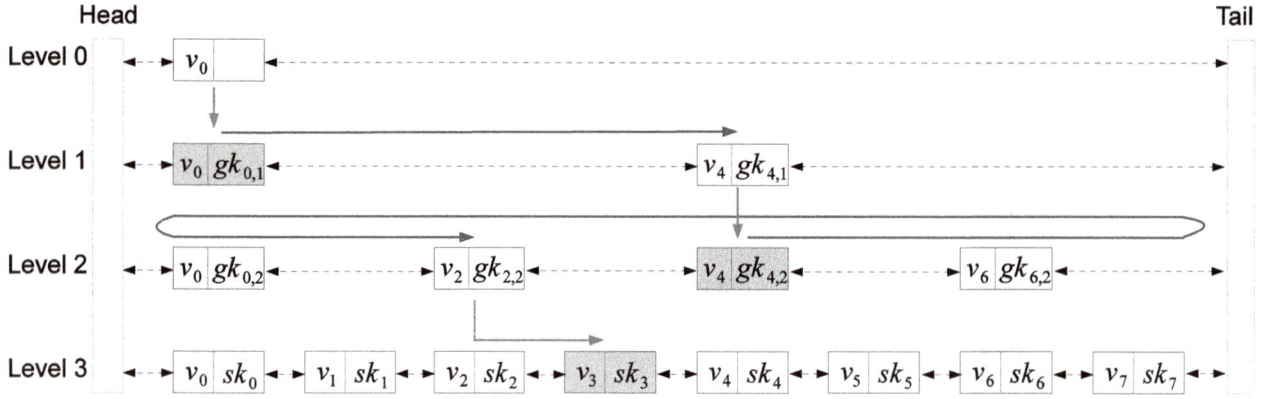

Figure 1: An example of authentication.

2. PRIVATE AUTHENTICATION PROTOCOL

2.1 Protocol Overview

In this paper, we propose Randomized Skip Lists-based Authentication (RSLA) which consists of four components: key issuing (initialization), private authentication, key-updating, and system maintenance.

In the key issuing process, the system generates skip lists. RF tags are randomly assigned to nodes in the lowest level list. A unique key and a set of group keys are assigned to each tag by traveling from a node at the bottom to the top level list. In the authentication, an RF reader scans group keys to narrow the search space of the corresponding unique key for a tag by traveling from the top list to the bottom list.

Due to the space constraint, we omit the discussion of key-updating and system maintenance in this paper. Interested readers please refer to our technical report [10] for detail.

2.2 Def nitions and Assumptions

In our assumptions, an RFID system consists of N tags and a reader, which is connected to the back-end server. For simplicity, it is assumed that the reader and the back-end server can securely communicate, and thus the reader is the final destination of a tag's data.

n_r and n_t represent nonce randomly selected by the reader and a tag, respectively. For a given key K and an input x, the hash function $H(x)$ is assumed to be collision resistant, and an encryption function $E(K, x)$ is implemented by low-cost cryptographic operations [4]. A reader is assumed to have enough computational power to run a decryption function $D(K, x)$ with a key K and an input x.

2.3 Construction of Skip Lists

To construct skip lists for key management, we modify the construction process as follows. Instead of randomly selecting nodes that appear at the list in the upper levels, we deterministically select nodes to keep the number of nodes at each level consistent.

Let L_i be the list at the i-th top level. Each list consists of a set of nodes. A node i, denoted as v_i, has pointers to left and right nodes in the same list, which are denoted by $v_i.left$ and $v_i.right$. The left pointer of the first node and the right pointer of the last node are null. In addition, the

pointers to the first and last nodes of list L_i are kept in $L_i.head$ and $L_i.tail$.

We generate skip lists that contain $\eta + 1$ lists. Each list L_i contains k^i nodes, where η is defined as $\lceil log_k(N) \rceil$ so that we can map all tags to the nodes in the lowest level list. Note that if there are more than N nodes, some nodes are not assigned a tag. Given the number of tags N and a balancing factor k, a list L_η with k^η nodes is first created. Then, node v_i is added into $L_{\eta-1}$ if $i \bmod k = 0$. For each level j, node v_i ($0 \leq j \leq \eta - 1$) is added into L_j if $i \bmod k^{\eta-j} = 0$. This process is repeated from η to 0. The top level list always has one node, i.e., $L_0 = \{v_0\}$, since the number of nodes at the lowest level list is k^η.

Each node in skip lists has a set of keys. We define $v_i.key[j]$ as the variable to store node v_i's key for Level j. If v_i does not appear in L_j, $v_i.key[j]$ is empty. Assuming Tag t is mapped to v_i, the unique key sk_t of Tag t is located at $v_i.key[\eta]$. Let us denote $gk_{i,j}$ the group key, which is stored at $v_i.key[j]$. Thus, all nodes in skip lists have a unique key in $v.key[\eta]$, and group keys for Level j ($1 \leq j \leq \eta - 1$) in $v.key[j]$ if v appears in L_j. We do not assign any key to the node in the top level list L_0, since L_0 has only one node. Thus, $v_0.key[0]$ is empty.

Since the construction of skip lists is deterministic, our skip lists with factor k work in similar fashion as a k-balanced tree. The reason why we employ skip lists instead of a balanced tree is that the link among the nodes in the same level is utilized for random rotation. Thus, we do not have

Table 1: Definition of notations.

Symbols	Definition
k	The balancing factor of skip lists
N	The number of tags in the system
η	The height of skip lists, $\lceil log_k N \rceil$
L_i	The list at Level i in skip lists ($0 \leq i \leq \eta$)
v_i	Node i in a list
sk_i	Tag i's unique secret key
GK_i	A set of group keys of Tag i, $\{gk_1, gk_2, ..., gk_{\eta-1}\}$
R_i	A set of random numbers of Tag i, $\{r_1, r_2, ..., r_{\eta-1}\}$
n_t, n_r	Nonces from a tag and a reader
β	Tag's reply, $\{\beta_1, \beta_2, .., \beta_\eta\}$
N_c	The number of compromised tags in the system
N_g	The number of compromised tags in a group
$E(.), D(.)$	The encryption and decryption functions
$H(.)$	The hash function
A	System anonymity
S_i	Anonymous set that Tag i belongs to

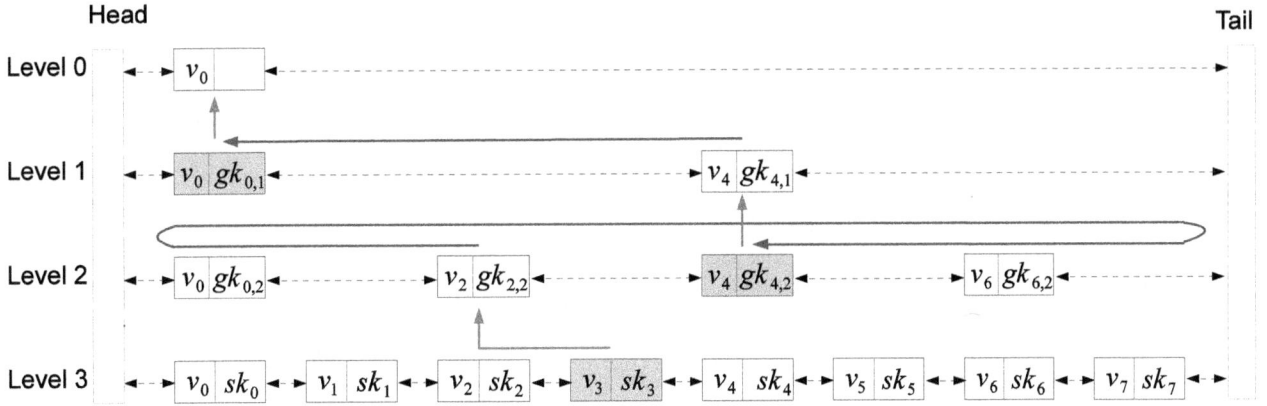

Figure 2: An example of key issuing.

2.4 Key Issuing

In RSLA, Tag t has three variables, the unique secret key sk_t, a set of group keys GK_t, and a set of random numbers R_t. Each tag t is randomly assigned to a node, say v_i, in the lowest level list L_η. Starting from v_i, the key issuer traverses to L_0 by shifting to the left for r_j nodes at each L_j $(1 \leq j \leq \eta - 1)$, where r_j is randomly chosen between 0 and $|L_j| - 1$ (i.e., $k^j - 1$). By doing this, the key of the selected node for each level is assigned to a tag.

At v_i in L_η, Tag t obtains the unique key from $v_i.key[\eta]$, and then the pointer moves to $L_{\eta-1}$. When v_i in L_η does not appear in $L_{\eta-1}$, the pointer first moves to node v_j where $j = i - i \bmod k$, and then moves to $L_{\eta-1}$. In general, for the current node v_i in L_m, the pointer moves to v_j where $j = i - i \bmod k^{\eta-m+1}$, and then goes to L_{m-1}. Thus, this can be seen as a parent and children relation of a k-balanced tree, i.e., v_j in L_{m-1} has k children v_i in L_m $(j \leq i \leq j + k - 1)$.

Every time, the pointer arrives at a upper level list, for instance L_j, the key issuer takes the left shift by r_j at L_j $(1 \leq j \leq \eta - 1)$. Here, r_j is randomly selected between 0 and $|L_j| - 1$, and added to set R_t. Note that the left shift is not taken at the L_η and L_0. The shifting can be done by moving the pointer via $v_i.left$. If $v_i.left = null$, i.e., v_i is the first node in L_j, the pointer moves to $L_j.tail$, i.e., the last node in L_j. Let v_i be the node in L_j after shifting. Tag t obtains the group key from $v_i.key[j]$. Then, the pointer moves to the upper level. This process continues until the key issuer reaches L_0.

At the end of this process, Tag t has one unique key, $\eta - 1$ group keys, and $\eta - 1$ random numbers. The pseudo code of the key issuing is given in Algorithm 1.

Example Consider an RFID system with 8 tags that uses skip lists with $k = 2$ and $\eta = 3$ for key assignment as shown in Figure 2. Tags are mapped to the t-th node in L_3. We illustrate how the key issuer assigns group keys and random numbers to a tag, for instance Tag 3. Starting from v_3, the key issuer traverses to the top level list. First, Tag 3 obtains sk_3 stored at $v_3.key[3]$, and the pointer moves to Level 2 via v_2. Because v_3 does not appear in L_2, the pointer goes to v_2 $(3 - 3 \bmod k = 2)$, and then moves to Level 2. Assume the key issuer randomly selects $r_2 = 3$ and the pointer shifts

---

Algorithm 1 Key Issue

1: /* Key Issuer does following */
2: Issuer locates all tags t to node v_i
3: /* For each tag i Key Issuer does following */
4: **for** each tag t in the system **do**
5:     KeyIssue(i, v_i)
6: **end for**
7: /* The function to assign keys to Tag t */
8: **KeyIssue(t, v_i)**
9: /* v_i is the current node */
10: $R_t = \phi$ /* Initialize the random numbers list */
11: $GK_t = \phi$ /* Initialize the grop keys list */
12: / * At the lowest level list L_η */
13: $sk_t \leftarrow v_i.key[\eta]$
14: $v_i \leftarrow v_m$ where $m = i - i \bmod k$
15: **for** (j from $\eta - 1$ to 1) **do**
16:     /* Random shifting by r and add a group key */
17:     $r \xleftarrow{uniform} [0, |L_j| - 1]$
18:     Add r to R_t
19:     $v_i \leftarrow$ shift to the left by r
20:     Add $v.key[j]$ to GK_t
21:     /* Move to upper level */
22:     $v_i \leftarrow v_m$ where $m = i - i \bmod k^{\eta-j+1}$
23:     $j = j - 1$
24: **end for**

---

to the left by 3. At the same time, 2 is added to R_3. The current pointer is now at v_4 in L_2. The issuer assigns $gk_{4,2}$ stored in $v_4.key[2]$ to Tag 3. This process continues until the issuer reaches L_0. Assume Tag 3 selects $r_1 = 1$ at Level 1. It obtains sk_3, $GK_3 = \{gk_{0,1}, gk_{4,2}\}$, and $R_3 = \{1, 3\}$.

2.5 Authentication

After issuing keys, the reader can securely communicate with tags. In RSLA authentication protocol, the reader first sends a query with nonce n_r, then a tag generates a reply message with nonce n_t, and then the reader decrypts the tag's reply.

Assume Tag t has one unique key sk_t, a set of group keys $GK_t = \{gk_1, gk_2, ..., gk_{\eta-1}\}$, and a set of random numbers $R_t = \{r_1, r_2, ..., r_{\eta-1}\}$. On receiving a query with nonce n_r from the reader, Tag t generates a reply message with nonce n_t. Let $\beta = \{\beta_1, \beta_2, ..., \beta_{\eta-1}\}$ be the reply message. Here, β_i $(i \leq 1 \leq \eta)$ consists of a hash value $\beta_i.hash$ and encrypted number $\beta_i.num$ at each level i. The hash value $\beta_i.hash$ is obtained by $H(gk_i||r_{i-1}||n_t||n_r)$ with the base

$r_0 = empty$. In other words, $\beta_1.hash = H(gk_1||n_t||n_r)$ because there is no rotation at L_0. The reason that we include the number at the previous level, i.e., r_{i-1} for $\beta_i.hash$, is to enforce dependency between the levels to keep high anonymity. The random number $\beta_i.num$ is encrypted by $E(gk_i, r_i)$. For the last element β_η, the hash value $\beta_\eta.hash$ is defined by $H(sk_t||r_{\eta-1}||n_t||n_r)$ where the unique key is used, and $\beta_\eta.num$ is empty. Finally, the tag sends n_t and β to the reader. Note that β contains η elements. One is computed by sk; the other $\eta-1$ are computed by gk. The pseudo code of the replying process is illustrated in Algorithm 2.

Algorithm 2 ReplyToReader(n_r)

1: /* Assume Tag t has sk_t, GK_t, and R_t */
2: /* where $GK_t = \{gk_1, gk_2, ..., gk_{\eta-1}\}$ */
3: /* and $R_t = \{r_1, r_2, ..., r_{\eta-1}\}$ */
4: Generate nonce n_t
5: **for** i from 1 to $\eta - 1$ **do**
6: $\quad \beta_i.hash \leftarrow H(gk_i||r_{i-1}||n_t||n_r)$ /* $r_0 = $ empty */
7: $\quad \beta_i.num \leftarrow E(gk_i, r_i)$
8: $\quad$ Add β_i to β
9: **end for**
10: $\beta_\eta.hash = H(sk_t||r_{\eta-1}||n_t||n_r)$
11: reply n_t and β

Algorithm 3 Authentication(n_r, n_t, β)

1: /* $\beta = \{\beta_1, \beta_2, ..., \beta_\eta\}$*/
2: $v_0 \leftarrow head$ /* the pointer to the current node */
3: **for** j from 1 to η **do**
4: $\quad$ /* Scan $v.key[j]$ for k nodes from v_i */
5: $\quad$ **for** m from 1 to k **do**
6: $\quad\quad$ /* Note that the base $r_0 = $ empty */
7: $\quad\quad$ **if** $H(v_i.key[j]||r_{j-1}||n_r||n_t) = \beta_j.hash$ **then**
8: $\quad\quad\quad$ **if** $j == \eta$ **then**
9: $\quad\quad\quad\quad$ Identify Tag t by the unique key $v_i.key[j]$
10: $\quad\quad\quad$ **else**
11: $\quad\quad\quad\quad$ $r \leftarrow D(v_i.key[j], \beta_j.num)$
12: $\quad\quad\quad\quad$ $v_i \leftarrow$ shift to the right by r
13: $\quad\quad\quad\quad$ $j \leftarrow j + 1$
14: $\quad\quad\quad$ **end if**
15: $\quad\quad$ **end if**
16: $\quad\quad$ $m \leftarrow m + 1$
17: $\quad$ **end for**
18: $\quad$ **if** The key is not found for L_j **then**
19: $\quad\quad$ return $FAIL$
20: $\quad$ **end if**
21: **end for**
22: return t

On receiving Tag i's reply, the reader scans group keys associated to nodes from the top level list. At the beginning, the pointer is at node v_0 in L_0. In L_1, there are k nodes, and one of them has the group key $v_i.key[1]$ $(v_i \in L_1)$ that matches the group key used for $\beta_1.hash$. After finding the corresponding key used for $\beta_1.hash$, the reader decrypts $\beta_1.num$ with the key. Then, we first move the pointer to L_1 form L_0, and shift the pointer to the right by $\beta_1.num$. If the pointer reaches the tail during shifting, it moves to the head of the same list. Note that the left shift was taken for key assignment by traveling from the lowest level, and on the contrary, the authentication process takes the right shift since the reader travels skip lists from the top. Assume v_i is the current node after shifting right by r_1. The list L_2 has k^2 nodes, but only k nodes v_j $(i \le j \le i + k)$ need to be scanned. This is because one of the k nodes has the group key for β_2. This process continues until the reader reaches

the bottom. Since the key at L_η is unique for a tag, the reader singulates the tag from β. The reader scans no more than k keys at each level $1 \le i \le \eta$, hence our skip lists imitate the search operation of a k-balanced tree. During this process, should the reader be unable to find a group key at any level, the tag's reply is invalid and the reader returns a $FAIL$ message. The pseudo code of the authentication process is provided in Algorithm 3.

3. CONCLUSION

Large-scale RFID systems always have tradeoffs between performance and security/privacy. The private authentication protocols proposed in the past are either slow or vulnerable to active attacks. In this paper, we propose RSLA which provides both high authentication efficiency and a strong privacy protection mechanism. RSLA relies on skip lists, a different data structure from the existing solutions. We believe the proposed skip lists-based approach is the most suitable authentication scheme for the next generation RFID systems.

4. REFERENCES

[1] G. Avoine, L. Buttyan, T. Holczer, and I. Vajda. Group-based Private Authentication. In *WoWMoM*, pages 1–6, 2007.

[2] T. Dimitriou. A Secure and Efficient RFID Protocol that cold make Big Brother (partially) Obsolete. In *PerCom*, pages 269–275, 2006.

[3] M. E. Hoque, F. Rahman, and S. I. Ahamed. AnonPri: An Efficient Anonymous Private Authentication Protocol. In *PerCom*, pages 102–110, 2011.

[4] A. Juels. Minimalist Cryptography for Low-Cost RFID Tags. In *SCN*, pages 149–164, 2004.

[5] W.-S. Ku, K. Sakai, and M.-T. Sun. The Optimal k-Covering Tag Deployment for RFID-Based Localization. *Special Issues of JNCA on RFID Technology, Systems, and Applications*, 34(3):914–924, 2011.

[6] L. Lu, J. Han, L. Hu, Y. Liu, and L. M. Ni. Dynamic Key-Updating: Privacy-Preserving Authentication for RFID Systems. In *PerCom*, pages 13–22, 2007.

[7] D. Molnar and D. Wagner. Privacy and Security in Library RFID Issues, Practices, and Architectures. In *CCS*, pages 210–219, 2004.

[8] W. Pugh. Skip Lists: a Probabilistic Alternative to Balanced Trees. *Comms. of the ACM*, 33(6):668–676, 1990.

[9] K. Sakai, W.-S. Ku, R. Zimmermann, and M.-T. Sun. Dynamic Bit Encoding for Privacy Protection against Correlation Attacks in RFID Backward Channel. *IEEE Trans. Computers*, 62(1):112–123, 2013.

[10] K. Sakai, M.-T. Sun, W.-S. Ku, T. H. Lai, and A. V. Vasilakos. Randomized Skip Lists-Based Private Authentication for Large-Scale RFID Systems. *OSU-CISRC-5/13-TR12*, 2013. ftp://ftp.cse.ohio-state.edu/pub/tech-report/2013/TR12.pdf.

[11] S. Wagner, M. Handte, M. Zuniga, and P. J. Marron. On Optimal Tag Placement for Indoor Localization. In *PerCom*, pages 162–170, 2012.

[12] J. Yu, W.-S. Ku, M.-T. Sun, and H. Lu. An RFID and particle filter-based indoor spatial query evaluation system. In *EDBT*, pages 263–274, 2013.

Energy-Efficient Fault-Tolerant Data Storage & Processing in Dynamic Networks

Chien-An Chen, Myounggyu Won,
Radu Stoleru
Department of Computer Science and
Engineering, Texas A&M University
{jaychen,mgwon,stoleru}@cse.tamu.edu

Geoffrey G. Xie
Department of Computer Science, Naval
Postgraduate School
xie@nps.edu

ABSTRACT

With the advance of mobile devices, cloud computing has enabled people to access data and computing resources without spatiotemporal constraints. A common assumption is that mobile devices are well connected to remote data centers and the data centers securely store and process data. However, for systems like mobile cloud deployed in infrastructureless dynamic networks (i.e., with frequent topology changes because of node failure/unavailability and mobility), reliability and energy efficiency remain largely unaddressed challenges. To address these issues, we develop the first "k-out-of-n computing" framework that ensures nodes retrieve or process data stored in mobile cloud with minimum energy consumption as long as k out of n storage/processing nodes are accessible. We demonstrate the feasibility and performance of our framework through both hardware implementation and extensive simulations.

Categories and Subject Descriptors

C.2 [**Networks**]: Mobile Networks, Mobile ad hoc networks, Dynamic Network; H.3 [**Information Systems**]: Distributed Storage

Keywords

Mobile Cloud, Distributed Storage, Distributed Data Processing

1. INTRODUCTION

With the explosion of personal mobile devices, cloud computing has gained popularity and become the norm in the past decade. However, due to limited resources of mobile devices (e.g., energy, processing, and memory), executing sophisticated applications (e.g., video and image storage and processing, or map-reduce type) on mobile devices remains challenging. To enable such applications on mobile devices, many commercial applications like Google Goggle and Siri rely on offloading data and computation-intensive

tasks to high-performance computers. Nevertheless, in situations such as disaster response or military operations, high-performance computers are not immediately available; instead, a quick and flexible ad hoc network is required.

In this paper, we propose the first framework to support fault-tolerant and energy-efficient remote storage & processing under a dynamic network topology. We integrate the k-out-of-n reliability mechanism into distributed computing in dynamic networks. The k-out-of-n system, a well-studied topic in reliability control, ensures that a system of n components operates correctly as long as k or more components work [5]. More specifically, we investigate how to store data as well as process the stored data with k-out-of-n reliability such that: 1) the energy consumption for retrieving distributed data is minimized; 2) the energy consumption for processing the distributed data is minimized; and 3) data and processing are distributed considering dynamic topology changes.

In our proposed framework, a data object is partitioned into n fragments and stored on n different nodes. As long as k or more of the n nodes are available, the data object can be recovered. Similarly, another set of n nodes are assigned tasks for processing the stored data and all tasks can be completed as long as k or more of the n nodes finish the assigned tasks. The parameters k and n determine the degree of reliability; system administrators select these parameters based on their reliability requirements. The contributions of this paper are summarized as follows:

- it presents a mathematical model for both optimizing energy consumption and meeting the fault tolerance requirements of data storage and processing under a dynamic network topology.

- it presents an efficient algorithm for estimating the communication cost in a dynamic network, where nodes fail or move, joining/leaving the network.

- it presents the evaluation of our proposed framework through real hardware implementation as well as extensive simulations particularly for large-scale networks.

2. RELATED WORK

Replicating data to multiple locations is a common technique for improving performance and reliability in distributed computing. Erasure coding has also been a widely used technique for enhancing data reliability [12]. Replicating data incurs storage overhead while achieving higher reliability; coding techniques reduce storage overhead while not performing well when nodes are not reliable, e.g., in dynamic networks [13].

Some researchers proposed solutions for achieving higher reliability in dynamic networks. Dimakis et al. proposed several erasure coding algorithms for maintaining a distributed storage system in a dynamic network [6]. Leong et al. proposed an algorithm for optimal data allocation that maximizes the recovery probability [9]. Aguilera et al. proposed a protocol to efficiently adopt erasure code for better reliability [1]. These solutions, however, focused only on system reliability but did not consider energy efficiency.

Cloud computing in a small-scale network with battery-powered devices has also gained attention recently. Cloudlet is a resource-rich cluster that is well-connected to the Internet and is available for use by nearby mobile devices [10]. A mobile device delivers a small Virtual Machine (VM) overlay to a cloudlet infrastructure and lets it take over the computation. Similar works that use VM migration are also done in CloneCloud [2] and ThinkAir [8]. MAUI uses code portability provided by Common Language Runtime to create two versions of an application: one runs locally on mobile devices and the other runs remotely [4]. MAUI determines which processes to be offloaded to remote servers based on their CPU usages. Serendipity considers using remote computational resource from other mobile devices [11]. Most of these works focus on minimizing the energy and do not address system reliability.

3. ARCHITECTURE

An overview of our proposed framework is depicted in Figure 1. Applications generate *data*; our framework stores the data in the network. For higher data reliability, each data is encoded and partitioned into *fragments*. These fragments then are distributed to a set of *storage nodes*. In order to process the data, applications provide *functions* that take the stored data as inputs. Each function is instantiated into multiple *tasks* that will process the data simultaneously. Nodes that execute these tasks are *processor nodes*. We call a set of tasks instantiated from one function as a *job*. *Client nodes* are the nodes that request data allocation or processing operations. A node can have any combination of roles from: storage node, processor node, or client node, and any node can retrieve data from storage nodes. Without loss of generality, we consider a single dataset – which is stored once and may be retrieved multiple times by any node. Our framework consists of five components: Topology Discovery and Monitoring, Failure Probability Estimation, Expected Transmission Time (ETT) Computation, k-out-of-n Data Allocation, and k-out-of-n Data Processing.

When a request for data allocation is received from applications, the *Topology Discovery and Monitoring component* provides network topology information and failure probabilities of nodes. The failure probability, which depends on the battery lifetime, network connectivity, and application-specific factors, is estimated by the *Failure Probability component* on each node. Based on the retrieved failure probabilities and network topology, the *ETT Computation component* computes the ETT matrix, which represents the expected energy consumption for communication between any two nodes in the network. The "expected" communication cost indicates that the estimated ETT between nodes considers the dynamically changing topology. Given the ETT matrix, our framework finds the locations for storing frag-

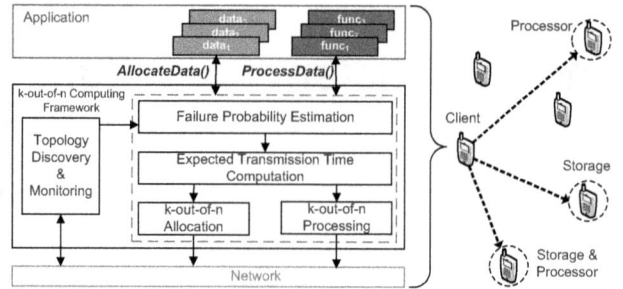

Figure 1: Architecture for integrating the k-out-of-n data allocation framework.

ments. The *k-out-of-n Data Allocation component* partitions a data object into n fragments and stores these fragments in the network such that the energy consumption for retrieving k fragments by any node is minimized. If an application needs to process the data, the *k-out-of-n Data Processing component* creates a job of M tasks and schedules the tasks on n processor nodes such that the energy consumption for processing these data is minimized. This component ensures that all tasks complete as long as k or more processor nodes finish their assigned tasks.

4. PROBLEM FORMULATION

Having explained the overall architecture of our framework, we now present design primitives for the k-out-of-n data allocation and k-out-of-n data processing. We consider a dynamic network with N nodes denoted by a set $V = \{v_1, v_2, ..., v_N\}$. We assume nodes are time synchronized. For convenience, we will use i and v_i interchangeably hereafter. The network is modeled as a graph $G = (V, E)$, where E is a set of edges. Each node has an associated *failure probability* $P[f_i]$ where f_i is the event that causes node v_i to fail.

Relationship Matrix R is a $N \times N$ matrix which defines the relationship between nodes and storage nodes. More precisely, each element R_{ij} is a binary variable – if R_{ij} is 0, node i will not retrieve data from storage node j; if R_{ij} is 1, node i may retrieve data from storage node j. *Storage node list X* is a binary vector containing storage nodes, i.e., $X_i = 1$ indicates that v_i is a storage node.

The *Expected Transmission Time Matrix D* is defined as a $N \times N$ matrix where element D_{ij} corresponds to the ETT for transmitting a fixed size packet from node i to node j considering the failure probabilities of nodes in the network, i.e., multiple possible paths between node i and node j. The ETT metric [3] has been widely used for estimating transmission time between two nodes in one hop. We assign each edge of graph G a positive estimated transmission time. Then, the path with the shortest transmission time between any two nodes can be found. However, the shortest path for any two nodes may change over time because of the dynamic topology. ETT, considering multiple paths due to nodes failures, represents the "expected" transmission time, or "expected" transmission energy between two nodes.

In general, each node may have different energy cost depending on their energy sources, e.g., nodes attached to a constant energy source may have zero energy cost while nodes powered by a battery may have relatively high energy cost. For simplicity, we assume the network is homogeneous and nodes consume the same amount of energy for proces-

$$R_{opt} = \arg\min_{R} \sum_{i=1}^{N}\sum_{j=1}^{N} D_{ij} R_{ij} \qquad (1)$$

$$\text{Subject to:} \quad \sum_{j=1}^{N} X_j = n \qquad (2)$$

$$\sum_{j=1}^{N} R_{ij} = k \ \forall i \qquad (3)$$

$$X_j - R_{ij} \geq 0 \ \forall i \qquad (4)$$

$$X_j \text{ and } R_{ij} \in \{0,1\} \ \forall i,j \qquad (5)$$

sing the same task. As a result, only the *transmission energy* affects the energy efficiency of the final solution.

4.1 k-out-of-n Data Allocation

In this problem, we are interested in finding n storage nodes denoted by $S = \{s_1, s_2, ...s_n\}, S \subseteq V$ such that the *total expected transmission cost* from any node to its k closest storage nodes – in terms of ETT – is minimized. We formulate this problem as an ILP as shown in Equations 1 - 5.

The first constraint (Eq 2) selects exactly n nodes as storage nodes; the second constraint (Eq 3) indicates that each node has access to k storage nodes; the third constraint (Eq 4) ensures that j^{th} column of R can have a non-zero element if only if X_j is 1; and constraint (Eq 5) is binary requirements for the decision variables.

4.2 k-out-of-n Data Processing

The objective of this problem is to find n nodes in V as processor nodes such that energy consumption for processing M tasks of a single job is minimized. In addition, all tasks must complete as long as k or more processor nodes can finish their assigned tasks.

In this problem, n nodes are selected as *processor nodes*; each processor node is assigned one or more tasks. Each task is replicated to $n - k + 1$ processor nodes. However, not all instances are processed – once an instance of the task completes, all other instances will be canceled. The task allocation can be formulated as an ILP as shown in Equations 6 - 10. In the formulation, $\overline{R_{ij}}$ is a $N \times M$ matrix which predefines the relationship between processor nodes and tasks; each element $\overline{R_{ij}}$ is a binary variable indicating whether task j is assigned to processor node i. $\overline{X}$ is a binary vector containing processor nodes, i.e., $\overline{X_i} = 1$ indicates that v_i is a processor node. The objective function minimizes the transmission time for n processor nodes to

$$\overline{R_{opt}} = \arg\min_{\overline{R}} \sum_{i=1}^{N}\sum_{j=1}^{M} T_{ij}^{r} \overline{R_{ij}} \qquad (6)$$

$$\text{Subject to:} \quad \sum_{i=1}^{N} \overline{X_i} = n \qquad (7)$$

$$\sum_{i=1}^{N} \overline{R_{ij}} = n - k + 1 \ \forall j \qquad (8)$$

$$\overline{X_i} - \overline{R_{ij}} \geq 0 \ \forall i \qquad (9)$$

$$\overline{X_j} \text{ and } \overline{R_{ij}} \in \{0,1\} \ \forall i,j \qquad (10)$$

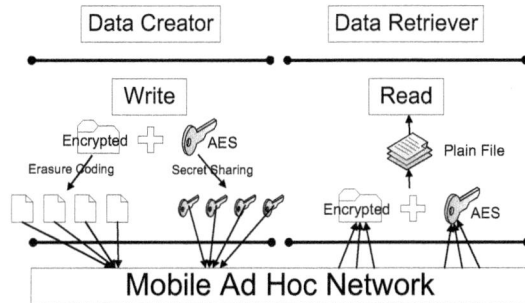

Figure 2: An overview of improved MDFS.

retrieve all tasks and their instances. The first constraint (Eq 7) indicates that n of the N nodes will be selected as processor nodes. The second constraint (Eq 8) replicates each task to $(n - k + 1)$ different processor nodes. The third constraint (Eq 9) ensures that the jth column of $\overline{R}$ can have a non-zero element if only if $\overline{X_j}$ is 1; and constraint (Eq 10) is binary requirements for the decision variables.

5. SYSTEM IMPLEMENTATION

This section investigates the feasibility of running our framework on real hardware. We implemented a mobile distributed file system (MDFS) on top of the k-out-of-n framework. Figure 2 shows an overview of our MDFS. As shown, each file is encrypted by a secret key and partitioned into n_1 file fragments. The secret key is also decomposed into n_2 key fragments by using Shamir's key sharing algorithm. The file and key fragments are distributed independently to the network. When a node needs to access a file, it must retrieve at least k_1 file fragments and k_2 key fragments. Our k-out-of-n data allocation allows nodes to optimally distribute the file and key fragments, when compared with the state-of-art MDFS that distributes file and key fragments uniformly in the network [7]. The ratio k_1/n_1 determines the reliability of a file and the ratio k_2/n_2 indicates the security of the file. Decreasing k_1/n_1 ratio improves the file reliability, while increasing k_2/n_2 ratio strengthens the security. Consequently, our k-out-of-n enabled MDFS to achieve higher reliability, energy efficiency, and security. On top of our MDFS, we also test our k-out-of-n data processing component by implementing a face recognition application that counts the number of faces appearing in a set of files stored in the network. Our k-out-of-n framework selects n processor nodes to retrieve and analyze the files in an energy-efficient and reliable way.

We implemented our system on HTC Evo 4G Smartphones, which run Android 2.3 OS using 1G Scorpion CPU, 512MB RAM, and a Wi-Fi 802.11 b/g interface. To enable the Wi-Fi AdHoc mode, we rooted the device and modified a config file – wpa_supplicant.conf. For this experiment we varied network size N and set $n_1 = \lceil 0.6N \rceil$, $k_1 = \lceil 0.3n_1 \rceil$ and $n_2 = \lceil 0.6N \rceil$, $k_2 = \lceil 0.6n_2 \rceil$. Figure 3 shows measured time of each component. As shown, distributing/retrieving fragments takes much longer time than computation, which affirms that our goal of minimizing communication energy is necessary. We also observe that larger network sizes incur longer distributing/retrieving time. This is because fragments are more sparsely distributed, resulting in more hops to distribute/retrieve fragments.

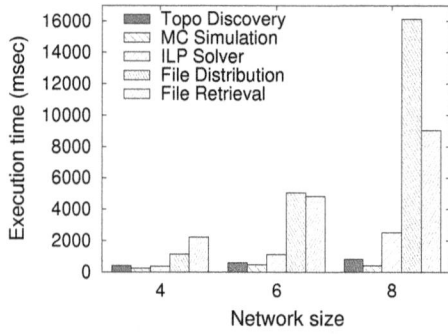

Figure 3: Execution time of different components.

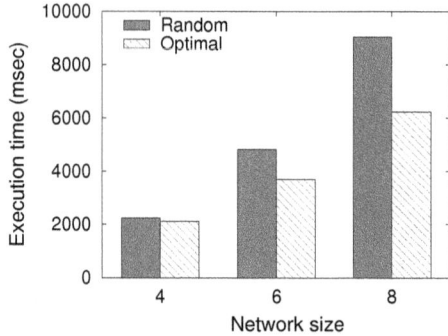

Figure 4: Data retrieval time.

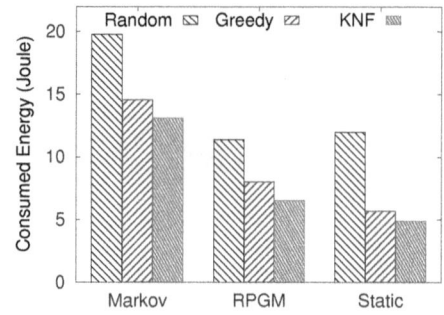

Figure 5: Effect of mobility.

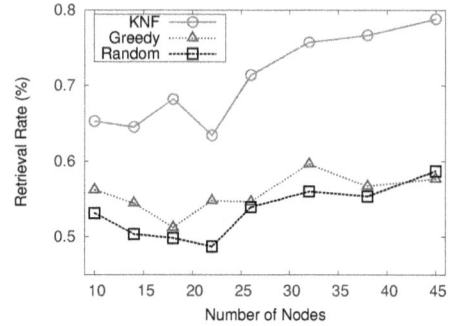

Figure 6: Effect of node density.

We then compared the file retrieval time between our allocation and random allocation. As shown in Figure 4, our framework achieves 15% to 25% smaller data retrieval time than Random. To validate the performance of our k-out-of-n data processing, we measured the completion rate of our face-recognition algorithm by varying the number of node failures. The face recognition algorithm had an average completion rate of 95% in our experimental setting.

6. SIMULATION RESULTS

We performed simulations to evaluate the performance of our k-out-of-n framework (denoted by KNF) in larger scale networks. We consider a network of $400 \times 400 \text{m}^2$ where up to 45 mobile nodes are randomly deployed. The communication range of a node is 130m. Two different mobility models are tested – Markovian Waypoint Model and Reference Point Group Mobility (RPGM). Markovian Waypoint is similar to Random Waypoint Model in which it randomly selects the waypoint of a node, but it accounts for the current waypoint when it determines the next waypoint. RPGM is a group mobility model where a set of leaders are selected; leaders move based on Markovian Waypoint; and other nodes follow their closest leaders.

We compare KNF with two other schemes – a greedy algorithm (Greedy) and a random placement algorithm (Random). Greedy selects nodes with the largest number of neighbors as storage nodes. Random selects storage nodes randomly. The goal is to evaluate how the selected storage nodes impact the performance. We measure the following metrics: consumed energy for retrieving data, consumed energy for processing a job, data retrieval rate, and completion rate of a job. A node may fail due to two independent factors: depleted energy and an application-dependent failure probability; specifically, the energy associated with a node

decreases as the time elapses. The lower the energy is, the higher the failure probability. Each node is assigned a constant application-dependent failure probability. In particular, for the simulations of our k-out-of-n data processing, we also artificially force various number of nodes to fail.

6.1 k-out-of-n data allocation

Figure 5 shows that mobility causes nodes to spend higher energy in retrieving data compared with the static network and the energy consumption for RPGM is smaller than that for Markov. It also shows that the energy consumption for RPGM is smaller than that for Markov. The reason is that a storage node usually serves the nodes in its proximity; thus when nodes move in a group, the impact of mobility is less severe than when all nodes move randomly. In all scenarios, KNF consumes lower energy than others.

Figure 6 shows that KNF achieves 15% to 25% higher retrieval rate than others because KNF takes the dynamic nature of the network into account when selecting storage nodes. We also observe that the retrieval rates increase with higher network density. The explanation is that, with higher network density, client nodes can find more reliable paths to storage nodes and the network is less likely to be partitioned due to node failures. Since the simulations run for 4 hours and several nodes fail in the last hour due to depleted energy, the highest retrieval rate of KNF is only 80%.

6.2 k-out-of-n data processing

This section investigates how the failures of processor nodes affect the energy efficiency and job completion rate. In Greedy, each task is replicated to $n-k+1$ processor nodes that have the lowest energy consumption for retrieving the task, and given a task. In Random, the processor nodes are selected randomly and each task is also replicated to $n-k+1$

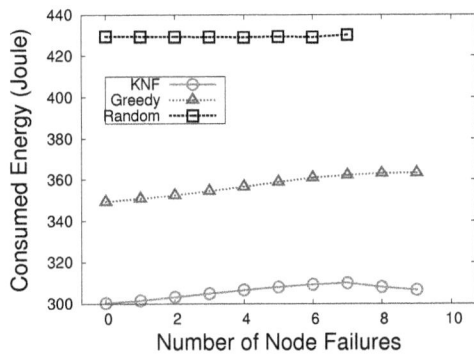

Figure 7: Effect of node failure on energy efficiency.

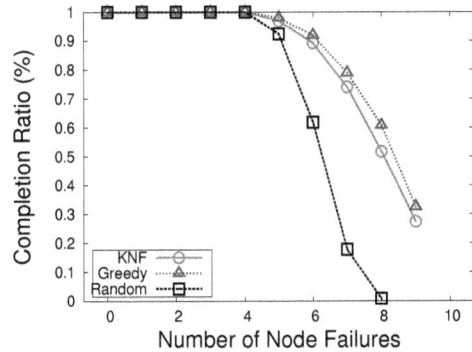

Figure 8: Effect of node failure on completion ratio.

processor nodes randomly. When a node fails, we assume none of the tasks assigned to it can be completed.

Figure 7 shows that KNF consumes 10% to 30% lower energy than Greedy and Random. We observe that the energy consumption is not sensitive to the number of node failures. When there is a node failure, a task may be executed on a less optimal processor node and causes higher energy consumption. However, this difference is small because given a task, it is replicated to n-k+1 processor nodes and failing an arbitrary processor may have no effect on the execution time of the job at all.

In Figure 8, we see that the completion ratio is 1 when no more than $n-k$ nodes fail. An interesting observation is that Greedy has the highest completion ratio. The main reason is because, in Greedy, the load on each node is highly uneven, meaning that some processor nodes may have many tasks but some may not have a task. This allocation strategy achieves high completion ratio because all tasks can complete as long as one of such high load processor nodes can finish all its assigned tasks. In our simulation, about 30% of processor nodes in Greedy are assigned all M tasks. Analytically, if three of the ten processor nodes contain all M tasks, the probability of completion when 9 processor nodes fail is $1 - \binom{7}{6}/\binom{10}{9} = 0.3$.

7. CONCLUSIONS

We presented the first k-out-of-n data allocation/processing framework that jointly addresses the energy-efficiency and fault-tolerance challenges in dynamic networks. It allocates data fragments to storage nodes such that other nodes retrieve data reliably with minimal energy consumption; when applications need to process data stored in the network, the framework selects a set of processor nodes to execute the job such that the job can be completed with minimal energy consumption and high completion rate. We demonstrated the effectiveness of our framework through both system implementation and simulations in large-scale networks.

Acknowledgment

This work was supported by Naval Postgraduate School under Grant No. N00244-12-1-0035.

8. REFERENCES

[1] M.K. Aguilera, R. Janakiraman, and Lihao Xu. Using erasure codes efficiently for storage in a distributed system. In *Proceedings of Dependable Systems and Networks*, 2005.

[2] Byung-Gon Chun, Sunghwan Ihm, Petros Maniatis, Mayur Naik, and Ashwin Patti. Clonecloud: elastic execution between mobile device and cloud. In *Proceedings of EuroSys*, 2011.

[3] Douglas S. J. De Couto. *High-Throughput Routing for Multi-Hop Wireless Networks*. PhD dissertation, MIT, 2004.

[4] Eduardo Cuervo, Aruna Balasubramanian, Dae-ki Cho, Alec Wolman, Stefan Saroiu, Ranveer Chandra, and Paramvir Bahl. Maui: making smartphones last longer with code offload. In *Proceedings of MobiSys*, 2010.

[5] Jiachen Liu David W. Coit. System reliability optimization with k-out-of-n subsystems. *Int. Journal of Reliability, Quality and Safety Engineering*, 7(2):129–142, 2000.

[6] A.G. Dimakis, K. Ramchandran, Y. Wu, and Changho Suh. A survey on network codes for distributed storage. *Proceedings of the IEEE*, 99:476–489, 2011.

[7] S. Huchton, G. Xie, and R Beverly. Building and evaluating a k-resilient mobile distributed file system resistant to device compromise. In *Proceedings of Military Communications Conference*, 2011.

[8] S. Kosta, A. Aucinas, Pan Hui, R. Mortier, and Xinwen Zhang. Thinkair: Dynamic resource allocation and parallel execution in the cloud for mobile code offloading. In *Proceedings of INFOCOM*, 2012.

[9] D. Leong, A.G. Dimakis, and Tracey Ho. Distributed storage allocation for high reliability. In *Proceedings of ICC*, 2010.

[10] M. Satyanarayanan, P. Bahl, R. Caceres, and N. Davies. The case for vm-based cloudlets in mobile computing. *Pervasive Computing, IEEE*, 8:14–23, 2009.

[11] Cong Shi, Vasileios Lakafosis, Mostafa H. Ammar, and Ellen W. Zegura. Serendipity: enabling remote computing among intermittently connected mobile devices. In *Proceedings of MobiHoc*, 2012.

[12] A Shokrollahi. Raptor codes. *Information Theory*, 52:2551–2567, 2006.

[13] Hakim Weatherspoon and John Kubiatowicz. Erasure coding vs. replication: A quantitative comparison. In *Proceedings of Peer-to-Peer Systems*, 2002.

Energy-Efficient Reliable Data Dissemination in Duty-Cycled Wireless Sensor Networks

Kai Han[†*]   Liu Xiang[†]   Jun Luo[†]   Mingjun Xiao[‡]   Liusheng Huang[‡]

[†]School of Computer Engineering, Nanyang Technological University, Singapore, 639798
[*]School of Computer Science, Zhongyuan University of Technology, China, 450007
[‡]School of Computer Science and Technology, University of Science and Technology of China, China, 230026
hankai@gmail.com, {xi0001iu, junluo}@ntu.edu.sg, {xiaomj, lshuang}@ustc.edu.cn

ABSTRACT

Because data dissemination is crucial to Wireless Sensor Networks (WSNs), its energy-efficiency and reliability are of paramount importance. While achieving these two goals together is highly non-trivial, the situation is exacerbated if WSN nodes are duty-cycled (DC) and their transmission power is adjustable. In this paper, we study the problem of minimizing the expected total transmission power for reliable data dissemination (multicast/broadcast) in DC-WSNs. Due to the NP-hardness of the problem, we design efficient approximation algorithms with provable performance bounds for it.

Categories and Subject Descriptors

C.2.1 [**Computer-Communication Networks**]: Network Architecture and Design—*Network topology*

Keywords

Wireless sensor networks, data dissemination, multicast, broadcast, duty-cycle, energy-efficiency, reliability

1. INTRODUCTION

Designing effective data dissemination mechanisms for *Wireless Sensor Networks* (WSNs) is of paramount importance, as WSNs rely on data dissemination to carry critical commands or code updates from a sink to a set of (or all) nodes in the networks [7,16]. *Reliability* and *energy-efficiency* are perhaps the most crucial requirements for data dissemination due to the notorious packet-losses in wireless communications and the limited power supply of sensor nodes.

Achieving reliability and energy-efficiency simultaneously is by no means trivial, and the problem gets even more challenging when modern WSNs' features such as duty-cycling and power-adjustability are taken into account. On one hand, in a *Duty-Cycled WSN* (DC-WSN) [5], nodes usually switch between active/dormant states periodically, which

imposes extra difficulty on attaining energy-efficiency even without regarding reliability [4,12,15]. On the other hand, nodes' capability of adjusting the transmission power (tx-power) to several levels brings out a tradeoff between reliability and energy-efficiency, as a node can increase its tx-power to improve link reliability at the cost of higher energy consumption [9]. All these entangled features make it extremely challenging to design energy-efficient reliable data dissemination mechanisms for DC-WSNs.

Despite major efforts on energy-efficient data dissemination in wireless networks, a large body of them have taken the over-optimistic assumption of error-free wireless transmissions [1, 4, 8, 12, 14, 15, 17]. Other proposals considering unreliable links roughly follow two lines. one line bears the common objective of minimizing the expected tx-power consumption under guaranteed reliability, but they are designed only for *Always-Active* Wireless Networks (AAWNs) [2,9]. This batch of work has unanimously concentrated on unicasting or employing unicasting to achieve multicasting [9–11] without leveraging the *Wireless Broadcast Advantage* (WBA). Another line aims at reducing the tx-power consumption for broadcasting in DC-WSNs without a stringent requirement on reliability. As a representative proposal in this line, [3] proposed a tree-based opportunistic flooding approach. However, the broadcast tree used in [3] is constructed for very low duty-cycle, hence per-hop transmissions are still done through unicasting. Pessimistically neglecting the WBA feature of wireless communication harms energy-efficiency. For example, suppose that nodes in Figure 1 all have a unit tx-power and A is the source node, then [9] creates the tree in Figure 1(b) for broadcast, whose expected tx-power is 19. Considering WBA leads to a better solution in Figure 1(c), whose expected tx-power is at most 12.9 (computed by **Lemma** 1 in Section 2).

In this paper, we present the first study on energy-efficient reliable data dissemination in DC-WSNs with guaranteed performance. We aim to minimize the **expected** total tx-power consumption in multicasting or broadcasting under guaranteed reliability, and we take into account WBA, unreliable links, power-adjustability and duty-cycling holistically. Given that the resulting *Energy efficient Reliable Data Dissemination* (ERDD) problem is NP-hard, we propose approximation algorithms for it with performance ratios of $\mathcal{O}(\ln \Delta \ln k)$, where Δ is the maximum node degree in the network and k is the total number of nodes involved in a multicast/broadcast session. Due to the page limits, all the proofs of our lemmas/theorems are omitted.

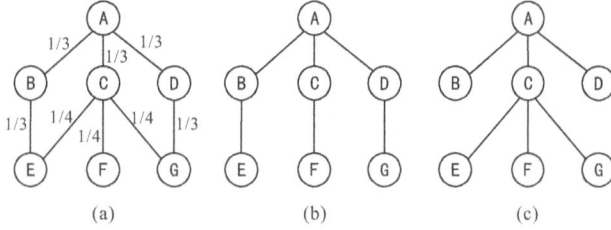

Figure 1: The original network topology is shown in (a) with link qualities marked beside the links. Both [9] and [3] create the same broadcast tree shown in (b), but (c) is a far better solution.

2. MODEL AND PROBLEM DEFINITION

We assume a set $\mathcal{V}$ of DC-WSN nodes; they switch between active/sleeping states periodically. The set of active time slots in a *working period* of any node $u \in \mathcal{V}$ is denoted by $\mathcal{A}(u) \subseteq \mathcal{I} = \{1, 2, ..., I\}$, where I is the length of the working period. Following [3, 4, 12, 15], we assume that all nodes in $\mathcal{V}$ are time synchronized, and *a node can wake up its transceiver to transmit data at any time slot, but can only receive data when it is active*. We also assume that each node $u \in \mathcal{V}$ can adjust its tx-power to several levels, which are $\varepsilon_{min} = \varepsilon_1 \leq \varepsilon_2 \leq ... \leq \varepsilon_d = \varepsilon_{max}$. Let $\mathcal{L} = \{\varepsilon_1, \varepsilon_2, ..., \varepsilon_d\}$. When u's tx-power is adjusted to $l \in \mathcal{L}$, let the link quality $p_{uv}(l) \in (0, 1]$ denote the success ratio of data transmission on link (u, v). Following [9], we assume that $p_{uv}(l)$ increases with $l \in \mathcal{L}$ and the link qualities considered in the network have a positive lower bound (denoted by Λ). Let $\mathcal{N}_u(l)$ denote the node set $\{v | v \in \mathcal{V} \backslash \{u\} \wedge p_{uv}(l) \geq \Lambda\}$.

In a data dissemination session, there exists a source node $\mathfrak{s} \in \mathcal{V}$ that needs to send data to a set of destination nodes $\mathcal{R} \subseteq \mathcal{V} - \{\mathfrak{s}\}$. Let $k = |\mathcal{R}| + 1$, hence the session is a broadcast if $k = |\mathcal{V}|$; otherwise a multicast. A *valid power assignment* for such a session is a function $L : \mathcal{V} \rightarrow \mathcal{L}$, such that when each node $u \in \mathcal{V}$ is adjusted to the tx-power level $L(u)$, there exists a data dissemination tree T spanning the nodes in $\mathcal{R} \cup \{\mathfrak{s}\}$; and the set of parent/children nodes of any node $v \in T$ must be contained in $\mathcal{N}_v(L(v))$ for v to send data and to receive control messages.

To conduct a data dissemination session in a DC-WSN, we need not only to find a valid power assignment L and a data dissemination tree T, but also to select the transmission time slots of the forwarding nodes in T, in order to avoid transmitting data to sleeping nodes. To this end, we provide the definition of *Viable Data-dissemination Solution* (VDS) in **Definition 1**, where we denote the set of non-leaf nodes in T by $\lambda(T)$, the set of children nodes of any node u in T by $\mathcal{C}_T(u)$, and the set $\{v | v \in \mathcal{N}_u(l) \wedge t \in \mathcal{A}(v)\}$ by $\partial_t(u, l)$:

Definition 1 (VDS). *Suppose that T is a data dissemination tree under a valid power assignment L, and S is a function that satisfies:*

1. *For any $t \in \mathcal{I}$ and any $u \in \lambda(T)$, $S(u, t) \subseteq \partial_t(u, L(u)) \cap \mathcal{C}_T(u)$,*
2. *For any $u \in \lambda(T)$, $\mathcal{C}_T(u) \subseteq \bigcup_{t \in \mathcal{I}} S(u, t)$,*

then S is called a Viable Transmission Schedule for T and $\langle L, T, S \rangle$ is called a Viable Data-dissemination Solution (VDS).

Basically, VDS requires a node $u \in T$ to be responsible for sending data to nodes in the set $S(u, t)$ at time slot t. Due to

the link unreliability, u may have to retransmit several times (in different working periods) for all the nodes in $S(u, t)$ to receive the data (if $S(u, t) \neq \emptyset$). To understand how many retransmissions are needed for a forward node, we introduce **Lemma 1**:

Lemma 1. *For any node $u \in \mathcal{V}$, any power level $l \in \mathcal{L}$ and any set $\mathcal{Q} \subseteq \mathcal{N}_u(l)$, let $\chi_u(l, \mathcal{Q})$ be the random variable that denotes the number of transmissions by u for all the nodes in $\mathcal{Q}$ to receive a data packet when u's tx-power is l. If $\mathcal{Q} \neq \emptyset$, then we have:*

$$\mathbb{E}[\chi_u(l, \mathcal{Q})] = \sum_{i=0}^{\infty} \left[1 - \prod_{v \in \mathcal{Q}} (1 - (1 - p_{uv}(l))^i) \right]$$

Our objective is to find a VDS such that the expected total transmission energy of the forward nodes is minimized. We introduce the formal definition of this problem by **Definition 2** and **Definition 3**:

Definition 2 (ENERGY-CONSUMPTION FUNCTION). *The energy-consumption function of a VDS $\langle L, T, S \rangle$ is*

$$\Psi(L, T, S) = \sum_{u \in \lambda(T)} \sum_{t \in \mathcal{I}} L(u) \cdot \chi_u(L(u), S(u, t))$$

Definition 3 (ERDD PROBLEM). *Given a set $\mathcal{V}$ of DC-WSN nodes, a source node $\mathfrak{s} \in \mathcal{V}$, and a set of receiver nodes $\mathcal{R} \subseteq \mathcal{V} - \{\mathfrak{s}\}$, the Energy-efficient Reliable Data Dissemination (ERDD) problem is to find a VDS $\langle L^*, T^*, S^* \rangle$ such that $\mathbb{E}[\Psi(L^*, T^*, S^*)]$ is minimized.*

Clearly, if we assume that all links are reliable and nodes are active at all time slots, then the ERDD problem degenerates to the min-energy broadcast/multicast problem in traditional AAWNs with perfect links, which is known to be NP-hard [1, 17]. Hence we have:

Theorem 1. *The ERDD problem is NP-hard.*

In the following section, we will propose approximation algorithms for the ERDD problem.

3. APPROXIMATION ALGORITHMS

A basic idea of our approximation algorithm design is to build a special data structure called the *Time-Reliability-Power* (TRP) Space, where data dimensions on time, reliability and power levels are all involved to facilitate our algorithm design and analysis. In this section, we will first introduce the concepts about the TRP space, and then present our algorithms in details.

3.1 Time-Reliability-Power Space

To build a TRP space, we first define a positive number $\gamma_{uv}(l)$ for any $u \in \mathcal{V}$, $l \in \mathcal{L}$ and $v \in \mathcal{N}_u(l)$ as follows:

$$\gamma_{uv}(l) = \begin{cases} 1 - 1/\ln(1 - p_{uv}(l)), & p_{uv}(l) \in (0, 1) \\ 1, & p_{uv}(l) = 1 \end{cases}$$

Note that $\gamma_{uv}(l)$ is no less than 1 and decreases when $p_{uv}(l)$ increases, hence a larger $\gamma_{uv}(l)$ indicates a poorer link. Besides, since link qualities have a constant lower bound in practice [9], $\gamma_{uv}(l)$ has a constant upper bound. Based on this definition, we introduce the concept of TRP space and a weight assignment method in **Definition 4** and **Definition 5**, respectively:

Definition 4 (TRP Space). *For any $u \in \mathcal{V}$, define the TRP Set of u as $\vartheta(u) = \{\langle u, t, r, l \rangle | t \in \mathcal{I} \wedge l \in \mathcal{L} \wedge r \in \bigcup_{v \in \mathcal{N}_u(l)} \{\gamma_{uv}(l)\}\}$. Define $\mathcal{U} = \bigcup_{u \in \mathcal{V}} \vartheta(u) \cup \mathcal{V}$. An access-relationship $\mathcal{W}$ is any set of ordered 2-tuples wherein each 2-tuple consists of two elements from $\mathcal{U}$. The tuple $\langle \mathcal{U}, \mathcal{W} \rangle$ is called a TRP space.*

Definition 5 (Weight Assignment). *Any $x \in \mathcal{U}$ has a weight $\varpi(x)$. If $x \in \mathcal{V}$ then $\varpi(x) = 0$. If $x = \langle u, t, r, l \rangle \in \mathcal{U} \backslash \mathcal{V}$, then $\varpi(x) = r \cdot l$. The weight of any $\mathcal{Y} \subseteq \mathcal{U}$ is also denoted by $\varpi(\mathcal{Y}) = \sum_{x \in \mathcal{Y}} \varpi(x)$.*

Intuitively, the elements in $\mathcal{U}$ can be viewed as weighted 4-dimensional (vector) nodes whose adjacent relationships are determined by an access-relationship $\mathcal{W}$, while $\mathcal{W}$ is defined by a given problem, as we shall do in Section 3.2 and 3.3. Based on above definitions, we introduce the concept of "TRP path" in **Definition** 6:

Definition 6 (TRP Path). *Given a TRP space $\langle \mathcal{U}, \mathcal{W} \rangle$, a sequence $h = \langle x_1, x_2, ..., x_m \rangle$ is called a TRP path from x_1 to x_m iff $(x_i, x_{i+1}) \in \mathcal{W}, \forall 1 \leq i \leq m - 1$. The length of h is defined as*

$$\ell(h) = \begin{cases} \sum_{i=2}^{m-1} \varpi(x_i) & m > 2 \\ 0 & \text{otherwise} \end{cases}$$

Based on the length function ℓ, the shortest TRP path from $x \in \mathcal{U}$ to $y \in \mathcal{U}$ is the one with the minimum length, which is denoted by $x \rightsquigarrow y$.

In our algorithm design, we will map the components in a TRP space $\langle \mathcal{U}, \mathcal{W} \rangle$ to a data-dissemination solution. Hence, we introduce some definitions for this mapping. The *node-image* of any $x \in \mathcal{U}$ is defined as the node $u \in \mathcal{V}$ such that $x \in \vartheta(u) \cup \{u\}$. The *edge-image* of any tuple $(x, y) \in \mathcal{W}$ is defined as the directed edge $(u, v) : u \in \mathcal{V}, v \in \mathcal{V}$ such that $x \in \vartheta(u) \cup \{u\}$ and $y \in \vartheta(v) \cup \{v\}$. Suppose that $\langle \mathcal{U}_1, \mathcal{W}_1 \rangle$ is a *subspace* of $\langle \mathcal{U}, \mathcal{W} \rangle$ (i.e., $\mathcal{U}_1 \subseteq \mathcal{U}$, $\mathcal{W}_1 \subseteq \mathcal{W}$), we define a mapping function $\mathfrak{F}$ such that $\mathfrak{F}(\mathcal{U}_1, \mathcal{W}_1)$ is a directed graph constructed by all node-images of the elements in $\mathcal{U}_1$ and all edge-images of the tuples in $\mathcal{W}_1$, i.e., all nodes in $\mathfrak{F}(\mathcal{U}_1, \mathcal{W}_1)$ are in $\mathcal{V}$.

For the convenience of description, we introduce some other notations/definitions about the TRP space here. Given a TRP space $\langle \mathcal{U}, \mathcal{W} \rangle$ and any $x, y \in \mathcal{U}$, we say x is *accessible* to y (or y is accessible from x) iff $(x, y) \in \mathcal{W}$. Given any $\mathcal{U}_1 \subseteq \mathcal{U}$, we define the *element closure* and *relationship closure* of $\mathcal{U}_1$ with respect to $\mathcal{W}$ as $\varrho_1(\mathcal{U}_1, \mathcal{W}) = \mathcal{U}_1 \cup \{y | x \in \mathcal{U}_1 \wedge (x, y) \in \mathcal{W}\}$ and $\varrho_2(\mathcal{U}_1, \mathcal{W}) = \{(x, y) | x \in \mathcal{U}_1 \wedge (x, y) \in \mathcal{W}\}$, respectively. For any TRP path $h = \langle x_1, ..., x_m \rangle$ in $\langle \mathcal{U}, \mathcal{W} \rangle$, we define $in(h) = \{x_i | 1 \leq i \leq m - 1\}$. If $\mathcal{U}_1$ contains all $x_i : 1 \leq i \leq m$, we say $\mathcal{U}_1$ embraces h. For any $u \in \mathcal{V}$, define $\gamma_{max}^u = \max\{\gamma_{uv}(l) | l \in \mathcal{L} \wedge v \in \mathcal{N}_u(l)\}$ and $\gamma_{min}^u = \min\{\gamma_{uv}(l) | l \in \mathcal{L} \wedge v \in \mathcal{N}_u(l)\}$. Define $\lambda = \max\{\frac{\gamma_{max}^u}{\gamma_{min}^u} | u \in \mathcal{V}\}$; it has a constant upper bound according to the definition of $\gamma_{uv}(l)$.

3.2 Solving ERDD for the Multicast Case

To design an approximation alorithm for the ERDD problem under multicast, we construct a TRP space $\langle \mathcal{U}, \mathcal{W}_M \rangle$ where $\mathcal{W}_M$ is defined by the following rules:

M1: Two elements $\langle u_1, t_1, r_1, l_1 \rangle$ and $\langle u_2, t_2, r_2, l_2 \rangle$ in $\mathcal{U} \backslash \mathcal{V}$ are accessible to each other iff either $u_1 = u_2$ or $[u_1 \in$

$\partial_{t_2}(u_2, l_2)] \wedge [u_2 \in \partial_{t_1}(u_1, l_1)] \wedge [r_1 \geq \gamma_{u_1 u_2}(l_1)] \wedge [r_2 \geq \gamma_{u_2 u_1}(l_2)]$ is true.

M2: Two elements $x = \langle u_1, t_1, r_1, l_1 \rangle \in \mathcal{U} \backslash \mathcal{V}$ and $u_2 \in \mathcal{V}$ are accessible to each other iff either $u_1 = u_2$ or there exists $y \in \vartheta(u_2)$ such that $(x, y) \in \mathcal{W}_M$.

Note that $\mathcal{W}_M$ is symmetric, i.e., $(x, y) \in \mathcal{W}_M \Rightarrow (y, x) \in \mathcal{W}_M$. Intuitively, **M1** embodies the covering relationship between two nodes under the time, reliability and power constraints, and **M2** is set up to avoid counting the weights of a multicast tree's leaf nodes in our algorithm. Based on this TRP space, we introduce an approximate algorithm MC-ERDD (**Algorithm 1**).

Algorithm 1: MC-ERDD$(\mathcal{V}, \mathfrak{s}, \mathcal{R}, \mathcal{I}, \mathcal{U}, \mathcal{W}_M, \mathcal{L}, \mathcal{A})$

1   Find a subspace $\langle \mathcal{U}', \mathcal{W}' \rangle$ of $\langle \mathcal{U}, \mathcal{W}_M \rangle$ containing $\mathcal{R} \cup \{\mathfrak{s}\}$ by the node weighted Steiner tree algorithm [6]; $\overline{\mathcal{U}} \leftarrow \mathcal{U}'$

2   Let T^m be an arbitrary spanning tree of $\mathfrak{F}(\mathcal{U}', \mathcal{W}')$

3   **foreach** $u \in \lambda(T^m)$ **do**

4   **foreach** $v \in \mathcal{C}_{T^m}(u)$ **do**

5   **if** $\{(u', v) \in \mathcal{W}_M | u' \in \mathcal{U}' \cap \vartheta(u)\} = \emptyset$ **then**

6   Find $\langle v, \bar{t}, \bar{r}, \bar{l} \rangle \in \mathcal{U}' \backslash \mathcal{V}$ such that $u \in \partial_{\bar{t}}(v, \bar{l})$ and $\bar{r} \geq \gamma_{vu}(\bar{l})$

7   Select an arbitrary time slot $\hat{t} \in \mathcal{A}(v)$ and add $\langle u, \hat{t}, \gamma_{uv}(\bar{l}), \bar{l} \rangle$ into $\overline{\mathcal{U}}$

8      $B \leftarrow \mathcal{C}_{T^m}(u)$

9   **foreach** $t \in \mathcal{I}$ **do**

10        $S^m(u, t) \leftarrow \emptyset$

11   **foreach** $l \in \mathcal{L}$ **do**

12           $\hat{r} \leftarrow \max\{r' | \langle u, t, r', l \rangle \in \overline{\mathcal{U}} \backslash \mathcal{V} \vee r' = 0\}$

13           $S^m(u, t) \leftarrow S^m(u, t) \cup \{v' | \gamma_{uv'}(l) \leq \hat{r} \wedge v' \in B \cap \partial_t(u, l)\}$

14        $B \leftarrow B - S^m(u, t)$

15   **foreach** *node u in T^m* **do**

16      $J \leftarrow$ the set of u's neighboring nodes in T^m

17      $\hat{l} \leftarrow \min\{l' | J \subseteq \mathcal{N}_u(l') \wedge l' \in \mathcal{L}\}$

18      $L^m(u) \leftarrow \max\{l' | \langle u, t', r', l' \rangle \in \overline{\mathcal{U}} \backslash \mathcal{V} \vee l' = \hat{l}\}$

19   **return** $\langle L^m, T^m, S^m \rangle$

In **Algorithm 1**, we first consider $\langle \mathcal{U}, \mathcal{W}_M \rangle$ as an undirected graph with node set $\mathcal{U}$ and edge set $\mathcal{W}_M$ and find an approximate Node Weighted Steiner Tree (NWST) with node set $\mathcal{U}'$ (line 1), then we map the NWST to an approximate solution $\langle L^m, T^m, S^m \rangle$ for ERDD (lines 2-18). Roughly speaking, the idea for doing this is that we can find the tx-power and transmission schedule of any $u \in \lambda(T^m)$ based on the NWST nodes in $\vartheta(u)$ (lines 8-18). To make this idea work correctly, we need to add some elements to $\mathcal{U}'$ (hence expand $\mathcal{U}'$ to $\overline{\mathcal{U}}$) (lines 4-7), because there may exist $u \in \lambda(T^m)$ such that $\mathcal{U}' \cap \vartheta(u) = \emptyset$ according to rule **M2**. The correctness and performance ratio of **Algorithm 1** will be proved in Section 4. The dominating running time of **Algorithm 1** is spent in line 1, which is determined by the time complexity of the NWST algorithm [6]. So we get:

Theorem 2. *The time complexity of the MC-ERDD algorithm is $\mathcal{O}(k^2 |\mathcal{V}|^2 \Delta^2)$.*

3.3 Solving ERDD for the Broadcast Case

Although **Algorithm 1** can also be used for broadcast, we introduce another algorithm with a better approximation ratio for the broadcast case in this section. Again, the first step is to construct a TRP space $\langle \mathcal{U}, \mathcal{W}_B \rangle$, but the access-relationship $\mathcal{W}_B$ is defined differently from $\mathcal{W}_M$ according to the following rules:

B1: An element $\langle u_1, t_1, r_1, l_1 \rangle \in \mathcal{U} \backslash \mathcal{V}$ is accessible to another element $\langle u_2, t_2, r_2, l_2 \rangle \in \mathcal{U} \backslash \mathcal{V}$ iff the boolean expression $[u_1 \neq u_2] \wedge [u_2 \in \partial_{t_1}(u_1, l_1)] \wedge [r_1 \geq \gamma_{u_1 u_2}(l_1)] \wedge [u_1 \in \mathcal{N}_{u_2}(l_2)]$ is true.

B2: An element $x = \langle u_1, t_1, r_1, l_1 \rangle \in \mathcal{U} \backslash \mathcal{V}$ is accessible to $u_2 \in \mathcal{V}$ iff $u_1 \neq u_2$ and there exists $y \in \vartheta(u_2)$ such that $(x, y) \in \mathcal{W}_B$.

B3: For any $u \in \mathcal{V}$ and any $y \in \vartheta(u)$, $(u, y) \in \mathcal{W}_B$.

Note that $\mathcal{W}_B$ is not symmetric and hence we cannot run **Algorithm 1** on $\langle \mathcal{U}, \mathcal{W}_B \rangle$. Actually, **B1-B3** are deliberately designed with desirable properties for our new approximation algorithm BC-ERDD (**Algorithm 2**), whose idea originates from the NWST algorithm in [6]. As we shall prove, by leveraging rules **B1-B3**, **Algorithm 2** yields a better performance ratio than **Algorithm 1** in the broadcast case.

Algorithm 2: BC-ERDD$(\mathcal{V}, \mathfrak{s}, \mathcal{I}, \mathcal{U}, \mathcal{W}_B, \mathcal{L}, \mathcal{A})$

---
1   $X \leftarrow \emptyset; \ Z \leftarrow \mathcal{V} - \{\mathfrak{s}\}$
2   **while** $|Z| > 0$ **do**
3   Find $a \in \mathcal{U}$ and a non-empty set $D \subseteq Z$ such that $(|D| \geq 2 \vee a = \mathfrak{s}) = true$ and $avl(a, D)$ is minimized. Let b be the node-image of a
4      $A \leftarrow \bigcup_{y \in D} in(a \rightsquigarrow y)$
5      $X \leftarrow A \cup X$
6      $Z \leftarrow Z - \varrho_1(A, \mathcal{W}_B)$
7   **if** $b \neq \mathfrak{s}$ **then** $Z \leftarrow Z \cup \{b\}$
8   Let T^b be an arbitrary directed spanning tree of $\mathfrak{F}(\varrho_1(X, \mathcal{W}_B), \varrho_2(X, \mathcal{W}_B))$ rooted at $\mathfrak{s}$
9   Set L^b, S^b for the nodes in T^b using the same method as lines 8-18 of **Algorithm 1**
10   **return** $\langle L^b, T^b, S^b \rangle$

In **Algorithm 2**, we use X to denote the subset of $\mathcal{U}$ which will be mapped to the forward nodes in broadcasting, and use Z to denote the set of nodes in $\mathcal{V}$ which are not accessible from any element in X. At first, X is initialized to $\emptyset$ and Z is initialized to $\mathcal{V} - \{\mathfrak{s}\}$. Then the algorithm uses a greedy strategy to expand X and reduce Z until $Z = \emptyset$ (lines 2-7). In each iteration, we greedily select an element $a \in \mathcal{U}$ and a subset D of $\mathcal{U}$ such that

$$avl(a, D) = \frac{\varpi(a) + \sum_{y \in D} \ell(a \rightsquigarrow y)}{|D|}$$

is minimized (line 3), and then update X and Z accordingly (lines 4-7). Intuitively, $avl(a, D)$ denotes the average cost for "covering" the nodes in D, which is analogous to the "cost effectiveness" measure in the greedy set-cover algorithm [13]. When X is finally determined, we use it to find an approximation solution $\langle L^b, T^b, S^b \rangle$ in lines 8-9 based on a similar mapping process as that in **Algorithm 1** (regarding X as $\bar{\mathcal{U}}$). The time complexity of **Algorithm 2** is given by:

Theorem 3. *The time complexity of the* BC-ERDD *algorithm is* $\mathcal{O}(k^4 \Delta^2)$.

4. PERFORMANCE ANALYSIS

In this section, we prove the correctness and performance ratios of the algorithms proposed in Section 3. We shall first introduce an analysis method which is used for analyzing both the MC-ERDD and the BC-ERDD algorithms, and then give the detailed performance analysis of each algorithm.

4.1 A Method for Performance Analysis

From **Lemma** 1 and **Definition** 2 we can see that, for any VDS $\langle L, T, S \rangle$, the expectation value $\mathbb{E}[\Psi(L, T, S)]$ is a summation of infinite series, which is hard to calculate. This makes it hard to find the performance ratios of our algorithms. To bypass this problem, we introduce a surrogate function ϕ, which is defined as follows:

$$\phi(L, T, S) = \sum_{u \in \mathcal{A}(T)} \sum_{t \in \mathcal{I}} L(u) \cdot \max\{\gamma_{uv}(L(u)) | v \in S(u, t)\}$$

It serves as an approximation to the expectation of Ψ. Let $\langle \bar{L}, \bar{T}, \bar{S} \rangle$ denote the output of **Algorithm 1** or **Algorithm 2**. We find the quantitative relationships between ϕ and Ψ by **Lemma 2** and **Lemma 3**:

Lemma 2. *We have* $\mathbb{E}[\Psi(\bar{L}, \bar{T}, \bar{S})] \leq (\ln \Delta + 1) \cdot \phi(\bar{L}, \bar{T}, \bar{S})$ *and* $\sigma[\Psi(\bar{L}, \bar{T}, \bar{S})] \leq \sqrt{2}(\ln \Delta + 1)\phi(\bar{L}, \bar{T}, \bar{S})$, *where* $\sigma(\cdot)$ *denotes the standard deviation.*

Lemma 3. $\phi(L^*, T^*, S^*) \leq (1 + \frac{1}{\ln 2})\lambda \mathbb{E}[\Psi(L^*, T^*, S^*)]$

Let $\langle \widehat{L^*}, \widehat{T^*}, \widehat{S^*} \rangle$ be a VDS such that $\phi(\widehat{L^*}, \widehat{T^*}, \widehat{S^*})$ is minimized. Intuitively, $\langle \widehat{L^*}, \widehat{T^*}, \widehat{S^*} \rangle$ is a solution in which each forwarding node selects the more reliable links for data transmission. Combining the above lemmas with the fact that $\phi(\widehat{L^*}, \widehat{T^*}, \widehat{S^*}) \leq \phi(L^*, T^*, S^*)$, we immediately get:

Theorem 4. *If* $\phi(\bar{L}, \bar{T}, \bar{S}) \leq \beta \cdot \phi(\widehat{L^*}, \widehat{T^*}, \widehat{S^*})$, *then we have* $\mathbb{E}[\Psi(\bar{L}, \bar{T}, \bar{S})] \leq \alpha\beta \cdot \mathbb{E}[\Psi(L^*, T^*, S^*)]$ *and* $\sigma[\Psi(\bar{L}, \bar{T}, \bar{S})] \leq \sqrt{2}\alpha\beta \cdot \mathbb{E}[\Psi(L^*, T^*, S^*)]$ *where* $\alpha = (1 + 1/\ln 2)\lambda(\ln \Delta + 1)$.

Theorem 4 actually suggests a method for analyzing our algorithms, i.e., to find the performance ratios of **Algorithm 1** and **Algorithm 2**, we only need to find their approximation ratios with respect to the surrogate function ϕ. In the following sections, we will analyze our algorithms based on this method.

4.2 Analyzing the MC-ERDD Algorithm

We first prove the correctness of **Algorithm 1** in **Lemma 4**, and then prove in **Lemma 5** that $\phi(L^m, T^m, S^m)$ is within constant times of $\varpi(\mathcal{U}')$, which is the weight of the NWST we found in **Algorithm 1**. The proofs of these lemmas are based on the construction rules of $\langle \mathcal{U}, \mathcal{W}_M \rangle$ as well as the mapping process employed in **Algorithm 1** that maps an NWST to a VDS.

Lemma 4. $\langle L^m, T^m, S^m \rangle$ *is a VDS for multicast.*

Lemma 5. $\phi(L^m, T^m, S^m) \leq (2\varepsilon_{max}/\varepsilon_{min}) \cdot \varpi(\mathcal{U}')$

Next, we reveal the quantitative relationship between $\varpi(\mathcal{U}')$ and $\phi(\widehat{L^*}, \widehat{T^*}, \widehat{S^*})$ by **Lemma 6**. The main idea behind **Lemma 6** is that, we can find a tree spanning $\mathcal{R} \cup \{\mathfrak{s}\}$

in $\langle \mathcal{U}, \mathcal{W}_B \rangle$ whose weight is at most $(1 + \lambda)\phi(\widehat{L}^*, \widehat{T}^*, \widehat{S}^*)$, whereas the NWST algorithm we used in **Algorithm 1** has an approximation ratio of $2 \ln k$.

Lemma 6. $\varpi(\mathcal{U}') \leq 2(1 + \lambda) \ln k \cdot \phi(\widehat{L}^*, \widehat{T}^*, \widehat{S}^*)$

Recall that $\varepsilon_{max}, \varepsilon_{min}$ are both pre-defined constants and λ has a constant upper bound. Hence, combing **Lemma** 5, **Lemma** 6, and **Theorem** 4 yields:

Theorem 5. *The expectation value and standard deviation of* $\Psi(L^m, T^m, S^m)$ *are within 9.8η and 13.9η times of* $\mathbb{E}[\Psi(L^*, T^*, S^*)]$, *respectively, where* $\eta = \frac{\varepsilon_{max}}{\varepsilon_{min}} \cdot \lambda(1 + \lambda)(\ln \Delta + 1) \ln k = \mathcal{O}(\ln \Delta \ln k)$.

4.3 Analyzing the BC-ERDD Algorithm

Based on similar reasoning to the proof of **Lemma 4**, we can also prove the correctness of **Algorithm 2** based on the definition of $\mathcal{W}_B$, as shown by **Lemma 7**:

Lemma 7. $\langle L^b, T^b, S^b \rangle$ *is a VDS for broadcast.*

Suppose that the while loop in lines 2-7 of **Algorithm 2** executes for q times. Let $Z_j = Z$ and $X_j = X$ after the jth while loop is executed. Let D_j be the set D found in the jth while loop. Clearly we have $\varpi(X_0) = 0$ and $|Z_q| = 0$. Indeed, the value $(\varpi(X_{j+1}) - \varpi(X_j))/|D_{j+1}|$ represents the "cost effectiveness" of the elements added to X in the $(j + 1)$th loop, which is bounded by **Lemma 8**:

Lemma 8. $(\varpi(X_{j+1}) - \varpi(X_j)) \cdot |Z_j| \leq |D_{j+1}| \cdot \phi(\widehat{L}^*, \widehat{T}^*, \widehat{S}^*)(0 \leq j \leq q - 1)$

Note that the elements in X_q are finally mapped to the transmission schedules of the non-leaf nodes in T^b. Hence, based on **Lemma 8** and the mapping process in **Algorithm 2**, we can get:

Lemma 9. $\phi(L^b, T^b, S^b) \leq (\varepsilon_{max}/\varepsilon_{min}) \cdot (2 \ln k + 1) \cdot \phi(\widehat{L}^*, \widehat{T}^*, \widehat{S}^*)$

Combing **Lemma 9** with **Theorem 4**, we get:

Theorem 6. *The expectation value and standard deviation of* $\Psi(L^b, T^b, S^b)$ *are within 2.5μ and 3.5μ times of* $\mathbb{E}[\Psi(L^*, T^*, S^*)]$, *respectively, where* $\mu = \frac{\varepsilon_{max}}{\varepsilon_{min}} \lambda(\ln \Delta + 1)(2 \ln k + 1) = \mathcal{O}(\ln \Delta \ln k)$.

5. CONCLUSION

We have studied the energy-efficient reliable data dissemination problem in DC-WSNs with unreliable links. We seek to minimize the total expected tx-power consumption for reliable multicasting/broadcasting. Due to the NP-hardness of the problem, we have proposed approximation algorithms with provable performance ratios. To the best of our knowledge, these algorithms are the first, one one hand, to holistically take into account various aspects including duty-cycling, wireless broadcast advantage, unreliable links and power-adjustability, and on the other hand, to provide guaranteed performance bounds for energy-efficient reliable data dissemination in DC-WSNs.

Acknowledgments

This work was supported in part by the AcRF Tier 2 Grant ARC15/11, the Start-up Grant of NTU, the National Natural Science Foundation of China (NSFC) No.61103007, the National Grand Fundamental Research 973 Program of China No.2011CB302905, and the National Science and Technology Major Project No. 2011ZX03005-004-04.

6. REFERENCES

[1] A. S. Ahluwalia and E. Modiano. On the complexity and distributed construction of energy-efficient broadcast trees in wireless ad hoc networks. *IEEE Transactions on Wireless Communications*, 4(5):2136–2147, 2005.

[2] S. Banerjee and A. Misra. Minimum energy paths for reliable communication in multi-hop wireless networks. In *Proc. ACM MobiHoc*, pages 146–156, 2002.

[3] S. Guo, Y. Gu, B. Jiang, and T. He. Opportunistic flooding in low-duty-cycle wireless sensor networks with unreliable links. In *Proc. ACM MobiCom*, pages 133–144, 2009.

[4] K. Han, Y. Liu, and J. Luo. Duty-cycle-aware minimum-energy multicasting in wireless sensor networks. *IEEE/ACM Transactions on Networking*, 21(3):910–923, 2013.

[5] K. Han, J. Luo, Y. Liu, and A. Vasilakos. Algorithm design for data communications in duty-cycled wireless sensor networks: a survey. *IEEE Communications Magazine*, 51(7), 2013.

[6] P. Klein and R. Ravi. A nearly best-possible approximation algorithm for node-weighted Steiner trees. *Journal of Algorithms*, 19(1):104–115, 1995.

[7] S. Kulkarni and L. Wang. Energy-efficient multihop reprogramming for sensor networks. *ACM Transactions on Sensor Networks*, 5(2):1–40, 2009.

[8] D. Li, X. Jia, and H. Liu. Energy efficient broadcast routing in static ad hoc wireless networks. *IEEE Trans. on Mobile Computing*, 3(2):144–151, 2004.

[9] X. Y. Li, Y. Wang, H. Chen, X. Chu, Y. Wu, and Y. Qi. Reliable and energy-efficient routing for static wireless ad hoc networks with unreliable links. *IEEE Transactions on Parallel and Distributed Systems*, 20(10):1408–1421, 2009.

[10] J. Luo, P. Eugster, and J.-P. Hubaux. Route driven gossip: probabilistic reliable multicast in ad hoc networks. In *Proc. of the 22nd IEEE INFOCOM*, pages 2229–2239, 2003.

[11] J. Luo, P. Eugster, and J.-P. Hubaux. Probabilistic reliable multicast in ad hoc networks. *Elsevier Ad Hoc Networks*, 2(4):369–386, 2004.

[12] L. Su, B. Ding, Y. Yang, T. F. Abdelzaher, G. Cao, and J. C. Hou. Ocast: optimal multicast routing protocol for wireless sensor networks. In *Proc. IEEE ICNP*, pages 151–160, 2009.

[13] V. V. Vazirani. *Approximation Algorithms.* Springer-Verlag, Berlin, 2001.

[14] P.-J. Wan, G. Calinescu, and C.-W. Yi. Minimum-power multicast routing in static ad hoc wireless networks. *IEEE/ACM Transactions on Networking*, 12(3):507–514, 2004.

[15] F. Wang and J. Liu. Duty-cycle-aware broadcast in wireless sensor networks. In *Proc. IEEE INFOCOM*, pages 468–476, 2009.

[16] Q. Wang, Y. Zhu, and L. Cheng. Reprogramming wireless sensor networks: challenges and approaches. *IEEE Network*, 20(3):48–55, 2006.

[17] J. E. Wieselthier, G. D. Nguyen, and A. Ephremides. On the construction of energy-efficient broadcast and multicast trees in wireless networks. In *Proc. IEEE INFOCOM*, pages 585–594, 2000.

Minimum Makespan Workload Dissemination in DTNs: Making Full Utilization of Computational Surplus Around

Sheng Zhang
State Key Lab. for Novel
Software Technology
Nanjing University, China
ynifbs215@gmail.com

Jie Wu
Department of Computer and
Information Sciences
Temple University, USA
jiewu@temple.edu

Sanglu Lu
State Key Lab. for Novel
Software Technology
Nanjing University, China
sanglu@nju.edu.cn

ABSTRACT

This paper poses the following problem: given a task that originates at some node in a Delay Tolerant Network (DTN), how are we to disseminate the workload during pairwise contacts to minimize the makespan? We first investigate the scenario in which each node has access to an oracle that knows global and future knowledge of node mobility, and we propose a centralized polynomial-time optimal algorithm. We then develop a distributed dissemination protocol, $D2$, which maintains r-hop neighborhood information at individual nodes. $D2$ makes dissemination decisions based on the estimations of the potential computational capacities and the future workloads of DTN nodes. Using trace-driven simulations, we show that, $D2$ with only 1-hop information is already near-optimal in a wide variety of environments, and the performance gap becomes smaller as the amount of information maintained at individual nodes increases.

Categories and Subject Descriptors

C.2.1 [**Network Architecture and Design**]: Wireless Communication

Keywords

Crowd computing; delay tolerant networks; makespan; workload dissemination

1. INTRODUCTION

The last few years have witnessed an explosive proliferation of personal wireless devices, whose communication ranges are much smaller than their roaming regions. Due to the unpredictable node mobility, devices can only contact each other opportunistically. Hence, the network around us is in essence intermittently-connected, and is a type of Delay Tolerant Network (DTN) [5]. Much existing DTN work has been devoted to message routing [2], content distribution [6], and cellular traffic offloading [8], however, little attention has been given to utilizing computational resources

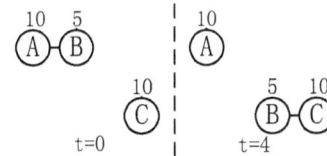

Figure 1: Two snapshots of a DTN with three nodes. The processing rate of each node is written next to the respective node that represents it.

in DTNs. Indeed, the computational capacity of a single device is generally quite small compared to the scale of tasks we have to handle at work or in scientific research; therefore, it would take a long time for us to complete a task. Inspired by crowd computing (e.g., the SETI@home [1] project exploits the massive idle computing resources across the Internet to analyze radio signal data), in this paper, we propose to disseminate the workload within the DTN around us, so that multiple devices can collaboratively finish a task, which may greatly reduce the makespan.

Compared with offloading parts of the workload to remote clouds, disseminating workload around us has some desirable properties. First and foremost, users do not need to pay any service fee to cloud providers, and can achieve economic efficiency through opportunistically sharing computational resources. Second, it is ad-hoc and can be used in cases where there are no infrastructure-based services, e.g., in a disaster. Third, the Internet is relieved to some extent, since uploading tasks and related data to clouds may take up a large amount of bandwidth.

This paper poses the following question: given a task that originates at some node in a DTN, how are we to disseminate the workload during pairwise contacts, so as to minimize the makespan? Unlike the classical minimum makespan scheduling problem [7], we do not have information about which devices would contribute to the completion of the task or from which time a device begins to participate in the collaboration. We use the example in Fig. 1 to illuminate the challenges in designing a distributed protocol. There are only two contacts in this DTN, i.e., A meets B at the beginning of slot 0, and B meets C at the beginning of slot 4. The processing rate of each node is written next to the respective node that represents it. For instance, A can finish 10 units of workload in one slot. Suppose that nodes A, B, and C have 300, 0, and 50 units of workload at the beginning of slot 0, respectively. We then show two possible dissemination schemes. Fig. 2(a) shows a naïve scheme, where the workload is split between two nodes in a contact

time node	0	0′	4	4′	completion time
A	300	200	160	160	20
B	0	100	80	30	10
C	50	50	10	60	10

(a) Naïve scheme: the makespan is 20 slots.

time node	0	0′	4	4′	completion time
A	300	145	105	105	14.5
B	0	155	135	48.3	13.7
C	50	50	10	96.7	13.7

(b) D2 protocol: the makespan is 14.5 slots.

Figure 2: Comparison of two schemes. The numbers indicate the amount of workload in each node before workload splitting ($t = 0, 4$) and after workload splitting ($t = 0', 4'$).

based on the ratio of their processing rates. That is, node A transfers 100 units of workload to B since, in this way, A and B would finish their respective workloads by the same time. The makespan of this scheme is 20 slots. Fig. 2(b) shows the protocol developed in this paper, where the impact of the future contact between B and C is taken into account when B meets A. To measure the impact, B needs to know not only when the contact will emerge, but also how many units of workload that C has. In our protocol, node A transfers 155 units of workload to B at the beginning of slot 0, which reduces the makespan to 14.5 slots.

In this paper, to gain a better understanding of the problem, we first investigate the scenario where each node has access to an oracle that knows global and future knowledge of node mobility, and we propose a centralized polynomial-time disseminating algorithm based on the Dijkstra shortest path algorithm [4]. The proposed centralized algorithm is proven to be optimal and serves as the comparison benchmark in extensive simulations.

With the insights obtained from the oracle case, we then develop a distributed dissemination protocol, D2, which enables each device to determine its disseminating strategy, such that all devices can collaboratively complete the task and achieve the minimal makespan. More specifically, in each individual node, D2 is comprised of four components: the workload queue manages operations (e.g., integration and splitting) on the actual workload; the r-hop neighborhood information manager stores and updates the contact rate, opportunistic path, and workload information for every r-hop neighbor of a node; the finish time estimator calculates the expected finish time of a given workload; and the future workload estimator predicts the expected workload in a node at a future time slot. We show through trace-driven simulations that the performance, in terms of makespan, of D2 with only 1-hop neighborhood information is already near-optimal in a wide variety of environments, and the performance gap becomes smaller as the amount of information maintained at individual nodes increases.

2. MODEL AND PROBLEM

Task Model. This paper considers a type of task that has the following two properties. One is that the output of (part of) a task is small, which makes it possible for participating devices to send their respective output to the task source through long distance communications. Thus, this paper mainly concentrates on the workload disseminating process

and does not care about the result gathering phase. The other property is that the workload of a task is fine-grained and can be parsed into arbitrarily small chunks. When two devices have a contact, their total workloads can be redistributed in any ratio between them.

These two properties are not made-up. For example, in the SETI@home [1] project, the observation data from the Arecibo radio telescope is divided into extremely small pieces, which are then assigned to the volunteers; a volunteer only has to report whether there are abnormal signals in a piece of data after analyzing it. Another example of this type of task could be an evaluation of the delivery ratio of a routing protocol on a data trace. To make the evaluation thoroughly, the routing protocol needs to be executed 100 times for each of the 1,000 different combinations of protocol parameters. In this example, we can see that the output (i.e., the delivery ratio) produced in a single execution is very small, and a single execution can be used as the smallest indivisible unit of workload.

Network Model. We model a delay tolerant network as a graph $G = (V, E)$. The vertex set V consists of all the mobile devices/users/nodes. Device $i \in V$ can process r_i units of workload in one time slot. We assume that the processing rate is constant for each node, and is not affected by the execution of the proposed protocols developed in this paper. The edge set E represents the stochastic contacts between devices. The inter-contact time between i and j is assumed to be exponentially distributed with the contact rate λ_{ij}. Each node is also assumed to contact its neighbors one by one, since a node does not frequently contact multiple neighbors at the same time. We do not consider storage and bandwidth constraints in this paper.

The Minimum Makespan Workload Dissemination Problem. As a starting point, this paper focuses on the problem of disseminating the workload from a single task in an empty DTN, where by "empty" we mean that: 1) when the single task originates at its source, all of the other nodes do not have any units of unfinished workload, and 2) before the single task is finished, there are no more tasks that would originate in the DTN.

To put it formally, the single task C consists of W units of workload; time is partitioned into slots of equal length. Denote the amount of workload in node i at the beginning of slot t (or slot t for short without causing confusion) as W_i^t. Without loss of generality, we assume that, the task C originates at its source $s \in V$ at slot 0. Since we consider the case where there is only one task, we have $W_s^0 = W$, and $\forall i \in V \setminus \{s\}$, $W_i^0 = 0$.

The completion time T, also called the makespan, is defined as the time difference between the origin time, i.e., slot 0, and the finish time, i.e., the time point when all nodes finish their respective workload that belongs to C. That is, $T = \min t$, subject to $W_i^t = 0, \forall i \in V$. Our problem is how to disseminate the workload during pairwise contacts so as to minimize the makespan T.

3. OPT: CENTRALIZED POLYNOMIAL TIME OPTIMAL ALGORITHM

In this section, we introduce the centralized polynomial time optimal algorithm for the case that each node has global and future knowledge. The basic idea is to utilize the computational capacity of each node as early as possible and

(a) Discrete contact graph   (b) Shortest delay tree

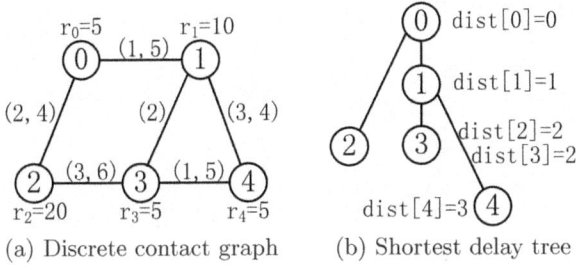

Figure 3: A numerical example of the OPT algorithm.

make sure that all of the participating nodes finish their respective workloads at the same time. Based on the Dijkstra shortest path algorithm, we can construct a shortest-delay tree, which encodes the shortest delays from the task source to all of the other nodes. Each node then makes workload dissemination decisions according to the tree. The details are omitted due to space limitations.

Fig. 3(a) shows a DTN that we will use throughout this paper. The processing rate of each node is labeled next to the respective node that represents it, and the contact opportunities of each pair of nodes are written next to the respective edge that represents it. For example, node 0 contacts node 1 at slots 1 and 5. Suppose that node 0 has 60 units of workload at slot 0, how are we to disseminate them?

As shown in Fig. 3(b), we first build the shortest delay tree; after simple calculations, we have $assignment[0] = 15$, $assignment[1] = 20$, $assignment[2] = 20$, $assignment[3] = 5$, and $assignment[4] = 0$. That is, at the beginning of slot 1, node 0 transfers 25 units of workload to node 1; at the beginning of slot 2, node 0 transfers 20 units of workload to node 2, and node 1 transfers 5 units of workload to node 3.

4. D2: DISTRIBUTED DISSEMINATION PROTOCOL

This section provides an overview of the D2 protocol. The details are omitted due to space limitations. Fig. 4 shows the architecture of the D2 protocol.

Workload queue. The workload queue manages operations on the actual workload in four aspects: 1) it accumulates the output of the finished workload and sends it to a source-specified server via long-distance communications at the proper time; 2) it stores the unfinished workload and is responsible for updating W_i^t; 3) when another node j transfers some workload to node i, it integrates the new workload into its own; 4) when node i goes to transfer a certain part of its workload to another node j, it splits the workload into two corresponding parts.

The r-hop neighborhood information manager. Denote the set of nodes that are within r hop(s) of node i as N_i^r, where $r \in [1, R]$, and R is the network diameter that represents the hop count of the longest shortest path among any two nodes in a DTN. For each node $k \in N_i^r$, this manager is responsible for storing and updating the contact rate λ_{ik}, the workload $\Phi_k(t)$, and the opportunistic path P_{ik}. When node i comes in contact with j, for each node $k \in N_i^r \setminus N_j^{r-1}$, node i keeps P_{ik} unchanged; for each node $k \in N_j^{r-1} \setminus N_i^r$, node i initializes P_{ik} by catenating i and P_{jk} (denoted as $i + P_{jk}$); for each node $k \in N_j^{r-1} \cap N_i^r$, node i updates P_{ik} to the path with the smaller weight among P_{ik} and $i + P_{jk}$.

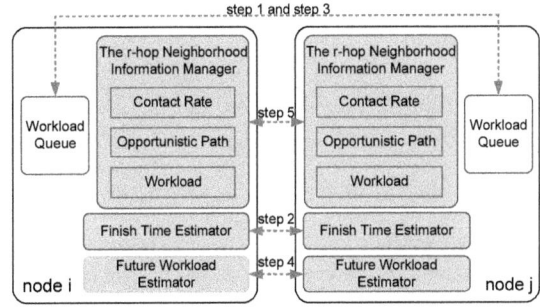

Figure 4: The architecture of D2.

Finish time estimator. Based on the information maintained by the r-hop manager, this component generates and updates a function $\mathcal{T}_i(w)$, which returns the expected time for node i to finish w units of workload. We note that it is non-trivial to obtain $\mathcal{T}_i(w)$, since the processing capacity and the amount of workload in all nodes that are within r hop(s) from node i should be taken into account. For example, suppose that node i has w units of workload, and has only one 1-hop neighbor j; if $\Phi_j(t)/r_j < w/r_i$, then we have:

$$\mathcal{T}_i(w) = \frac{1}{r_i + r_j}\left(w + \Phi_j(t) + \frac{r_j}{\lambda_{ij}}\left(e^{-\frac{\lambda_{ij}}{r_j}\Phi_j(t)} - e^{-\frac{\lambda_{ij}}{r_i}w}\right)\right)$$

Future workload estimator. This component generates and updates a function $\Phi_i(t)$, which returns the expected workload in node i in a future point t in time, i.e., the domain of this function is $\{t|t \geq t_0\}$, where t_0 is the generating time of this function. Since node i may transfer/receive some workload to/from other nodes during future contacts, it is also non-trivial to obtain $\Phi_i(t)$.

Take Fig. 4 for example; we illustrate how D2 works when nodes i and j meet at time slot t: (1) After neighbor discovery, i and j exchange the information about the amount of their respective workloads, i.e., W_i^t and W_j^t. (2) Node i generates $\mathcal{T}_i(w)$ based on the information (i.e., λ_{ik}, r_k, and $\Phi_k(t)$) of all nodes in $N_i^r \setminus \{j\}$; node j generates $\mathcal{T}_j(w)$ based on the information (i.e., λ_{jh}, r_h, and $\Phi_h(t)$) of all nodes in $N_j^r \setminus \{i\}$. Without loss of generality, suppose that $W_i^t < W_j^t$, then node j sends the function $\mathcal{T}_j(w)$ to i. (3) In order to locally minimize the makespan, their total workloads should be re-distributed in such a way that two nodes finish their separate parts by the same time. Specifically, denote the amounts of workload that nodes i and j would get after workload splitting by x_i and x_j, respectively; then, x_i and x_j should satisfy:

$$x_i + x_j = W_i^t + W_j^t \quad \text{and} \quad \mathcal{T}_i(x_i) = \mathcal{T}_j(x_j)$$

Node i returns the results to node j. The workload queues then complete the necessary workload transfers. (4) Nodes i and j generate $\Phi_i(t)$ and $\Phi_j(t)$, respectively. Node i updates its maintained copy $\Phi_j(t)$ to be the new one generated by j, and node j does the same in a similar way. (5) For each node $k \in N_i^{r-1}$, node i updates $\Phi_k(t)$ and P_{ik}; node j updates the corresponding information in a similar way.

Applying D2 to Fig. 3(a), the makespan is 3.34 slots, while the makespan of the Naïve scheme is 3.53 slots.

5. PERFORMANCE EVALUATION

This section evaluates the performance of D2, which is compared with the following dissemination algorithms: OP-

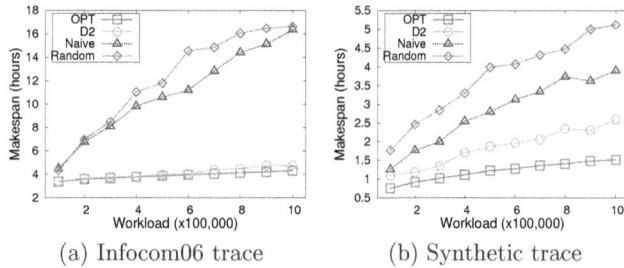

(a) Infocom06 trace (b) Synthetic trace

Figure 5: Performance comparison under different workloads while keeping $MaxCapacity = 10$.

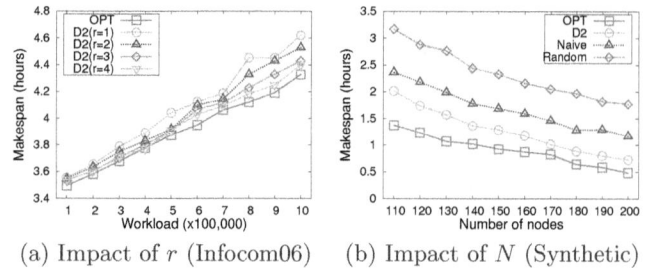

(a) Impact of r (Infocom06) (b) Impact of N (Synthetic)

Figure 6: Sensitivity results.

T, the polynomial-time optimal algorithm that has access to global and future knowledge; *Naïve*, the total workload is split between two nodes in a contact based on the ratio of their processing rates; and *Random*, the total workload is randomly split between two nodes in a contact.

Simulation Setup. Our evaluations are conducted on three realistic traces and a synthetic trace. Due to space limitations, we only present results on the Infocom06 [3] and synthetic traces. In the Infocom06 trace, mobile users with Bluetooth-enabled devices periodically detect their peers nearby and record contacts over several days. After some preprocessing on the raw data, we find that there are few contacts made during the nighttime. In order to have a meaningful and usable DTN, we only use trace data that was collected during the daytime.

Since the scale of these traces is relatively small and cannot be flexibly tuned, we also generate a synthetic trace with N nodes: the number of 1-hop neighbors of a node is uniformly generated from the range $[AvgDeg - 5, AvgDeg + 5]$; the inter-contact time between two nodes follows exponential distribution, with λ being uniformly generated from the range $[1/2 \cdot AvgLambda, 3/2 \cdot AvgLambda]$. Based on the observations from the realistic traces, the defaults are set as $AvgDeg = 10$, $AvgLambda = 0.0001$, and $N = 100$. D2 maintains $r = 1$ hop information at individual nodes—unless otherwise noted. In each trace, the processing rates of mobile devices are uniformly generated from the range $[1, MaxCapacity]$, and W units of workload originate at a randomly selected node at the beginning of the partial trace that we use. The result is averaged over multiple running times for statistical convergence.

Simulation Results. Fig. 5 shows the comparison results under different workloads in two traces, while keeping $MaxCapacity = 10$. In general, D2 achieves a near-optimal performance and outperforms Naïve and Random. Specifically, the makespan in D2 is within 113% and 172% of that in OPT in the two traces, respectively. With various workloads, D2 maintains a 10%-70% performance advantage over Naïve and Random. We also notice that the advantage becomes greater when the amount of workload goes up. The reason behind this phenomenon is that D2 makes decisions by estimating the expected finish time and the future workload of each node, and the estimation becomes more accurate when the computation task lasts for a longer time.

We are also interested in evaluating the impact of the scope of network information maintained at individual nodes. Fig. 6(a) shows the simulation results. Generally, D2 performs better when more information is maintained at individual nodes. Surprisingly, we notice that D2 with 1-hop

information has already received most of the performance gain under various workloads, and the marginal benefit of maintaining additional information is small. This is because each node does not maintain the information between its r-hop neighbors; when r increases, the estimation accuracy of D2 would be reduced. This insight can be used to trade-off between the performance and the overhead of D2.

Fig. 6(b) shows the impact of the number of nodes on the makespan, when there are more mobile nodes in a DTN, the makespan in all four algorithms goes down since more nodes provide more computational capacity.

In summary, our trace-driven simulations show that D2 performs well in a variety of settings. In future work, we believe a more sophisticated estimation of the potential computational capacity and the future workload of each node will improve our results, and perhaps bringing us closer to guarantees of performance.

6. CONCLUSION

In this paper, we propose making full utilization of computational surplus around us, and we design a centralized optimal algorithm, OPT, and a distributed protocol, D2, for the minimum makespan workload dissemination problem in DTNs. Extensive simulations show that D2 achieves a near-optimal performance in a wide variety of settings.

Acknowledgements. This work was supported in part by NSFC (No. 61073028, No. 61202113, and No. 61021062), Key Project of Jiangsu Research Program (No. BE2010179), Jiangsu NSF(No. BK2011510), China 973 Program (No. 2009CB320705), Research and innovation project of Jiangsu (No. CXZZ12_0055), Program A for outstanding PhD candidate of Nanjing University, and US NSF (ECCS 1128209, CNS 1065444, CCF 1028167, CNS 0948184, and CCF 0830289).

7. REFERENCES

[1] SETI@home. *http://setiathome.berkeley.edu/*.
[2] A. Balasubramanian, B. Levine, and A. Venkataramani. DTN routing as a resource allocation problem. In *ACM SIGCOMM 2007*.
[3] A. Chaintreau, P. Hui, J. Scott, R. Gass, J. Crowcroft, and C. Diot. Impact of human mobility on opportunistic forwarding algorithms. *IEEE TMC 2007*.
[4] T. H. Cormen, C. E. Leiserson, R. L. Rivest, and C. Stein. *Introduction to Algorithms*. The MIT Press, 2nd revised edition edition.
[5] K. Fall. A delay-tolerant network architecture for challenged internets. In *ACM SIGCOMM 2003*.
[6] W. Gao and G. Cao. User-centric data dissemination in disruption tolerant networks. In *IEEE INFOCOM 2011*.
[7] V. Vazirani. *Approximation algorithms*. Springer, 2004.
[8] X. Zhuo, W. Gao, G. Cao, and Y. Dai. Win-Coupon: An incentive framework for 3G traffic offloading. In *IEEE ICNP 2011*.

Adaptive Channel Bonding in Multicarrier Wireless Networks

Pei Huang
Department of Computer
Science and Engineering
Michigan State University
East Lansing, MI 48824, USA
huangpe3@cse.msu.edu

Xi Yang
Department of Computer
Science and Engineering
Michigan State University
East Lansing, MI 48824, USA
yangxi1@cse.msu.edu

Li Xiao
Department of Computer
Science and Engineering
Michigan State University
East Lansing, MI 48824, USA
lxiao@cse.msu.edu

ABSTRACT

To support high data rate applications such as multimedia streaming, the ongoing standardization of the next generation Wi-Fi increases the channel bonding from 40 MHz in 802.11n to 80, and even 160 MHz under certain conditions in 802.11ac. However, inefficiency and unfairness issues arise when devices that use different channel widths coexist in a contention domain. A device with channel bonding has to wait until all bonded channels to be idle to commence a transmission while narrow channel interferers have more channel access opportunities. To address the inefficiency and unfairness issues in channel bonding, we propose an adaptive channel bonding (ACB) protocol in which a node is allowed to initiate a transmission as long as a narrow channel is available and gradually increase channel width during transmission whenever a new narrow channel becomes available. ACB aggregates all available narrow channels as one wide channel, removing the need of setting guard bands between contiguous narrow channels. A challenge in the design is the communication over uncertain channels. To enable fast spectrum agreement between transmitter and receiver, a partial spectrum correlation method is introduced. ACB also considers the severe contention in a wide band of spectrum. When new channels become available, multiple nodes may contend for them. A compound preamble is designed to make collisions detectable in the frequency domain and a parallel bitwise arbitration mechanism is introduced to quickly resolve the collisions in the time domain. We implemented and evaluated the ACB through the GNU Radio/USRP platform. Experimental results show that ACB can well address the inefficiency and unfairness issues caused by heterogeneous radio coexistence.

Categories and Subject Descriptors

C.2.1 [**Computer-Communication Networks**]: Network Architecture and Design—*Wireless Communication*

General Terms

Design, Experimentation

Keywords

Channel Bonding; NC-OFDM; Medium Access; Testbed

1. INTRODUCTION

Applications such as video conferencing and multimedia streaming demand low-lag gigabit speeds. To support high speed wireless communication, 802.11 standards are being driven to improve channel bonding, an efficient method that increases data rate regardless of other technologies in use. However, inefficiency and unfairness issues arise when heterogeneous radios coexist in a contention domain. When a device operates in a wide channel that spans across multiple narrow channels, it has to defer its transmission to the time when all of the narrow channels are vacant. This is inefficient because the device cannot utilize the other narrow channels when only one of them is occupied. Further, when there are narrow channel devices in more than one narrow channel, the devices with channel bonding are harder to win medium access opportunities because narrow channel devices may work independently in the non-overlapping narrow channels.

The inefficiency and unfairness issues can be alleviated if a device does not need to wait for all narrow channels to be idle to initiate a transmission. Recent spectrum-agile designs [5] [6] [7] have shown that it is practical to aggregate noncontiguous narrow channels as one channel. However, current spectrum-agile designs are frame-based. Therefore, when a new narrow channel becomes available, it cannot be utilized until the next frame. An unfairness issue arises as not all nodes get contention opportunities for the new channel. The new channel is invisible to nodes that are in transmission. Some nodes may never be able to acquire the channel.

Current spectrum-agile designs are also lack of an efficient mechanism to address the multiple access issue. When a narrow channel becomes available, several nodes may attempt to acquire it at the same time. Leaving contention resolution to the carrier sense multiple access with collision avoidance (CSMA/CA) may require a large contention window to reduce collisions, reducing communication efficiency.

After a transmitter has won some narrow channels, it needs to inform the receiver of used spectrum; otherwise, the receiver is unable to decode packets sent by the transmitter. Using control packets to achieve spectrum agreement can introduce high overhead because of medium access contention and physical layer convergence procedure. A better spectrum agreement method is desired.

To address the aforementioned issues, this paper introduces an adaptive channel bonding (ACB) protocol in which a node is allowed to initiate a transmission as long as there exist some idle narrow channels and the node gradually increases channel width during transmission when new narrow channels are sensed to be idle. The proposed ACB employs self-interference cancellation techniques to ensure that any idle channel is available to all nodes. As many nodes may contend for narrow channels that become available, a compound preamble is designed to probe collisions in all channels at the same time and a bitwise arbitration mechanism is introduced to resolve the collisions in parallel. After a transmitter wins some narrow channels, a partial spectrum correlation method is introduced to achieve fast spectrum agreement between the transmitter and the target receiver.

This paper integrates all the components as a complete system and prototypes it on the GNU Radio/USRP platform to validate its effectiveness. Experimental results show that the proposed ACB is practical and the design significantly improves the efficiency of wide band spectrum sharing.

2. ADAPTIVE CHANNEL BONDING

Adaptive channel bonding (ACB) imposes three challenges: *how to detect newly available channels during transmission, how to contend for the channels, and how to achieve spectrum agreement between transmitter and receiver.* In this section, we present our design that well addresses them.

2.1 Overview

We assume that a wide band of spectrum is divided into n narrow channels of equal size for independent evaluation (e.g., 6 MHz channels in the TV bands). Before commencing a transmission, the transmitter checks what narrow channels are available in the target wide band. The clear channel assessment (CCA) can be done through checking the power spectral density (PSD) of the wide band [6] [7]. Once available channels are identified, the transmitter needs to contend for these channels. Because there may exist many contending transmitters in a wide band of spectrum, medium access contention based on carrier sense multiple access with collision avoidance (CSMA/CA) may require a large contention window (CW) to ensure low collision probabilities. However, a large contention window leads to a long average backoff time, significantly reducing the communication efficiency. We thus introduce a parallel bitwise arbitration mechanism to reduce the medium access overhead for achieving low collision probabilities. In addition, the bitwise arbitration allows a node to contend for different channels with different priorities simultaneously. A node is unlikely to lose all channels in a contention.

After a transmitter wins some channels, it informs the receiver of the occupied channels through a partial spectrum correlation method introduced in the paper. Once a node starts transmission, it was unable to receive simultaneously with the same RF front end and antenna because the self-interference is so strong that it buries other signals. However, recent advances in self-interference cancellation [4] [3] show that new radio designs will allow simultaneous transmission and reception even with the same antenna. Therefore, a transmitter can monitor the medium state even when it is transmitting. This allows all nodes to perceive a newly available channel. All nodes get equal chances to acquire the new channel. The contention is resolved through bitwise arbitration. The winner piggy-backs the new spectrum use via data to inform the receiver of increased channel width and then bonds more channels for subsequent data transmission.

2.2 Spectrum Contention

In ACB, nodes do not need to back off randomly for collision avoidance. Instead, a node transmits a compound preamble immediately when some narrow channels are identified as available. The compound preamble is designed with unique frequency features for collision detection. If any collision is detected, a bitwise arbitration mechanism is activated to resolve the collision.

ACB employs non-contiguous orthogonal frequency division multiplexing (NC-OFDM) to opportunistically utilize spectrum fragments in a wide band of spectrum. When a node has data to send, it first checks what narrow channels are available. If there are available narrow channels, it transmits a compound preamble that occupies all of the channels but the preamble is designed to set some subcarriers to inactive. First, some narrow channels are busy, hence all subcarriers in those channels are set to inactive. Second, we intentionally deactivate some subcarriers in available narrow channels for collision detection.

2.2.1 Compound Preamble Design

In the frequency domain, a compound preamble occupies the entire wide band with some subcarriers set to inactive. In the time domain, a compound preamble is one OFDM symbol that consists of N samples when a N-point IFFT is used. Suppose the target wide band is composed of n narrow channels. Each channel contains $l = N/n$ subcarriers. A unique sequence of l symbols $\{p_1 p_2 p_3 ... p_l\}$ is assigned to each node. For collision detection, the actual sequence is of no meaning. We thus postpone the discussion about the sequence to spectrum agreement. Here we only consider how to create high magnitudes at some but not all subcarriers.

If a node repeats its unique sequence by n times ($l \times n = N$ symbols) and performs N-point IFFT, it generates a compound preamble that causes high magnitudes at all N subcarriers. To deactivate some subcarriers for collision detection, a node feeds some 0s instead of modulation symbols to the IFFT algorithm, leading to zero power at corresponding subcarriers. To achieve this, a node draws a random number from a uniform distribution over the interval $(0, 2^k)$ for each available channel. For busy channels, the number 0 is used. Because each number can be represented by a k bit binary code, the binary code is bitwise AND with the node's unique sequence. As we feed some 0s to the IFFT, not all subcarriers in each available channel are used and all subcarriers in busy channels are set to inactive. The binary representation of each number is essentially a map of active and inactive subcarriers in the corresponding channel with '1' indicating active and '0' indicating inactive.

With the compound preamble design, a collision is detectable if a node observes high magnitudes at subcarriers that are deactivated by it. The collision detection fails only when two nodes choose the same random number, but the probability is very low as it decreases exponentially along with the increased subcarrier number l. For example, assuming a 64-point IFFT is performed on a 20 MHz band

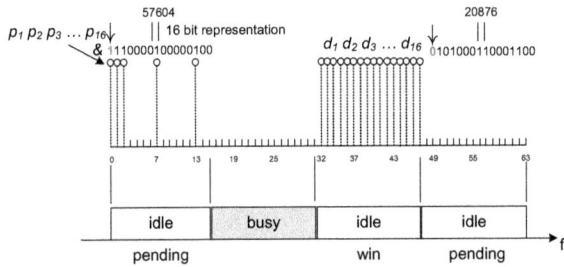

Figure 1: An example of the 1st compound preamble in the arbitration phase.

that comprises four 5 MHz channels, each channel contains 16 subcarriers. With the binary mapping, in total $2^{16} - 1$ random numbers can be represented (0 is excluded) in each channel. The probability that two nodes choose the same number for a channel is very low. To get the same low collision probability in CSMA/CA, the contention window (CW) will be too large to be practical.

Because collisions are detectable in the frequency domain, instead of deferring nodes' transmissions randomly for collision avoidance, we allow a node to transmit immediately when some channels are identified as available. To check whether the bold attempt will cause a collision, a node performs spectrum analysis on received signal while it is transmitting its compound preamble. The simultaneous transmission and reception has been demonstrated to be practical in recent studies in self-interference cancellation [3] [4].

2.2.2 Medium Access Procedure

As collisions are detectable, it is important to determine which node is the winner of a particular narrow channel in a contention. The medium access procedure in ACB has two phases: *collision probe* and *bitwise arbitration*. When a node has data to send, it checks what channels are available. Once the CCA is done, the node transmits its first compound preamble to check whether there is a collision in any of the contending channels. We call this phase *collision probe*.

If there is no collision in a channel, the node wins the channel immediately. It then uses all subcarriers in the channel for the destination's unique sequence. If there is no collision in all channels, the bitwise arbitration is skipped and the node starts data transmission. However, if there is a collision in any of the channels, the bitwise arbitration is triggered.

In the bitwise arbitration phase, the compound preamble of a node is updated to consider three conditions as shown in Fig. 1. First, for busy channels, all subcarriers are set to inactive. Second, for a channel that is won by the node, the destination's unique sequence is mapped to all subcarriers. Third, for a channel in which the winner has not been determined, the ith bit of the corresponding binary code is checked for constructing the ith compound preamble in the bitwise arbitration phase.

The compound preamble is used to contend for multiple channels at the same time. If we focus on one arbitrary channel, a node traverses the corresponding binary code starting from the most significant bit. If the ith bit is 0, all subcarriers in the channel are set to inactive in the ith compound preamble. If the ith bit is 1, the node maps its unique sequence to active subcarriers according to the binary code. While a node is transmitting its compound preamble, it also performs spectrum analysis on received signal. If it sets all subcarriers in a channel to inactive but observes high

magnitudes at subcarriers in the channel, it loses the channel. On the contrary, if the channel is idle, the winner has not been determined, it proceeds to check the next bit. If a node uses some subcarriers in a channel in a compound preamble and the collision still exists (detecting high magnitude at subcarriers that are deactivated by it), it proceeds to check the next bit. On the contrary, if the frequency spectrum of the channel matches its corresponding binary code, the collision has been resolved and the node wins the channel. Once a node wins a channel, all subcarriers in the channel are used for the destination's unique sequence. The node shall occupy the channel to prevent other nodes from taking it. When collisions in all channels are resolved, the node initiates the spectrum agreement.

2.3 Spectrum Agreement

After the spectrum contention, a transmitter has won some narrow channels. However, the receiver is unable to decode any packet sent by the transmitter if the occupied spectrum is unknown. To achieve spectrum agreement, we introduce partial spectrum correlation.

Each node has been assigned a unique sequence as its signature. The signature is known to its neighbors through neighbor discovery (routing also requires neighbor information). To indicate channels that are used for the receiver, the transmitter encodes the receiver's signature in the frequency domain. A node maps the receiver's signature to subcarriers of each channel that is occupied by it. Suppose a narrow channel contains l subcarriers. We design a signature sequence of length l for each node. A key feature of the signature is that a node's signature cannot result in high correlation values with other nodes' signatures. Some polyphase codes possess such correlation properties and we use the Zadoff-Chu (ZC) sequences introduced in [2]. The transmitter uses the signature of the receiver to modulate all l subcarriers in each obtained channel while in other channels it feeds 0s to the corresponding subcarriers.

The receiver calculates the expected IFFT results assuming its signature is present in each of the n channels. A sliding window is used to extract samples from received signal and the receiver correlates the extracted samples with all expected IFFT results in parallel. When the sliding window is aligned with the right OFDM symbol, a correlation peak will be observed for each used channel. Although the received signal is the sum of all modulated subcarriers, the orthogonality means that the cross-talk between subcarriers sums up to zero.

3. PERFORMANCE EVALUATION

We prototype the ACB on the GNU radio/USRP platform. The first step is to evaluate the collision detection and the bitwise arbitration. We compare them with CSMA/CA to demonstrate that the bitwise arbitration is more efficient in medium access contention. We then evaluated the spectrum agreement design, verifying that the receiver can be informed of used channels through partial spectrum correlation.

When a node is transmitting, the noise floor is increased. If a node measures the noise floor during the interval when there is no transmission and uses the noise floor to determine whether there is a signal at a subcarrier, it may incorrectly regard that there is a collision. To reduce the false alarm

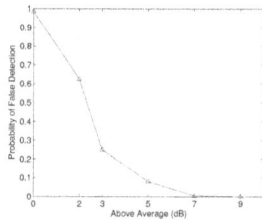

Figure 2: Setting noise threshold x dB above the noise floor.

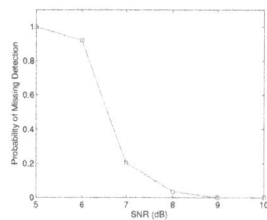

Figure 3: Collision detection performance.

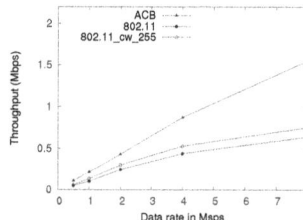

Figure 4: Throughput gain over 802.11.

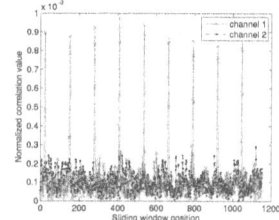

Figure 5: Partial spectrum correlation under 10 dB interference.

rate, we examine how much we need to increase the noise threshold.

We let one USRP E110 with WBX transceiver operate in the full-duplex mode. It measures the noise floor when the channel is idle and then starts transmission. When only one node is transmitting, it should not identify inactive subcarriers as active. We set the noise threshold x dB above the measured noise floor. A subcarrier is considered active if its magnitude is above the threshold. The false collision detection rate is presented in Fig. 2. If the noise threshold is increased by 7 dB from the measured noise floor, the probability of mistaking an inactive subcarrier as active is low enough.

Once the noise threshold is determined, we study whether a node can detect the collision from a weak signal. We start another USRP E110 and gradually increase the transmission power. Fig. 3 shows that a node is unlikely to miss the detection of another node's compound preamble if the SNR is above 9 dB.

The bitwise arbitration reduces the medium access overhead, leading to higher throughput. We make three USRP E110 keep sending packets to a E110. To compare with CSMA/CA, we use the parameters defined in 802.11n [1]. Fig. 4 shows the improvement of ACB over 802.11n. When there are contending transmitters, CSMA/CA may need a larger contention window to reduce collisions.

In ACB, both the medium access overhead and the collision probability are reduced. The throughput improvement of ACB over 802.11 is more significant at high speeds as shown in Fig. 4. At high speeds, the medium access contention is a large portion of a transmission where the actual data transmission is completed in a short time. The overhead of medium access contention in 802.11 is invariable to the physical layer (PHY) data rate. The throughput is thus constrained by the medium access overhead. On the contrary, in the bitwise arbitration, the time used to collect samples is reduced with higher data rates. The collision detection and bitwise arbitration is upper bounded by $(1 + k)$ FFT frames when k bit binary codes are used for medium access contention. Suppose each FFT frame consumes a vector of 64 samples. When the data rate is 4 Msps, each FFT frame takes 16 μs. When the data rate is increased to 8 Msps, each FFT frame takes only 8 μs. The medium access overhead is thus reduced with higher data rates. The high PHY data rate is translated to high throughput.

To validate the effectiveness of partial spectrum correlation, we use 128 point IFFT/FFT on a band of 4 MHz. The band is divided into four channels of 1 MHz and each contains 32 subcarriers. The transmitter encodes the receiver's signature in channel 1 and 3. Fig. 5 shows the correlation results when the receiver expects its signature to be present

in channel 1 and channel 2. Because the receiver's signature is indeed present in channel 1, each time the sliding window is aligned with the signature, the receiver identifies a correlation peak. On the contrary, the receiver's signature is not encoded in channel 2. There is no significant correlation peak at any time. Experiments prove that although the time domain samples are the sum of all modulated subcarriers, the receiver is still able to identify whether a subset of subcarriers are modulated with a known sequence.

4. CONCLUSION

In this paper we presented an adaptive channel bonding (ACB) protocol to address the inefficiency and unfairness issues caused by coexistence of heterogeneous radios. The ACB introduces a compound preamble design for collision detection and a parallel bitwise arbitration mechanism to address the severe contention in wide band spectrum sharing. A partial spectrum correlation method is also introduced to achieve fast spectrum agreement between transmitter and receiver. Experiments on testbed validate the effectiveness of ACB.

5. REFERENCES

[1] IEEE Standard for Information technology–Telecommunications and information exchange between systems Local and metropolitan area networks–Specific requirements Part 11: Wireless LAN Medium Access Control (MAC) and Physical Layer (PHY) Specifications. *IEEE Std 802.11-2012 (Revision of IEEE Std 802.11-2007)*, pages 1 –2793, 2012.

[2] D. Chu. Polyphase codes with good periodic correlation properties (corresp.). *IEEE Transactions on Information Theory*, 18(4):531–532, 1972.

[3] S. S. Hong, J. Mehlman, and S. Katti. Picasso: flexible RF and spectrum slicing. In *Proc. SIGCOMM*, pages 37–48, 2012.

[4] M. Jain, J. I. Choi, T. Kim, D. Bharadia, S. Seth, K. Srinivasan, P. Levis, S. Katti, and P. Sinha. Practical, real-time, full duplex wireless. In *Proc. MobiCom*, pages 301–312, 2011.

[5] H. Rahul, N. Kushman, D. Katabi, C. Sodini, and F. Edalat. Learning to share: narrowband-friendly wideband networks. In *Proc. SIGCOMM*, pages 147–158, 2008.

[6] L. Yang, W. Hou, L. Cao, B. Y. Zhao, and H. Zheng. Supporting demanding wireless applications with frequency-agile radios. In *Proc. NSDI*, pages 5–5, 2010.

[7] X. Zhang and K. Shin. Adaptive subcarrier nulling: Enabling partial spectrum sharing in wireless LANs. In *Proc. ICNP*, pages 311–320, 2011.

Connecting, Scaling and Securing RS code and TD Based KPDs in WSNs: Deterministic Merging

Pinaki Sarkar[*]
Assistant Professor,
Department of Mathematics,
NIST, Behrampur, India
pinakisark@gmail.com

Brijesh Kumar Rai
EEE Department,
IITGuwahati,
Guwahati, India.
bkrai@iitg.ernet.in

Aritra Dhar CS–IT
Department,
IIITDelhi,
New Delhi, India.
aritra1204@iiitd.ac.in

ABSTRACT

Key management, one of the most challenging problems in Wireless Sensor Network (WSN) has been efficiently addressed using Key Predistribution (KPD) schemes. This paper analyzes a localized KPD based on the Transversal Design (TD) design or Reed Solomon (RS) codes schemes; later two shown to be similar. They lack full direct communications among their constituent nodes and so, rely on multi-hop involving other nodes reducing the overall efficiency of the system. The communication issue for TD or RS and hence the localized KPD gets resolved by Deterministic Merging of exactly two nodes. The weakness of 'selective node attack' of the merged designs, similar to their original KPDs, is overcome by invoking the novel trick of Sarkar *et al.* Simulation results confirm that the various network parameters of the proposed schemes improves significantly over a random counterpart among other existing schemes.

Categories and Subject Descriptors

C.2.m, [**Computer Systems Organization**]: COMPUTER -COMMUNICATION NETWORKS—*Security*; G.2.1,G.2.3 [**Mathematics of Computing**]: DISCRETE MATHEMAT-ICS—*Combinatorics,Applications*

General Terms

Theory

Keywords

WSN Security; Communication; Deterministic Merging; Designs; Codes; Connectivity; Selective node attack.

1. INTRODUCTION

Any Wireless Sensor Network (WSN) consists of numerous tiny low cost nodes spread across wide adversarial area.

---

[*]Dr. Pinaki Sarkar is the main contributor of this work.

One or a few powerful Base Station (BS) connects the network to the user. At times, some special node called Cluster Heads (CH) are deployed to subdivide the network into clusters. The presence or absence of CH makes the network Hierarchical (HWSN) or Distributed (DSWN) respectively.

Message exchange in the network should be achieved securely as they deal with sensitive data. Due to resource constraint in the nodes, encryption and decryption of messages are done using symmetric key cryptology. (Symmetric) Key management is a challenging problem and has been well answered by the standard Key Predistribution (KPD) techniques, comprising of two steps: (i) *Preloading of the keys into nodes before deployment;* and (ii) *Key establishment.*

After some KPD [4] proposals, where the preloading and establishment of keys are random, the focus shifted toward deterministic schemes [2, 5, 6, 8, 7, 1] that utilized Mathematical tools (refer to [5, 6, 7] and references therein). There are schemes like [3] which are *hybrid* of both approach while localized schemes [8, 7] can be built on some basic KPD [5, 6] etc. Deterministic (Det.) schemes are preferred over random or hybrid KPD due to better *key establishment.*

1.1 Terminology:Contribution:Organization

Following notation and symbols are reserved for the terms mentioned here. Any more will be defined before their use.

- 'Key predistribution scheme(s)' and 'Communication scheme(s)' mean the same and are denoted by 'KPD'.

- Connectivity between nodes: If two nodes within the 'communication radius' of each other and share a common key, they are said to be connected.

- Full–communication of the network (security): If any two given nodes of the network share cryptographic KPD keys, the network is said to have full communication. The individual node's 'communication radius' is not taken into account as the deployment is random.

- k_{min} :=min. no. of keys required in each node forming a block for *full−inter* & *intra−nodal* communication.

- For notational simplicity, in this paper, $\mathbb{F}_{p^r}$ will be associated with $\mathbb{Z}_{p^r} = \{0, 1, \ldots, p^r-1\}$ under the map:

$$f : \mathbb{F}_{p^r} \longrightarrow \mathbb{Z}_{p^r} = \mathbb{Z}_q, (a_i \in \mathbb{F}_p = \{\overline{0}, \overline{1}, \overline{2}, \ldots, \overline{p-1}\}.)$$

$$\bigoplus_{i=0}^{r-1}{}_p \, a_i\overline{x^i} = (a_{r-1}, a_{r-2}, \ldots, a_1, a_0) \longmapsto \sum_{i=0}^{r-1} a_i p^i,$$

The operations are those in $\mathbb{F}_{p^r}$, and not of $\mathbb{Z}_{p^r}$.

1.2 Basics of Coding and Combinatorics

One is referred to [5, 6] for basic concepts needed for the understanding and thereby comparing the chosen schemes. Further, they provide good overview of coding and combinatorial tools needed for other designs. *Group-divisible design (GDD) (type g^u, block size k)* – is a special type of *combinatorial design* where a triplet $(\mathcal{X}, \mathcal{H}, \mathcal{A})$ with the following properties is considered:

1. $\mathcal{X}$ is a finite set with $|\mathcal{X}| = gu$.

2. $\mathcal{H}$ is a partition of $\mathcal{X}$ into u parts. That is, $\mathcal{H} = \{H_1, H_2, \ldots, H_u\}$ with $\mathcal{X} = H_1 \cup H_2 \cup H_3 \cup \ldots \cup H_u$, $|\mathcal{H}_i| = g \ \forall \ 1 \leq i \leq u$ & $\mathcal{H}_i \cap \mathcal{H}_j = \phi \ \forall \ 1 \leq i \neq j \leq u$.

3. $\mathcal{A}$ is the collection of blocks of $\mathcal{X}$ having the following properties: $|H \cap A| \leq 1 \ \forall \ H \in \mathcal{H}, \ \forall \ A \in \mathcal{A}$, given any pair of varieties $x \in H_i, y \in H_j$ with $i \neq j \ \exists$ unique $A \in \mathcal{A}$ such that $x, y \in A$.

2. RS CODE AND EXTENDED TD KPD

This section compares the $TD(k, p^r)$ KPD, extension of [5] with the RS code KPD [6] using the GDD representation. Note that the $TD(k, p)$ extends readily to $TD(k, p^r)$ by considering the construction of the design over $\mathbb{F}_{p^r}$ in place of $\mathbb{F}_p$. For the extended $TD(k, p^r)$ KPD, the polynomial corresponding to the node $N_{(a,b)} = N_m = N_{aq+b}$ is $p_m = ay + b, 0 \leq a, b \leq p^r - 1 \implies 0 \leq m (= aq + b) \leq q^2 - 1$.

The RS code KPD [6] uses evaluation at non–zero points, i.e., $k = 1, 2, \ldots, p^r - 1$ and tuples are taken in reverse order as opposed to $TD(k, p^r)$. So the key ids in [6] become $(a + by, y)$ instead of $(y, ay + b)$, considered in [5]. For unification of notation, the convention of the scheme [5] will be imitated here. The similarities between both the schemes help in fixing some common weakness.

2.1 Comparison: Weakness

Table 1 depicts the key sharing of a sample network having $p^{r2} = 4^2 = 16$ nodes built over $\mathbb{F}_4, (p = 2, r = 2, p^r = 2^2 = 4)$, for $k = 3, 4$ using $TD(k, p^r)$ KPD (extended [5]) and the RS code KPD [6]. V/C denoted the distinct *Variety Classes*–H_0, H_1, H_2, H_3, where $H_d = \{(d, j) : 0 \leq j \leq 3\}$ for $d = 0, 1, 2, 3$. Selectively choice of V/C results in the alternative GDD form of representation demonstrated below:

- Considering the rows H_0, H_1 & H_2 yields $TD(k, p^r)$ KPD, extension of $TD(k, p)$ [5] with $k = p^r - 1 = 3, p = 2, r = 2$. The V/C $H_3 = H_{p^r - 1}$ is considered.

- Rows H_1, H_2 & H_3 corresponds to *RS code* KPD [6] with $k = p^r - 1, p = 2, r = 2$. H_0 is not considered.

These above observations readily generalizes for other values of $p^r, r \in \mathbb{N}, p$ is a prime. Clearly the RS code and $TD(p^r - 1, p^r)$ KPDs are similar, both *subsets* of $TD(p^r, p^r)$. Lack of direct full communication among nodes and feasibility of 'selective node attack' are two prominent drawbacks of $TD(k, p^r)$ and RS code KPD. The lack of communication weakness can be observed by noting N_i, N_{i+1}, N_{i+2} and $N_{i+3}, \forall i = 0, 2, 3, \ldots p^r - 1$ do not communicate directly as they do not share any common key in both the KPDs [5, 6]. This results in multi-hop communications increasing the overall communication cost and making the system vulnerable to adversarial attacks. Thus both security and energy efficiency of message exchange gets affected.

'Selective node attack' occurs when an attacker overseeing the wireless channel can tap the unencrypted node ids when they are exchanged during key establishment. Then they equate the node ids to find out the clique of each key in a manner similar to the nodes. Knowledge of the exact key sharing enables the attacker to selectively attack nodes (sharing same key) during actual message exchange.

3. REMEDY: DETERMINISTIC MERGING

The schemes under consideration are a $p^r(p^r - 1)$–CID (for $k = p^r$ in [5] or $(p^r - 1)(p^r - 2)$–CID (for $k = p^r - 1$ in extended [5, 6]). Thus for nodes not sharing any key, there are enough nodes which can play the role of the intermediate node in multi-hop (2-hop) process. This encourages one to search for a deterministic design with exactly two merged nodes per big (new) block yielding full communication among the blocks. While proposing the Deterministic (Det.) Merging Block (MB) Designs for the above mentioned KPDs, the following criteria are given priority:

- The essence of the merging concept is to ensure that any message received by either of the two nodes of a block passes down securely to the other node. So, blocks must have intra–nodal communication.

- Random deployment of blocks instead of nodes ensures the clubbed nodes fall within 'communication range' of each other. Inner–block connectivity assured.

- One tries to evaluate at as much less value of $1 \leq y \leq q - 2$ as possible, where $ay + b$ is any node $N_{(a,b)}$'s polynomial. This helps analyze the min. value of $k := k_{min}$ for which 2-merging is feasible.

- The RS code KPD [6] misses evaluation at $y = 0$, but is otherwise similar to $TD(p^r - 1, p^r)$ extension of [5]. So for generalization to [6], evaluation at $y = 0$ is avoided.

- Establishing the keys for the nodes ensures key establishment for the merged blocks.

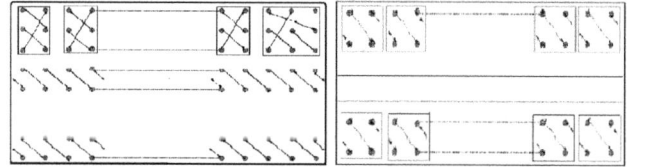

(a) General odd prime.   (b) Even prime power.

Figure 1: Det. MB Strategy on $\mathbf{TD(k, p^r)}$, extended [5] and RS code KPD [5] for $\mathcal{N} = p^2 \text{or} 2^{2r}, r \in \mathbb{N}$ nodes.

3.1 Connectivity Model: SNA

'Selective Node Attack' (SNA) for any Det. KPD [2, 1, 5, 6, 8] can be addressed by applying the novel connectivity model of Sarkar *et al.* [7] after suitable modifications. The number of CHs in the MB designs can be taken as $\frac{q-1}{2}$ each having $q + 1$ nodes as opposed to q CHs each having q nodes below them in the original networks. The deployment is group–wise or cluster–wise random. The key establishment technique of [7] can be imitated for the Det. MB designs.

Table 1: Polynomials, Node and Key identifiers for $4^2 = 16$ nodes, (max.) $k = p^r = 4$ (i.e. $x = 0, 1, 2, 3$). Table constructed according to [5].

Nodes V − C	N_0 $0y+0$	N_1 1	N_2 2	N_3 3	N_4 y	N_5 $y+1$	N_6 $y+2$	N_7 $y+3$	N_8 $2y$	N_9 $2y+1$	N_{10} $2y+2$	N_{11} $2y+3$	N_{12} $3y$	N_{13} $3y+1$	N_{14} $3y+2$	N_{15} $3y+3$
H_0	(0,0)	(0,1)	(0,2)	(0,3)	(0,0)	(0,1)	(0,2)	(0,3)	(0,0)	(0,1)	(0,2)	(0,3)	(0,0)	(0,1)	(0,2)	(0,3)
H_1	(1,0)	(1,1)	(1,2)	(1,3)	(1,1)	(1,0)	(1,3)	(1,2)	(1,2)	(1,3)	(1,0)	(1,1)	(1,3)	(1,2)	(1,1)	(1,0)
H_2	(2,0)	(2,1)	(2,2)	(2,3)	(2,2)	(2,3)	(2,0)	(2,1)	(2,3)	(2,2)	(2,1)	(2,0)	(2,1)	(2,0)	(2,3)	(2,2)
H_3	(3,0)	(3,1)	(3,2)	(3,3)	(3,3)	(3,2)	(3,1)	(3,0)	(3,1)	(3,0)	(3,3)	(3,2)	(3,2)	(3,3)	(3,0)	(3,1)

1 The entire network consisting of $p^r \times p^r$ can be split into $p^{r-1} \times p^{r-1}$ clusters of size $p \times p$ (see Figure 2);

2 Merging of nodes to form blocks in each cluster is along the backward diagonal excluding the top/bottom 3 rows. This is to facilitate common intra-nodal key evaluation at $y = 1, 2, \frac{p-1}{2}$;

3 Visualize a sample $p \times p$ square grid (bottom–left cluster of Figure 2, explicitly shown in Figure 1). The merging in the chosen cluster is as follows:;

4 The cluster is split into $2 \times p$ sub–grids yielding $\frac{p-3}{2}$ pair of rows barring the top 3 rows (in this case);

5 Merging of $\frac{p-3}{2}$ many pair of rows is similar to the case of $p = 2^r$ and is as follows;

6 **for** $i = 0$ to $(p-5), i = i+2$ **do**

7   **for** $j = 0$ to $(p-1), j = j+1$ **do**

8   MERGE $N_{(i,j+1)}$ & $N_{(i+1,j)}$. Common key is the result of evaluation at $y = 1$;

9   MERGE $N_{(i,0)}$ & $N_{(i+1,p-1)}$. Common key is the result of evaluation at $y = 1$;

   end

end

10 For the remain three rows, i.e., the $3 \times p$ sub–grid, the merging is done is two steps;

12 1 The first $p-3$ columns of the last 3 rows , i.e., $3 \times p-3$ sub–grid is merged as follows;

13 **for** $ctr = 0$ to $(p-5), ctr = ctr+2$ **do**

14   **for** $i = (p-3)$ to $(p-1), i = i+1$ **do**

15   **for** $j = ctr$ to $(ctr+1), j = j+1$ **do**

16   MERGE $N_{(p-3,j)}$ & $N_{(p-1,j+1)}$. Common key evaluated at $y = \frac{p-1}{2}$;

17   MERGE $N_{(p-3,j+1)}$ & $N_{(p-2,j)}$. Common key evaluated at $y = 1$;

18   MERGE $N_{(p-2,j+1)}$ & $N_{(p-1,j)}$. Common key evaluated at $y = 1$;

     end

   end

end

20 2 Merge the remaining 3×3 sub–grid as follows;

21 MERGE $N_{(p-3,p-3)}$ & $N_{(p-1,p-2)}$. Common key evaluated at $y = \frac{p-1}{2}$;

22 MERGE $N_{(p-3,p-2)}$ & $N_{(p-2,p-3)}$. Common key evaluated at $y = 1$;

23 MERGE $N_{(p-3,p-1)}$ & $N_{(p-2,p-2)}$. Common key evaluated at $y = 1$;

24 MERGE $N_{(p-2,p-1)}$ & $N_{(p-1,p-3)}$. Common key evaluated at $y = 2$;

25 Various similar merging patterns can be constructed for the remaining clusters by choosing 3 rows from top/bottom and leaving one node non–merged;

Algorithm 1: Deterministic Merging Strategy: General odd prime power $q = p^r, r \in \mathbb{N}$.

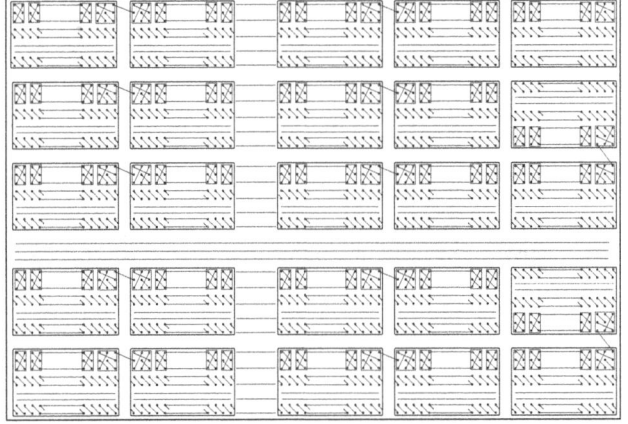

Figure 2: Deterministic Merging for general prime power case: $p^r, r \in \mathbb{N} \setminus \{1\} \implies \mathcal{N} = p^{r2} = p^{2r} = (p \times p) \times p^{2(r-1)}$ of $TD(k, p^r)$ or RS code **KPD**. The network split into $p^{r-1} \times p^{r-1}$ grids of p^2 nodes.

4. SIMULATION RESULTS: COMPARING

Communication Probability or Connectivity of the network is defined to be the probability that any given pair of nodes can communicate directly. From the constructions of the MB designs over extended KPD [5, 6], clearly the *Connectivity is 1 for $k \geq k_{min}$*. Figure 3 compares the connectivity of $TD(k, p^r)$ (or RS code), the random merging proposals of [3] v/s current Det. MB designs. The plotting in Figure 4 is done with no. of nodes in the network along x-axis v/s the resiliency coefficient $E(s)$ along the y-axis, which measures the ratio of links broken on compromise of s nodes to total no. of links in original network. It compares the Det. MB designs with some existing KPDs with the parameters:

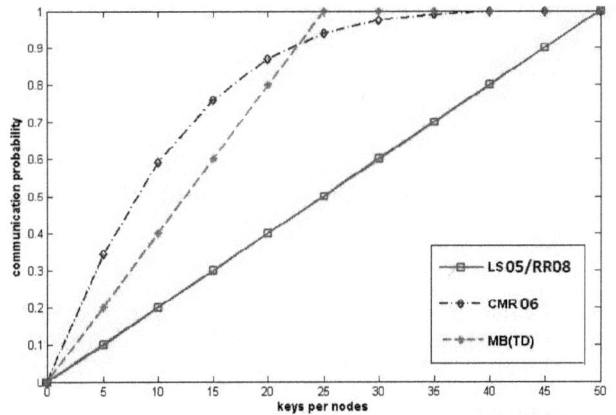

Figure 3: Connectivity: Det. MB(TD) vs. random (CMR), original (RS code or TD(k, p^r)) KPDs.

1. Simonova *et al.* (SLW) [8], parameters: $k = 16, p = 11$ & $m = 2$ over $TD(4, 11)$ KPD [5];

2. Lee Stinson (LS) $TD(k, q))$ [5], parameter: $k = 101$; same as Ruj Roy (RS code), parameter: $k = 100$;

3. Du *et al.* [4] (DDHV), $k = 200, \omega = 11, \tau = 2$;

4. Chakrabarti Seberry [2] ((i) CS06-1 (CY & CY) (both tiers based on [1]), parameter: $k = 12$;

5. (ii) CS-2 (TD & CY) (KPD [5] for lower tier, projective planes KPD [1] for upper), parameter: $k = 11$; and

6. the present designs: MB(TD), parameter: $k = 149$; MB(RS), parameter: $k = 148$;

7. MB (Scalable), parameters: $p = 47, k = 48 = 47 + 1$.

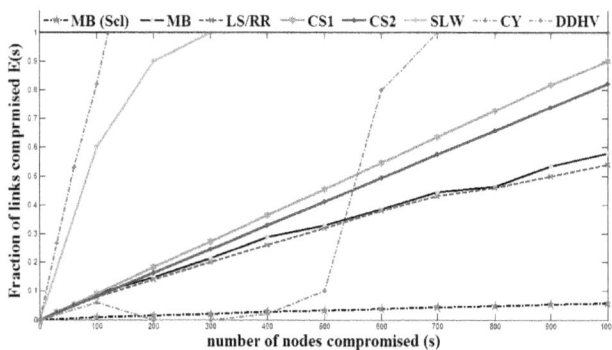

Figure 4: Comparative study of resiliency coefficient (E(s)) Det. MB schemes vs. various existing KPD.

5. CONCLUSION: FUTURE WORK

This work focuses on rectifying two pertinent weakness viz. 'lack of full direct communication' and 'selective node attack' (SNA) of prominent deterministic KPD. The former weakness is overcome by deterministically merging exactly 2 nodes of the $TD(k, p^r)$ KPD, extension of KPD [5] as well as the RS code KPD [6]. The network size is halved as compared to 'z' times reduction in the random model [3] which even failed to guarantee full communication.

On the other hand, SNA gets eradicated by applying the connectivity model of [7] after suitable modifications to both the original deterministic KPD as well as their MB designs. Some important features of the MB Designs as well as the connectivity model are as follows:

- The MB designs, suggested here provides full communication among the blocks through at least 1 and at most 4 common keys between any given pair of blocks.

- Evaluation at $y = 1, 2, \frac{p-1}{2}$ to assure full inner–nodal communication in a block for the MB designs.

- Minimum number of keys MB Design to provide full–intra–inter block communication for $TD(k, p^r)$, extension of [5] and [6] $= k_{min} = max(\frac{q+6}{4}, \frac{p-1}{2})$.

- Full communication can not be assured when nodes are merged randomly to form blocks as the exact key rings of the constituent nodes are not known. For instance $N_{(0,0)} \cup N_{(0,1)}$ and $N_{(0,2)} \cup N_{(0,3)}$ share no common key.

- Merging does not mean that the nodes physically combine to become one big node. Just that they are to be treated as one unit, termed block.

- The total number of links in the merged scheme is same as that of the original Transversal Design or Reed Solomon based KPD in [5, 6] respectively.

- The connectivity model uses repeated encryption by introducing an hierarchy in the system through Cluster Heads (CHs). This results in improved resiliency.

There are some immediately future research directions which stems out of the current work. The optimality of k_{min} needs to be established more prominently. The connectivity model may malfunction under CH jamming/capture. Scalability is an issue as with most existing KPD. Devising schemes which allow CH replacement/addition under security environment may be an interesting problem.

Acknowledgement

Special thanks is due towards Sarbajit Mukherjee, BESU, INDIA and Aniket Roy, Kalyani Govt. Eng. College for their assistance during the final drafting of the paper.

6. REFERENCES

[1] S. A. Çamtepe and B. Yener. Combinatorial design of key distribution mechanisms for wireless sensor networks. In *ESORICS 2004, LNCS*, volume 3193, pages 293–308, 2004.

[2] D. Chakrabarti and J. Seberry. Combinatorial structures for design of wireless sensor networks. In *ACNS 2006, LNCS*, volume 3989, pages 365–374, 2006.

[3] M. S. Chakrabarti, D. and B. Roy. A key pre-distribution scheme for wireless sensor networks: merging blocks in combinatorial design. *Inter. Jour. of Infor. Sec.*, 5(2):105–114, 2006.

[4] D. J. H. Y. S. Du, W. and P. K. Varshney. A key predistribution scheme for sensor networks using deployment knowledge. *IEEE Trans. Dependable Sec. Comput.*, 3(1):62–77, 2006.

[5] J. Y. Lee and D. R. Stinson. A combinatorial approach to key predistribution for distributed sensor networks. In *WCNC 2005, IEEE*, pages 1200–1205, 2005.

[6] S. Ruj and B. Roy. Key predistribution schemes using codes in wireless sensor networks. In *Inscrypt 2008, LNCS*, volume 5487, pages 275–288, 2008.

[7] S. A. Sarkar, P. and M. U. Chowdhury. Secure connectivity model in Wireless Sensor Networks (WSN) using first order Reed-Muller codes. In *MASS*, pages 507–512, 2010.

[8] L. A. C. H. Simonova, K. and X. S. Wang. Location-aware key predistribution scheme for wide area wireless sensor networks. In *SASN 2006, ACM*, pages 157–168, 2006.

SMASHER: A Strategy-Proof Combinatorial Auction Mechanism for Heterogeneous Channel Redistribution[*]

Zhenzhe Zheng, Fan Wu[†], and Guihai Chen
Shanghai Key Laboratory of Scalable Computing and Systems
Department of Computer Science and Engineering
Shanghai Jiao Tong University, China
zhengzhenzhe@sjtu.edu.cn, {fwu,gchen}@cs.sjtu.edu.cn

ABSTRACT

Auction is believed to be an effective way to solve or relieve the problem of radio spectrum shortage, by dynamically redistributing idle wireless channels of primary users to secondary users. However, to design a practical channel auction mechanism, we have to consider five challenges, including strategy-proofness, channel spatial reusability, channel heterogeneity, bid diversity, and social welfare maximization. Unfortunately, none of the existing works fully considered the five design challenges. In this paper, we present the first in-depth study on the problem of dynamic channel redistribution by jointly considering the five design challenges, and present SMASHER, which is a Strategy-proof coMbinatorial Auction mechaniSm for HEterogeneous channel Redistribution. Our analyses show that SMASHER achieves both strategy-proofness and approximately efficient social welfare.

Categories and Subject Descriptors

C.2.1 [**Computer-Communication Networks**]: Network Architecture and Design – *Wireless Communication*

General Terms

Algorithm, Design, Economic

Keywords

Channel Allocation; Combinatorial Auction

[†]F. Wu is the corresponding author.
[*]This work was supported in part by the State Key Development Program for Basic Research of China (Grant No. 2012CB316201), in part by China NSF grant 61272443, 61133006, 61073152, and in part by Shanghai Science and Technology fund 12PJ1404900 and 12ZR1414900. The opinions, findings, conclusions, and recommendations expressed in this paper are those of the authors and do not necessarily reflect the views of the funding agencies or the government.

1. INTRODUCTION

The last two decades have witnessed a rapid development of wireless communication technology. Unfortunately, naturally limited radio spectrum is becoming a more and more serious bottleneck of the ongoing growth of wireless applications and services. Most of the countries have specific departments to regulate spectrum usage, *e.g.*, Federal Communications Commission (FCC) in the US and Radio Administration Bureau (RAB) in China. They statically allocate spectrum to wireless application service providers on a long term basis for large geographical regions. Such static management leads to low spectrum utilization in the spatial and temporal dimensions. Large chunks of radio spectrum are left idle most of the time at a lot of places, while new wireless applications are starving for the radio spectrum. Therefore, an open and market-based framework is highly needed to dynamically redistribute the radio spectrum, and thus improve the utilization of the radio spectrum.

Auctions are the most well-known market-based mechanisms to redistribute resources [3]. Since 1994, FCC has conducted a series of auctions for the licenses of radio spectrum. While FCC auctions target at large wireless service providers, our focus is on small wireless applications, such as community wireless networks or home wireless networks.

There exist many challenges in designing a practical channel auction mechanism. We list five major challenges:

- *Strategy-Proofness*: In strategy-proof auction mechanisms, simply submitting truthful channel demands (*e.g.*, valuation of the channels) maximizes each participant's utility. Since the participants are normally rational and selfish, they always tend to strategically manipulate the auction, if doing so can increase their utilities. Therefore, it discourages truthfully behaving participants from joining the auction, if strategy-proofness is not guaranteed.

- *Spatial Reusability*: Spatial reusability differentiates the wireless channels from conventional goods. Two wireless users can use the same wireless channel simultaneously, if they are well-separated.

- *Channel Heterogeneity*: Channel heterogeneity comes from both *spatial heterogeneity* and *frequency heterogeneity*. On one hand, the availability and quality of a channel vary at different locations. On the other hand, channels with different central frequency may have different propagation and penetration characteristics.

- *Bid Diversity*: Wireless devices may be equipped with multiple radios, each of which can work on a distinguished channel at the same time. Consequently, a wireless user may request multiple bundles of channels, according to her quality of service requirement. Buyers have higher opportunities to obtain channels by submitting diverse bids, which makes the channel redistribution more flexible. Therefore, it is necessary to allow users to express diverse demands for channels.

- *Social Welfare*: The objective of any auction is to maximize social welfare, which is the sum of the auction winners' valuations of the allocated goods.

A number of related works (*e.g.*, [1, 2, 4–6, 8–10]) exist in the literature. Unfortunately, none of these works fully consider the five design challenges. Some of strategy-proof channel auction mechanisms (*e.g.*, VEARITAS [8], TRUST [9], SMALL [5]) consider channel spatial reusability, but only work when the trading channels are homogenous. Two recent works TAHES [2] and CRWDP [1] consider channels' heterogeneity, but TAHES restricts each user to bid for a single channel while CRWDP ignores the spatial reusability of channels.

In this paper, we conduct an in-depth study on the problem of dynamic channel redistribution by jointly considering the five design challenges, and present SMASHER, which is a <u>S</u>trategy-proof co<u>M</u>binatorial <u>A</u>uction mechani<u>S</u>m for <u>HE</u>terogeneous channel <u>R</u>edistribution. SMASHER is a novel combinatorial auction mechanism for indivisible heterogeneous channel redistribution, and achieves both strategy-proofness and approximately efficient social welfare.

We make the following contributions in this paper:

- First, we present a general model of combinatorial auction for heterogeneous channel redistribution. The auction model is powerful enough to express channel spatial reusability and heterogeneity, as well as bid diversity.

- Second, we introduce the concept of *virtual channel* to capture the conflict of channel usage among different auction participants. By using virtual channels, we transform the problem of heterogeneous channel allocation to a classic multi-unit combinatorial auction.

- Third, we propose SMASHER, which is a combinatorial auction mechanism for heterogeneous channel redistribution, achieving both strategy-proofness and approximately efficient social welfare.

2. PRELIMINARIES AND PROBLEM FORMULATION

In this section, we present the auction model for the problem of heterogeneous channel allocation.

2.1 Auction Model

We consider a static scenario, in which there is a primary spectrum user, called "seller", who wants to lease out her temporarily unused wireless channels, and some secondary users (*e.g.*, WiFi access points), called "buyers", who want to lease channels to provide services to their customers at certain quality of service (QoS). We consider that the channels for leasing are *heterogeneous*, and thus the buyers have their own preference over the channels due to spatial variance (*e.g.*, background noise, temperature, and landform). Since wireless devices can be equipped with multiple radios, the buyers may request more than one channel according to their requirements of QoS. Considering the diversity of

QoS demand and heterogeneity of the channels, we allow the buyers to submit multiple channel requests, among which one of the requests can be granted. We assume that the buyers have uniform valuation over any of their channel requests, because the buyers' requirement of QoS can be satisfied if one of their requested bundles is allocated. Different from the allocation of traditional goods, wireless channels can be spatially reused, meaning that well-separated buyers can work on the same channel simultaneously, if they do not have interference between each other.

We model the process of heterogeneous channel redistribution as a sealed-bid combinatorial auction, in which buyers simultaneously submit their demands for channels to a trustworthy auctioneer, such that no buyer can know other participants' information. The auctioneer makes the decision on channel allocation and the charge to each winner. We denote the set of orthogonal and heterogeneous channels for leasing by $\mathbb{C} \triangleq \{c_1, c_2, \ldots, c_m\}$, and the set of buyers by $\mathbb{N} \triangleq \{1, 2, \ldots, n\}$. We list useful notations in our model of combinatorial channel auction as follows:

Channel Request R_i: Each buyer $i \in \mathbb{N}$ submits a vector of requested channel bundles $R_i \triangleq \left(S_i^1, S_i^2, \ldots, S_i^K\right)$ to the auctioneer. Any channel bundle $S_i^j \subseteq \mathbb{C}, 1 \leq j \leq K$ can satisfy her QoS. We assume that buyer's request is strict, meaning that the buyer is only interested in winning a whole bundle S_i^j in her request vector. We call a buyer, who submits a request vector of K channel bundles, and is interested in winning one of the bundles, as K-minded buyer. If $K = 1$, then the buyer is single-minded. Note that our auction model is a generalization of existing models with single-minded buyers (*e.g.*, [1, 2]). We denote the channel request vector $\vec{\mathbb{R}}$ of all the buyers as $\vec{\mathbb{R}} \triangleq (R_1, R_2, \ldots, R_n)$.

Valuation v_i: Each buyer $i \in \mathbb{N}$ has an uniform valuation v_i over any requested channel bundles in R_i. Here, v_i is the private information of the buyer i. This is also known as *type* in mechanism design. We denote the valuation vector $\vec{\mathbb{V}}$ of all the buyers as $\vec{\mathbb{V}} \triangleq (v_1, v_2, \ldots, v_n)$.

Bid b_i: Each buyer $i \in \mathbb{N}$ submits a bid b_i to the auctioneer, meaning that if she wins any channel bundle S_i^j, she would like to pay no more than b_i for it. Here, the bid b_i may not necessarily be equal to her valuation v_i. Let vector $\mathbb{B}$ represent the bids of all the buyers $\mathbb{B} \triangleq (b_1, b_2, \ldots, b_n)$.

Clearing price p_i: The auctioneer charges each winning buyer $i \in \mathbb{N}$ a clearing price p_i. The loser in the auction is free of any charge. We use vector $\mathbb{P} \triangleq (p_1, p_2, \ldots, p_n)$ to represent the clearing prices of all the buyers.

Utility u_i: The utility of a buyer $i \in \mathbb{N}$ in the auction is defined as the difference between her valuation on the bundle of channels she wins and her clearing price p_i:

$$u_i \triangleq v_i - p_i. \quad (1)$$

We consider that the buyers are rational and selfish, thus their goals are to maximize their own utilities. In contrast to the buyers, the auctioneer's objective is to maximize *social welfare*. Here social welfare is defined as follows:

DEFINITION 1 (SOCIAL WELFARE). *The social welfare in a channel auction is the sum of winning buyers' valuations on their allocated bundles of channels, i.e.,*

$$SW \triangleq \sum_{i \in W} v_i, \quad (2)$$

where W is the set of winners.

In this paper, we assume that buyers do not collude with each other and do not cheat about their channel bundles, while leaving these problems to our future works.

3. MULTI-UNIT COMBINATORIAL CHANNEL AUCTION

Different from existing works on strategy-proof channel allocation, we introduce a novel concept of *virtual channel* to represent the conflict of channel usage among the buyers. By introducing virtual channels, we transform the problem of heterogeneous channel allocation to a classic multi-unit combinatorial auction.

3.1 Virtual Channel

We introduce *virtual channel* to capture the interference among the buyers on different channels. Specifically, a virtual channel $vc_{i,j}^k$ denotes that the buyer i and the buyer j may cause interference between each other on channel c_k, and thus they cannot work on channel c_k simultaneously. Since virtual channel $vc_{i,j}^k$ represents the exclusive usage of channel c_k between the buyer i and j, its quantity is set to 1. When virtual channel $vc_{i,j}^k$ is added to the requested bundle(s) that contains channel c_k from the buyer i and j, at most one of the requests containing channel c_k from the two buyers can be granted. Consequently, the exclusive usage of channel c_k between the buyer i and j is guaranteed. We present the definition of virtual channel as follows.

DEFINITION 2 (VIRTUAL CHANNEL). *There is a virtual channel $vc_{i,j}^k$, if the buyer i and buyer j are within the interference range of each other on channel c_k.*

In most of existing works on channel auction, a single conflict graph is used to represent the interference among buyers. However, in the case of heterogeneous channels, each channel may have a distinctive conflict graph. Let $G_k \triangleq (O_k, E_k)$ denote the conflict graph on channel c_k, where $O_k \subseteq \mathbb{N}$ is the set of buyers who can access channel c_k, and each edge $(i, j) \in E_k$ represents the interference between the buyer i and j on channel c_k.

Since conflict graph is commonly assumed to be available in wireless networks, we construct the virtual channel from conflict graph. We create a virtual channel $vc_{i,j}^k$, if there is an edge between the buyer i and j in conflict graph G_k, and append $vc_{i,j}^k$ to the requested bundle(s) containing channel c_k from the buyer i and j, while remaining the corresponding bids unchanged. Let $\mathbb{VC}$ be the set of virtual channels and $\vec{\mathbb{R}}'$ be the vector of updated requests with virtual channels.

3.2 Multi-Unit Combinatorial Auction

Given the virtual channel introduced in last section, we are ready to transform the problem of heterogenous channel allocation to a classic multi-unit combinatorial auction.

The goods in the multi-unit combinatorial auction are the channels and virtual channels. The quantities of each channel $c_k \in \mathbb{C}$ and virtual channel $vc_{i,j}^k \in \mathbb{VC}$ are n and 1, respectively. Let $x\left(i, S_i'^j\right) = 1$ denote that the channel set $S_i'^j$ is granted to the buyer i; otherwise, $x\left(i, S_i'^j\right) = 0$. The process of winner determination can be modeled as a binary program. The objective is to maximize the social welfare. We use b_i, instead of v_i, because the strategy-proof mechanism shown in later sections will guarantee that bidding truthfully is the dominate strategy of each buyer $i \in \mathbb{N}$.

Objective:

$$Maximize \quad \sum_{i \in N} \sum_{j=1}^{K} x(i, S_i'^j) \times b_i$$

Subject to:

$$\sum_{i \in N} \sum_{S_i'^j \in \mathbb{R}_i', S_i'^j \ni c_k} x\left(i, S_i'^j\right) \leq n \qquad \forall c_k \in \mathbb{C} \qquad (3)$$

$$\sum_{i \in N} \sum_{S_i'^j \in \mathbb{R}_i', S_i'^j \ni vc_k} x\left(i, S_i'^j\right) \leq 1 \qquad \forall vc_k \in \mathbb{VC} \qquad (4)$$

$$\sum_{j=1}^{K} x\left(i, S_i'^j\right) \leq 1 \qquad \forall i \in \mathbb{N} \qquad (5)$$

$$x\left(i, S_i'^j\right) \in \{0, 1\} \quad \forall i \in \mathbb{N}, 1 \leq j \leq K \qquad (6)$$

If the optimal social welfare can be achieved by solving the above binary program, then the celebrated VCG mechanism can be applied to calculate the clearing price that can ensure the strategy-proofness of the auction mechanism. Unfortunately, the above winner determination problem can be proven to be NP-hard by reducing to the *exact cover* problem. Considering the computational intractability of the winner determination problem, we integrate a greedy allocation algorithm with a novel pricing mechanism to provide a strategy-proof and approximately efficient combinatorial auction mechanism for heterogeneous channel redistribution in next section.

4. HETEROGENEOUS CHANNEL REDISTRIBUTION

As shown in Section 3.2, finding the optimal auction decision is computationally intractable. In this section, we present SMASHER, which is a strategy-proof and approximately efficient combinatorial auction mechanism for heterogenous channel redistribution.

4.1 Design of SMASHER

SMASHER consists of the following three major components: virtual channel generation, winner determination, and clearing price calculation.

4.1.1 Virtual Channel Generation

The process of virtual channel generation is the same as the method discussed in Section 3.1, except that we add one more virtual channel vc_i with unit quantity to each requested bundle of buyer $i \in N$. Virtual channel vc_i is used to ensure that at most one of the requested bundles from the buyer i can be granted.

$$S_i'^j = S_i'^j \cup \{vc_i\}, i \in N, 1 \leq j \leq K,$$

where $S_i'^j$ is updated bundle with virtual channels.

4.1.2 Winner Determination

Before presenting the approximation algorithm for winner determination, we introduce *virtual bid*. The uniform virtual bid $\tilde{b}_i$ over any of requested bundles from the buyer i is defined as

$$\tilde{b}_i \triangleq \frac{b_i}{\max_{1 \leq l \leq K} \left(\sqrt{|S_i'^l|}\right)}. \qquad (7)$$

Algorithm 1: Approximation Algorithm for Winner Determination

Input: Vector of updated channel requests $\vec{\mathbb{R}}'$, vector of bids $\vec{\mathbb{B}}$.

Output: A pair of sets of winning buyers and allocated bundles of channels $(\mathbb{W}, \mathbb{S})$.

1   $(\mathbb{W}, \mathbb{S}) \leftarrow (\varnothing, \varnothing); \mathcal{V} \leftarrow \varnothing;$

2   **foreach** $i \in \mathbb{N}$ **do**

3     $\tilde{b}_i \leftarrow b_i / \max_{1 \leq l \leq K} \left(\sqrt{|S_i'^l|} \right);$

4   **end**

5   Sort $\tilde{b}_i$ in non-increasing order: $\mathbb{L}_1 : \tilde{b}_1 \geq \tilde{b}_2 \geq \ldots \geq \tilde{b}_n;$

6   **for** $i = 1, \ldots, n$ **do**

7   Sort $S_i'^j$ in non-decreasing order of bundle size: $\mathbb{L}_2 : |S_i'^1| \leq |S_i'^2| \leq \ldots \leq |S_i'^K|;$

8   **for** $j = 1, \ldots, K$ **do**

9   **if** $S_i'^j \cap \mathcal{V} = \varnothing$ **then**

10   Add virtual channels in $S_i'^j$ to $\mathcal{V}$;

11        $(\mathbb{W}, \mathbb{S}) \leftarrow (\mathbb{W} \cup \{i\}, \mathbb{S} \cup \{S_i'^j\});$

12   **break**;

13   **end**

14   **end**

15   **end**

16   **return** $(\mathbb{W}, \mathbb{S})$;

SMASHER sorts all the buyers according to their virtual bids in non-increasing order:

$$\mathbb{L}_1 : \tilde{b}_1 \geq \tilde{b}_2 \geq \ldots \geq \tilde{b}_n.$$

In case of a tie, SMASHER breaks the tie following a bid-independent rule, such as lexicographic order of buyers' ID and channel number.

Following the order in $\mathbb{L}_1$, SMASHER greedily grants the smallest channel bundle, in which no virtual channel has already been allocated, to each buyer.

Algorithm 1 shows the pseudo-code of above winner determination process. In practice, the number of buyers n is much larger than K, thus the time complexity of Algorithm 1 is $\mathcal{O}(n \log n)$.

4.1.3 Clearing Price Calculation

The clearing price is calculated based on *critical virtual bid*.

DEFINITION 3 (CRITICAL VIRTUAL BID). *The critical virtual bid $cr(i) \in \mathbb{L}_1$ of buyer $i \in \mathbb{N}$ is the minimum virtual bid that the buyer i must exceed to be allocated one of her channel bundles.*

We note that according to the definition of critical virtual bid, no matter which request of the buyer i is granted in the auction, the critical virtual bid $cr(i)$ is always the same.

The critical virtual bid of the buyer $i \in \mathbb{N}$ can be calculated by the following procedure. Given other buyers' requests and bids $\left(\vec{\mathbb{R}}'_{-i}, \vec{\mathbb{B}}_{-i} \right)$, we greedily select virtual bid by rerunning Algorithm 1 until none of the buyer i's requests can be satisfied. The threshold virtual bid $cr(i)$ we select finally is regarded as the critical virtual bid of the buyer i. We now show the method of calculating the clearing price of the buyer i by distinguishing two cases:

1. If the buyer i loses the auction or $cr(i)$ does not exist (denoted by $cr(i) = 0$), then her clearing price is 0.

2. If the buyer i is granted channel bundle $\hat{S}_i'^j$ and there exists a critical virtual bid $cr(i)$, the clearing price p_i of the buyer i is set to

$$p_i \triangleq cr(i) \times \max_{1 \leq l \leq K} \left(\sqrt{|S_i'^l|} \right). \qquad (8)$$

4.2 Analysis

We prove the strategy-proofness and analyze the approximation ratio of SMASHER in this section.

THEOREM 1. *SMASHER is a strategy-proof combinatorial auction mechanism for heterogeneous indivisible channel redistribution.*

THEOREM 2. *The approximation ratio of SMASHER is $\mathcal{O}(n\sqrt{m})$, where n is the number of buyers, m is the number of channels.*

We leave the detailed proofs in our technical report [7].

5. CONCLUSION

In this paper, we have made an in-depth study on channel redistribution problem by jointly considering the five design challenges. We have presented a strategy-proof combinatorial auction mechanism for dynamic heterogeneous channel redistribution, namely SMASHER. Our analyses show that SMASHER achieves strategy-proofness and approximately efficient social welfare.

6. REFERENCES

[1] M. Dong, G. Sun, X. Wang, and Q. Zhang. Combinatorial auction with time-frequency flexibility in cognitive radio networks. In *INFOCOM*, 2012.

[2] X. Feng, Y. Chen, J. Zhang, Q. Zhang, and B. Li. TAHES: Truthful double auction for heterogeneous spectrums. In *INFOCOM*, 2012.

[3] J. Huang, R. A. Berry, and M. L. Honig. Auction-based spectrum sharing. *Mobile Networks and Applications*, 11(3):405–418, 2006.

[4] J. Jia, Q. Zhang, Q. Zhang, and M. Liu. Revenue generation for truthful spectrum auction in dynamic spectrum access. In *MobiHoc*, 2009.

[5] F. Wu and N. Vaidya. SMALL: A strategy-proof mechanism for radio spectrum allocation. In *INFOCOM*, 2011.

[6] P. Xu, S. Wang, and X.-Y. Li. SALSA: Strategyproof online spectrum admissions for wireless networks. *IEEE Transactions on Computers*, 59(12):1691 –1702, 2010.

[7] Z. Zheng, F. Wu, and G. Chen. SMASHER: Strategy-proof combinatorial auction mechanisms for heterogeneous channel redistribution. Technical report, available at http://www.cs.sjtu.edu.cn/~fwu/res/Paper/ZWC12TRSMASHER.pdf, 2012.

[8] X. Zhou, S. Gandhi, S. Suri, and H. Zheng. eBay in the sky: Strategy-proof wireless spectrum auctions. In *MobiCom*, 2008.

[9] X. Zhou and H. Zheng. TRUST: A general framework for truthful double spectrum auctions. In *INFOCOM*, 2009.

[10] X. Zhou and H. Zheng. Breaking bidder collusion in large-scale spectrum auctions. In *MobiHoc*, 2010.

Author Index

www.ingramcontent.com/pod-product-compliance
Lightning Source LLC
Chambersburg PA
CBHW080930220326
41598CB00034B/5742